CRC HANDBOOK SERIES IN NUTRITION AND FOOD

Miloslav Rechcigl, Jr.
Editor-in-Chief

SECTION OUTLINE

SECTION A: Science of Nutrition and Food
Minimum of 1 volume projected.
Nomenclature, Nutrition Literature, Nutrition Societies, Foundations, and Historical Milestones in Nutrition.

SECTION B: The Living Organisms, Their Chemical Constitution, Feeding and Digestive System, and Ecological Aspects
Minimum of 3 volumes projected.
Taxonomy, Distribution of Organisms, Ecology — Symbiosis, Feeding and Digestive System, Chemical Constitution of Organisms, and Biological Productivity.

SECTION C: The Nutrients and Their Metabolism
Minimum of 1 volume projected.
Nutrients and Growth Regulators, Antinutrients and Antimetabolites, Naturally Occurring Food Toxicants, Regulatory Aspects of Nutrition, and Availability of Nutrients.

SECTION D: Nutritional Requirements
Minimum of 4 volumes projected.
Comparative Requirements of Organisms, Qualitative Requirements of Specific Organisms, Quantitative Requirements (Nutritional Standards), and Nutritional Requirements for Specific Processes and Functions: Animals, Microorganisms, Plants.

SECTION E: Nutritional Disorders
Minimum of 2 volumes projected.
Nutritional Disorders in Living Organisms, Effect of Specific Nutrient Deficiencies and Toxicities, and Nutritional Disorders in Specific Tissues.

SECTION F: Food Composition
Minimum of 1 volume projected.
Nutrient Content and Energy Value of Food, Factors Affecting Nutrient Composition of Food, and Utilization and Biological Value of Food.

SECTION G: Diets, Culture Media, and Food Supplements
Minimum of 4 volumes projected.
Diets, Culture Media, and Food and Feed Supplements.

SECTION H: The State of World Food and Nutrition
Minimum of 1 volume projected.
World Population, Natural and Food Resources, Food Production, Food Losses, World Food Usage and Consumption, Geographic Distribution of Nutritional Diseases, Nutritional Requirements — Current and Projected, Agricultural Inputs — Current and Projected, Food Aid, Food Marketing and Distribution, Socioeconomic, Cultural, and Psychological Factors Affecting Nutrition.

SECTION I: Food Safety, Food Spoilage, Food Preservation
Minimum of 1 volume projected.
Food Contamination, Food Spoilage, Foodborne Diseases, Food Sanitation and Preservation, Food Laws, and Nutrition Labeling.

SECTION J: Production, Utilization, and Nutritive Value of Foods
Minimum of 1 volume projected.

SECTION K: Nutrition and Food Methodology
Minimum of 1 volume projected.
Assessment of Nutritional Status and Measuring
Nutritive Value of Food.

CRC Handbook Series in Nutrition and Food

Miloslav Rechcigl, Jr., Editor-in-Chief

Nutrition Advisor and Director
Interregional Research Staff
Agency for International Development
U.S. Department of State

Section G: Diets, Culture Media, Food Supplements

Volume III

Culture Media for Microorganisms and Plants

CRC PRESS, Inc.
18901 Cranwood Parkway · Cleveland, Ohio 44128

Library of Congress Cataloging in Publication Data

Main entry under title:

Culture media for microorganisms and plants.

    (Diets, culture media, food supplements, v. 3)
(CRC handbook series in nutrition and food; section G)
    Bibliography: p.
    Includes index.
    1. Microbiology—Cultures and culture media.
2. Plant growing media. I. Series. II. Series:
CRC handbook series in nutrition and food; section G.
QH519.D54 vol. 3 [QR66.3] 591'.13s [576'.028] 77-16838
ISBN 0-8493-2738-5

© 1978 by CRC Press, Inc.

International Standard Book Number 0-8493-2700-8 (Complete Set)
International Standard Book Number 0-8493-2735-0 (Section G)
International Standard Book Number 0-8493-2736-9 (Volume I)
International Standard Book Number 0-8493-2737-7 (Volume II)
International Standard Book Number 0-8493-2738-5 (Volume III)
International Standard Book Number 0-8493-2739-3 (Volume IV)

Library of Congress Card Number 77-16838
Printed in the United States

PUBLISHER'S PREFACE

In 1913, when the First Edition of the *Handbook of Chemistry and Physics* appeared, scientific progress, particularly in chemistry and physics, had produced an extensive literature but its utility was seriously handicapped because it was fragmented and unorganized. The simple but invaluable contribution of the *Handbook of Chemistry and Physics* was to provide a systematic compilation of the most useful and reliable scientific data within the covers of a single volume. Referred to as the "bible," the Handbook soon became a universal and essential reference source for the scientific community. The 57th Edition, published in the bicentennial year of 1976, represents more than 63 years of continuous service to millions of professional scientists and students throughout the world.

In the years following World War II, scientific information expanded at an explosive rate due to the tremendous growth of research facilities and sophisticated analytical instrumentation. The single-volume Handbook concept, although providing a high level of convenience, was not adequate for the reference requirements of many of the newer scientific disciplines. Due to the sheer quantity of useful and reliable data being generated, it was no longer feasible or desirable to select only that information which could be contained in a single volume and arbitrarily to reject the remainder. **Comprehensiveness** had become as essential as **convenience**.

By the late 1960's, it was apparent that the solution to the problem was the development of the multi-volume Handbook. This answer arose out of necessity during the editorial processing of the *Handbook of Environmental Control*. A hybrid discipline or, to be more precise, an interdisciplinary field such as Environmental Science could be logically structured into major subject areas. This permitted individual volumes to be developed for each major subject. The individual volumes, published either simultaneously or by some predetermined sequence, collectively became a multi-volume Handbook series.

The logic of this new approach was irrefutable and the concept was promptly accepted by both the scientist and science librarian. It became the format of a growing number of CRC Handbook Series in fields such as Materials Science, Laboratory Animal Science, and Marine Science.

Within a few years, however, it was clear that even the multi-volume Handbook concept was not sufficient. It was necessary to create an information structure more compatible with the dynamic character of scientific information, and flexible enough to accommodate continuous but unpredictable growth, regardless of quantity or direction. This became the objective of a "third generation" Handbook concept.

This latest concept utilizes each major subject within an information field as a "Section" rather than the equivalent of a single volume. Each Section, therefore, may include as many volumes as the quantity and quality of available information will justify. The structure achieves permanent flexibility because it can, in effect, expand "vertically" and "horizontally." Any section can continue to grow (vertically) in number of volumes, and new sections can be added (horizontally) as and when required by the information field itself. A key innovation which makes this massive and complex information base almost as convenient to use as a single-volume Handbook is the utilization of computer technology to produce up-dated, cumulative index volumes.

The *Handbook Series in Nutrition and Food* is a notable example of the "sectionalized, multi-volume Handbook series." Currently underway are additional information programs based on the same organizational design. These include information fields such as Energy and Agricultural Science which are of critical importance not only to scientific progress but to the advancement of the total quality of life.

We are confident that the "third generation" CRC Handbook comprises a worthy contribution to both information science and the scientific community. We are equally certain that it does not represent the ultimate reference source. We predict that the most dramatic progress in the management of scientific information remains to be achieved.

B. J. Starkoff
President
CRC Press, Inc.

PREFACE
CRC HANDBOOK SERIES IN NUTRITION AND FOOD

Nutrition means different things to different people, and no other field of endeavor crosses the boundaries of so many different disciplines and abounds with such diverse dimensions. The growth of the field of nutrition, particularly in the last two decades, has been phenomenal, the nutritional data being scattered literally in thousands and thousands of not always accessible periodicals and monographs, many of which, furthermore, are not normally identified with nutrition.

To remedy this situation, we have undertaken an ambitious and monumental task of assembling in one publication all the critical data relevant in the field of nutrition.

The *CRC Handbook Series in Nutrition and Food* is intended to serve as a ready reference source of current information on experimental and applied human, animal, microbial, and plant nutrition presented in concise tabular, graphical, or narrative form and indexed for ease of use. It is hoped that this projected open-ended multivolume set will become for the nutritionist what the *CRC Handbook of Chemistry and Physics* has become for the chemist and physicist.

Apart from supplying specific data, the comprehensive, interdisciplinary, and comparative nature of the *CRC Handbook Series in Nutrition and Food* will provide the user with an easy overview of the state of the art, pinpointing the gaps in nutritional knowledge and providing a basis for further research. In addition, the *Handbook* will enable the researcher to analyze the data in various living systems for commonality or basic differences. On the other hand, an applied scientist or technician will be afforded the opportunity of evaluating a given problem and its solutions from the broadest possible point of view, including the aspects of agronomy, crop science, animal husbandry, aquaculture and fisheries, veterinary medicine, clinical medicine, pathology, parasitology, toxicology, pharmacology, therapeutics, dietetics, food science and technology, physiology, zoology, botany, biochemistry, developmental and cell biology, microbiology, sanitation, pest control, economics, marketing, sociology, anthropology, natural resources, ecology, environmental science, population, law, politics, nutritional and food methodology, and others.

To make more facile use of the *Handbook,* the publication has been divided into sections of one or more volumes each. In this manner the particular sections of the *Handbook* can be continuously updated by publishing additional volumes of new data as they become available.

The Editor wishes to thank the numerous contributors, many of whom have undertaken their assignment in pioneering spirit, and the Advisory Board members for their continuous counsel and cooperation. Last but not least, he wishes to express his sincere appreciation to the members of the CRC editorial and production staffs, particularly President Bernard J. Starkoff, Mrs. Kathryn H. Harter, Mr. Robert Datz, Ms. Pamela Dworken, and Mr. Paul R. Gottehrer, for their encouragement and support.

We invite comments and criticism regarding format and selection of subject matter, as well as specific suggestions for new data (and additional contributors) which might be included in subsequent editions. We should also appreciate it if the readers would bring to the attention of the Editor any errors or omissions that might appear in the publication.

Miloslav Rechcigl, Jr.
Editor-in-Chief
August 1977

PREFACE
SECTION G: DIETS, CULTURE MEDIA, FOOD SUPPLEMENTS

This section furnishes information relating to food habits and dietary regimens of various types of animals, including livestock, laboratory, and zoo animals, as well as animals living in the wild. Data are provided on natural as well as semipurified and purified diets, listing individual ingredients, their concentrations, and their sources. In the case of natural ingredients or mixtures, chemical composition is also given when known.

Separate subsections are devoted to media used in cultivation of microorganisms and plants, while others give attention to culture media for cells, tissues, organs, and embryos of animals and plants. Whenever available, information is presented on both the agnotobiotic and the axenic types of media.

Miloslav Rechcigl, Jr.
Editor
August 1977

MILOSLAV RECHCIGL, JR., EDITOR

Miloslav Rechcigl, Jr. is Nutrition Advisor and Director of the Interregional Research Staff in the Agency for International Development, U.S. Department of State.

He has a B.S. in Biochemistry (1954), a Master of Nutritional Science degree (1955), and a Ph.D. in nutrition, biochemistry, and physiology (1958), all from Cornell University. He was formerly a Research Biochemist in the National Cancer Institute, National Institutes of Health and subsequently served as Special Assistant for Nutrition and Health in the Health Services and Mental Health Administration, U.S. Department of Health, Education, and Welfare.

Dr. Rechcigl is a member of some 30 scientific and professional societies, including being a Fellow of the American Association for the Advancement of Science, Fellow of the Washington Academy of Sciences, Fellow of the American Institute of Chemists, and Fellow of the International College of Applied Nutrition. He holds membership in the Cosmos Club, the Honorary Society of Phi Kappa Pi, and the Society of Sigma Xi, and is recipient of numerous honors, including an honorary membership certificate from the International Social Science Honor Society Delta Tau Kappa. In 1969, he was a delegate to the White House Conference on Food, Nutrition, and Health and in the last two years served as President of the District of Columbia Institute of Chemists and a Councilor of the American Institute of Chemists.

His bibliography extends over 100 publications, including contributions to books, articles in periodicals, and monographs in the fields of nutrition, biochemistry, physiology, pathology, enzymology, and molecular biology. Most recently he authored and edited *World Food Problem: A Selective Bibliography of Reviews* (CRC Press, 1975), *Man, Food, and Nutrition: Strategies and Technological Measures for Alleviating the World Food Problem* (CRC Press, 1973), *Food, Nutrition and Health: A Multidisciplinary Treatise Addressed to the Major Nutrition Problems from a World Wide Perspective* (Karger, 1973), following his earlier pioneering treatise on *Enzyme Synthesis and Degradation in Mammalian Systems* (Karger, 1971), and that on *Microbodies and Related Particles. Morphology, Biochemistry and Physiology* (Academic Press, 1969). Dr. Rechcigl also has initiated and edits a new series on Comparative Animal Nutrition and is Associated Editor of *Nutrition Reports International.*

CONTRIBUTORS
SECTION G: DIETS, CULTURE MEDIA, FOOD SUPPLEMENTS
VOLUME III

Vernon Ahmadjian
Department of Biology
Clark University
Worcester, Massachusetts

C. J. Asher
Department of Agriculture
University of Queensland
Brisbane, Australia

Dominick V. Basile
Herbert H. Lehman College
City University of New York
Bronx, New York

E. Y. Bridson
Oxoid Limited
Basingstoke, Hampshire
England

James D. Caponetti
Department of Botany
The University of Tennessee
Knoxville, Tennessee

N. G. Carr
Department of Biochemistry
The University of Liverpool
Liverpool, England

M. R. Droop
Scottish Marine Biological Association
Dunstaffnage Marine Research Laboratory
Oban, Argyll
Scotland

Victor S. C. Fan
Health Science Center
University of Louisville
Louisville, Kentucky

E. A. Freundt
FAO/WHO Collaborating Centre for Animal
Mycoplasmas
Institute of Medical Microbiology
University of Aarhus
Aarhus, Denmark

Robert K. Gerloff
Department of Health, Education, and
Welfare
National Institute of Allergy and Infectious
Diseases
Rocky Mountain Laboratory
Hamilton, Montana

D. Kay
Sir William Dunn School of Pathology
University of Oxford
Oxford, England

L. H. Mattman
Department of Biology
Wayne State University
Detroit, Michigan

David McLaughlin
Department of Zoology
Howard University
Washington, D.C.

Harry H. Murakishi
Department of Botany and Plant
Pathology
Michigan State University
East Lansing, Michigan

J. S. Porterfield
National Institute of Medical Research
Medical Research Council
London, England

John Tuite
Department of Botany and Plant Pathology
Purdue University
West Lafayette, Indiana

M. F. Turner
Scottish Marine Biological Association
Dunstaffnage Marine Research Laboratory
Oban, Argyll
Scotland

To my inspiring teachers at Cornell: Harold H. Williams, John K. Loosli, Richard H. Barnes, the late Clive M. McCay, and the late Leonard A. Maynard.
And to my supportive and beloved family: Eva, Jack, and Karen.

TABLE OF CONTENTS
SECTION G: DIETS, CULTURE MEDIA, FOOD SUPPLEMENTS
VOLUME III

Microorganisms

METHODS FOR THE GROWTH OF BACTERIOPHAGES, CYANOPHAGES, AND MYCOPHAGES

D. Kay

INTRODUCTION

The aim of this chapter is to provide a selection of methods for the growth of bacterial viruses or bacteriophages, often abbreviated as "phage." Also included are methods for cyanophages, specific for the blue-green algae and mycophages active on certain fungi. Bacteriophages were discovered in 1915 by Twort[139] in cultures of staphylococci, and for many years studies were confined to phages active on bacteria of medical interest, e.g., Salmonellae, Shigellae, staphylococci, and *Vibrio cholerae*. In 1945, Demerec and Fano[140] assembled a collection of seven phages active on *Escherichia coli* strain B. These phages were readily grown on this relatively harmless host organism, and during subsequent years the "T" phages, as they were known, became the subject of much of the research into the bacteriophage phenomenon.

In recent years our knowledge of the range of bacteria that possess phages has widened greatly, and phages have also been found for the actinomycetes, the blue-green algae, and some fungi. No attempt has been made in this chapter to prepare a comprehensive list of all the known phages, but attention is drawn to the following publications where large numbers of phages are listed and described in varying degrees of detail. Eisenstark[141] described over 500 phages listed by their host organisms, which were arranged in order according to Bergey's classification (7th edition, 1957). The details given covered the phage designation, morphology, and key references with an indication of the type of information to be found in them. Tikhonenko[142] listed a large number of phages classified according to morphology. Numerous references are given but the bulk of the information is morphological. Frankel-Conrat[143] listed over 300 viruses active on protists in alphabetical and numerical order. The details given for each phage range from comprehensive to negligible, where not even a reference is quoted. Bradley,[144] in a comprehensive review of phage structure, proposed a classification scheme for bacteriophages in which they were divided into six groups, A through F, according to morphology. Bradley's classification was accepted by Topley and Wilson,[145] and it will be used in this chapter.

Methods for growing phages generally resolve themselves into methods for growing the host organism rapidly and exponentially on a sufficient scale, as it is usually found that phage multiplies to the highest titer under these conditions. Occasionally, special conditions are needed for phage growth. These may be particular concentrations of mono- and divalent cations or the presence of the adsorption cofactor, L-tryptophan, for the growth of coliphage T4. Generally, rich complex media allowing rapid host multiplication are chosen for the growth of phages, but defined media enriched with amino acids in the form of hydrolyzed casein and other growth-promoting substances are equally suitable. Where the host organism can grow on a minimal salts medium with glucose (or other carbohydrate), e.g., *E. coli* B, a medium such as M9 (Reference M46) may be used with success for growing phages.

Adams[147] gives a comprehensive account of the methods used for the study of the T phages active on *E. coli* B and which are applicable to phages generally.

All aerobic host organisms show an increased rate of growth when an adequate supply of air is provided. Also, growth will be prolonged to a higher cell density, provided the medium is adequate, when the rate of oxygen solution is maintained to match the demand. Consequently, numerous workers have used stirred aeration apparatus, known as fermentors, to grow the host organisms to high cell densities and obtain very high phage titers. This method can suffer from severe frothing problems when cellular lysis occurs, and steps must be taken to control this.

After lysis has taken place, it is usually necessary to concentrate and purify the phage. A universally applicable method for concentration of phage has been described by Yamamoto and Alberts.[146] This should be followed by density gradient equilibrium centrifugation in CsCl. Phage densities are listed in Table 1.

METHODS FOR THE GROWTH OF BACTERIOPHAGES, CYANOPHAGES, AND MYCOPHAGES

Glossary of Terms

Aeration — The supply of oxygen to liquid cultures of microorganisms. Usually done by bubbling sterile air into the culture but also by shaking the culture on a reciprocating or gyratory shaker. Stirred culture vessels into which air is blown, called fermentors, are used when the highest degree of aeration is required.

Host — The organism in which the virus multiplies. As phages are usually species- and often strain-specific, the host is described precisely by its generic name and its code number, e.g., ATCC or NTCC number or its laboratory code number.

Induction — The treatment of a lysogenic strain of bacteria with ultraviolet light or mitomycin C or another agent in order to permit the normally repressed phage genome to replicate and give rise to progeny phage in all the cells of the culture.

Lysis — The dissolution of the host cells at the end of phage multiplication accompanying the release of progeny phage.

Lyzate — A host culture whose cells have suffered lysis. Contains bacterial debris and unlysed cells in addition to phage.

Medium — An aqueous solution of inorganic salts and organic compounds, usually buffered to a suitable pH value, which supports the growth of a microorganism. Complex media are made from enzymic digests of meat and extracts of yeast. Their composition is not precisely known. Defined media are composed of substances of known chemical composition.

Sterilization — Liquid media must be heated to 125°C for 30 min in an autoclave and dry apparatus to 165°C for 30 min to kill all bacteria, spores, and viruses.

Temperate phage — Also known as lysogenic phage. A phage that can be carried indefinitely in a strain of bacteria in a form known as a prophage, in which the phage genome is integrated into and multiplies concomitantly with the host chromosome.

Lysis of Host Organism in Liquid Media

The quantity of phage required should be carefully considered in order to determine the volume of liquid culture to be grown. The size and mass of the phage particle can be estimated from the data in Table 1,* as can the amount of nucleic acid. A small culture of, say, 10 mℓ, should be grown, and from its titer the volume of culture required can be estimated. Where the phage is not mentioned in the table, the value for nucleic acid content may be estimated by comparison with a phage of similar size.

Where the volume is small (10 to 100 mℓ), the culture can be grown in a conical flask of ten times greater volume. It can be aerated by shaking on a reciprocating or gyratory shaker. In the latter case, special flasks with three to four indentations in the walls should be used; these break up the fluid and greatly increase the aeration. They are very useful for growing mycelial organisms.

*Tables are found at the end of the text.

Volumes from 100 ml to one liter may be grown in flasks or bottles fitted with aeration tubes furnished with fritted glass discs passed into the vessels through rubber bungs in their necks. The air should be freed from oil droplets and airborne organisms by filtration through cotton wool, making sure that the filters are large enough for the purpose. It is also important to filter the effluent air from the cultures to prevent release of phage into the atmosphere. Some phages are resistant to drying in air and remain infective for a long period. The airborne phage particles can cause spontaneous lysis of cultures. If the host organism is pathogenic, special precautions must be taken to ensure safety.

When volumes greater than one liter are needed, it is necessary either to prepare several small cultures or to use a fermentor in which volumes up to 10 l or more can be grown. These devices are commercially available and are fitted with air inlets and outlets, baffled stirring fitments, and controls for the measurement and adjustment of temperature, pH, and the addition of antifoaming agents.

After deciding the volume of the phage lysate required and selecting a suitable apparatus, an appropriate culture medium should be prepared, sterilized, and loaded into the apparatus. Usually the medium can be sterilized in the culture apparatus but if this is not convenient it must be loaded aseptically into the sterile apparatus. Media suitable for the phages listed in Table 1 are given in Table 2. The temperature of the culture must be maintained at the correct level (see Table 1) by setting up the apparatus in a controlled temperature room or incubator or in a thermostatically controlled water bath.

Two variables that must be considered are the cell density at the time of infection and the multiplicity of infection (MOI). Generally, the MOI should be about five phage particles per bacterium, but with rapidly proliferating phages this can be lowered to as little as 0.01. The cell density at which the infecting phage is added to the culture should be decided by experiment. If the phage is added at too high a cell density, complete lysis may not occur and the progeny phage titer will be low. On the other hand, if the phage is added at too low a cell density, lysis may occur before the optimum cell density has been reached, again resulting in a lower progeny titer.

When all the conditions have been decided on, the procedure for growing phage is as follows: A culture of the host organism, about one tenth the volume of the main culture, is prepared by growth in the same medium to a density of about 5×10^8 cells per milliliter. This is used to inoculate the main culture, which should be prewarmed to the growing temperature. The culture is aerated by whatever means have been chosen, and the growth of the culture monitored by optical density measurements. When the appropriate cell density has been reached, the phage inoculum should be added to give the chosen MOI. Incubation with aeration should be continued until lysis occurs. The vast majority of phages cause lysis, but some, e.g., the filamentous phages fd, f1, ZJ/2, and Xf, are released without cellular lysis. Certain mycoviruses have to be released from the mycelium by grinding or homogenization.

The culture should be stopped when lysis is complete and stored at 4°C overnight for processing or, if convenient, may be processed immediately.

Where cultures are aerated by bubbling air in, trouble may be experienced from foaming when lysis takes place. This can be controlled by the addition of an antifoam agent, e.g., Dow Corning® Antifoam B, or can be minimized by reducing the air flow. It is advisable to use a culture vessel at least twice the volume of the culture to contain the foam, which is unavoidable.

Sometimes phage-infected bacteria do not lyse readily and the progeny phage are not released. Lysis can be effected by the addition of about 1% of chloroform, followed by gentle shaking. This procedure is unsuitable for the filamentous phages that are degraded by chloroform and of course for the lipid-containing phages.

Lysis of Host Organism on Solid Media

This method is very suitable to prepare small amounts of high-titer phage. The culture medium is solidified by the addition of 1.5% of agar. A bacterial culture is prepared, mixed with phage, and spread onto the surface of the solid media. The proportion of phage to bacteria has to be carefully gauged so that all the cells are lysed. Another method is to mix bacteria and phage with sloppy nutrient agar (about 0.6%) and pour this onto the surface of plates containing firm (1.5%) nutrient agar. After incubation and complete lysis the phage is recovered from the plate by extraction with a small volume of medium or buffer for a period ranging from a few minutes to overnight.

Methods for Growing Temperate Phages

Lysis of lysogenic bacterial cultures can be brought about by induction with ultraviolet light or with mitomycin C. A typical example is that given in Reference 50. A strain of *E. coli* lysogenic for phage lambda was grown in medium M20 up to 2×10^8 cells per milliliter. The cells were centrifuged and resuspended in half the volume of UV buffer (Reference M47). The suspension was irradiated for 30 sec in a layer 1 to 2 mm deep in a dish 60 cm below a UV lamp. The energy of the lamp was such that after 14 sec irradiation there were 10^{-1} survivors in a suspension of phage $T4r^+$ placed at the same distance. After irradiation, the bacteria were diluted with an equal volume of double-strength medium M20 plus $10^{-3} M$ $MgSO_4$ and incubated at $37°C$ for 50 min without aeration followed by incubation with aeration until lysis occurred. All procedures after irradiation were carried out in the dark or under a red safe lamp to avoid photoreactivation.

In another procedure suitable for large volumes of culture,[44] a lysogenic strain of *E. coli* was grown in M19 to 5×10^8 per milliliter and then cooled by adding one-third volume of ice. The culture was allowed to flow at 0.16 l/min through a quartz tube 2 mm in internal diameter placed 3 cm from a 15-W UV lamp. Seventy percent of the cells were induced. Yeast extract (3 g) and tryptone (4 g) were then added and aeration at $37°C$ continued until lysis occurred. Chloroform (1 ml/l) and DNAse (1 μg/ml) was added and the culture stored at $37°C$ for 1 hr. A titer of 1 to 2×10^8 per milliliter was obtained.

Growth of Isotopically Labeled Phage

In order to label phage efficiently with an isotope or a particular element, whether radioactive or not, it is necessary to grow the host organism in a medium that contains that element (or compound) in a form that can be utilized by the organism in an amount just sufficient for growth to the required density. In the case of a nonradioactive isotope, e.g., ^{15}N, the element in the form of $^{15}NH_4Cl$ can be substituted for the normal $^{14}NH_4Cl$ present in the medium (e.g., M46). In the case of radioactive elements, complete substitution is rarely required, and all that is necessary is to reduce the nonradioactive element present in the medium to as low a level as practicable for the growth of the host. This usually means reducing the inorganic phosphorus when labeling with ^{32}P, the sulfate when labeling with ^{35}S, or the amino acids when using either ^{14}C- or ^3H-amino acids. The use of thymine-requiring strains of bacteria is advantageous when labeling the nucleic acid of phage. Tritiated thymidine is the label of choice.

A medium low in phosphorus precludes the use of phosphate buffer, so recourse is usually made to tris(hydroxymethyl)ammonium chloride (Tris) as a buffer, which is effective at neutral pH and compatible with the growth of bacteria. The procedure is to grow the host bacteria in a defined medium containing the minimum amount of the unlabeled element or compound to be used up to the required cell density. The phage inoculum is then added, followed either immediately or after 10 min delay by the isotope or labeled compound. Lysis should occur, but the titer is usually not so high as with a complex medium.

Table 1
HOST ORGANISMS, MORPHOLOGY, NUCLEIC ACID COMPOSITION, AND CULTURAL CONDITIONS FOR THE GROWTH OF BACTERIOPHAGES, CYANOPHAGES, AND MYCOPHAGES

Key

Column 1 — Host organism and ATCC or other number

Column 2 — Phage designation, laboratory number (NN = no number given). "Temp" indicates temperate phage and usually means that it has been grown by induction; def: defective particle. Second entry: density of phage particle in cesium chloride (g/cm^3). Third entry: particle weight in daltons $\times$ 10^{-6}.

Column 3 — First entry: Bradley classification[144] A through F (Env = envelope particle). Second entry: head dimensions in nm, length $\times$ width; single figure means isometric capsid. Third entry: tail dimensions in nm, length $\times$ width or length alone.

Column 4 — First entry: 2-DNA = duplex DNA; 1-DNA = single-strand DNA; 2-RNA = double-stranded RNA; 1-RNA = single-stranded RNA. Second entry: percentage w/w of nucleic acid in phage particle. Third entry: buoyant density of phage nucleic acid in CsCl.

Column 5 — First entry: contour length of phage nucleic acid in μm. Second entry: percentage of guanine plus cytosine. Third entry: molecular weight of phage nucleic acid in daltons $\times$ 10^{-6}.

Column 6 — Medium (see Table 2) in which the phage was grown in reference in Column 2.

Column 7 — First entry: temperature of incubation in °C. Second entry: aeration method (S, shaken; G, gyratory; Ae, bubbled air; F, fermentor; P, surface plate culture). Third entry: time for lysis (O/N = culture incubated overnight).

Column 8 — Special notes or extra information.

All numbers in parentheses are references. References to media appear as (Mxx). A dash indicates that no information was available for that entry or was not applicable.

Table 1 (continued)
HOST ORGANISMS, MORPHOLOGY, NUCLEIC ACID COMPOSITION, AND CULTURAL CONDITIONS FOR THE GROWTH OF BACTERIOPHAGES, CYANOPHAGES, AND MYCOPHAGES

1	2	3	4	5	6	7	8
Agrobacterium radiobacter 8149	P8149 temp (1)	C 50	— —	— —	YEB (M1)	28 G 8 hr	(2)
tumefaciens B2A	PB2A temp (1)	B 80 × 71 280 × 18	2-DNA —	22 —	YEB (M1)	28 G 8 hr	(2)
B6	PB6(ω) temp (1)	B as PB2A	2-DNA —	22 —	YEB (M1)	28 G 8 hr	(2)
925	P925 def (1)	A 67 × 61 150 × 21	— —	— —	YEB (M1)	28 G 8 hr	(2)
Ancalomicrobium adetum No. 18	Sp (3)	B 58 104–170	— —	— —	MMB (M2)	21	
Aspergillus foetidus IMI 41871	AfV-s (4)	C 40–42	2-RNA —	—	CS (M3)	27 F, S 48 hr	
	AfV-f (4)	C 40–42	2-RNA (5)	—	(M3)	27	
	1.351–1.381		—	—		F, S 48 hr	
Bacillus anthracis CN 18-74 *cur*	AP 50 1.307 1.198 (sucrose) (7)	E 80 (6)	RNA	—	YEP (M4)	37	(7a)

Table 1 (continued)

HOST ORGANISMS, MORPHOLOGY, NUCLEIC ACID COMPOSITION, AND CULTURAL CONDITIONS FOR THE GROWTH OF BACTERIOPHAGES, CYANOPHAGES, AND MYCOPHAGES

1	2	3	4	5	6	7	8
cereus	CP51	B	2-DNA	–	NBY	37	(9)
569	1.527	90	56.7	43	(M5)	P	
	102	160 × 20	1.746	57.9		20 hr	
	(8)						
	CP53	B	2-DNA	–	NBY	37	(10)
	1.486	66	51.9	37	(M5)	S	
	28.8	276 × 13	1.697	16.5			
	(8)						
megaterium	CS-1	B	2-DNA	36–38.5	NB	–	(12)
ATCC 19213	1.49	55	–	41.7	(M6)		
	(11)	201 × 10		13.9			
stearothermophilus	TP-84	B	2-DNA	42	T	62–65	(14)
10	1.508	53	1.705	22.6	(M7)	G	
	50	130 × 5		(14a)			
	(13)						
ATCC 8005	Tφ3	B	2-DNA	12.1	TYNGC	60	(16a)
	1.526	57	–	44	(M8)	S	
	(15)	125	1.702	23.2		Ae	
			(16)				
subtilis	PBS1	A	2-DNA	82	NB+	37	(19)
A26	1.426	120	–	28	(M9)		
		200	1.722	190			
		(17a)	(18)				
168	temp				PEN	37	(20)
	(17)				(M10)	S	
H	φ29	B	2-DNA	5.7–5.9	PB	37	(22)
	–	41.5 × 31.5	–	34(23)	(M11)	S	
	(21)	32.5	1.654	10.9–11.3			
3610	φe		2-DNA	–	LB	37	(25)
(ATCC 6051)	(24)		–	–	(M12)	S	
S Hg W	AR9	A	–	–	–	37	(27)
	(26)	150 × 270	–				
		290	–				

Table 1 (continued)
HOST ORGANISMS, MORPHOLOGY, NUCLEIC ACID COMPOSITION, AND CULTURAL CONDITIONS FOR THE GROWTH OF BACTERIOPHAGES, CYANOPHAGES, AND MYCOPHAGES

1	2	3	4	5	6	7	8
Caulobacter crescentus	φCbK	B	2-DNA	—	PYE	30	(29)
CB13B1a	1.53	64 × 195	57	62.3	(M13)	G	
	(28)	275				7—12 hr	
CB15	φCb13	B	DNA	—	PYE	P	(31)
	(30)	50 × 170	—	—	(M13)		
		250					
CB13B1a	φCd1	C	2-DNA	—	PYE	30	(33)
	1.51	60		61	(M14)	G	
	(32)			29			
CB1a	φ101	B	2-DNA	—	PYE	65—120 min	(35)
	temp	10—12	1.721		(M15)	30	
	(34)	55	2-DNA			24 hr	
		117	—				
Caulobacter fusiformis 27	φCv-23r	E	RNA	—	PYE	P	(31)
	(30)	21—23			(M13)		
vibrioides CV6	φ6	B	DNA	—	PYE	30	(37)
	(36)	—	—		(M13)		
		—	—				
Clostridium tetani	NN	B	—	—	FTM	37	(40)
ATCC 10779		65			(M16)	Ae	
	(38)	100 × 5				7—8 hr	
ATCC 453	NN	A	—	—	FTM	37	(40)
	(38)	65			(M16)	Ae	
		100 × 10				7—8 hr	
		(39)					
Diplococcus pneumoniae R36A	Dp1	B	2-DNA	—	C		
	(41)	50	—		(M17)		
		100					

Table 1 (continued)

HOST ORGANISMS, MORPHOLOGY, NUCLEIC ACID COMPOSITION, AND CULTURAL CONDITIONS FOR THE GROWTH OF BACTERIOPHAGES, CYANOPHAGES, AND MYCOPHAGES

1	2	3	4	5	6	7	8
Escherichia coli K12, Hfr (3300/1)	fd 1.29 (42) 14.6 ± 0.6 (43)	F 900 × 6	1-DNA 11.3 1.716	1.25 ± 5% — 1.89 ± 0.07	Tryptone (M18)	38 Ae 6 hr	
W3104	λ (44) temp 1.50 (45) 57 (47)	B 65 157 (46)	2-DNA 60 1.7093	17.2 49 33–36 (48)	(M19)	37 Ae	(49) (49a)
W3110 (λ)	λ (50) temp		2-DNA		Tryptone (M20)	37 S	(51)
K12F+ (112-12)	M13 (52)	F See fd	See fd 1.7224 (23)		3XD (M21)	37 S 6–10 hr	(53)
159F+thy−	M13 (52a)	See fd	See fd		M9+ (M22)	37 S	(53)
K10 (Hfr)	MS-2 1.422 3.6 (54)	E 25	1-RNA — 1.626 (55a)	— 1.1	L (M23)	37 Ae 6–8 hr	(55)
C600	Mu-1 (56) 1.454 (57)	B 49 115 (61)	2-DNA — 1.706	12.9 ± 0.07 — 22–23	T broth (M24)	37 G 2 hr	(58)
C (BTCC 122)	φX-174 1.40 6.2 (59)	D 24.3–25.2 (60) 22.5 (61)	1-DNA 25.5 1.72	1.77 ± 0.13 (61a) 42.6 1.7	3XD (M21)	37 Ae 45 min	(62)

Table 1 (continued)

HOST ORGANISMS, MORPHOLOGY, NUCLEIC ACID COMPOSITION,
AND CULTURAL CONDITIONS FOR THE GROWTH
OF BACTERIOPHAGES, CYANOPHAGES, AND MYCOPHAGES

1	2	3	4	5	6	7	8
K10 (Hfr)	Qβ 1.439 4.2 (54)	E 25	1-RNA ≃30 —	— — —	L medium (M23)	37 Ae 6–8 hr	(64)
K12 (Hfr1)	R17 1.46 3.6 ± 0.3 (65)	E 25	1-RNA 31.7 1.630 (65a)	1.07 — 1.1	(M26)	37 Ae 3 hr	(66)
B	T1 (67)	B 56 170 (68)	2-DNA — 1.705 (68a)	15.9 ± 0.3 48 31–32	TRIS (M27)	37 Ae 45–60 min	(67a) —
B	T2r⁺ ≃1.45 (69) (69a)	A (70)	2-DNA ≃60 1.702 (72)	56 ± 2 (76) 33 130 (72)	3XD (M21)	37 Ae	(73) (74) (75)
B	T4 (69a)	A 115 × 85 95 × 18 (71)	2-DNA ≃60 1.7005 (72)	— 34.4	3XD (M21)	37 Ae	(73) (73a)
B	T3 — 49 (77)	C	2-DNA — 1.711	11.93 ± 0.47 (78) 50 25	Peptone (M28)	37 Ae 1 hr	(79)
F	T5 (80) 1.539 (81, 82)	B 75 180	2-DNA — 1.701 (83)	38.3 ± 0.9 (84) 39 (84a) 76, 82.5 (85, 84)	NB	37 Ae	(83)
B	T7 1.50 38 (86)	C 62.6 20	2-DNA 50 1.710	12.2 48 26	3XD (M21)	37 Ae	(87) (88)

Table 1 (continued)
HOST ORGANISMS, MORPHOLOGY, NUCLEIC ACID COMPOSITION, AND CULTURAL CONDITIONS FOR THE GROWTH OF BACTERIOPHAGES, CYANOPHAGES, AND MYCOPHAGES

1	2	3	4	5	6	7	8
Capsulated strains of *E. coli* and *Klebsiella* 28-1, 28-2, 38, 29, 31 and *Klebsiella* 7	28-1	A 87.5 × 56 95	—	—	P (M40)	37	
	7	B 56 × 50 168	—	—	P (M40)	37	
	26 1.45–1.51 (89)	C 65 × 61 (90)	—	—	P (M40)	37	
Lactobacillus casei ATCC 27092	PL-1 (91)	B	—	—	MR (M29)	37	
Mycobacterium phlei	NN 1.51 (92)	B 63 158	2-DNA 34 –	—	Sauton (M30)	37 S O/N	(93)
butyricum	NN 1.51 (92)	B 63 213	2-DNA 42 –	—	Sauton (M30)	37 S O/N	(94)
Mycoplasma Acholeplasma laidlawii BN1	MV-L1 (95, 98)	F 90 × 16	DNA –	—	GS (M31)	37 24 hr	
M1305/68	MV-L2 (96)	E 80	DNA –	—	GS (M31)	37 P	
M1305/68	MV-L3 1.477 (97)	C 57 × 61 25	2-DNA 35.2 –	—	GS (M31)	37 48 hr	

Table 1 (continued)

HOST ORGANISMS, MORPHOLOGY, NUCLEIC ACID COMPOSITION, AND CULTURAL CONDITIONS FOR THE GROWTH OF BACTERIOPHAGES, CYANOPHAGES, AND MYCOPHAGES

1	2	3	4	5	6	7	8
Myxococcus xanthus, A *fulvus*, M *virescens*, V2	MX-1 1.531 (99)	A 75 100 (100)	2-DNA — —	— 55 —	CT (M32)	30	(101) (102)
Nocardia canicruria 3, 57	φC (103)	B 52 ± 2 192 ± 8 × 10	— — —	— — —	PY (M33)	30	
	φEC (103)	B 52 197 × 10	— —	— —	PY (M33)	30	
Nostoc muscorum	N-1 1.498 (104)	A 55 110	2-DNA — 1.696	17.2 37 38	Chu 10 (M34)		(105)
Penicillium stoloniferum ATCC 14586 NRRL 5267	PsV-s 1.36 (106) 5.97 PsV-f (106, 109)	E 30–33 E 31–34	2-RNA 15.5 2-RNA — —	— 0.94 and 1.11 — (108)	Corn steep (M35)	27 F, S 2–5 days	(107)
Plectonema boryanum IU 594	LPP-1 1.48 51 ± 3 (110, 112)	C 60 20	2-DNA — 1.714	13.2 55 27	Chu 10 (M34)	RT Ae 2 days	(111)
Pseudomonas aeruginosa PA0D	B33 temp 1.473 (113, 114)	B 50 135 × 15	2-DNA — 1.718	11.8 60 24.5	NA (M36)	P	

Table 1 (continued)
HOST ORGANISMS, MORPHOLOGY, NUCLEIC ACID COMPOSITION, AND CULTURAL CONDITIONS FOR THE GROWTH OF BACTERIOPHAGES, CYANOPHAGES, AND MYCOPHAGES

1	2	3	4	5	6	7	8
phaseolicola HB10Y	φ6 1.26 (115, 116)	Env 60–70	2-RNA 13	–	SSM (M37)	26 F 7 hr	
Unnamed BAL-31	PM2 1.28 (118, 119)	Env 61.4	2-DNA 14 –	(117) – 15	AMS (M38)	28 Ae 3 hr	(120)
Salmonella abortus equii NCTC 5727	x 1.48 90 (121)	B 65 220 × 14	2-DNA 46 1.715	21.7 57 41.6	CAAG (M39)	36 S	(122) (123)
typhi Ci 23	ViIII 1.485 (124)	C 47 (125)	– –	– –	P (M40)	37 F 30 min	
typhimurim	P22 (126)	C	2-DNA	15	M9CAA	37	
LT2(P22Tc-10)	1.45 (127)	60 (128)	– 1.710	50 28 ± 0.9	(M41)	(129) Ae	
LT2	temp P22 (130)				(M42)	37 Ae 40 min	
Staphylococcus aureus 8510	3A, 7 1.460 (131, 132)	B 80 × 60 300 (133)	–	–	CY (M43)	35 P	
8511	29, 53 1.502 (131, 134)	B	–	–	CY (M43)	35 P	
Xanthomonas oryzae	Xf 1.272 (135)	F 976 × 8.4	1-DNA	–	PS (M44)	28	136
Vibrio cholerae NIH 41	VA-1 temp (137)	A 58 × 50 87 × 26	–	–	L broth (M45)		138

Table 2
CULTURE MEDIA USED FOR THE GROWTH OF THE BACTERIOPHAGES
LISTED IN TABLE 1

M1: YEB. Reference 1. Peptone (Difco®), 0.5%; yeast extract, 0.1%; beef extract, 0.5%; sucrose, 0.5%; $MgSO_4$, 0.002 M; pH 7.6.

M2: MMB. Reference 3. Peptone, 0.015%; yeast extract, 0.015%; ammonium sulfate, 0.025%; glucose, 0.1%; plus mineral salts (Van Ert, M. and Staley, J. T., *J. Bacteriol.*, 108, 236—240, 1971.

M3: CS. References 4 and 5. Glucose, 2%; peptone, 0.1%; corn steep liquor, 5%; yeast autolysate, 0.1%; K_2HPO_4, 0.5%; pH 7.0.

M4: YEP. Yeast extract-peptone (Csiszar, K. and Ivanovics, G., *Acta Microbiol. Acad. Sci. Hung.*, 12, 73—89, 1965).

M5: NBY. Nutrient broth (Difco), 0.8%; yeast extract, 0.3% (Thorne, C. B., *J. Virol.*, 2, 657—662, 1968).

M6: NB. Standard Difco nutrient broth.

M7: T broth. Reference 13. Trypticase, 20 g; yeast extract, 3 g; NaCl, 8.5 g; $CaCl_2 \cdot 2H_2O$, 1.47 g; $FeCl_3 \cdot 6H_2O$, 7 mg; $MgSO_4 \cdot 7H_2O$, 15 mg; $MnCl_2 \cdot 4H_2O$, 1 mg; water, one liter; pH 7.4.

M8: Reference 15. TYNGC. Difco tryptone, 10 g; Difco yeast extract, 5 g; NaCl, 10 g; water, one liter. After autoclaving add glucose, 1 g, and $CaCl_2$, 2×10^{-3} M.

M9: Difco nutrient broth plus uracil at 5 μg/ml.

M10: Difco Penassay broth.

M11: Difco Antibiotic Medium No. 3.

M12: LB broth. Bacto® tryptone, 1%; yeast extract, 0.5%; NaCl, 0.5%; pH 7.0 (Hall, D. W., *Proc. Natl. Acad. Sci. U.S.A.*, 58, 584—591, 1967.

M13: PYE. Reference 27. Peptone, 0.2%; yeast extract, 0.1%; $MgSO_4 \cdot 7H_2O$, 0.02% made up in tap water.

M14: M13 made up in deionized water.

M15: M13 made up in distilled water plus 1% Hutner's salts (Cohen-Bazire, G., Sistron, W. R., and Stanier, R. Y., *J. Cell. Comp. Physiol.*, 49, 24—68, 1957).

M16: Reference 38. Difco fluid thioglycollate medium.

M18: Reference 42. Difco Bacto tryptone, 10 g; Difco yeast extract, 1 g; NaCl, 8 g; glucose, 1 g; Tris, 0.5 g; water, one liter; plus 2 ml sterile M-$CaCl_2$; pH 7.4.

M19: Reference 44. $NaH_2PO_4 \cdot H_2O$, 4.6 g; K_2HPO_4, 11.6 g; NH_4Cl, 1.73 g; NaCl, 4.6 g; water, one liter; plus glucose, 2 g, and $MgSO_4 \cdot 7H_2O$, 0.25 g autoclaved separately.

M20: Tryptone. Difco tryptone, 10 g; NaCl, 5 g; water, one liter (Sollar, A., Levine, L., and Epstein, H. T., *Virology*, 26, 708—714, 1965).

M21: 3XD. Na_2HPO_4, 10.5 g; KH_2PO_4, 4.5 g; NH_4Cl, 1.0 g; $MgSO_4 \cdot 7H_2O$, 0.3 g; M-$CaCl_2$, 0.3 ml; gelatin, 1%, 1.0 ml; casamino acids, 15 g; glycerol, 30 g; water, one liter. Add the $CaCl_2$ separately (Fraser, D. and Jerrell, E. A., *J. Biol. Chem.*, 205, 291—295, 1953).

M22: Reference 52a. M9+CAA. Na_2HPO_4, 7 g; KH_2PO_4, 3 g; NH_4Cl, 1 g; NaCl, 0.5 g; water, one liter. Autoclave and add M-$MgSO_4$, 1 ml; 0.01 M-$FeCl_3$, 0.3 ml; glucose, 0.2%; and vitamin-free casamino acids, 0.5%. Thymidine at 2—40 μg/ml was added for the growth of *E. coli* 159 F$^+$ (thy$^-$).

M23: Reference 54. L medium. Glucose, 1.0 g; Difco tryptone, 10.0 g; yeast extract, 5 g; NaCl, 10 g; water, one liter. Add 1 ml of 2M-$CaCl_2$ per liter.

M24: Reference 56. T broth. Difco casamino acids, 10 g; Tris, 2.42 g; NaCl, 1 g; KCl, 0.5 g; L-tryptophan, 20 mg; pH 7.4. Autoclave, then add separately $MgSO_4$ to 4 mM; geatin, 0.01%; glucose, 0.5%; thiamine, 0.2 μg/ml; and at time of infection with phage $CaCl_2$ to 0.6 mM.

M26: Reference 65. Difco casamino acids, 10 g; NaCl, 5 g; NH_4Cl, 1 g; glycerol, 4 ml; Na_2HPO_4, 0.1 mmol; $FeCl_3$, 10 μmol; $MgSO_4$, 1 mmol; Tris, 1.21 g. Adjust pH to 7.4. Add $CaCl_2$, 2 mmol.

M27: Glucose (sterilized separately), 2.0 g; NaCl, 5.4 g; KCl, 3.0 g; NH_4Cl, 1.1 g; $CaCl_2$ (hydrated), 0.011 g; $MgCl_2$, 0.095 g; Tris, 12.1 g; KH_2PO_4, 0.087 g; Na_2SO_4, 0.023 g; $FeCl_3$, 0.00016 g. Adjust pH to 7.4 and volume to one liter (Hershey, A. D., *Virology*, 1, 108—127, 1955).

M28: Reference 77. Proteose peptone, 1.5 g; liver digest, 0.25 g; Oxoid yeast extract, 0.4 g; NaCl, 0.5 g; water 100 ml.

M29: Reference 91. Mr. Pancreatic casein digest, 10 g; glucose, 10 g; sodium acetate, 10 g; yeast extract, 3 g; beef extract, 3 g; NaCl, 1 g; $MgSO_4 \cdot 7H_2O$, 0.2 g; $MnSO_4 \cdot H_2O$, 0.01 g; $FeSO_4 \cdot 7H_2O$, 0.001 g; water, one liter, pH 6.0.

M30: Reference 92. Sauton medium. Asparagine, 4 g; $MgSO_4 \cdot 7H_2O$, 0.5 g; ferric ammonium sulfate, 0.5 g; KH_2PO_4, 0.5 g; glycerol, 6%; water to one liter. Adjust pH to 7.4 with NaOH. Add 0.001 M $CaCl_2$.

M31: GS. Difco PPLO broth plus 10% fetal calf serum. Solidified with 0.5% agarose as needed (Gourlay, R. N. and Wyld, S. G., *J. Gen. Virol.*, 14, 15—23, 1972).

M32: Reference 99. CT medium. Difco Bacto casitone, 1%; yeast extract, 0.5%.

M33: Reference 103. PY medium. Peptone, 0.5%; yeast extract, 0.3%; $Ca(NO_3)_2$, 0.01 M.

M34: Chu No 10. $Ca(NO_3)_2 \cdot 4H_2O$, 0.232 g; K_2HPO_4, 0.01 g; $MgSO_4 \cdot 7H_2O$, 0.025 g; Na_2CO_3, 0.02 g; ferric citrate, 0.0035 g; citric acid, 0.035 g; water, one liter plus; 1 ml/l of Hoagland's micronutrients (described in *California Agric. Exp. Stn. Circ.*, No. 347, 1950).

Table 2 (continued)
**CULTURE MEDIA USED FOR THE GROWTH OF THE BACTERIOPHAGES
LISTED IN TABLE 1**

M35:   Corn steep medium. U.S. Patent No. 3,108,047.

M36:   Reference 113. Nutrient agar. Difco Nutrient Broth, 1.6%; NaCl, 0.5%. Solidified with agar, 1%.

M37:   Reference 115. SSM medium. $MgSO_4 \cdot 7H_2O$, 0.2 g; Na_2HPO_4, 6.0 g; KH_2PO_4, 3.0 g; water, 850 ml. Add 50 ml glucose (10%) and 100 ml of a 20% solution of acid-hydrolyzed casein, sterilized by filtration. Casein from Nutritional Biochemicals Corporation gave better phage stability than Difco casein hydrolysate.

M38:   Reference 118. AMS medium. Nutrient broth, 8 g; KCl, 0.7 g; $CaCl_2 \cdot 2H_2O$, 1.5 g; $MgSO_4 \cdot 7H_2O$, 12 g; NaCl, 26 g; water, one liter.

M39:   Reference 121. CAAG. Potassium phosphate buffer, 0.1 M, pH 7.0; ammonium sulfate, 0.015 M; magnesium sulfate, 0.001 M; ferric sulfate, $10^{-6}M$; Difco casamino acids, 1%; glycerol, 0.2%; thiamine, 1 μg/ml.

M40:   P medium. Solution A – 25 g glucose in one liter; Solution B – Difco casamino acids, 6.25 g; L-tryptophan, 0.4 g; L-cysteine, 0.3 g; KH_2PO_4, 2.5 g; Na_2HPO_4, 15.6 g; NH_4Cl, 1.3 g; gelatin, 0.01 g in one liter. Solution C – $MgSO_4 \cdot 7H_2O$, 5 g; $FeSO_4$, 0.01 g in 100 ml. Solution D – $CaCl_2$, 0.5 g in 10 ml. Solutions A, B, C, and D sterilized separately and added in proportion 20:80:1:0.1. pH 7.2.

M41:   M9CAA. KH_2PO_4, 0.022 M; Na_2HPO_4, 0.042 M; NH_4Cl, 0.018 M; $MgSO_4$, 2.5×10^{-3} M; NaCl, 8.5×10^{-3} M; glucose, 0.2% Difco casamino acids decolorized with Norit$^\circledR$ A, 1.5% (Smith, H. O. and Levine, M., *Proc. Natl. Acad. Sci. U.S.A.*, 52, 356–363, 1964).

M42:   Tris buffer, 5×10^{-2} M, pH 7.2; NaCl, 8.5×10^{-3} M; NH_4Cl, 10^{-3} M; $MgCl_2$, 10^{-3} M; $CaCl_2$, 10^{-4} M; glucose, 0.1%; Difco casamino acids, 0.15%; Difco Bacto peptone, 0.05% (Rhoades, M., McHattie, L. A., and Thomas, C. A., *J. Mol. Biol.*, 37, 21–40, 1968).

M43:   CY medium. β-Glycerophosphate, 0.06 M; $MgSO_4$, 0.001 M; $CaCl_2$, 0.004 M (added after autoclaving); trace metals, 0.02 ml/l of $CuSO_4$, $ZnSO_4$, $FeSO_4$, 0.5%, $MnCl_2$, 0.2% in 10% HCl; yeast extract, 0.3%; acid-hydrolyzed casein, 0.3% in deionized water (Novick, R. P., *Virology*, 33, 121–136, 1963).

M44:   Reference 135. P medium. Potato, 200 g; peptone, 5 g; sucrose, 15 g; $Ca(NO_3)_2 \cdot 4H_2O$, 0.5 g; $Na_2HPO_4 \cdot 12H_2O$, 2.0 g; water, one liter. pH 6.8.

M45:   L Broth. Bacto tryptone, 1%; yeast extract, 0.5%; NaCl, 1%; glucose, 0.1%. pH to 7.0 with NaOH (Bertani, G., *J. Bacteriol.*, 62, 293–300, 1951).

M46:   M9 medium. NH_4Cl, 1.0 g; $MgSO_4 \cdot 7H_2O$, 0.49 g; KH_2PO_4, 3.0 g; Na_2HPO_4, 6.0 g; glucose, 4.0 g; water, one liter. $MgSO_4$ and glucose must be sterilized separately.

M47:   UV buffer. Phosphate buffer, 0.067 M (pH 7.0); NaCl, 0.85%; gelatin, 0.001%; $MgSO_4$, 10^{-3} M.

REFERENCES AND NOTES

1. **Vervliet, G., Holsters, M., Teuchy, H., Montagu, M. V., and Schell, J.,** *J. Gen. Virol.,* 26, 33–48, 1975.

2. Induced with mitomycin C (1 μg/ml) for 5 min.

3. **Stanley, P. M.,** *J. Gen. Virol.,* 32, 37–43, 1976.

4. **Banks, G. T., Buck, K. W., Chain, E. B., Darbyshire, J. E., Himmelweit, F., Ratti, G., Sharpe, T. J., and Planterose, D. N.,** *Nature,* 227, 505–507, 1970.

5. **Ratti, G. and Buck, K. W.,** *J. Gen. Virol.,* 14, 165–175, 1972.

6. Degraded by prolonged suspension in CsCl, contains phospholipids (see Reference 7).

7. **Nagy, E., Pragai, B., and Ivanovics, G.,** *J. Gen. Virol.,* 32, 129–132, 1976.

7a. Contains 14.3% lipid.

8. **Yelton, D. B. and Thorne, C. B.,** *J. Virol.,* 8, 242–253, 1971.

9. Transducing phage. Contains hydroxymethyl uracil.

10. UV induced.

11. **Cooney, P. H., Jacob, R. J., and Slepecky, R. A.,** *J. Gen Virol.,* 26, 131–134, 1975.

12. Add 0.8 mM Ca or 1.8 mM Mg for maximum plaque count.

13. **Sanders, G. F. and Campbell, L. L.,** *J. Bacteriol.,* 91, 340–348, 1966.

14. For phage particles, $S_{20,w}^{50}$ = 436 S; for phage DNA, T_m = 86.4°.

14a. **Sanders, G. F. and Campbell, L. L.,** *Biochemistry,* 4, 2836–2844, 1965.

15. **Egbert, L. N. and Mitchell, H. K.,** *J. Virol.,* 1, 610–616, 1967.

16. **Egbert, L.,** *J. Virol.,* 3, 528–532, 1969.

16a. For phage DNA, T_m = 87.5° corresponding to G+C = 44.4%.

17. **Takahashi, I.,** *J. Gen. Microbiol.,* 32, 211–217, 1963.

17a. **Eiserling, F. A.,** *J. Ultrastruct. Res.,* 17, 342–347, 1967.

18. **Hunter, B. I., Yamagishi, H., and Takahashi, I.,** *J. Virol.,* 841–842, 1967.

19. Contains deoxyuridylic acid instead of thymidylic acid. The bouyant density of the DNA is higher and the T_m lower than expected for the G+C content.
20. Transducing phage.
21. **Anderson, D. L., Hickman, D. D., and Reilly, B. E.,** *J. Bacteriol.,* 91, 2081–2089, 1966.
22. Complex morphology.
23. **Szybalski, W.,** *Methods Enzymol.,* 12B, 330, 1968.
24. **Sonenschein, A. L. and Roscoe, D. H.,** *Virology,* 39, 265–276, 1969.
25. Contains hydroxymethyluracil instead of thymine.
26. **Belyaeva, N. N. and Azybekian, R. R.,** *Virology,* 34, 176–179, 1968.
27. Transducing phage with complex tail morphology.
28. **Agabian-Keshishian, N. and Shapiro, L.,** *J. Virol.,* 5, 795–800, 1970.
29. For phage DNA, $S_{20,w} = 63.5$ S.
30. **Schmidt, J. M. and Stanier, R. Y.,** *J. Gen. Microbiol.,* 39, 95–107, 1965.
31. One of 23 strains from sewage.
32. **West, D., Lagenaur, C., and Agabian, N.,** *J. Virol.,* 17, 568–575, 1976.
33. For phage DNA, $T_m = 78.5°$.
34. **Driggers, L. J. and Schmidt, J. M.,** *J. Gen. Virol.,* 6, 421–427, 1970.
35. Induced with 1.0 μg/ml mitomycin C for 15 min or by UV treatment.
36. **Jollick, J. D. and Wright, B. L.,** *J. Gen. Virol.,* 22, 197–205, 1974.
37. Flagellar specific.
38. **Prescott, L. M. and Altenbern, R. A.,** *J. Virol.,* 1, 1085–1086, 1967.
39. Particles show contracted tail sheaths, 50 × 20 nm.
40. Induced by 3 to 5 μg/ml mitomycin C. Particles are inactive.
41. **McDonnell, M., Ronda-Lain, C., and Tomasz, A.,** *Virology,* 63, 577–582, 1975.
42. **Marvin, D. A. and Schaller, H.,** *J. Mol. Biol.,* 15, 1–7, 1966.
43. **Berkowitz, S. A. and Day, L. A.,** *J. Mol. Biol.,* 102, 531–547, 1976.
44. **Dyson, R. D. and van Holde, K. E.,** *Virology,* 33, 559–566, 1967.
45. **Kaiser, A. D. and Hogness, D. S.,** *J. Mol. Biol.,* 2, 392–415, 1960.
46. **Eiserling, F. A. and Boy de la Tour, E.,** *Pathol. Microbiol.,* 28, 175–180, 1965.
47. **Cummings, D. J., Chabram, V. A., and De Long, S. S.,** *J. Mol. Biol.,* 14, 418–422, 1965.
48. **MacHattie, L. A. and Thomas, C. A.,** *Science,* 144, 1142–1144, 1964.
49. UV induced (Reference 44).
49a. **Ledinko, N.,** *J. Mol. Biol.,* 9, 834–835, 1964.
50. **Sollar, A. and Epstein, H. T.,** *Virology,* 26, 715–725, 1965.
51. UV induced (Reference 50).
52. **Salivar, W. O., Tzagoloff, H., and Pratt, D.,** *Virology,* 24, 359–371, 1964.
52a. **Pratt, D. and Erdahl, W. S.,** *J. Mol. Biol.,* 37, 181–200, 1968.
53. Filamentous phages M13, fd, f1, and ZJ/2 are released from the host without cellular lysis.
54. **Overby, L. R., Barlow, G. H., Doi, R. H., Jacob, M., and Spiegelman, S.,** *J. Bacteriol.,* 91, 442–448, 1966.
55. For phage particles, $S_{20,w} = 79$.
55a. **Burdon, R. H., Billeter, M. A., Weissman, C., Warner, R. C., Ochoa, S., and Knight, C. A.,** *Proc. Natl. Acad. Sci. U.S.A.,* 52, 768–775, 1964.
56. **Martuschelli, J., Taylor, A. L., Cummings, D. J., Chapman, V. A., De Long, S. S., and Canedo, L.,** *J. Virol.,* 8, 551–563, 1971.
57. **Torti, F., Barksdale, C., and Abelson, J.,** *Virology,* 41, 567–568, 1970.
58. Length of phage DNA also given as 14.50 ± 0.67 μm and molecular weight as 28 × 10⁶ daltons (Reference 57).
59. **Sinsheimer, R. L.,** *J. Mol. Biol.,* 1, 37–42, 1959.
60. **Hall, C. E., Maclean, E. C., and Tessman, I.,** *J. Mol. Biol.,* 1, 192–194, 1959.
61. **Tromans, W. J. and Horne, R. W.,** *Virology,* 15, 1–7, 1961.
61a. **Freifelder, D., Kleinschmidt, A. K., and Sinsheimer, R. L.,** *Science,* 146, 254–255, 1964.
62. Phage DNA is circular.
63. Large-scale production method.
64. For phage particles, $S_{20,w} = 84$ S.
65. **Gesteland, R. F. and Bodtker, H.,** *J. Mol. Biol.,* 8, 497–507, 1964.
65a. **Erikson, R. L. and Szybalski, W.,** *Virology,* 22, 111–124, 1964.
66. For phage particles, $S_{20,w} = 79$ S.
67. **Bresler, S. E., Kiselev, N. A., Mayakov, V. F., Mosevitsky, M. I., and Timkovsky, A. L.,** *Virology,* 33, 1–9, 1967.
67a. The coliphages T1 through T7 can be grown in *E. coli* B on medium M9 (Reference M46).
68. Scaled from **Trouwborst, T., Kuyper, S., and Teppema, J. S.,** *J. Gen. Virol.,* 25, 75–81, 1974.

68a. Mandel, M., *J. Mol. Biol.,* 5, 435, 1962.

69. Cummings, D. J., Chapman, V. A., and De Long, S. S., *Virology,* 37, 94–108, 1969.

69a. Brenner, S., Streisinger, G., Horne, R. W., Champe, S. P., Barnett, L., Benzer, S., and Rees, H. W., *J. Mol. Biol.,* 1, 281–292, 1959.

70. Phages T2, T4, and T6 are identical in size (Reference 71).

71. Kellenberger, E., Boy de la Tour, E., Epstein, R. H., Franklin, N. C., Jerne, N. K., Reale-Scafati, A., and Sechaud, J., *Virology,* 26, 419–440, 1965.

72. Marmur, J. and Doty, P., *J. Mol. Biol.,* 5, 109–118, 1962.

73. Phage contains hydroxymethyl cytosine instead of cytosine.

73a. Phage T4 requires L-tryptophan (10 μg/ml) for growth.

74. For phage DNA, T_m = 83°.

75. For phage T2 particle, $S_{20,w}$ = 1000 S (Reference 69).

76. Thomas, C. A. and MacHattie, L. A., *Proc. Natl. Acad. Sci. U.S.A.,* 52, 1297–1301, 1964.

77. Barnett, W. A., Morrod, R. S., Spragg, S. P., and Flewett, T. H., *J. Gen. Virol.,* 25, 187–196, 1974.

78. Lang, D. and Coates, P., *J. Mol. Biol.,* 36, 137–151, 1968.

79. For phage particles, $S_{20,w}$ = 580 S (Reference 77); for phage DNA, T_m = 90° (Reference 72).

80. Zweig, M. and Cummings, D. J., *Virology,* 51, 443–453, 1973.

81. Hertel, R., Marchi, L., and Muller, K., *Virology,* 18, 576–581, 1962.

82. For particles of phage T5 st$^+$, density is 1.554 (Reference 83).

83. For particles of phage T5 st$^+$, $S_{20,w}$ = 600 ± 18 S, Rubinstein, I., *Virology,* 36, 356–376, 1968.

84. Bujard, H., *Proc. Natl. Acad. Sci. U.S.A.,* 62, 1167–1174, 1969.

84a. Wyatt, G. R. and Cohen, S. S., *Biochem. J.,* 55, 773, 1953.

85. Abelson, J. and Thomas, C. A., *J. Mol. Biol.,* 18, 262–291, 1966.

86. Luftig, R. and Haselkorn, R., *Virology,* 34, 664–674, 1968.

87. For phage particles, $S_{20,w}$ = 487 ± 5 S (Davison, P. F. and Freifelder, D., *J. Mol. Biol.,* 5, 635–642, 1962).

88. Studier, F. W., *Virology,* 39, 562–574, 1969, gives molecular weight of phage DNA as 26 × 10^6; Misra, D. N., Sinha, R. K., and das Gupta, N. N., *Virology,* 39, 183–193, 1969, give its length as 13.1 μm.

89. Stirm, S. and Freund-Molbert, E., *J. Virol.,* 8, 330–342, 1971.

90. Phage C has a base plate attached to the head.

91. Watanabe, K. and Takesu, S., *J. Gen. Virol.,* 20, 319–326, 1973.

92. Somogyi, P. A., Petrovsky, G. V., and Grigorev, V. B., *J. Gen. Virol.,* 29, 235–238, 1975.

93. Phage particle $S_{20,w}$ = 410 S.

94. Phage particles $S_{20,w}$ = 490 S.

95. Gourlay, R. N., *Nature,* 225, 1165, 1970.

96. Gourlay, R. N., *J. Gen. Virol.,* 12, 65–67, 1971.

97. Garwes, D. J., Pike, B. V., Wyld, S. G., Pocock, D. H., and Gourlay, R. N., *J. Gen. Virol.,* 29, 11–24, 1975.

98. The viruses MV-L1, L2, and L3 are called Mycoplasmatales virus-laidlawii.

99. Tsopankis, C. and Parish, J. H., *J. Gen. Virol.,* 30, 99–112, 1976.

100. Burchard, R. P. and Dworkin, M., *J. Bacteriol.,* 91, 1305–1313, 1966.

101. Phage MX-1 requires 10^{-3} M Ca^{++} and 10^{-2} M monovalent ions for best adsorption.

102. For phage particles, $S_{20,w}$ = 1145 S.

103. Brownell, G. H., Adams, J. N., and Bradley, S. G., *J. Gen. Microbiol.,* 47, 247–256, 1967.

104. Adolph, K. W. and Haselkorn, R., *Virology,* 46, 200–208, 1971.

105. *Nostoc muscorum* (blue-green alga) fixes nitrogen and requires light for growth and phage production. For phage particles, $S_{20,w}$ = 539.

106. Buck, K. W. and Kempson-Jones, G. F., *J. Gen. Virol.,* 18, 223–235, 1973. Refers to the L particle of virus "s." These viruses are termed Mycoviruses.

107. Virus released by homogenizing the mycelium.

108. 2-RNA of three molecular weight classes was found (0.99, 0.89, and 0.27 × 10^6), together with some 1-RNA.

109. PsV-s and f are slow- and fast-moving particles under electrophoresis. The s particles can infect protoplasts.

110. See Reference 86.

111. Twelve-liter culture vessel surrounded by ring fluorescent lamps.

112. Phages specific for blue-green algae are termed "cyanophages."

113. Morgan, T. M. and Stanisch, V. A., *J. Gen. Virol.,* 30, 73–80, 1976.

114. Phage B33 is temperate and female specific.

115. Vidaver, A. K., Koski, R. K., and Van Etten, J. L., *J. Virol.,* 11, 799–805, 1973.
116. Phage contains 25% lipid.
117. Contains three pieces of RNA (Semancik, J. S., Vidaver, A. K., and Van Etten, J. L., *J. Mol. Biol.,* 78, 617–625, 1973.)
118. Espejo, R. T. and Canelo, E. S., *Virology,* 34, 738–747, 1968.
119. Phage contains 10% lipid.
120. Phage DNA is circular.
121. Schade, S. Z. and Adler, J., *J. Virol.,* 1, 591–598, 1967.
122. For phage DNA, T_m = 92.1°.
123. Active on motile strains of salmonellae, serratia, or coli. Phage grown on *E. coli* AW 313, a derivative of K12.
124. Kwiatkowski, B., Beilharz, H., and Stirm, S., *J. Gen. Virol.,* 29, 267–280, 1975.
125. Phage particle has base plate with spikes attached to head.
126. Chan, R. K., Botstein, D., Watanabe, T., and Ogata, Y., *Virology,* 50, 883–898, 1972.
127. Sheppard, D. E., *Virology,* 17, 212–214, 1962.
128. Base plate attached to head by a short neck.
129. Induced by UV light for 20 sec at 30 cm from lamp. Aerated in dark for 25 hr (Watanabe, T., Ogata, Y., Chan, R. K., and Botstein, D., *Virology,* 50, 874–882, 1972).
130. Rhoades, M., MacHattie, L. A., and Thomas, C. A., *J. Mol. Biol.,* 37, 21–40, 1968.
131. Rosenblum, E. and Tyrone, S., *J. Bacteriol.,* 88, 1737–1742, 1964.
132. Phages 3 and 7 are specific for bacteria of serological group A.
133. Bradley, D. E., *J. Ultrastruct. Res.,* 8, 552–565, 1963.
134. These phages are specific for bacteria in serological group B.
135. Kao, T. T., Huang, T. C., and Chow, T. Y., *Virology,* 39, 548–555, 1969.
136. This phage is released without lysis of the host.
137. Weston, L., Drexler, H., and Richardson, S. H., *J. Gen. Virol.,* 21, 155–158, 1973.
138. Indicator strain is RV 79.
139. Twort, F. W., *Lancet,* ii, 189, 1251–1252, 1915.
140. Demerec, M. and Fano, U., *Genetics,* 30, 119–136, 1945.
141. Eisenstark, A., in *Methods in Virology,* Vol. 1, Academic Press, New York, 449–524, 1967.
142. Tikhonenko, A. S., *Ultrastructure of Bacterial Viruses,* Plenum, New York, 1970.
143. Frankel-Conrat, H., in *Comprehensive Virology,* Vol. 1, Plenum, New York, 1974.
144. Bradley, D. E., *Bacteriol. Rev.,* 31, 230–314, 1967.
145. Topley, W. W. C. and Wilson, G. S., *The Principles of Bacteriology and Immunity,* 6th ed., Arnold, London, 1975.
146. Yamamoto, K. R., Alberts, B. M., Benzinger, R., Lawthorne, L., and Treiber, G., *Virology,* 40, 734–744, 1970.
147. Adams, M. H., *Bacteriophages,* Interscience, New York, 1959.

PLANT TISSUE CULTURES FOR TOBACCO MOSAIC VIRUS INFECTION

H. H. Murakishi

Tobacco mosaic virus (TMV) infection of tobacco leaves usually results in systemic mosaic symptoms or in localized necrotic lesions. Crystalline viral inclusions are often associated with systemic infection and provide cytological evidence of virus infection in plant leaf hair cells,[1] in tissue cultures derived from previously infected plants,[2-4] in cells microinjected[5,6] with TMV, and in tissue culture cells inoculated in vitro.[7-12] The in vitro method has resulted in moderate to high virus titer and crystalline viral inclusions in callus cells derived from *Nicotiana tabacum* var. Havana-38. Local lesions are produced in callus cells[13,14] obtained from the hypersensitive tobacco Xanthi-nc, which contains the N gene for virus localization.[15] For both types of reaction, systemic or localized, the method of inoculation is the same.

PROCEDURE

Callus Cultures

The culture medium is a combination of the media described by White[16] (W) and by Murashige and Skoog[17] (MS). The W and MS media are prepared separately, and each is brought to pH 5.9; both media are then combined in a ratio of 3 volumes of MS to 2 volumes of W. The combined medium is supplemented with 48 ml of coconut water (first treated to remove precipitates), 30 μg of kinetin, 240 μg of 2,4-dichlorophenoxyacetic acid (2,4-D), and 40 μg of naphthalene acetic acid (NAA) per liter. Fresh coconut water will form precipitates during autoclaving. To prevent this, separately add 2N NaOH to fresh coconut water to adjust the pH to 10.5, then allow to set overnight in the cold. Siphon off the clear supernatant fluid from the precipitates at the bottom, readjust the pH to 5.9, then add the fluid to the rest of medium. The complete medium (MS-W) is sterilized by autoclaving at 15 lb pressure for 20 min. Solid medium is prepared by adding agar (1% w/v).

Stem pith cores (explants) of *N. tabacum* var. Havana-38 and of *N. tabacum* var. Xanthi-nc are removed aseptically with a cork borer, placed on solidified MS-W medium, and grown under fluorescent lamps (80 fc). Callus growth usually becomes green-pigmented within 2 weeks and is maintained by monthly transfers on solid medium. To prepare cells for inoculation, 1 g of agar-grown cells is transferred to 20 ml of liquid medium in 125-ml Erlenmeyer flasks and placed on a rotary shaker (120 rpm). After 6 to 9 days, 1 to 2 ml of the resulting cell suspension is transferred with a large-mouth pipet (3-mm opening) to fresh liquid medium. After 6 to 9 days of additional growth, the cells develop into small aggregates, 20 to 75 cells per aggregate, and are ready for inoculation.

Virus Inoculation

Virus is purified by differential centrifugation[18] of leaf sap extracted from Havana-38 tobacco inoculated 2 to 4 weeks previously. The purified virus is freed of microorganisms by passage through an ultrafine sintered glass filter or Millipore filter.

Using aseptic technique, 300 to 400 mg of cell aggregates are transferred to 40-ml test tubes, each containing TMV in 3 ml of MS-W medium. Different concentrations of TMV may be used. We have obtained good infection with as little as 12 μg/ml, but better infection is obtained when we use 150 μg/ml. The tubes are vibrated at 800 rpm for 20 sec on a Vortex mixer, such as Model S8220 (Scientific Products, Evanston, Ill.). This dissociates the cell clumps into smaller aggregates. It is thought that during this inoculation process the connections between the cells are broken, exposing openings in the cell walls to virus entry and infection of the protoplast. The contents of each tube are emptied into a 50-mm funnel lined with Miracloth or Whatman No. 4 filter paper, and washed with 15 mℓ of fresh MS-W medium. The cells are then transferred with a microspatula to MS-W agar medium in petri plates and incubated at 24°C under fluorescent lights (80 fc).

In Havana-38 tobacco callus, viral inclusions are first visible at 44 hr, but become more abundant in 3 to 6 days. In Xanthi-nc tobacco callus, local lesions are first visible about 40 hr after inoculation, but become more distinct and increase in number after 4 days. Use of continuously subcultured callus cells usually results in lower efficiency of infection. When this occurs, new explants should be made to start fresh cultures.

COMPOSITION AND PREPARATION OF THE TISSUE CULTURE MEDIUM FOR TOBACCO MOSAIC VIRUS INFECTION

MS-W medium is a mixture of Murashige and Skoog[17] Medium (MS) and Modified White's[16] Medium (W-9). We find it convenient to prepare it by making up MS and W-9 separately, then combining the two in a 3:2 ratio. The preparation of MS-W medium is described in Table 1. Tables 2 and 3 respectively list the ingredients for stock solutions MS and W-9.

Table 1
PREPARATION OF MS-W MEDIUM[14]

MS, 3 Liters

1. MS stock solution I, minerals	300 ml
2. MS stock solution II, vitamins	30 ml
3. MS stock solution III, iron	15 ml
4. Sucrose	90 g
5. Water, glass-distilled, to make final volume	3 l

Adjust pH to 5.85 before mixing with W-9

W-9, 2 Liters

1. White's stock solution I, minerals	200 ml
2. White's stock solution II, vitamins	20 ml
3. White's stock solution III, growth regulators	
2,4-D	0.4 ml
Ca pantothenate	4.0 ml
NAA	0.4 ml
4. White's stock solution IV, iron	5 ml
5. Coconut Water[a]	240 ml
6. Glucose	40 g
7. Water, glass-distilled, to make final volume	2 l

Adjust pH to 5.85 before mixing with MS

Preparation of Liquid and Solid Media

To make MS-W liquid medium, combine 3 l of MS medium with 2 l of W-9 medium. Distribute MS-W medium into 125-ml Erlenmeyer flasks (25 ml each) for suspension cultures, into prescription bottles for washing cells, and into large test tubes (3 ml each) for inoculation medium. If MS-W agar medium is desired, adjust the pH of the liquid to 6.2, add 1% Difco Bacto agar, then steam to melt the agar. After autoclaving at 15 lb pressure for 20 min, the final pH should be around 5.9 for both agar and liquid media.

[a] Treat fresh coconut water with 2N NaOH to remove insoluble precipitates before autoclaving (see text).

Compiled by H. H. Murakishi.

Table 2
PREPARATION OF MS STOCK SOLUTIONS

MS Stock Solution I, Minerals, 10X, in 1 Liter of Glass-distilled Water

NH_4NO_3	16.5 g	H_3BO_3	62.0 mg
KNO_3	19.0 g	$MnSO_4 \cdot H_2O$	169.0 mg
$CaCl_2 \cdot 2H_2O$	4.4 g	$ZnSO_4 \cdot 7H_2O$	106.0 mg
$MgSO_4 \cdot 7H_2O$	3.7 g	KI	8.3 mg
KH_2PO_4	1.7 g		

MS Stock Solution II, Vitamins, 10X, in 100 ml of Glass-distilled Water

Glycine anhydride	20 mg	Nicotinic acid	5 mg
Indoleacetic acid	10 mg	Pyridoxine·HCl	5 mg
Kinetin	0.5 mg	Thiamine HCl	1 mg
Myoinositol	1,000 mg		

Distribute in 10-ml lots into vials, cap, and store in deep-freeze.

MS Stock Solution III, Iron, in 100 ml of Glass-distilled Water

$Fe(SO_4)_3 \cdot 7H_2O$ 557 mg Na_2-EDTA   745 mg

Store in a 125-ml Erlenmeyer flask under refrigeration.

From Murashige, T. and Skoog, F., *Physiol. Plant.*, 15, 473–497, 1962. With permission.

Table 3

PREPARATION OF W-9 STOCK SOLUTIONS

W-9 Stock Solution I, Minerals

Set out four 1-liter flasks, each containing 1 liter of glass-distilled water, and label them 1, 2, 3, and 4.

To flask 1 add:		To flask 3 add:	
$Ca(NO_3)_2 \cdot 4H_2O$	12.0 g	$NaH_2PO_4 \cdot H_2O$	0.76 g
KNO_3	3.2 g	To flask 4 add:	
KCl	2.6 g	$MnSO_4 \cdot 4H_2O$	0.20 g
To flask 2 add:		$ZnSO_4 \cdot 7H_2O$	0.12 g
$MgSO_4 \cdot 7H_2O$	30.0 g	H_3BO_3	0.06 g
Na_2SO_4	8.0 g	KI	0.03 g

When the contents of the four flasks are dissolved, they are mixed slowly by transfer into a 4-liter bottle to provide stock solution I, which is then autoclaved.

W-9 Stock Solution II, Vitamins, 10X

Glycine (anhydride)	30 mg	Pyridoxine·HCl base	1 mg
Nicotinic acid	5 mg	Water, glass-distilled	100 ml
Thiamine HCl	1 mg		

Distribute in 10-ml lots into pyrex test tubes, cap, and store in freezer.

W-9 Stock Solution III, Growth Regulators[a]

Calcium pantothenate, 25 mg/20 ml H_2O (1.25 mg/ml); refrigerate, Naphthalene-acetic acid (NAA), 10 mg/20 ml H_2O (0.5 mg/ml); refrigerate. 2,4-Dichlorophenoxyacetic acid (2,4-D), 60 mg/ml in 30% EtOH (3 mg/ml); store at room temperature.

W-9 Stock Solution IV, Iron

$Fe_2(SO_4)_3 \cdot 7H_2O$	52.6 mg
Na_2-EDTA	80.0 mg
Water, glass-distilled	100 ml

Dissolve in a 125-ml Erlenmeyer flask (screw-capped) and store under refrigeration.

[a] Store each in a separate screw-capped tube.

From White, P. R., *The Cultivation of Animal and Plant Cells,* 2nd ed., © Ronald Press Co., New York, 1963, pp. 57–63. With permission.

REFERENCES

1. McWhorter, F. P., *Annu. Rev. Phytopathol.*, 3, 287–312, 1965.
2. Ball, E., *Bull. Torrey Bot. Club*, 93, 244–258, 1966.
3. Chandra, N. and Hildebrandt, A. C., *Virology*, 31, 414–421, 1967.
4. Singh, M. and Hildebrandt, A. C., *Virology*, 30, 134–142, 1966.
5. Nims, R. C., Halliwell, R. S., and Rosberg, D. W., *Cytologia*, 32, 224–235, 1967.
6. Russell, T. E. and Halliwell, R. S., *Phytopathology*, 64, 1520–1526, 1974.
7. Motoyoshi, F. and Oshima, N., *Jpn. J. Microbiol.*, 12, 317–320, 1968.
8. Murakishi, H. H., *Phytopathology*, 58, 993–996, 1968.
9. Murakishi, H. H., Hartmann, J. X., Pelcher, L. E., and Beachy, R. N., *Virology*, 41, 365–367, 1970.
10. Murakishi, H. H., Hartmann, J. X., Beachy, R. N., and Pelcher, L. E., *Virology*, 43, 62–68, 1971.
11. Pelcher, L. E., Murakishi, H. H., and Hartmann, J. X., *Virology*, 47, 787–796, 1972.
12. Hartmann, J. X. and Murakishi, H. H., *In Vitro*, 6, 373–374, 1971.
13. Beachy, R. N. and Murakishi, H. H., *Phytopathology*, 61, 877–878, 1971.
14. Beachy, R. N. and Murakishi, H. H., *Virology*, 55, 320–328, 1973.
15. Holmes, F. O., *Phytopathology*, 28, 533–561, 1938.
16. White, P. R., *The Cultivation of Animal and Plant Cells*, 2nd ed., Ronald Press Co., New York, 1963, pp. 57–63.
17. Murashige, T. and Skoog, F., *Physiol. Plant.*, 15, 473–497, 1962.
18. Knight, C. A., in *Biochemical Preparations*, Vol. 9, Coon, M. J., Ed., Wiley & Sons, New York, 1962, pp. 132–136.

TISSUE CULTURES FOR ANIMAL (VERTEBRATE) VIRUSES

J. S. Porterfield

Improved tissue culture methods have contributed substantially to recent developments in animal virology. From the large number of tissue culture media now available, the virologist is able to choose those best suited to his needs, which may differ significantly from those of the cell biologist, although the two disciplines have much in common.

Tissue culture media may be divided into three categories: natural media, partially defined media, and chemically defined media.

Natural culture media — These are made up of natural substances, such as serum and tissue extracts, diluted in balanced salt solutions. A list of balanced salt solutions is presented in Table 1.

Partially purified culture media — The great majority of media currently used in *in vitro* studies on animal cells and animal viruses fall into this category, since they usually contain a proportion of serum, serum dialysate, or other complex natural product. With the development of improved chemically defined culture media, an increasing number of cells can now be grown — or at least maintained for longer periods — on serum-free media; the most commonly used of these are described in Table 2.

Chemically defined culture media — Table 2 sets out the constituents of commonly used chemically defined culture media. The tabulation is designed to allow comparisons between different media. Several media are available in a variety of different modifications not listed. Reference should be made to the original publication for further details of the formulation of the media, and to the important Reference 2. All quantities are expressed as mg/l; antibiotics are omitted. Note that although some of the media will support the growth of some cells without supplementation, the addition of serum is necessary with other cells.

Table 1

BALANCED SALT SOLUTIONS

	Dulbecco[1]	Earle[2]	Gey[3] (slides)	Gey[3] (tubes)	Hanks[4]	Locke[5]	Puck "F"[6]	Puck "G"[6]	Ringer[7] (mammalian)	Rappaport[8]	Simms[9]	Tyrode[10]
$CaCl_2$	100	200	170	170	140	200	16[a]	16[a]	250	420	150[a]	200
KCl	200	400	370	370	400	200	285	400	420	400	200	200
KH_2PO_4	200		30	30	60		83	150				
$MgCl_2 \cdot 6H_2O$	100		210	210						100		100
$MgSO_4 \cdot 7H_2O$		100	70	70	200		154	154		100	200	
$NaCl$	8,000	6,800	8,000	7,000	8,000	9,500	7,400	8,000	9,000	8,000	8,000	8,000
$NaHCO_3$		2,200	227	2,270	350	200	1,200			780	1,000	1,000
$NaH_2PO_4 \cdot 2H_2O$		125								20		
$Na_2HPO_4 \cdot 2H_2O$	1,150		150	150	60		290[b]	290[b]		80	240	50
Glucose	1,000	1,000	1,000	1,000	1,000	1,000	1,100	1,100		2,000	1,000	1,000
Gas phase	Air	CO$_2$[c]	CO$_2$[c]	CO$_2$[c]	Air	Air	CO$_2$[c]	Air	Air	CO$_2$[c]	CO$_2$[d]	Air[e]

[a] $2H_2O$.
[b] $7H_2O$.
[c] 5% CO_2 in air.
[d] 2% CO_2 in air.
[e] Generally used in a closed system.

Compiled by J. S. Porterfield.

REFERENCES

1.   Dulbecco, R. and Vogt, M., J. Exp. Med., 99, 167–182, 1954.
2.   Earle, W. R., J. Natl. Cancer Inst., 4, 165–212, 1943.
3.   Gey, G. O. and Gey, M. K., Am. J. Cancer, 27, 45–76, 1936.
4.   Hanks, J. H. and Wallace, R. E., Proc. Soc. Exp. Biol. Med., 71, 196–200, 1949.
5.   Locke, F. S., Cent. Physiol., 14, 670–672, 1900.
6.   Puck, T. T., Cieciura, S. J., and Robinson, A. J., J. Exp. Med., 108, 945–956, 1958.
7.   Ringer, S. and Buxton, D. W., J. Physiol. London, 7, 288–295, 1886.
8.   Rappaport, C., Proc. Soc. Exp. Biol. Med., 91, 464–470, 1956.
9.   Simms, H. S. and Sanders, M., Arch. Pathol., 33, 619–635, 1942.
10. Tyrode, M. V., Arch. Int. Pharmacodyn. Ther., 20, 205–223, 1910.

Table 2
CHEMICALLY DEFINED CULTURE MEDIA

Constituents, mg/l	Eagle's Basal[1,2]	Eagle's MEM[2-4]	Dulbecco[2,4]	Joklik[5]	Glasgow[6]
Inorganic salts					
(1) $CaCl_2$	111	200	200		200
(2) $CoCl_2 \cdot 6H_2O$					
(3) $CuSO_4 \cdot 5H_2O$					
(4) $FeSO_4 \cdot 7H_2O$			0.1[b]		0.1[b]
(5) KCl	373	400	400	400	400
(6) KH_2PO_4					
(7) $MgCl_2 \cdot 6H_2O$	102	200		200	
(8) $MgSO_4 \cdot 7H_2O$			200		200
(9) $MnSO_4 \cdot H_2O$					
(10) NaCl	5,845	6,800	6,400	6,500	6,400
(11) $NaHCO_3$	1,680	2,000	3,700	2,000	2,750
(12) Na_2HPO_4					
(13) $NaH_2PO_4 \cdot H_2O$	138	150[g]	125[g]	1,327	124
(14) $(NH_4)_6Mo_7O_{24} \cdot 4H_2O$					
(15) $ZnSO_4 \cdot 7H_2O$					
Amino acids					
(16) L-Alanine					
(17) L-α-Aminobutyric acid					
(18) L-Arginine	17.5	105	84[k]	105[k]	42[k]
(19) L-Asparagine					
(20) L-Aspartic acid					
(21) L-Cysteine					
(22) L-Cystine	12	24	48	32.4[m]	24
(23) L-Glutamic acid					
(24) L-Glutamine	292	292	584	294	584
(25) Glycine			30		
(26) L-Histidine	7.75	31	42[kl]	31	16
(27) L-Hydroxyproline					
(28) L-Isoleucine	26	52	104.8	52	52
(29) L-Leucine	26	52	104.8	52	52
(30) L-Lysine	29	58	146[k]	58	74[k]
(31) L-Methionine	7.5	15	30	15	15
(32) L-Ornithine					
(33) L-Phenylalanine	16	32	66	32	33
(34) L-Proline					
(35) L-Serine			42		
(36) L-Taurine					
(37) L-Threonine	24	48	95	48	48
(38) L-Tryptophan	4	10	16	10	8
(39) L-Tyrosine	18	36	72	47[n]	36
(40) L-Valine	23	46	94	46	47
Vitamins					
(41) p-Aminobenzoic acid					
(42) Ascorbic acid					
(43) D-Biotin	0.24				
(44) Choline chloride	0.12	1	4	1	2
(45) Folic acid	0.44	1	4	1	2
(46) Folinic acid					

[b] $Fe(NO_3)_3 \cdot 9H_2O$
[g] $2H_2O$.
[k] HCl.
[l] H_2O.
[m] 2HCl.

Table 2 (continued)
CHEMICALLY DEFINED CULTURE MEDIA

Constituents, mg/l	Eagle's Basal[1,2]	Eagle's MEM[2-4]	Dulbecco[2,4]	Joklik[5]	Glasgow[6]
Vitamins (continued)					
(47) *i*-Inositol		2	7	2	3.6
(48) Lipoic acid					
(49) Niacin					
(50) Nicotinamide	0.12	1	4	1	2
(51) Pantothenate (Ca)	0.2 (acid)	1	4	1	2
(52) Pyridoxal HCl	0.2	1	4	1	2
(53) Pyridoxine HCl					
(54) Riboflavine	0.04	0.1	0.4	0.1	0.2
(55) Thiamine HCl	0.34	1	4	1	2
(56) Vitamin A					
(57) Vitamin B_{12}					
(58) Vitamin D_2					
(59) Vitamin E (toco-pherol)					
(60) Vitamin K_3 (menadione)					
Other constituents					
(61) Adenine sulfate					
(62) Adenine 5′-triphosphate					
(63) Adenylic acid					
(64) Cholesterol					
(65) Cocarboxylase					
(66) Coenzyme A					
(67) Deoxyadenosine					
(68) Deoxycytidine					
(69) Deoxyguanosine					
(70) 2-Deoxy-D-ribose					
(71) Ethanol					
(72) Flavine adenine dinucleotide					
(73) D-Galactose					
(74) D-Glucosamine HCl					
(75) D-Glucose	900	1,000	1,000	2,000	4,500
(76) D-Glucuronolactone					
(77) Glutathione					
(78) Guanine					
(79) Hypoxanthine					
(80) Insulin					
(81) α-Ketoglutaric acid					
(82) Linoleic acid					
(83) Methyl cellulose					
(84) 5-Methyl cytosine					
(85) 5-Methyl deoxycytidine					
(86) Methyl oleate					
(87) Nicotine adenine dinucleotide					
(88) Nicotine adenine trinucleotide phosphate					
(89) Phenol red	5	10	15	10	15
(90) Polysorbate 80					
(91) Polyvinyl pyrrolidone					
(92) Putrescine·2HCl					
(93) D-Ribose					
(94) Salmine sulfate					

Table 2 (continued)
CHEMICALLY DEFINED CULTURE MEDIA

Constituents, mg/l	Eagle's Basal[1,2]	Eagle's MEM[2-4]	Dulbecco[2,4]	Joklik[5]	Glasgow[6]
	Other constituents (continued)				
(95) Sodium acetate·$3H_2O$					
(96) Sodium glucuronate					
(97) Sodium mucate					
(98) Sodium pyruvate			110		
(99) Thymidine					
(100) Thymine					
(101) Uracil					
(102) Uridine 5'-triphosphate					
(103) Xanthine					

Constituents, mg/l	α-MEM[7]	199[2,4,8]	CMRL 1066[4,9]	CMRL 1415[10]	CMRL 1969[11]
	Inorganic salts				
(1) $CaCl_2$	200	200	200	140	140
(2) $CoCl_2 \cdot 6H_2O$					
(3) $CuSO_4 \cdot 5H_2O$					
(4) $FeSO_4 \cdot 7H_2O$		0.2[b]			
(5) KCl	400	400	400	400	400
(6) KH_2PO_4					
(7) $MgCl_2 \cdot 6H_2O$					
(8) $MgSO_4 \cdot 7H_2O$	200	200	200	240	200
(9) $MnSO_4 \cdot H_2O$					
(10) NaCl	6,800	6,800	6,800	6,800	8,000
(11) $NaHCO_3$	2,000	2,200	2,200	1,000	560
(12) Na_2HPO_4				200	180
(13) $NaH_2PO_4 \cdot H_2O$	140	140	140	50	70
(14) $(NH_4)_6Mo_7O_{24} \cdot 4H_2O$					
(15) $ZnSO_4 \cdot 7H_2O$					
	Amino acids				
(16) L-Alanine	25	50[o]	25	30	25
(17) L-α-Aminobutyric acid					
(18) L-Arginine	105[k]	70[k]	70	500	58
(19) L-Asparagine	25[l]				
(20) L-Aspartic acid	30	50[o]	30	10	30
(21) L-Cysteine	100[kl]	0.1[k]	260	179	0.1[k]
(22) L-Cystine	24	20	20	24	20
(23) L-Glutamic acid	75	150[ol]	75	10	67
(24) L-Glutamine	292	100	100	292	200
(25) Glycine	50	50	50	17	50
(26) L-Histidine	42[kl]	20[k]	20[kl]	3.1	16
(27) L-Hydroxyproline		10	10	10	10
(28) L-Isoleucine	52	40[o]	20	52	20
(29) L-Leucine	52	120[o]	60	52	60
(30) L-Lysine	58	70[k]	70[k]	47	70
(31) L-Methionine	15	30[o]	15	15	15
(32) L-Ornithine					
(33) L-Phenylalanine	33	50[o]	25	32	25
(34) L-Proline	40	40	40	30	40
(35) L-Serine	25	50[o]	25	12	25
(36) L-Taurine					

[n] Na salt.
[o] DL.

Table 2 (continued)
CHEMICALLY DEFINED CULTURE MEDIA

Constituents, mg/l	α-MEM[7]	199[2,4,8]	CMRL 1066[4,9]	CMRL 1415[10]	CMRL 1969[11]
Amino acids (continued)					
(37) L-Threonine	48	60°	30	48	30
(38) L-Tryptophan	10	20°	10	10	10
(39) L-Tyrosine	36	40	40	36	40
(40) L-Valine	47				
Vitamins		50°	25	46	25
(41) *p*-Aminobenzoic acid		0.05	0.05		0.05
(42) Ascorbic acid	50	0.05	50	50	0.05
(43) D-Biotin	0.1	0.01	0.01	1	1
(44) Choline chloride	1	0.5	0.5	1	1
(45) Folic acid	1	0.01	0.01	1	1
(46) Folinic acid					
(47) *i*-Inositol	2	0.05	0.05[m]	2 (NF)	2[m]
(48) Lipoic acid	0.2				
(49) Niacin		0.025	0.025		
(50) Nicotinamide	1	0.025	0.025		1
(51) Pantothenate (Ca)	1	0.01	0.01	0.5	1
(52) Pyridoxal HCl	1	0.025	0.025		1
(53) Pyridoxine HCl		0.025	0.025		
(54) Riboflavine	0.1	0.01	0.01		0.1
(55) Thiamine HCl	1	0.01	0.01		1
(56) Vitamin A		0.14			
(57) Vitamin B$_{12}$					
(58) Vitamin D$_2$		0.1			
(59) Vitamin E (toco-pherol)	1.36	0.01			
(60) Vitamin K$_3$ (menadi-one)					
Other constituents		0.01			
(61) Adenine sulfate		10			
(62) Adenine 5′-triphos-phate		1			
(63) Adenylic acid		0.2			
(64) Cholesterol		0.2	0.2		
(65) Cocarboxylase			1	1	
(66) Coenzyme A			2.5	1	
(67) Deoxyadenosine			10	10	
(68) Deoxycytidine			10	10	
(69) Deoxyguanosine			10	10	
(70) 2-Deoxy-D-ribose		0.5			
(71) Ethanol			16		
(72) Flavine adenine dinu-cleotide			1	1	
(73) D-Galactose				500	
(74) D-Glucosamine HCl					
(75) D-Glucose	1,000		1,000	500	
(76) D-Glucuronolactone					
(77) Glutathione		0.05	10	10	
(78) Guanine		0.3			
(79) Hypoxanthine		0.3			
(80) Insulin					
(81) α-Ketoglutaric acid					
(82) Linoleic acid					
(83) Methyl cellulose					
(84) 5-Methyl cytosine					
(85) 5-Methyl deoxycytidine			0.1	0.1	

Table 2 (continued)
CHEMICALLY DEFINED CULTURE MEDIA

Constituents, mg/l	α-MEM[7]	199[2,4,8]	CMRL 1066[4,9]	CMRL 1415[10]	CMRL 1969[11]
Other constituents (continued)					
(86) Methyl oleate					
(87) Nicotine adenine dinucleotide			7	7	
(88) Nicotine adenine trinucleotide phosphate			1	1	
(89) Phenol red	10	15	20	20	
(90) Polysorbate 80		20	5		
(91) Polyvinyl pyrrolidone					
(92) Putrescine·2HCl					
(93) D-Ribose		0.5			
(94) Salmine sulfate					
(95) Sodium acetate·3H$_2$O		83	83		
(96) Sodium glucuronate			4.2		
(97) Sodium mucate					
(98) Sodium pyruvate	110			225	
(99) Thymidine			10	10	
(100) Thymine		0.3			
(101) Uracil		0.3			
(102) Uridine 5'-triphosphate			1	1	
(103) Xanthine		0.3			

Complete α-MEM medium also contains the following deoxyribosides and ribosides, all at 10 mg/l: thymidine, deoxyadenosine, deoxycytidine, deoxyguanosine, uridine, adenosine, cytidine, and guanosine.

Constituents, mg/l	McCoy 5A[2,4,12]	RPMI 1603[2,13]	RPMI 1634[2,14]	RPMI 1640[2,15]	NCTC 109 and NCTC 135[2,16]
Inorganic salts					
(1) CaCl$_2$	200	200[h]	100[h]	100[h]	200
(2) CoCl$_2$·6H$_2$O					
(3) CuSO$_4$·5H$_2$O					
(4) FeSO$_4$·7H$_2$O		1			
(5) KCl	400	400	400	400	400
(6) KH$_2$PO$_4$					
(7) MgCl$_2$·6H$_2$O		1			
(8) MgSO$_4$·7H$_2$O	200	200	100	100	100[a]
(9) MnSO$_4$·H$_2$O					
(10) NaCl	6,460	6,000	6,000	6,000	6,800
(11) NaHCO$_3$	2,200	1,500		2,000	2,200
(12) Na$_2$HPO$_4$		1,512[e]	2,835[e]	1,512[e]	
(13) NaH$_2$PO$_4$·H$_2$O	140	230			140
(14) (NH$_4$)$_6$Mo$_7$O$_{24}$·4H$_2$O					
(15) ZnSO$_4$·7H$_2$O		2			

[a] Anhydrous.

[d] 0.05 Fe(NO$_3$)$_3$·9H$_2$O + 0.26 FeSO$_4$ anhydrous.

[e] 7H$_2$O.

[h] Ca(NO$_3$)$_2$·4H$_2$O.

Table 2 (continued)
CHEMICALLY DEFINED CULTURE MEDIA

Constituents, mg/l	McCoy 5A[2,4,12]	RPMI 1603[2,13]	RPMI 1634[2,14]	RPMI 1640[2,15]	NCTC 109 and NCTC 135[2,16]
Amino acids					
(16) L-Alanine	13				31.5
(17) L-α-Aminobutyric acid					5.5
(18) L-Arginine	42[k]	200	100	200	31[k]
(19) L-Asparagine	45[l]	50	30	50	9[l]
(20) L-Aspartic acid	20		30	20	10
(21) L-Cysteine	24				26
(22) L-Cystine		50	100	50	10.5
(23) L-Glutamic acid	22	15	80	20	8
(24) L-Glutamine	219	500	300	300	136
(25) L-Glycine	7.5	15	15	10	13.5
(26) L-Histidine	21[kl]	20	35	15	27[kl]
(27) L-Hydroxyproline	20			20	4
(28) L-Isoleucine	39	80	50	50	18
(29) L-Leucine	39	80	50	50	20
(30) L-Lysine	36.5[k]	25	75[k]	40[k]	38[k]
(31) L-Methionine	15	30	15	15	4.5
(32) L-Ornithine					9[k]
(33) L-Phenylalanine	16.5	20	30	15	16.5
(34) L-Proline	17	10	30	20	6
(35) L-Serine	26	100	50	30	11
(36) L-Taurine					4
(37) L-Threonine	18	35	50	20	19
(38) L-Tryptophan	3	20	10	5	17.5
(39) L-Tyrosine	18	20	30	20	16
(40) L-Valine	17.5	10	40	20	25
Vitamins					
(41) *p*-Aminobenzoic acid	1	0.5		1	0.125
(42) Ascorbic acid	0.5				50
(43) D-Biotin	0.2	0.05	0.1	0.2	0.025
(44) Choline chloride	5	2	3	3	1.25
(45) Folic acid	0.2	0.01	1	1	0.025
(46) Folinic acid		0.01			
(47) *i*-Inositol	36	5	15	35	0.125
(48) Lipoic acid					
(49) Niacin	0.5	0.01			0.0625
(50) Nicotinamide	0.5	0.2	2.5	1	0.0625
(51) Pantothenate (Ca)	0.2	0.25	0.25	0.25	0.025
(52) Pyridoxal HCl	0.5				0.0625
(53) Pyridoxine HCl	0.5	1	2	1	0.0625
(54) Riboflavine	0.2	0.1	0.25	0.2	0.025
(55) Thiamine HCl	0.2	0.25	0.25	1	0.025
(56) Vitamin A				0.005	0.25
(57) Vitamin B_{12}	0.00075	0.05	0.1		10
(58) Vitamin D_2					0.25
(59) Vitamin E (tocopherol)					0.025
(60) Vitamin K_3 (menadione)					0.025

Table 2 (continued)
CHEMICALLY DEFINED CULTURE MEDIA

Constituents, mg/l	McCoy 5A[2,4,12]	RPMI 1603[2,13]	RPMI 1634[2,14]	RPMI 1640[2,16]	NCTC 109 and NCTC 135[2,16]
Other constituents					
(61) Adenine sulfate					
(62) Adenine 5'-triphosphate					
(63) Adenylic acid					
(64) Cholesterol					
(65) Cocarboxylase					1
(66) Coenzyme A					2.5
(67) Deoxyadenosine					10
(68) Deoxycytidine					10
(69) Deoxyguanosine					10
(70) 2-Deoxy-D-ribose					
(71) Ethanol					40
(72) Flavine adenine dinucleotide					1
(73) D-Galactose					
(74) D-Glucosamine HCl					3.85
(75) D-Glucose	3,000	2,500	2,000	2,000	1,000
(76) D-Glucuronolactone					1.8
(77) Glutathione	0.5		10	1	10
(78) Guanine					
(79) Hypoxanthine					
(80) Insulin					
(81) α-Ketoglutaric acid					
(82) Linoleic acid					
(83) Methyl cellulose					
(84) 5-Methyl cytosine					0.1
(85) 5-Methyl deoxycytidine					
(86) Methyl oleate					
(87) Nicotine adenine dinucleotide					7
(88) Nicotine adenine trinucleotide phosphate					1
(89) Phenol red	5	5	5	5	20
(90) Polysorbate 80					12.5
(91) Polyvinyl pyrrolidone					
(92) Putrescine·2HCl					
(93) D-Ribose					
(94) Salmine sulfate					
(95) Sodium acetate·3H$_2$O					50
(96) Sodium glucuronate					1.8
(97) Sodium mucate					
(98) Sodium pyruvate					
(99) Thymidine					10
(100) Thymine					
(101) Uracil					
(102) Uridine 5'-triphosphate					1
(103) Xanthine					No cysteine in NCTC 135

Table 2 (continued)
CHEMICALLY DEFINED CULTURE MEDIA

Constituents, mg/l	Ham F10[2,4,17]	Ham F12[2,4,18]	Neuman and Tytell[19]	N16[2,20]	MAB 87/3[21]
Inorganic salts					
(1) $CaCl_2$	44[g]	44[g]	200	16[g]	120[g]
(2) $CoCl_2 \cdot 6H_2O$					0.022
(3) $CuSO_4 \cdot 5H_2O$	0.0025	0.0025			
(4) $FeSO_4 \cdot 7H_2O$	0.834	0.834			0.5[c]
(5) KCl	285	224	400	285	150
(6) KH_2PO_4	83			83	208
(7) $MgCl_2 \cdot 6H_2O$		122			240
(8) $MgSO_4 \cdot 7H_2O$	153		200	154	100
(9) $MnSO_4 \cdot H_2O$					0.016[j]
(10) NaCl	7,400	7,600	6,460	7,400	6,000
(11) $NaHCO_3$	1,200	1,176	2,200	1,200	2,240
(12) Na_2HPO_4	290[e]	268[e]	235[e]	290[e]	300
(13) $NaH_2PO_4 \cdot H_2O$					
(14) $(NH_4)_6Mo_7O_{24} \cdot 4H_2O$					0.025
(15) $ZnSO_4 \cdot 7H_2O$	0.029	0.863			0.03
Amino acids					
(16) L-Alanine	9	9	9		11
(17) L-α-Aminobutyric acid					
(18) L-Arginine	211[k]	211[k]	42	37.5[k]	75[k]
(19) L-Asparagine	15[l]	15[l]	40		24
(20) L-Aspartic acid	13	13	13	30	60
(21) L-Cysteine	25	35[kl]	24		90[k]
(22) L-Cystine				7.5	15
(23) L-Glutamic acid	15	15	15	75	150
(24) L-Glutamine	146	146	219	200	350
(25) Glycine	7.5	7.5	8	100	50
(26) L-Histidine	21[k]	21[k]	21	37.5[k]	150[k]
(27) L-Hydroxyproline			13		
(28) L-Isoleucine	2.5	4	39	12.5	25
(29) L-Leucine	13	13	39	25	50
(30) L-Lysine	29	37	37	80[k]	240[k]
(31) L-Methionine	4.5	4.5	15	25	50
(32) L-Ornithine					
(33) L-Phenylalanine	5	5	17	25	50
(34) L-Proline	11.5	34.5	12	25	50
(35) L-Serine	10.5	10.5	26		12.5
(36) L-Taurine					
(37) L-Threonine	4	12	12	37.5	75
(38) L-Tryptophan	0.5	2	6	20	40
(39) L-Tyrosine	2	5	18	40	40
(40) L-Valine	3.5	2	12	25	65
Vitamins					
(41) p-Aminobenzoic acid					
(42) Ascorbic acid			0.2		17.5
(43) D-Biotin	0.024	0.007	0.2	0.1	0.02
(44) Choline chloride	0.7	14	5	3	250
(45) Folic acid	1.3	1.3	2	0.1	0.5
(46) Folinic acid			0.1		

[c] 0.05 $Fe(NO_3)_3 \cdot 9H_2O$ + 0.45
    $FeSO_4$ anhydrous.
[j] $4H_2O$.

Table 2 (continued)
CHEMICALLY DEFINED CULTURE MEDIA

Constituents, mg/l	Ham F10[2,4,17]	Ham F12[2,4,18]	Neuman and Tytell[19]	N16[2,20]	MAB 87/3[21]
Vitamins (continued)					
(47) *i*-Inositol	0.54[m]	18[m]	9	1[m]	1 (M)
(48) Lipoic acid	0.2	0.2			
(49) Niacin					
(50) Nicotinamide	0.6	0.04	0.5	3	1
(51) Pantothenate (Ca)	0.7	0.24	0.2	3	1
(52) Pyridoxal HCl			0.5		
(53) Pyridoxine HCl	0.2	0.06	0.5	0.5	1
(54) Riboflavine	0.4	0.04	0.2	0.5	1
(55) Thiamine HCl	1	0.3		5	10
(56) Vitamin A					
(57) Vitamin B_{12}	1.4	1.4	0.00075		0.2
(58) Vitamin D_2					
(59) Vitamin E (toco-pherol)					
(60) Vitamin K3 (menadione)					
Other constituents					
(61) Adenine sulfate					
(62) Adenine 5′-triphosphate					
(63) Adenylic acid					
(64) Cholesterol					
(65) Cocarboxylase					
(66) Coenzyme A					
(67) Deoxyadenosine					
(68) Deoxycytidine					
(69) Deoxyguanosine					
(70) 2-Deoxy-D-ribose					
(71) Ethanol					
(72) Flavine adenine dinucleotide					
(73) D-Galactose					
(74) D-Glucosamine HCl					
(75) D-Glucose	1,100	1,800	3,000	1,100	5,000
(76) D-Glucuronolactone					
(77) Glutathione			0.5		15
(78) Guanine					
(79) Hypoxanthine	4	4		25	25
(80) Insulin					8
(81) α-Ketoglutaric acid					
(82) Linoleic acid		0.08			
(83) Methyl cellulose					
(84) 5-Methyl cytosine					
(85) 5-Methyl deoxycytidine					
(86) Methyl oleate			15		
(87) Nicotine adenine dinucleotide					
(88) Nicotine adenine trinucleotide phosphate					
(89) Phenol red	1.2	1.2	2.5	5	10
(90) Polysorbate 80					
(91) Polyvinyl pyrrolidone					
(92) Putrescine·2HCl		0.16			

Table 2 (continued)
CHEMICALLY DEFINED CULTURE MEDIA

Constituents, mg/l	Ham F10[2,4,17]	Ham F12[2,4,18]	Neuman and Tytell[19]	N16[2,20]	MAB 87/3[21]
Other constituents (continued)					
(93) D-Ribose					
(94) Salmine sulfate			5		
(95) Sodium acetate·3H$_2$O					
(96) Sodium glucuronate					
(97) Sodium mucate			23		
(98) Sodium pyruvate	110	110	110		
(99) Thymidine	0.7	0.7			8
(100) Thymine					
(101) Uracil					
(102) Uridine 5'-triphosphate					
(103) Xanthine			Also 2,000 mg/l of lactalbumin hydrolysate		

Constituents, mg/l	MB 752/1[22]	MD 705/1[23]	L15[2,4,8]	Birch and Pirt[25]	DM 20[26]
Inorganic salts					
(1) CaCl$_2$	120^g	120^g	140	1^g	264
(2) CoCl$_2$·6H$_2$O		0.11			
(3) CuSO$_4$·5H$_2$O		0.25		0.5	0.04−0.06
(4) FeSO$_4$·7H$_2$O		0.26		1	0.104−0.108
(5) KCl	150	150	400	400	200
(6) KH$_2$PO$_4$	80	80	60		177
(7) MgCl$_2$·6H$_2$O	240	240	200	200	100
(8) MgSO$_4$·7H$_2$O	200	100	200		
(9) MnSO$_4$·H$_2$O		0.08		0.1^i	0.03
(10) NaCl	6,000	6,000	8,000	5,900	8,000
(11) NaHCO$_3$	2,240	2,240		2,500	1,000
(12) Na$_2$HPO$_4$	300	300	50		35^f
(13) NaH$_2$PO$_4$·H$_2$O				500^g	
(14) (NH$_4$)$_6$Mo$_7$O$_{24}$·4H$_2$O		0.12			
(15) ZnSO$_4$·7H$_2$O		0.15		1	
Amino acids					
(16) L-Alanine			450°	90	400
(17) L-α-Aminobutyric acid					
(18) L-Arginine	75^k	75^k	500	450	100
(19) L-Asparagine			250		25
(20) L-Aspartic acid	60	60			
(21) L-Cysteine	90^k	90^k	120		80
(22) L-Cystine	15	15		75	
(23) L-Glutamic acid	150	150		1,530	150
(24) L-Glutamine	350	350	300	100	
(25) Glycine	50	50	200	15	15
(26) L-Histidine	150^k	150^k	250	99	30
(27) L-Hydroxyproline					
(28) L-Isoleucine	25	25	25	180	150
(29) L-Leucine	50	50	125	180	400

^{f} 12H$_2$O.
^{i} MnCl$_2$·4H$_2$O.

Table 2 (continued)
CHEMICALLY DEFINED CULTURE MEDIA

Constituents, mg/l	MB 752/1[22]	MD 705/1[23]	L15[2,4,8]	Birch and Pirt[25]	DM 20[26]
Amino acids (continued)					
(30) L-Lysine	240[k]	240[k]	75	175	100
(31) L-Methionine	50	50	150[o]	30	80
(32) L-Ornithine					
(33) L-Phenylalanine	50	50	250[o]	70	80
(34) L-Proline	50	50			
(35) L-Serine			200	20	20
(36) L-Taurine					
(37) L-Threonine	75	75	600[o]	100	100
(38) L-Tryptophan	40	40	20	20	40
(39) L-Tyrosine	40	40	300	70	50
(40) L-Valine	65	65	200	150	85
Vitamins					
(41) p-Aminobenzoic acid					
(42) Ascorbic acid	17.5	17.5			40
(43) D-Biotin	0.02	0.02		0.1	0.002
(44) Choline chloride	250	240	1	20	250
(45) Folic acid	0.4	0.5	1	1	0.01
(46) Folinic acid		0.4			
(47) i-Inositol	1 (M)	1 (M)	2	2 (M)	
(48) Lipoic acid					
(49) Niacin					
(50) Nicotinamide	1	1	1	1.2	5
(51) Pantothenate (Ca)	1	1	1	1.2	1
(52) Pyridoxal HCl					
(53) Pyridoxine HCl	1	1	1	1	1
(54) Riboflavine	1	1	0.1	0.2	1
(55) Thiamine HCl	10	10	1[p]	2	10
(56) Vitamin A					
(57) Vitamin B_{12}	0.2	0.2		1	0.005
(58) Vitamin D_2					
(59) Vitamin E (toco-pherol)					
(60) Vitamin K3 (menadione)					
Other constituents					
(61) Adenine sulfate					
(62) Adenine 5'-triphosphate					
(63) Adenylic acid					
(64) Cholesterol					
(65) Cocarboxylase					
(66) Coenzyme A					
(67) Deoxyadenosine					
(68) Deoxycytidine					
(69) Deoxyguanosine					
(70) 2-Deoxy-D-ribose					
(71) Ethanol					
(72) Flavine adenine dinucleotide					
(73) D-Galactose			900		
(74) D-Glucosamine HCl					
(75) D-Glucose	5,000	5,000		2,000	1,000
(76) D-Glucuronolactone					

[p] Monophosphate

Table 2 (continued)
CHEMICALLY DEFINED CULTURE MEDIA

Constituents, mg/l	MB 752/1[22]	MD 705/1[23]	L 15[2,4,5]	Birch and Pirt[25]	DM 20[26]
Other constituents (continued)					
(77) Glutathione	15	15			
(78) Guanine					
(79) Hypoxanthine	25	25		10	
(80) Insulin					
(81) α-Ketoglutaric acid				80	
(82) Linoleic acid					
(83) Methyl cellulose				1,000	
(84) 5-Methyl cytosine					
(85) 5-Methyl deoxycytidine					
(86) Methyl oleate					
(87) Nicotine adenine di-nucleotide					
(88) Nicotine adenine tri-nucleotide phosphate					
(89) Phenol red	10	10	10		
(90) Polysorbate 80					
(91) Polyvinyl pyrrolidone				1,000	
(92) Putrescine·2HCl					
(93) D-Ribose					
(94) Salmine sulfate					
(95) Sodium acetate·3H$_2$O					
(96) Sodium glucuronate					
(97) Sodium mucate					
(98) Sodium pyruvate			550	220	
(99) Thymidine					
(100) Thymine					
(101) Uracil					
(102) Uridine 5'-triphosphate					
(103) Xanthine					

REFERENCES

1. Eagle, H., *Science,* 122, 501–504, 1955.
2. Morton, H. J., *In Vitro,* 6, 89–108, 1970.
3. Eagle, H., *Science,* 130, 432–437, 1959.
4. Rutzky, L. P. and Pumper, R. W., *In Vitro,* 9, 468–469, 1975.
5. Joklik, W. K., quoted in GIBCO BIO-CULT Catalog, 1974–75, as personal communication.
6. Macpherson, I. A. and Stoker, M. G. P., *Virology,* 16, 147–151, 1962.
7. Stanners, C. P., Eliceira, G. L., and Green, H., *Nature New Biol.,* 230, 52–53, 1971.
8. Morgan, J. F., Morton, H. J., and Parker, R. C., *Proc. Soc. Exp. Biol. Med.,* 73, 1–8, 1950.
9. Parker, R. C., in *Methods of Tissue Culture,* 3rd ed., Harper and Row, New York, 1961, pp. 74–75.
10. Healy, G. M. and Parker, R. C., *J. Cell Biol.,* 30, 531–538.
11. Healy, G. M., Teleki, S., von Seefried, A., Walton, M. J., and Macmorine, H. G., *Appl. Microbiol.,* 21, 1–5, 1971.
12. McCoy, T. A., Maxwell, M., and Kruse, P. F., *Proc. Soc. Exp. Biol. Med.,* 100, 115–118, 1959.
13. Moore, G. E. and Kitamura, H., *N.Y. State J. Med.,* 68, 2054–2060.
14. Moore, G. E., Gerner, R. E., and Minowada, J., in *The Proliferation and Spread of Neoplastic Cells,* Williams & Wilkins Co., Baltimore, 1968, pp. 41–60.
15. Moore, G. E., Gerner, R. E., and Franklin, H. A., *J. Am. Med. Assoc.,* 199, 519–524, 1967.
16. Evans, V. J., Bryant, J. C., Kerr, H. A., and Schilling, E. L., *Exp. Cell Res.,* 36, 439–474, 1964.
17. Ham, R. G., *Exp. Cell Res.,* 29, 515–526, 1963.
18. Ham, R. G., *Proc. Natl. Acad. Sci. U.S.A.,* 53, 288–293, 1965.
19. Neuman, R. E. and Tytell, A. A., *Proc. Soc. Exp. Biol. Med.,* 104, 252–256, 1960.
20. Puck, T. T., Cieciura, S. J., and Robinson, A., *J. Exp. Med.,* 108, 945–956, 1958.
21. Gorham, L. W. and Waymouth, C., *Proc. Soc. Exp. Biol. Med.,* 119, 287–290, 1965.
22. Waymouth, C., *J. Natl. Cancer Inst.,* 22, 1003–1017, 1959.
23. Kitos, P. A., Sinclair, R., and Waymouth, C., *Exp. Cell Res.,* 27, 307–316, 1962.
24. Leibovitz, A., *Am. J. Hyg.,* 78, 173–180, 1963.
25. Birch, J. R. and Pirt, S. J., *J. Cell Sci.,* 7, 661–670, 1970.
26. Takaoka, T. and Katsuta, H., *Exp. Cell Res.,* 67, 295–304, 1971.

TISSUE CULTURES AND CULTURE MEDIA FOR
MICROORGANISMS OF THE TRIBE *RICKETTSIEAE*

R. K. Gerloff

The composition of the order *Rickettsiales,* according to the latest edition of *Bergey's Manual,*[1] and the highlights of requirements for growth of species within the order are discussed in this Handbook by R. A. Ormsbee in his chapter, "Qualitative Requirements and Utilization of Nutrients: Rickettsiales." The purpose here will be to consider tissue cultures and cell-free media which are suitable for propagating organisms of the tribe *Rickettsieae,* all of which are pathogenic for man. They are known collectively as the rickettsias, a term comparable to bacteria, viruses, and fungi, and belong to the genera *Rickettsia, Coxiella,* and *Rochalimaea.* These are the three principal genera of the order *Rickettsiales* to which tissue culture techniques have been applied. Such techniques have also had some application in the study of rickettsia-like organisms of the genus *Ehrlichia,*[2] tribe *Ehrlicheae,* and of the genera *Wolbachia*[1] and *Rickettsiella,*[1,3] tribe *Wolbachieae.*

Rochalimaea quintana, formerly called *Rickettsia quintana,* is the etiologic agent of trench fever. It differs from other rickettsias inasmuch as it resides in an extracellular environment in its arthropod host, the human body louse, and is able to grow on cell-free media. Cultivation of the organism on a cell-free medium was first reported in 1961.[4] The medium consisted of blood-agar base, 6% inactivated horse serum, and 4% lysed horse erythrocytes. Growth in a liquid medium was first reported in 1967.[5] The media used in early studies contained lysate of red blood cells as an essential ingredient. Subsequently, much effort was directed toward finding substitutes for this lysate,[6,7] a more precise definition of the growth requirements of the organism,[7,8] and development of a transparent liquid medium which would permit rickettsial growth to be quantitated turbidimetrically.[9] Crystalline hemoglobin or hemin can be used in place of lysed erythrocytes, and bovine albumin or a colloidal substance such as starch or charcoal can substitute for serum.[7] Fetal calf serum, but not calf serum, can substitute for hemoglobin in a liquid medium.[9]

All other rickettsias are obligate intracellular parasites. The advent of present-day tissue culture techniques has given the rickettsiologist an important new tool for cultivating these organisms. Previously, experimental work was confined to laboratory animals, yolk sacs of embryonated eggs, and natural arthropod vectors, including ticks, fleas, mites, and lice. The kinds of cells in which rickettsias have been studied and the media used for maintaining such cells after infection will be discussed in this chapter. Because certain technical procedures and environmental factors often play an important part in successful cultivation of rickettsias in tissue cultures, as compared with cultivation of free-living bacteria in bacteriological media, some of these other aspects will be considered here also.

Cultivation of rickettsias in tissue culture dates back to work by Wolbach and Schlesinger in 1923.[10] Other early workers in the field included Zinsser, Batchelder, Fitzpatrick, and Wei,[11,12] Pinkerton and Haas,[13] Nigg and Landsteiner,[14] Kligler and Aschner,[15] Burnet,[16] and Cox and Bell.[17] These pioneer investigators were successful in cultivating rickettsias, principally those of typhus, Rocky Mountain spotted fever, and Q fever, in minced tissues such as guinea pig tunica or chick embryo and in cells that migrated from explants of these tissues and formed monolayers. Although their contributions to rickettsiology are not to be discounted, further details of their methods and findings will not be included in this chapter because, for the most part, the techniques and media that they used are now considered obsolete. Very few publications dealing with rickettsias in tissue culture appeared during the years 1940 through 1954. By the end of this period, propagation of viruses, notably poliovirus, was becoming a very popular subject, and methods, media, and cells that were useful to the virologist soon became equally useful to the rickettsiologist. The remainder of this chapter will deal principally with information accumulated since 1954 and published in the English language.

Table 1 provides a guide to studies that have been performed in tissue culture with the various strains of each rickettsial species. It is readily apparent that more work has been done with *Coxiella burnetii, Rickettsia rickettsii, Rickettsia prowazekii,* and *Rickettsia tsutsugamushi,* as representatives of the four rickettsial biotypes[1] and as agents of the four most important rickettsial diseases, than with other species.

Table 2 provides a guide to all studies done with rickettsias of the various species in each of several kinds of host cell, whether cultivation of the organisms in the designated cell was successful or not. In this table and subsequent ones, rickettsial species are divided into the four groups or biotypes: Q fever, spotted fever, typhus, and scrub typhus. It is evident from Table 2 that a wide variety of cells have been employed, including primary and secondary passage cells, various lines of heteroploid cells, strains of diploid cells, and tissues of arthropod origin. Primary chick embryo fibroblasts and mouse L cells, including clones derived from L cells, have been most commonly used.

Table 3 provides an index to the literature on various topics related to growth of rickettsias in tissue cultures. Topics were selected in an effort to aid the reader in understanding such aspects as the mechanics of rickettsial infection; mode of growth in susceptible cells; means for detecting rickettsias in the cells; morphology and structural details of rickettsias; use of tissue culture for assaying rickettsial infectivity; production of rickettsias in tissue cultures as the starting material for biologicals; and environmental factors, both physical and chemical, that affect the growth of rickettsias in tissue cultures.

In cultivation of rickettsias for purposes other than the plaque test, which will be discussed separately, the medium on the cells may be any of several mixtures. Table 4 gives examples of combinations that have been used successfully. The variation shown is probably due not so much to broad differences in the requirements of rickettsial species as it is to differences in routine cell culture maintenance procedures which develop within laboratories. Whatever the medium used, it should be capable, preferably without replacement over several days, of maintaining control monolayers in a healthy condition during the time needed for the desired rickettsial growth to occur in infected monolayers. The pH should remain in the range that is optimal for the cell system, usually 6.8 to 7.4 but as low as 6.4 for some insect cells. Control of pH is important in rickettsial production. With *Rickettsia typhi,* for example, infected cells lower the pH even more rapidly than uninfected cells, and the acidity eventually causes the metabolic activity of both cells and rickettsias to fall irreversibly to negligible levels.[66] Similarly, accelerated acid production by L cells after they are infected with *R. rickettsii* has been observed repeatedly by the author.

In general, a tissue culture medium consists of a base to which serum and, in some cases, other nutrients are added. The two most commonly used bases are Medium 199, developed by Morgan, Morton, and Parker,[94] and minimum essential medium (MEM), developed by Eagle.[95] The formulae for preparing these media are shown in Table 5. These and several other commercially available tissue culture media are included in a review by Morton.[96]

Eagle's basal medium[97] preceded MEM. They differ in that MEM contains larger concentrations of most of the amino acids, and their proportions are more nearly what they are in the proteins of cultured human cells. Thus, MEM permits cell cultures to be kept for somewhat longer periods without refeeding. Either medium requires the addition of 5 to 10% serum, either whole or dialyzed. Medium 199 is a complex mixture containing some ingredients that have subsequently been proven nonessential for cells. In the absence of serum it will keep cells alive only for relatively short periods. Most investigators supplement it with 2 to 5% serum, although bovine plasma albumin (Fraction V; Armour Pharmaceutical Co., Chicago, Illinois) in a final concentration of 0.5 to 1% has been used with equal success.

It is evident from Table 4 that the kinds of sera used in tissue culture media for rickettsial production have been rather limited. They have included horse, calf, fetal calf, human, guinea pig, and monkey sera. The two bovine sera, collected from young calves and calf fetuses, are now the most popular kinds of sera used, partly because such sera are readily available as by-products of the meat industry, and partly because they contain proteins which aid cells in attaching to and spreading over glass and plastic surfaces. Whatever the kind of serum, it is generally added to medium bases in concentrations of 2 to 10%. Some investigators prefer to inactivate the serum first (56° C for 30 min) in order to inactivate *Mycoplasma* contaminants[98] and to lessen the chances for contamination by viruses which might have been present in the animal blood and which escaped removal by filtration. Yunker and Cory, in working with insect cells, found that better growth was obtained if serum for medium was inactivated at 56°C for 1 hr, presumably because some toxic component was thereby destroyed. They now routinely treat all serum in this manner, whether it is used for insect or mammalian cells in viral or rickettsial studies.[99,111]

Serum is usually filtered through 0.22-μm membrane filters, preferably twice, before it is added to a medium base. The purpose is to remove microbial contaminants, including mold spores, yeasts, bacteria, larger viruses, and most, if not all, mycoplasmas. The last-named organisms, because of their insidious nature, represent a notorious enemy of the tissue culturist. It is very difficult to eliminate them once they are established in a line of cells; hence, the best solution to the problem of mycoplasma contamination is to prevent their entry.

Of the miscellaneous organic nutrient additives, lactalbumin hydrolysate is probably the most common. Others include Yeastolate (Difco Laboratories, Detroit, Michigan), Fraction V of bovine plasma albumin, tryptose phosphate broth, the so-called nonessential amino acids, and L-glutamine in quantity beyond that provided by the medium base itself. Most tissue culture media contain sodium bicarbonate, added for its buffering effect. Inclusion of it necessitates culturing cells in gas-tight vessels; otherwise, escape of CO_2 from the medium is unimpeded and the resulting rise in pH of the medium will eventually be deleterious to the cells. If the culture vessel cannot be sealed (e.g., a Petri dish used in plaque tests), it must be kept in an incubator which provides a humidified atmosphere containing 5% CO_2. Transparent gas-impermeable plastic bags (Anaerobag, Cedanco Co., Framingham, Massachusetts), which can be sealed with heat after introduction of the desired gas mixture,[100] have also been used successfully in place of more costly CO_2 incubators.[112]

The amount of $NaHCO_3$ in a medium can be altered according to the concentration of cells in the culture. In Eagle MEM, for example, it may be reduced from 2000 to 400 mg/l in order to minimize the alkalization of the medium in the early stages of cell growth. However, in production of rickettsias in tissue cultures, mature monolayers are ordinarily used, so the emphasis is more on achieving maximum buffering capacity and preventing the medium from becoming too acid.

Leibovitz L-15 medium[101] permits growth and maintenance of cell cultures in free gas exchange with the atmosphere. It is a medium in which the free base amino acids of L-arginine, L-histidine, and L-cysteine are used in place of bicarbonate as the buffer, and D(+) galactose, sodium pyruvate, and DL-α-alanine are substituted for glucose. The organic buffer *N*-2-hydroxyethylpiperazine-*N'*-2-ethane sulfonic acid (HEPES) has also been included in media,[32,66,72] and more effective buffering has been obtained with it than with sodium bicarbonate.

Whether or not it is specifically mentioned in formulae, phenol red is nearly always a constituent of tissue culture media and serves as an acid-base indicator. The concentration is usually 0.001 to 0.002%. This dye, which is red or red-purple in an alkaline environment, orange at neutrality, and yellow in an acid environment, serves as a convenient indicator of the general condition of a tissue culture. A strongly alkaline condition might suggest a leak in the vessel which is allowing CO_2 to escape, some chemical contaminant in the vessel, and/or dead, degenerating cells. A decrease in pH of the medium is a normal consequence of cellular metabolism and is desirable within limits. However, the pH can reach a level so low that it is harmful to cells. This level, which varies with different kinds of cells, indicates that the medium should be replaced or at least its pH should be adjusted upward.

A final consideration in tissue culture media is that of antibiotics for inhibition of microbial contaminants. Because rickettsias are susceptible to many antibiotics, the kind and concentration of the latter must be chosen with care. This restriction is in contrast to the relatively free use of antibiotics that is permissible with either normal or virus-infected cells. Penicillin and streptomycin, the most commonly used antibiotics in tissue culture, affect rickettsias in varying degrees, depending on the species. They are frequently included in media for rickettsial studies, yet many investigators prefer to omit antibiotics completely lest they have some effect, however small, on rickettsial growth.

Table 6 reviews rickettsial studies in which penicillin and streptomycin have been used. If 100 units of penicillin and 100 μg of streptomycin/ml are considered concentrations that provide reasonable protection against bacterial contaminants, data in the table indicate that this degree of protection cannot be utilized without some risk of affecting rickettsial growth, except in studies of *R. tsutsugamushi*. However, the time of adding the antibiotics is also an important factor. They have less effect if they are introduced after the interval necessary for rickettsias to enter the cells. Hence, stock cell cultures can be maintained in medium containing antibiotics without affecting the future susceptibility of subcultures to rickettsial infection.

A relatively new antibiotic named gentamycin (Schering Corporation, Bloomfield, New Jersey) has appealed to tissue culturists because of its unusual properties. It withstands autoclaving, is stable over a wide range of pH, inhibits the growth of *Mycoplasma* as well as a broad spectrum of bacteria, and has no effect on replication of many viruses.[102] The recommended concentration for inhibiting bacterial growth in tissue culture media is 50 μg/ml, yet concentrations up to 20 times this great are nontoxic to many cell lines. However, gentamycin is not generally suitable for work with rickettsias. Wisseman et al.[76] found that only 10 μg/ml of gentamycin in agar overlay completely inhibited plaque formation by *R. prowazekii*. Ormsbee and Peacock,[103] using chick embryo fibroblasts, obtained similar results with this antibiotic against *Rickettsia canada*, *Rickettsia conorii*, *R. typhi*, and *R. rickettsii*.

One of the chief advantages of tissue culture in rickettsiology has been the provision of a more definitive method for infectivity assay via the plaque test. Plaques also permit establishment of rickettsial clones which are useful in genetic studies and which help to rule out the possibility of having rickettsias with multiple genetic backgrounds in a given strain. In addition, plaque tests have been useful for studying the susceptibility of rickettsias to antibiotics[51,76] and the neutralization of rickettsias by homologous antisera.[64]

Mechanics of the plaque test for rickettsias are similar to those for viral plaque tests. Incubation times are generally longer, however, and temperatures lower. Table 7 cities studies in which various rickettsias have been successfully plaqued in one or more kinds of cells. Primary chick embryo fibroblasts have been used most widely and serve as a universal host for plaquing rickettsias of all four biotypes. However, several spotted fever (SF) and typhus group rickettsias also form plaques in Vero cells,[32] and *R. conorii* produces plaques in a variety of cells.

More difficulty has been encountered in plaquing *C. burnetii* than any other rickettsia. In some studies plaques either did not form[30,32] or were very small (1 mm) and required 8 to 10 days to appear, yielding an assay method with sensitivity lower than that of standard assay methods.[42] With *C. burnetii*, 9 Mile strain, no difference in plaquing ability was observed between phase I and II organisms.[31] In general, plaquing is most satisfactory with members of the SF group. Plaques are larger, better defined with sharper margins, and become visible earlier. They usually acquire a diameter of 2 mm (occasionally up to 3 mm) and appear in 5 to 7 days. In contrast, members of the typhus group generally produce plaques 1 mm or less in diameter, and these require 10 to 14 days to reach countable size and clarity. Plaques produced by *R. tsutsugamushi* are similar to *C. burnetii* plaques in size but require an additional week for maturity.[42]

Table 8 lists the major alternative conditions that have been used in plaque procedures and provides references. In any procedure it is customary to remove the growth medium from the confluent cell monolayer, add the rickettsial inoculum in a small volume (usually 0.1 ml), and provide some means, either manual or mechanical, for keeping the inoculum spread over the monolayer during an adsorption period. Subsequently, the nutrient overlay is added and the cultures are incubated. With any rickettsia, various environmental factors affect the size and time of appearance of plaques.[31,32,50,56,68] These include the kind of host cell, number of cells planted initially in each vessel, the solution used for diluting the rickettsias prior to inoculation of monolayers, medium used in the agar overlay, and temperature of incubation. Temperatures for best plaque formation in all studies cited were nearly always between 32 and 35°C. In a few instances,[32] temperatures above and below this range were preferable for rickettsias of the typhus group.

The choice of diluent for the rickettsial inoculum appears to be a very important factor. Among several comparisons that have been made,[32,50,56] brain heart infusion (BHI)[56] was found to be superior to all other media tested in preventing both a significant decrease in plaque-forming units and a delay in plaque formation, as regards *R. rickettsii* and *R. typhi*. Other investigators[50] found Medium 199 with 5% calf serum as satisfactory as BHI. In the third study cited,[32] phosphate-buffered saline with bovine plasma albumin (Fraction V) was found superior to BHI for three members of the SF group, as shown by number and size of plaques formed.

Antibiotics are usually omitted in performing plaque tests, for reasons indicated in Table 6. In one study,[30] inclusion of 200 units/ml of penicillin and 250 µg/ml of streptomycin in the nutrient overlay completely inhibited plaque formation by *Rickettsia akari*, *R. prowazekii*, *R. typhi*, and *R. conorii*. Decreasing the quantities to 100 units/ml and 50 µg/ml, respectively, severely affected the size and number of plaques but did not completely inhibit them.[50] Still smaller quantities incorporated in paper discs placed atop the agar surface had no effect.[51] The organic buffer, HEPES, has been employed in agar overlay medium without effect on plaque formation.[32]

Infection of cells in monolayers can be enhanced by centrifuging rickettsias onto the monolayers. This increase in infection rate is due, at least in part, to the fact that the organisms are brought into close contact with cells before they undergo thermal inactivation.[45] Logically, a greater increase would be expected for rickettsias that are relatively thermolabile. Conversely, with *C. burnetii*, which is very stable at 37°C, centrifugation is less advantageous since practically the same amount of adsorption will eventually occur in stationary cultures.

Centrifugation of monolayers presents certain technical problems, since adapters for all types of vessels are not commercially available and must be improvised. The most convenient vessel is a 16- × 100-mm flat-bottomed tube containing a 12-mm coverslip on which a cell monolayer is established. Such tubes can be centrifuged in a horizontal rotor.[45] Modified basket-type rotors with special adapters have been used to centrifuge plastic tissue culture flasks.[31] Different conditions which have been used for centrifugation and the advantages gained are summarized in Table 9.

Use of irradiated cells also produces better rickettsial growth than can be obtained with nonirradiated cells. The effects of a wide range of radiation dosages on host cells themselves and on growth of *R. prowazekii* were described by Weiss and Dressler.[79] In subsequent studies in their laboratory[55,65,66,81] and in the laboratory of Wisseman and associates,[72,76,77] dosages of 3000 to 5000 R, as either gamma radiation from cobalt[60] or X-radiation from a radiation therapy unit, were used. The cells, either LM_3, L-929, primary chick embryo, or primary duck embryo, were irradiated in concentrated suspension and then planted in tissue culture vessels. Monolayers formed which consisted of unusually large cells which were viable but incapable of multiplying. They were subsequently infected.

Cells which grow and are maintained as a monodispersed suspension have also been used in rickettsial studies. Cohn et al.[85] and Bozeman et al.[47] used strain MB III mouse lymphoblasts in this manner. The medium consisted of 40% Gey's balanced salt solution, 10% beef embryo extract, and 50% horse serum. Cells were kept in suspension by placing the vessels in a roller drum. Hopps et al.[86,87] used the L-929 strain of mouse fibroblasts in a similar manner except that the medium was altered to contain 20% chick embryo extract and 40% horse serum. In all four studies cited, growth of *R. tsutsugamushi* was predominantly studied, but it was also determined that *R. rickettsii, R. prowazekii, R. typhi*, and *C. burnetii* would infect MB III cells under these conditions.

Burton et al. used suspended L-929 cells as host for *C. burnetii.*[37] The cell suspensions were agitated by placing them in Erlenmeyer flasks on a rotary shaker. The medium was Eagle MEM (spinner modified, meaning that calcium was omitted to help prevent cells from clumping) containing 5% calf serum and 0.1% methyl cellulose.

For additional information on the general aspects of tissue culture and of working with rickettsias, the reader is referred to any of several textbooks on these subjects.[104-109] A review of the use of arthropod tissue cultures in the study of rickettsias was provided by Yunker,[92] and Weiss reviewed the overall subject of growth and physiology of rickettsias.[110]

Table 1
GUIDE TO LITERATURE ON STUDIES WITH VARIOUS SPECIES AND STRAINS OF RICKETTSIAS IN TISSUE CULTURES

Species and strain	Ref.
Coxiella burnetii	
9 Mile	18–38
Geschwandtner	39
L-35	35, 39
AD (California bovine)	40–42
California	43–45
Henzerling	35, 46, 47
Derrick	48
Florian	35
Vanco	35
Australian QD	38
Rickettsia diaporica[a]	17
SND[b]	49
Rickettsia rickettsii	
Bitter Root (R)	32, 47, 50–59
Sheila Smith	31, 52, 60–62
SND[b] (from naturally infected ticks)	63
SND[b]	64
K43	49
Rickettsia akari	
MK	51, 60, 65, 66
Toger	29, 30, 67
SND[b]	31
Rickettsia conorii	
Simko	32, 53
M1	29, 30, 67
Malish	31
R42	49
SND[b]	60, 68
Rickettsia rhipicephali[c]	
SND[b]	69
Rickettsia sibirica	
SND[b]	31, 60
Rickettsia australis	
SND[b]	60, 64
Rickettsia montana[d]	
M/5-6B	32

[a] *R. diaporica* was an early name for the rickettsia of Q fever. Presumably, the agent used in this study was the one later designated as *C. burnetii*, 9 Mile strain.
[b] The strain used was not designated in the paper cited.
[c] Proposed name[90] for this spotted fever group agent from *Rhipicephalus sanguineus* ticks.[69]
[d] Described as a new species [91] but not recognized as such by *Bergey's Manual*.[1]

Table 1 (continued)
GUIDE TO LITERATURE ON STUDIES WITH VARIOUS
SPECIES AND STRAINS OF RICKETTSIAS IN TISSUE CULTURES

Species and strain	Ref.
Rickettsia prowazekii	
Breinl	29–32, 47, 49, 51, 54, 60, 67, 70–78
Madrid E	45, 49, 51, 54, 60, 71, 76, 77, 79
Cairo	31
ZRS	32
W2	67
Rickettsia typhi	
Wilmington	31, 32, 51, 57, 60, 66, 71, 73, 80–82
Mexican	29, 67
Warszawa	30
Rickettsia canada	
SND[b]	31, 32
Rickettsia tsutsugamushi	
Karp	42, 47, 54, 57, 73, 82–88
Gilliam	42, 47, 49, 55, 73, 82, 83
Kato	42, 73, 83
Rickettsia sennetsu[e]	
SND[b]	54, 89
Rickettsia quintana	
Fuller	4
Ossejik	47

[e] *R. sennetsu* is regarded as a *species incertae sedis* in *Bergey's Manual.*[1]

Table 2
GUIDE TO LITERATURE ON RICKETTSIAL STUDIES IN DIFFERENT KINDS OF HOST CELLS AS TISSUE CULTURES

| Host cell | | | Rickettsial species[a] | | | | | | | | | | | | | |
Cell designation	Species	Origin — Organ or tissue	Q fever — bur	Spotted fever group — rick	ak	con	sib	aus	mont	rhip[b]	Typhus group — prow	typ	can	Miscellaneous — tsu	sen	quin
L	Mouse	Connective tissue (areolar and adipose)	20,[c] 22, 33, 40, 41, 46, 52	52, 58, 59						69	70, 74, 75, 78			85		
L-929 (clone of L)	Mouse	Same as L	34, 36, 37		65	68										
LM₂, derived from L-929	Mouse	Same as L			66	68								86, 87		
HeLa	Human	Cervical epitheloid carcinoma	18, 39							69		66, 81				
KB	Human	Oral epidermoid carcinoma	18													
HEp-2	Human	Epidermoid carcinoma of larynx	18, 39													4
Detroit-6	Human	Sternal bone marrow	18, 21, 39			68										
WI-38	Human	Normal embryonic lung		52												
McCoy	Human		41		65											
I1	Human	Normal amnion	39													
Vero	Monkey (African green)	Kidney	32	32, 50, 52, 53, 93		32, 53			32	69	32	32	32		89	
BS-C-1	Monkey (African green)	Kidney		54							54, 73	73, 82		54, 73, 82, 83	54	
LLC-MK₂	Monkey (Rhesus)	Kidney			65											
OL	Monkey (Rhesus)	Kidney														
DBS-FRhL-2	Monkey (Rhesus)	Kidney	24, 27, 28			68										
HK	Hamster	Kidney		50												
	Cat	Kidney		50												
MBIII (lymphoblasts)	Mouse	Lymphosarcoma	47	47		47					47	47		47, 85		
Primary chick embryo	Chicken	Embryo	22, 25, 30, 31, 35, 42	31, 32, 50–52, 56, 60, 62–64	30, 51, 60, 65	30, 31, 60, 68				69	30–32, 51, 60, 72, 76	30–32, 51, 56, 60	31, 32	42, 84		47

[a] Meaning of species abbreviations: bur = *Coxiella burnetii*, rick = *Rickettsia rickettsii*, ak = *R. akari*, con = *R. conorii*, sib = *R. sibirica*, aus = *R. australis*, mont = *R. montana*, prow = *R. prowazekii*, typ = *R. typhi*, can = *R. canada*, tsu = *R. tsutsugamushi*, sen = *R. sennetsu*, quin = *Rochalimaea quintana*.

[b] Rickettsia from *Rhipicephalus sanguineus* ticks.[69] The name *Rickettsia rhipicephali* has been proposed.[90]

[c] Numbers indicate reference citations.

Table 2 (continued)
GUIDE TO LITERATURE ON RICKETTSIAL STUDIES IN DIFFERENT KINDS OF HOST CELLS AS TISSUE CULTURES

Rickettsial species[a]

Cell designation	Species	Organ or tissue	Q fever	Spotted fever group							Typhus group			Miscellaneous		
			bur	rick	ak	con	sib	aus	mont	rhip[b]	prow	typ	can	tsu	sen	quin
Secondary chick embryo	Chicken	Embryo	19, 26								76, 77					
Chick embryo yolk sac	Chicken	Yolk sac	43—45								45, 79					
Explants from avascular entoderm	Chicken	Embryo														
Chick embryo, Maitland-type tissue culture	Chicken	Embryo	48													
Chick embryo, modified Maitland-type tissue culture	Chicken	Embryo	17													
Primary duck embryo	Duck	Embryo		52, 55										55		
Primary bone marrow cells	Guinea pig	Bone marrow		61, 62												
Primary circulating monocytes	Guinea pig	Blood cells		61, 62												
Primary circulating monocytes	Human	Blood cells									71	71, 80				
14pf strain of fibroblasts	Rat	Subcutaneous areolar tissue		57										57		
Peritoneal macrophages	Meriones, mouse, guinea pig, rabbit, rat, hamster	Peritoneal macrophages	49	49		49					49			49		
Primary liver and kidney	Colubrid snakes	Liver and kidney	29		29	29					29	29				
Singh's Aedes albopictus cell line	Mosquito	Larval tissues	38	53, 93		53								88		
Hemocytes from Hyalomma dromedarii	Tick	Hemocytes	19													
Tissues from Hyalomma asiaticum asiaticum	Tick	Nymphal tissues	23													
Singh's Aedes aegypti cell line	Mosquito	Whole larvae												88		
Hsu's Culex quinquefasciatus cell line	Mosquito	Ovaries from adults												88		
Grace's Antheraea eucalypti cell line	Emperor gumworm	Ovaries from diapausing pupae	38	92							92			88		
Tissues from Hyalomma dromedarii	Tick	Nymphal tissues			67	67					67	67				

Table 3

GUIDE TO THE LITERATURE ON TOPICS RELATED TO GROWTH OF RICKETTSIAS IN TISSUE CULTURES

Specific subject	Ref.
Mere demonstration of the susceptibility or resistance of a particular kind of cell to rickettsial infection and growth	18, 21−24, 26, 29, 48, 49, 67, 69, 88
Light microscopy	
Giemsa stain	18, 24−26, 28, 29, 43−46, 54, 65, 70, 74, 77, 79, 83, 85, 88
Macchiavello stain	4, 22, 26, 44
Acridine orange stain (ultraviolet light)	46, 78
Giménez stain	38, 52, 61−63, 71, 76, 77, 81
Phase contrast microscopy	18, 57
Electron microscopy	21, 34, 37, 54, 75
Morphology of rickettsias; mechanism of infection and multiplication	18, 21, 22, 26, 29, 34, 39, 46, 57, 67, 70, 74, 75, 77, 78, 85
Environmental factors that affect penetration of rickettsias into host cells	47, 85, 86
Effect of tissue culture medium on rickettsial growth	43
Use of irradiated cells as host	55, 65, 66, 72, 76, 77, 79, 81
Enhancement of infection by centrifugation	31, 40, 45, 65
Growth in suspended cells	37, 38, 47, 85−87
Growth in roller cell cultures	38
Metabolism of rickettsias	55, 66
Effect of strain virulence on growth of typhus rickettsias	71, 77
Formation of plaques and their use for assay of infectivity	30−32, 42, 50, 51, 53, 56, 60, 63, 64, 68, 76
Infectivity assays other than plaque method	40, 43, 44, 83
Rickettsial counts per cell	71, 77
Ultrastructure of rickettsias	37, 54, 75
Development of rickettsial antigen	19
Use of fluorescent antibody staining	19, 22, 28, 33, 40, 46, 61−63, 70
Rickettsial plaque reduction with homologous antiserum and antiglobulin	64
Neutralization of rickettsias; infectivity of rickettsia-antibody complex	40, 50, 64, 83
Effect of immune serum on phagocytosis and intracellular destruction of rickettsias	72, 80
Susceptibility of rickettsias to antibiotics	22, 47, 50, 51, 71, 76, 82, 85, 87
Effect of metabolic inhibitors on rickettsial growth	87
Interferon	
Production by rickettsias	41, 84
Effect on rickettsial growth	25, 65
Interference of viral infection with subsequent rickettsial infection	25
Interference of rickettsial infection with subsequent viral infection	35
Mass cultivation of rickettsias for	
Diagnostic antigen preparation	38, 69, 73
Vaccine preparation	52
Purification studies	58, 59, 81
Use of tissue cultures for rickettsial isolation	61−63, 93
Phase variation in *Coxiella burnetii;* comparative infectivity and growth of phase I and II	20, 33, 36, 39
Rickettsial strain differentiation by type of cytopathic effect	83
Persistence of rickettsial infection in serially passaged cells	27−29, 89

Table 4
KINDS OF MEDIA USED FOR TISSUE CULTURES AFTER INFECTION WITH RICKETTSIAS[a]

Base	Serum			Other additives	Ref.
	% in final medium	Kind			
Medium 199	2	Calf			*35*,[b] 58
	5	Calf			40, 41
	10	Calf			*34*, 36
	10	Fetal calf			72, 76
	2	Calf		0.5% LaH[c]	*19, 46*
	10	Calf		0.5% LaH	*70*
	20	Fetal calf		0.1% Yeastolate (Difco)	54
Hanks BSS[c]	10	Horse			23, *24*
	10	Horse		0.5% LaH	*18, 21, 39*
	2	Bovine		0.5% LaH	89
Medium 199 and Eagle basal medium No. 2 in 1:1 ratio	4	Fetal calf		5% tryptose phosphate broth	82
	2 or 4	Calf			*73*
	2	Calf		5% tryptose phosphate broth and 2 m*M* glutamine	*83*
Eagle MEM[c]	2	Calf			52
	2	Calf		1% glutamate	*65*
	10	Horse		10% Hanks BSS	22
	30	Guinea pig		0.1 *M* L-glutamine	61, 62
	30	Monkey		0.1 *M* L-glutamine	62
Eagle medium (Dulbecco's modification)	10	Fetal calf			77
BSS and beef embryo extract in 5:1 ratio	40	Horse			57
Earle BSS and medium 199 in 2.4:1 ratio	15	Human			28
Eagle basal medium	30	Human			71

[a] This table does not consider kinds of tissue culture used in rickettsial plaque tests; see Table 7 for such information.
[b] Italicized reference number indicates that the investigators used heated serum (usually 56°C for 30 min).
[c] LaH = lactalbumin hydrolysate, BSS = balanced salt solution, MEM = minimum essential medium.

Table 5
FORMULAE OF TWO TISSUE CULTURE MEDIA COMMONLY USED FOR MAINTAINING CELLS AFTER INFECTION WITH RICKETTSIAS

Medium 199

Compound	Milligrams/liter
Amino acids	
L-Arginine	70.0
L-Histidine	20.0
L-Lysine	70.0
L-Tyrosine	40.0
DL-Tryptophane	20.0
DL-Phenylalanine	50.0
L-Cystine	20.0
DL-Methionine	30.0
DL-Serine	50.0
DL-Threonine	60.0
DL-Leucine	120.0
DL-Isoleucine	40.0
DL-Valine	50.0
DL-Glutamic acid	150.0
DL-Aspartic acid	60.0
DL-Alanine	50.0
L-Proline	40.0
L-Hydroxyproline	10.0
Glycine	50.0
L-Glutamine	100.0
Vitamins	
p-Aminobenzoic acid	0.05
Biotin	0.01
Calcium pantothenate	0.01
Choline	0.50
Folic acid	0.01
Inositol	0.05
Niacin	0.025
Niacinamide	0.025
Pyridoxal	0.025
Pyridoxine	0.025
Riboflavin	0.01
Thiamin	0.01
Vitamin A	0.10
Calciferol (vitamin D)	0.10
α-Tocopherol phosphate (vitamin E)	0.01
Menadione (vitamin K)	0.01
Coenzymes	
Adenosine triphosphate	10.0
Reducing agents	
Ascorbic acid	0.05
Cysteine	0.1
Glutathione	0.05
Nucleic acid derivatives	
Adenine	10.0
Guanine	0.3
Hypoxanthine	0.3
Thymine	0.3
Uracil	0.3
Xanthine	0.3
Adenylic acid	0.2

Table 5 (continued)
FORMULAE OF TWO TISSUE CULTURE MEDIA
COMMONLY USED FOR MAINTAINING CELLS
AFTER INFECTION WITH RICKETTSIAS

Compound	Milligrams/liter
Lipid sources	
Cholesterol	0.2
Tween® 80 (oleic acid)	20.0
Carbohydrate sources (other than glucose)	
Ribose	0.5
Desoxyribose	0.5
Inorganic salts	
NaCl	6800.0
KCl	400.0
$CaCl_2$	200.0
$MgSO_4 \cdot 7H_2O$	200.0
NaH_2PO_4	140.0
$NaHCO_3$	2200.0
$Fe(NO_3)_3 \cdot 9H_2O$	0.1
Sodium acetate	50.0
Glucose	1000.0

Eagle Minimum Essential Medium

Compound	mM	Milligrams/liter
L-Amino acids		
Arginine	0.6	105
Cystine	0.1	24
Glutamine	2.0	292
Histidine	0.2	31
Isoleucine	0.4	52
Leucine	0.4	52
Lysine	0.4	58
Methionine	0.1	15
Phenylalanine	0.2	32
Threonine	0.4	48
Tryptophan	0.05	10
Tyrosine	0.2	36
Valine	0.4	46
Vitamins		
Choline		1
Folic acid		1
Inositol		2
Nicotinamide		1
Pantothenate		1
Pyridoxal		1
Riboflavin		0.1
Thiamine		1
Carbohydrate		
Glucose	5.5	1000
Salts		
NaCl	116	6800
KCl	5.4	400
$CaCl_2$	1.8	200
$MgCl_2 \cdot 6H_2O$	1.0	200
$NaH_2PO_4 \cdot 2H_2O$	1.1	150
$NaHCO_3$	23.8	2000

Note: For use on cells, these media are supplemented with serum in 2 to 10% concentration.

Table 6
EFFECT OF PENICILLIN AND STREPTOMYCIN ON RICKETTSIAL INFECTION AND GROWTH IN TISSUE CULTURES

Rickettsia	Antibiotic concentration		Relative time for introducing rickettsias and antibiotics[b]	Observations	Ref.
	Penicillin (units or µg/ml[a])	Streptomycin (µg/ml)			
Coxiella burnetii	300		After	RGO[c]	19
	200		After	RGO	36
		100	After	Number of infected cells reduced 50%	22
		500	Together	Complete inhibition of infection and growth	22
	100	20	Together	Marked effect on number of infected cells in suspended culture	47
	200	200	After	RGO	39
	200	200	After	RGO	18
	200	200	Not stated	RGO	21
Rickettsia rickettsii	100	20	Together	Marked effect on number of infected cells in suspended culture	47
	100	50	After	Both number and size of plaques greatly reduced	50
	5		Together	RGO	58
Rickettsia akari	200	250	Before and after	Complete inhibition of plaques	30
Rickettsia conorii	200	250	Before and after	Complete inhibition of plaques	30
Rickettsia prowazekii	20—50 µg		After	In plaque tests, MIC[d] was >20 and <50 µg/ml.	76
		100	After	In plaque tests number of plaques was reduced to 63% of that in controls.	76
	100	20	Together	Marked effect on number of infected cells in suspended culture	47
	200	250	Before and after	Complete inhibition of plaques	30

[a] Quantity of penicillin is expressed in units unless otherwise specified.

[b] Before = rickettsial inoculum treated with antibiotics before addition to cells; together = rickettsias and antibiotics added to cells simultaneously or antibiotics introduced immediately after rickettsias; after = antibiotics introduced after a rickettsial adsorption period of 15 to 120 min.

[c] RGO = rickettsial growth occurred but comparative data on the amount of growth with and without antibiotics were not provided.

[d] MIC = minimal inhibitory concentration.

Table 6 (continued)
EFFECT OF PENICILLIN AND STREPTOMYCIN ON RICKETTSIAL INFECTION AND GROWTH IN TISSUE CULTURES

Rickettsia	Antibiotic concentration		Relative time for introducing rickettsias and antibiotics[b]	Observations	Ref.
	Penicillin (units or μg/ml[a])	Streptomycin (μg/ml)			
Rickettsia typhi	200	250	Before and after	Complete inhibition of plaques	30
	100	20	Together	Marked effect on number of infected cells in suspended culture	47
	100 μg	500	After	MIC for cytopathic effects = 100 μg/ml.	82
		20	After	No influence on cytopathic effects	82
	100	20	Together	RGO	73
	100	20	Together	RGO	57
	100	20	After	No influence on infectious titer of resulting rickettsial preparations	83
Rickettsia tsutsugamushi	500 μg	500	After	No influence on cytopathic effects	82
		20	After	No influence on cytopathic effects	82
	100	20	Together	Influence on growth very minor, if any, in suspended cells	47

Table 7

REFERENCES TO STUDIES IN WHICH RICKETTSIAL SPECIES HAVE BEEN SUCCESSFULLY PLAQUED

Host cell	Rickettsial species[a]										
	bur	rick	ak	con	sib	aus	mont[b]	prow	typ	can	tsu
Primary			30	30				30	30		
Chick embryo		50									
		60	60	60	60	60		60	60		
	42										42
				68							
	31	31	31	31	31			31	31	31	
		56							56		
		63									
		32						32	32	32	
		64				64					
								76			
Vero		50									
		32		32			32	32	32	32	
WI-38				68							
L-929				68							
HeLa				68							
DBS-FRhL-2				68							
Singh's											
A. albopictus		53		53							

[a] Meaning of species abbreviations: bur = *Coxiella burnetii*, rick = *Rickettsia rickettsii*, ak = *R. akari*, con = *R. conorii*, sib = *R. sibirica*, aus = *R. australis*, mont = *R. montana*, prow = *R. prowazekii*, typ = *R. typhi*, can = *R. canada*, tsu = *R. tsutsugamushi*.

[b] Described as a new species[91] but not recognized as such by *Bergey's Manual*.[1]

Table 8

VARIATIONS USED IN THE PERFORMANCE OF RICKETTSIAL PLAQUE TESTS

Alternative condition	Ref.
Culture vessel	
30-ml tissue culture flask (25-cm² monolayer area)	31, 32, 50
60-mm plastic Petri plate (incubated in 5% CO_2)	68
Plastic trays, 24 wells, sealed with self-adhesive plastic sheets	32
Solution for diluting rickettsias (either used solely or found most satisfactory in comparative studies)	
Brain heart infusion (Difco)	42, 50, 56, 63, 68
Medium 199 with 5% calf serum	50
SPG:[a] Hanks BSS with 10% calf serum (1:1)	64
0.15 *M* phosphate-buffered saline with 0.75% bovine plasma albumin, Fraction V (Armour Pharmaceutical Co.)	32, 53
Adsorption period at room temperature	
15 min	31, 42, 50, 56, 60
30 min	64
60 min	32, 68, 76

[a] Sucrose-potassium-glutamate solution.

Table 8 (continued)
VARIATIONS USED IN THE PERFORMANCE OF RICKETTSIAL PLAQUE TESTS

Alternative condition	Ref.
Nutrient overlay	
Base medium	
Medium 199	30—32, 50
Eagle minimum essential medium	32, 68
Dulbecco modification of Eagle medium	68
Leibovitz (L-15) medium	32
Serum	
5% fetal calf	31, 32
10% fetal calf	68
5% calf	30, 50
Solidifying agent	
Noble agar (Difco), 0.15 or 0.3%	30, 50
Agarose (Mann Research Laboratories), 0.5%	42, 50
Agarose (Marine Colloids, Inc.), 0.5%	31, 64, 68
Agarose (Marine Colloids, Inc.), 1%	32
Miscellaneous additives	
10% tryptose phosphate broth (Difco)	32
HEPES buffer (*N*-2-hydroxyethylpiperazine-*N'*-2-ethane sulfonic acid; Calbiochem)	32
Dye for enhancing visibility of plaques	
0.01% neutral red in second nutrient overlay	31, 32, 64, 68
Toluidine blue spread over cells after fixation with Bouin's solution	30
1% neutral red spread over cell monolayer after removal of semisolid agar, or added to surface of solid agar and allowed to diffuse through to the cells	50

Table 9
USE OF CENTRIFUGATION TO ENHANCE INFECTION OF CELL MONOLAYERS BY RICKETTSIAS

Species of rickettsia[a]	Centrifugal force (× G)	Time (min)	Temperature (°C)	Vessel	Authors' conclusions	Ref.
rick	1600—1800	60	20	16- × 100-mm tubes	10,000-fold increase in infected cells	45
bur	1600—1800	60	20	16- × 100-mm tubes	Only slight increase in infected cells	45
ak	1600	60	20	16- × 100-mm tubes	None	65
bur	500	15	23—25	18- × 100-mm tubes	Increase of force to 1000 G or time to 30 min offered no advantage. With centrifugation, 97% of the rickettsias were adsorbed in 15 min as compared to 63% in 2 hr without centrifugation.	40
rick	600	15	Room temp	30-ml tissue culture flasks	Number of plaques was increased by factor of 7.2.	31
typ	600	15	Room temp	30-ml tissue culture flasks	Number of plaques was increased by factor of 5.9.	31

[a] Meaning of species abbreviations: rick = *Rickettsia rickettsii*, bur = *Coxiella burnetii*, ak = *Rickettsia akari*, typ = *R. typhi*.

REFERENCES

1. **Moulder, J. W., Weiss, E., Philip, C. B., Brooks, M. A., Weinman, D., Ristic, M., and Kreier, J. P.**, in *Bergey's Manual of Determinative Bacteriology*, 8th ed., Buchanan, R. E. and Gibbons, N. E., Eds., Williams and Wilkins, Baltimore, 1973, 882–914.

2. **Nyindo, M. B. A., Ristic, M., Huxsoll, D. L., and Smith, A. R.**, Tropical canine pancytopenia: *in vitro* cultivation of the causative agent – *Ehrlichia canis, Am. J. Vet. Res.*, 32, 1651–1658, 1971.

3. **Suitor, E. C., Jr.**, Propagation of *Rickettsiella popilliae* (Dutky and Gooden) Philip and *Rickettsiella melolonthae* (Krieg) Philip in cell cultures, *J. Insect Pathol.*, 6, 31–40, 1964.

4. **Vinson, J. W. and Fuller, H. S.**, Studies on trench fever. I. Propagation of rickettsia-like microorganisms from a patient's blood, *Pathol. Microbiol. Suppl.*, 24, 152–166, 1961.

5. **Huang, K.-Y.**, Metabolic activity of the trench fever rickettsia, *Rickettsia quintana, J. Bacteriol.*, 93, 853–859, 1967.

6. **Vinson, J. W.**, In vitro cultivation of the rickettsial agent of trench fever, *Bull. WHO*, 35, 155–164, 1966.

7. **Myers, W. F., Cutler, L. D., and Wisseman, C. L., Jr.**, Role of erythrocytes and serum in the nutrition of *Rickettsia quintana, J. Bacteriol.*, 97, 663–666, 1969.

8. **Myers, W. F., Osterman, J. V., and Wisseman, C. L., Jr.**, Nutritional studies on *Rickettsia quintana:* nature of the hematin requirement, *J. Bacteriol.*, 109, 89–95, 1972.

9. **Mason, R. A.**, Propagation and growth cycle of *Rickettsia quintana* in a new liquid medium, *J. Bacteriol.*, 103, 184–190, 1970.

10. **Wolbach, S. B. and Schlesinger, M. J.**, The cultivation of the micro-organisms of Rocky Mountain spotted fever (*Dermacentroxenus rickettsi*) and of typhus (*Rickettsia prowazeki*) in tissue plasma cultures, *J. Med. Res.*, 44, 231–256, 1923.

11. **Zinsser, H. and Batchelder, A. P.**, Studies on Mexican typhus fever, *J. Exp. Med.*, 51, 847–858, 1930.

12. **Zinsser, H., Fitzpatrick, F., and Wei, H.**, A study of rickettsiae grown on agar tissue cultures, *J. Exp. Med.*, 69, 179–190, 1939.

13. **Pinkerton, H. and Hass, G. M.**, Typhus fever. III. The behavior of *Rickettsia prowazeki* in tissue cultures, *J. Exp. Med.*, 54, 307–314, 1931.

14. **Nigg, C. and Landsteiner, K.**, Studies on the cultivation of the typhus fever rickettsia in the presence of live tissue, *J. Exp. Med.*, 55, 563–576, 1932.

15. **Kligler, I. J. and Aschner, M.**, Immunization of animals with formalized tissue cultures of *Rickettsia* from European and Mediterranean typhus, *Br. J. Exp. Pathol.*, 15, 337–346, 1934.

16. **Burnet, F. M.**, Tissue culture of the rickettsia of Q fever, *Aust. J. Exp. Biol. Med.*, 16, 219–224, 1938.

17. **Cox, H. R. and Bell, E. J.**, The cultivation of *Rickettsia diaporica* in tissue culture and in the tissues of developing chick embryos, *Public Health Rep.*, 54, 2171–2178, 1939.

18. **Kordová, N. and Kvicala, P.**, *Coxiella burneti* in tissue cultures, studied by the optic microscope and in phase contrast, *Folia Microbiol. Prague*, 7, 89–92, 1962.

19. **Kordová, N. and Kováčová, E.**, Appearance of antigens in tissue culture cells inoculated with filterable particles of *Coxiella burneti* as revealed by fluorescent antibodies, *Acta Virol.*, 12, 460–463, 1968.

20. **Kordová, N.**, Complementation of *Coxiella burneti* in mixed infection of heat-inactivated "Phase" II and host-dependent "Phase" I organisms in L cells, *Acta Virol.*, 14, 78–81, 1970.

21. **Rosenberg, M. and Kordová, N.**, Multiplication of *Coxiella burneti* in Detroit-6 cultures, *Acta Virol.*, 6, 176–179, 1962.

22. **Roberts, A. N. and Downs, C. M.**, Study of the growth of *Coxiella burnetii* in the L strain mouse fibroblast and the chick fibroblast, *J. Bacteriol.*, 77, 194–204, 1959.

23. **Rehaček, J. and Brezina, R.**, Propagation of *Coxiella burneti* in tick tissue cultures, *Acta Virol.*, 8, 380, 1964.

24. **Pospišil, V. F.**, Propagation of *Coxiella burneti* in cultures of a monkey kidney stable cell line, *Acta Virol.*, 9, 188–189, 1965.

25. **Kazár, J.**, Multiplication of *Coxiella burneti* in virus-infected and interferon-treated cell cultures, *Acta Virol.*, 13, 346–348, 1969.

26. **Kordová, N.**, Microscopic study of tissue cultures infected with filtrable *Coxiella burneti* particles, *Folia Microbiol. Prague*, 4, 237–239, 1959.

27. **Pospišil, V. F.**, Persistent infection of monkey kidney stable cells with *Coxiella burneti, Acta Virol.*, 10, 176–177, 1966.

28. **Pospišil, V. F.**, Persistent infection of cell cultures with *Coxiella burneti, Acta Virol.*, 10, 542–548, 1966.

29. **Kazár, J., Řeháček, J., and Brezina, R.,** Cultivation of rickettsiae in Colubrid snake tissue cultures, *J. Hyg. Epidemiol. Microbiol. Immunol.,* 10, 240–245, 1966.

30. **Kordová, N.,** Plaque assay of rickettsiae, *Acta Virol.,* 10, 278, 1966.

31. **Wike, D. A., Tallent, G., Peacock, M. G., and Ormsbee, R. A.,** Studies of the rickettsial plaque assay technique, *Infect. Immun.,* 5, 715–722, 1972.

32. **Cory, J., Yunker, C. W., Ormsbee, R. A., Peacock, M., Meibos, H., and Tallent, G.,** Plaque assay of rickettsiae in a mammalian cell line, *Appl. Microbiol.,* 27, 1157–1161, 1974.

33. **Kordová, N., Kováčová, E., and Wilt, J. C.,** Mixed infections of *Coxiella burnetii* in restrictive host cells: an immunofluorescent study, *Can. J. Microbiol.,* 16, 561–566, 1970.

34. **Burton, P. R., Kordová, N., and Paretsky, D.,** Electron microscopic studies of the rickettsia *Coxiella burneti:* entry, lysosomal response, and fate of rickettsial DNA in L-cells, *Can. J. Microbiol.,* 17, 143–150, 1971.

35. **Kazár, J.,** Interference between *Coxiella burneti* and cytopathic viruses in chick embryo cell cultures. I. Establishment of the interference, *Acta Virol.,* 13, 124–134, 1969.

36. **Kordová, N., Burton, P. R., Downs, C. M., Paretsky, D., and Kováčová, E.,** The interaction of *Coxiella burnetii* phase I and phase II in Earle's cells, *Can. J. Microbiol.,* 16, 125–133, 1970.

37. **Burton, P. R., Stueckemann, J., and Paretsky, D.,** Electron microscopy studies of the limiting layers of the rickettsia *Coxiella burneti, J. Bacteriol.,* 122, 316–324, 1975.

38. **Yunker, C. E., Ormsbee, R. A., Cory, J., and Peacock, M. G.,** Infection of insect cells with *Coxiella burneti, Acta Virol.,* 14, 383–392, 1970.

39. **Kordová, N. and Brezina, R.,** Multiplication dynamics of phase I and II *Coxiella burneti* in different cell cultures, *Acta Virol.,* 7, 84–87, 1963.

40. **Hahon, N. and Cooke, K. O.,** Assay of *Coxiella burnetii* by enumeration of immunofluorescent infected cells, *J. Immunol.,* 97, 492–497, 1966.

41. **Hahon, N. and Kozikowski, E. H.,** Induction of interferon by *Coxiella burneti* in cell cultures, *J. Gen. Virol.,* 3, 125–127, 1968.

42. **McDade, J. E. and Gerone, P. J.,** Plaque assay for Q fever and scrub typhus rickettsiae, *Appl. Microbiol.,* 19, 963–965, 1970.

43. **Blackford, V. L.,** Influence of various metabolites on growth of *Coxiella burnetii* in monolayer cultures of chick embryo entodermal cells, *J. Bacteriol.,* 81, 747–754, 1961.

44. **Weiss, E. and Pietryk, H. C.,** Growth of *Coxiella burnetii* in monolayer cultures of chick embryo entodermal cells, *J. Bacteriol.,* 72, 235–241, 1956.

45. **Weiss, E. and Dressler, H. R.,** Centrifugation of rickettsiae and viruses onto cells and its effect on infection, *Proc. Soc. Exp. Biol. Med.,* 103, 691–695, 1960.

46. **Kordová, N. and Kováčová, E.,** Histochemical and fluorescent antibody studies on the early stages of infection of L cells with *Coxiella burneti, Acta Virol.,* 12, 23–30, 1968.

47. **Bozeman, F. M., Hopps, H. E., Danauskas, J. X., Jackson, E. B., and Smadel, J. E.,** Study on the growth of rickettsiae. I. A tissue culture system for quantitative estimations of *Rickettsia tsutsugamushi, J. Immunol.,* 76, 475–488, 1956.

48. **Burnet, F. M.,** Tissue culture of the rickettsia of Q fever, *Aust. J. Exp. Biol. Med.,* 16, 219–224, 1938.

49. **Capponi, M.,** Rickettsies et Macrophages *in vitro, C.R. Soc. Biol.,* 167, 904–907, 1973.

50. **Weinberg, E. H., Stakebake, J. R., and Gerone, P. J.,** Plaque assay for *Rickettsia rickettsii, J. Bacteriol.,* 98, 398–402, 1969.

51. **McDade, J. E.,** Determination of antibiotic susceptibility of *Rickettsia* by the plaque assay technique, *Appl. Microbiol.,* 18, 133–135, 1969.

52. **Kenyon, R. H., Acree, W. M., Wright, G. G., and Melchior, F. W., Jr.,** Preparation of vaccines for Rocky Mountain spotted fever from rickettsiae propagated in cell culture, *J. Infect. Dis.,* 125, 146–152, 1972.

53. **Cory, J. and Yunker, C. E.,** Rickettsial plaques in mosquito cell monolayers, *Acta Virol.,* 18, 512–513, 1974.

54. **Anderson, D. R., Hopps, H. E., Barile, M. F., and Bernheim, B. C.,** Comparison of the ultrastructure of several rickettsiae, ornithosis virus, and Mycoplasma in tissue culture, *J. Bacteriol.,* 90, 1387–1403, 1965.

55. **Weiss, E., Green, A. E., Grays, R., and Newman, L. M.,** Metabolism of *Rickettsia tsutsugamushi* and *Rickettsia rickettsi* in irradiated host cells, *Infect. Immun.,* 8, 4–7, 1973.

56. **Wike, D. A., Ormsbee, R. A., Tallent, G., and Peacock, M. G.,** Effects of various suspending media on plaque formation by rickettsiae in tissue culture, *Infect. Immun.,* 6, 550–556, 1972.

57. **Schaechter, M., Bozeman, F. M., and Smadel, J. E.,** Study on the growth of rickettsiae. II. Morphologic observations of living rickettsiae in tissue culture cells, *Virology,* 3, 160–172, 1957.

58. Anacker, R. L., Gerloff, R. K., Thomas, L. A., Mann, R. E., Brown, W. R., and Bickel, W. D., Purification of *Rickettsia rickettsi* by density-gradient zonal centrifugation, *Can. J. Microbiol.,* 20, 1523–1527, 1974.

59. Anacker, R. L., Gerloff, R. K., Thomas, L. A., Mann, R. E., and Bickel, W. D., Immunological properties of *Rickettsia rickettsii* purified by zonal centrifugation, *Infect. Immun.,* 11, 1203–1209, 1975.

60. McDade, J. E., Stakebake, J. R., and Gerone, P. J., Plaque assay system for several species of *Rickettsia, J. Bacteriol.,* 99, 910–912, 1969.

61. Buhles, W. C., Jr., Huxsoll, D. L., and Elisberg, B. L., Isolations of *Rickettsia rickettsi* in primary bone marrow cell and circulating monocyte cultures derived from experimentally infected guinea pigs, *Infect. Immun.,* 7, 1003–1005, 1973.

62. Buhles, W., Huxsoll, D. L., Ruch, G., Kenyon, R. H., and Elisberg, B. L., Evaluation of primary blood monocyte and bone marrow cell culture for the isolation of *Rickettsia rickettsii, Infect. Immun.,* 12, 1457–1463, 1975.

63. Wike, D. A. and Burgdorfer, W., Plaque formation in tissue cultures by *Rickettsia rickettsii* isolated directly from whole blood and tick hemolymph, *Infect. Immun.,* 6, 736–738, 1972.

64. Kenyon, R. H. and McManus, A. T., Rickettsial infectious antibody complexes: detection by antiglobulin plaque reduction technique, *Infect. Immun.,* 9, 966–968, 1974.

65. Kazár, J., Krautwurst, P. A., and Gordon, F. B., Effect of interferon and interferon inducers on infections with a nonviral intracellular microorganism, *Rickettsia akari, Infect. Immun.,* 3, 819–824, 1971.

66. Weiss, E., Newman, L. W., Grays, R., and Green, A. E., Metabolism of *Rickettsia typhi* and *Rickettsia akari* in irradiated L cells, *Infect. Immun.,* 6, 50–57, 1972.

67. Řeháček, J., Brezina, R., and Majerska, M., Multiplication of rickettsiae in tick cells in vitro, *Acta Virol.,* 12, 41–43, 1968.

68. Osterman, J. V. and Parr, R. P., Plaque formation by *Rickettsia conori* in WI-38, DBS-FRhL-2, L-929, HeLa, and chicken embryo cells, *Infect. Immun.,* 10, 1152–1155, 1974.

69. Burgdorfer, W., Sexton, D. J., Gerloff, R. K., Anacker, R. L., Philip, R. N., and Thomas, L. A., *Rhipicephalus sanguineus:* vector of a new spotted fever group rickettsia in the United States, *Infect. Immun.,* 12, 205–210, 1975.

70. Kordová, N. and Kováčová, E., Replication of *Rickettsia prowazeki* in L-cells as revealed by immunofluorescence, *Acta Virol.,* 11, 252–255, 1967.

71. Gambrill, M. R. and Wisseman, C. L., Jr., Mechanisms of immunity in typhus infections. II. Multiplication of typhus rickettsiae in human macrophage cell cultures in the nonimmune system: influence of virulence of rickettsial strains and of chloramphenicol, *Infect. Immun.,* 8, 519–527, 1973.

72. Wisseman, C. L., Jr., Waddell, A. D., and Walsh, W. T., Mechanisms of immunity in typhus infections. IV. Failure of chicken embryo cells in culture to restrict growth of antibody-sensitized *Rickettsia prowazeki, Infect. Immun.,* 9, 571–575, 1974.

73. Barker, L. F. and Patt, J. K., Production of rickettsial complement-fixing antigens in tissue culture, *J. Immunol.,* Vol. 100, 821–824, 1968.

74. Kordová, N., Die Vermehrung der *Rickettsia prowazeki* in L-Zellen. Lichtmikroskopische Untersuchungen, *Arch. Gesamte Virusforsch.,* 15, 697–706, 1965.

75. Kordová, N., Rosenberg, M., and Mrena, E., Die Vermehrung der *Rickettsia prowazeki* in L-Zellen. Elektronenmikroskopische Untersuchungen an infizierten Gewebezellen in Dünn-schnitten, *Arch. Gesamte Virusforsch.,* 15, 707–720, 1965.

76. Wisseman, C. L., Jr., Waddell, A. D., and Walsh, W. T., In vitro studies of the action of antibiotics on *Rickettsia prowazeki* by two basic methods of cell culture, *J. Infect. Dis.,* 130, 564–574, 1974.

77. Wisseman, C. L., Jr. and Waddell, A. D., In vitro studies on rickettsia-host cell interactions: intracellular growth cycle of virulent and attenuated *Rickettsia prowazeki* in chicken embryo cells in slide chamber cultures, *Infect. Immun.,* 11, 1391–1401, 1975.

78. Kováčová, E. and Kordová, N., Die Vermehrung der *Rickettsia prowazeki* in L-Zellen. III. Acridin-orange-fluoreszenz und Ultraviolett-Mikroskopie, *Arch. Gesamte Virusforsch.,* 19, 57–62, 1966.

79. Weiss, E. and Dressler, H. R., Growth of *Rickettsia prowazeki* in irradiated monolayer cultures of chick embryo entodermal cells, *J. Bacteriol.,* 75, 544–552, 1958.

80. Gambrill, M. R. and Wisseman, C. L., Jr., Mechanisms of immunity in typhus infections. III. Influence of human immune serum and complement on the fate of *Rickettsia mooseri* within human macrophages, *Infect. Immun.,* 8, 631–640, 1973.

81. Weiss, E., Collbaugh, J. C., and Williams, J. C., Separation of viable *Rickettsia typhi* from yolk sac and L cell host components by renografin density gradient centrifugation, *Appl. Microbiol.,* 30, 456–463, 1975.

82. **Barker, L. F.,** Determination of antibiotic suspectibility of rickettsiae and chlamydiae in BS-C-1 cell cultures, *Antimicrob. Agents Chemother.,* 8, 425–428, 1968.

83. **Barker, L. F., Patt, J. K., and Hopps, H. E.,** Titration and neutralization of *Rickettsia tsutsugamushi* in tissue culture, *J. Immunol.,* 100, 825–830, 1968.

84. **Hopps, H. E., Kohno, S., Kohno, M., and Smadel, J. E.,** Production of interferon in tissue cultures infected with *Rickettsia tsutsugamushi, Bacteriol. Proc.,* pp. 115–116, 1964.

86. **Cohn, Z. A., Bozeman, F. M., Campbell, J. M., Humphries, J. W., and Sawyer, T. K.,** Study on growth of rickettsiae. V. Penetration of *Rickettsia tsutsugamushi* into mammalian cells in vitro, *J. Exp. Med.,* 109, 271–292, 1959.

86. **Hopps, H. E., Jackson, E. B., Danauskas, J. X., and Smadel, J. E.,** Study on the growth of rickettsiae. III. Influence of extracellular environment on the growth of *Rickettsia tsutsugamushi* in tissue culture cells, *J. Immunol.,* 82, 161–171, 1959.

87. **Hopps, H. E., Jackson, E. B., Danauskas, J. X., and Smadel, J. E.,** Study on the growth of rickettsiae. IV. Effect of chloramphenicol and several metabolic inhibitors on the multiplications of *Rickettsia tsutsugamushi* in tissue culture cells, *J. Immunol.,* 82, 172–181, 1959.

88. **Yunker, C. E. and Cory, J.,** Growth of *Rickettsia tsutsugamushi* in insect cells, in *Proc. 3rd Int. Colloq. Invertebrate Tissue Culture,* pp. 451–457, 1973.

89. **Minamishima, Y. and Mori, R.,** Persistent infection of FL cells by *Rickettsia sennetsu (S. todai), J. Bacteriol.,* 88, 1195–1196, 1964.

90. **Burgdorfer, W., Krinsky, W. L., Brinton, L. P., and Philip, R. N.,** *Rickettsia rhipicephali* n. sp.: a new rickettsia from the brown dog tick, *Rhipicephalus sanguineus,* in 2nd Int. Symp. Rickettsiae and Rickettsial Diseases, Bratislava, Czechoslovakia, June 21–25, 1976.

91. **Lackman, D. B., Bell, E. J., Stoenner, H. G., and Pickens, E. G.,** The Rocky Mountain spotted fever group of rickettsias, *Health Lab. Sci.,* 2, 135–141, 1965.

92. **Yunker, C. E.,** Arthropod tissue culture in the study of arboviruses and rickettsiae: a review, *Curr. Top. Microbiol. Immunol.,* 55, 113–126, 1971.

93. **Cory, J., Yunker, C. E., Howarth, J. A., Hokama, Y., Hughes, L. E., Thomas, L. A., and Clifford, C. M.,** Isolation of spotted fever group and Wolbachia-like agents from field-collected materials by means of plaque formation in mammalian and mosquito cells, *Acta Virol.,* 19, 443–445, 1975.

94. **Morgan, J. F., Morton, H. J., and Parker, R. C.,** Nutrition of animal cells in tissue culture. I. Initial studies on a synthetic medium, *Proc. Soc. Exp. Biol. Med.,* 73, 1–8, 1950.

95. **Eagle, H.,** Amino acid metabolism in mammalian cell cultures, *Science,* 130, 432–437, 1959.

96. **Morton, H. J.,** A survey of commercially available tissue culture media, *In Vitro,* 6, 89–108, 1970.

97. **Eagle, H.,** Nutrition needs of mammalian cells in tissue culture, *Science,* 122, 501–504, 1955.

98. **Edward, D. G.,** The pleuropneumonia group of organisms: a review together with some new observations, *J. Gen. Microbiol.,* 10, 27–64, 1954.

99. **Yunker, C. E. and Cory, J.,** Plaque production by arboviruses in Singh's *Aedes albopictus* cells, *Appl. Microbiol.,* 29, 81–89, 1975.

100. **Rosenblatt, J. E. and Stewart, P. R.,** Anaerobic bag culture method, *J. Clin. Microbiol.,* 1, 527–530, 1975.

101. **Leibovitz, A.,** The growth and maintenance of tissue-cell cultures in free gas exchange with the atmosphere, *Am. J. Hyg.,* 78, 173–180, 1963.

102. **Schafer, T. W., Pascale, A., Shimonaski, G., and Came, P. E.,** Evaluation of gentamycin for use in virology and tissue culture, *Appl. Microbiol.,* 23, 565–570, 1972.

103. **Ormsbee, R. A. and Peacock, M. G.,** unpublished data, 1975.

104. **Paul, J.,** *Cell and Tissue Culture,* 4th ed., Churchill Livingstone, London, 1973.

105. **Whitaker, A. M.,** *Tissue and Cell Culture,* Williams and Wilkins, Baltimore, 1972.

106. **Wasley, G. D., Ed.,** *Animal Tissue Culture,* Williams and Wilkins, Baltimore, 1973.

107. **Kruse, P. F., Jr. and Patterson, M. K., Jr.,** *Tissue Culture Methods and Applications,* Academic Press, New York, 1973.

108. **Vago, C., Ed.,** *Invertebrate Tissue Culture,* Vol. 1, Academic Press, New York, 1971.

109. **Elisberg, B. L. and Bozeman, F. M.,** Rickettsiae, in *Diagnostic Procedures for Viral and Rickettsial Infections,* 4th ed., Lennette, E. H. and Schmidt, N. J., Eds., American Public Health Association, New York, 1969, 826–868.

110. **Weiss, E.,** Growth and physiology of rickettsiae, *Bacteriol. Proc.,* pp. 259–283, 1973.

111. **Yunker, C. E. and Cory, J.,** personal communication.

112. **Cory, J.,** personal communication.

CULTURE MEDIA FOR CHLAMYDIAE

V. S. C. Fan

INTRODUCTION

Chlamydiae are nonmotile, spherical, Gram-negative bacteria. They are obligatory intracellular organisms.

The developmental cycle of Chlamydiae begins with the "elementary body," a small, spherical (0.2 to 0.4 μm), electron-dense structure containing nucleus material and ribosomes and surrounded by a rigid cell wall. After adsorption and uptake by host cells, the elementary bodies gradually reorganize into the "initial body," a large (0.8 to 1.5 μm) reticulated form bounded by a thin cell wall. The initial bodies undergo active fission within the cytoplasmic vacuoles of host cells. Daughter cells continue to multiply and eventually transform into elementary bodies. The site of multiplication can easily be detected by histological staining (Giemsa-May-Grunwald) and is termed "inclusion." The initial body is the metabolically active form of the organism, while the elementary body is the infectious form.

Chlamydial organisms have been found to have limited metabolic activities. They appear to have some endogenous metabolism similar to that of some bacteria, but they fail to synthesize high-energy compounds (e. g., adenosine triphosphate). Chlamydiae are known as "energy parasites." A living cell capable of providing sufficient metabolic requirements to sustain chlamydial growth is a requisite in chlamydial development. To date, no chlamydial propagation outside a living cell has been achieved.

Two species of chlamydial organisms are recognized. They are *Chlamydia trachomatis* and *C. psittaci.*

CULTURE MEDIA FOR CHLAMYDIA

Embryonated Egg Culture

In the past, embryonated eggs have been used widely in the isolation and cultivation of chlamydial organisms. In recent years the use of embryonated eggs has been gradually replaced by tissue culture. There are three methods of cultivating chlamydiae in embryonated egg.

Chlorioallantoic membrane (CAM) — This method for cultivation of chlamydiae was first used by Burnet and Rountree[16] and Fortner and Pfaffenburg.[34] They demonstrated the development of Chlaymdiae in CAM and used the membrane lesions produced as a means for infectivity titration. Larzarus and Meyer[63] were able to cultivate *C. psittaci* strains for an extended period of 38 months without loss of infectivity for mice. CAM was used by Litwin[65] to study the growth cycle of the organism. However, not all strains of chlamydia can be successfully cultivated by this method. Marriott and Storz[69] showed that strains of ewe and bovine abortion organisms and an enteric isolate were able to grow continuously in CAM, while isolates of polyarthritis and guinea pig inclusion conjunctivitis organisms failed.

Allantoic cavity — Williams[128] reported the cultivation of psittacosis agents in allantoic cavity. The yield of organisms from the allantoic cavity was higher than from the yolk sac. Francis and Gorden[35] showed that feline pneumonitis, meningopneumonitis, human pneumonitis (SF), and psittacosis can be cultivated in allantoic cavity. The advantage of this method over others is that it yields relatively clean organisms without extraneous contamination from egg materials. This method was used by Colon and Moulder,[20] Moulder et al.,[84] and Weiss[126] for preparing clean organisms for biochemical studies.

Yolk sac inoculation — In 1940, Rake et al.[101] found that yolk sac trophoblasts were susceptible to chlamydia and yielded high numbers of organisms. This procedure became a popular method for the isolation and cultivation of Chlamydiae. Shaffer et al.[106] demonstrated the cultivation of lymphogranuloma venerum (LGV) organisms in yolk sacs. Hamre and Rake[51] used yolk sacs for the isolation of feline pneumonitis agent. Golub[42] demonstrated the growth of psittacosis (6BC), meningopneumonitis (Cal-10) and human pneumonitis (Borg) agents in yolk sacs. Trachoma organism (TE-55) was isolated by Tang et al.[118] via yolk sac cultivation. This method is found to be successful for most Chlamydiae and is widely used for isolation and routine cultivation. Fertile chicken eggs are usually obtained from chicken flocks on antibiotic-free diets. Eggs are incubated at 38°C for 6 to 8 days before use. The development of the embryo is regularly monitored by candling. Methods used for egg inoculation are similar to that used for virus preparation. These methods are described by Blaskovic and Styk.[14]

Organ Culture

Organ culture technique has been used in studying the infection of chlamydial organisms in vitro. The method provides an opportunity to study chlamydial infection in an organized cell arrangement, which is not possible with monolayer or suspension cell cultures.

The sources of tissue for organ culture used were limited to conjunctival tissues of monkeys,[7] baboons,[96] and guinea pigs. The culture medium used by Barron et al.[7] consisted of Eagle's minimum essential medium (MEM)[27] with double the prescribed amount of amino acids, vitamins, and L-glutamine in Earle's saline solution. Human γ-globulin-free serum (Grand Island Biological Company, Grand Island, N.Y.) was added at 20% of the final medium. For infection studies, the Eagle's MEM contained the doubled amount of amino acids, vitamins, and L-glutamine. Nonessential amino acids were also supplemented. Newborn calf serum was used at 5%. The glucose concentration of the medium was doubled to 2.0 mg/ml. Bicarbonate concentration was 1.9 mg/ml. Streptomycin and vancomycin were used in both media at 250 μg/ml and 100 μg/ml, respectively.

C. trachomatis strains SA-2, TE-55,[7] and MRC-4f[96] and *C. psittaci* stains meningopneumonitis[7] and guinea pig inclusion conjunctivitis[96] have been cultivated.

Cell Cultures

Cell cultures have been used extensively in chlamydial research involving the use of a variety of cells and medium formulations. Some of the cells and media used in chlamydial studies are listed in Table 1.

LABORATORY ANIMALS EMPLOYED IN CHLAMYDIAL STUDIES

Common laboratory animals including mice, guinea pigs, hamsters, cats, rabbits, monkeys, pigeons, parakeets, and ricebirds have been used in chlamydial research (Table 2).

THE NATURAL HOSTS OF CHLAMYDIAE

Natural distribution of Chlamydiae exists in man, where they cause oculogenital and respiratory diseases; birds, where respiratory and generalized infections are known; and mammals (except primates), where organisms cause respiratory, encephalomyelitic, enteric, arthritic, and placental diseases.

The known natural hosts and the animals that show signs of chlamydial infections are listed in Table 3. Some arthropods that associate with mammals and birds have been found to contain chlamydial organisms. The exact function of these arthropods in carrying chlamydial organisms is poorly understood at the moment. Table 4 lists arthropods that are associated with Chlamidiae or from which chlamydial organisms have been isolated.

Table 1
CELLS AND CULTURE MEDIA USED IN CULTIVATION OF CHLAMYDIAL AGENTS

Cell	Medium	Organism	Ref.
Human HeLa (epitheloid carcinoma, cervix)	Hank's[a] balanced salt solution (BSS), 80%; yeast-olate (Difco®), 0.08%; human serum, 20%	C. psittaci strain meningopneumonitis (MN)	68
	Hank's BSS, 81.1%; tryptic digest broth, 5%; lactal-bumin hydrolysate (LAH), 0.5%; horse serum, 10%; rabbit serum, 2%; sodium bicarbonate, 1.4%	C. trachomatis strain TE-55	36
	Hank's BSS, 84.5%; LAH, 0.5%; human serum, 10%; calf serum, 5%	C. trachomatis inclusion blennorrhora strain LB1	37
	Gey's BSS,[b] 89.5%; LAH, 0.5%; human serum, 10%	C. trachomatis Tang strain	2
	1.75 × Hank's BSS, 40%; Medium 199,[c] 0.5%; LAH, 40%; human serum, 20%	C. psittaci strain "New Jersey" turkey ornithosis	21
	Hank's BSS, 99.4%; yeast extract, 0.1%; LAH, 0.5%	C. Trachomatis strains Mita, Bour, Kami, Kondo, and Shim	80
	Hank's BSS, 97.9%; yeast extract, 0.1%; bovine serum, 2%		81
	Eagle's minimal essential medium[27] (MEM), 90%; calf serum, 10%	C. trachomatis strain Cal-1, Cal-2, pK2f	102
	Eagle's MEM, 90%; calf serum, 10%	C. trachomatis strains Bour and TW-3 (no growth on HeLa S3 or HeLa-calf)	56
	Hank's BSS, 94.5%; LAH, 0.5%; veal serum, 5%	C. psittaci strains EAE/A_1 enzootic abortion of ewes, and 01010 pigeon ornithosis; C. trachomatis strains lymphogranuloma venereum (LGV) JH, iso-lates TN2, TN26, TN34, and TN38	119
	Eagle's MEM, 90%; calf serum, 10%	C. trachomatis genital isolates D, E, F; ocular isolates A, B, C, LGV I and II, mouse pneumonitis (Mopn), C. psittaci (MN)	123
	Eagle's BSS, 76.8%; L-glutamine, 3.2%; human serum, 20%	C. psittaci ovine abortion strain B57	9
	Eagle's MEM, 95%; fetal calf serum, 5%	C. trachomatis strain TW-3 LGV440N; C. psittaci strain MN	58
	Eagle's MEM, 90%; calf serum, 10%	C. trachomatis strains G17, TW-5, and Mopn); C. psittaci strains Cal-10 and MN	62
FL cells (amnion)	Medium 199, 94.5%; LAH, 0.5%; calf serum, 5%	C. trachomatis strain SA2	99
	Earle's BSS, 79.4%; LAH, 0.5%; yeast extract, 0.1%; human serum, 20%	C. trachomatis strain TE-55	12
		C. trachomatis strains Tang and Bour Camal	13
	Eagle's MEM, 90%; calf serum, 10%	C. trachomatis strains TN3 and Bour; no growth	56

a   Hank, J. and Wallace, R. E., *Proc. Soc. Exp. Biol. Med.*, 71, 196–200, 1949.
b   Gey, G. O. and Gey, M. K.. *Am. J. Cancer*, 27, 45–76, 1936.
c   Morgan, J. F., Morton, H. J., and Parker, R. C., *Proc. Soc. Exp. Biol. Med.*, 73, 1–8, 1950.

Table 1 (continued)
CELLS AND CULTURE MEDIA USED IN CULTIVATION OF CHLAMYDIAL AGENTS

Cell	Medium	Organism	Ref.
FL cells (amnion) (continued)	Medium 199, 92.2%; fetal calf serum, 5%; sodium bicarbonate, 2.8%	C. trachomatis strains MU-16, HAR-2, and HAR-2f	50
Human prepuce (Hp-1)	Eagle's MEM, 90%; calf serum, 10%	C. trachomatis strains Bour and TW-3 (poor growth)	56
Human amnion cells	Medium 199, 85%; LAH, 10%; calf serum, 5%	C. psittaci turkey strain TTS	117
	Medium 199, 40%; Tyrode's[d] solution, 40%; human umbilical cord serum, 20%	C. psittaci "New Jersey" turkey strain	21
WI-38 (human diploid)	Eagle's MEM, 97%; calf serum, 3%	C. psittaci strains 6BC, SF, and Borg	116
Human embryonic lung	Eagle's MEM, 97%; calf serum, 3%	C. psittaci strains 6BC, SF, and Borg	95
Chang liver (CHL)	Hank's BSS, 89.5%; LAH, 0.5%; calf serum, 10%	C. psittaci strains 6BC and Borg	89
HEp-2 (epidermoid carcinoma, larynx)	Hank's BSS, 89.5%; LAH, 0.5%; calf serum, 10%	C. trachomatis strains MN, GBC, and Borg	66
	Eagle's MEM, 90%; calf serum, 10%	C. trachomatis strains Cal-9 and Cal-1	102
	Medium 199, 90%; calf serum, 10%	C. psittaci ovine strain HE	52
Cells of mouse origin			
MaCoy — a human synovial cell in origin, but shown to be a mouse cell contaminant[49]	Medium 199; 94.5%; LAH, 0.5%; calf serum, 5%	C. trachomatis strain SA-2	99
	Medium 199, 99.5%; LAH, 0.5%	C. psittaci turkey strain TTF	117
	Medium 199, 89.5%; LAH, 0.5%; calf serum, 10%	C. psittaci strain Borg	61
Irradiated cells	Eagle's MEM, 90%; horse serum, 10% (glucose 35.6 mM)	C. trachomatis strains LGV, TW-6, and MRC-1/0T	45
		C. trachomatis isolates from Iran	22
		C. psittaci strain MN; C. trachomatis strain MRC-1/G and LGV strains NMR25, NMR26, and NMR41	47
		C. psittaci strain 6BC; C. trachomatis strains Mopn and MRC-1/G	44
L cells (connective tissue)	Hank's BSS; LAH, 0.2%; yeast extract (YE), 0.07%; amino acids; vitamins; horse serum, 25%	C. psittaci strain 6BC	3
	Earle's BSS, 89.4%; LAH, 0.5%; YE, 0.1%; bovine serum, 10%	C. psittaci strain Cal-10	53
	Hank's BSS, 77.3%; LAH, 2%; YE, 0.07%; horse serum, 20%	C. psittaci strain 6BC	4
	L cell clone 5b (L5B); Medium 199, 90%; fetal calf serum, 10%	C. psittaci strains 6BC and MN; C. trachomatis strain Mopn	121
	Eagle's MEM, 90%; fetal calf serum, 10%	C. psittaci strains MN: turkey — T575, New Jersey, Calif., and Minn.; pigeon — M46, Calif., and Carlson pigeon; parakeet — Arizona, Calif., and 6BC	5

d   Tyrode, M. V., Arch. Int. Pharmacodyn. Ther., 20, 205—223, 1910.

Table 1 (continued)
CELLS AND CULTURE MEDIA USED IN CULTIVATION OF CHLAMYDIAL AGENTS

Cell	Medium	Organism	Ref.
L cells (connective tissue) (continued)	L5b in suspension culture; Waymouth medium,[e] 99%; oleate, 5 μg/ml; serum albumin (fatty acid free), 2 mg/ml	*C. psittaci* strain MN	83
Cells of avian origin			
Chicken embryonic fibroblast cell	(a) Simm's[f] salt solution, 77.8%; chicken embryo extract, 11.1%; rabbit serum, 11.1%	*C. psittaci* strain 6BC	82
	(b) Simm's salt solution, 66.7%; rabbit serum, 33.3%		
	(a) Hank's BSS, 75%; chicken serum, 25%	*C. psittaci* strain feline pneumonitis (Fepn)	125
	(b) Hank's BSS, 90%; bovine plasma albumin, 10%		
	(c) Hank's BSS, 65%; bovine serum ultrafiltrate, 35%		
	Hank's BSS, 90%; beef embryo extract, 10%	*C. psittaci* strain 6BC	109
	Hank's BSS, 75%; chicken serum, 25%	*C. trachomatis* strain TW-10	48
	Medium 199, 80%; tryptose phosphate broth (Difco), 10%; calf serum, 8%; bicarbonate, 2%	*C. psittaci* strain 6BC	97
Chicken ectodermal cell (CE)	Hank's BSS, 75%; chicken serum, 25%	*C. trachomatis* strains TE-55, Mopn, Cal-1, LGV, and MRC-1/0T	46
Cells of primate origin			
BSC-1 cell (kidney, African green)	Eagle's MEM, 90%; calf serum, 10%	*C. trachomatis* strain Tang; *C. psittaci* strain Bour	11
LLC-Mk2 (kidney, rhesus)	Eagle's MEM, 95%; fetal calf serum, 5%	*C. trachomatis* strain Bour	57
	Eagle's MEM, 95%; fetal calf serum, 5%	*C. psittaci* strain 6BC	67
	Eagle's MEM, 95%; fetal calf serum, 5%	*C. trachomatis* strain LGV440N	33
Others			
BHK-21 (kidney, baby Syrian hamster)	Eagle's MEM, 89.9%; tryptose phosphate broth, 10%; bicarbonate, 0.035%	*C. trachomatis* strains HAF-2f and MRC-4f	15
Turtle kidney cells (*Testuda graeca*)	NaCl, 6.5 g/l; KCl, 0.14 g/l; CaCl₂, 0.12 g/l; MgSO₄, 0.2 g/l; 1% phenol red, 2 ml/l; LAH, 0.25%; calf serum, 5%; pH 7.2, 35°C	*C. psittaci* onithosis SK strain	108

[e] Waymouth, C., *J. Natl. Cancer Inst.*, 22, 1003—1017, 1959.
[f] Simm, H. S. and Sanders, M., *Arch. Pathol.*, 33, 619—635, 1942.

Table 2

SUSCEPTIBILITY OF COMMON LABORATORY ANIMALS USED IN CHLAMYDIAL STUDIES[a]

	Embryonated yolk sac	Mouse	Guinea pig	Hamster	Cat	Rabbit	Monkey	Pigeon	Parakeet	Ricebird
Lymphogranuloma venereum	+	+		±			+			
Trachoma	+	+					+			
Human pneumonitis	+	+	±	+						
Meningopneumonitis	+	+		+						
Mouse pneumonitis	+	±	−	±				−	−	−
Feline pneumonitis	+	±	+	+	+	−		−	−	−
Sheep pneumonitis	+	+				+				
Ewe, enzootic abortion	+	+								
Opossum A encephalitis	+	+								
Bovine enteric	+		+							
Bovine encephalomyelitis	+	+	+							
Psittacosis	+	+	+	+			+		+	+
Pigeon ornithosis	+	+		+				+		+
Chicken ornithosis	+	+		+				+	+	
Turkey ornithosis	+	+	+						+	+
Fulmar ornithosis	+	+								
Egret ornithosis	+	+	+						+	+

[a] +, animal showed signs of infection; ±, animal showed irregular response; −, animal showed no sign of infection.

Table 3
NATURAL AND EXPERIMENTAL HOSTS OF CHLAMYDIAL ORGANISMS

Hosts	Type of infection	Ref.
Class Mammalia		
Order Marsupialia		
Didelphis paraguayensis (common South American opossum)	*C. psittaci* strain opossum A isolated	103
Caluromys laniger (woody opossum)	*C. psittaci* strain opossum A isolated	103
Metachirus nudicaudatus (brown-marked opossum)	*C. psittaci* strain opossum B of Roca-Garcia	127
Order Primate	No natural chlamydia infection known	
Family Cebidae		
Cebus capucinus (white throated capuchin)	Experimental LGV infection	64
Cebus apella (hooded capuchin)	Experimental LGV infection	74
Family Cercopithecidae (Old World monkeys)		
Macaca irus (crab-eating macaque)	Experimental trachoma and conjunctivitis	23
Macaca mulatta (Rhesus monkey)	Experimental trachoma and conjunctivitis	23
Macaca cyclopis (Taiwan monkey)	Experimental trachoma infections	122
Macaca brevicaudatus (Island of Hainan macaque)		74
Macaca sylvanus (Barbary ape)	Experimental trachoma infections	87
Cercocebus torquatus (capped mangabey)	Experimental LGV infections	74
Cercopithecus aethiops	Experimental LGV infections	74
Chaeropithecus cynocephalus (yellow baboon)	Experimental LGV infections	74
Chaeropithecus ursinus (chacma baboon)	Experimental trachoma infection	38
Papio papio (Western baboon)	Experimental trachoma infection	19
Comopithecus hamadryas	Experimental trachoma infection	19
Family Pongidae		
Pan troglodytes (chimpanzee)	Experimental trachoma infection	87
Family Hominidae		
Homo sapiens	*C. trachomatis* infection	
	Trachoma	118
	Inclusion conjunctivitis	58a
	Lymphogranuloma venereum	101
	C. psittaci infections	
	Psittacosis	64
	Lymphogranuloma venereum (rare)	105a
	Laboratory infection of enzootic abortion of ewe;	
	Pneumonitis with alvealor capillary block after exposure to epizootic bovine abortion	6
	Laboratory infection with bovine encephalomyelitis	8
Order Lagomorpha		
Lepus americanus (snowshoe hare)	*C. psittaci* isolated	111
Lepus californicus (jackrabbit)	*C. psittaci* infection	110
Sylvilagus audubonii (cottontail rabbit)	*C. psittaci* infection	110
Oryetolagus cuniculus (domestic rabbit)	*C. psittaci* infection	43
Order Rodentia		
Family Sciuridae		
Citellus beecheyi (California ground squirrel)	Experimental *C. psittaci* infection	54
Citellus leucurus (ground squirrel)	Natural Chlamydia infection	110
Citellus rariegatus (rock squirrel)	Natural Chlamydia infection	110

Table 3 (continued)
NATURAL AND EXPERIMENTAL HOSTS OF CHLAMYDIAL ORGANISMS

Hosts	Type of infection	Ref.
Family Geomyoidae		
Thomomys bottae (valley pocket gopher)	Experimental *C. psittaci* infection	54
Family Heteromyidae		
Dipodomys ordii (kangaroo rat)	Natural Chlamydia infection	110
Dipodomys microps (chisel-toothed kangaroo rat)	Natural Chlamydia infection	110
Family Cricetidae		
Nectomys squamipes (neotropical water rat)	Chlamydia infection	74
Peromyscus truli (white-footed mouse)	Chlamydia isolated	74
Peromyseus maniculatus (deer mouse)	Natural infection of chlamydia	110
	Laboratory infection of hamster pneumonitis	59
Onychomys leucogaster (grasshopper mouse)	Natural Chlamydia infection	110
Reithrodontomys megalotis (Western harvest mouse)	Natural Chlamydia infection	110
Neotoma fuscipes (dusty-footed wood rat)	Natural Chlamydia infection	110
Neotoma lepida (desert wood rat)	Natural Chlamydia infection	110
Sigmondon sp. (cotton rat)	Experimental Chlamydia infection	74
Mesocricetus auratus (golden hamster)	Experimental Chlamydia infection	28
Cricetus auratus (Syrian hamster)	Pneumonia	94
Family Microtinae		
Ondatra zibethicus (muskrat)	*C. psittaci* isolated	111
Family Muridae		
Rattus norvegicus (Norway rat)	*C. trachomatis* (LGV) laboratory infection	28
	C. psittaci laboratory infection	18
Mus musculus (laboratory mouse)	*C. trachomatis* mouse pneumonitis	88
Family Caviidae		
Cavia porcellus (guinea pig)	*C. psittaci* guinea pig conjunctivitis	85
Family Feloidae		
Felis catus (domestic cat)	Keratoconjunctivitis	17
	Pneumonitis	4a
Felis concolor (mountain lion)	Natural Chlamydia infection	110
Order Pinnipedia		
Callorhinus ursinus (Northern fur seal)	Chlamydia isolated	30
Order Perissodactyla		
Family Equidae		
Equus sp. (horse)	Encephalitic reaction, perivascular hemorrhage	105
Order Artiodactyla		
Family Suidae		
Sus scrofa (domestic pig)	Arthritis	71
	Pericarditis	129
	Conjunctivitis	93
Family Cervidae		
Odocileus hemionus (deer)	Natural Chlamydia infection	110
Family Bovidae		
Bos taurus (domestic cattle)	Encephalomyelitis	73
	Enteritis	131
	Pneumonitis	91
	Epizootic abortion	113
	Polyarthritis	114
Capra hircus (domestic goat)	Pneumonitis	55
	Abortion	25
	Fecal isolation	90

Table 3 (continued)
NATURAL AND EXPERIMENTAL HOSTS OF CHLAMYDIAL ORGANISMS

Hosts	Type of infection	Ref.
Family Bovidae		
(continued)		
Ovis aries (domestic sheep)	Abortion	92
	Pneumonitis	26
	Fecal isolation	130
	Polyarthritis	72
	Conjunctivokeratitis	24
Class Aves	*C. psittaci* ornithosis	
Order Podicipediformes (grebes)		74
Order Procellariiformes		
Fulmarus glacialis (fulmar)		79
Puffinus tenuirostrus (mutton bird)		86
Order Ciconiiformes		
Leucophoyx thula (snowy egret)		104
Order Anseriformes		
Anas platyrhynchos (domestic duck)		60
Order Falconiformes		74
Order Galliformes		
Gallus gallus (domestic fowl)		75
Phasianus colchicus (pheasant)		124
Meleagris gallopavo (turkey)		77
Order Charadriiformes		
Larus ridibundus (black head gull)		115
Larus argentatus (herring gull)		79
Larus atricilla (laughing gull)		98
Catoptrophorus semipalmatus (willet)		98
Larus fuscus (lesser black gull)		98
Sterna albiforns (least tern)		98
Sterna hirundo (common tern)		98
Calidris alba (sanderling)		98
Rhynchops nigra (skimmer)		98
Order Psittaciformes		
Family Psittacide		
Trichoglossus chlorolepidotus		107
Trichoglossus moluccanus		107
Kakatoe galerita (sulfur-crested cockatoo)		107
Kakatoe roseicapilla (galah)		107
Ara ararauna (macaw)		107
Amazona aestiva		107
Amazona barbadensis		107
Amazona oratrix		100
Palaoernis torquatus		107
Psittacula conspicillata		107
Psittacula spengeli		107
Psephotus haematonotus (grass parrot)		107
Purpureicephalus spurius		10
Platycerus elegans		10
Platycerus eximius		10
Agapornis sp.		107
Melopsittaccus undulatus (budgerigras)		107
Order Columbiformes		
Columbia livia (wild pigeon)		76
Streptopelia sp. (doves)		76
Gallicolumba luzonica (bleeding heart dove)		76
Gallicolumba cruenta		76
Goura oristata (goura pigeon)		1

Table 3 (continued)
NATURAL AND EXPERIMENTAL HOSTS OF CHLAMYDIAL ORGANISMS

Hosts	Type of infection	Ref.
Order Cuculiformes		74
Order Apodiformes		74
Order Passeriformes		
Family Piridae		
Parus major (titmouse)		107
Family Sturnidae (starling)		112
Family Ploceidae		
Passer domesticus (English sparrow)		10
Family Estrildidae		
Padda oryzirora (Java sparrow)		70
Lagonosticta seuegala (firefinch)		107
Poephila mirabilis (lady goldfinch)		107
Family Carduelidae		
Serinus canaria (canary)		107
Carduelis carduelis (goldfinch)		107
Loxia curvirostra (crossbill)		107
Spinus pinus (siskin)		107
Pyrrhula pyrrhula (bullfinch)		107
Zonotrichia sp. (crowned sparrow)		70
Family Icteridae		
Turdus meurla (blackbird)		107
Family Muscicapidae		
Leiothrix lutea (pekin robin)		107

Table 4
ARTHROPODS ASSOCIATED WITH CHLAMYDIA ORGANISMS

Hosts	Sources of isolation	Ref.
Class Arthropoda		
Order Acari		
Family Parasitidae	Litter from infected birds, U.S.A.	29
Family Laelapidae	Litter from infected birds, U.S.A.	29
Family Dermanyssidae	Litter from infected birds, U.S.A.	
Ornithonyssus sylviarum (Northern fowl mite)	Domestic fowl, U.S.A.	29
Family Cheyletidae	Litter from infected birds, U.S.A.	29
Cheyletus eruditis	Straw from horse stables, Calif.	29
Family Trombiculidae		
Chigger	Shear water (*Puffinus baroli*), Russia	120
Schoutedenichia penetrans	Cattle, Africa	39
Family Tyroglyphidae		
Glycyphagus domesticus	Turkey farm, U.S.A.	78
Glycyphagus cadaverum	Straw from horse stables, Calif.	29
Glycyphagus michaeli	Straw from horse stables, Calif.	29
Order Acarina		
Super family Argasidae (soft ticks)		
Ornithodoros moubata	Africa	39
Ornithodoros coriaceus	Rodent, ground, or vegetation, Calif.	31
Agas arboreus	White-necked cormorant, Ethiopia	32
Super Family Ixodidae (hard ticks)		
Ixodes ricinus	Cattle, France	41
Ixodes pacificus	Rodent, ground, or vegetation, Calif.	31
Ixodes sp.	Rodent, ground, or vegetation, Calif.	31
Dermacentor occidentalis	Rodent, ground, or vegetation, Calif.	31
Dermacentor marginatus	France	40
Haemaphysalis leachii	Source not specified	74
Hyalomma transiens	Cattle, Africa	41
Hyalomma anatolicum	Source not specified	74
Hyalomma excuvatum	Cattle, Africa	41
Hyalomma rufipes	Goats, Africa	41
Boophilus annulatus	Cattle, Africa	74
Boophilus decoloratus (blue tick)	Goats, Africa	41
Order Mallophaga		
Family Menoponidae		
Manopen gallinae (soft louse)	Chicken, Iowa hatchery	29
Order Anoplura (sucking louse)		
Family Hoplopleuridae		
Polyplax serrata (suckling louse)	Laboratory mice	74
Order Siphonaptera (flea)		
Family Pulicidae		
Pulex simulans	Rodents, Calif.	74
Unidentified	Rodents, Calif.	31

REFERENCES

1. Andrianne, V. F., *Ann. Med. Vet.,* 97, 63–72, 1953.
2. Armstrong, J. A.,, Valentine, R. C., and Fildes, C., *J. Gen. Microbiol.,* 30, 59–73, 1963.
3. Bader, J. P. and Morgan, H. R., *J. Exp. Med.,* 113, 271–281, 1958.
4. Bader, J. P. and Morgan, H. R., *Proc. Soc. Exp. Biol. Med.,* 106, 311–313, 1961.
4a. Baker, J. A., *J. Exp. Med.,* 79, 159–172, 1944.
5. Banks, J., Eddie, B., Schachter, J., and Meyer, K. F., *Infect. Immun.,* 1, 259–262, 1970.
6. Barnes, M. G. and Brainerd, H., *N. Engl. J. Med.,* 271, 981–985, 1964.
7. Barron, A. L., Mount, D. T., Cornell, D. S., and Brody, H., in *Trachoma and Related Disorders Caused By Chlamydial Agents,* Nichols, R. L., Ed., Excerpta Medica, Amsterdam, 1971, 71–78.
8. Barwell, C. F., *Lancet,* 2, 1369–1371, 1955.
9. Becerra, V. M. and Storz, J., *Zentralbl. Bakteriol. Parasitenkd. Infektionskr. Hyg. Abt. 1 Orig.,* 214, 250–258, 1970.
10. Beech, M. D. and Miles, J. A. R., *Aust. J. Exp. Biol. Med. Sci.,* 31, 473–480, 1953.
11. Bernkopf, H., *Am. J. Ophthalmol.,* 63, 1206–1207, 1967.
12. Bernkopf, H. and Mashiah, P., *J. Immunol.,* 188, 570–571, 1962.
13. Bernkopf, H., Treu, G., and Maythar, G., *Arch. Ophthalmol.,* 71, 693–700, 1964.
14. Blaskovic, D. and Styk, B., in *Methods in Virology,* Vol. 1, Maramorosch, K. and Koprowski, H., Eds., Academic Press, New York, 1967, 163–255.
15. Blyth, W. A., Tarvern, J., and Garrett, A. J., in *Trachoma and Related Disorders Caused by Chlamydial Agents,* Nichols, R. L., Ed., Excerpta Medica, Amsterdam, 1971, 79–87.
16. Burnet, I. M. and Rountree, P. M., *J. Pathol. Bacteriol.,* 40, 471–481, 1935.
17. Cello, R. M., *Am. J. Ophthalmol.,* 63, 1270–1273, 1967.
18. Chialov, R. J. and Parodi, A. S., *Rev. Inst. Bacteriol. Dep. Nac. Hig. Argent.,* 12, 337–346, 1944.
19. Collier, L. H., *Ann. N.Y. Acad. Sci.,* 98, 188–196, 1962.
20. Colon, J. I. and Moulder, J. W., *J. Infect. Dis.,* 103, 109–119, 1958.
21. Croker, T. T. and Eastwood, J. M., *Virology,* 19, 23–31, 1963.
22. Darougar, S., Kinnison, J. R., and Jones, B. R., in *Trachoma and Related Disorders Caused by Chlamydial Agents,* Nichols, R. L., Ed., Excerpta Medica, Amsterdam, 1971, 63–70.
23. Dawson, C. R., Mordhorst, C. H., and Thygeson, P., *Ann. N.Y. Acad. Sci.,* 98, 167–176, 1962.
24. Dickinson, L. and Cooper, B. S., *J. Pathol. Bacteriol.,* 78, 257–266, 1959.
25. Dragonas, P. N., *Bull. Soc. Pathol. Exot.,* 56, 17–21, 1963.
26. Dungworth, D. L., *J. Comp. Pathol. Ther.,* 73, 68–75, 1963.
27. Eagle, H., *Science,* 130, 432–437, 1959.
28. Eaton, M. D., Martin, W. P., and Beck, M. D., *J. Exp. Med.,* 75, 21–33, 1942.
29. Eddie, B., Meger, K. F., Lambrecht, F. L., and Furman, D. P., *J. Infect. Dis.,* 110, 231–237, 1962.
30. Eddie, B., Sladen, W. J., Sladen, B. K., and Meyer, K. I., *Am. J. Epidemiol.,* 84, 405–410, 1966.
31. Eddie, B., Radovsky, F. J., Stiller, D., and Kumade, N., *Am. J. Epidemiol.,* 90, 449–460, 1969.
32. Eddie, B., Foster, W. A., Radovsky, F. J., and Stiller, D., *J. Med. Entomol.,* 7, 745–746, 1971.
33. Fan, V. S. C. and Jenkin, H. M., *Infect. Immun.,* 10, 464–470, 1974.
34. Fortner, J. and Pfaffenburg, R., *Z. Hyg. Infektionskr.,* 117, 286–297, 1935.
35. Francis, R. D. and Gorden, F. B., *Proc. Soc. Exp. Biol. Med.,* 59, 270–272, 1945.
36. Furness, D. M., Graham, P., and Reeve, J., *J. Gen. Microbiol.,* 23, 613–619, 1960.
37. Furness, D. M. and Fraser, E., *J. Gen. Microbiol.,* 27, 299–304, 1962.
38. Gear, J., Cuthbertson, E., and Ryan, J., *Ann. N.Y. Acad. Sci.* 98, 197–200, 1962.
39. Giroud, P. and Jadin, J., *Bull. Soc. Pathol. Exot.,* 47, 578–588, 1954.
40. Giroud, P. and Colas-Belcour, J., *Bull. Soc. Pathol. Exot.,* 50, 194–197, 1957.
41. Giroud, P., Colas-Belcour, J., Pfister, R., and Morel, P., *Bull. Soc. Pathol. Exot.,* 50, 529–532, 1957.
42. Golub, O. J., *J. Immunol.,* 59, 71–82, 1948.
43. Gorden, N. H., *Lancet,* 1, 1174–1177, 1930.
44. Gorden, F. B. and Gillmore, J. D., *Aerosp. Med.,* 45, 257–262, 1974.
45. Gorden, F. B. and Quan, A. L., *Proc. Soc. Exp. Biol. Med.,* 118, 354–363, 1965.
46. Gorden, F. B. and Quan, A. L., *J. Infect. Dis.,* 115, 186–196, 1965.
47. Gorden, F. B. and Quan, A. L., *Antimicrob. Agents Chemother.,* 2, 242–244, 1972.
48. Gorden, F. B., Quan, A. L., and Trimmer, R. W., *Science,* 131, 733–734, 1960.
49. Gorden, F. B., Dressler, H. R., Quan, A. L., McQuilkin, W. T., and Thomas, J. L., *Appl. Microbiol.,* 23, 123–129, 1972.

50. **Graham, D. M. and Layton, J. E.,** in *Trachoma and Related Disorders Caused By Chlamydial Agents,* Nichols, R. L., Ed., Excerpta Medica, Amsterdam, 1971, 145–157.
51. **Hamre, D. and Rake, G.,** *J. Infect. Dis.,* 74, 206–211, 1944.
52. **Harrison, M. J.** *Aust. J. Exp. Biol. Med. Sci.,* 48, 207–213, 1970.
53. **Higashi, N. and Tamura, A.,** *Virology,* 12, 578–588, 1960.
54. **Hoge, V. M.,** *Public Health Rep.,* 49, 1415–1419, 1934.
55. **Ishitani, R., Sugawa, Y., Shibata, D., and Omuri, T.,** *Exp. Rep. Gov. Exp. Stn. Anim. Hyg. Jpn.,* 27, 121–130, 1953.
56. **Jenkin, H. M.,** *J. Infect. Dis.,* 116, 390–399, 1966.
57. **Jenkin, H. M. and Lu, Y. K.,** *Am. J. Ophthalmol.,* 63, 1110–1115, 1967.
58. **Jenkin, H. M. and Fan, V. S. C.,** in *Trachoma and Related Disorders Caused By Chlamydial Agents,* Nichols, R. L., Ed., Excerpta Medica, Amsterdam, 1971, 52–59.
58a. **Jones, B. R., Collier, L. H., and Smith, C. H.,** *Lancet,* 1, 902–905, 1959.
59. **Kemp, A. H., Wheeler, A. H., and Nungester, W. J.,** *J. Infect. Dis.,* 76, 135–143, 1945.
60. **Kerns, R. F.,** in *Psittacosis: Diagnosis, Epidemiology and Control,* Beaudette, F. R., Ed., Rutgers University Press, New Brunswick, N.J., 1955, 80–89.
61. **Kozikowski, E. H. and Hanon, N.,** *J. Bacteriol.,* 88, 533–534, 1964.
62. **Kuo, C. C., Kenny, G. E., and Wang, S. P.,** in *Trachoma and Related Disorders Caused By Chlamydial Agents,* Nichols, R. L., Ed., Excerpta Medica, Amsterdam, 1971, 113–123.
63. **Larzarus, A. S. and Meyer, K. F.,** *J. Bacteriol.,* 38, 121–151, 1939.
64. **Levinthal, W.,** *Klin. Wochenschr.,* 9, 654–659, 1930.
65. **Litwin, J.,** *J. Infect. Dis.,* 105, 129–160, 1959.
66. **Litwin, J., Officer, J. Z., Brown, A., and Moulder, J. W.,** *J. Infect. Dis.,* 109, 251–279, 1961.
67. **Makino, S., Jenkin, H. M., Yu, H. M., and Townsend, D.,** *J. Bacteriol.,* 103, 62–70, 1970.
68. **Manire, G. P. and Galasso, G. J.,** *J. Immunol.,* 83, 529–533, 1959.
69. **Marriott, M. E. and Storz, J.,** *Zentralbl. Bakteriol. Parasitenkd. Infektionskr. Hyg. Abt. 1 Orig.,* 200, 304–323, 1966.
70. **Matumoto, M., Inaba, O. Y., Ishitani, R., Kurogi, H., Morimoto, T., and Ishii, S.,** *Bull. Nat. Inst. Anim. Health* (Tokyo), 35, 67–81, 1958.
71. **McNutt, S. H., Leith, T. S., and Underbjerg, G. K.,** *Am. J. Nat. Res.,* 6, 247–251, 1945.
72. **Mendlowski, B. and Segre, D.,** *Am. J. Vet. Res.,* 21, 68–73, 1960.
73. **Menges, R. W., Harshfield, G. S., and Wenner, H. A.,** *Am. J. Hyg.,* 57, 1–14, 1953.
74. **Meyer, K. F.,** *Am. J. Ophthalmol.,* 63, 1225–1246, 1967.
75. **Meyer, K. F. and Eddie, B.,** *Proc. Soc. Exp. Biol. Med.,* 49, 522–525, 1942.
76. **Meyer, K. F., Eddie, B., and Yanamura, H. Y.,** *Proc. Soc. Exp. Biol. Med.,* 49, 609–615, 1942.
77. **Meyer, K. F. and Eddie, B.,** *Proc. Soc. Exp. Biol. Med.,* 83, 99–101, 1953.
78. **Meyer, K. F. and Eddie, B.,** *Science,* 132, 300, 1960.
79. **Miles, J. A. R. and Shrivastav, J. B.,** *J. Anim. Ecol.,* 20, 195–200, 1951.
80. **Mitsui, Y., Kitamuro, T., Endo, K., and Matsumura, K.,** *Science,* 145, 715–716, 1964.
81. **Mitsui, Y., Kitamuro, T., and Fujimoto, M.,** *Am. J. Ophthalmol.,* 63, 1191–1205, 1967.
82. **Morgan, H. R. and Wiseman, R. W.,** *J. Infect. Dis.,* 79, 131–133, 1946.
83. **Morrison, S. J. and Jenkin, H. M.,** *In Vitro,* 8, 94–100, 1972.
84. **Moulder, J. W., Novosel, D. L., and Tribby, I. C.** *J. Bacteriol.,* 85, 701–706, 1963.
85. **Murray, E. S.,** *J. Infect. Dis.,* 114, 1–12, 1964.
86. **Mykytowycz, R., Dane, D. S., and Beech, M.,** *Aust. J. Exp. Med. Biol. Sci.,* 33, 629–636, 1955.
87. **Nicolle, C., Cue'nod, A., and Blaizot, L.,** *Ann. Ocul.,* 156, 405–414, 1911.
88. **Nigg, C.,** *Science,* 95, 49–50, 1942.
89. **Officer, J. E. and Brown, A.,** *J. Infect. Dis.,* 107, 283–299, 1960.
90. **Omuri, T., Morimoto, T., Harada, K., Inaba, Y., Ishii, S., and Matumoto, M.,** *Jpn. J. Exp. Med.,* 27, 131–143, 1957.
91. **Palotay, J. L. and Christinessen, N. R.,** *J. Am. Vet. Med. Assoc.,* 134, 220–230, 1963.
92. **Parker, H. D.,** *Am. J. Vet. Res.,* 21, 243–250, 1960.
93. **Parlov, P., Milanov, M., and Tchilev, D.,** *Ann. Inst. Pasteur Paris,* 105, 450–454, 1963.
94. **Pearson, H. E. and Eaton, M. D.,** *Proc. Soc. Exp. Biol. Med.,* 45, 677–679, 1940.
95. **Pearson, J. W., Duff, J. T., Gearinger, N. F., and Robbins, M. L.,** *J. Infect. Dis.,* 115, 49–58, 1965.
96. **Pearce, J. H. and Lowrie, D. B.,** in *Microbiolgical Pathogenicity in Man and Animals,* Smith, H. and Pearce, J. H., Eds., Cambridge University Press, London, 1972, 193–216.
97. **Piraino, F. and Abel, C.,** *J. Bacteriol.,* 87, 1503–1511, 1964.
98. **Pollard, M.,** *Proc. Soc. Exp. Biol. Med.,* 64, 200–202, 1947.
99. **Pollard, M., Starr, T. J., Tanami, Y., and Moore, R. W.,** *Proc. Soc. Exp. Biol. Med.,* 104, 223–225, 1960.
100. **Pollard, M.,** *Tex. Rep. Biol. Med.,* 17, 186–193, 1959.

101. Rake, G., McKee, C., and Shaffer, M. F., *Proc. Soc. Exp. Biol. Med.,* 43, 332–334, 1940.

102. Reeve, P. and Tarvern, J., *Am. J. Ophthalmol.,* 63, 1167–1173, 1967.

103. Roca-Garcia, M., *J. Infect. Dis.,* 85, 275–289, 1949.

104. Rubin, H., Kissling, R. E., Chamberlain, R. W., and Eidson, M. E., *Proc. Soc. Exp. Biol. Med.,* 78, 696–698, 1951.

105. Sanchez, B. R., *Rev. Patronato Biol. Anim.,* 15, 349–367, 1971.

105a. Schachter, J., *Am. J. Ophthalmol.,* 63, 1049/23–1053/27, 1967.

106. Shaffer, M. F., Jones H., Grace, A. W., Hamre, D. M., and Rake, G., *J. Infect. Dis.,* 75, 109–112, 1944.

107. Shaughnessy, H. J., in *Diseases Transmitted from Animals to Man,* Hull, T. G., Ed. Charles C Thomas, Springfield, Ill., 1963, 350–373.

108. Shindarov, L., Runevski, N., and Vassileva, V., *Zentralbl. Bakteriol. Parasitenkd. Infektionskr. Hyg. Abt. 1 Orig.,* 216, 9–14, 1971.

109. Schiøtt, C. R. and Morgan, H. R., *Proc. Soc. Exp. Biol. Med.,* 96, 647–648, 1957.

110. Sidwell, R. W., Lundgren, D. L., and Thorpe, B. D., *Am. J. Trop. Med. Hyg.,* 13, 591–594, 1964.

111. Spalatin, J., Iverson, J. V., and Hanson, R. P., *Can. J. Microbiol.,* 17, 935–942, 1971.

112. Sperling, G. F., *Pa. Bull. Vet. Ext. Q.,* 47, 156–158, 1957.

113. Storz, J. and McKercher, P. G., *Zentralbl. Veterinarmed.,* 9, 520–541, 1962.

114. Storz, J., Smart, R. A., Marriott, M. E., and Davis, R. V., *Am. J. Vet. Res.,* 27, 633–641, 1966.

115. Strauss, J., Bednar̆, B., and Serý, V., *J. Hyg. Epidemiol. Microbiol. Immunol.,* 1, 230, 1957.

116. Swack, N. S. and Duff, J. T., *J. Infect. Dis.,* 118, 468–472, 1968.

117. Tanami, Y., Dollard, M., Starr, T. J., and Moore, R. W., *Tex. Rep. Biol. Med.,* 18, 515–522, 1960.

118. Tang, F. F., Chang, H. L., Huang, Y. T., and Wang, K. C., *Chin. Med. J.,* 75, 429–447, 1957.

119. Tarrizo, M. L. and Nabli, B., *Am. J. Ophthalmol.,* 63, 1215–1225, 1967.

120. Terskikh, I. I., Cheltsov-Bebutov, A. M., Kubovina, L. N., and Keleinikov, A. A., *Vopr. Virusol.,* 6, 131–135, 1961.

121. Treuhaft, M. W. and Moulder, J. W., *J. Bacteriol.,* 96, 2004–2011, 1968.

122. Wang, S. P. and Grayston, J. T., *Ann. N.Y. Acad. Sci.,* 98, 177–187, 1962.

123. Wang, S. P. and Grayston, J. T., *Am. J. Ophthalmol.,* 70, 367–380, 1970.

124. Ward, G. C. and Birge, J. P., *JAMA,* 150, 217–219, 1952.

125. Weiss, E. and Huang, J. S., *J. Infect. Dis.,* 94, 107–125, 1954.

126. Weiss, E., *J. Bacteriol.,* 90, 243–253, 1965.

127. Wenner, H. A., *Adv. Virus Res.,* 5, 39–93, 1958.

128. Williams, S. E., *Aust. J. Exp. Biol. Med. Sci.,* 22, 205–208, 1944.

129. Willigan, D. A. and Beamer, P. D., *J. Am. Vet. Med. Assoc.,* 126, 118–122, 1955.

130. Wilson, M. R. and Dungworth, D. L., *J. Comp. Pathol. Ther.,* 73, 277–284, 1963.

131. York, C. J. and Baker, J. A., *J. Exp. Med.,* 93, 587–603, 1951.

CULTURE MEDIA FOR CELL WALL-DEFICIENT MICROORGANISMS

L. H. Mattman

All incubation is done in 2% carbon dioxide. If an anaerobe is suspected, a duplicate culture is incubated in an anaerobic atmosphere containing CO_2.

CULTURE OF WALL-DEFICIENT VARIANTS PRODUCED IN VITRO

Medium with High Concentration of NaCl

Distilled water, 90 ml
Brain Heart Infusion (BHI) (Difco®), 2.7 g
NaCl, 5 g
Sterile horse serum or swine serum, 10 ml.

The BHI broth is autoclaved at 120°C for 20 min and then held in a 50°C water bath until the serum is added. A gamma horse or swine serum may be substituted for whole serum. The serum is heat inactivated at 56°C for 30 min and added to the autoclaved medium. If serum is added to the agar when it is too hot, an opacity results which makes detection of colonies difficult. "Sterile" commercial serum often contains wall-deficient bacteria, but we have encountered no contaminants which grow in 5% NaCl. The final pH of the medium should be 7.4 to 7.6. If pH adjustment is required it is done with NaOH.

For a solid medium purified agar (Difco) is added to give a concentration of 0.9%. Thus, 0.9 g agar is added to 90 ml of broth-NaCl solution before autoclaving.

Medium with Sucrose as Osmotic Stabilizer*

NIH Thioglycollate broth (Difco) 3.0 g
Noble's Agar (Difco), 0.8 g
Sucrose, 15 g
Distilled water, 100 ml
The pH is adjusted to 7.1 to 7.3.

The agar and thioglycollate broth are autoclaved together in 50 ml of the water. The sucrose is autoclaved separately in 50 ml of distilled water. Horse or swine serum is added to give a final concentration of from 1 to 20%, varying with the requirements of the bacterial strain. If the serum is used in a concentration above 2%, it will probably require inactivation at 56°C for 30 min. The serum and other solutions should be at 50°C when mixed before pouring the petri plates. If more than 5% serum is used, correspondingly less water is employed to dilute the broth and agar. After a heavy inoculum is streaked over the surface of the plates, a sterile glass slide is placed on the agar. Growth may be initiated only under or at the edges of the slide.

In practice, the media for in vitro-induced variants is usually used with the same concentration of the same stabilizing agent present when variation was induced by 100 to 1000 units of penicillin per milliliter. The preference for NaCl or sucrose as a stabilizer can be a characteristic of either species or strain.

*Slightly modified from Brem and Eveland.[1]

Synthetic-type Medium for the L-type Colonies of Proteus Induced In Vitro*

The percent composition of the basal synthetic medium is

Glucose	1.0
Sodium lactate	2.0
H_2HPO_4	0.9
KH_2PO_4	0.1
Nicotinamide	0.1
Salt mixture	0.2

The salt mixture contains the following per liter:

$MgSO_4 \cdot 7H_2O$	40.0 g
NaCl	2.0 g
$FeSO_4 \cdot 7H_2O$	2.0 g
$MnSO_4 \cdot 4H_2O$	8.0 g
Casamino acids	5.0 mg

An amino acid mixture of the following composition (mg/10 ml) may be substituted for the vitamin-free casamino acids:

Glycine	0.2
L-Leucine	2.5
L-Tyrosine	3.0
L-Proline	4.0
L-Histidine	1.0
L-Arginine	2.5
DL-Alanine	1.0
DL-Serine	0.4
DL-Valine	4.0
DL-Isoleucine	2.5
DL-Phenylalanine	2.0
DL-Aspartic acid	2.0
DL-Lysine	4.0
L-Glutamic acid	10.0
DL-Methionine	2.0

This synthetic-type medium without serum gives large L-phase colonies of *Proteus mirabilis* and *P. vulgaris* within 24 hr. Reversion is usually complete by 48 hr if penicillin was the inducing agent. However, for cell wall-deficient variants of most genera this medium needs an osmotic stabilizer and serum.

CULTURE OF WALL-DEFICIENT MICROORGANISMS FROM IN VIVO SOURCES

Veal Infusion Agar for Propagation of Cell Wall-deficient Forms from Blood Cultures Not Showing Growth by Routine Methods**

Veal Infusion Broth, dehydrated (Difco), 2.2 g
Soluble Starch (Difco), 0.5%
Purified agar (Difco), 1.2 g
Distilled water, 100 ml
Sucrose, 20.0 g
Supplement C (Difco), 1.0 ml
The final pH should be 7.6 to 7.8 adjusted with NaOH.

---

*More media which may be employed for in vitro induced wall-deficient variants from in vivo sources are described in Mattman.[2]

**This subculture may be made after the blood cultures has been incubated for 2 days.

The sucrose is autoclaved separately in half of the water. All autoclaving is done at least 5 days before use of the medium. Immediately before use, the broth-agar-starch mixture is melted by boiling in a water bath.

The Supplement C, although free of classical bacteria, has not been free of wall-deficient variants. It is filtered through a membrane of 0.25 μ porosity or through a Seitz pad of sterilizing grade. Large volumes can be filtered and stored in aliquots at $-70°C$. The Supplement C can be added to individual petri dishes or to the flask of medium.

Many essential nutrients are contributed by the inoculum, which consists of 2 ml of blood culture sediment. These erythrocytes of the patient are lysed by freeze-thawing three times in stainless steel tubes. The lysed erythrocytes are placed in a petri dish and mixed into the agar medium poured over them. From 2 to 20 classical colonies may develop. If a disc containing 1% potassium tellurite is placed on the surface of the medium at the time of inoculation, the tellurite will be reduced, giving a definite blackening around the discs if the culture contains wall-deficient *Staphylococci*. The tellurite reduction occurs even if no reverted colonies appear on the plate.[3-5] The wall-deficient colonies in the pour plates may be very difficult to detect without a marker such as the reduction of tellurite.

These pour plates may be used for testing with standard antibiotic sensitivity discs. At times, reversion will occur in a ring at the periphery of a circle of inhibition.

Horse Muscle Broth for Growing Large Numbers of Wall-deficient Bacteria and Fungi from Clinical Samples

Thaw a 12 oz package of horse meat (Topper Dog and Cat Food, packed by Lewis Fur Farms, Rodney, Mich.) in warm water (this requires approximately 30 min);

Pack the horse meat in 20-ml tubes to 1.5 cm in height;

Prepare Difco veal infusion broth according to supplier's directions;

Dispense 15 ml of the broth in each tube containing horse meat;

Autoclave the tubes for 10 min at 122°C;

Using a disposable pipette, discard the top fatty layer completely from the tubes while still hot;

Reautoclave the broth for 30 min at 122°C;

Store the tubes at 0 to 4°C for at least 5 days for elimination of inhibitors before using;

At the time of use, add Supplement C (Difco) to make a concentration of 1.5% in the medium.

Inoculate the above medium with 0.5 ml of the blood of the patient (or synovial fluid, spinal fluid, exudate, or whatever constitutes the specimen). One half milliliter of a 48-hr blood culture may be used.

This horse muscle broth will foster growth of large numbers of wall-deficient fungi and bacteria. However, the artifacts in the horse muscle make characterization of the growth difficult unless clues to the identity of the variant are known and fluorescent antibody can be applied.[6] With many but not all isolates, 0.2 ml of a 48-hr horse muscle culture will yield colonies in pour plates of the veal infusion agar.

REFERENCES

1. **Brem, A. M. and Eveland, W. C.,** L-forms of *Listeria monocytogenes, J. Infect. Dis.,* 118, 181–187, 1968.
2. **Mattman, L. H.,** *Cell Wall Deficient Forms,* CRC Press, Cleveland, 1974.
3. **Whalen, M.,** Biochemical Reactivity of *Staphylococci* in the Cell Wall Deficient Stage, M.S. thesis, Wayne State University, Detroit, 1975.
4. **Zafar, R.,** Quantitation and Identification of Cell Wall Deficient Microorganisms Occurring in Pre-therapy Blood Cultures, M.S. thesis, Wayne State University, Detroit, 1975.
5. **Abbassian, R.,** M.S. thesis, Wayne State University, Detroit, anticipated.
6. **Motwani, N. G.,** Identification of Cell Wall Deficient Bacteria by Immunofluorescence and Polyacrylamide Gel Electrophoresis, Ph.D. thesis, Wayne State University, Detroit, 1976.

CULTURE MEDIA (NATURAL AND SYNTHETIC): MYCOPLASMAS

E. A. Freundt

INTRODUCTION

Mycoplasmas are prokaryotic microorganisms which vary in shape from spheres about 250 to 300 nm in diameter to long filamentous structures. At present, more than 50 species are known, many of which are important pathogens causing disease in man, animals, and plants.

The mycoplasmas are assigned to a separate class, the Mollicutes, which differs from other prokaryotes in lacking a true cell wall. The class contains one order, the Mycoplasmatales, which consists of three families: the Mycoplasmataceae (with two genera, *Mycoplasma* and *Ureaplasma*), Acholeplasmataceae (genus *Acholeplasma*), and Spiroplasmataceae (genus *Spiroplasma*). The classification into families and genera is based mostly on differences in nutritional requirements and metabolism. Species of *Mycoplasma, Ureaplasma,* and *Spiroplasma* are distinguished by requiring cholesterol or certain other sterols for growth, in which respect they differ from *Acholeplasma* species and from bacteria.[6,13,15,17] *Ureaplasma* is unique in its capability to catabolize urea.

A recently proposed new genus, *Anaeroplasma*,[14] cannot be placed within any of the already established families because it contains sterol-requiring as well as sterol-nonrequiring strains.

Thermoplasma is another genus that is recognized as a member of the Mollicutes, but its precise affiliation is uncertain.

For detailed information about the mycoplasmas the reader is referred to recent reviews.[8,10,12] Most of the species referred to in the following sections are described in Reference 8.

NUTRITION

The mycoplasmas are the smallest organisms that are able to replicate in artificial cell-free media. As mentioned earlier, the Mycoplasmataceae, which include the vast majority of mycoplasma species, and Spiroplasmataceae depend on cholesterol or related sterols as nutritional factors. In the laboratory, this demand is usually complied with by adding horse or some other animal serum to the growth medium. Yeast extract stimulates the growth of a great many mycoplasmas. A few species have special requirements, as, for example, the requirement shown by *M. synoviae* for β-nicotinamide dinucleotide in the reduced form. Glucose seems to enhance the growth rate of glucose-utilizing organisms, and L-arginine the growth rate of organisms which hydrolyze this amino acid. The growth of *M. synoviae, M. buccale, M. lipophilum,* and *U. urealyticum* is significantly stimulated by cysteine hydrochloride. With the said exceptions, very little is known of the amino acid requirements of mycoplasmas. The addition of L-histidine to the growth medium prolongs the stationary phase and survival (buffering effect), but does not affect logarithmic growth of *U. urealyticum.*[1] The growth of *M. bovigenitalium* and possibly some other species is markedly enhanced by deoxyribonucleic acid (DNA).

It is still a matter of controversy whether *U. urealyticum* has an absolute requirement for urea or whether this substance can be replaced by related compounds such as allantoin.[11] The demonstration that urea is quantitatively metabolized to NH_3 and CO_2 and that no ^{14}C is detectable in organisms grown in ^{14}C labeled urea[7] indicates that urea is not utilized by *U. urealyticum* as a source of nitrogen or carbon. Neither is there any known pathway by which urea can be utilized as a source of energy.

The very particular conditions needed for the growth of *Spiroplasma* and *Thermoplasma* species will be dealt with in the pertinent sections.

Most mycoplasma species are facultative anaerobes, but growth is frequently stimulated in atmospheric air containing 5% added CO_2. Some mycoplasmas of human and nonhuman primate sources prefer anaerobic conditions (95% N_2 and 5% CO_2), at least on primary cultivation and during the first culture passages. *Anaeroplasma* species are strict anaerobes.

The pH optimum is about 7.3 to 7.8 for most mycoplasmas, although it is significantly lower for *U. urealyticum,* namely, 6.0 ± 0.5. The pH optimum for *T. acidophilum* is about 1 to 2.

The temperature optimum is about 37°C for almost all *Mycoplasma* and *Ureaplasma* species. Most *Acholeplasma* species grow at a wide range of temperatures (22 to 37 °C), but optimum temperatures have not been defined precisely. The optimum temperatures for *Spiroplasma* vary from 28 to 32°C, while the optimum for *Thermoplasma* is about 56 to 60°C.

FORMULAE OF GROWTH MEDIA

In the formulae presented in the following, information on commercial sources of chemicals and other medium ingredients is given in some cases. The preparations of a commercial firm specified by name are those with which the FAO/WHO Collaborating Centre for Animal Mycoplasmas has experience. This does not, of course, necessarily imply that preparations of equal quality may not be obtainable from other firms as well.

Media designated by capital letters are those used at the FAO/WHO Collaborating Centre for Animal Mycoplasmas.

The modifications of medium compositions described are also those of the Collaborating Centre.

Sterilization Methods

Autoclavable (121°C for 20 min) medium components include broth bases, glucose and other carbohydrates, yeast extract, DNA, Hanks balanced salt solution, and other salt solutions.

Animal sera are sterilized by filtration through Seitz® Pore Filter K5 (prefiltration), followed by filtration through Seitz Filter Sheets EK, at 15 psi pressure. Unless otherwise stated, animal sera are not heat-inactivated.

The remaining medium constituents are sterilized by filtration through Millipore® or Gelman® membrane filters, pore size 200 nm.

MEDIA FOR *MYCOPLASMA* SPECIES

Medium B (Hayflick's Medium, Modified)

This is a modification of the medium originally devised by Hayflick[3] for the growth of *M. pneumoniae.* It is a suitable medium for the cultivation of the vast majority of *Mycoplasma* species.

Liquid Medium

Heart infusion broth (Difco)®	2.85 g
Distilled water	90.0 ml
Sterilize by autoclaving (121°C for 20 min)	

Horse serum	20.0 ml
Yeast extract, 25% (w/v) solution[18]	10.0 ml
Thallium acetate, 1% (w/v) solution	1.0 ml
Penicillin G, 200,000 IU/ml	0.25 ml
DNA, 0.2% (w/v) solution[a]	1.2 ml
Adjust pH to 7.8.	

Note: Thallium acetate and penicillin are included as inhibitors of bacterial growth; they do not inhibit the growth of most mycoplasmas in the concentrations indicated.

[a] From calf thymus, Type 1, Sigma Chemical Company, St. Louis, Missouri; Catalogue No. 1501.

Solid Medium

This is prepared by replacing heart infusion broth with 3.6 heart infusion agar (Difco).

Medium N

This is a medium devised at the FAO/WHO Collaborating Centre for Animal Mycoplasmas for strains that, although not requiring very complex media, are not easily grown on the B medium. The following species are known to prefer the N medium: *M. anatis, M. bovigenitalium, M. edwardii, M. felis, M. maculosum, M. meleagridis, M. spumans,* and *M. verecundum.* For the growth of unknown strains it has been proved advantageous to include both media in a first attempt at cultivation. If no growth occurs, trials are made with more special media.

Liquid Medium

Bacto® brain heart infusion (Difco)	3.7 g
Yeast extract (Difco)	0.5 g
Distilled water	100.0 ml
Sterilize by autoclaving (121°C for 20 min).	

Horse serum	20.0 ml
Yeast extract, 25% (w/v) solution[18]	10.0 ml
Thallium acetate, 1% (w/v) solution	1.0 ml
Penicillin G, 200,000 IU/ml	0.25 ml
DNA, Sigma®, 0.2% (w/v) solution	1.3 ml
Glucose, 50% (w/v) solution	2.0 ml
Adjust pH to 7.8.	

Solid Medium

This is prepared by adding 1.7 g Noble (Difco) special agar before autoclaving.

Medium BACY (for *M. faucium* and *M. lipophilum*)
Liquid Medium

Heart infusion broth (Difco)	2.25 g
Distilled water	90.0 ml
Sterilize by autoclaving (121°C for 20 min).	

Horse serum	20.0 ml
Yeast extract, 25% (w/v) solution[18]	10.0 ml
DNA, Sigma, 0.2% (w/v) solution	1.3 ml
Cysteine hydrochloride (10% (w/v) solution)	1.24 ml
L-Arginine, 30% (w/v) solution	4.25 ml
Phenol red, 0.06% (w/v) solution	5.0 ml
Thallium acetate, 1.0% (w/v) solution	1.0 ml
Penicillin G, 200,000 IU/ml	0.25 ml
Adjust pH to 7.3.	

Solid Medium

Replace heart infusion broth with 3.6 g heart infusion agar (Difco).

Medium F (Frey's Medium; for *M. synoviae*)

Liquid Medium

Bacto PPLO broth w/o CV (Difco)	2.25 g
Distilled water	90.0 ml

Sterilize by autoclaving (121°C for 20 min).

Eagle's essential vitamins (× 100)[19]	0.025 ml
Glucose, 50% (w/v) solution	2.0 ml
Swine serum (heated 56°C/30 min)	12.0 ml
β-NADH, 1% (w/v) solution[a]	1.0 ml
Cysteine hydrochloride, 1% (w/v) solution	1.0 ml
Phenol red, 0.06% (w/v) solution	5.0 ml
Penicillin G (200,000 IU/ml)	0.25 ml

Note: The solutions of β-NADH and cysteine hydrochloride are mixed, and after 10 min added to the other ingredients. Adjust pH to 7.8.

[a] β-Nicotinamide dinucleotide, reduced form, Sigma, Grade III.

Solid Medium

This is prepared by adding 1.4 g Noble (Difco) special agar before autoclaving.

Medium FF74 (Friis' Medium,[9] Modified 1974; for *M. hyopneumoniae, M. flocculare*, and *M. dispar*)*

Liquid Medium

Brain heart infusion broth (Difco)	0.82 g
Double distilled water	75.0 ml
Bacto PPLO broth w/o CV (Difco)	0.87 g
Hanks balanced salt solution (modified; see below)	50.0 ml

Sterilize by autoclaving (121°C for 2 to 5 min).

Fresh yeast extract (FG, see below)	6.0 ml
Horse serum	15.0 ml
Swine serum (from SPF pigs), heated at 56°C for 60 min	15.0 ml
Phenol red, 0.06% (w/v) solution	5.0 ml
Bacitracin, 2.5% (w/v) solution	1.0 ml
Meticillin, 2.5% (w/v) solution	1.0 ml

Solid Medium

Special agar, Noble (Difco)	1.3 g
DEAE-dextran	0.017 g
Hanks balanced salt solution (modified, see below)	16.0 ml

Sterilize by autoclaving (121°C for 2 to 5 min). After cooling below 100°C, this solution is added to one portion of preheated FF74 broth.

Hanks Balanced Salt Solution (HBSS), Modified

The solution is prepared from stock solutions A and B.

* Dr. N. F. Friis, State Veterinary Serum Laboratory, Copenhagen, Denmark.

A. NaCl, 80.0 g; KCl, 4.0 g; MgSO$_4$ · 7H$_2$O, 1.0 g; MgCl$_2$ · 6H$_2$O, 1.0 g. Dissolve in 400 ml water, add 1.4 g CaCl$_2$ (anhydrous) and water to 500 ml.

B. Na$_2$HPO$_4$ · 12H$_2$O, 1.5 g in 400 ml water; add 0.6 g KH$_2$PO$_4$ and water to 500 ml.

To prepare HBSS, 25 ml of A is added to 400 ml of water, then mixed with 25 ml of B and 50 ml of water.

Yeast Extract FG

Fleischmann's® pure dry yeast, type 2040	125 g
Double distilled water	750 ml

The suspension is heated to 37°C for 20 min, thereafter heated to 90 to 100°C for 5 min. After cooling, it is centrifuged at 100 × g for 30 min. The supernatant is dispensed in appropriate volumes and autoclaved at 115°C for 2 to 5 min. The medium is stored at −20°C.

MEDIA FOR *ACHOLEPLASMA* SPECIES

Acholeplasma species, which do not require cholesterol for growth, can be grown in Medium B without horse serum, this being replaced by broth supplemented with 1% serum fraction (Difco, Code 0441).

MEDIA FOR *UREAPLASMA* SPECIES

Medium S (Shepard's Medium, Modified by Black;[2] for *U. urealyticum*)

Liquid Medium

Trypticase® soy broth (Baltimore Biological Laboratories)	3.7 g
Distilled water	80.0 ml
Sterilize by autoclaving (121°C for 20 min)	
Horse serum	22.0 ml
Yeast extract, 25% (w/v) solution[18]	10.0 ml
Urea, 40% (w/v) solution	1.6 ml
Phenol red, 0.06% (w/v) solution	5.0 ml
Penicillin G, 150,000 IU/ml	1.0 ml
Adjust pH to 6.0	

Solid Medium

This is prepared by adding 1.7 g Noble agar (Difco) before autoclaving.

MEDIA FOR *SPIROPLASMA* SPECIES

Medium P (Saglio's Medium (SMC), Slightly Modified from Saglio et al.;[15] for Growing *S. citri*)

Although the P medium or like hypertonic media are required for primary isolation of *S. citri* from plant or insect (leafhopper) tissues, this organism may be adapted to conventional media, e.g., the B medium. The optimal temperature for growth of *S. citri* is 32°C, significantly less growth being obtained at 29 and 35°C.

Liquid Medium

Heart infusion broth (Difco)	2.0 g
Distilled water	80.0 ml
Tryptone (Difco)	1.0 g
Yeast extract (Difco)	1.0 g
Sterilize by autoclaving (121°C for 20 min).	

Sorbitol, 50% (w/v) solution	14.0 ml
Sucrose, 50% (w/v) solution	2.0 ml
Fructose, 50% (w/v) solution	0.2 ml
Glucose, 50% (w/v) solution	0.2 ml
Phenol red, 0.06% (w/v) solution	5.0 ml
Horse serum	20.0 ml
Penicillin G (200,000 IU/ml)	0.6 ml
Adjust pH to 7.4.	

Solid Medium

This is prepared by replacing heart infusion broth with 3.4 g heart infusion agar (Difco).

Medium M1-A (Williamson and Whitcomb's Medium,[20] Revised[23])

This medium was devised for the isolation of the Corn Stunt spiroplasma from plant or insect (*Drosophila*) tissues, but recent observations[23] indicate that it is an even better medium for *S. citri*. The optimal temperature for growth of the Corn Stunt organism is about 29°C.

Liquid Medium

Autoclavable Fraction

Distilled water	70.0 ml
α-D-Glucose	0.1 g
D-Fructose	0.1 g
Sucrose	1.0 g
Bacto tryptone (Difco)	1.0 g
Sorbitol	7.0 g
PPLO broth base (Difco)	2.1 g
Bacto peptone (Difco)	0.8 g

Adjust pH to 7.8 with 1 *N* NaOH. Autoclave 30 min at 15 psi pressure. Cool to room temperature. Add the following *nonautoclavable components:* 160 ml Schneider's Drosophila Medium (revised, Grand Island Biological Co.), 50 ml fetal bovine serum (Flow Laboratories, "mycoplasma free"; heat-inactivate 1 hr at 56°C), 10 ml fresh yeast extract (Microbiological Associates), 1.2 ml phenol red (0.5% w/v solution), and 2.5 ml potassium penicillin G (100,000 IU/ml).

Solid Medium

Add 5.0 g Noble (Difco) special agar to the autoclavable fraction of the liquid medium. Preheat Schneider's Drosophila Medium to 37°C; if a precipitate forms, let it settle and ignore it. Equilibrate the autoclavable medium fraction at 56°C and treat the serum for 1 hr at that temperature. Equilibrate Schneider's Medium quickly to 37°C and mix the components rapidly. Finally, add penicillin and pour immediately.

Medium C-3 (Chen and Liao's Medium)

This medium was shown independently by Chen and Liao[4] to support growth of the Corn Stunt spiroplasma. It consists of Medium 199 (1x), 1 ml; Schneider's

Drosophila Medium, 0.5 ml; CMRL-1066 medium, 0.5 ml; Difco PPLO broth base, 1.5 g; sucrose, 16 g; horse serum, 20 ml; fresh yeast extract, 10 ml; phenol red, 10 μg/ml; and water to 100 ml. The pH is adjusted to 7.4.

Very recently, Liao and Chen[22] reported that all the nutritional and physical requirements of the Corn Stunt spiroplasma can be met by an extremely simple medium containing only 20 ml horse serum, 1.5 g Difco PPLO broth base, 16 g sucrose, and 74 ml distilled water.

This medium, as well as Chen and Liao's Medium C-3, have also been found to be excellent media for *S. citri*.[23]

MEDIA FOR *THERMOPLASMA* SPECIES

Medium TA

The description given below of the composition and preparation of the liquid medium for growing *T. acidophilum*[5] is based on a prescription provided by Dr. T. D. Brock.[21] The optimal temperature for growth of *T. acidophilum* is about 56°C. With a large inoculum (about 2% v/v), growth is visible in 2 to 3 days.

Liquid Medium

Ingredients for salt solution A:

$FeCl_3 \cdot 6H_2O$	1.93 g
$MnCl_2 \cdot 4H_2O$	0.18 g
$Na_2B_4O_7 \cdot 10H_2O$	0.45 g
$ZnSO_4 \cdot 7H_2O$	0.022 g
$CuCl_2 \cdot 2H_2O$	0.005 g
$NaMoO_4 \cdot 2H_2O$	0.003 g
$VOSO_4 \cdot 2H_2O$	0.003 g
$CoSO_4$	0.001 g
Water	1000 ml

Ingredients for salt solution B:

$(NH_4)_2SO_4$	13.2 g
KH_2PO_4	3.72 g
$MgSO_4 \cdot 7H_2O$	2.47 g
$CaCl_2 \cdot 2H_2O$	0.74 g
Water	1000 ml

Mix 10 ml of solution A and 100 ml of solution B into about 800 ml water, adjust the pH to 2.0 with 10 N H_2SO_4, make to one liter, and autoclave. If a precipitate develops, ignore it. Add one volume yeast extract solution (10.0 g yeast extract and 100.0 ml distilled water, autoclaved at 121°C for 20 min) and two volumes 50% (w/v) glucose solution (40.0 ml) to 100 volumes salt solution in sterile containers. For small volumes, use 5 ml medium in 16- X 125-mm stainless steel or screw-capped tubes. For larger volumes, use screw-capped bottles or flasks with one fifth of the volume as culture medium.

Solid Medium

This is prepared according to a prescription by T. Sander, pharmacist, Institute of Medical Microbiology, University of Aarhus, Denmark.

Combine 5.0 ml salt solution A, 50.0 ml salt solution B, and 180.0 ml distilled water. Adjust the pH to 2.0 with 10 N H_2SO_4, make to 250 ml, dispense aliquots of 50 ml into 100-ml Sovirel® screw-capped bottles, and autoclave. Add 1.0 ml autoclaved 10% (w/v) yeast extract solution and 2.0 ml 50% (w/v) glucose solution to 50.0 ml of the double-concentrated salt solution.

Combine 1.2 g Noble agar (Difco) and 50.0 ml distilled water; autoclave at 121°C for 20 min and cool to about 50°C. Add this to the above, which should be heated to about 50°C. Dispense immediately into petri dishes.

REFERENCES

1. Ajello, F. and Romano, N., *Appl. Microbiol.,* 29, 293–294, 1975.
2. Black, F. T., *Appl. Microbiol.,* 25, 528–533, 1973.
3. Chanock, R. M., Hayflick, L., and Barile, M. F., *Proc. Natl. Acad. Sci. U.S.A.,* 48, 41–49, 1962.
4. Chen, T. A. and Liao, C. H., *Science,* 188, 1015–1017, 1975.
5. Darland, G., Brock, T. D., Samsonoff, W., and Conti, S. F., *Science,* 170, 1416–1418, 1970.
6. Edward, D. G. ff. and Freundt, E. A., *J. Gen. Microbiol.,* 62, 1–2, 1970.
7. Ford, D. K., McCandlish, K. L., and Gronlund, A. F., *J. Bacteriol.,* 102, 605–606, 1970.
8. Freundt, E. A., The mycoplasmas, in *Bergey's Manual of Determinative Bacteriology,* 8th ed., Buchanan, R. E. and Gibbons, N. E., Eds., Williams & Wilkins, Baltimore, 1974, 929.
9. Friis, N. F., *Acta Vet. Scand.,* 12, 120–121, 1971.
10. Hayflick, L., Ed., *The Mycoplasmatales and the L-Phase of Bacteria,* Appleton-Century-Crofts, New York, 1969.
11. Masover, G. K. and Hayflick, L., *Ann. N.Y. Acad. Sci.,* 225, 118–130, 1973.
12. Razin, S., in *Handbook of Microbiology,* Vol. 1, Laskin, A. I. and Lechevalier, H. A., Eds., CRC Press, Cleveland, 1973, 105.
13. Razin, S. and Tully, J. G., *J. Bacteriol.,* 102, 306–370, 1970.
14. Robinson, I. M., Allison, M. J., and Hartman, P. A., *Int. J. Syst. Bacteriol.,* 25, 173–181, 1975.
15. Saglio, P., Lhospital, D., Laflèche, G., Dupont, G., Bové, J. M., Tully, J. G., and Freundt, E. A., *Int. J. Syst. Bacteriol.,* 23, 191–204, 1973.
16. Schneider, I., *J. Exp. Zool.,* 156, 91–104, 1964.
17. Shepard, M. C., Lunceford, C. D., Ford, D. K., Purcell, R. H., Taylor-Robinson, D., Razin, S., and Black, F. T., *Int. J. Syst. Bacteriol.,* 24, 160–171, 1974.
18. Taylor-Robinson, D., Somerson, N. L., Turner, H. C., and Chanock, R. M., *J. Bacteriol.,* 85, 1261–1273, 1963.
19. Therkelsen, A. J., *Biochem. Pharmacol.,* 8, 269–279, 1961.
20. Williamson, D. L. and Whitcomb, R. F., *Science,* 188, 1018–1020, 1975.
21. Brock, T. D., Dept. of Microbiology, Indiana University, Bloomington, April 1971, personal communication.
22. Liao, C. and Chen, T. C., *Phytopathology,* in press, 1976.
23. Whitcomb, R. F., Agricultural Research Center, U.S. Dept. of Agriculture, Beltsville, Md., May 1976, personal communication.

NATURAL AND SYNTHETIC CULTURE MEDIA FOR BACTERIA

E. Y. Bridson

I. INTRODUCTION

All living matter is endowed with the capacity for multiplication and for coming to terms with its environment; these fundamental properties are exhibited by even the simplest type of organism, the single cell, of which the bacterium is the commonest example.[1]

Bacterial cells contain macromolecules such as proteins, nucleic acids, and polysaccharides, which are built up from simple materials such as ammonia and three-carbon atom compounds. These macromolecules, together with mixed compounds such as mucopeptides and lipopolysaccharides, are the most important high-molecular-weight constituents of bacteria. They are all polycondensation products — proteins from 24 amino acids, nucleic acids from 4 nucleotides, and polysaccharides from approximately 20 monosaccharides.

The raw materials that bacteria demand for growth vary widely from the few simple carbon, nitrogen, and inorganic compounds required by oligotrophic organisms to the very elaborate growth factors (glutamine, hemin, cocarboxylase, etc.) demanded by some exacting strains, without which growth will not occur. Although many culture media have been devised (over 2000 were listed in 1930 and the current estimate is probably approximately 6000 to 7000 formulations), they are all variations on a relatively few essential nutritive ingredients, supplemented with a large number of selective, elective, or indicative compounds. The nutrients may be inorganic salt solutions, from which the organism can scavenge sufficient carbon and nitrogen, or they may be very elaborate hydrolysates of protein plus the heat-labile factors donated by blood, serum, or other tissue fluids. Such elaborate media may be mandatory to grow pathogenic organisms isolated from clinical specimens.

However, the primary distinction is between defined media prepared from known chemical constituents (referred to here as synthetic media) and undefined media containing unknown natural mixtures of substances (referred to here as natural media). The preference for use of synthetic or natural media depends largely on the type of bacteria being investigated. There are obvious advantages to using growth substrates that are both defined and reproducible. Most bacteriologists would prefer to use such media for all their purposes. Unfortunately, all the hitherto published formulations are limited to restricted groups of organisms. Growth of even some of these organisms may be atypical in rate, pigment formation, by-product metabolism, etc., when compared with their growth on natural media. Moreover, the synthesis of complex solutions of known chemicals, unless carried out with great care, may result in unstable interaction products which can have inhibitory effects and may themselves represent a nonreproducible component in the medium.

As is well known, the natural media contain unknown factors which, in spite of quantitative analyses, give uncontrolled additives to the medium. A full analysis of all the components in many natural nutrients is mainfestly impossible. The purpose of Sections II to IV is to describe the background of many of the ingredients, including manufacturing and processing data where pertinent, so that the bacteriologist is aware of the problems and limitations of the constituents of his media.

The factors involved in culture media may be intrinsic (i.e., concerned with the ingredients) or extrinsic (i.e., concerned with what is done to the medium in the form of sterilization, storage, incubation, or gaseous environment). Sections II to VII concern intrinsic factors. Section VIII concerns extrinsic factors. The classification of culture media under the names of organisms described in the 8th edition of *Bergey's Manual of Determinative Bacteriology*[2] has little in its defense except ease of reference for seekers of information. Almost inevitably some overlap and duplication occurs, although where this is obvious cross-references have been introduced.

II. PROTEINS AND HYDROLYSATES

A. Proteins

The role of proteins in culture media is largely supportive, protective, or indicative. More commonly hydrolyzed protein is added as a source of amino nitrogen in the form of peptides or amino acids. The disadvantage of protein additives is that they denature on heating and form a coagulum in the medium. Occasionally, advantage is taken of this fact in the form of "inspissated media." Loeffler serum slopes, used for the cultivation of *Corynebacterium* species and for storage of organisms, are formed from horse serum plus glucose broth, heated in small volumes at 80 to 85°C for 2 hr. The serum forms a solid coagulum, on the surface of which organisms can be smeared. The egg media of Lowenstein are similarly formed, with whole egg forming the protein coagulum. The addition of chopped meat, coagulated egg, soya protein, etc., to broth to encourage the growth of anaerobic organisms is also carried out. In these media low eH potentials occur after heating and the inoculated organisms can establish themselves in micro niches in the protein surfaces and create an environment suitable for multiplication.

Proteins may be added as indicator substrates to detect proteolytic enzymes produced by certain organisms. The addition of gelatin to broth is a common example, where the production of gelatinase is indicated by failure of the gelatin to gel after cooling to low ambient temperature. When it is added to agar media, it is necessary to pour a denaturing solution (20% w/v aqueous solution of sulfosalicylic acid) over the surface. The medium will become opaque except for clear areas around gelatinase-producing colonies.[3] Egg yolk emulsion added to agar media may yield an opaque medium, depending on pH and salt content. After inoculation and incubation of the medium, zones of clearing around the colonies indicate a proteolytic reaction; a subsequent halo of opacity in the clear zone indicates lipase activity.[4]

A number of culture media formulations call for the addition of serum or tissue fluids to encourage the growth of fastidious organisms. The role of these substances is not clear; they may be contributing heat-labile factors in the form of key enzymes or precursor compounds. The serum proteins may have a protective effect on the organisms, absorbing toxic material. The effect of serum in glucose serum broth is to act as a buffering agent. It can be replaced with sodium carbonate when growing *Streptococcus pneumoniae*.[5] It is known that most species of mycoplasma require native protein for growth, and media for the detection of mycoplasma require supplementation with serum.[6] Omission of the serum reduces the level of mycoplasma metabolism.[7] Serum for use in culture media can be collected at the abattoir and need not be sterile if it is promptly frozen after separation from the blood. At a later stage it can be filtered to remove bacterial contaminants with the aid of cellulose-acetate membranes or asbestos pads.

Naturally clotted blood can be stored overnight in a refrigerator and the extruded serum carefully collected, or blood may be collected into vessels containing 10 ml of a 10% w/v aqueous solution of neutral potassium oxalate for each liter of blood required. Blood treated in this way will not clot. After standing overnight in the refrigerator, the supernatant plasma may be separated from the cells. The separated plasma will then require warming to 37°C and 23 ml of 4% w/v aqueous solution of calcium chloride added per liter of serum. After shaking, the fibrin separates and the serum can be filtered off. This process yields higher volumes of serum than natural clot serum.

After filtration through a sterilizing grade of filter, the sterile product may be held in sterile containers at 4°C or, preferably, frozen at -20°C. Serum taken from young animals may have high calcium and lipid contents. On storage, such serum may become opaque through the formation of calcium-lipid salts, and further filtration may be required.

Whole blood is the most widely used protein additive to culture media in clinical microbiology. The product must be collected with full aseptic precautions from healthy animals because there is no satisfactory way of sterilizing whole blood. Because the use of blood in culture media includes diagnostic reactions on the erythrocytes present, there is a limited length of time for which it may be kept. The maximum time that whole blood for culture media should be stored is 3 to 4 weeks at 4°C. Any degree of blackening or deep hemolysis of the supernatant plasma indicates that the blood is unsuitable for use.

Blood taken from horses, sheep, cows, or rabbits is collected into sterile vessels which may contain an oxalate or citrate anticoagulant. The excess of anticoagulant (which is essentially a calcium chelate) present will be transferred to the culture medium and may inhibit the growth of certain organisms.[8] To overcome this problem, a more common method of preventing collected blood clotting is to shake it vigorously while it is still liquid; the fibrin clot separates, leaving a liquid suspension of erythrocytes in serum. This process has the added advantage that the small degree of cell damage occurring on shaking releases growth-activating substances into the blood. Blood treated in this way is described as defibrinated blood.

Blood can be deliberately lysed by freezing and thawing several times or by the addition of a sterile solution of white saponin (0.1% w/v concentration in blood). The shelf-life of this product is much longer, and it is used in media for organisms that demand hematin or where enzymes from the erythrocytes are used to improve the performance of antibiotic sensitivity media (see section on peptones).

Heated blood agar (chocolate agar) is used to grow organisms that require the release of diphosphopyridine nucleotide from the red cells, e.g., *Haemophilus influenzae*. It has been shown that blood from different species of animals may have different specific effects on the growth of this organism.[9-11] The digestion of whole blood with pepsin gives comparable results with *H. influenzae*,[12] but the resulting growth with strains of *Neisseria gonorrhoeae* is poorer than on chocolate agar. It is possible that the heated blood is acting as an absorbent of toxic materials in a manner similar to that of starch.[13]

When whole blood is added to culture media, the most important diagnostic reaction observed is the production of hemolysins. These substances may completely lyse the erythrocytes, giving clear, transparent halos on agar media (β hemolysis) or total lysis in blood-broth tubes, or they may partially lyse the cells, giving greenish, semitranslucent, hemolysis on agar plates (α hemolysis). A full description of the changes produced in blood-containing media is given by Wilson and Miles.[14]

B. Hydrolysates

As chemoorganotrophic bacteria must secrete proteolytic exoenzymes to hydrolyze protein before they can transfer the protein fractions back into the cell, it follows that all these organisms will utilize prehydrolyzed protein more effectively.

Natural or undefined media contain a mixture of the breakdown products of hydrolyzed protein. A wide variety of protein sources may be used to provide protein hydrolysates, including animal, plant, and single-cell proteins. The following list shows the wide variety available:

> Meat (fresh, frozen, dried)
> Fish (fresh, dried)
> Casein
> Gelatin
> Keratin (horn, hair, feathers)
> Ground nuts
> Soya protein
> Cottonseed
> Sunflower seed
> Microorganisms (yeasts, algae, bacteria)
> Guar protein
> Blood meal
> Corn gluten
> Egg albumin

The quality of the hydrolysate will be no better than that of the original protein; it follows, therefore, that quality specifications should be observed in the selection of the protein. Thus, meat and offal (heart and liver) should be fresh from an abattoir or properly frozen and stored. Dried meat is commonly used in the form of meat meal. Drying such products is a compromise between overheating and putrefaction. Temperatures below 50°C should be used, and drying under negative pressure prevents overheating. Caramelization of sugars (Maillard reaction) and inactivation of B-group vitamins occurs on overheating.

Fish proteins contain highly unsaturated fats that are readily oxidized to give rancid, fishy odors. Solvent extraction of these oils or steam distillation reduces the level to an acceptable amount. Dried fish meal is the most common form of fish protein, although the proteolysis of fresh fish and fish waste is being advocated.[15]

Milk proteins such as casein and lactalbumin rank next to fresh meat in terms of quality of the peptone produced by enzymic hydrolysis. The manufacture of casein varies and may involve fermentation of the lactose of skim milk or the addition of acid to reduce the pH to 4.6. Separation and washing of the curd is carried out before drying it to produce "isoelectric" casein. Alternatively, the wet curd is dispersed in alkali to give a pH of 7.0 and spray-dried as sodium caseinate. The whey proteins left behind after separation of the curd are heat-denatured at 76°C and precipitated at pH 4.8 to 5.3 to form lactalbumin. The whey proteins are considered to be nutritionally superior to casein,[16] but they contain substantial amounts of lactose that can interfere with its use as a peptone.

Gelatin, extracted from collagen, contains few sulfur-containing amino acids and is high in proline and hydroxyproline. It is easily hydrolyzed by steam as well as acids and enzymes to yield peptones that have relatively poor growth-promoting effects. Keratin, contained in wool, hair, nails, hooves, feathers, etc., is high in proline and cystine, but deficient in lysine. The strong disulfide bonding makes the protein resistant to enzyme hydrolysis. These bonds must be broken by heating or chemical reduction before enzyme treatment. A peptone produced from chicken feathers was reported as perfectly satisfactory when treated in this way.[17]

Soya meal, groundnut meal, cottonseed, and sunflower seed are primarily grown for their oil content. After pressing to extract the oil, the residue may contain up to 40% protein. Digestion of the crude meal is usually carried out, and the resulting peptone may contain high levels of fermentable carbohydrate. Soya protein, which, after digestion, is free of this sometimes objectionable characteristic, is available. The peptones obtained from plant proteins may be low in lysine, depending on the processing temperatures.

Microbial proteins may be obtained from many sources, including the following broad divisions:

1. Single-cell protein from hydrocarbons using bacteria grown on methane, bacteria grown on methanol, yeasts grown on either pure N-alkanes or the N-alkanes in a middle distillate fraction of petroleum products.

2. Algae which can be grown on open lagoons (such as *Spirulina platensis* which has up to 70% of its dry weight as protein and can yield 10 tons of dry protein per acre of lagoon).

3. Fungal proteins which can be grown in fermenters and can produce around 45% of the dry weight as protein.

C. Principal Methods of Hydrolysis

Hydrolysis of proteins may be effected by

1. Boiling the protein with mineral acids or strong alkalis at atmospheric pressure or under increased pressure to raise the temperature of the reaction.

2. Digestion with proteolytic enzymes.

3. Acid hydrolysis, which is commonly carried out on casein and gelatin.

Casein requires boiling with five to ten times its weight of 6 N hydrochloric acid or 8 N sulfuric acid for 6 to 24 hr. At increased pressure and the temperature raised to 120°C, 45 min is sufficient for adequate hydrolysis. Special hydrolysis vessels are required to prevent corrosion and the transfer of unwanted materials into the peptone. There is complete destruction of tryptophan, a severe loss of cystine, and minor losses of serine and threonine. Alkaline hydrolysis is used to prevent the destruction of tryptophan, but it results in the partial or complete destruction of cysteine, cystine, and arginine. The hydrolysate is initially deep black, and decolorization is performed with activated charcoal. This treatment alters the nutritional properties of the hydrolysate by causing absorption of vitamins.[18] Neutralization of the acid with alkalis considerably increases the salt content (up to 30 to 40% w/w peptone).

Enzymic hydrolysis of proteins by pepsin, pancreatin, trypsin, or papain is carried out industrially to produce peptones. Microbial proteases obtained from molds or bacteria are of increasing interest, as are the plant enzymes ficin and bromelain. All these latter enzymes resemble papain in their activity and peptide-producing spectrum. The enzymes used are either crude preparations such as pancreatic extract (pancreatin) or partially purified as high-activity papain preparations. The product of protein digestion by enzymes is an ill-defined mixture of peptides of various sizes known as peptones. The polypeptides ranging in size from 5000 mol wt down to dipeptides and single amino acids.[19] The upper limit in size is imposed by the condition that the polypeptide must remain in solution after heating at 120°C for 15 min at a pH range of 6.0 to 9.0. Although such peptones are undefined, by controlling the process of digestion it is possible to standardize the products so that growth characteristics are reproducible from batch to batch. Large-scale manufacture with opportunities of blending material overcomes many of the problems previously associated with small-lot laboratory manufacture. Laboratory-scale preparation of casein and muscle digests are described in Part IV of this paper. Industrial scale processes are described by Bridson and Brecker.[20]

1. Enzymes Commonly Used to Prepare Protein Digests

Papain (chymopapain) — The most widely used enzyme for proteolysis; approximately

350 t of crude papain are exported annually from production areas.[21] It is obtained by tapping the green, unripe fruit and collecting and drying the resulting latex exudate which contains the enzyme. The quality of the dried product is extremely variable, and its color varies from cream-white to red-brown, depending on the degree of oxidation of the latex or degree of burning during drying. It is also unstable and, unless stored at low temperatures, there is loss of enzymatic activity. Refined papain is prepared by water extraction of the latex using reducing and chelating agents or salt precipitation and solvent extraction. Filtration and spray-drying are the finishing operations. Papain is a sulfhydryl enzyme and requires activation by reducing agents such as cysteine, bisulfite, or hydrogen sulfide. It is rapidly inactivated with oxidizing compounds. It shows remarkable stability at high temperatures, and digestion can be carried out at 70°C. Maximum enzyme activity is in the range of pH 5.0 to 7.5. It has a broad hydrolytic action similar to that of pepsin. Chymopapain is also found in papaya latex. It has great similarity to papain, but it is relatively more stable, especially at low pH values.[22]

Pancreatin (trypsin, chymotrypsin) — Generally prepared from pig pancreas. It has amylase and lipase activities as well as proteolytic properties. The fresh or frozen glands are pulverized with duodenal tissue; this activates the zymogens trysinogen and chymotrypsinogen to their respective enzymes by autocatalysis. After drying *in vacuo*, defatting is carried out by extraction with petroleum ether. After further drying the glands are powdered and packed. The active proteolytic enzyme in pancreatin is trypsin, which is an endopeptidase having high specificity for certain peptide bonds which are linkages between the carboxyl groups of lysine and arginine to the amino group of any other amino acid. Therefore, the relatively large peptide fragments present after hydrolysis have terminal lysine or arginine groupings. Trypsin is used in the pH range 7 to 9, but due to its strong tendency to autolyse it will not remain active at high pH values for long periods of time. The autolysis of the enzyme is retarded considerably in the presence of calcium ions.[23] Although not inactivated by the common sulfhydryl reagents or mild oxidation, trypsin is inactivated by some naturally occurring proteins from soya bean, lima bean, wheat, and ovomucoid.[24] The presence of the inhibitor in soya flour prevents tryptic digestion unless it is neutralized. Chymotrypsin is also present in crude pancreatic preparations. It is more stable at neutral pH values and has specific activity for peptides of amino acids with aromatic rings, such as tyrosine, phenylalanine, and tryptophan.[25]

Pepsin — Manufactured from the mucosa of the stomach. The minced fundus is held for many hours at pH 2 (after the addition of 2 to 3 volumes of dilute hydrochloric or phosphoric acid) in order to change the pepsinogen to pepsin. After filtering off the undigested tissue, the filtrate is dried *in vacuo*. An enzyme preparation of higher activity can be prepared by precipitation with ethanol and filtering and drying at low temperature. Commercially obtainable pepsin is a mixture of enzymes. It is stable at pH 5 to 6, but denatured above pH 6. The optimal activity pH for proteins is 1.8, but it is not stable at this pH because of autodigestion. The enzyme attacks peptide bonds only and, although it shows somewhat broad specificity, it favors those with adjacent aromatic amino acids.

Bromelain — Present in the pineapple plant *Ananas comosus*. The enzyme-bearing juice is pressed from the stems of the plant and recovered by acetone precipitation. The enzyme is similar to papain, with an optimum pH activity at 6 to 8. However, it is not as heat stable, and its activity drops off at temperatures slightly above 55°C.[26] Bromelain is activated by cysteine and KCN, but not by EDTA.

Ficin — Obtained from the latex of tropical fig trees of the genus *Ficus* and extracted by solvent precipitation. The tissue-dissolving properties of concentrated ficin demand that it is handled with care. It is similar to papain in many respects. Ficin shows optimal stability at pH 6 to 8 and reasonably good stability over the pH range 3.5 to 9.0.[27] The optimum temperature for proteolytic activity is 63°C and it is completely inactivated at 80°C.

Microbial proteases — These are generally extracellular enzymes and are classified as acid, neutral, and alkaline proteases. Acid proteases are primarily of fungal origin and are similar to pepsin. They are active in the pH range 2 to 5 and have maximum stability in the range 2 to 6. Hydrolysis of a wide range of peptide bonds is characteristic of these enzymes.[28] Neutral proteases are widespread in bacteria and fungi. They show maximum activity and stability in the pH range 7 to 9, but their stability decreases rapidly on either side of these values; however, they can be stabilized with calcium ions. Some specific activity toward peptide bonds adjacent or close to hydrophobic residues (e.g., phenylalanine and leucine) has been shown.[29,30] Alkaline proteases are widespread in bacteria and fungi, showing resemblance to trypsin and chymotrypsin. They are generally stable between pH 5 to 10, but are rapidly inactivated at 65°C. The maximum activity occurs in the range pH 9 to 11.

D. Protein Hydrolysates — Peptones (Peptides and Amino Acids)

The degradation products of protein hydrolysis, commonly called peptones, are mixtures of polypeptides, oligopeptides, amino acids, organic nitrogen bases, salts, and trace elements. It is extremely difficult to obtain a complete analysis of a peptone because of the heterogeneous nature of its components. Depending on the protein substrate (casein, meat, or soya) and the enzyme used (trypsin, pepsin, or papain), the nitrogen-containing compounds exist in differing qualitative and quantitative relations. Therefore, the control of peptone preparation must concentrate on the nitrogen fractions present which are necessary to grow microorganisms and take care that such fractions are present in sufficient quantity. The classical analyses of peptones are restricted to the determination of individual types of combined nitrogen, together with an analysis of amino acids present. These analyses are shown in Tables 1 to 4, which have been obtained with permission from the major manufacturers of culture media. Definitions of the terms used and the methods of analysis are given by Bridson and Brecker[20] and Herbert et al.[31]

The ratio of amino nitrogen to total nitrogen measures the degree of degradation of the peptone and this measurement, together with the classical data, characterizes peptones to some degree, but little indication is given of the ratio of polypeptides, peptides, and amino acids present in the peptone.

Ziska[19] produced a table (Table 5) of nitrogen fractions present in commercial peptones, expressed as percentages. These values were later expressed as amino nitrogen:total nitrogen ratios in footnotes to histograms of peptide distribution in the peptones. Ziska[32] described the method whereby he was able to fractionate peptones using a gel filtration (molecular exclusion) process. Using a 144 × 2.2 cm column of Sephadex® G25 fine beads, he dissolved 200 to 300 mg of peptone in 5 ml of 0.2 M acetic acid and applied it to the column in the normal way. The peptone was then eluted from the column with 0.2 M acetic acid solution. Previous experience showed that the first 200 ml could be discarded, and fractions were collected and later pooled. The separate fractions and pools were evaluated by two different methods:

1. Determination of amino acids and peptides before and after acid hydrolysis using a photometric ninhydrin method.
2. Estimation of the total nitrogen using the Kjeldahl technique.

Both methods gave similar results and all Ziska's further work used the total nitrogen method.

Table 1
APPROXIMATE COMPOSITION OF BBL® PEPTONES

Catalog number	11842	11869	11879	11892	11905	11918	11920	11928
Name	Acidicase® Peptone	Gelysate® Peptone	Lactalysate® Peptone	Myosate® Peptone	Phytone® Peptone	Thiotone® Peptone	Trypticase® Peptone	Yeast Extract
Source	Casein	Gelatin	Lactalbumin	Cardiac muscle	Soy meal	Animal tissues	Casein	Yeast cells
Hydrolysis	HCl	Pancreatic	Pancreatic	Pancreatic	Papaic	Peptic	Pancreatic	Autolytic enzymes
Nitrogen (%)								
Total	8.0	16	11.9	12.3	9.2	12.8	11.7	10.3
Amino	6.4	1	6.9	4.8	1.8	3.2	3.5	5.5
NaCl	37.2	0.6	1.5	1.2	4.4	2.3	0.5	0.5
Ca	0.05	Trace	0.19	0.002	0.05	0.05	0.35	0.06
Fe	0.0045	0.006	0.056	0.016	0.02	0.02	0.03	0.20
K	0.4	0.06	0.32	2.0	3.99	1.4	0.24	3.4
Mg	0.003	0.001	0.005	0.06	0.19	0.05	0.03	0.07
P	0.32	0.07	0.34	0.63	0.38	0.6	0.65	1.16
S	0.066	0.35	0.43	1.0	0.39	1.0	0.73	
Carbohydrates (%)	0.0	0.0	+a	+	37	+	0.0	16.6
Amino acids (%)								
Arginine	1.4	8.0	3.1	4.7	4.6	5.0	2.6	3.5
Aspartic acid	3.7	3.9	8.2	7.0	5.8	5.9	5.1	
Cystine	0.3	0.1	2.1	0.4	0.5	0.6	0.3	1.6
Glycine	1.0	20.7	1.7	5.2	2.8	9.3	1.8	
Glutamic acid	14.2	9.1	10.6	12.9	9.3	10.0	17.0	
Histidine	0.7	0.9	1.9	1.6	1.6	1.8	2.4	1.5
Isoleucine	2.7	1.6	5.4	4.0	2.5	3.3	5.0	4.7
Leucine	3.5	2.8	10.9	5.8	3.2	6.0	7.1	6.4
Lysine	3.7	4.3	10.0	5.2	3.6	5.5	5.3	6.5
Methionine	1.7	1.0	2.5	1.8	0.6	1.6	2.4	2.0
Phenylalanine	0.7	1.9	3.4	3.2	3.6	3.3	3.8	3.5
Proline	4.0	14.0	6.3	4.9	3.4	7.3	11.5	
Threonine	2.5	1.8	3.8	3.0	1.8	3.5	3.5	3.3
Tryptophan	0.0	0.2	2.0	1.2	0.7	0.7	0.9	1.0
Tyrosine	3.1	0.8	3.3	1.2	1.9	2.1	2.3	4.0
Valine	4.1	2.2	4.1	4.0	2.0	4.6	5.6	4.8

a +, carbohydrate detected.

Courtesy of Becton, Dickinson and Company, Cockeysville, Md.

Table 2
TYPICAL ANALYSIS OF DIFCO PEPTONES AND HYDROLYSATES

Percent	Peptone	Proteose Peptone	Propeose Peptone No. 3	Tryptone	Tryptose	Neopeptone	Protone	Casitone	Casamino acids, technical	Casamino acids	Yeast extract
Ash	3.53	9.61	4.90	7.28	8.44	3.90	2.50	6.66	30.8	3.64	10.1
Ether-soluble extract	0.37	0.32		0.30	0.31	0.30	0.31				
Total N	16.16	14.37	13.06	13.14	13.76	14.33	15.41	13.00	7.85	11.15	9.18
Primary proteose N	0.06	0.60		0.20	0.40	0.46	5.36				
Secondary proteose N	0.68	4.03		1.63	2.83	3.03	7.60				
Peptone	15.38	9.74		11.29	10.52	10.72	2.40				
Ammonia N	0.04	0.00		0.02	0.01	0.12	0.05				
Free amino N (Van Slyke)	3.20	2.66		4.73	3.70	2.82	1.86				
Amide N	0.49	0.94		1.11	1.03	1.23					
Monoamino N	9.42	7.61		7.31	7.46	7.56					
Diamino N	4.07	4.51		3.45	3.98	4.43					
Arginine	8.0	6.8	5.9	3.3	5.05	4.7	3.9	3.2	1.9	3.8	0.78
Aspartic acid	5.9	7.4	6.6	6.4	6.9	6.7	10.8	6.5	4.0	0.49	5.1
Cystine (Sullivan)	0.22	0.56		0.19	0.38	0.39	0.27				
Glutamic acid	11.0	12.0	11.2	18.9	15.4	15.2	8.1	20.0	12.6	5.1	6.5
Glycine	23.0	11.6	8.9	2.4	7.0	6.3	5.0	2.5	1.3	1.1	2.4
Histidine	0.96	1.7	1.7	2.0	1.8	2.3	5.9	2.1	1.4	2.3	0.94
Isoleucine	2.0	3.3	3.3	4.8	4.0	4.3	0.71	5.0	2.9	4.6	2.9
Leucine	3.5	6.4	6.0	3.5	7.4	8.4	13.6	8.2	4.0	9.9	3.6
Lysine	4.3	5.3	5.1	6.8	6.0	6.4	10.3	7.0	4.4	6.7	4.0
Methionine	0.83	2.0	1.8	2.4	2.2	2.4	1.9	2.6	1.08	2.2	0.79
Phenylalanine	2.3	3.3	3.1	4.1	3.7	4.3	6.8	4.3	2.0	4.0	2.2
Threonine	1.6	3.5	3.2	3.1	3.3	3.7	4.6	4.2	2.2	3.9	3.4
Tryptophan	0.42	0.72	0.85	1.45	1.08	1.01	1.65	1.38	Nil	0.8	0.88
Tyrosine	2.3	3.4	0.36	7.1	5.2	5.3	3.0	2.8	0.52	1.9	0.60
Valine	3.2	4.4	4.0	6.3	5.3	6.0	10.1	6.3	3.8	7.2	3.4
Organic sulfur	0.33	0.60		0.53	0.57	0.63	0.45				
Inorganic sulfur	0.29	0.04		0.04	0.04	0.09	0.16				
Phosphorus	0.079	0.24	0.46	0.75	0.49	0.112	0.15	0.72	0.29	0.35	0.89
Iron	0.0023	0.0038	0.0044	0.0071	0.0054	0.0021	0.0099	0.0039	0.0101	0.0006	0.028
SiO_2	0.042	0.078	0.019	0.090	0.084	0.18	0.52	0.073	0.022	0.053	0.052
Potassium	0.22	0.70	0.21	0.30	0.50	0.85	0.06	0.12	0.16	0.88	0.042
Sodium	1.08	2.84	0.033	2.69	2.76	0.45	0.30	0.24	1.05	0.77	0.32
Magnesium	0.056	0.118	0.00048	0.045	0.081	0.051	0.057	0.00060	0.0039	0.0032	0.030
Calcium	0.058	0.137	0.0396	0.096	0.116	0.198	0.263	0.0913	0.0538	0.0025	0.040
Chlorine	0.27	3.95		0.29	2.77	0.84	0.38				

Table 2 (continued)
TYPICAL ANALYSIS OF DIFCO PEPTONES AND HYDROLYSATES

Percent	Peptone	Proteose Peptone	Propeose Peptone No. 3	Tryptone	Tryptose	Neopeptone	Protone	Casitone	Casamino acids, technical	Casamino acids	Yeast extract
Chloride	0.27	3.95	4.15	0.29	2.12	0.84	0.38	0.425	21.34	11.2	0.190
PPM											
Manganese	8.6	5.3	7.8	13.2	9.2	5.8	6.0	9.7	5.7	7.6	7.8
Lead	15.00	5.00	3.00	6.00	5.50	5.00	9.00	5.00	3.00	4.00	16.00
Arsenic	0.09	0.25	0.00	0.07	0.16	0.37	0.46	0.32	0.00	0.50	0.11
Copper	17.00	31.00	9.00	16.00	23.50	19.00		10.00	8.00	10.00	19.00
Zinc	18.00	44.00	37.00	30.00	37.00	2.00	13.00	10.00	14.00	8.00	88.00
Reaction pH	7.0	6.8		7.2	7.3						

Courtesy of Difco Laboratories, Detroit.

Table 3
AMINO ACID CONTENT OF CULTURE MEDIUM BASES

(Micromoles/10 mg)

Product	Catalogue no.	Asp	Thr	Ser	Pro	Glu	Gly	Ala	Val	Met	Ile	Leu	Tyr	Phe	Lys	His	Arg	Cys
Agar-agar	1614	0.37	0.162	0.237	0.197	0.372	0.380	0.375	0.225	—	0.122	0.227	0.182	0.110	0.115	0.047	0.090	—
Casein hydrolysate, acid hydrolyzed	2245	3.660	1.680	3.10	5.540	8.080	1.60	1.880	2.460	0.180	1.260	2.640	0.940	1.140	2.440	0.660	0.690	—
Casein hydrolysate, pancreatic	2239	6.860	4.020	6.420	7.680	17.10	3.920	5.10	6.680	1.720	4.340	8.120	2.440	3.20	5.560	1.680	3.760	—
Meat extract	3979	7.30	3.375	3.425	4.30	10.0	9.975	7.30	4.250	0.350	3.150	5.450	1.40	2.150	6.150	2.125	3.125	—
Gelatin	4070	6.450	2.80	4.750	11.40	8.60	38.10	13.60	4.10	0.750	2.050	4.10	0.950	1.950	4.150	1.20	5.20	—
Yeast extract	3753	5.240	2.520	2.540	2.180	5.220	4.60	4.760	3.280	—	2.240	3.160	1.920	1.60	4.060	0.740	0.920	Traces
Liver powder	5347	5.460	2.720	2.940	2.930	6.50	6.350	5.610	4.260	0.860	2.60	5.780	1.350	2.340	3.910	0.990	2.330	—
Peptone from meat, peptic	7224	6.90	3.550	3.60	4.250	8.050	9.550	7.70	4.80	0.90	3.350	6.20	1.650	2.550	5.050	1.950	3.550	—
Peptone from meat, tryptic	7214	6.120	1.680	2.640	2.620	7.340	7.10	5.520	3.20	—	2.460	3.960	0.260	0.80	4.960	1.720	2.660	—
Peptone from casein, tryptic	7213	7.70	4.10	5.975	10.0	19.925	3.10	4.325	6.60	0.0875	4.225	7.650	1.625	3.350	6.650	1.475	2.10	—
Peptone from soybean meal, papaic	7212	6.960	2.560	3.940	3.020	9.0	5.180	3.820	3.020	0.260	2.160	3.660	1.540	2.040	3.520	0.780	2.660	—

Courtesy of E. Merck, Darmstadt, W. Ger.

Table 4
AVERAGE ANALYSES OF OXOID PEPTONES

	Casein hydrolysate (acid) L41	Lactalbumin hydrolysate L48	Liver digest L27	Peptone, mycological L40	Peptone, bacteriological L37	Peptone, bacteriological neutralized L34	Peptone P L49	Peptonized milk L32	Proteose peptone L46	Special peptone L72	Peptone soya L44	Tryptone L42	Tryptone T L43	Tryptose L47
General (% w/w)														
Moisture	3.0	5.9	5.2	4.4	4.4	3.8	3.2	5.4	6.0	3.5	4.1	3.6	3.6	4.7
Ash	36.6	5.5	13.1	3.8	6.7	11.3	16.3	9.4	9.5	12.1	13.7	8.1	13.9	12.4
Chloride (as NaCl)	29.0	1.8	4.6	0.6	2.3	5.5	10.3	2.0	4.3	4.6	3.3	0.4	4.8	5.7
Phosphate (as P_2O_5)	1.3	1.4	2.3	1.5	1.3	1.2	0.7	2.7	1.6	1.7	0.7	1.9	2.5	1.2
Total nitrogen	8.3	12.5	11.4	13.6	14.3	13.3	12.8	5.7	13.6	12.4	9.3	13.2	12.7	12.7
Amino nitrogen	4.8	6.1	3.2	3.4	2.5	2.3	2.8	1.7	3.6	3.9	2.2	4.8	3.7	3.7
Amino N/total N	57.8	49.0	28.1	25.0	17.5	17.3	21.9	29.8	26.4	31.5	23.7	36.4	29.1	29.1
Lipids	<0.1	<0.1	<0.1	<0.1	<0.1	<0.1	<0.1	<0.1	<0.1	<0.1	<0.1	<0.1	<0.1	<0.1
Ammonia	0.62	1.29	0.92	1.13	0.70	0.47	0.99	0.43	0.92	0.66	0.96	0.75	0.97	0.84
Lactose (by difference)	—							43.4						
Carbohydrate (as dextrose)	—										13.9			
pH of 2% sol. (after autoclaving)	7.0	6.3	7.0	5.4	6.2	7.0	7.0	5.4	7.0	7.0	7.0	7.0	7.0	7.0
Amino acids (% w/w)														
Alanine	1.64	4.58	3.42	5.15	6.08	5.00	6.94	1.93	4.95	3.72	2.52	2.6	2.20	3.53
Arginine	1.68	3.07	3.43	4.45	5.43	3.69	6.24	0.40	3.43	3.16	5.03	3.1	0.83	2.71
Aspartic acid	4.04	10.49	5.07	6.79	8.29	7.65	8.28	2.78	6.82	6.47	7.94	6.8	4.72	6.15
Cystine	0.13	2.60	0.84	0.67	0.46	0.48	0.33	0.17	0.59	0.53	0.80	0.3	0.17	0.40
Glutamic acid	15.29	15.94	9.48	12.14	8.76	12.59	14.62	12.49	12.29	16.91	14.68	17.3	17.32	15.37
Glycine	1.09	1.84	4.62	5.98	5.55	8.20	15.84	0.90	7.59	3.19	2.98	1.8	1.36	4.49
Histidine	0.42	1.78	1.53	1.81	1.66	1.51	1.41	1.22	1.53	1.88	1.70	2.1	1.65	1.68
Isoleucine	2.50	4.97	2.15	2.83	2.76	2.50	2.46	2.11	2.15	3.47	2.50	3.8	3.65	2.72
Leucine	4.33	10.70	4.70	6.03	5.83	4.46	5.13	3.76	4.70	6.04	4.46	6.9	6.42	5.05
Lysine	5.14	11.25	4.52	6.11	6.60	4.67	4.82	2.53	5.65	5.52	3.92	7.9	4.90	6.17
Methionine	1.85	1.63	1.22	1.68	1.71	1.11	1.36	0.95	1.22	1.88	0.55	1.8	1.44	1.22
Phenylalanine	2.89	3.08	2.55	3.26	2.64	2.98	2.75	1.82	2.55	3.18	3.01	3.6	3.49	2.83
Proline	5.54	4.06	2.87	4.61	6.90	5.54	7.47	4.10	4.36	7.43	3.64	8.5	6.79	5.19
Serine	0.93	1.70	0.75	0.73	2.28	2.03	1.46	1.11	0.68	2.38	0.68	1.7	0.88	0.86
Threonine	1.47	3.24	1.41	1.87	2.59	1.86	1.95	1.25	1.70	2.36	1.38	2.3	1.69	1.66
Tryptophan	—	1.32	0.80	0.72	0.52	0.53	0.33	0.55	0.61	0.68	0.69	1.0	1.11	0.86
Tyrosine	2.08	0.67	2.31	2.01	2.31	1.95	1.95	0.84	2.13	1.07	2.36	1.4	2.16	1.78
Valine	3.54	3.92	3.33	4.36	3.96	3.40	3.70	2.92	3.04	4.59	3.10	5.1	4.94	3.75

Note: The analysis of materials containing soluble degradation products of proteins presents considerable difficulties in the interpretation of results. Because of the overlap of nitrogen containing fractions such as primary and secondary proteoses, peptones, polypeptides, and the non-specific nature of most analytical techniques — the results obtained are highly empirical and mean very little unless the nature of the sample and the method of analysis is known.

Table 4 (continued)
AVERAGE ANALYSES OF OXOID PEPTONES

	Casein hydrolysate (acid) L41	Lact-albumin hydrolysate L48	Liver digest L27	Peptone, mycological L40	Peptone, bacteriological L37	Peptone, bacteriological neutralized L34	Peptone P L49	Peptonized milk L32	Proteose peptone L46	Special peptone L72	Peptone soya L44	Tryptone L42	Tryptone T L43	Tryptose L47
Metals (w/w)														
Calcium	243 ppm	5100 ppm	1800 ppm	2300 ppm	1330 ppm	143 ppm	109 ppm	1.35%	980 ppm	2000 ppm	,930 ppm	3460 ppm	87 ppm	2220 ppm
Copper	1 ppm	1.3 ppm	49 ppm	13 ppm	2 ppm	1.1 ppm	1.6 ppm	4.2 ppm	3.5 ppm	5.6 ppm	3.7 ppm	1.0 ppm	1.4 ppm	2.25 ppm
Iron	56 ppm	110 ppm	405 ppm	210 ppm	71 ppm	79 ppm	31 ppm	102 ppm	68 ppm	97 ppm	207 ppm	69 ppm	22 ppm	68 ppm
Lead	<2 ppm	3.9 ppm	<2 ppm	<2 ppm	<2 ppm	<2 ppm	<2 ppm	<2 ppm	<2 ppm	<2 ppm	<2 ppm	<2 ppm	<2 ppm	<2 ppm
Magnesium	143 ppm	480 ppm	1050 ppm	1230 ppm	790 ppm	200 ppm	98 ppm	1600 ppm	1130 ppm	710 ppm	2630 ppm	283 ppm	26 ppm	706 ppm
Manganese	0.3 ppm	3.9 ppm	0.2 ppm	0.2 ppm	0.2 ppm	0.2 ppm	0.2 ppm	0.2 ppm	0.2 ppm	0.2 ppm	0.2 ppm	0.3 ppm	<0.2 ppm	0.2 ppm
Potassium	0.27%	0.85%	1.45%	1.17%	1.23%	1.23%	1.43%	1.30%	1.46%	1.10%	2.90%	0.20%	1.40%	0.83%
Sodium	13.5%	0.92%	0.40%	0.58%	1.60%	2.46%	5.16%	2.10%	1.34%	4.40%	1.10%	3.20%	4.50%	2.27%
Tin	<20 ppm	86 ppm	<20 ppm	<20 ppm	<20 ppm	<20 ppm	<20 ppm	<20 ppm	<20 ppm	<20 ppm	<20 ppm	<20 ppm	<20 ppm	<20 ppm
Zinc	25 ppm	9.8 ppm	120 ppm	70 ppm	26 ppm	27 ppm	15 ppm	62 ppm	56 ppm	69 ppm	27 ppm	49 ppm	36 ppm	53 ppm

From *The Oxoid Manual of Culture Media, Ingredients, and Other Laboratory Services*, 3rd ed. (rev.), Oxoid Limited, Basingstoke, England, 1976, 214–215. With permission.

Table 5
PERCENTAGES OF NITROGEN FRACTIONS IN COMMERCIAL PEPTONES

	Total	Amino	Ammonium	Amide	Protein	Non-protein	Poly-peptide
Bactotryptone, Difco	13.08	3.70	0.05	0.70	3.67	9.41	5.66
Bactotryptose, Difco	12.28	3.52	0.13	0.62	3.53	8.75	5.10
Bactocasitone, Difco	12.01	3.56	0.04	0.87	1.70	10.31	6.71
Thiotone, BBL	13.20	4.97	0.11	0.49	2.02	11.18	6.10
Trypticase, BBL	11.67	4.01	0.08	0.87	0.74	10.93	6.84
Polypeptones, BBL	12.43	4.96	0.12	0.67	1.48	11.95	6.88
Bacteriological peptone, Oxoid	14.20	2.69	0.24	0.48	3.76	10.44	7.51
Tryptose, Oxoid	12.46	4.18	0.02	0.63	0.75	11.71	7.51
Tryptone, Oxoid	12.26	4.31	0.02	0.61	0.91	11.35	7.02
Tryptone T, Oxoid	11.20	4.88	0.07	0.59	0.07	11.13	6.18
Casein peptone, Merck	14.42	4.76	1.57	0.87	4.53	9.89	3.56
Meat peptone, Merck	13.50	4.96	0.14	0.68	3.75	9.75	4.65
Casein peptone, IFS	11.89	3.47	0.32	0.71	1.83	10.06	6.25
Meat peptone, IFS	15.17	3.97	1.91	0.43	4.21	10.95	4.06
Peptone ST, Brunnengräber	11.75	3.72	0.60	0.45	2.97	8.78	4.46
Casein peptone CAT, Brunnengräber	13.75	3.63	0.22	1.06	3.92	9.83	5.98

From Ziska, P., *Arch. Hyg.*, 152, 73–76, 1968. With permission.

From experiments with amino acids and synthetic peptides, Ziska had shown that the molecular weights present in the separate fractions were as follows:

Eluate volume (ml)	Molecular weight	Fraction 1
210–260	5000–2500	1
260–310	2500–1500	2
310–360	1500–600	3
360–410	600–250	4
410–800	250–100	5

The percentage distribution of total nitrogen in 16 peptones was expressed by Ziska in the form of bars, where each fraction represented a 10 X 5-ml fraction pool. He showed that the results for protein N, nonprotein N, and polypeptide N showed no relationship with the nitrogen values obtained by gel filtration, but he was able to divide his peptones into the following two groups:

1. Peptones containing over 30% total N as peptides of mean molecular weight 1500 to 5000 (Difco bactopeptones, proteose peptones, soytones, bactoprotones, Orthana meat peptone, Oxoid peptones, BBL gelysate).
2. Peptones containing over 45% total N as peptides of mean molecular weight 250 to 1000 (Difco bactoprotones, proteose-peptones, neopeptones, Oxoid soya, proteose, bacteriological and mycological peptones, BBL phytone).

This valuable technique of analyzing peptones using a modification of the Ziska technique was reported by Bridson and Brecker.[20] The eluate from the column was fed through a continuously recording absorptiometer at 280 nm. Aromatic amino acids absorb strongly in this part of the spectrum. Assuming that these amino acids are randomly distributed among the various fractions present in the peptone, quantitative estimations of the peptide content can be made. Free aromatic amino acids (phenylalanine, tyrosine, and tryptophan) tend to be adsorbed onto the resin and are eluted significantly later than the other amino acids. Later work[19a] demonstrated that Sephadex® G15 resin gave finer resolution of peaks with this method.

From these studies, Bridson and Brecker[20] were able to show that the "peptide profile" of peptones was strongly influenced by the enzyme used in hydrolysis. Figure 1 shows the different results obtained with three enzymes and four protein substrates. By carrying out such tests during enzyme hydrolysis, it is possible to determine the efficiency and speed of the enzymic process.

Nekvasilova et al.[33] used the technique of gel filtration of casein digest to isolate the essential peptides in the peptone required for growth and toxinogenesis of *Clostridium perfringens*. The authors drew attention to the fact that quantitative cation distribution among the peptide fractions should not be ignored. In this work, the distribution of Fe was shown to be of particular significance.

Further identification of the fractions isolated from peptones may be accomplished by paper chromatography[34] and high-voltage electrophoresis.[35] Peptide separation by two-dimensional chromatography and electrophoresis has also been described.[36] By using a medium that is deficient in nitrogen and seeding it with test organisms, it is possible to detect biologically active components in paper chromatography separations by laying strips cut from the sheets on the preseeded agar. Nitrogen fractions required by the organisms are exhibited by growth in appropriate zones after incubation.[37]

A technique of bioautography using a preseeded layer of agar medium poured onto the surface of a chromatography plate was described.[38] After incubation for 24 hr, the agar surface was flooded with a 1% w/v solution of neotetrazolium and the plate was kept in the dark. Reduction of the tetrazolium salt occurred where growth had taken place.

The analyses of peptones contain data on ash, sodium chloride, phosphate, metals, and pH, as well as nitrogen-containing substances. Full analyses of some of the commercially available peptones are shown in Tables 6 to 8. It must be emphasized that these probably represent typical analyses, with mean figures quoted. Deviations from these figures will occur from batch to batch, but reputable manufacturers would restrict the deviation to within ±1.96 SD, i.e., 95% confidence limits.

The ash content of peptone contains salt, phosphates, sulfates, silicates, and metal oxides. Acid-insoluble ash is usually composed of silicates only.

The salt content (as NaCl) is related to the ash figure and, although variable, it is chiefly influenced by any substantial pH change during processing. Thus, acid digests (e.g., pepsin) and acid hydrolysates require pH adjustment in order to obtain a neutral product, and such peptones have high salt contents.

Phosphate acts as both a buffer and a source of phosphorous for organisms. Highly buffered peptones may be unsuitable for media to which carbohydrates and indicators are added to detect small changes in pH during fermentation or oxidation. However, phosphate plays a role in gas production, and the amount of gas produced by an organism can sometimes be enhanced by the deliberate addition of sodium phosphate.

The presence of metals in both macro- and microquantities is a major influence on the quality of peptone and its suitability for specific functions. The role of metals and minerals in culture media is discussed later, but it can be mentioned here that the role of Fe^{++} in peptone is very significant in the production of toxins by *Clostridia* and *Corynebacteria*.[39] The varying concentrations of trace metals and minerals are probably

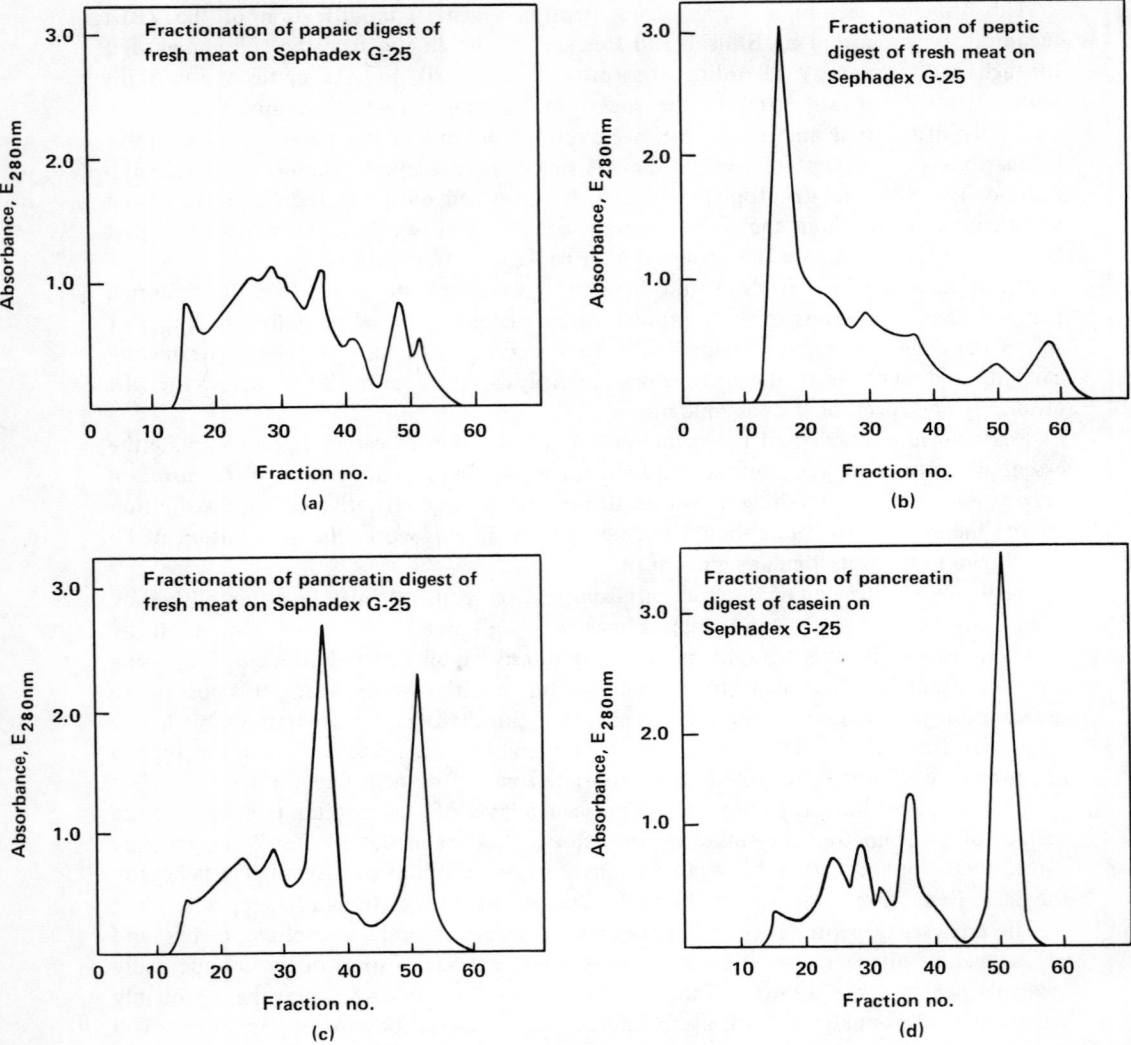

FIGURE 1.   Fractionation on Sephadex® G-25 at 280 nm: (a) Papaic digest of fresh meat. (b) Peptic digest of fresh meat. (c) Pancreatic digest of fresh meat. (d) Pancreatic digest of casein. (From Bridson, E. Y. and Brecker A., in *Methods in Microbiology,* Vol. 3A, Norris, J. R. and Ribbons, D. W., Eds., Academic Press, London, 1970, 229–295. With permission.)

major growth factors in the microbiological assessment of peptones. In 1956[40] it was stated that it did not seem practicable to establish standards for the content of these substances in peptone because

1. The amount and variety of growth factors required by the different groups of bacteria are so great that the practical control of a few may be misleading.
2. The proportions of the different factors required by different organisms may vary.
3. Other growth factors as yet unknown may be of great importance.
4. Inhibitory substances may remain undetected or confuse the results of the test.

Now, twenty years later, there is little more that can be added to this list or to understanding the growth factor problems of complex media.

The principal microbiological applications of peptones are for packaged microbiological media — the production of vaccines, antibiotics, enzymes, toxins, and other substances of commercial or chemical value. It would be impossible to measure the performance of peptones in all these applications and simpler, more reproducible tests must be devised.

The practical approach to measuring the criteria of peptone quality is to determine its ability to support the growth of certain strains of bacteria that are known to be demanding in their nutritional requirements or to measure the production of by-products of bacterial metabolism such as gas production or indol formation. However, it is recognized that a peptone suitable for the growth of an organism often may not be satisfactory for the production of a desired end product by that organism. The U.S. Pharmacopeia[41] lists the desirable criteria of a bacteriological peptone under the title "Peptic Digest of Animal Tissue." Similar criteria for pancreatic digest of casein are reproduced in full from a memorandum issued by the U.S. Department of Health, Education, and Welfare (1955).

Degree of digestion — Dissolve 1 g of peptone in 10 ml of distilled water. (a) Onto approximately 1 ml of the solution, layer a few drops of 10% acetic acid in 50% alcohol. No ring of precipitate forms at the junction of the two fluids, and when shaken no turbidity results, indicating the absence of undigested casein. (b) Mix 1 ml of the peptone solution with 4 ml of a saturated solution of zinc sulfate. A moderate amount of precipitate is formed, indicating the presence of proteoses. (c) To 1 ml of filtrate from (b) add 3 ml of distilled water, and follow with 4 drops of saturated bromine water. A violet-red color is produced, indicating that digestion has been carried beyond the stage of proteose production but that tryptophan has not been destroyed.

Loss at 100°C — Weigh accurately approximately 2 g of peptone and dry to constant weight at 100°C. The loss in weight is not more than 7%.

Nitrogen content — Determine by the Kjeldahl method the nitrogen content of a sample of the peptone previously dried to constant weight at 100°C. Not less than 10% of nitrogen is found.

Residue on ignition — Weigh accurately approximately 0.5 g of the peptone previously dried to constant weight at 100°C and heat slowly until thoroughly charred. Cool, add 1 ml of sulfuric acid, and ignite to constant weight. The weight of the residue corresponds to not more than 15%.

Freedom from nitrites — To 5 ml of a 2% solution of peptone add (1) a few drops of sulfanilic acid reagent (sulfanilic acid, 0.8 g; sulfuric acid of sp gr 1.84, 5 ml; distilled water, 100 ml and (2) a few drops of dimethyl-α-naphthylamine reagent (dimethyl-α-naphthylamine, 0.6 ml; glacial acetic acid, 30 ml; distilled water, 70 ml). Mix and allow to stand for 15 min. No pink or red color develops, indicating absence of nitrite.

Bacteriological tests — Prepare media of the following compositions:

1. 2% peptone, 0.5% sodium chloride in distilled water;
2. 0.1% peptone, 0.5% sodium chloride in distilled water;
3. 1% peptone, 0.5% sodium chloride, 0.5% dextrose in distilled water;
4. 1% peptone, 0.5% sodium chloride in distilled water;
5. 2% peptone, 0.5% sodium chloride, 1.5% agar in distilled water.

Adjust all media to pH of 7.2 to 7.4.

Freedom from fermentable carbohydrate — To medium (a) add sufficient phenol red indicator to give a readable color, tube in Durham fermentation tubes, and autoclave. Inoculate with a loopful of a 24-hr culture of *Escherichia coli*. No acid or only a trace in the inner tube and no gas is produced during incubation for 48 hr.

Production of indole — Inoculate 5 ml of medium (b) with *E. coli*, incubate for 24 hr at 35 to 37°C, and test by the addition of approximately 0.5 ml of indole reagent (*p*-dimethylaminobenzaldehyde, 1 g; ethyl alcohol, 95 ml; hydrochloric acid of sp gr 1.18, 20 ml). A distinct pink or red color which is soluble in chloroform is produced.

Production of acetyl methyl carbinol — Inoculate 5 ml of medium (c) with *Aerobacter aerogenes* and incubate for 24 hr. Test by adding to the culture an equal volume of 10% solution of sodium potassium hydroxide; shake and allow to stand at room temperature for several hours. A pink color is produced, indicating the presence of acetyl methyl carbinol.

Production of hydrogen sulfide — Inoculate 5 ml of medium (d) with *Salmonella typhi*. Place a strip or loop of lead acetate paper between the cotton plug and the mouth of the test tube so that it hangs approximately 2 in. above the medium. After incubation for 24 hr, the lower tip of the lead acetate paper shows little, if any, darkening; after 48 hr it shows an appreciable amount of brownish blackening (lead sulfide).

Growth-supporting properties — In the above tests the media support good growth of *E. coli, A. aerogenes,* and *S. typhi*. Medium (e) stab-inoculated with a stock culture of *Brucella abortus* shows good growth in the line of the stab after incubation for 48 hr. Slants of medium (e) inoculated with *E. coli, A. aerogenes, S. Pseudomonas aeruginosa, Staphylococcus aureus,* and *S. albus,* show characteristic growth after incubation for 24 hr. Medium (e), to which approximately 5% of rabbit blood has been added, inoculated, and poured into Petri dishes, shows characteristic α or β zones about colonies of *Diplococcus pneumoniae* and β-hemolytic streptococci (serological groups A and B), recognizable within 24 hr and fully developed after incubation for 48 hr. Medium (e), to which ca. 10% of blood has been added, then heated to 80 to 90°C until the blood turns chocolate brown, permits the growth of *Neisseria gonorrhoeae* colonies within 48 hr when incubated in an atmosphere containing approximately 10% carbon dioxide.*

A method for the microbiological evaluation of peptone was presented by Kheshgi and Saunders.[43] The method employed eight carefully selected aerobic and anaerobic organisms. Only the optical density of the resulting growth was measured. Therefore, the test determines the nutritive value of the peptone for supporting the growth of microorganisms, but does not measure the products of their metabolism. They discovered that such tests could be made to differ significantly by the addition of trace amounts of certain salts. In order to prevent their tests from merely revealing differences in peptones which varied in trace metal content, they prepared the following basal medium of mineral salts:

NaCl	2 g
K_2HPO_4	2 g
$MgSO_4 \cdot 7H_2O$	0.2 g
$FeSO_4 \cdot 7H_2O$	0.1 g
$MnSO_4 \cdot 4H_2O$	0.1 g
Water	1 liter
pH 7.0	

For certain test organisms, dextrose and yeast extract were added. The unknown peptones were added at 2% w/v level, and the pH was checked at 7.0.

The authors did not specify the conditions under which the aerobic organisms should grow, which, in fact, should be in shaken (or air-sparged) cultures. The first growth factor to be exhausted in a closed vial of culture medium is the dissolved oxygen. The higher the temperature of incubation, the lower the content of dissolved oxygen.

Another precaution not specified in any of the above papers is the necessity of putting up known standard peptones in parallel with the unknown tests. This is necessary to

* From Bridson, E. Y. and Brecker, A., in *Methods in Microbiology*, Vol. 3A, Norris, J. R. and Ribbons, D. W., Eds., Academic Press, London, 1970, 229—295.

control normal biological variation; the results can be read as 90% of standard or 110% of standard, etc.

The vitamin content of peptones is variable and depends on the quality of the protein substrate, the manufacturing processes used, and the subsequent storage of the finished peptone. Typical analyses of commercially available peptones are reproduced in Tables 6—8 by courtesy of the companies mentioned.

III. EXTRACTS OF MEAT, YEAST, PLANTS, AND SOIL

A. Meat Extract

Meat extract was probably the first nutrient solution deliberately used by early microbiologists for the cultivation of organisms. The bouillons of the French school of bacteriology were derived directly from the stews and soups of the kitchen. The aqueous extract obtained from boiling meat contains most of the growth factors required by bacteria, but it normally requires supplementation with protein digest in order to increase the amino nitrogen content for adequate biomass development.

In commercial practice, "meat extract" is a by-product from the manufacture of corned beef, although meat extract was originally produced by the Liebig process and corned beef was developed to use up the extracted meat. Chopped beef is immersed in water and heated for a prescribed time, usually 20 min at 100°C. During this infusion process the water-soluble fractions pass out of the meat into the aqueous solution.

Although standard textbooks on bacteriology contain instructions on the preparation of meat extract in the laboratory, the use of commercial concentrates is more convenient and is widely practiced. One such product, Lab-Lemco®, is now synonymous with meat infusion and is prescribed in the formulation of culture media.

Meat extract can be described as an aqueous solution of peptides, amino acids, nucleic acid fractions, organic acids, minerals, and vitamins. It functions in culture media by contributing metals, phosphates, energy sources (carbon compounds), and essential factors that may be lacking in peptone. The most extensive analytical work on meat extracts was that carried out by Wood and Bender[44] and Bender et al.[45] A summary of their findings, together with data presented in the Society for General Microbiology Supplement[40] is presented in Table 9.

Freshly extracted meat infusion is a pale yellow solution. Darkening occurs during concentration of the commercial product, due primarily to Maillard reactions between the amino acids and reducing sugar present in the fresh extract. The water-soluble vitamins remain in the concentrated extract, but the thiamine level is reduced. The high level of organic acid present as sodium lactate should be noted. This substance is the major available carbon energy source for chemorganotrophic organisms.

Meat extract for bacteriological use should be low in copper, lead, and zinc. It must be free from fermentable carbohydrate if it is to be used as a nutrient base in indicator-carbohydrate media. The high phosphate content of meat extract may act as a buffer in such media and interfere with the pH value unless the carbohydrate added to the medium is in high concentration. It is normally used at 0.3 to 0.5% w/v in nutrient media. This weight usually refers to the amount of concentrate taken, which is approximately 70% w/w total solids. Dehydrated meat extracts can be used at lower concentrations.

Liver infusion or extract is widely used in the preparation of media for the growth of *Brucella* or *Clostridia*. Huddleston et al.[47] used a liver preparation for the growth of *Brucella abortus*. Huddleston[48] described the preparation of liver infusion as follows: "Mince fresh ox liver and then pulp it in a mortar. Mix with distilled water in the ratio 500 g of liver to 500 ml of water and keep in the cold for 24 hours. Steam for 1½ hours and filter through a 60 mesh wire gauge." The fragments of liver tissue play a significant role in the cultivation of microorganisms, and liver infusion is usually a cloudy product rather than clear. The infusion process may be carried out at room temperature after steaming.[49]

Table 6
APPROXIMATE COMPOSITION OF BBL® PEPTONES

Catalog number Name	11842 Acidicase® Peptone	11869 Gelysate® Peptone	11879 Lactalysate® Peptone	11892 Myosate® Peptone	11905 Phytone® Peptone	11918 Thiotone® Peptone	11920 Trypticase® Peptone	11928 Yeast extract
Source	Casein	Gelatin	Lactalbumin	Cardiac muscle	Soy meal	Animal tissues	Casein	Yeast cells
Hydrolysis	HCl	Pancreatic	Pancreatic	Pancreatic	Papaic	Peptic	Pancreatic	Autolytic enzymes
Vitamins mcg/g								
Biotin	0.018	0.014	0.062	0.10	0.35	0.18	0.083	4
Choline	0.0	0.0	1.15	4250	3.05	1980	0.0	2000
Cyanocobalamin	0.00006	0.13	0.0039	0.37	0.00115	0.5	0.45	0.0
Folic acid	0.0057	0.052	0.038	1.06	0.81	1.17	0.67	20
Niacin	0.10	2	6.3	390	33	212	8	400
Pantothenic acid	0.26	1.76	4.0	1.76	7.6	8.9	2.2	100
Pyridoxine	0.024	1.7	0.7	11.9	4.0	3.2	0.06	30
Riboflavin	0.10	4	0.69	37	4.3	19	5.75	50
Thiamine	0.105	160	0.44	96	1.9	+	+	100
PABA		0.164	0.08		5.4	0.66	0.21	24

Note: Absence of a figure means absence of data, not of the component.

Courtesy of Becton, Dickinson and Company, Cockeysville, Md.

Table 7
VITAMIN CONTENTS OF PEPTONES AND HYDROLYSATES

Micrograms per gram	Peptone	Proteose peptone	Propeose peptone No. 3	Tryptone	Tryptose	Yeast extract
Pyridoxine	2.5	3.0	4.1	2.6	2.8	20.0
Biotin	0.32	0.43	0.24	0.36	0.39	1.4
Thiamine	0.50	3.0	2.7	0.33	1.66	3.2
Nicotinic acid	35.00	131.00	169.00	11.00	71.00	279.00
Riboflavin	4.00	11.00	13.00	0.18	5.59	19.00
Reaction pH	7.0	6.8		7.2	7.3	

	Neopeptone	Protone	Casitone	Casamino acids, technical	Casamino acids
Pyridoxine	5.0	0.24	1.1	0.025	0.073
Biotin	0.73	0.0021	0.34	0.050	0.102
Thiamine	3.4	0.17	0.48	0.02	0.12
Nicotinic acid	134.00	2.1	24.00	2.5	2.7
Riboflavin	11.4	0.046	0.68	0.019	0.03

Courtesy of Difco Laboratories, Detroit.

Table 8
VITAMIN CONTENT OF CULTURE MEDIUM BASES

(Micrograms per Gram)

Product	Catalogue no.	B_1	B_2	B_6	B_{12}	Biotin	Nicotinic acid	Folic acid	Pantothenic acid (Ca)
Casein hydrolysate, acid-hydrolyzed, vitamin-free	2238	2.0	5.16	<1.0	11.8	0.12	15.5	<0.1	<0.6
Casein hydrolysate, acid-hydrolyzed	2245	0.86	4.1	4.0	12.8	0.18	5.3	0.48	6.0
Casein hydrolysate, pancreatic, free from sulfonamide antagonists	2239	38.5	8.86	16.8	26.3	0.27	3.8	0.51	6.0
Meat extract	3979	2.0	10.32	7.4	98.7	0.24	102.0	0.50	18.0
Gelatin	4070	2.0	3.94	1.0	11.6	0.11	0.1	0.55	6.0
Yeast extract	3753	72.4	49.3	28.8	32.1	1.6	1128.3	1.83	60.0
Liver hydrolysate, enzymatic	5402	0.8	45.6	15.4	30.6	2.85	34.7	3.03	36.6
Liver powder	5347	0.78	52.6	7.8	84.6	2.5	312.2	3.69	60.0
Malt extract	5391	4.0	4.33	8.6	16.7	0.19	38.6	0.95	11.2
Ox bile, dried	3756	2.0	3.74	2.6	22.6	1.05	113.5	0.1	15.0
Peptone from casein, tryptic	7213	2.0	3.76	3.2	34.5	0.18	8.7	0.47	6.0
Peptone from meat, peptic	7224	4.02	2.76	16.6	19.1	0.69	36.0	0.51	10.2
Peptone from meat, tryptic	7214	27.0	11.95	11.32	86.8	0.59	121.0	1.34	15.0
Peptone from soybean meal, papaic	7212	5.0	10.66	7.0	16.4	0.72	155.2	0.96	60.0
Proteose peptone	7229	2.0	6.5	6.2	134.4	0.20	90.8	0.51	21.0
Tryptose	10213	2.0	3.35	4.1	100.0	0.18	44.2	0.50	26.2

Courtesy of E. Merck, Darmstadt, W. Ger.

Table 9
ANALYSES OF MEAT EXTRACTS[40,44,45]

Analysis of Ox-Muscle Extracts

Component	% fresh extract	% commercial extract
Amino acids		
α-Alanine	1.27	1.48
Aspartic acid	0.25	
Glutamic acid	0.78	
Glycine	0.25	
Isoleucine	0.24	0.09
Leucine	0.63	0.09
Lysine	0.07	
Methionine	0.21	0.01
Phenylalanine	0.40	
Serine, threonine, asparagine	2.59	0.11
Tyrosine	0.21	
Taurine	1.67	0.36
Valine	0.39	
Peptides		
Carnosine	4.86	4.15
Anserine	1.12	0.84
Imidazole peptide	Trace	2.10
Guanidines		
Creatine	9.45	5.40
Creatinine	0.75	6.18
Methylguanidine	Absent	~0.1
Guanidine	Absent	~0.1
Purines		
Hypoxanthine (free and combined)	1.46	2.57
Organic acids		
Lactic	23.04	16.40
Glycollic	2.34	1.10
Succinic	0.88	1.42
Carnitine	3.45	3.7
Urea	0.68	0.12
Ammonia	0.37	0.47
Inorganic matter (8.95% K, 7.3% P_2O_5)	27.50	33.0
Coloring matter	15.80	20.55
Reducing sugar (as glucose)	2.10	Absent

Vitamins of the B Group in Meat Extract

Vitamin (μg/g)	Fresh beef (μg/g)	Meat extract (μg/g)
Cynanocobalamin (Vitamin B_{12})	0.3–1.0	0.2
Thiamine (Vitamin B_1)	0.9–3.0	0–1
Riboflavin (Vitamin B_2)	1.8–3.5	30–35
Nicotinic acid	24–102	1000–1200
Pyridoxine (Vitamin B_6)	0.77–4.0	5
Panthothenic acid	4.9–15	25
Choline	760	1500

Table 9 (continued)
ANALYSES OF MEAT EXTRACTS[40,44,45]

Typical Analysis of Meat Extract Paste

Total solids	76.9%
Ash	19.1%
Chlorides as NaCl	6.2%
Phosphate as P_2O_5	4.3%
Total nitrogen	9.2%
Amino nitrogen	1.5%
Ether-soluble extract	0.1%
Calcium	0.05%
Magnesium	0.2%
Sodium	4.9%
Potassium	5.8%
Iron	51 ppm
Copper	15 ppm
Lead	2 ppm
Tin	10—20 ppm
Zinc	25 ppm

Why liver infusion is of such value in bacteriology is not clear, except that the tissue is very rich in glycogen and probably has a high content of vitamins and trace metals such as cobalt, as well as all the other components present in meat infusion. The high fermentable-carbohydrate content helps in the production of low redox potentials for the successful cultivation of anaerobic organisms.[49] The analysis of liver tissue is as follows:

Analysis of Liver[50]

Water	70.9% w/w
Protein	19.8%
Fats	4.2%
Carbohydrate (largely glycogen)	3.6%

Salts (mg/100 g)

P	373
K	298
S	251
Na	130
Mg	22
Fe	12.1
Ca	8
Cu	2.08
Mn	0.3

Vitamins (mg/100 g)

A	19,200 (IU/100 g)
B_1	0.27
B_2	2.80
Nicotinic acid	16.1
C	31.0
B_6	1.0
B_{12}	0.03—0.1
Pantothenic acid	5.0
p-aminobenzoic acid	0.2

B. Yeast Extract

Yeast extract consists of a concentrated solution of hydrolyzed yeast protein. It is produced by the autolytic reaction of the proteolytic enzymes of the cell on yeast protein. It should have a meaty, soup-like flavor; be free from bitter, burnt, acid or other off-flavors; and dissolve in hot water to give a clear solution. Most of the yeast extract manufactured is required for the culinary market, where flavor-enhancing substances are added. The raw material used is waste brewing yeast, and a high salt content with a dark color is an advantage.

Yeast extract for microbiological purposes should be light in color and have a relatively low salt content (1 to 2% w/w). The use of fermenter-grown baker's yeast ensures a more reproducible source of yeast protein free from hop resins which are inhibitory to the growth of most bacteria. If brewing waste yeasts are used, they should be washed in alkaline solution. A sodium carbonate solution at pH 6.5 to 7.0 is suitable.

Autolysis of the cells is initiated by raising the temperature of the suspension to 50°C. To speed up the reaction, plasmolysis may be obtained by adding sodium chloride. For bacteriological purposes it is more usual to add organic solvents such as ethyl acetate or isopropanol. The times and temperatures used are based on the final analysis required in the extract or on particular flavor characteristics. Normally, 24 to 48 hr is sufficient to obtain the maximum amount of diffusion products in the final filtrate. The extract is separated from cell wall debris by centrifugation and then filtration. Concentration or drying must follow very quickly or the extract will deteriorate as the microbial content increases to spoilage levels. Spray-drying is the usual method of finally processing the extract before packaging. Quality control tests include chemical analysis and microbiological growth tests.

Yeast extract is a mixture of amino acids, peptides, water-soluble vitamins, nucleic acid fractions, and carbon compounds. The carbohydrates of yeast are primarily glycogen and trehalose, but these substances undergo enzymic hydrolysis during the extraction process. A typical analysis of powdered yeast extract is

Moisture	4.1%
Ash	11.5%
Salt (as NaCl)	1.3%
Phosphate (as P_2O_5)	3.2%
Total nitrogen	10.5%
Amino nitrogen	4.8%
Vitamins ($\mu g/g$)	
Thiamine (B_1)	20–70
Riboflavine (B_2)	55–100
Pyridoxine (B_6)	15
Niacin	250–700
Pantothenic acid	90
Inositol	3800
Folic acid	30–40
Amino acids (g/100 g of extract)	
Alanine	3.6
Arginine	1.0
Aspartic acid	5.1
Cystine	0.5
Glutamic acid	6.0
Glycine	2.7
Histidine	1.2
Isoleucine	2.3
Leucine	3.5

Amino acids (g/100 g of extract)
(continued)

Lysine	3.5
Methionine	0.8
Phenylalanine	1.9
Proline	2.5
Serine	2.3
Threonine	1.0
Tryptophan	0.8
Tyrosine	1.5
Valine	3.0

Malt Extract (wort) is a water-soluble extract of malted barley. The malting process is the selective liberation of the cereal starch from its reserve status to an intermediate preparedness for conversion into sugars.[51] The effectiveness of the extraction process is measured by calculating the gross percentage of solubilizable material. The art and technique of malting is now a scientific process that has evolved from the largely empirical methods previously practised in the brewing industry.

A brief description of the steeping, aeration cycle is as follows:*

Steep	6 hr at 34% moisture at 20—22°C
Aerate	24 hr at 15—16°C until 90% germination
Steep	2 hr at 42% moisture at 16°C
Aerate	24 hr at 15—16°C until 100% germination and some growth
Soak	3 hr at 48% moisture at 35°C to check rootlet growth
Aerate	60 hr at 15—16°C to complete conversion of cereal starch

The malted barley, after drying by kilning, is put through a mill and then extracted with warm water. The extract is filtered and concentrated to a thick syrup (80% total solids) having a smell and taste characteristic of malt extract. During the process of germination, proteolytic enzymes act on the insoluble proteins and produce soluble amino nitrogen compounds (peptones). There is some loss of these products in the steeping water. Piendl and Wagner[52] discussed the amino acid content of malt and the effect of malting on the concentrations of the end products of proteolysis.

The most extensive review of malt is that written by MacWilliam,[53] who showed that the major constituents of malt extract are carbohydrates (90% of total solids) and nitrogenous materials (4% of total solids). The minor constituents include lipids, sulfur compounds, and inorganic constitutes. The major carbohydrate components are glucose, fructose, maltose, sucrose, maltotriose, and other linear and branched glucose polymers (dextrin fractions with 4 to 20 glucose units). The concentrations of each sugar present depend on the malting process, but average figures are as follows:

Sugar	g/100 ml of wort
Glucose	0.22
Fructose	0.96
Sucrose	0.35
Maltose	5.34
Maltotriose	1.44
Dextrins	3.12

* From Griffin, O. T., *Process Biochem.*, 7, 17—20, 1972. With permission.

The nitrogenous material is a mixture of compounds ranging from ammonia to polypeptides. The concentrations of amino acids in malt extract vary with the nitrogen content of the malt used and average values are given below:

Amino acids	mg/100 ml
Alanine	6.9
α-Aminobutyric acid	6.0
Arginine	10.0
Aspartic acid	6.3
Cystine	Trace
Glutamic acid	3.6
Glycine	2.6
Histidine	3.8
Isoleucine	4.0
Leucine	9.0
Lysine	6.9
Methionine	2.5
Phenylalanine	8.2
Proline	25.8
Threonine	4.5
Tryptophan	4.5
Tyrosine	6.0
Valine	7.6
Serine + asparagine	15.0
Ammonia	2.5

The vitamins present in malt extract have been quantified by MacWilliam[53] and the average result is as follows:

Vitamin	μg/100 ml
Biotin	0.96
Folic acid	11.00
Inositol	5.00
Nicotinic acid	1.40
Pantothenic acid	63.00
Pyridoxine	89.00
Riboflavin	50.00
Thiamine	49.00

The total ash content of malt extract is 1.5 to 2.0% of the solids. In addition to phosphates, chlorides, sulfates, and fluorides, it contains the metallic ions Na, K, Ca, Mg, Fe, Cu, and Zn. Malt extract is often presented as maltose, but the presence of diastase in most extracts, together with the undefined quantity of hexoses present, means that great care must be exercised if it is used for fermentation studies.

Malt extract is widely used in media for the cultivation of molds and yeasts. An average analysis is as follows:

	%
Maltose	52.2
Dextrose	19.1
Sucrose	1.8
Dextrin	15.0
Other carbohydrates	3.8
Protein	4.6
Ash	1.5
Moisture	2.0
pH (10% solution)	5.5

It is incorporated into culture media as malt extract, as in the following example:

Malt extract broth	g/l
(Oxoid CM57)	
Malt extract	17
Mycological peptone,	3
pH 5.4	

Wort agar	g/l
(Oxoid CM247)	
Malt extract	15.0
Peptone	0.78
Maltose	12.75
Dextrin	2.75
Glycerol	2.35
Dipotassium phosphate	1.0
Ammonium chloride	1.0
Agar	15.0
pH 4.8	

C. Potato Extract

Potato extract may be prepared by steaming 500 g of peeled, chopped potatoes in 1 liter of distilled water for 3 to 4 hr and filtering off the liquid. A concentrate may be prepared by extraction with ethanol.[54] Jensen and Spencer[55] supplemented their medium with potato extract and were able to stimulate the growth of small inocula of *Clostridium butyricum*. Potato-containing media can be used for counting nitrogen-fixing *Clostridia*.[56] An example of the composition of a potato-containing medium is

Potato dextrose agar	
(Oxoid CM139)	(Oxoid CM139)
Potato extract	4 g
Dextrose	20 g
Agar	15 g
Water	1 liter
pH 5.6	

Table 10
PREPARATION OF SOIL EXTRACTS

	Fischer	Löhnis	Smith and Worden	Fred. Watesman	Lockhead et al.
Soil (g)	1000	1000	500	1000	1000
Water (ml)	1000	800	1200	1000	1000
Precipitant[a]	Na_2CO_3	–	–	$CaCO_3$	$CaCO_3$
Final volume	1000	800	1000	1000	1000
K_2HPO_4 (g)	2.0	–	0.5	0.5	0.2
Agar (g)	12.0	15.0	20.0	12.5	15.0

[a] The function of the precipitant is to help remove colloidal material left in suspension after heating soil in water at 121°C.

From James, N., *Can. J. Microbiol.*, 4, 363–370, 1958. With permission.

Potato dextrose agar can be used to count yeasts and molds in butter.[57] The addition of sterile 10% tartaric acid solution to the autoclaved medium to reduce the pH to 3.5 improves its selective properties for yeasts and molds.

D. Soil Extract

From the beginning of the study of soil microbiology, aqueous extracts of soil have been used as the main source of nutrients and as additives to media containing known organic and inorganic materials. The use of soil extract for media required to grow soil microorganisms is a rational extension of the knowledge that, as it supports a teeming population, soil must contain all the nutritional substances required for their growth.

Lockhead and Chase[58] reported that 63 of 332 cultures isolated from soil on media containing soil extract would not grow on an inorganic base medium containing glucose, 10 amino acids, and 7 growth factors or with yeast extract as a substitute for the amino acids and growth factors. It was later reported that vitamin B_{12} could replace soil extract for some of these cultures.[59] Further work showed that, for some organisms, other unknown essential growth factors were present in soil extract which could not be supplanted by vitamin B_{12} or yeast extract.[60]

James[61] conducted further evaluations of the use of soil extract in culture media. He drew attention to the variation in procedures for producing soil extract agar described by previous workers (Table 10). He showed that dilution of the soil extract gave lower counts, that unheated extract was no better (indeed, a little worse than heated extract), and that potassium phosphate significantly increased the number of organisms recovered. He also stated that if care is taken to use 500 g of fertile soil and add that volume of water which will yield 1000 ml of extract after heating, the maximum amount of extractives will be obtained. The preparation of soil extract agar is as follows:[61]

1. Use 500 g soil from a field of high fertility.
2. Add approximately 1500 ml of tap water (or that amount to yield 1000 ml of extract).
3. Autoclave for 30 min at 121°C.
4. Filter through cloth or paper; refilter the first portion if cloudy.
5. Add 0.5 g $CaSO_4$ or $CaCO_3$ while hot; mix, and let stand for 5 min before refiltering.
6. Add 0.2 g K_2HPO_4 to every 1000 ml of filtrate.
7. Add 15 g of agar per liter and heat to dissolve the agar completely.

8. Adjust the pH to 6.8; filter again through cotton wool or muslin, if necessary, and dispense in 100 ml quantities.

9. Sterilize at 121°C for 20 min.

In spite of the largely unknown composition of soil extract, it continues to give higher counts of soil organisms than substitute materials and it will continue to be used in soil microbiology.

Augier[62] found that soil extract improved the yield of *Azotobacter* in his medium. It also improved the yield of nitrogen-fixing *Clostridia.*[63]

IV. BILE DERIVATIVES

Bile is a product of the liver. The composition of the yellow to green fluid varies according to its animal source and state of preservation. It contains bile pigments, bile salts, fatty acids and their salts, cholesterol, mucin, lecithin, inorganic salts, ethereal sulfates, glycuronic acids, urea, and porphyrins.[64] Fresh bile contains approximately 10% w/v total solids. An analysis of ox bile expressed as percentage of total solids is as follows:[65]

Glycocholic acid	37.3
Taurocholic acid	28.9
Taurodihydroxycholanic acids	7.1
Glycodihydroxycholanic acids	6.1
Unconjugated bile acids	Trace

A bile salt is usually considered to be the sodium salt of a conjugated bile acid. A conjugated bile acid is a peptide formed from a bile acid and glycine (H_2NCH_2COOH) or taurine ($H_2NCH_2CH_2SO_3H$). As far as is known, all bile acids present in normal (fresh) animal bile are conjugated. The structure of bile acids may be represented as follows:

Acid	R^1	R^2	R^3	R^4
Lithocholic	OH	H	H	H
Hyodeoxycholic	OH	OH	H	H
Chenodeoxycholic	OH	H	OH	H
Deoxycholic	OH	H	H	OH
Hyocholic	OH	OH	OH	H
Cholic	OH	H	OH	OH

The principal acids in ox and sheep bile are cholic and deoxycholic, whereas in pig bile they are hyodeoxycholic and hyocholic. Human bile contains large amounts of chenodeoxycholic acid. Acidification of an aqueous solution of bile salt will precipitate the conjugated bile acid. Alkaline hydrolysis of a bile salt or conjugated bile acid followed by acidification precipitates the bile acid.

A. Bile Desiccant

Freshly collected or preserved animal bile is collected at the abattoir, filtered, concentrated at low temperatures, and dried. When reconstituted it should form clear solutions with a brown-green color at 10% w/v concentration. This was the first bile additive used in culture media.[66] Bile is a nutritive environment for a number of organisms and putrefaction quickly occurs in the abattoir unless it is chilled or preserved with chemical solutions. The bacterial spoilage leads to deconjugation of the bile salts and releases varying amounts of free bile acids.[67,68] As these acids are more toxic to bacteria than the conjugates, there is considerable scope for variation in the toxicity of bile in culture media. The presence of bile pigments, which may cause confusion in media with indicator dyes, plus the variation in quality caused the search for more refined products.

Burman[69] described tests to select suitable grades of bile salts and introduce standardization. He was not aware of the variable effect of conjugates and free bile acids and relied on screening microbiological tests using river water flora. Bile was commonly used in diagnostic tests to distinguish groups of *Streptococci* as bile-tolerant or bile-sensitive, or to distinguish between *Streptococci* and *Pneumococci* using the fact that *Pneumococci* will completely lyse in bile solutions. A 10% solution of sodium deoxycholate is now more commonly used.[70]

B. Bile Salts

A further refinement to the use of bile is to precipitate the bile salts and separate them from the bile pigments and other substances present in crude bile. Hydrochloric acid is added to the bile solution until the pH is 4.5. The temperature may be raised to 70°C, and carbon can be added to reduce the color of the bile. The bile salts are still in solution at this pH and the bile can be filtered, neutralized with caustic soda, and the filtrate concentrated to a syrup. Above 60% total solids the bile salts will begin to crystallize out. Alternatively, the pH of the filtered solution can be lowered by adding more acid and the precipitated bile salts filtered off, washed, and dried.

Although the process described yields bile salts, the product obtained may be a mixture of conjugates and free bile acids, depending on the quality of the bile used in the process. Northolt[71] described the mixture of conjugates and their derivatives found in commercial bile salt preparations. Using thin-layer absorption chromatography (TLC) he obtained the following results:

Bile acids	Oxoid bile salts[a]	Difco bile salts[a]	Difco sodium taurocholate[a]
Cholic acid	+++	−	−
Unknown acid	−	+	−
Hyodeoxycholic acid	−	−	−
Chenodeoxycholic acid	−	+	−
Deoxycholic acid	+	−	−
Lithocholic acid	−	+	−
Taurocholic acid	+	−	+++
Taurodeoxycholic acid	+	+	++
Glycocholic acid	+	−	++
Unknown	−	+	−
Glycochenodeoxycholic acid	+	+++	+

[a] −, no spot; +, small spot; ++, moderate spot; +++, large spot.

Many papers have been published on TLC methods for the separation of free and conjugated bile acids. Workers who wish to follow up the technique are directed to References 72 to 78.

The presence of more than 0.02% w/w deoxycholic acid in bile or bile salts indicates a degree of toxicity that may be unacceptable in some culture media. Work carried out by Oxoid Laboratories and Mossel[78a] showed that the bile used in EE Broth (Oxoid CM317) must be free of bile acids, deoxycholic acid in particular. The criterion for this medium was that it should be able to demonstrate the presence of one cell of enterobacteria present in the enrichment sample, and a purified ox bile was prepared for the medium. Northolt[70a] demonstrated the results he found with TLC technique as follows:

	EE Broth[a]			
Bile acid	Oxoid (purified bile)	BBL	Merck	Difco
Cholic	–	+	+++	+
Unknown	–	–	+	–
Hyodeoxycholic	–	–	–	–
Chenodeoxycholic	–	–	+	–
Deoxycholic	–	–	+	+
Taurocholic	+	++	+	+
Taurodeoxycholic	–	+	+	+
Glycodeoxycholic	+	+	+	++
Glycochenodeoxycholic	–	–	–	–

[a] –, no spot; +, small spot; ++, moderate spot; +++, large spot.

Although bile-containing media are formulated to suppress Gram-positive organisms and to select the Gram-negatives, bile salts (free from bile acids) will allow *Staphylococci* and *Streptococci* to grow. Particular advantage is taken of this characteristic in the U.K., where MacConkey Agar (e.g., Oxoid CM7) is used as a general-purpose medium for clinical bacteriology. The growth of Group D *Streptococci* and *Staphylococcus aureus* would be looked for on this medium when it is used for culture of urine, neonatal feces, and purulent material obtained from patients. To prevent the growth of such organisms and Gram-positive sporing rods, aniline dyes are often added to bile-containing media (e.g., Violet-Red Bile Agar, Brilliant Green Bile Broth, and Brilliant Green MacConkey Agar). Bile salts should be purchased as a fine, white powder and, when reconstituted as a 2% w/v solution in distilled water, should be a clear, bright, slightly yellow solution at pH 7.0. When incorporated into culture media, it should not affect the color of the indicator dyes or their subsequent change in color. It should not form a surface scum or deposit in the medium after storage.

C. Hydrolyzed Bile Acids

It has been stressed that free bile acids may be considered too toxic for some organisms; nevertheless, there are circumstances in which free bile acids are deliberately chosen for inclusion in culture media. Such media are designed to be very selective and they may receive large quantities of commensal organisms in the inoculum. Bile Salts No. 3, bile salts mixture, etc. contain cholic and deoxycholic acid. Northolt[70a] demonstrated this with a TLC analysis as follows:

Bile acids	Oxoid bile salts No. 3[a]	Difco bile salts No. 3[a]	BBL bile salts mixture[a]
Cholic	++++	+++	++++
Hyodeoxycholic	–	–	–
Chenodeoxycholic	–	–	–
Deoxycholic	+++	+++	++++
Lithocholic	–	–	–
Conjugated	–	–	–

[a] –, no spot; +, small spot; ++, moderate spot; +++, large spot.

As opposed to normal concentrations of bile salts in culture media at 5 g/l, hydrolyzed bile salts are used at 1.5 g/l or less. It would be usual practice to titrate the bile acids in the medium, using the growth of test/control microorganisms as an indicator.

A characteristic of bile acids in culture media is the precipitation of the acids as an opaque halo around the colony if the pH drops below a critical value. The growth of *E. coli* on Violet Red Bile Agar is indicated by the presence of red colonies (indicating lactose fermentation) and a white halo (which indicates a greater production of organic acids than red colonies without the halo).

Although microorganisms can deconjugate bile compounds with great facility, the chemical process requires temperatures of 120 to 160°C for 6 to 8 hr in strongly alkaline solutions. The hydrolyzed free bile acids are then precipitated with acid, washed, redissolved, neutralized, and dried. The resulting product is composed of approximately equal parts of sodium cholate and sodium deoxycholate. Sodium deoxycholate may be separated from such a mixture and used as a relatively pure salt. Leifson[79] showed that the constituent of bile which had the greatest effect on bacterial growth was deoxycholic acid. In a detailed study he showed the following factors influenced its performance:

pH – Gram-positive bacteria could not grow below pH 7.5, but various groups of organisms grew above pH 7.5.

NaCl – Sodium chloride enhanced its inhibitory action. Thus, at 2 to 3% NaCl, intestinal pathogens were only slightly inhibited, while other coliform organisms were greatly inhibited.

Organic acids – Acetates, propionates, butyrates, and especially citrates in concentrations of 1 to 2% strongly inhibited coliform organisms, but not intestinal pathogens.

Nutrients – The action of deoxycholate varied with the source of peptone or meat extract.

Oxygen tension – The inhibitory effect of deoxycholate was less pronounced on coliform organisms under low redox potentials.

Fats, fatty acids, lipids, proteins – Reduced the activity of deoxycholic acid, possibly by combining with it. Gram-positive bacteria will grow in the presence of sufficient of these materials.

One of the most important factors in the role of deoxycholate citrate in culture media is the effect of magnesium. The activity of deoxycholate is enhanced in the absence of magnesium, and the role of citrate is to act as a chelate. The addition of magnesium to deoxycholate-citrate medium reduces the inhibitory effect and improves the growth of *Shigellae*.

It can be seen from these variable factors that the production of standard bile salt media is a difficult process. In attempting to improve this situation, defined chemicals have been tried in place of bile derivatives. Use of wetting agents such as Teepol®[80-82] and Tergitol® 7[83,84] was tried with varying success. Ricinoleate was used in Lactose Ricinoleate Broth[85] as a replacement for bile. Glutamic acid was successfully used in the medium devised by Gray,[86] which was later modified to improve the mineral content.[87] The most recent substitution is that by Mossel et al.,[88] in which purified sodium dodecylsulfate Analytical Reagent Grade has been substituted for ox bile in EE Broth.[89] It should be anticipated that the search for further substitutes for bile, bile salts, or their derivatives will continue.

V. GELLING AGENTS

A. Agar

Agar is a complex mixture of polysaccharides extracted from species of the red algae known as agarophytes (*Gelidium, Gracilaria, Pterocladia, Acanthopeltis* and *Ahnfeltia* species). It is a sulfuric acid ester of a linear galactan, insoluble in cold water but soluble in hot water. A 1.5% w/v solution should set at between 32 and 39°C and not melt below 85°C. Araki and Araki[90] and Araki and Hirase[91] suggested that agar yielded two polysaccharide fractions:

1. A virtually neutral polymer, agarose — $(1 \rightarrow 4)$-linked 3,6-anhydro-α-L-galactose alternating with $(1 \rightarrow 3)$-linked β-D-galactose.
2. A charged polymer, agaropectin, having the same repeating unit as agarose, with some of the 3,6-anhydro-L-galactose residues replaced with L-galactose sulfate residues, together with partial replacement of the D-galactose residues with pyruvic acid acetal 4,6-0-(1-carboxyethylidene)-D-galactose.

Agarose is the component responsible for the high-strength gelling properties of agar, whereas agaropectin provides the viscous component. Thus, the gel strength of an agar is related to the percentage of agarose present in relation to agaropectin. However, Duckworth and Yaphne,[92] using DEAE Sephadex® fractionation, have shown that agar is not made up of one neutral and one charged polysaccharide, but is composed of a complex series of related polysaccharides which range from a virtually neutral molecule to a highly charged galactan.

The characteristic property of agar to form high-strength gels that are thermally reversible with a hysteresis cycle or lag over a range of 40°C is due to three equatorial hydrogen atoms on the 3,6-anhydro-L-galactose residues which constrain the molecule so as to form a helix with a threefold screw axis. The interaction of these helices causes gel formation.[93] Agar is easily hydrolyzed with acids because the 3,6-anhydro-α-L-galactoside linkage is very sensitive to acid cleavage.

Agar is manufactured in many parts of the world, although the industry must be geographically located near suitable weed beds. The major countries producing agar are Japan, the United States, Spain, Portugal, Morocco, Korea, and New Zealand. Apart from a supply of suitable algae, the most important factor is a suitable population prepared to gather weed in often uncomfortable circumstances. Some degree of care is required in pulling weed from the tidal edge in order to ensure that the bed is not destroyed in one harvest. The commercial production of agar is outlined by Bridson and Brecker[20] and fully described by Chapman.[94] It is important to remember that agar not only varies according to the source of weed but also according to the method of manufacture. The effect of extraction temperature, time and pH, the clarification and bleaching processes used, the conditions under which the gel is frozen or pressed to remove water, and the subsequent drying and milling are all important.

Table 11
AGAR ANALYSES

	American	Portuguese	Spanish	New Zealand	Moroccan	Japanese
Ash	3.9%	2.2%	2.5%	1.5%	4.52%	2.95%
Acid-insoluble ash	0.46%	0.1%	0.23%	0.11%	0.17%	0.29%
Sulfate	2.0%	1.4%	1.2%	0.7%	1.74%	1.82%
Chloride	0.21%	Trace	0.31%	0.06%	0.2%	Trace
Calcium	0.18%	0.5%	0.34%	0.59%	0.8%	0.68%
Magnesium	0.07%	0.1%	0.16%	0.08%	0.22%	0.11%
Total nitrogen	0.09%	0.1%	0.24%	0.09%	0.15%	0.02%
Iron	21 ppm	170 ppm	233 ppm	161 ppm	324 ppm	256 ppm
Grade strength	125	155	155	182	156	135
Recommended concentration	1.5%	1.2%	1.2%	1.0%	1.2%	1.5%

From Bridson, E. Y. and Brecker, A., in *Methods in Microbiology,* Vol. 3A, Norris, J. R. and Ribbons, D. W., Eds., Academic Press, London, 1970, 229–295. With permission.

For every ton of dried agar, up to 100 tons of water are used in the process. Thus, the concentration of calcium and magnesium in the agar will often reflect the quality of the local water. Table 11 shows some of the differences to be found in agars of different geographical origins.

The presence in agar of metals (Ca, Mg, Fe, etc.) that can react with phosphates present in culture media to form insoluble precipitates or hazes is undesirable. Filtering agar-containing culture media is not an easy task. On the other hand, the total elimination of these essential metals or a drastic reduction in phosphate level could lead to poor growth-promoting properties in the medium. Various "purified" agars are marketed that are particularly low in Ca and Mg and these can be used with confidence. Most bacteriological agars are compatible with the normal ingredients of culture media, but prolonged heating or storage of the molten medium or exceptionally high phosphate levels at pH values above 7.0 will eventually lead to formation of precipitates. The effect on culture media of adventitious metals present in agar is shown in the work carried out by Barth-Reller et al.,[95] where a large difference was observed between the performance of the medium with and without agar.

All gels have a tendency to shrink and expel water containing solutes. This is known as syneresis, and it varies directly with the concentration of agar in the gel. There is also a general rule that the higher the gel strength of the agar, the more water is expelled. It is possible that the strength of the gel is determined by the interlocking of the helices as described above, and that this slow knitting together of the molecules forces water to the surface of the gel. Although agar loses water with relative ease, it is difficult to rehydrate without raising the temperature to the liquefying point. Examples can be seen where, because of cycling temperatures between 4°C and high ambient temperatures, water has been lost from agar slopes or plates in the form of condensation on the walls of the container. After a period of time, dehydrated films of agar can be observed in closed vessels containing an excess of condensate. It appears that agar becomes hardened on the agar/air interface and will not take up water. This point must especially be borne in mind when drying the surface of agar media in petri dishes (see section on storage of culture media).

The fact that agar remains in the gel form until the temperature reaches 80°C means that agar-containing media can be used to grow thermophilic organisms, which require incubation at temperatures up to 75°C. Equally, the fact that agar remains liquid until the temperature drops to 39°C means that heat-sensitive additives such as blood can be added and mixed in the gel. Microorganisms can also be added and thoroughly distributed at temperatures which will not harm them. Such procedures are used in preparing "pour plates" for counting bacteria in dilutions of water, soil, foodstuffs, etc. Agar also resists liquefaction by the organisms growing on or in the medium. This proved to be a major problem with gelling agents such as gelatin. Some agar-liquefying organisms (notably, *Vibrio* species) have been isolated from sea water, but these are comparatively rare and few microbiologists will encounter them.

As the primary purpose of agar in culture media is to form a solid gel, considerable attention is paid to measuring this characteristic. Agars commonly produced will form firm gel at concentrations varying from 0.9% w/v to 1.6% w/v. Although culture media formulas fix the agar concentration at prescribed grams per liter, it is wise to consider the value of raising or reducing this level to obtain comparable gel strengths in the final prepared media. There are two standard tests to measure the gel (or grade) strength of agar: (1) the Kobe method using a "Nikan" apparatus and (2) the BFMIRA method using a rotating blade.

The Kobe method measures the compressive force acting for 20 sec to break an agar gel of 1.5% w/v concentration that is 3 cm thick and stabilized at 20°C. The method consists of carefully weighing 3 g of agar (±0.01 g) and pouring it into 600 ml of distilled water contained in a 1 liter flask that can be connected to a reflux condenser. A magnetic stirrer is placed in the flask and the flask placed on a hot plate/stirrer apparatus. The agar is boiled to ensure that it goes completely into solution and cold water is run through the condenser to reduce water loss during the process. Foaming must be avoided. The molten agar is then poured into a box measuring 6 × 30 × 45 cm to give a layer of agar 3 cm thick. The agar gel is allowed to stand at 20°C for 15 hr; it is inverted to avoid evaporation of surface moisture. The "Nikan" apparatus is composed of a small, cylindrical rod of 1 cm² cross-section, on which weights may be added. The apparatus is sold by Kuja Seisakusho Ltd., 50 Komagomo Oiwake-Cho, Bunkyo, Japan. The cylindrical rod (piston) weighs 100 g and further weights are added as necessary. The piston is located on the surface of the agar and the time is noted. The piston should penetrate the agar surface after 20 sec of contact. If it occurs sooner than this, the weight is reduced and the test repeated on another location in the box. If the agar resists penetration after 20 sec have elapsed, the weights should be increased again after changing the location to another part of the box. The gel strength is expressed in grams per square centimeter, and the normal range for bacteriological agar lies between 500 to 700 g/cm².

The BFMIRA method measures the gel rigidity against a tortional stress produced by a rotating blade placed in the agar. The stress is increased by filling a reservoir with water at a standard rate. Using the same precautions for dissolving the agar described as for the Kobe method, 3.5 g (±0.01 g) of agar is added to 500 ml of distilled water. The molten agar is poured into three 2-in. cubical boxes, filling them to the very top. The tops of the filled cubes are covered with squares of polyethylene film to reduce evaporation. The gel is allowed to stabilize for a standard period of time, usually overnight at 20 or 4°C. The apparatus is supplied by H. A. Gaydon & Co. Ltd., 93 Lansdowne Road, Croydon, Surrey, U.K. To carry out the test, the stabilized gel is placed on the platform of the apparatus, and the platform is raised so that the rotating blade (previously set to zero) is plunged into the center of the agar cube. A flow of water, previously calibrated at 100 ml/min, is allowed to flow into the reservoir, and the increasing weight induces the blade to press against the rigid agar. The stress is allowed to continue until the blade has rotated 20°. The time taken is measured (and the water volume is noted), and this gives the grade strength of the agar. The test may be continued until the blade rotates 90°; this measures the breaking strength of the agar. The gel strength is expressed as Grade Strength and bacteriological agars are in the range 145 to 190.

It is clear that the Kobe test and the **BFMIRA** test measure different physical characteristics of the agar. Ultimately, both measurements must be converted into grams per liter for culture media manufacture, and the tests detect the amount of variation in gel strength of the raw material.

The clarity of agar gels in culture media is an important characteristic. Ideally, the molten solution should be water-white, clear without any deposit. The gel is always less transparent; nevertheless, in 4-mm thick layers it should appear to be transparent. Haze in agar may be caused by mineral incompatibilities or minute fragments of debris which have passed through the filters during manufacture. Sand, diatoms, plant cells, and other debris may be present. Onöz and Hoffmann[96] described a method whereby agar can be clarified by treatment with sodium phosphate. In this method a phosphate "floc" is induced in the molten agar which acts as a filter aid and also removes excess calcium and magnesium salts. von Kirschninck[97] discussed the various steps that must be taken to obtain agars of better clarity. The common technique of gelling agar, cutting it into small cubes, and washing in running demineralized water for 2 to 3 days reduces the water-soluble adulterants, but does little to reduce tightly bound chemicals or insoluble debris that are present in the gel.

The nature of toxic agents in agar is not clear. Ley and Mueller[13] reported inhibitors for *Neisseria gonorrhoea* which they could extract with methanol or neutralize with starch. They suggested that the toxic substances were fatty acids. Other workers[99-102] have indicated the presence of these substances in media. Jacobs and Harris[103-105] reported that culture media showing inhibitory effects towards phenol-treated organisms could be improved by treatment with charcoal or ferric chloride. In a later study[106] they were able to demonstrate that this toxic effect came from the agar component of the medium. Starch did not influence the results, and the authors concluded that the toxic effects they observed were due to the presence of heavy metals rather than fatty acids.

The high viscosity of agar makes it a very useful substance in semisolid media. Media such as Thioglycollate Medium (USP) and Stuart's Transport Medium are prepared with agar concentrations of between 2 and 5 g/l. The function of the agar is to reduce convection in the medium during cooling and thus prevent the diffusion of oxygen into the medium. These prereduced media are used to encourage the growth of anaerobic or microaerophilic organisms. The addition of redox dyes such as resazurin and methylene blue enable the user to recognize how much oxidized medium is present in the container. Not more than one third of the depth of the medium should show the oxidized color of the dye. Should the color show deeper, the medium can be reheated to 100°C, but not more than one reheating should be allowed.

Not all agars are suitable for use at low concentrations. A phenomenon shown by some high-gel-strength agars is the retraction of the soft agar to form a hard "blob" of agar floating in the liquid broth. It is preferable to deliberately choose low-gel-strength agar or use other gelling agents.

Another characteristic of agar is its ability to allow diffusion of compounds although locking water into a rigid gel. This property is exploited in antibiotic sensitivity test agar, in which the measured rate of diffusion of antibiotics from a central reservoir indicates the susceptibility of the test organism to the antibiotic. Cooper[107] explained the formation of zones of inhibition based on mathematical models of rate of diffusion vs. speed of multiplication of the organism. Under standard conditions the phenomenon is sufficiently constant to confidently measure the minimal inhibitory concentration (MIC) of antibiotic that will prevent the multiplication of bacteria. The diffusion of large-molecule antibiotics through agar can be impeded by the mineral content of the agar. The purer the polysaccharide, the better the diffusion (see section on agarose). It can be seen that the use of unsatisfactory agars can lead to small zones of inhibition and, therefore, cause erroneous results in sensitivity testing.

Bechtle and Scherr[108] described a test to distinguish agars that would allow satisfactory diffusion from agars that would not. A plate of agar gel 4 mm thick is poured and allowed to set. Holes 6 mm in diameter are cut into the agar using a cork borer. A few drops of a large-molecule dye, saffranin, are placed into the wells and allowed to diffuse into the agar at room temperature. The halo of color is measured after 2 and 4 hr. If the plates are left at room temperature overnight, the dye will spread uniformly through satisfactory agar, whereas unsatisfactory agar causes the dye to remain concentrated around the perimeter of the well.

When preparing nitrogen-free media containing agar, the effect of nitrogen donated by the agar is sometimes important. Skinner[54] produced the following data on the total nitrogen content of different agar samples:

Agar type	N_2 content
Oxoid Agar No. 1	0.060
Oxoid Ionagar No. 2	0.072
Difco Noble®	0.077
Difco Bacto®	0.098

Thus, an agar containing 0.098% combined nitrogen used at 1.5% w/v in a medium donates only 14.7 mg of combined nitrogen per liter of medium. He considered this amount to be too low to effect the selectivity of the media for nitrogen-fixing organisms.

B. Agarose

As previously described, agarose is a separated polysaccharide fraction of agar which has two primary uses: (1) as a gel medium for electrophoresis and immunodiffusion techniques and (2) as a granulated medium for molecular exclusion chromatography. Agarose has several advantages for immunodiffusion and the electrophoretic separation of antigens, antibodies, and other proteins. It gives a clear, transparent gel which exhibits less background staining than agar. Due to the high gel strength, the mechanical properties of these gels are superior to those of agar. In the gel diffusion analysis of basic antigens, agarose gives more distinct lines of precipitation, without halo effects.

Agarose is prepared by a procedure which results in its having a very low sulfur content (e.g., <0.02%), at which level no adsorption of the basic substances crystal violet and cytochrome C occurs. These tests are used to assess the efficiency of purification of agar.[109] As a result of the low sulfur content, agarose does not react with serum as does agar, so that artifacts previously observed in gel precipitation reactions between antigens and antibodies do not occur, even when the antigens carry strong negative charges. Similarly, agarose gels do not exhibit adsorption phenomena or ion-exchange effects, and the dissociation of acidic groups during electrophoresis is negligible.[110-112]

The suitability of agarose for immunoelectrophoresis can be tested by measuring the relative electroendosmosis value (m_r) of a gel. This is defined as the relative mobility of dextran, which is assumed to have zero mobility, to that of crystalline human albumin. The m_r values quoted vary from 0.10 (very good) to 0.39 (very poor).

Agarose granules may be used for gel filtration, as little, if any, nonspecific adsorption of molecules occurs. On cooling an agarose solution a strong gel is formed, even at low concentration. This gel will have an ordered porosity and its pore size will depend on the agarose concentration.[113] For agarose concentrations of greater than 5%, the porosity of the gel is roughly comparable to cross-linked dextrans and polyacrylamide gels in that, while separation of smaller molecules can be achieved, it is not possible to effect separation of molecules of molecular weight greater than 200,000. However, agarose is notable for the fact that gels having sufficient strength for gel filtration uses are obtained even when the concentration of agarose is lower than 5%. As the concentration decreases, the effective pore size of the gel increases and, since agarose can even be used at concentrations as low as 0.5%, molecular sorting of high-molecular-weight molecules can be effected. Thus, separation of macromolecules is possible with agarose granules. Granules may be prepared from agarose by various methods, including those recorded by Egorov et al.[114] and Hjerten.[115] Such agarose columns have been used for the purification of papain;[116] the separation of cells from ribosomes, proteins, dyes, salts, etc.;[117] and chromatography of other subcellular particles. A typical analysis of agarose is the following:

pH	5.9
Moisture	9.0%
Ash	<0.3% w/w
Sulfate	<0.02% w/w
Pyruvic acid	None detected
Gel strength	BFMIRA 189 g/cm^2 at 0.75% gel
	Kobe 1000 g/cm^2 at 1.5% gel
Melting point	87°C
Gel point	33°C
m_r	0.2

C. Gelatin

Gelatin is a collagenous extract prepared from boiled hides, hooves, and other collagen-rich materials. It forms "gels" at concentrations of 12 to 15% w/v and was the first gelling agent used in bacteriology. It has been entirely displaced by agar because of its low melting point (30 to 35°C). Gelatin plates can only be incubated at relatively low ambient temperatures of around 20 to 25°C. As it is a proteinatious material, it is readily utilized by many bacteria. Liquefaction of the plates because of gelatinase production was a constant hazard to microbiologists. The sole use of gelatin in microbiology today is as a substrate to detect gelatinase formation. The classical technique of preparing nutrient gelatin butts and stabbing the inoculum into the center of the butt has been superceded by more modern methods. The reason for the decline in popularity of the technique lies in the problems of having to cool the gelatin after incubation at 37°C to determine if proteolysis has taken place. If the gelatin stabs are incubated at room temperature, long periods of time are required (3 to 5 days). The characteristic liquefaction around the stab inoculum, described and illustrated in great detail in the older text books, is no longer required to aid in bacterial identification.

The addition of gelatin to an agar medium enabled plates of culture medium to be incubated at 37°C. Detection of gelatinase action around the colonies is obtained by flooding the surface of the medium with a protein-denaturing solution such as sulfosalicylic acid.[3] Formalized gelatin does not melt at 37°C, but is still susceptible to gelatinase. Kohn[118] and Lautrop[119] described the addition of finely divided charcoal to formalized gelatin as an indicator of proteolysis. The release of charcoal from the gelatin disc after incubation on an agar slope or in broth demonstrated that gelatinase had been produced. Other microtests, using photographic film (either black and white or color), have been described by Hartman.[120]

Tolle et al.[121] used gelatin in the cultivation of organisms in milk. The microcolonies formed in nutrient gelatin are fixed with acid-formaldehyde, which liquefies the gelatin at the same time. The molten gelatin containing a suspension of fixed colonies is then passed through a Coulter counter. Rapid and very accurate counts are obtainable by this technique. However, it is essential that the gelatin used is of the highest clarity to prevent background "noise" interfering with the counter.

D. Carrageenan and Furcellaran

Carrageenan is a polysaccharide complex extracted from *Chondrus* species (Irish moss). The hydrocolloid, extracted with boiling water, is composed of two components: lambda- and kappa-carrageenan.[122] Although both fractions are high-molecular-weight, strongly negatively charged polymers, only kappa-carrageenans will gel, and then only in the presence of potassium ions.

There has been interest in the use of carrageenans as gelling agents for bacteriological purposes. Marine Colloids Inc. of the U.S. manufacture a range of extractives with gel strengths rising to 900 g/cm^2. The high mineral content (15% w/w) has been a deterrent, as the criterion for gelling agents in culture media is that they should have no influence on the medium apart from providing a support gel.

The same comments can be made about furcellaran, sometimes called Danish agar. It is a high-gel-strength colloid extracted from *Furcellaria fastigiata*. It also requires the addition of potassium salt to give a firm gel. Furcellaran is probably a mixture of carrageen and agar.[123] It is less soluble than carrageenan in cold water and has fewer sulfate groups. Again, the high mineral content means that the gelling agent will have significant effect on the metabolism of organisms grown in its presence.

E. Silica Gel

The forms of silica gel most familiar to the microbiologist are the dried crystals used for dehydration. However, a true gel of silicon dioxide (SiO_2) can be prepared by passing a 10% w/v solution of sodium silicate through an ion-exchange column (Amberlite® 1R120 Zeo-Karb® 215) to yield a clear, colorless solution containing approximately 3% w/v SiO_2. The solution has a pH of 3.0, is stable for 1 to 3 weeks, and can be sterilized by autoclaving at 121°C for 15 min.[124]

A highly purified colloidal silica preparation is marketed as Ludox® (DuPont).[125] It is a 30% aqueous solution of colloidal silica and it may be diluted to 10%, nutrients dissolved, pH adjusted, and the medium autoclaved. Gelation occurs during sterilization; therefore, the medium must be prepared in its final container. It forms a cloudy, translucent gel rather than the clear gels normally associated with silica gel. Kingsbury and Barghoorn[126] modified the Ludox preparation to give clear gels by using a combination of strong-base exchange resin (Amberlite 1R-120) and weak-acid exchange resin (Amberlite 1R-45). Firm, brilliantly clear gels were formed using substituted silica esters in place of sodium silicate,[127] but the procedure is complicated by the necessity to centrifuge to obtain a clear gel.

Pramer[128] reviewed several silicate preparations and showed the effects of varying the silica content, pH, temperature, and the ions used for gelling. A gel containing 1.5% SiO_2 and 30 M NaCl formed the best gel at pH 6.0 to 7.0 and in the shortest time. Skerman[129] described the preparation of silica gel plates for cultivating *Nitrosomas* and *Nitrobacter* species.

Although at first sight the concept of using silica gel as a completely inert gelling agent for bacterial culture media is highly attractive, the problems of actually handling the material are daunting. The various descriptions given by the cited workers involve mixing the nutrients and silica preparation with substantial quantities of acid and buffer before autoclaving the medium in its final container, this process often requiring the use of glass petri dishes; or the sterilization of silica colloid is carried out separately and then mixed with double-strength nutrients, poured into dishes, and a gelling agent (strong NaCl solutions) added. A further alternative process is to prepare dishes containing silica gel and then pour nutrients over the surface and allow them to diffuse into the gel.

Syneresis is a constant problem, with the release of greater volumes of water from the gel than normally occurs with agar. Following syneresis, cracking of the surface invariably occurs on storage and adds to the difficulty of inoculating the medium. Nevertheless, for strict testing of carbon metabolism, silica gel media may be mandatory and further developments of this method may eventually lead to its wider use in bacteriology.

F. Other Gelling Agents

The range of water-soluble polymers that can be used in culture media has not been properly explored. The limitations are that a rigid gel must be formed at relatively low concentrations and such gels should be formed at pH values acceptable to the organisms to be inoculated. They must not be inhibitory to microbial growth or form inhibitory substances during the growth of microorganisms.

Lorian and Gray[130] described the use of polyacrylamide. By adding 0.5% w/v agar to a 1% w/v solution of Separan® NP10 (Dow Chemical Co.), a gel was formed that was inert to bacterial growth, but too soft for inoculation with a platinum loop.

Myrvik et al.[131] described the use of carboxymethylcellulose (CMC) as an agar expander. Using 1% w/v Type 7 CMC (Hercules Powder Co. Inc.) plus 0.6% agar gave a gel equivalent in strength to 1.5% agar alone.

There must be other compounds that are inert, water soluble, and may be used in conjunction with agar, but they will not be acceptable unless the method of preparation is no more troublesome than that of agar alone or, preferably, if they improve the preparation of agar media.

VI. DEFINED INGREDIENTS OF CULTURE MEDIA

Microbiologists tend to think of culture media as progressing upward from ill-defined nutrients to better-defined organic mixtures to known mixtures of amino acids, carbohydrates, nucleic acid fractions, vitamins, and minerals. Rose[132] defined three stages in the isolation of a microorganism from its natural habitat and the definition of its nutritional requirements: (1) growth in enrichment culture, (2) growth in a chemically complex, nondefined medium, and (3) growth in a medium containing only known chemical compounds. However, Lichstein[133] emphasized that the adequacy of nutrition cannot solely be measured by growth rather than function. Equal growth does not imply equal function. For example, synthetic media were shown to give good growth of *Clostridium perfringens* without any α-toxin production.[134,135] Bacteria, on the other hand, vary widely in their demands for essential nutrients. Their demands vary from the photolithotrophic organisms to the chemorganotrophs. Some of the latter group of organisms have yet to be isolated on artificial media, defined or otherwise.

The components of culture media may act as energy substrates, biosynthetic raw materials, selective agents, or indicators of metabolism. There is obviously considerable scope for an exchange of roles between energy substrates and biosynthetic materials, and it would be difficult to be precise about the exact function of any of these ingredients during the various phases of growth and metabolism. In the formulation of defined culture media, particular attention must be paid to the question of whether or not the compound included can be transported into the cell. Payne[136] described the mechanisms of bacterial peptide transport and showed that peptide shape, size, and the presence of specific binding proteins controlled the uptake of specific nutrients. Requirements for particular nutrients may not be constant, but may vary, depending in part on the nature and concentration of other chemical components in the medium.[133]

A. Nitrogen Sources

The element nitrogen must be present in the medium in a form that the organism can use to synthesize amino acids, proteins, nucleotides, and vitamins. The atom exists in various states of oxidation, from 0 (N_2 gas) to +6 (NO_3) or -3 (NH_4^+).

Nitrogen fixation is now clearly shown to be a function that many bacteria share.

Once apparently confined to *Rhizobium* and *Azotobacter* species, it is now seen to be widely distributed among microorganisms. *Clostridia, Achromobacter, Aerobacter, Nocardia,* and *Pseudomonas* species also fix nitrogen. To test the ability of organisms to use nitrogen gas as the sole source of N_2 requires the defined medium to be prepared free from any nitrogen-containing substrate. Alternately, the isotope $^{15}N_2$ may be used in the atmosphere and looked for in the organism.

Nitrates (NO_3^-), although readily assimilated by algae and fungi, are less extensively used by bacteria and yeasts.[132] Ammonium salts (NH_4^+) are the most preferred source of inorganic nitrogen for bacteria. This would be expected, as nitrogen is incorporated into organic compounds in the same form. Davis and Mingioli[137] used $(NH_4)_2SO_4$ at 1 g/l. Tatum and Lederberg[138] used NH_4Cl at 5 g/l and NH_4NO_3 at 1 g/l in their synthetic media to test the role of carbon sources in nutrition. Wolin et al.[139] showed that NH_4Cl could be used as the sole nitrogen source for the growth of *Streptococcus bovis.*

Amino acids are the usual organic nitrogen additives to culture media. Sodium glutamate, urea, biuret, and similar organic nitrogen compounds are more commonly used as energy sources. The oxidative assimilation of glutamate is recorded for very nutritionally demanding organisms such as *Pasteurella talarensis,*[140] *Haemophilus pertussis,*[141] and *Shigella flexneri.*[142] Where preliminary tests suggest that amino acid supplementation is required, the organism may be tested with the individual compounds or against mixtures grouped to include most of the common multiple and alternative requirements occurring in biosynthesis. Meynell and Meynell[124] suggest the following combinations:

> Isoleucine + valine
> Phenylalanine + tyrosine (+ tryptophan + PAB)
> Adenine + thiamine
> Methionine + lysine

at between 10 to 50 $\mu g/ml$ of medium. Aaronson[143] suggests the following groupings:

> I.   Aspartic acid + glutamic acid + alanine + glycine
> II. Arginine + histidine + lysine
> III. Isoleucine + phenylalanine + leucine + tyrosine
> IV. Methionine + threonine + tryptophan
> V.   Proline + serine + valine

Stock solutions (0.1 to 1.0 g/100 ml) may be prepared in distilled water for most of the amino acids. Cystine and tryptophan may require 0.2 to 0.4 N HCl to give complete solution, whereas aspartic acid, tyrosine, and glutamic acid require 0.2 to 0.4 N NaOH. As a general rule, the L isomer is preferred, although the DL form is usually half the price and often suffices. The D form is usually antagonistic except for those organisms that require D-alanine for the synthesis of glycopeptides.

If, when testing with amino acids, it is found that improved growth is obtained with small quantities of undefined material (yeast extract, meat extract, etc.), it is likely that an amino acid imbalance is present, requiring peptides to reverse it. Blocking of the amino-acid pathway, as suggested by Kihara and Snell[144] using *Lactobacillus casei,* could be overcome by using a peptide containing the blocked amino acid. The concept of separate transport systems for amino acids and peptides was discussed by Payne.[136] Demain and Hendlin,[145] using a mutant strain of *Bacillus subtilis,* were able to show that the growth-stimulating factor in yeast extract added to their synthetic glucose/amino acid/nucleotide/vitamin/mineral medium was glycine peptide. It appeared that histidine prevented the uptake of glycine into the cell. Leach and Snell[146] showed that glycine peptides are accumulated at a faster rate than glycine as a free acid.

The investigation of peptides as growth factors in bacteria is a daunting subject, although the fact that there is a limiting size of peptide that can be utilized helps. Payne and Gilvarg,[147] using *E. coli*, showed that the nutritional effectiveness of the oligopeptides ceased abruptly with tetrapeptides. The pentapeptides and higher homologues were all unable to support growth. However, the varying effect of such peptides when the identical acids are rearranged in the chain suggests that the problems are very great.[148]

B. Carbon Sources

The most simple form of carbon assimilable by microorganisms is CO_2. Organisms capable of utilizing CO_2 as the sole source of carbon form the autotrophic group. The heterotrophs require organic carbon to supplement their requirements, although they have the ability to utilize some CO_2. The range of organic compounds that can be used as a source of carbon by microorganisms is very large. A general rule exists that if an organic compound is made by one form of life, another form of life uses it for carbon and energy. Usually the last of the chain of organisms breaking down organic carbon are the ubiquitous bacteria and fungi. Among the more bizarre carbon feed-stock used by bacteria are methane, oil slicks, and industrial effluents. In the normal range of carbon materials that microorganisms use and that are deliberately added to culture media, the following are the most prominent:

Carbohydrates — The following are utilized by microorganisms:

Pentoses	Arabinose, xylose, rhamnose
Hexoses	Glucose (dextrose), fructose (laevulose), mannose, galactose
Disaccharides	Sucrose (saccharose), maltose, lactose, trehalose
Trisaccharides	Raffinose
Polysaccharides	Starch, inulin, dextrin, glycogen
Alcohols	Glycerol, erythritol, adonitol, mannitol, dulcitol, sorbitol
Glucosides	Salicin, esculin

Some microorganisms utilize a very wide range of carbohydrates while others are much more selective or have definite preferences for certain carbohydrates. These latter organisms can be identified by their fermentation or oxidative assimilation of test carbohydrates added to the medium. A large number of media formulations in which carbohydrates and an indicator are included to detect specific enzyme reactions are used in diagnostic bacteriology. A very useful contribution on the structure and metabolism of carbohydrates is that by Berkeley and Hedges.[149]

Organic acids — Acetate, malate, lactate, citrate, formate, succinate. Microorganisms are often impermeable to organic acids; therefore, these compounds cannot be used as the sole source of carbon. However, they are utilized and a useful role in culture media is that the utilization of neutralized organic acids will cause a rise in pH value and counteract any fall caused by the fermentation of other carbohydrates. Marshall and Kelsey[150] used potassium lactate to counteract the fermentation of glucose. It should also be borne in mind that some organic acids are powerful chelates for calcium and magnesium. The effect of adding citrates or tartrates to culture media could cause a deficiency of these alkali earth metals.

Amino acids — Glutamate, glycine, and other amino acids are often added to culture media to supply carbon and nitrogen. Most organisms can readily use amino acids as carbon sources; this is demonstrated by the fact that they grow so well in nutrient broth peptone solutions. (See also Section VI.A.)

Fatty acids — Lipids can be utilized as sources of carbon although their relative insolubility makes them less available. Microorganisms produce powerful lipases which yield glycerol and fatty acids which can be used.

C. Nucleic Acid Fractions

The significance of the role of nucleic acid fractions in culture media was only appreciated by studies of genetic mutants that required specific fractions to be added to media to enable them to grow. Such mutants produced in vitro were of interest in studying the synthesis of nucleic acids. The impact of chemotherapy has been the development of antibacterial compounds that interfere with the replication, transcription, or translation of the nucleic acids. The use of these compounds has shown that microorganisms can be selected that are deficient in their normal synthetic pathways and require pyrimidines from their environment. Much interest is now being directed toward thymine/thymidine-dependent organisms which can only be isolated on media containing these compounds at a sufficiently high level.[151,152]

When investigating nucleic acid requirements, the effect of adding an alkaline hydrolysate of nucleic acid can be measured. After this a mixture of purine-pyrimidine compounds can be tried at 10 to 40 μg/ml concentration. Separate trials of the purines and pyrimidines may then be carried out before the isolated fraction is finally tested. Figure 2 shows the products of nucleic acid hydrolysis that can be added separately to media. When organisms cannot incorporate exogenous purines and pyrimidines into nucleotides, preformed nucleosides or nucleotides must be supplied.[153,154]

D. Vitamins

The investigation of growth factor requirements for microorganisms always includes measurement of the vitamin requirement for optimum growth or metabolism. In undefined media, supplementation with 0.3% w/v yeast extract will usually supply the vitamins required, along with amino acids, peptides, and nucleic acid factors. In defined media the vitamins are added at levels of approximately 1 μg/ml of medium. With the exception of menadione (Vitamin K_3), the vitamins essentially required by microorganisms are in the water-soluble B group. It is usual to start by adding a complete vitamin mixture and then testing the effect of smaller subgroups of vitamins to determine the few key supplementary compounds required by the organisms under test.

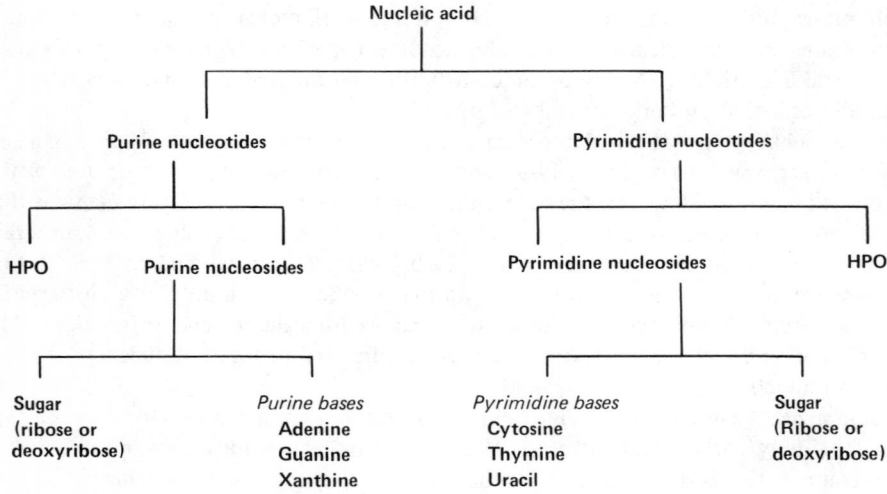

FIGURE 2.   Nucleic acid factions.

The metabolic functions of vitamins in microorganisms as coenzymes or parts of coenzymes are well documented in the standard textbooks of bacteriology. Thiamine and biotin are the most commonly required vitamins, the latter being active in exceedingly small amounts (0.2 ng/ml). Lactic acid bacteria are among the most demanding organisms for vitamin requirements in their growth media and may require five or six compounds before exhibiting growth in defined media that is equal to their growth in complex undefined media.

In the determination of vitamin requirements it is essential to remember that organisms may carry reserves of the essential compounds. Several subcultures to known vitamin-free media may be required before true dependence will be determined.

The practical details of vitamin assay and the preparation of vitamin-free growth supplements have been described by Barton-Wright.[155] The author emphasizes the care required to eliminate vitamin contamination of distilled water and the scrupulous care required in the preparation of glassware. Aaronson[143] commented that laboratory dust, finger grease, or "chemically-pure" organic molecules of biological origin may contain sufficient vitamin to induce growth.

When preparing solutions of vitamins for culture media, Meynell and Meynell[124] offered the following tentative concentrations:

1 μg/ml
   Thiamine, pantothenate, riboflavin, nicotinic acid, choline and pyridoxamine
0.05 μg/ml
   Folic acid and *p*-aminobenzoic acid
0.005 μg/ml
   Biotin

Stock solutions may be prepared in water, except for riboflavin (0.02 N acetic acid) and folic acid (0.01 N NaOH). Such stock solutions are stable for at least 1 month at 4°C. Experience has shown that the stock solutions remain free from infection if the solutions are prepared in sterile distilled water and quickly pasteurized by immersion in boiling water for 3 min.[124]

Vitamins may be added to culture media as filter-sterilized solutions or, less effectively, prior to heat-processing. Table 12, taken from Ten Ham,[156] shows the stability of the vitamins to various deleterious effects.

Table 13 shows the upper limit of vitamin supplementation that the specified organisms require for maximum growth assuming that all other growth requirements have been satisfied.

E. Lipids and Sterols

Lipids are almost insoluble in water, but are soluble in fat solvents. The sterols, as precursors of the fat-soluble vitamins D, E, and K, may play significant roles in bacterial metabolism as well as cell structure. Therefore, they overlap into the previous discussion on vitamins. Lipids may be classified as

1. Fatty acids and triglycerides.
2. Phospholipids (including lecithin).
3. Carotenoids.

The last-named group is of significance in that they are lipid pigments (of orange or red color) produced by many bacterial species. The pigmentation of *Staphylococci* or *Micrococci* can be enhanced by the inclusion of lipid substances.[157]

The great insolubility of lipids in water has hampered investigation into their role in microbial metabolism. It was not until gas-liquid chromatography (GLC) techniques became available that the large numbers of complex related structures could be properly analyzed. Such techniques are now widely used in the analysis of lipid byproducts of metabolizing bacteria.[158]

The most important requirement by bacteria for lipids and sterols is in the synthesis of cell-wall and cell-membrane structures. Lipopolysaccharides produced by microorganisms are used for specific immunodiagnostic tests, and these substances also form essential parts of endotoxic compounds produced by Gram-negative bacteria. The role of lipids and sterols as carbon energy sources does not have the significance that it does for animal nutrition. It is only as a primary supply of essential skeletal structures which the organism can directly utilize in its synthesis of cell structure that these compounds are important in microbial nutrition. The protozoa, mycobacteria, and mycoplasma are three specific groups of organisms that have special demands for lipids and sterols.

Table 12

SOME CHEMOPHYSICAL PROPERTIES OF COMMON VITAMINS
AS RELATED TO THEIR STABILITY[a]

	pH		Ultraviolet light	Oxidation (peroxides)	Reducing substances	Heavy metals	Heat
	<7	>7					
Fat-soluble Vitamins							
Vitamin A esters	--	---	---	---	---	---	0
Vitamin D (D$_2$ and D$_3$)	--	-	---	---	-	---	0
Vitamin E (tocopherol)	0	0	---		-	---	+
Vitamin A acetate	+	-	0	0		---	0
Vitamin K$_3$ (menadione)	--	---	---	-	--	---	0
Vitamin K$_3$ bisulfite	--	---	---	-			
Water-soluble Vitamins							
Vitamin B$_1$ HCl	+	--	---	---	--	---	0
Vitamin B$_2$	+	--	--	0	--	---	0
Vitamin B$_6$ HCl	+	--	--	--	-	---	0
Vitamin B$_{12}$	+	--	--	--	--	---	0
Vitamin C	+	--	--	--	-	---	0
Nicotinic acid	0	0	0	0	0	0	0
Nicotinamide	-	--	0	0	0	0	-
Calcium pantothenate	--	--	0	--	0	0	0
Biotin	--	--	---	--	---	---	0
Folic acid	-	-	---	---	---	---	0
p-Aminobenzoic acid	0	0	0	--	0	---	0
Choline	0	-	0	0	0	+	0

[a] Stability: ++, very good; +, good; 0, reasonable; -, moderate; --, bad.

From Ten Ham, E. J., in *Effects of Sterilization on Components in Nutrient Media*, Van Bragt, J., Mossel, D. A. A., Pierik, R. L. M., and Veldstra, H., Eds., Miscellaneous Papers 9, Landbouwhogeschod, Wageningen, The Netherlands, 1971, 121—123. With permission.

Table 13
THE AMOUNTS OF VITAMINS AND GROWTH FACTORS REQUIRED BY VARIOUS MICROORGANISMS

Vitamin	Examples of dependent organisms	Growth-limiting concentrations[a] (ng/ml)
p-Aminobenzoic acid	Acetobacter suboxydans	0–1.0
	Clostridium acetobutylicum	0–0.2
Folic acid	Lactobacillus casei	0–0.15
	Streptococcus faecalis	0–0.8
Biotin	Lactobacillus arabinosus	0–0.2
	Rhizobium trifolii	0–0.1
	Streptococcus faecalis	0–0.2
Nicotinic acid	Lactobacillus arabinosus	0–40
	Proteus vulgaris	0–20
	Shigella paradysenteriae	0–25
Pantothenic acid	Lactobacillus casei	0–20
	Proteus morganii	0–0.5
Riboflavin	Lactobacillus casei	0–25
	Clostridium tetani	0–100
Thiamine	Lactobacillus fermenti	0–5
	Staphylococcus aureus	0–0.5
Vitamin B_6		
Pyridoxal	Lactobacillus casei	0–0.7
Pyridoxal or pyridoxamine	Streptococcus faecalis	0–0.4
Vitamin B_{12}	Lactobacillus lactis	0–0.026
	Lactobacillus leichmannii	0–0.025
Inositol	Saccharomyces carlsbergensis	0–100
	Schizosaccharomyces pombe	0–1000
Lipoic acid	Streptococcus lactis	0–0.003
	Tetrahymena geleii	0–0.3
Heme	Hemophilus influenzae	0–200
Ferrichrome	Arthrobacter JG-9	0–10
Coprogen	Pilobolus kleinii	0–5
Terregens factor	Arthrobacter terregens	0–20[b]
Vitamin K	Mycobacterium paratuberculosis	0–1000
	Fusiformis nigrescens	0–1000
Choline	Pneumococcus, type III	0–6000
	Neurospora crassa, cholineless	0–2000

[a] The higher concentration is that at which growth approaches maximum.
[b] Nanograms per milliliter of a concentrate.

From Guirard, B. M. and Snell, E. E., in The Bacteria, Vol. 4, Gunsalus, I. C. and Stanier, R. Y., Eds., Academic Press, London, 1962, 40. With permission.

There is a very wide range of compounds which must be considered when studying the nutritional requirements for lipids and sterols from acetate (as a universal precursor for many lipids) to steroids and other cholesterol compounds, incorporating fatty acids, inositol, choline, vitamin K, and complex phospholipids and waxes. Goren[159] reviewed the lipid requirements of mycobacteria. Adler and Shifline[160] and Fallon and Whittlestone[161] reviewed the lipid nutritional requirements for mycoplasma.

Problems in the investigation of fatty acid requirements of bacteria are (1) the often narrow margin between stimulation and inhibition exhibited by substances such as oleic acid and (2) the almost ubiquitous character of fatty acids as contaminants of organic materials such as agar and inorganic materials such as charcoal. The inhibitory effects of fatty acids in bacteriological materials are mentioned by Meynell and Meynell.[124] Agar, cotton, wool, and distilled water may all exhibit inhibitory effects on susceptible organisms. The addition of absorbents such as serum, bovine albumin, or charcoal to the medium may be necessary to absorb the toxic effects of dissociated fatty acids.

F. Metals and Minerals

The inorganic nutrients present in media are the mineral elements required for the growth of all microorganisms. These elements may be macro inorganic nutrients, or major elements that are required in relatively high concentrations (grams per 100 ml or 10^{-3} to 10^{-4} M), and are deliberately added to nutrient media. Micro inorganic nutrients, or minor elements, are required in much lower concentrations (milli- or micrograms per 100 ml or 10^{-6} to 10^{-8} M) and are usually present as contaminants in the medium constituents. Examples of the inorganic elements involved in culture media are as follows:

Macro inorganic nutrients
  (g/100 ml)
        Na   as $NaCl$, Na_2SO_4, Na_3PO_4
        K   as KCl, K_2SO_4, K_3PO_4
        Ca   as $CaCl_2$, $CaCO_3$
        P   as Na or K phosphates
        S   as Na or Mg sulfates
        Mg   as $MgCl_2$, $MgSO_4$, $MgCO_3$
        Cl   as Na, K, Ca, or NH_4 salts
Microinorganic nutrients
  (mg/100 ml)
        Fe   as $FeCl_3$, $Fe(NH_4)_2SO_4$, ferric citrate
        Zn   as SO_4 or Cl salts
        Mn   as SO_4 or Cl salts
        Br   as Na, K, Ca, or NH_4 salts
        B   as H_3BO_3
Micro inorganic nutrients
  (μg/100 ml)
        Cu   as SO_4 or Cl salts
        Co   as SO_4 or Cl salts (or Vitamin B_{12})
        Mo   as Na or NH_4 molybdate salts
        V   as Na_3VO_4
        Sr   as SO_4 or Cl salts
        Al   as SO_4 or Cl salts
        Rb   as SO_4 or Cl salts
        Li   as SO_4 or Cl salts
        I   as Na, K, Ca, or NH_4 salts

Defined media are usually formulated to contain Na^+, K^+, Mg^{2+}, Mn^{2+}, $Fe^{2+\ or\ 3+}$, PO_4^{3-}, SO_4^{2}, and Cl ions. Unless specific trace elements are required, it is normally assumed that natural contamination of these compounds will supply the micronutrients. Complex media often have high inorganic levels provided by the digests and extracts used in the formulation. Kempner[164] described the trace-metal content of nutrient broth. Bovallius and Zacharias[165] analyzed the Mg^{2+}, Fe^{2+}, Ca^{2+}, Zn^{2+}, Cu^{2+}, and Mn^{2+} contents of several commercially available media. The concentrations of the cation content were shown to vary among manufacturers and batches of the same medium. The authors were able to show that the rate-limiting factor in their complex media was magnesium.

Marshall and Kelsey[150] found that the addition of Mg, Fe, or Mn improved the growth of several organisms on their complex medium. The following inorganic salts solution was added at 5 ml per liter of the final medium:

$MgSO_4 \cdot 7H_2O$	4.0 g
$MnSO_4 \cdot 4H_2O$	0.4 g
$FeSO_4 \cdot 7H_2O$	0.4 g
Distilled water acidified with	100 ml
2 drops of 10 N H_2SO_4	

The concentration of inorganic elements (in milligrams per liter) in the final medium was quoted as follows:

Ca	6.0
Cl	340.0
Cu	0.06
Fe	5.7
Mg	30.0
Mn	5.0
P	1180.0
K	2040.0
Na	1670.0
S	46.0

Shankar and Bard[167] showed that metal-deficient basal medium failed to grow *Clostridium perfringens*. When the ashed constituents of tryptone and yeast extract were added to the basal medium, almost optimal growth was obtained. Although some 30 metallic salts were tested as replacements for the ash, only Mg^{2+}, Fe^{2+}, and Ca^{2+} played a significant role in stimulating growth. Final concentrations of 7.5 $\mu g/ml$ Mg^{2+}, 2.5 $\mu g/ml$ Ca^{2+}, and 2 $\mu g/ml$ Fe^{2+} were required in the basal metal-deficient medium.

Mineral elements function in microbial metabolism primarily as activators of various enzymes. Many of these functions are specific for certain metals, although some nonspecific effects have been observed. Thus, isocitrate lyase of *Pseudomonas aeruginosa* is activated by Mg^{2+}, Mn^{2+}, Fe^{2+}, or Co^{2+}.[168] The role of Fe in microbial metabolism has been especially studied. Gram-negative bacteria require 0.3 to 1.8 μM and Gram-positive bacteria 0.4 to 4 μM iron. Defined media contain 0.5 to 3.0 μM iron as a contaminant of sugar and phosphate salts. Complex culture media contain 3.0 to 12.0 μM iron.[169] Toxin production by *Clostridia* and *Corynebacteria* is stimulated or inhibited by Fe^{2+} concentrations between 3.0 to 30 μM.[170] Macham[171] emphasized the problems of insoluble metals such as iron where Fe^{3+} compounds have a solubility of 10^{-18} M at pH 7.0. Therefore, in aerobic environments of neutral pH, inorganic iron is effectively insoluble. Mild chelating salts such as citrate are often used to hold iron compounds in solution at these pH values. The role of metal ions in the sporulation of bacteria is discussed by Murrell.[172]

The importance of metals and minerals in media designed for antibiotic sensitivity testing has been well documented.[173-190] Ca^{2+} and Mg^{2+} ions play a major role; NaCl also has influence on the bacteriostatic properties of antibiotics.

The production of anthrocyanin pigments by organisms such as *Pseudomonas* is also controlled by metal content of the medium, with Mg^{2+} having a key role in the production of typical green pigmentation.

The incompatibility of certain metals with phosphate salts present in the medium can prove troublesome. Autoclaving media at neutral pH values often produces insoluble phosphate "flocs" which are Mg^{2+} and Ca^{2+} polyphosphate complexes. Iron will form similar complexes or will precipitate with alkaline earth/polyphosphate complexes.[191] Phosphate-salt precipitation was described as a method of treating culture media to remove metals.[192]

It is clear that critical studies on metal dependence can only be carried out on metal-depleted media, as many of the constituents of culture media contain contaminating metals. The methods used for the removal of metallic ions from bacteriological media may be divided into four categories:

1. Use of chelating agents (see Section VI.H).
2. Removal by precipitation.[192,193]
3. Previous microbial growth in the medium.[194]
4. Use of ion-exchange substances.[195-197]

In spite of every care taken in such procedures, it is extremely difficult to produce culture media that will demonstrate dependence on trace elements required in minute quantities.

G. Buffers

Buffer salts are added to culture media to initially poise the pH value at a desired level and, hopefully, to maintain the pH level during subsequent sterilization and incubation. Changes of pH inevitably occur during microbial growth. The presence of peptones causes a rise in pH because of deamination of amino acids. A fall in pH value occurs in the presence of fermentable sugars. The inclusion of buffering substances in the medium limits the change in pH. Older culture media formulations contained chalk as finely precipitated calcium carbonate to control pH changes. Clear zones produced around the colonies in such opaque media indicated the production of acid by the organism.

The metabolism by many species of lactate, formates, or succinates to release Na^+ is also used to help neutralize acid production. Marshall and Kelsey[150] incorporated potassium lactate (10 ml of a 50% w/w solution per 1 liter) in their medium to help counteract the effect of the fermentation of 2 g of glucose per liter of medium.

Most buffers are effective over a range of 2 pH units, which is approximately the pK value ±1. At the pK value the buffer salt is 50% dissociated and is at its most effective ionic state. The choice of a suitable buffer is limited by the following criteria given by Munro:[198]

1. The pK should be close to the desired pH value.
2. The buffer should be very soluble in water and nonvolatile.
3. The medium concentration, temperature, and ionic composition should have a minimum effect on the dissociation of the buffer.
4. The buffer should not complex with cations. If so, these should be soluble complexes and the binding constants of such complexes known.
5. The buffer should not produce unwanted salt effects or exhibit toxicity.
6. The buffer should be nonmetabolizable, resist enzymatic degradation, and should not act as an analogue inhibitor.

Phosphate is the most commonly used buffering substance, with a pK_1 value of 6.8. The fact that it interacts with cations and is metabolizable has not prevented its wide use, although other buffering compounds are now used. Hutner et al.[199] described the use of the following compounds:

Tris(hydroxymethyl)aminomethane
Quadrol(N,N,N^1,N^1-tetrakis(2-hydroxypropyl)ethylenediamine)
Pyromellitic acid (Benzene-1,2,4,5-tetracarboxylic acid)
Succinic acid

Tris buffer (pK = 8.2), although commonly used, tends to antagonize potassium competitively, and provision of ample K^+ should be made in the medium. It also exhibits a significant temperature variation. Thus, a Tris buffer adjusted to pH 7.8 at room temperature will have a pH of 8.4 at 4°C and a pH of 7.4 at 37°C.[198] Lewis[200] reported the use of other Tris compounds, "mono-tris" and "bis-tris," with pK values of 7.8 and 6.5 respectively. These compounds showed advantages over conventional Tris buffer.

Good et al.[201] examined a range of zwitterion buffers which are water soluble, have low binding capacities for divalent cations, and have pK values of from 6 to 8. However, they are susceptible to bacterial degradation. The following two compounds were found to be superior to the other tested buffers:

N-Tris(hydroxymethyl)methyl-2-aminoethane sulfonic acid (TES), pK = 7.14.
N^1,2-Hydroxymethylpiperazine-N^1-ethane sulfonic acid (HEPES), pK = 7.31.

Williamson and Cox[202] found HEPES buffer to be better than TES buffer in overall performance.

Mallette[203] examined a range of carboxylic acid compounds available as buffers for bacteriological systems. The pK values of 23 compounds were listed, with some compounds having as many as 6 pK values. The most suitable compound for studying bacterial metabolism was 3,6-endo-methylene-1,2,3,6-tetrahydrophthalic acid (EMTA). Table 14 lists buffer compounds other than phosphate salts that can be used in culture media, with recommended concentrations and pH ranges over which the buffer acts.

In an effort to overcome the disadvantage of phosphate buffers sequestering alkaline earth metals and forming insoluble complexes, glycerophosphate is widely used. Douglas[204] reviewed the use of glycerophosphates in microbiological media. The relatively high solubilities of the glycerophosphates of the heavy metals as compared to those of their phosphate salts enables high concentrations of metals to be added without precipitation. The pK value of glycerophosphates (around 6.8) makes them very useful for most media. The author stressed that α-glycerophosphate compounds can be metabolized by microorganisms and, although providing a source of phosphate and glycerol to the organisms, will also cause precipitation to occur during the terminal growth phase. The β-glycerophosphate isomer cannot be metabolized and is, therefore, preferable as a buffering agent.

Many tables of buffer compounds are available in the published literature and no attempt will be made to reproduce them here. Bates[205] has prepared tables of buffer and indicator dyes used for pH measurement.

H. Chelates

Metal solutions which tend to precipitate may be held in solution by soluble sequestering agents or chelates. The degree of binding (or avidity) of such substances varies and tends to alter the properties and effects of metals in the culture medium. There are many natural sequestrants, including the macromolecular porphyrins, peptides, and proteins, as well as amino acids, carboxylic acids, and polyphosphates.

Table 14
BUFFERS

Compound	Concentration	Best pH range
Tris(hydroxymethyl)-aminomethane	0.01–0.1%	7.5–8.5
Triethanolamine	0.03%	7.0–8.0
L-Histidine	0.03%	6.0–7.0
Glycylglycine	0.08%	7.0–8.2
Succinic acid	0.03%	6.0–7.0
Malic acid	0.03%	6.0–7.0
N,N-Bis(2-hydroxyethyl)-glycine	0.1 M	8.0–9.0
N-Tris(hydroxymethyl)-glycine	0.1 M	7.5–8.5
N-2-Hydroxyethyl-piperazine-N^1-2-ethane sulfonic acid	0.1 M	7.0–8.0
N-Tris(hydroxymethyl)-methyl-2-amino-ethanesulfonic acid	0.1 M	7.0–8.0
2-(N-Morpholine)ethane sulfonic acid	0.1 M	6.0–7.0

From Aaronson, S., *Experimental Microbial Ecology*, Academic Press, London, 1970, 62. With permission.

Ethylenediamine tetracetic acid (EDTA) is probably the best known and most widely used synthetic chelate. Nitrilo-triacetic acid (NTA) is also used as a less powerful and less toxic chelate in culture media. Further examples of these synthetic compounds are ethylenediamine (*o*-hydroxyphenyl)acetic acid; 1,2-diaminocyclohexane tetraacetic acid; diethylenetriamine pentaacetic acid; ethylene glycol bis-(aminoethyl ether)tetraacetic acid; and dihydroxyethylglycine.

Citric acid is the most commonly used chelate, either in the form of metal citrate salts or addition of sodium citrate to solutions containing metal salts. Its avidity is such that it will prevent the formation of metal-phosphate complexes when phosphate buffers are present in metal-containing media. A specific example of the chelate effect of citric acid is its inclusion in deoxycholate media. Media containing this bile fraction are more toxic to commensal flora of the intestine when sodium citrate is added to the formulation. This increased toxicity can be reduced by adding Mg^{2+} to the medium; after saturating the citrate it will alow commensal flora to grow. Fe^{3+} salts in particular are commonly added to culture media in the citrate form to ensure that they stay in solution.

Other organic acids such as succinic, tartaric, malic, or acetic are also chelates to Ca^{2+}, Mg^{2+}, and Fe^{3+}. Cystine, histidine, and glycine are examples of amino acids that will act as cation chelating agents in culture media.[124]

Sodium hexametaphosphate and sodium glycerophosphate are examples of phosphates that will form soluble complexes with metals. Douglas[204] quoted the following solubilities of glycerophosphate salts (in parts per 100 ml of water):

Ca^{2+}	2
Fe^{3+}	50
Mg^{2+}	2
Mn^{2+}	1

Chelates may be deliberately added to culture media to form metal-depleted media.[192] Specific metals can then be added to measure the effect on the growth of microorganisms. Albert[206] reviewed the metal-binding agents available for biological work. He pointed out that many of them are toxic to cells solely because their metal-binding capacity was greater than the chelating proteins on the surfaces of cells. The stability constants of such chelates can only be determined within strictly defined pH systems, and it is difficult to draw general conclusions to predict the performance of such chelates in other complex environments. Neither is it possible to predict the effect of sequestrants on individual microorganisms without carefully testing each strain.

I. Antibiotics and Dyes

1. Antibiotics

Selective agents such as antibiotics, dyes, and other inhibitory chemicals have been incorporated into culture media since the time of Ehrlich (dyes) and Fleming (antibiotics). In his original paper on the discovery of penicillin in 1929, Fleming[207] advocated the use of penicillin as a selective substance to aid in the isolation of *Haemophilus influenzae*. Since that time antibiotics have been used successfully for the selective cultivation of microorganisms in many media.

As new antibiotics have appeared they have been incorporated into culture media. The criteria for their use has been stated as follows by Goldberg:[208]

1. Stability.
2. Solubility.
3. A highly specific antimicrobial spectrum.
4. Freedom from toxicity for the organisms being selected.

Stability in the medium is a very important factor, especially as the antibiotic may have to be added to molten agar held at 50°C before pouring plates or slants. Some antibiotics, such as the aminoglycosides (neomycin, kanamycin, and streptomycin), chloramphenicol, and cycloheximide, are sufficiently stable to withstand autoclaving. Such antibiotics may be added to the medium in the preparation stage and then sterilized by autoclaving. Most antibiotics, especially the β-lactam group (penicillins — natural and synthetic and cephalosporins), must be added to the molten medium after it has been sterilized.

Vials of antibiotic prepared for intramuscular or intravenous injection are most suitable for aseptic additions to culture media. Sufficient volume of sterile distilled water is added aseptically into the vial to produce a concentration per milliliter suitable for the medium. Antibiotic concentrations are always quoted in micrograms or units per milliliter of the final medium; therefore, the mathematics involved are very simple. The true potency of the antibiotic may have to be taken into account, especially if it is below 80% of the theoretical level. Sometimes antibiotic levels are quoted in amounts of the base molecule. Thus, streptomycin may be quoted in micrograms per milliliter of base, and the use of a dihydrostreptomycin salt (sulfate) means that the amount must be correspondingly increased. Table 15 lists the activities of accepted international preparations that are held in the various national antibiotic reference laboratories. Antibiotics weighed out from a vial may be dissolved in buffer solution (see the *U.S. Pharmacopoeia* for appropriate buffer solvents) and filtered through a membrane filter. A Swinney filter attached to a syringe is usually appropriate for the small volumes involved.

The stability of antibiotics in culture media varies with the constituents of the medium, the pH, the temperature and duration of storage, and the length of incubation at 37°C or similar temperatures. Unless the pH of the medium is wholly unsuitable for the antibiotic, it is common to find that the stability is greater in the medium than in the aqueous solution used in the preparation of the antibiotic. This is especially evident in the penicillin group. Prolonged incubation eventually leads to total destruction of the

Table 15
ACTIVITIES OF INTERNATIONAL
PREPARATIONS

	IU/mg	$\equiv \mu g$ of
Amphotericin B	960	
Bacitracin	55	
Chlortetracycline (HCl)	1000	HCl
Dihydrostreptomycin (SO_4)	760	Base
Erythromycin (base)	950	Base
Kanamycin (SO_4)	812	Base
Neomycin (SO_4)	680	Base
Novobiocin (Na salt)	835	Free acid
Oleandomycin	845	Base
Oxytetracycline (base)	900	Anhydrous base
Penicillin G (Na salt)	1670	
Polymyxin B	7874	
Streptomycin (SO_4)	780	Base
Vancomycin	1007	
Viomycin (SO_4)	730	Base

Adapted from Kavanagh, F., in *Analytical Microbiology*, Kavanagh, F., Ed., Academic Press, New York, 1963, 256–259. With permission.

antibiotic; but, providing it is active in early incubation, this does not significantly alter the incidence of contaminated cultures.

It would seem to be a natural criterion that an antibiotic must be water soluble to be active in culture media. Some antibiotics are almost insoluble at the concentrations required for some formulations (e.g., nystatin, nitrofurandantin, and chloramphenicol). The use of suitable organic solvents to help disperse them into the broth or agar media seems to overcome this problem. Examples of such solvents and antibiotic solubilities are given in an extensive table by Kavanagh.[209]

The antimicrobial spectrum of the antibiotics used is most important. Ideally, such substances should totally inhibit unwanted organisms and allow the multiplication of not even one cell of the selected organisms. In a nonideal situation, compromises must be sought. For example, in the selection of Gram-positive organisms, an antibiotic is chosen that inhibits the majority of Gram-negatives. Unfortunately, strains of bacteria do not react uniformly to antibiotics, and care must be taken not to assume that recovery of the selected organism will always take place in the presence of the particular antibiotic chosen.

Finegold et al.[210] used the following antibiotics in blood agar to isolate anaerobic organisms of human origin:

Neomycin (100 μg/ml) for the isolation of anaerobic cocci (other than *Veillonella*) and *Clostridia*.
Kanamycin (75 μg/ml) for *Bifidobacterium*.
Kanamycin (75 μg/ml) plus Vancomycin (7.5 μg/ml) for *Bacteriodes* spp.
Neomycin (100 μg/ml) plus Vancomycin (7.5 μg/ml) for *Sphaerophorus* spp., *Fusobacterium fusiforme*, and *Veillonella* spp.
Rifampin (50 μg/ml) for *Sphaerophorus mortiferus* and *S. varius*.

Selective media permit earlier recovery and identification of organisms, especially when isolated from complex mixtures of organisms found in clinical specimens or soil. It is possible to construct media so selective that the presence of bacterial growth of the right morphological features provides reliable tentative identification. This situation occurred with the Thayer-Martin medium used to isolate *Neisseria gonorrhoeae*.[211] This very fastidious organism required a highly nutritive medium plus incubation for 48 hr in a moist CO_2 atmosphere. Commensal organisms grow very freely in such circumstances, and the dishes of heated blood agar can be grossly overgrown with unwanted organisms. The addition of antibiotics (vancomycin, 3 μg/ml; colistin, 7.5 μg/ml; and nystatin, 12.5 μg/ml) cleared the dish of most Gram-negative and all Gram-positive organisms plus yeasts. The occasional overgrowth by swarming *Proteus* spp. was overcome by adding trimethoprim (5 μg/ml) to the antibiotic cocktail.[212-214]

Another example of isolating difficult organisms from heavily contaminated environments is the isolation of *Listeria monocytogenes* from soil, manure, slaughterhouse waste, etc. The use of nalidixic acid blood agar considerably improved the isolation rate.[215]

A computer search of the *Index Medicus* from January 1970 to January 1973 revealed 43 published papers containing specific reference to the use of antibiotic-containing media for the isolation of microorganisms (Medlars Pilot Search No. 11103). Examples of the use of modified antibiotic-containing media can be found in the literature for most of the genera of bacteria that must be isolated from heavily contaminated flora.

2. Dyes

Natural and synthetic dyes were used in histology in the late 19th century and were observed to inhibit bacterial growth.[216] Churchman[217] extensively studied the selective action of gentian violet; this was followed by the work of Krumwiede and Pratt.[218] However, long before this, Endo[219] had incorporated basic fuchsin into a medium that had been decolorized with sodium sulfite. This medium was used for the isolation of coliform organisms and has been extensively modified since.

Hartman et al.[220] reviewed the use of crystal violet (gentian violet) incorporated into culture media for the isolation of *Streptococci*. They also listed the formulations of 26 media containing dyes for the isolation of *Enterococci*. Bryan[221] and Kline[222] tested brilliant green and found it similar in activity to crystal violet, confirming Churchman's[217] original generalization that basic dyes are more effective than acid dyes. Quastel and Wheatley[223] were also able to show, by the Thunberg technique, that basic dyes were more toxic than acid dyes when studying multienzyme systems.

The number of dyes tested was extended by Petroff and Gump,[224] who tested 130 dyes against 7 bacterial genera for bacteriostatic and bacteriocidal effects. They concluded that they could not be used for chemotherapeutic purposes, and it remained for other workers to test the effects of such dyes in culture media.

A problem in the study of dyes for bacteriological purposes is that the dyestuff industry produces compounds that vary considerably in active constituents. Conn[225] and Gurr[226] have given specifications for each dye and included an absorbance graph (between 325 and 700 nm) which measures purity of the dye to some extent. Therefore, it is essential that the dye used meet a specification of purity, as dyes are made for coloring fabrics and it would not be surprising if some batches could not meet strict specifications. Fung and Miller[227] screened 30 species of bacteria against 42 different dyes in different concentrations in order to test their inhibitory and differential properties. Using 19 acidic, 20 basic, and 3 neutral dyes, they showed that Gram-negative organisms have greater resistance to dyes than Gram-positive organisms, that basic dyes were more inhibitory than acidic or neutral dyes at the same concentration, and that only brilliant green, crystal violet and methyl violet showed good activity at the greatest dilution used (10^{-5}).

One of the most important dye sensitivity tests is that used to distinguish species of *Brucellae*. Originally, the dyes thionin, basic fuchsin, methyl violet, and pyronine were used at two concentrations.[228] However, only thionin and basic fuchsin are currently used, at three and two concentrations respectively.[229]

A comprehensive description of pH and eH indicator dyes, their pH ranges, color changes, and preparation is given by Bates.[205]

VII. FORMULATION, DESIGN, INTERACTION, AND INCOMPATIBILITIES

The very large number of culture media described in the published literature indicates that optimum growth of separated strains of microorganisms requires special environments. These special environments include specific basic nutrients that may be required and specialized enrichment or selective substances that must be added to complete the formulation. There are few rules that control the construction of successful media. The following parts of the media must be supplied, although much overlapping occurs:

> Protein parts (amino acids, peptides)
> Gene parts (nucleic acid fractions)
> Minerals (metals and inorganic molecules)
> Vitamins
> Energy source (photosynthetic, chemolithotropic, or chemorganotrophic)
> Selective agents (inhibitory chemicals)
> Enrichment factors (blood, serum, etc.)

Microorganisms vary enormously in their synthetic ability and range from those organisms that can build all their requirements from a source of light and very simple chemicals, to those which require many of the cell parts preformed because their parasitic evolution seems to have left them poor synthetic abilities.

Many culture media formulations are built on preceding formulas and often contain one-step changes that increase the isolation success or provide greater biomass. Culture media that are developed to demonstrate biochemical reactions of specific enzymes that the organism may possess by including substrates and indicators of fermentation, hydrolysis, or utilization are especially numerous. Reference should be made to the appropriate review to obtain further information (see References 230 to 232).

The design of culture media would require statistical evaluation of each ingredient and measurement of the interaction of each ingredient on the other. This procedure could only be applied if an objective evaluation were being made (e.g., recovery counts in which measurement was made of the number of colonies formed from a known inoculum of cells). The classical design study was that reported in 1954[233] and 1963[234] and completed 1972,[235] in which a medium was designed that would yield standard counts of organisms from milk. However, many of the criteria looked for in bacterial colonies, are highly subjective (e.g., pigment, appearance, hemolytic properties, smell). Even colony size is a highly variable characteristic unless care is taken to space the colonies equidistant from one another. Fermentation media, in which measurable biomass or by-product are produced, are more often designed to determine maximum economy of nitrogen or carbon energy in the ultimate yields.[236]

The first rule in the design of a culture medium for a specific organism is to survey the literature and determine what has been used before for the same or similar organisms. The most difficult situation is the isolation of organisms from soil, mud, or other environments in which a teeming ecology of organisms all lend support to each other. Unless there is clear microscopic evidence, it is very difficult to know if all the organisms present in the original sample are, in fact, isolated. The addition of such material in increasing dilutions to a relatively inert gelling agent, followed by incubation under varying conditions is often the first useful step that can be taken.

The physiological conditions of growth, light, presence of oxygen, temperature, and pH may affect nutrition. Therefore, it is important that these factors be closely controlled during the subsequent stages when organic nutrients or defined chemicals are substituted for the natural substrate. Aaronson[143] describes the stages of refinement of the culture media ingredients in the progress toward totally defined ingredients. Useful tables of organic mixtures are given which may later be tested by reducing the size of the mixture in the attempt to identify the minimal nutrients required to give optimal growth.

The interaction of culture media ingredients is to be expected, especially if the preparation stage involves heating to a high temperature. The great majority of these interactions are unnoticed and go unrecorded. Interactions that result in changes in physical appearance of the medium (e.g., color, clarity, smell, or alteration of the pH value) are often associated with inhibition of normal growth and must be avoided. Gross precipitation of phosphates or metal complexes are clear indications that the formulation is incorrect. The addition of chelating agents or pH changes may be called for. Color changes may be caused by Maillard reactions (amino acid/glucose interaction) or glucose/phosphate "caramelization." If the ingredients cannot be altered, lower temperatures must be used in the preparation of the medium or the pH must be moved to the acid side of neutrality. Primary Maillard complexes and their decomposition products are not without effect on microorganisms.[237] Hydroxymethylfurfural compounds formed during heat-induced caramelization of sugars have pronounced antimicrobial properties.[238,239]

Dyes added to culture media may function as selective agents or as pH/eH indicators. When used as selective agents, interaction may occur between the dye and the other components in the formulation. Thus, the same sample of brilliant green may show different inhibition titers against *E. coli* when tested in different peptone solutions. Bile salts interact with brilliant green and reduce the toxicity of the dye. This can be shown by adding toxic levels of brilliant green to MacConkey bile-salt medium where profuse growth of *E. coli* will occur. Similar differences will be found in brilliant green media, depending on whether the dye is added before sterilization or after sterilization, the latter process giving higher titers of activity. It is known that the aniline dyes are more toxic when fully oxidized. Therefore, it is possible that reduction of the dye takes place when it is sterilized in the presence of peptone or broth.

Indicator dyes show similar interactions. In particular, bromcresol purple reacts with peptone on storage to form a brownish-red compound that no longer reacts to changes in pH. Other indicator dyes such as phenol red and methylene may undergo changes as a result of microbial metabolism, which prevents their functioning as pH or eH indicators.

If the agar used for media preparation is high in minerals, incompatibilities with the broth constituents commonly occur. Precipitation may occur on sterilization or when the molten agar is held at 50°C. In the latter circumstance, it is probable that partially chelated metals are gradually released from the agar to react with phosphates in the broth. This problem can only be overcome by using higher quality agar (i.e., lower in mineral content) or adding chelating agents.

Adventitious agents may be added in the water used to prepare the media. The quality of distilled water varies with the raw water used and the type of apparatus. Demineralized water may be a potent source of problems if the ion-exchange columns are not properly prepared and the resistance of the water constantly checked.

The glassware used may also donate molecules or elements that are not desirable. Highly alkaline washing solutions may remain on the surface. Untreated "soda" glass is alkaline and will affect the pH of solutions stored. Toxic chemicals such as bile, tellurite, selenite, etc., may remain on the surface of glass vessels after inadequate processing. Later they will leach out into aqueous solutions stored in these same containers.

In summary, although there are many causes for unwanted interaction and incompatibilities in culture media, most of the causes are identifiable. It is essential that careful attention be paid to the method of media preparation. Accurate instructions should be established on the mode of dissolving dry ingredients, with special attention to the order of preparation. Water and glassware must be of a high standard. Particular attention should be paid to the method by which labile components are dissolved, sterilized, and added. A high level of quality in media preparation will overcome many of the problems commonly faced in the microbiology laboratory.

VIII. STERILIZATION AND STORAGE OF CULTURE MEDIA

Sterility is an absolute concept — a medium is either sterile or it is not. In practice, however, the detection of viable microorganisms depends on the testing of samples or aliquots of the medium at various temperatures and periods of time. Therefore, the probability of sterility is measured as a statistical concept. The process of sterilization is stated as the probability of a certain procedure yielding a sterile product. If the 12D concept for *Clostridium botulinum* is followed, the sterilization process must reduce 10^{12} spores to <1 spore. This gives a considerable margin of safety because it follows that if 100 ml bottles contain 10^3 spores per milliliter (i.e., 10^5 spores per bottle), there is approximately only 1 bottle likely to contain viable spores out of every 10^7 bottles processed.

Sterilization of culture media is usually carried out by two processes: (1) heat treatment (usually steam under pressure) and (2) cold treatment (filtration, irradiation, or chemical additives).

A. Heat Treatment

With the exception only of special heat-labile material, culture media should be sterilized by steam under pressure. This has proven to be the most reliable and universally applicable method of sterilization. Temperature of the steam and time of exposure are interconnected, with a higher temperature reading requiring a shorter time.

It has been calculated[240] that the temperature coefficient for each 10°C increase (Q_{10} spores) lies between 8 and 10 in the range 100 to 135°C. Theil et al.[241] calculated the equivalent sterilization times as follows:

Temperature (°C)	100	110	115	121	125	130
Time	20 hr	2.5 hr	51 min	15 min	6.4 min	2.4 min

The Medical Research Council[242] recommended the following exposure times:

Temperature (°C)	126	126	134
Time (min)	15	10	3

It is important to remember that these heat-process times relate to the containers of media reaching the prescribed temperature; it is not the time interval after the autoclave chamber reaches the temperature.

The sterilization cycle can be divided into four stages.

1. Autoclave chamber heat-up time.
2. Fluid heat-penetration time for the container.
3. Exposure time at the prescribed temperature.
4. Cooling-down time.

The autoclave chamber heat-up and cool-down times may be long and may unnecessarily add to the total heat input into the medium. In such cases it may help if this heat input is calculated into the sterilization time, thus shortening the exposure to the peak temperature. The use of water sprays inside the chamber helps to rapidly reduce the temperature; but, unless the containers are well protected, the sprayed water can enter them.

The fluid heat-penetration time is usually premeasured using thermocouples placed in the center of the containers. The following time intervals for media in glass bottles to reach 121°C in an autoclave chamber at that temperature have been recorded:

Size bottle (ml)	Time (min)
500	18
1000	22
2000	27
5000	37

These times will vary according to the size of the load in the autoclave. Sykes[243] stated that a single 100-ml bottle requiring 12 min to reach 121°C required 19 min when placed in a crate with other 100-ml bottles and 30 min when placed in stacked crates.

Agar-containing media should be presteamed and mixed to completely dissolve the agar before sterilization. Elliott and Georgala[244] described the equipment required in a small laboratory for the preparation of agar media.

It should be borne in mind that the temperature conditions inside an autoclave are not homogeneous. Therefore, two bottles of medium in the same autoclave load do not receive the same heat treatment. Autoclaves may be tested for this variation by filling bottles with a 2% w/v glucose solution in 2% w/v Na_2HPO_4 solution. The effect of heat treatment is to produce "browning" reactions in the buffered glucose solution. The degree of "browning" varies with the heat treatment, and bottles stacked near the steam inlet will be browner than bottles stacked over the bottom outlet (which is the coolest part of the autoclave). Similarly, variation in color will be seen between bottles stacked against the door or in the center of the bottles and those stacked against the heated or insulated walls.

Berhagel[245] described the effect of sterilization on pH. He considered the following reactions that could occur in heat sterilization of culture media and stated that all were pH dependent and resulted in a change of pH value:

1. Precipitation or dissolution of salts.
2. Neutralization or esterification of acids.
3. Hydrolysis of proteins, polysaccharides, and esters.
4. Reactions between carboxyl and amino groups.
5. Polymerization or depolymerization of alcohols, aldehydes, and saccharides.

Berhagel concluded that the pH values of media should be measured before and after sterilization. Not only does this measure the amount of pH drift occurring on heating, but it may also disclose an incorrect formulation, as in these circumstances the drift may be abnormally large.

B. Cold Treatment

Heat-labile constituents of culture media cannot be sterilized by heat; therefore, process of cold sterilization is normally carried out. The most common way of carrying this out is by membrane filtration using cellulose acetate/nitrate discs. Mulvany[246] and Grubert[247] gave complete details of the process of membrane filtration. Under certain circumstances chemical sterilization may be carried out. Needless to say, the sterilants used are not the normal biocidal agents, but chemicals with very reactive side groups that have short half-lives and hydrolyze to relatively inert substances. Two examples of useful substances are β-propiolactone (BPL) and pyrocarbonic acid diethylester (PKE) (Bayer).

β-Propiolactone — BPL is a corrosive, colorless liquid that must not be allowed to touch the skin. Its specific gravity is 1.146, vapor pressure 10 mm Hg at 51°C, molecular weight 51, boiling point 155°C, and freezing point -33.4°C. It is normally stored at 4°C to prevent formation of harmful vapor which is inflammable in air at concentrations greater than 8% v/v and intensely lachrymatory.[248] It is soluble in water to 3% v/v where the lactone ring of this alkylating agent opens on hydrolysis and β-substituted propionic acids are formed.

$$CH_2-CH_2-C=O + HOH \rightarrow HO-CH_2-CH_2-\overset{\overset{\textstyle O}{\|}}{C}-OH$$
$$\underset{\displaystyle O}{\rule{0pt}{0pt}}$$

β-Propiolactone

Biocidal activity occurs during the hydrolysis phase and varies in time according to the temperature. Hoffman and Warshowskey[249] quoted the following hydrolysis times:

Temp. (°C)	Half-life of hydrolysis (min)
10	1080
25	210
50	20
75	5

Although concentrated BPL solutions are stable for long periods of time, polymerization may occur at 4°C, and the resulting complex is less active. It also reacts exothermically with hydroxyl, carbonyl, sulfhydryl, amino, and phenolic groups. The method of use for culture media is to add BPL to normally prepared media at 0.2% v/v concentration. The β-propionic acid formed after hydrolysis usually lowers the pH, and subsequent correction must be made.[250] The activity of BPL against microorganisms was reported by Taquet et al.[251] Himmelfarb[252] used BPL to sterilize culture media containing carbohydrates. After the complete hydrolysis of BPL the culture medium is as vulnerable to microbial infections, as is any sterilized medium.

Pyrocarbonic acid diethylester — PKE is manufactured by Bayer Ltd. under the trade name BAYCOVIN®. When hydrolyzed in water it forms alcohol and carbon dioxide

$$C_2H_5O-CO-OCO-OC_2H_5 + HOH \rightarrow 2C_2H_5OH + 2CO_2$$

The biocidal activity again occurs during hydrolysis, which lasts approximately 8 hr at 20°C and 16 hr at 10°C. It is a corrosive, colorless liquid that is soluble in alcohol and most organic solvents, but only 0.6% v/v soluble in water. Molin et al.[253] reported their work with PKE on yeasts and fungi. Messrs. Bayer Ltd. will provide data and an extensive bibliography of work in the German literature which should be consulted before using BAYCOVIN® for sterilization.

Ethylene oxide — This gaseous agent has limited use in sterilizing culture media. The formation of the toxic substance ethylene chlorhydrin prevents its use in most circumstances. Kayser[254] and Ernst[255] reviewed the use of this sterilizing agent.

Radiosterilization — This is a relative newcomer to the scene of cold sterilization of culture media. The source of radiant energy is usually ^{60}Co or ^{137}Cs isotopes with half-lives of 5.27 and 30 years, respectively. Becking[256] has written a comprehensive review of radiosterilization. Although the concept of manufacturing containers of media which can then be sterilized *in toto* is very attractive, there are problems. The process is not without destructive effect on the medium components; carbohydrates, proteins, lipids, and vitamins are all degraded to some extent, depending on the irradiation dose. Microorganisms vary in their susceptibility to irradiation, and this is made worse by the presence of protective SH molecules in culture media. The construction of irradiation sources is very expensive and most laboratories would wish to hire space in commercial irradiation centers. Therefore, the product must undergo transport and storage before the process. Therefore, the standard technique is to prepare media in as sterile a form as possible and then send it for irradiation to ensure greater protection.

Radiopasteurization — This process is more applicable in these circumstances. Doses of 0.2 to 0.5 Mrd are sufficient to prevent microbial growth, but, at the same time, cause minimal damage to the constituents of the medium.

Pasteurization — This refers to the short, low-heat process used on solutions that cannot withstand high-temperature processing. An exposure of 30 min at 60°C ensures destruction of most vegetative organisms (except thermophiles or thermoduric organisms), but has little effect on spores.

Tyndallization — This process uses the same principle on 3 consecutive days. The philosophy behind it is that, providing the solution is nutritive, after the first heating any spores present will germinate and become susceptible to the second or third heating. The times and temperatures vary considerably; 100°C for 20 min is employed for nonprotein-containing media.

Inspissation — This process is used for protein-containing media (e.g., egg or serum protein), which may be heated to provide solid slopes of coagulated protein. The process is carried out at 75°C for 1 hr with the tubes or bottles sloped at an appropriate angle to provide the maximum surface area. It is essential to remember that this is not a sterilizing process and, therefore, the liquid medium should be free from infection.

C. Storage of Culture Media

It is probably true to say that any storage of culture media is deleterious. The more intensive the search of the tests on stored vs. freshly prepared media, the more convincing the answers. However, the demands of most laboratories can only be met by holding media "on the shelf" and drawing from stock as required. Therefore, storage will inevitably lead to changes in the medium. These must be kept in check by providing optimum conditions.

Once agar-containing culture media have been poured into dishes and stored at 4°C, loss of moisture commences. Freshly prepared media have a_w values of approximately 0.99. Refrigerators may have humidities of 0.85, and room temperature storage may be 0.60 humidity or lower.[257] When the surface of the agar plate becomes dry, organisms requiring high a_w environments will grow poorly or not at all. Wrapping the dishes in plastic film helps, but the transfer of moisture that can occur between the surface of the agar and the dish or plastic wrapper may affect the growth of susceptible organisms.

Agar is relatively easily dehydrated down to 60% of its water content, but it cannot be rehydrated without heating. Examples of this anomaly can be seen in bottles containing agar slopes. If the bottle is stored under cycling temperature fluctuations, the wall of the bottle acts as a condenser and a relatively dry agar slope will exist in a sealed bottle containing free water. Martin[258] reviewed the subject of storage of microbial culture media and concluded that precautions can be taken to protect media during storage. Elliott and Georgala[244] considered that the storage life of prepared media must be decided by the bacteriologist who is going to use it, but they considered that 6 months at room temperature was a maximum period. Optimal conditions of storage are at 2 to 4°C, but a cool, dark cupboard would be satisfactory for media that do not contain antibiotics, blood, or other thermolabile additives. Monthly checks should be made on the visual appearance of all batches of medium for contamination with molds or bacteria, precipitation, change in color, etc.

Certain inhibitory or indicator additives to media other than antibiotics are sensitive to storage, including dyes, azide, tellurite, selenite, and sulfur compounds. A number of dyes are light sensitive, and media containing them must be stored away from light. Brilliant green will fade and the leuco base of fuchsin in Endo Agar will become pink when exposed to bright light. Tetrazolium compounds are susceptible to light and will form red, insoluble formazan compounds in the stored media. Tellurite and selenite undergo slow changes on storage and may produce blackening or orange pigment, respectively, in the medium. Azide can form hydrozoic acid, which is more toxic than the original compound.

The chemical instability and reactivity of biologically important derivatives were clearly explained by Postgate.[259] Sulfite easily oxidizes to sulfate. Metabisulfite hydrolyzes to sulfite and bisulfite. Thiosulfate is reasonably stable at pH 7.0, but it oxidizes to sulfate in aerated alkaline conditions. It will also react with thiol groups (both in culture media and microorganisms) to yield free H_2S and organic sulfonic acids. Tetrathionate reacts like thiosulfate with thiol groups to give sulfite in place of H_2S. Dithionite (hydrosulfite) is very unstable, and autooxidation to sulfite takes place rapidly. Cysteine will autooxidize to cystine at 4°C over approximately a week at acid pH values.

The action of light on culture media was dramatically demonstrated by Waterworth,[260] who demonstrated the inhibitory effect of sunlight shining on the surfaces of dishes of nutrient agar. Hydrogen peroxide was suspected as being the lethal chemical formed in the medium, as cups containing catalase could reverse the lethality of short exposure to sunlight in the immediately surrounding media.

IX. COMMENTS

Armed with the background information on culture media given above, it is hoped that the culture media formulas which follow will appear less confusing. The common threads of nutrient requirements which run throughout the formulas should be distinguishable from the numerous variations of individual ingredients chosen by various authors at different times.

X. CULTURE MEDIA FORMULAS

A. General Basic Media (Complex)
Meat Infusion Broth

Lean meat	500 g
Peptone	10 g
Sodium chloride	5 g
Water	1 liter

Use fresh, fat-free meat, finely minced and extract in the water at 4°C for 24 hr. Skim off fat, boil for 15 min. Filter. Make up volume to 1 liter, add peptone and salt. Heat to dissolve and filter again, if necessary. Adjust pH to 7.5 with sodium hydroxide. Sterilize by autoclaving at 121°C for 15 min. Final pH should be 7.4.

Meat Extract Broth

Peptone	10 g
Meat extract (Lab-Lemco)	10 g
Sodium chloride	5 g
Water	1 liter

Mix all the ingredients in water and dissolve by heating. Adjust the pH to 8.0 and heat at 100°C for 30 min. Filter off alkaline-earth metal-phosphate precipitate and adjust pH to 7.5 with hydrochloric acid. Sterilize at 121°C for 15 min. Final pH should be 7.4.

Papaic Digest Broth

Defatted, minced ox heart	500 g
Water	1 liter
Papain	1 g

Digest in a water bath at 60°C for 18 hr. Add 5 g of sodium chloride. Heat at 100°C for 90 min and filter. Adjust the pH to 8.0 with sodium hydroxide and heat at 100°C for 30 min. Filter and adjust pH to 7.5 with hydrochloric acid. Sterilize at 121°C for 15 min. Final pH should be 7.4. The amino-nitrogen content of a good broth should be approximately 10 mg/100 ml.

Tryptic Digest Broth

Defatted minced ox heart	1500 g
Water	2500 g
Na_2Co_3 (0.8% w/v) solution	2500 g
Pancreatic extract	50 ml
Chloroform	50 ml

Mix together and digest at 45°C for 3 to 6 hr. Reduce the pH by adding 40 ml of 10 N HCl. Heat at 100°C for 30 min and filter. Raise pH to 8.0 with sodium hydroxide and heat at 100°C until phosphates precipitate. Filter, cool, and adjust pH to 7.5 with hydrochloric acid. Sterilize at 121°C for 15 min. Final pH should be 7.4.

Peptone Media

Peptone	10–20 g
Sodium chloride	5 g
Water	1 liter

Dissolve the peptone and salt. Adjust pH to 7.4 and filter if necessary. Sterilize at 121°C for 15 min. Final pH should be 7.3. To improve growth of organisms, small quantities of dextrose may be added (0.25 to 0.5 g/l). Higher quantities of fermentable sugar will require the addition of phosphate buffer to the formulation.

Dextrose Peptone Broth

Peptone	20 g
Dextrose	10 g
Sodium chloride	5 g
Distilled water	1 liter
pH 7.2 (approximately)	

Tryptone Soya Broth — Soybean Casein Digest Medium USP

Pancreatic digest of casein	17.0 g
Papaic digest of soybean meal	3.0 g
Sodium chloride	5.0 g
Dibasic potassium phosphate	2.5 g
Dextrose	2.5 g
Distilled water	1 liter

pH 7.3 (approximate)

For fermentation studies, indicator dyes can be added. Carbohydrate solutions are often added separately after sterilization. The peptone must be free from fermentable carbohydrates.[261]

Dextrose Tryptone Broth

Tryptone	10.0 g
Dextrose	5.0 g
Bromcresol purple	0.04 g
Distilled water	1 liter

pH 6.9 (approximate)

The addition of specific infusions increases the ability of the medium to support the growth of fastidious organisms.

Brain Heart Infusion

Calf brain infusion solids	12.5 g
Beef heart infusion solids	5.0 g
Proteose peptone	10.0 g
Dextrose	2.0 g
Sodium chloride	5.0 g
Disodium phosphate	2.5 g
Distilled water	1 liter

pH 7.4 (approximate)

The addition of starch provides protection against the production of toxic substances during growth.

Mueller Hinton Broth[262]

Beef, infusion from	300.0 g
Casein hydrolysate	17.5 g
Starch	1.5 g
Distilled water	1 liter

pH 7.4 (approximate)

The above media can be converted into solid media by the addition of agar. The concentration of agar varies according to its source. High-strength European agars are added at concentrations of 10 to 12 g per liter of broth. U.S. agars are normally added at 15 g/l. If the agar contains free minerals there may be insolubles produced when it is autoclaved with broth (see section on media incompatibilities).

The use of mixed peptones ("special" peptones) often improves the recovery of organisms requiring a variety of growth substances.

Columbia Blood Agar Base[263]

Special peptone	23 g
Starch	1 g
Sodium chloride	5 g
Agar	10 g
Distilled water	1 liter

pH 7.3 (approximate)

Any of the following additional enrichment may be added to the above nutrient media:

1. Whole or lysed blood at concentrations varying from 2 to 5% v/v. The blood must be added aseptically after the base medium has been sterilized and carefully mixed to ensure a homogeneous suspension free from air bubbles. To reduce the opacity of the medium, a thin layer of nutrient agar (7 ml) may be poured at the bottom of the dish. A similar volume of blood agar can then be poured over the solid bottom layer. The temperature of the blood medium can be raised to 75°C for a short period of time to form "chocolate" agar. Such a heated medium shows enhanced properties of growing *Haemophilus* spp. and *Neisseria* spp.

2. A peptic digest of blood (sheep or horse) (Fildes extract) added to nutrient agar at 2 to 5% v/v stimulates the growth of *Haemophilus influenzae, Clostridium tetani*, and *Clostridium perfringens.*[12]

3. Serum added at 10% v/v aseptically to nutrient media can be used to grow the more exacting pathogenic organisms. Hayward[265,266] incorporated it into a medium for the Nagler plate identification of *Clostridium welchii.*

4. Egg yolk was later substituted[267-272] for serum, as it gives more consistent and stronger reactions with *Clostridia*.

5. Egg yolk emulsion added to nutrient broth has also been used as an enrichment/indicator additive for the *Bacillus*[273-277] and *Staphylococcus*[278-292] species.

6. The addition of tissue fragments to nutrient broth to reduce the eH of the medium and encourage the growth of anaerobic organisms is commonly practiced in microbiology.

Cooked Meat Medium

Peptone	10 g
Lab-Lemco powder	10 g
Neutral heart tissue	30 g
Sodium chloride	5 g
Distilled water	1 liter

pH 7.4 (approximate)

Liver Broth

Infusion from fresh liver	23 g
Peptone	10 g
Potassium phosphate	1 g
Extracted liver tissue	30 g
Distilled water	1 liter

pH 6.8 (approximate)

The added protein need not be meat based.

Cooked Meat Medium (Synthetic)

Peptone	10 g
Lab-Lemco powder	10 g
Textured soya protein	30 g
Sodium chloride	5 g
Distilled water	1 liter
pH 7.0 (approximate)	

Agar may also be added to help maintain anaerobic conditions.[293]

Liver Agar (Shapton)

Infusion from fresh liver	23.0 g
Peptone	10.0 g
Dipotassium phosphate	1.0 g
Liver fiber	47.0 g
Agar	15.0 g
Distilled water	1 liter
pH 6.9 (approximate)	

Agar and reducing agents (thioglycollate, cystine) are used in sterility test media which must be able to recover both aerobic and anaerobic organisms.

Thioglycollate Medium (Brewer)[294]

Lab-Lemco powder	1.0 g
Yeast extract	2.0 g
Peptone	5.0 g
Dextrose	5.0 g
Sodium chloride	5.0 g
Sodium thioglycollate	1.1 g
Methylene blue	0.002 g
Agar	1.0 g
Distilled water	1 liter
pH 7.2 (approximate)	

Thioglycollate Medium USP[295]

Yeast extract powder	5.0 g
Tryptone	15.0 g
Dextrose	5.5 g
Sodium thioglycollate	0.5 g
Sodium chloride	2.5 g
L-Cystine	0.5 g
Resazurin	0.001 g
Agar	0.5 g
Distilled water	1 liter
pH 7.1 (approximate)	

Synthetic Basal Media

Skerman[296] doubted the plausibility for the very large number of different formulas given for the preparation of inorganic parts of synthetic media. He thought that in many cases the majority of constituents provided trace elements, but they were present in quantities far in excess of requirements. Skerman also believed that the phosphate content of most formulations could be considerably reduced and thus prevent precipitation in the medium. The following tables have been reproduced together with his description.

Skerman's Basal Mineral Salts Medium

Group A

Solution	Amount	Procedure
1. N NaOH	1 l	Sterilize at 121°C for 20 min
2. 0.074 M H$_3$PO$_4$	1 l	Sterilize as for Solution 1 (above)
3. Solution from 2, above	200 ml	Dilute to 2000 ml (0.0074 M); sterilize as for Solution 1
4. Solution from 3, above	1 l	Neutralize with use of N NaOH; sterilize as for Solution 1
5. NaHCO$_3$ in 100 ml of water	8.333 g	Sterilize as for Solution 1
6. CaCl$_2$ in 100 ml of water	5.0 g	Sterilize as for Solution 1
7. NaNO$_2$ in 100 ml of water	50.0 g	Sterilize as for Solution 1
8. Glucose in 100 ml of water	10.0 g	Sterilize at 110°C for 25 min
9. Mannitol in 100 ml of water	10.0 g	Sterilize as for Solution 8
10. Sucrose in 100 ml of water	10.0 g	Sterilize as for Solution 8
11. Sodium citrate in 100 ml of water	2 g anhydrous (or 2.77 g hydrated)	Sterilize at 121°C for 20 min
12. Phenol in 100 ml of water	10.0 g	Sterilize as for Solution 11
13. Na$_2$S$_2$O$_3$ in 100 ml of water	10.0 g	Sterilize as for Solution 11
14. 0.5 N HCl	100 ml	Sterilize as for Solution 11
15. 0.0167 M H$_3$PO$_4$	1 l	Neutralize 500 ml with N NaOH; sterilize as for Solution 11
16. Monoethylamine hydrochloride in 100 ml of water	5 ml	Sterilize by filtration

Group B

Solution	Amount (per 100 ml solvent)	Solvent required	Final concentration (μg/l medium)
1. NaCl	3.0 g	0.0074 M H$_3$PO$_4$	300,000
2. (NH$_4$)$_2$SO$_4$	6.6 g	0.074 M H$_3$PO$_4$	660,000
3. LiCl$_2$	21.0 mg	0.0074 M H$_3$PO$_4$	21
4. CuSO$_4$·5H$_2$O	80.0 mg	0.0074 M H$_3$PO$_4$	80
5. ZnSO$_4$·7H$_2$O	106.0 mg	0.0074 M H$_3$PO$_4$	106
6. H$_3$BO$_4$	600.0 mg	0.0074 M H$_3$PO$_4$	600
7. Al$_2$(SO$_4$)$_3$·18H$_2$O	123.0 mg	0.0074 M H$_3$PO$_4$	123
8. NiCl$_2$·6H$_2$O	110.0 mg	0.0074 M H$_3$PO$_4$	110
9. CoSO$_4$·7H$_2$O	109.0 mg	0.0074 M H$_3$PO$_4$	109
10. TiCl$_4$	60.0 mg	0.074 M H$_3$PO$_4$	60
11. KBr	30.0 mg	Water	30
12. KI	30.0 mg	Water	30
13. MnCl$_2$·4H$_2$O	629.0 mg	0.074 M H$_3$PO$_4$	629
14. MgSO$_4$·7H$_2$O	1.4 g	Water	140,000
15. SnCl$_2$·2H$_2$O	36.0 mg	Water	36
16. FeSO$_4$·7H$_2$O	300.0 mg	Water	300

From Skerman, V. B. D., *A Guide to the Identification of the Genera of Bacteria,* 2nd ed., Williams & Wilkins, Baltimore, 1967, 213. With permission.

For the preparation of the various mineral salts media, the solutions described above are required. They should all be prepared with glass-distilled water and acid-cleaned glassware and using the following steps:

1. Pipette into a 1-liter standard flask the following amounts of solutions from Group B: 10.0 ml of Solutions 1 and 2 and 0.1 ml of each of Solutions 3 through 10.

2. Add approximately 600 ml of 0.0074 M H_3PO_4 (Solution 3, Group A) and 210 ml of water.

3. Adjust the pH to 7.0 with N NaOH (Solution 1, Group A).

4. Add 0.1 ml of Solutions 11 and 12 from Group B.

5. Take 0.1 ml of the $MnCl_2$ solution (Solution 13, Group B), add 9.9 ml of 0.074 M H_3PO_4 (Solution 2, Group A), and adjust the pH to 7.0. Autoclave and filter. Add the filtrate to the medium.

6. Add 10 ml of Solution 14, Group B and 0.1 ml of Solutions 15 and 16, Group B.

7. Using the neutralized 0.0074 M H_3PO_4 (Solution 4, Group A), make the final volume to 1 liter.

8. Sterilize at 121°C for 20 min.

This solution is crystal clear and will remain so for long periods if kept in acid-washed glassware. It provides a complex mineral salts base with ammonium-N, which has been found suitable, after addition of specific components, for a wide range of autotrophic and exacting heterotrophic bacteria. With the omission of the ammonium sulfate, it is a suitable base for nitrogen-fixing bacteria. It may not be the best base, but in all cases tested it has been found satisfactory for taxonomic purposes.

A range of media for autotrophic organisms is described by Skerman[296] using the above basal mineral mix at single- and double-strength solutions.

Davis and Mingioli Phosphate-buffered Salts Medium[137]

K_2HPO_4	7.0 g
KH_2PO_4	3.0 g
$MgSO_4 \cdot 7H_2O$	0.1 g
$(NH_4)_2SO_4$	1.0 g
Na citrate·$3H_2O$	0.5 g
Distilled water	1 liter
Glucose	2.0 g
Agar (if required)	15.0 g
pH 7.0	

The glucose and mineral salts solution are sterilized separately at 121°C for 15 min.

Hershey's Tris-buffered Salts Medium[298]

Glucose	2.0 g
Tris buffer (0.1 M)	12.1 g
NaCl	5.4 g
KCl	3.0 g
NH_4Cl	1.1 g
$CaCl_2$	0.011 g
$MgCl_2$	0.095 g
KH_2PO_4	0.087 g
Na_2SO_4	0.023 g
$FeCl_3$	0.00016 g
Distilled water	1 liter
pH to 7.4 with HCl solution	

Sterilize the glucose and mineral salts solutions separately at 121°C for 15 min.

Powell and Errington's Medium[299]

Solution 1

$NaH_2PO_4 \cdot 2H_2O$	31.0 g (0.2 M)
$(NH_4)_2HPO_4$	238.0 g (1.8 M)
K_2SO_4	70.0 g (0.4 M)
Distilled water	1 liter

Solution 2

HCl (conc.)	50 ml (0.58 M)
MgO	10.0 g (0.25 M)
$CaCO_3$	2.0 g (0.02 M)
$FeCl_3 \cdot 6H_2O$	5.4 g (0.02 M)
$ZnSO_4 \cdot 7H_2O$	1.44 g (0.005 M)
$MnSO_4 \cdot 4H_2O$	1.11 g (0.005 M)
$CuSO_4 \cdot 5H_2O$	0.25 g (0.001 M)
$CoSO_4 \cdot 7H_2O$	0.28 g (0.001 M)
H_3BO_4	0.062 g (0.001 M)
$Na_2MoO_4 \cdot 2H_2O$	0.49 g (0.002 M)
Distilled water	1 liter

Mix 50 ml of Solution 1 with 5 ml of Solution 2 and made up to 1 liter with distilled water. Sterilize by autoclaving at 121°C for 15 min.

Solution 3

Citric acid	4.2 g (0.02 M)
Glucose	3.6 g (0.02 M)
L-Glutamic acid	2.94 g (0.02 M)
Succinic acid	1.18 g (0.01 M)

Dissolve in 50 ml of distilled water and adjust pH to 7.0 with NaOH. Filter, sterilize, and add 1.24 g (0.005 M) $Na_2S_2O_3 \cdot 5H_2O$ to the 1-liter solution prepared above.

B. Phototrophic Bacteria
Yeast Malate Medium (Rhodopseudomonas viridis)[300]

Yeast extract	5.0 g
Sodium malate	1.0 g
Distilled water (or sea water)	1 liter
pH 7.0	

Yeast Peptone Broth (Rhodopseudomonas sp.)[300]

Yeast extract	2.5 g
Peptone	2.5 g
Distilled water (or sea water)	1 liter
pH 7.2	

Chloropseudomonas Medium[301]

Solution 1	
EDTA	59.0 g
$FeSO_4 \cdot 7H_2O$	24.9 g
Distilled water	1 liter
Solution 2	
$FeCl_3 \cdot 6H_2O$	1.6 g
$Na_2B_4O_7 \cdot 10H_2O$	0.88 g
$ZnSO_4 \cdot 7H_2O$	0.44 g
$CoSO_4 \cdot 7H_2O$	0.24 g
$CuCl_2 \cdot 2H_2O$	0.0135 g
$MnSO_4 \cdot 4H_2O$	0.0165 g

Solution 1	110 ml
Distilled water	to 1 liter
Solution 3	
KH_2PO_4	2 g
NH_4Cl	2 g
$MgCl_2 \cdot 6H_2O$	10 g
NaCl	40 g
$CaCl_2$	0.08 g
Solution 2	2 ml
Distilled water	998 ml
Solution 4	
Solution 3	500 ml
Distilled water	450 ml
$Na_2S \cdot 9H_2O$	0.2 g
$NaHCO_3$	4.0 g
$FeSO_4 \cdot 7H_2O$	2.5 mg in 0.3 M HCl
Ethylalcohol	3 ml
pH to 7.3 with HCl	

Sterilize the Na_2S, $NaHCO_3$, and $FeSO_4$ solutions separately by autoclaving at 121°C for 15 min. Mix the ingredients aseptically.

Chromatium, Chlorobium, and Thiopedia Medium[300]

Trace elements solution[a]	
$FeCl_3 \cdot 6H_2O$	2.7 g
H_3BO_3	0.1 g
$ZnSO_4 \cdot 7H_2O$	0.1 g
$Co(NO_3)_2 \cdot 6H_2O$	0.05 g
$CuSO_4 \cdot 5H_2O$	0.005 g
$MnCl_2 \cdot 6H_2O$	0.005 g
Distilled water	1 liter
Basal medium	
KH_2PO_4	1 g
NH_4Cl	1 g
$MgCl_2 \cdot 6H_2O$	0.5 g
NaCl	10 g
Distilled water	1 liter
Trace elements solution	1 ml
Sodium bicarbonate	
10% w/v in distilled water	
Sodium sulfide ($Na_2S \cdot 9H_2O$)	
10% w/v in distilled water	
Sodium thiosulfate ($Na_2S_2O_3 \cdot 5H_2O$)	
10% w/v in distilled water	
Sodium malate	
10% w/v in distilled water	

Autoclave all solutions separately at 121°C for 15 min

[a] A precipitate of iron oxide forms, but does not appear to interfere with the success of the medium.

Chromatium sp. (small-celled, such as Strain D)

> For every 10 ml of freshly boiled basal medium add aseptically 0.2 ml bicarbonate, 0.02 ml sulfide, 0.1 ml thiosulfate, and 0.1 ml malate

Thiopedia

> Follow same procedure as for *Chromatium*

Chlorobium limicola

> For every 10 ml of freshly boiled basal medium add aseptically 0.2 ml bicarbonate and 0.1 ml sulfide; adjust the pH of the complete medium to 7.0—7.2 with sterile phosphoric acid

Chlorobium thiosulfatophilum

> For every 10 ml of freshly boiled basal medium add aseptically 0.2 ml bicarbonate, 0.02 ml of sulfide, and 0.1 ml of thiosulfate; adjust the pH of the complete medium to 7.0—7.2 with sterile phosphoric acid;
>
> certain strains of *Chlorobium* are known to require vitamin B_{12} or related compounds; for these it is necessary to add the vitamin to the growth medium;
>
> if an agar medium is required, e.g., for stab cultures or purification procedures, 1% w/v Ionagar® No. 2 may be added to the basal medium

From Lapage, S. P., Shelton, J. E., and Mitchell, T. G., in *Methods in Microbiology,* Vol. 3A, Norris, J. R. and Ribbons, D. W., Eds., Academic Press, London, 1970, 1—228. With permission.

Chromatium Species Medium

Basal medium

NH_4Cl	0.1 g
KH_2PO_4	0.1 g
$MgCl_2$	0.05 g
NaCl (for marine species)	30.0 g
Agar	20.0 g
Tap water	925.0 ml

Dissolve the mineral salts in the tap water; add the agar and then heat to 121°C for 15 min to dissolve the agar; distribute in the required quantities in test tubes or screw-capped bottles; sterilize at 121°C for 15 min; for a liquid medium, omit the agar

Sterile sodium bicarbonate

$NaHCO_3$	5 g
H_2O	100 ml
Sterilize at 121°C for 15 min	

Sterile sodium sulfide

$Na_2S \cdot 9H_2O$	5 g
H_2O	100 ml
Sterilize at 121°C for 15 min	

Note: For use of the complete medium, melt the required number of tubes of basal medium and cool to 45°C. To each 10-ml amount of basal medium add aseptically 0.4 ml $NaHCO_3$ solution and 0.2 ml $Na_2S \cdot 9H_2O$ solution.

From Skerman, V. B. D., *A Guide to the Identification of the Genera of Bacteria,* 2nd ed., Williams & Wilkins, Baltimore, 1967, 213. With permission.

Rhodomicrobium Species Medium[302]

$NaHCO_3$	5.0 g
NaCl	2.0 g
$(NH_4)_2SO_4$	1.0 g
K_2HPO_4	0.5 g
$MgSO_4 \cdot 7H_2O$	0.1 g
$Na_2S \cdot 9H_2O$	0.1 g
Ethanol	2 ml
Yeast autolysate	2 ml
Distilled water	1 liter

Sterilize the $NaHCO_3$, ethanol, and Na_2S solutions separately by filtration and add them to the autoclaved yeast-basal salts medium. Adjust pH to 7.0 with H_3PO_4. For a solid medium, agar is added at 15 g/l.

Chromatiaceae Media

Medium T

KH_2PO_4	1 g
NH_4Cl	1 g
$MgCl_2 \cdot 6H_2O$	0.5 g
NaCl	10 g
Tap water	1 liter

(where necessary, make allowance in this volume
for the additions given below)

Add 1 ml of the following solution (sterilized separately) to each liter of basic Medium T:

Trace elements

$FeCl_3 \cdot 6H_2O$	1.5 mg
$CuCl_2 \cdot 2H_2O$	1 mg
$NiCl_4 \cdot 6H_2O$	2 mg
$Na_2MoO_4 \cdot Co(NO_3)_2 \cdot 6H_2O$	0.05 g
$CuSO_4 \cdot 5H_2O$	0.005 g
$MnCl_2 \cdot 6H_2O$	0.005 g
Distilled water	1 liter

The following solutions should be made up more concentrated, sterilized separately by filtration, and added to the basic Medium T immediately before use:

1. $NaHCO_3$ — 2 g/l of basic medium for all species. Sterilize just before use.
2. $Na_2S \cdot 9H_2O$ — 1 g/l for *Chlorobium limicola*; 0.2 g/l for *Chlorobium thiosulfatophilum, Chromatium, Thiopedia*.
3. $Na_2S_2O_3 \cdot 5H_2O$ — 1 g/l for *Chlorobium thiosulfatophilum, Chromatium, Thiopedia*.
4. Sodium malate — 1 g/l for *Chromatium, Thiopedia*.
5. H_3PO_4 to pH values of 7.0 to 7.2 for *Chlorobium* spp. and 8.0 to 8.4 for *Chromatium, Thiopedia*.

Chlorobium limicola grows better if the complete medium is prepared the day before inoculation and kept overnight in a refrigerator (in filled bottles); immediately before inoculation make a further addition of $Na_2S \cdot 9H_2O$ (1 drop of 10% $Na_2S \cdot 9H_2O$ per 10 ml of medium.)

Medium P

Solution 1

$CaCl_3$	0.4 g Sterilize in autoclave
Distilled water	1 liter

Solution 2(a)

EDTA	0.5 g Dissolve the EDTA first
$FeSO_4 \cdot 7H_2O$	0.2 g
$ZnSO_4 \cdot 7H_2O$	10 mg
$MnCl_2 \cdot 4H_2O$	3 mg
H_3BO_3	30 mg
$CoCl_2 \cdot 6H_2O$	20 mg
$CuCl_2 \cdot 2H_2O$	1 mg
$NiCl_2 \cdot 6H_2O$	2 mg
$Na_2MoO_4 \cdot 2H_2O$	3 mg
Distilled water	1 liter

Solution 2(b)

Cyanocobalamin	2 mg
Distilled water	100 ml

Solution 2(c)

KH_2PO_4	1 g	Dissolved in 70 ml dis-
KCl	1 g	tilled water plus 30 ml
NH_4Cl	1 g	of Solution 2(a) plus 3
$MgCl_2 \cdot 6H_2O$	1 g	ml of Solution 2(b)

Solution 3

$NaHCO_3$	4.5 g	Saturate with CO_2 for
Distilled water	900 ml	30 min by bubbling; mix with Solutions 2 and filter-sterilize under pressure of CO_2

Solution 4

Distillled water	3 g	May be autoclaved;
$Na_2S \cdot 9H_2O$	200 ml	partly neutralize with sterile H_2SO_4 before use

Mix Solutions 2 + 3 with Solution 1 in the ratio 2:1. Add 2.4 to 4.8% (v/v) of Solution 4 according to the sulfide requirement of the organisms. Use pH 6.7 to 6.9 for green bacteria, 6.9 to 7.2 for purple. Tightly closed filled stoppered bottles of Medium P may be stored in the dark for several months.

From Postgate, J. R., *Lab. Pract.*, 15, 1239–1244, 1966. With permission.

Rhodopseudomonas spp. Medium[304]

Peptone	10.0 g
Yeast extract	5.0 g
NaCl	5.0 g
Water	1 liter
Agar	15 g
pH 7.0	

Sterilize by autoclaving at 121°C for 15 min.

Rhodospirillum spp. Medium[304]

Yeast extract	0.1 g
NaCl	2.0 g
NH_4Cl	1.0 g
$MgCl_2$	0.2 g
$NaHCO_3$	5.0 g
Ethanol	2.0 ml
pH to 7.0 with sterile 0.1 NH_3PO_4	

Sterilize the $NaHCO_3$ and ethanol solutions separately by filtration. Add to the previously autoclaved basal medium (121°C for 15 min).

Chlorobium spp. Medium[305]

NH_4Cl	1.0 g
KH_2PO_4	1.0 g
$Na_2S \cdot 9H_2O$	0.5 g
$MgCl_2$	2.0 g
$NaHCO_3$	As required
NaCl	As required
Tap water	1 liter

Growth was much improved if the following trace elements were supplied (per liter of basal medium):

$CaCl_2$	0.1 g
$FeCl_3$	1600 µg
B	100 µg
Zn	100 µg
Co	50 µg
Cu	5 µg
Mn	5 µg

Rhodospirillaceae Medium[306]

Proteose peptone	1.5—20.0 g
$(NH_4)SO_4$	1.0 g
K_2HPO_4	0.5 g
$MgSO_4 \cdot 7H_2O$	0.2 g
NaCl	2.0 g
$NaHCO_3$	5.0 g
Distilled water	1 liter
pH 7.1—7.2	

Sterilize the $NaHCO_3$ solution separately by filtration and add aseptically to the previously sterilized (121°C for 15 min) basal medium.

Rhodomicrobium spp. Medium[143]

Enrichment medium

KH_2PO_4	0.136 g
$(NH_4)_2SO_4$	0.05 g
$CaCl \cdot 2H_2O$	1.0 mg
$MnSO_4 \cdot 4H_2O$	0.25 mg
Na_2HPO_4	0.213 g
$MgSO_4 \cdot 7H_2O$	0.02 g
$FeSO_4 \cdot 7H_2O$	0.5 mg
$Na_2MoO_4 \cdot 2H_2O$	0.25 mg
Agar	1.9 g
Water to	100 ml
pH (adjust with NaOH)	7.2

Isolation medium

KH_2PO_4	0.15 g
$(NH_4)_2SO_4$	0.1 g
NaCl	0.2 g
$MgSO_4 \cdot 7H_2O$	0.01 g
Yeast extract	0.1 g
Water to	79.0 ml

Sterilize the preceding solution by filtration and add aseptically

$NaHCO_3$ (5%, w/v)	10.0 ml
$Na_2S \cdot 9H_2O$ (3%, w/v)	1.0 ml
Sodium acetate (2%, w/v)	10.0 ml
pH (adjust with NaOH)	7.4

Fill a tall sterile glass beaker (400 ml) with sterile enrichment medium almost to the top; add 1 to 5 g of soil sample (or 5 to 10 ml of water sample); cover the top with aluminum foil and incubate on a window sill at room temperature.

From Aaronson, S., *Experimental Microbial Ecology*, Academic Press, London, 1970, 62. With permission.

Chromatiaceae Medium[307]

NH_4Cl	1.0 g
K_2HPO_4	0.5 g
$MgCl_2$	0.2 g
$NaHCO_3$	1.0 g
$Na_2S \cdot 9H_2O$	1.0 g
Distilled water	1 liter
pH 8.0—8.5	

Sterilize the $NaHCO_3$ and Na_2S solutions separately by filtration. Add to the basal medium which has been sterilized at 121°C for 15 min.

Growth Media for the Purple Bacteria

	Rhodopseudomonas spheroides	Rhodospirillum rubrum	Rhodopseudomonas palustris	Rhodomicrobium vannielii
KH_2PO_4	0.500	0.600	—	0.500
K_2HPO_4	0.500	0.900	—	0.500
$(NH_4)_2HPO_4$	0.800	—	—	0.800
$(NH_4)_2SO_4$	—	0.50—1.25	0.040	—
$NaHCO_3$	—	—	—	1.00
$MgSO_4 \cdot 7H_2O$	0.200	0.200	0.600	0.200
$MnSO_4 \cdot 4H_2O$	2.23 mg	—	—	2.23 mg
$CaCl_2 \cdot 2H_2O$	0.040	0.075	0.150	0.040
$FeSO_4 \cdot 7H_2O$	—	0.012	—	—
Iron citrate,[a] 4 mM	1 ml	—	—	1 ml
Glycerophosphate (Na)	—	—	2.00	—
Citrate (K)	—	—	0.500	—
Glutamate (Na)	1.90	—	—	—
L-Glutamic acid	—	—	3.00	—
DL-Malic acid	2.70	6.00	4.00[b]	—
Acetate (Na)	—	—	0.200	—
Propanol	—	—	—	0.50
L-Histidine	—	—	2.00	—
Tyrosine	—	—	0.100	—
Monobutyrin	—	—	0.400	—
Homocysteine thiolactone HCl	—	—	0.200	—
EDTA	—	0.020	—	—
Microelements	—	1 ml[c]	—	—
Nicotinic acid	1 mg	—	—	—
Thiamine HCl	1 mg	—	—	—
Biotin	10 µg	15 µg	—	—
p-Aminobenzoic acid	—	—	200 µg	—
pH	6.8	6.8	6.2—6.5	—

Note: Except where indicated, all quantities are in grams per liter of distilled water.

[a] Iron citrate: $(NH_4)_2 Fe(SO_4)_2 \cdot 6H_2O$, 157 mg and sodium citrate (dihydrate), 236 mg dissolved in 100 ml distilled water and stored at 0°C.

[b] Neutralized separately with KOH.

[c] Per 100 ml deionized water: H_3BO_3, 280 mg; $MnSO_4 \cdot 4H_2O$, 210 mg; $Na_2MoO_4 \cdot 2H_2O$, 75 mg; $ZnSO_4 \cdot 7H_2O$, 24 mg; $Cu(NO_3)_2 \cdot 3H_2O$, 4 mg.

[d] A mix containing (mg/l) Fe, 18; Mn, 14; Zn, 9.0; Mo, 1.8; Cn, 0.9; Co, 0.18; B, 0.18; V, 0.18; I, 0.18; Se, 0.036.

From Carr, N. G., in *Methods in Microbiology,* Vol. 3B, Norris, J. R. and Ribbons, D. W., Eds., Academic Press, London, 1969, 53—77. With permission.

Growth Media for the Purple Sulfur Bacteria

	Tap water, heterotrophic (g/l)	Distilled water, autotrophic (g/l)	Distilled water, heterotrophic (g/l)	Pfennig
Sodium malate	2.4			
$Na_2S_2O_3$	1.6			
$(NH_4)_2SO_4$	1.0			
NaCl	20.0			
K_2HPO_4	0.5			
$MgSO_4$	0.2			
pH	7.4			
*10% Na_2S	18 ml			
*10% Na_2CO_3	9 ml			
*10% H_3PO_3	6 ml			

Distilled water, autotrophic (g/l)

Solution A
- NaCl — 10.0
- NH_4Cl — 2.0
- KH_2PO_4 — 1.0
- $MgCl_2 \cdot 6H_2O$ — 1.0
- CaCl — 0.2
- Trace elements[a]
- Indigo carmine — 0.01
- Conc. HCl — 9.0

Solution B
- Na_2CO_3 — 12.6
- $Na_2S_2O_3 \cdot 5H_2O$ — 6.0
- $Na_2S \cdot 9H_2O$ — 0.1
- EDTA (di Na) — 0.2

*Added after sterilization

Solution A and B are sterilized separately and mixed in equal volumes after cooling, giving a final pH of 8.0

Distilled water, heterotrophic (g/l)

Stock solution
- $CaCl_2$ — 0.25
- NH_4Cl — 1.20
- $MgSO_4 \cdot 7H_2O$ — 0.66
- NaCl — 20.0
- Trace elements[b] — 1 ml

Solution I — Stock salt solution (120 ml) diluted to 800 ml

Solution II — M Tris, pH 7.5, 50 ml

Solution III — K_2HPO_4 (1 g) in 50 ml

Solution IV — $Na_2S \cdot 9H_2O$ (0.2 g) in 50 ml

Solution V — 5.5 g organic carbon source (neutralized) in 50 ml

Solutions I–V were autoclaved separately and mixed when cool; pH adjusted to 7.5 with conc. HCl

Pfennig

Solution 1 Distilled water — 5000 ml
- $CaCl_2 \cdot 2H_2O$ — 2.0 g

Solution 2 Distilled water — 40 ml
- Heavy metal solution[c] — 60 ml
- Vitamin B_{12} — 0.12 mg
- KH_2PO_4 — 2.0 g
- KCl — 2.0 g
- NH_4Cl — 2.0 g
- $MgCl_2 \cdot 6H_2O$ — 2.0 g

Solution 3 Distilled water — 900 ml
- $NaHCO_3$ — 9 g

Solution 4 Distilled water — 400 ml
- $Na_2S \cdot 9H_2O$ — 6 g

The sterilization and addition of these solutions are described below[d]

a Larsen's trace elements.

b Dissolve in 750 ml distilled water, EDTA, 50 g; $ZnSO_4 \cdot 7H_2O$, 22 g; H_3BO_3, 11.4 g; $MnCl_2 \cdot 4H_2O$, 5.1 g; $FeSO_4 \cdot 7H_2O$, 5.0 g; $CuSO_4 \cdot 5H_2O$, 1.6 g; $(NH_4)_6 Mo_7 O_{24} \cdot 4H_2O$, 1.1 g. Boil and after cooling bring to pH 6.5 to 6.8 with KOH; make to 1 liter final volume.

c Heavy metal solution. To 1 liter distilled water containing EDTA disodium salt (500 mg) were added $FeSO_4 \cdot 6H_2O$, 200 mg; $ZnSO_4 \cdot 7H_2O$, 10 mg; $MnCl_2 \cdot 4H_2O$, 3 mg; H_3BO_3, 30 mg; $CoCl_2 \cdot 6H_2O$, 20 mg; $CuCl_2 \cdot 2H_2O$, 1 mg; $NiCl_2 \cdot 6H_2O$, 2 mg; $NaMoO_4 \cdot 2H_2O$, 3 mg.

d Solution 1 (105 ml) was placed in a 135-ml screw-capped flask and autoclaved with loose caps. Solution 3 was flushed with CO_2 for at least 30 min while being vigorously stirred, until the pH was below 7.0, and then mixed with Solution 2. The mixture was sterilized by passage through a washed Seitz filter under a pressure of 121° C CO_2, and aseptically added (22 ml aliquots) to the sterile cold Solution A. Solution 4, after autoclaving, was added to the flasks in 4- (giving pH 7.0) to 6-ml (giving pH 7.2) amounts. The flasks were filled with sterile 0.04% $CaCl_2$ solution.

From Carr, N. G., in *Methods in Microbiology*, Vol. 3B, Norris, J. R. and Ribbons, D. W., Eds., Academic Press, London, 1969, 53–77. With permission.

Growth Media for Green Sulfur Bacteria

Chlorobium thiosulfatophilum and *Chlorobium limicola*		*Chloropseudomonas ethylicum*	
(g/l)			
		Basal salts (per 500 ml)	
NH_4Cl	1.0	KH_2PO_4	2.0 g
KH_2PO_4	1.0	NH_4Cl	2.0 g
$MgCl_2$	0.5	$MgCl_2 \cdot 6H_2O$	10.0 g
NaCl	10.0	NaCl	40.0 g
$NaHCO_3$ [a]	2.0	$CaCl_2$	80 mg
$Na_2S_2O_3$ or $Na_2S \cdot 9H_2O$ [a]	1.0	Trace elements	2.0 ml
		Trace elements (per liter)	
$CaCl_2$	0.1	$FeCl_3$	1.6 g
Fe ($FeCl_3 \cdot 6H_2O$) [a]	500 µg	$Na_2B_4O_7 \cdot 10H_2O$	880 mg
B (H_3BO_3)	100 µg	$ZnSO_4 \cdot 7H_2O$	440 mg
Zn ($ZnSO_4 \cdot 7H_2O$)	100 µg	$CoSO_4 \cdot 7H_2O$	240 mg
Co ($Co(NO_3)_2 \cdot 6H_2O$)	50 µg	$CuCl_2 \cdot 2H_2O$	13.5 mg
Cu ($CuSO_4 \cdot 5H_2O$)	5 µg	$MnSO_4 \cdot H_2O$	16.5 mg
Mn ($MnCl_2 \cdot 4H_2O$)	5 µg	Versenol iron solution [b]	110 ml
		Mix [c] I. Basal salts	500 ml
		Distilled water	450 ml
		II. 10% $Na_2S \cdot 9H_2O$	2 ml
		III. 10% $NaHCO_3$	40 ml
		IV. 0.05% $FeSO \cdot 7H_2O$ in 0.3 N HCl	5 ml
		V. 70% ethanol	3 ml

[a] Sterilized separately.

[b] Versenol iron solution: 59 g EDTA are dissolved in 500 ml distilled water and $FeSO_4 \cdot 7H_2O$ (24.9 g) added. Dilution to one liter and aeration overnight yields a final pH of 9.7.

[c] Solutions I to IV were sterilized separately and mixed after cooling. Ethanol was sterilized by membrane filtration and the complete medium adjusted with concentrated HCl to pH 7.3.

From Carr, N. G., in *Methods in Microbiology*, Vol. 3B, Norris, J. R. and Ribbons, D. W., Eds., Academic Press, London, 1969, 53–77. With permission.

C. "Higher" Bacteria

1. Gliding Bacteria
Beggiatoa spp. Medium[309]

Yeast extract	2 g
$CaCl_2$	0.1 g
Sodium acetate	0.5 g
Agar powder	2 g
Tap water	1 liter
Catalase	10 sigma units/ml medium

(One sigma unit decomposes 1 µmol H_2O_2 per minute at pH 7.0, 25°C).

Adjust pH to 7.0 and sterilize at 121°C for 15 min. Add the catalase enzyme after sterilization.

Myxococcus spp. Medium[300]

Potatoes, peeled and diced	10 g
Bakers yeast (fresh)	10 g
Agar	12 g
Distilled water	1 liter

Steam the potato in water for 30 min and filter. Make up volume to one liter and add the yeast and agar. Adjust pH to 7.2. Sterilize the medium at 121°C for 15 min.

Cytophaga spp. Media[300]

No. 1

Yeast extract	10 g
NaCl	20 g
K_2HPO_4	0.2 g
$MgSO_4$	0.5 g
NH_4Cl	1 g
$FeCl_3 \cdot 6H_2O$	0.25 mg
Distilled water	1 liter

Dissolve the ingredients in the water and adjust the pH to 7.0 to 7.4. After distribution in bottles or tubes, autoclave the medium at 121°C for 15 min.

No. 2

Tryptone	0.5 g
Yeast extract	0.5 g
Sodium acetate	0.2 g
Beef extract	0.2 g
Agar	9 g
Distilled water	1 liter

Dissolve all the ingredients except the agar in the water and adjust the pH to 7.2 to 7.4. Add the agar, steam the medium to dissolve, then distribute as required. Sterilize by autoclaving at 121°C for 15 min.

No. 3

Nutrient broth, dehydrated (Difco)	1 g
Yeast extract	1 g
Corn-steep liquor	1 ml
$NaHCO_3$	0.5 g
NaCl	30 g
Agar powder	10 g
Distilled water	1 liter

Mix all the ingredients except the agar with the water and adjust the pH to 7.2. Add the agar and steam the complete medium. Sterilize by autoclaving at 121°C for 15 min.

No. 4

Yeast extract	2.5 g
Agar	10 g
Distilled water	1 liter

Dissolve the yeast extract in the water and adjust the pH to 7.2 to 7.4. Add the agar, steam to dissolve, distribute in bottles, and autoclave at 121°C for 15 min.

Myxococcus spp. Medium[300]

Casitone (Difco)	1 g
Meat extract	1 g
Glucose	1 g
Agar powder	12 g
Distilled water	1 liter

Dissolve the glucose in a little of the water and autoclave separately. Dissolve the meat extract and casitone in the remaining water and adjust the pH to 7.2. Add the agar and steam to dissolve. Autoclave the medium at 121°C for 15 min and add the sterile glucose. Pour the medium into Petri dishes or into sterile bottles for slopes.

Flexibacteria Media[310]

No. 1 — *Saprospira grandis*

Tryptone	5 g
Yeast extract	5 g
$Ca(NO_3)_2 \cdot 4H_2O$	100 mg
K_2HPO_4	20 mg
Trace elements	See below
Filtered sea water	1 liter
pH 7.0	

No. 2 — *Flexibacter litorale* and *F. marinum*

Yeast extract	1 g
$Ca(No_3)_2 \cdot 4H_2O$	100 mg
K_2HPO_4	20 mg
Glucose (autoclaved separately)	1 g
Trace elements	See below
Filtered sea water	1 liter
pH 7.0	

No. 3 — *S. thermalis, F. elegans,* and *F. rubrum*

Yeast extract	1 g
Tris buffer	1 g
Glucose (autoclaved separately)	1 g
$MgSO_4 \cdot 7H_2O$	100 mg
KCl	100 mg
$NaNO_3$	100 mg
$CaCl_2 \cdot 2H_2O$	100 mg
Na glycerophosphate	100 mg
Cobalamin	1 µg
Trace elements	See below
Distilled water	1 liter
pH 7.5	

No. 4 — *S. albida*

As No. 3, but glycerol autoclaved in the medium, replacing glucose

Trace elements	mg/l
Fe	0.5
Zn	0.3
B	0.1
Co	0.1
Cu	0.1
Mn	0.1
Mo	0.1

Media for Freshwater Myxobacters[311]

No. 1 — Peptonized Milk Agar (PMA)

Peptonized Milk	1.0 g
Agar	15.0 g
Distilled water	1 liter

A liquid medium (PMS) was also made by omitting the agar.

No. 2 — Mineral Salt Peptonized Milk Agar (SPMA)

Peptonized milk	1.0 g
Agar	15.0 g
Distilled water	1 liter

Autoclave at 121°C for 15 min. Cool to 50°C and add the following from separately sterilized solutions:

$(NH_4)_2SO_4$	0.10 g
$MgSO_4 \cdot 7H_2O$	0.50 g
$FeCl_3 \cdot 6H_2O$	0.01 g
$CaCl_2$	0.25 g
$MnCl_2$	0.0001 g
K_2HPO_4	0.25 g

No. 3 — Crystal Violet PMA (CVPMA)

Peptonized milk	4.0 g
Agar	60.0 g
Distilled water	4 liters

Separate into four 1-liter volumes. Add the following volumes of a saturated alcoholic solution of crystal violet:

Flask 1	0.1 ml
Flask 2	0.005 ml
Flask 3	0.001 ml
Flask 4	0.0001 ml

Sterilize by autoclaving at 121°C for 15 min.

No. 4 — Dilute Dung Decoction Agar (DBA)

Rabbit dung (dry)	10.0 g
Distilled water	1 liter

Boil and allow to stand for 24 hr at 27°C. Filter through gauze and restore volume to 1 liter.

Agar	15.0 g

Adjust pH to 6.8. Sterilize by autoclaving at 121°C for 30 min.

No. 5 — Oatmeal Infusion Agar (OIA)

Oatmeal	100 g
Distilled water	1 liter

Boil for 5 min, filter hot through gauze. Restore volume to 1 liter and autoclave at 121°C for 15 min. Hold at 10°C for 5 days or until three distinct layers form in the flask. Draw off 50 ml of the gelatinous middle layer and add to 1 liter of 1.5% w/v agar. Sterilize by autoclaving at 121°C for 15 min.

No. 6 — Bacterial Cell Agar (BCA)

Bacto Tryptose Blood Agar base w/o agar	33.0 g
Distilled water	1 liter

Sterilize at 121°C for 15 min. Inoculate with a culture of *Aeromonas hydrophila*. Incubate at 30°C for 72 hr, shaking occasionally. Distribute liquid culture in 40-ml volumes and centrifuge. Wash the cells four times. Resuspend in 25 ml of distilled water and heat at 121°C for 15 min. Add the 25-ml volumes of washed cells at 250 ml of sterilized 1.5% w/v agar in distilled water.

Cytophaga fermentans (agar digesting)[312]

NaCl	30.0 g
KH_2PO_4	1.0 g
NH_4Cl	1.0 g
$MgCl_2 \cdot 6H_2O$	0.5 g
$CaCl_2$	0.04 g
$NaHCO_3$	5 g
$Na_2S \cdot 9H_2O$	0.1 g
Fe citrate ($M/250$)	5 ml
Yeast extract	0.3 g
Agar	5 g
Trace element solution	2 ml
Distilled water	1 liter

Adjust pH to 7.0.

Trace element solution

H_3BO_3	0.28 g
$MnSO_4 \cdot 6H_2O$	0.21 g
$Cu(NO_3)_2 \cdot 3H_2O$	0.02 g
$Na_2MoO_4 \cdot 2H_2O$	0.075 g
$CoCl_2 \cdot 6H_2O$	0.02 g
$Zn(NO_3)_2 \cdot 6H_2O$	0.025 g
Distilled water	100 ml

2. Sheathed Bacteria

Isolation Medium for Sphaerotilus spp.[313]

Pancreatic digest of casein	5.0 g
Glycerol	10.0 g
Yeast extract	1.0 g
Distilled water	1 liter
Agar (optional)	15 g
pH 7.0	

Sterilize by autoclaving at 121°C for 15 min.

Defined Medium for Sphaerotilus spp.[313]

Glycerol	5.0 g
Glutamic acid	0.9 g
$MgSO_4 \cdot 7H_2O$	0.1 g
$FeSO_4 \cdot 7H_2O$	0.5 g
$CaCl_2 \cdot 2H_2O$	0.03 g
$ZnSO_4 \cdot 7H_2O$	0.03 g
Distilled water	900 ml
Agar (optional)	15 g
pH 7.0	

Sterilize by autoclaving at 121°C for 10 min. Prepare a separate phosphate solution:

K_2HPO_4	5.7 g
KH_2PO_4	2.3 g
Distilled water	500 ml

Sterilize at 121°C for 15 min. Add 1 vol sterile phosphate buffer to 9 vol sterile basal medium. For some strains of Sphaerotilus spp. the addition of 0.1% casein digest or 0.01% yeast extract is required.

Sphaerotilus discophorus Medium[314]

Peptone		5 g
$MgSO_4 \cdot 7H_2O$		0.2 g
$CaCl_2$		0.05 g
$MnSO_4 \cdot H_2O$		0.05 g
Agar		12 g
Tap water		900 ml
Ferric NH_4 citrate	Prepared as 100-ml	0.5 g
$FeCl_3 \cdot 6H_2O$	separate solution	0.01 g
pH 7.0		

Sterilize the two solutions separately at 121°C for 15 min. Add them aseptically after sterilization.

Enrichment Medium for Sphaerotilus spp. and Leptothrix spp.[307]

Peptone	1.0 g
Glucose	1.0 g
$MgSO_4 \cdot 7H_2O$	0.2 g
$CaCl_2$	0.05 g
$FeCl_3 \cdot 6H_2O$	0.1 g
Distilled water	1 liter
pH 7.0	

Sterilize by autoclaving at 121°C for 15 min. Agar may be added to the medium at 10 to 15 g/l to isolate filamentous iron bacteria.

Leptothrix ochracea Medium[307]

Manganous acetate	0.1 g
Agar	10.0 g
Distilled water	1 liter
or	
$NaHCO_3$	0.01 g
$(NH_4)_2SO_4$	0.01 g
KH_2PO_4	Trace
$MgSO_4$	Trace

Add to 1 liter distilled water containing 100 ml of a saturated solution of manganese bicarbonate (prepared by passing CO_2 through a suspension of manganese carbonate and filtering).

3. Budding and Appendaged Bacteria
Enrichment and Isolation of the Stalked Bacteria Caulobacter and Hyphomicrobium[143]

Enrichment medium	
KH_2PO_4	0.136 g
$(NH_4)_2SO_4$	0.05 g
$CaCl_2 \cdot 2H_2O$	1.0 mg
$MnSO_4 \cdot 7H_2O$	0.25 mg
Na_2HPO_4	0.213 g
$MgSO_4 \cdot 7H_2O$	0.02 g
$FeSO_4 \cdot 7H_2O$	0.5 mg
$Na_2MoO_4 \cdot 2H_2O$	0.25 mg
Water to	100 ml
pH (adjust with NaOH)	7.2
(Agar 1.9% may be used to solidify)	
Isolation medium for *Caulobacter*	
Peptone	1.0 g
Yeast extract	0.3 g
Agar	1.8 g
Water to	100 ml
pH (adjust with NaOH)	7.2
Isolation medium for *Hyphomicrobium*	

Enrichment medium plus methylamine HCl, 0.675%

From Aaronson, S., *Experimental Microbial Ecology*, Academic Press, London, 1970, 62. With permission.

Enrichment for Budding Bacteria, Hyphomicrobium

$Na_2HPO_4 \cdot 7H_2O$	0.02 g
KNO_3	0.04 g
$MgSO_4 \cdot 7H_2O$	0.48 mg
$MnCl_2 \cdot 4H_2O$	0.01 mg
$FeCl_3 \cdot 6H_2O$	0.02 mg
Water to	100 ml
pH (adjust with HCl or NaOH)	7.2

From Aaronson, S., *Experimental Microbial Ecology*, Academic Press, London, 1970, 62. With permission.

Sterilize by autoclaving at 121°C for 15 min.

Hyphomicrobium Medium[300]

KH_2PO_4	1.36 g
Na_2HPO_4	2.13 g
$MgSO_4 \cdot 7H_2O$	0.2 g
$CaCl_2 \cdot 2H_2O$	9.95 mg
$FeSO_4 \cdot 7H_2O$	5 mg
$MnSO_4 \cdot 4H_2O$	2.5 mg
$Na_2MoO_4 \cdot 2H_2O$	2.5 mg
Agar powder	15 g
Distilled water	970 ml
Methanol	4 ml
Urea (20% w/v solution in distilled water)	30 ml

Sterilize the methanol and urea separately by filtration. Dissolve the salts in the water and add the agar. Steam the medium and autoclave it in bulk at 121°C for 15 min. Add the methanol and 30 ml of the urea solution to the molten agar and dispense the complete medium aseptically into tubes or bottles.

Caulobacter spp. Media[315]

Fresh water isolates

Peptone	2 g
Yeast extract	1 g
$MgSO_4 \cdot 7H_2O$	0.2 g
Agar	10 g
Tap water	1 liter
pH 7.0	

Some strains may require riboflavin (1000 μg/l). Sterilize by autoclaving at 121°C for 15 min.

Marine isolates

Peptone	0.5 g
Agar	10.0 g
Filtered sea water	1 liter
pH 7.0	

Sterilize by autoclaving at 121°C for 15 min.

Galliona ferruginea Medium[316]

The source of iron is an essential part of this medium. Ferrous sulfide is prepared by boiling equal molar quantities of ferrous ammonium sulfate and sodium sulfide in distilled water. The precipitated FeS is washed four times with boiling distilled water, avoiding oxidation of the iron.

NH_4Cl	1 g
K_2HPO_4	0.5 g
$MgSO_4 \cdot 7H_2O$	0.2 g
Distilled water	1 liter

Separately sterilize the above constituents in water and mix aseptically. Bubble carbon dioxide through the medium filled out in stoppered test tubes for approximately 5 sec. Add the FeS precipitate slowly from a pipette and allow it to settle at the bottom of the tube.

Metallogenium symbioticum Medium[317]

Manganous acetate	0.1 g
Agar	10 g
Distilled water	1 liter

Dissolve the manganous acetate in the water and check the pH to approximately 7.0. Add the agar and steam the medium. Dispense the medium into screw-capped bottles and autoclave at 121°C for 15 min.

D. Spirochetes
Spirochaeta aurantia Medium[318]

Isolation Medium

Hay extract is prepared by boiling 5 g dried hay in 1 liter distilled water for 10 min. Filter off the hay residue.

Hay extract	500 ml
Peptone	1 g
Yeast extract	1 g
pH 6.5	

Sterilize by autoclaving at 121°C for 15 min.

Growth Medium

Maltose	2 g
Peptone	2 g
Yeast extract	4 g
pH 7.5	

Sterilize by autoclaving at 121°C for 15 min and add sterile $M/1$ K_2HPO_4 buffer (pH 7.0) to a final concentration of 0.01 M. Final pH, 7.2.

Treponema zuelzerae Medium[300]

Basal Medium
KH_2PO_4	1 g
NH_4Cl	1 g
$MgCl_2 \cdot 6H_2O$	0.5 g
$CaCl_2$	0.04 g
$FeCl_3 \cdot 6H_2O$	0.0025 g
Trace element solution (see below)	2 ml
Agar powder	15 g
Distilled water	1 liter

Trace element solution
H_3BO_3	56 mg
$ZnSO_4 \cdot 7H_2O$	44 mg
$CaCl_2 \cdot 6H_2O$	20 mg
$CuSO_4 \cdot 5H_2O$	0.2 mg
$MnCl_2$	2 mg
$Na_2MoO_4 \cdot 2H_2O$	75 mg
Distilled water	100 ml

Additives
    $NaHCO_3$ 10% w/v in distilled water
    $Na_2S \cdot 9H_2O$ 10% w/v in distilled water
    Glucose 10% w/v in distilled water
    Yeast extract 10% w/v in distilled water

Prepare the basal medium and the additives separately and autoclave them at 121°C for 15 min in separate containers. For use, to 1 liter basal medium add aseptically 10 ml bicarbonate, 5 ml sulfide, 10 ml glucose, and 10 ml yeast extract as detailed above.

From Lapage, S. P., Shelton, J. E., and Mitchell, T. G., in *Methods in Microbiology,* Vol. 3A, Norris, J. R. and Ribbons, D. W., Eds., Academic Press, London, 1970, 1—228. With permission.

Spirolate Broth (BBL) for Treponema reiteri[319]

Trypticase peptone	15.0 g
Dextrose	5.0 g
Yeast extract	5.0 g
NaCl	2.5 g
Na thioglycollate	0.5 g
Cysteine	1.0 g
Distilled water	1 liter
pH 7.1	

Add 0.25 g/l of TEM 4T (diacetyl tartaric acid ester of monoglycerides of animal origin; T.M. Hachmeister Inc.). Boil to dissolve the medium. Fill out into tubes and sterilize at 121°C for 15 min. Cool and add 10% v/v sterile inactivated serum (rabbit, sheep, or bovine.)

Nonpathogenic Treponema Medium (Reiter, Kazan, Nichols, Noguchu)[320]

Thioglycollate Medium USP	100 ml
Rabbit serum (inactivated at 56—60°C for 1 hr)	10 ml

Chemically Defined Medium for Leptospira spp.[321]

Component	Conc. (mg/l)	Component	Conc. (mg/l)
NaH_2PO_4	531.0	Tween® 80	50.0
KH_2PO_4	69.0	Tween® 60	200.0
$MgSO_4 \cdot 7H_2O$	150.0	Glycerol	200.0
$(NH_4)_2[Fe(SO_4)_2] \cdot 6H_2O$	6.0	L-Asparagine	500.0
Ca^{++} b	4.0	Vitamin B_{12}	10^{-3}
EDTA	10.0	Thiamine HClc	1.0

a The medium was heated to boiling temperature, filtered, and sterilized by autoclaving at 121°C for 15 min. The final pH was 7.4 to 7.6. The basal medium was similar in composition except for the following changes: glycerol and Tween® 80 were omitted; asparagine was replaced by ammonium chloride (200 mg/l); and the concentration of Tween® 60 was reduced to 160 mg/l. (In some experiments, Tween® 60 was replaced with palmitic acid (12 mg/l) or stearic acid (14 mg/l).)

b Ca^{++} was weighed as $CaCO_3$ and dissolved in dilute HCl.

c Thiamine HCl was added aseptically after sterilization.

From Shenberg, E., *J. Bacteriol.*, 93, 1598—1606, 1967. With permission.

Korthof's Medium for Leptospira spp.[322]

Peptone	0.8 g
NaCl	1.4 g
$NaHCO_3$	0.02 g
KCl	0.04 g
$CaCl_2$	0.04 g
KH_2PO_4	0.24 g
$Na_2HPO_4 \cdot 2H_2O$	0.88 g
Distilled water	1 liter
pH 7.2	

Steam the ingredients, filter, and autoclave 100-ml volumes at 115°C for 15 min. To 100 ml peptone solution, add 8 ml of sterile inactivated rabbit serum and 0.8 ml of a filter-sterilized "hemoglobin" solution (rabbit blood clot with equal volume of distilled water, lysed by freezing and thawing).

Solid Medium for Growth of Leptospira spp.

Tryptose phosphate broth	0.2 g
Agar	1.0 g
Distilled water	90.0 ml
pH 7.5	

Sterilize by autoclaving at 121°C for 15 min. Cool to 50°C and add 10 ml sterile rabbit serum and 1 ml "hemoglobin" solution (rabbit cells lysed in distilled water, centrifuged, and filter sterilized). Place complete medium in a water bath at 56°C for 30 min before pouring plates.

E. Spiral and Curved Bacteria
Spirillum spp. Medium[300]

Peptone	5 g
Beef extract	3 g
Yeast extract	3 g
Calcium lactate	10 g
Distilled water	1 liter

Dissolve the constituents in the water and adjust the pH to 7.0. Dispense the medium as required into bottles or tubes and sterilize by autoclaving at 116°C for 20 min. The precipitate that forms is not removed.

Marine Spirillum Medium

Prepare the medium in the same way as *Spirillum* medium, except replace 750 ml of the distilled water by an equal volume of aged sea water.

Spirillum volutans Medium

Place a grain of wheat in a 20 × 180-mm test tube and cover it with 1 in. of garden soil. Add water to a height of 2 in. in the tube. After plugging the mouth of the tube, autoclave at 121°C for 15 min. Check the pH and adjust if necessary to 6.5 to 7.5.

Peptone-Succinate-Salts Broth (PSS) for Spirillum spp.[323]

Peptone	1.0 g
Succinic acid	0.1 g
$(NH_4)_2SO_4$	0.1 g
$MgSO_4 \cdot 7H_2O$	0.1 g
$FeCl_3 \cdot 6H_2O$	0.0002 g
$MnSO_4 \cdot H_2O$	0.0002 g
Distilled water	100 ml
pH 6.8	

Use KOH to adjust pH. Sterilize by autoclaving at 121°C for 15 min.

Marine Spirilla

Substitute synthetic sea water for distilled water.

NaCl	2.75 g
$MgCl_2$	0.50 g
$MgSO_4$	0.20 g
$CaCl_2$	0.05 g
KCl	0.10 g
$FeSO_4$	Trace
Distilled water	100 ml

The addition of 0.15 g % w/v agar to the **PSS** medium gives a semisolid medium that can be used to store cultures at 30°C.

Spirillum gracile Medium[324]

Peptone	5.0 g
Yeast extract	0.5 g
Tween® 80	0.02 g
K_2HPO_4	0.1 g
Tap water	1 liter
pH 7.2	

Campylobacter (Vibrio) fetus Medium[325]

Albimi brucella broth	28.0 g
Agar	5.0 g
Ox bile (fresh)	40.0 ml
Ethyl violet to give final concentration	1:800,000
Distilled water	1 liter

Sterilize by autoclaving at 121°C for 15 min. Aseptically add

Bacitracin	25,000 units
Polymyxin B sulfate	5,000 units
Actidione	100 mg

Dispense aseptically in tubes to a depth of 3 to 4 cm.

Bdellovibrio spp. Medium[326]

YDC
Yeast extract	10.0 g
Glucose	20.0 g
$CaCO_3$	20.0 g
Agar	15.0 g
Distilled water	1 liter
pH 7.2	

NBA
Nutrient broth	8.0 g
Casamino acids	5.0 g
Yeast extract	1.0 g
Distilled water	1 liter
pH 6.8	

Sterilize by autoclaving at 121°C for 15 min.

F. Aerobic Gram-negative Rods and Cocci

1. Pseudomonadaceae

Most of the species studied, including the parasitic ones, require no growth factors and can develop on defined mineral media with a single source of organic carbon. Acetate is the principle carbon source that can be utilized, but over 100 compounds have been described in test media.

Some strains of these species may require growth factors, and it does not follow that cultures requiring amino acids or vitamins belong to the three species known to require supplementation (*Pseudomonas multophila, P. vesicularis,* and *P. diminuta*). *Pseudomonas* spp. will grow on complex peptone/meat extract agar, and selection is made by incorporating selective agents in the formulas. The enhancement of pigment formation is a diagnostic characteristic that is developed on some media. A number of media have been specifically designed to isolate *P. aeruginosa* because of its importance in clinical medicine.

Medium A — for Pyocyanin Production[327]

Peptone	20 g
Glycerol	10 g
K_2SO_4	10 g
$MgCl_2$	1.4 g
Agar	15 g
Distilled water	1 liter
pH 7.2	

Sterilize by autoclaving at 115°C for 10 min.

Medium B — for Fluorescin Production

Peptone	20 g
Glycerol	10 g
K_2HPO_4	1.5 g
$MgSO_4 \cdot 7H_2O$	1.5 g
Agar	15 g
Distilled water	1 liter
pH 7.2	

Sterilize phosphate solution separately and add it to the sterile, cooled base.

Basal Mineral Medium for Pseudomonas spp.[328]

K_2HPO_4	12.5 g
KH_2PO_4	3.8 g
$(NH_4)_2SO_4$	1.0 g
$MgSO_4 \cdot 7H_2O$	0.1 g
Trace element solution	5 ml
Distilled water	1 liter
pH 7.2	

Sterilize by autoclaving at 121°C for 15 min. Add filter-sterilized carbon source at a final concentration of 0.08 M. Trace element solution contains (mg/l)

$FeSO_4(NH_4)_2SO_4 \cdot 6H_2O$	116.0
HBO_3	232.0
$CoSO_4 \cdot 7H_2O$	95.6
$CuSO_4 \cdot 5H_2O$	8.0
$MnSO_4 \cdot 4H_2O$	8.0
$(NH_4)_6Mo_7O_{24} \cdot 4H_2O$	22.0
$ZnSO_4 \cdot 7H_2O$	174.0

Pseudomonas saccharophila Medium[329]

KH_2PO_4-Na_2HPO_4, $M/30$, pH 6.64	1 liter
NH_4Cl	1 g
$MgSO_4 \cdot 7H_2O$	0.5 g
$FeCl_3 \cdot 6H_2O$	0.05 g
$CaCl_2 \cdot 2H_2O$	0.01 g
Sucrose	2 g

Dissolve the salts and sucrose in the buffer solution and distribute in bottles or tubes. Sterilize by autoclaving at 121°C for 15 min.

Malachite Green Broth for P. aeruginosa [330]

Peptone	15 g
Lab-Lemco®	9 g
Malachite Green	1:100,000 (final concentration)
Distilled water	1 liter
pH 7.3	

Sterilize by autoclaving at 121°C for 15 min.

Cetrimide Agar for P. aeruginosa[331]

Peptone	10.0 g
Lab-Lemco®	10.0 g
NaCl	5.0 g
Cetrimide	0.3 g
Agar	15.0 g
Distilled water	1 liter
pH 7.4	

Sterilize the cetrimide separately by filtration of a 2% w/v solution. After autoclaving the base medium at 121°C for 15 min, cool to 45°C and add the cetrimide.

Modified Cetrimide Agar for P. aeruginosa
Add 0.003 g % w/v cetrimide to 100 ml of Medium A.[327]

Low-peptone Cetrimide Agar for P. aeruginosa[332]

Peptone	0.2 g
K_2SO_4	10.0 g
$MgCl_2 \cdot 6H_2O$	1.4 g
Glycerol	5.0 g
D-Mannitol	5.0 g
Cetrimide	0.3 g
Acetimide	10.0 g
Phenol red	0.012 g
Distilled water	1 liter
pH 7.0	

Sterilize the acetimide/phenol red by filtration as a separate solution. Add it to the base medium sterilized at 121°C for 15 min.

Other Selective Agents

A review[333] of ten selective media for the detection of *P. aeruginosa* listed the following compounds:

Cetrimide
Nalidixic acid
China acid/penicillin
Triphenyl tetrazolium chloride
Dettol®
Ampholyte (Boehringer)
Cadmium sulfate

Media for Xanthomonas spp.

This group of plant pathogens are very closely related to *Pseudomonas* spp., but their minimal growth requirements are complex. A yellow, nondiffusable pigment is formed by most species, especially on nutrient agar containing 2% w/v dextrose, sucrose, or other utilizable carbohydrates. Malt extract enhances the production of xanthan gum in the developing colonies.

Malt Extract Agar

Malt extract	30.0 g
Peptone	5.0 g
Agar	15.0 g
Distilled water	1 liter
pH 7.0	

Sterilize by autoclaving at 115°C for 20 min.

2. Azotobacter

Glucose, Nitrogen-free Salt Solution[334]

Glucose	10.0 g
K_2HPO_4	1.0 g
$MgSO_4 \cdot 7H_2O$	0.2 g
$CaCO_3$	1.0 g
NaCl	0.2 g
$FeSO_4 \cdot 7H_2O$	0.1 g
$Na_2MoO_4 \cdot 2H_2O$	0.005 g
Distilled water	1 liter
pH 7.0	

Agar can be added at 15 g/l, and the glucose can be replaced by other energy sources. Sterilize the carbohydrate separately by filtration and add it to the basal medium which has been autoclaved at 121°C for 15 min.

Isolation of Azotobacter spp. from Water

Pipette 25-ml volumes of water into four sterile 100-ml flasks and add the following sterile nutrients to the given concentration:

1. 0.02% K_2HPO_4 plus 1% mannitol.
2. 0.02% K_2HPO_4 plus 0.5% mannitol plus 1% sodium benzoate.
3. 0.02% K_2HPO_4 plus 1% ethanol.
4. 0.02% K_2HPO_4 plus 1% ethanol plus 1% sodium benzoate.

Incubate at 25°C and subculture to glucose, nitrogen-free agar.

Silica Gel Media for Azotobacter spp.

Silica gel plates are impregnated with glucose, nitrogen-free salt solution and particles of soil are placed on the surface.

Soil-mannitol Buffer Plates

Azotobacter spp. will grow on the surface of sieved soil containing 1% w/w mannitol that has been mixed into a smooth paste with sterile buffer solution.

Buffer solution	
K_2HPO_4	0.67 g
KH_2PO_4	0.33 g
Distilled water	1 liter

3. Rhizobiaceae
Rhizobium Medium 1[300]

Yeast extract	10 g
K_2HPO_4	0.5 g
$MgSO_4 \cdot 7H_2O$	0.2 g
NaCl	0.2 g
$FeCl_3 \cdot 6H_2O$	0.002 g
Agar powder	15 g
Distilled water	1 liter

Dissolve all the ingredients except the agar in the water and adjust the pH to 7.2. Add the agar and steam the medium. After distribution into bottles or tubes, autoclave the medium at 121°C for 15 min.

Rhizobium Medium 2[300]

Glycerol	4.6 g
L-Arabinose	1 g
Yeast extract	1 g
K_2HPO_4	1 g
$CaSO_4$	1.3 g
KNO_3	0.7 g
$FeCl_3 \cdot 6H_2O$	0.004 g
$MgSO_4 \cdot 7H_2O$	0.36 g
Agar powder	15 g
Distilled water	1 liter

Mix all the ingredients except the agar with the water and adjust the pH to 7.2. Add the agar and steam the medium. After distribution into bottles or tubes, autoclave the medium at 121°C for 15 min.

Yeast Extract Mannitol Agar[335]

Mannitol	10.0 g
K_2HPO_4	0.5 g
$MgSO_4 \cdot 7H_2O$	0.2 g
$CaCO_3$	4.0 g
NaCl	0.1 g

Yeast Extract Mannitol Agar[335]
 (continued)

Yeast extract	0.4 g
Agar	15.0 g
Distilled water	1 liter
pH 6.8–7.0	

Sterilize by autoclaving at 121°C for 15 min. The $CaCO_3$ may be omitted when a clear medium is required.

Legume Extract Agar for Rhizobium[336]

Cut up 35 g of well-washed alfalfa roots and mix with 10 g of soybean meal. Add three times the volume of water, steam gently for 1 hr, and allow to stand overnight. Dilute to 1000 ml and filter through paper pulp. Add the following (g):

K_2HPO_4	1.0
$MgSO_4 \cdot 7H_2O$	0.2
NaCl	0.1
$CaCl_2$	0.1
$FeCl_3$	0.001
Agar	20.0

Heat to 121°C for 20 min, filter, and add (g)

$CaCO_3$	5
Sucrose	10
Glucose	5

When the sugars have dissolved, dispense as required and sterilize at 115°C for 20 min.

Glucose Peptone Agar for Agrobacterium spp.[335]

Glucose	10.0 g
Peptone	20.0 g
NaCl	5.0 g
Agar	15.0 g
Distilled water	1 liter
pH 7.2	

Sterilize by autoclaving at 121°C for 15 min. Most *Rhizobia* grow poorly or not at all on this medium.

4. Methylmonadaceae
Autotrophic Medium for Methanomonas[337]

$NaNO_3$	2.0 g
$MgSO_4 \cdot 7H_2O$	0.2 g
$FeSO_4 \cdot 7H_2O$	0.001 g
Na_2HPO_4	0.21 g
NaH_2PO_4	0.09 g
$CuSO_4 \cdot 5H_2O$	200.0 μg
H_3BO_3	60.0 μg

Autotrophic Medium for Methanomonas[337]
(continued)

$MnSO_4 \cdot H_2O$	30.0 µg
$ZnSO_4 \cdot 7H_2O$	300.0 µg
MoO_3	15.0 µg
KCl	0.04 g
$CaCl_2$	0.015 g
Distilled water	1000 ml

Dissolve the salts and sterilize. Incubate under an atmosphere of 50% methane and 50% air.

Methylococcus Medium[338]

$NaNO_3$	2 g
$MgSO_4 \cdot 7H_2O$	0.2 g
KCl	0.04 g
$CaCl_2$	0.015 g
Na_2HPO_4	0.21 g
NaH_2PO_4	0.09 g
$FeSO_4 \cdot 7H_2O$	1 mg
$CuSO_4 \cdot 5H_2O$	5 µg
H_3BO_3	10 µg
$MnSO_4 \cdot 5H_2O$	10 µg
$ZnSO_4 \cdot 7H_2O$	70 µg
$Na_2MoO_4 \cdot 2H_2O$	10 µg
Deionized water	1 liter

Dissolve the constituents in the water and dispense the medium in bottles or flasks. Autoclave the medium at 121°C for 15 min.

5. Halobacteriaceae
Halophile Medium[339]

Casamino acids	7.5 g
Yeast extract	10 g
Trisodium citrate	3 g
KCl	2 g
$MgSO_4 \cdot 7H_2O$	20 g
$FeSO_4 \cdot 7H_2O$	0.05 g
$MnSO_4 \cdot 4H_2O$	0.25 mg
Agar powder	20 g
NaCl	250 g
Distilled water	1 liter

Dissolve all ingredients except the agar in the water and adjust the pH to 7.4. Add the agar, steam the medium, and distribute as required. Autoclave at 121°C for 15 min.

Synthetic Medium for Halophilic Bacteria[340]

DL-Alanine	43.0 mg	Cytidylic acid	10 0	mg
L-Arginine	40.0 mg	NaCl	25.0	g
DL-Aspartic acid	45.0 mg	$MgSO_4 \cdot 7H_2O$	2.0	g
L-Cystine	5.0 mg	KNO_3	10.0	mg
L-Glutamic acid	130.0 mg	K_2HPO_4	5.0	mg
Glycine	6.0 mg	KH_2PO_4	5.0	mg
DL-Histidine	30.0 mg	Sodium citrate	50.0	mg
DL-Isoleucine	44.0 mg	$FeCl_2$	0.23	mg
L-Leucine	80.0 mg	$CaCl_2 \cdot 7H_2O$	0.7	mg
L-Lysine	85.0 mg	$MnSO_4 \cdot 5H_2O$	0.03	mg
DL-Methionine	37.0 mg	$ZnSO_4 \cdot 7H_2O$	0.044	mg
DL-Phenylalanine	26.0 mg	$CuSO_4 \cdot 5H_2O$	5.0	μg
L-Proline	5.0 mg	Glycerol	0.1	g
DL-Serine	61.0 mg	Tween® 40	50.0	mg
DL-Threonine	50.0 mg	Biotin	0.1	μg
L-Tyrosine	20.0 mg	Folic acid	10.0	μg
DL-Tryptophan	5.0 mg	Vitamin B_{12}	0.02	μg
DL-Valine	100.0 mg	Water to	100	ml
Adenylic acid	10.0 mg	pH (adjusted with KOH)	6.2	
Guanylic acid	10.0 mg			
Uridylic acid	10.0 mg			

From Aaronson, S., *Experimental Microbial Ecology,* Academic Press, London, 1970, 62. With permission.

Sterilization is not necessary with this medium.

Synthetic Medium for Halophilic Bacteria[341]

Basal nutrient medium
Glucose	0.1 g
KNO_3	0.05 g
$FePO_4$	0.01 g
Artificial seawater to	100 ml

Artificial seawater
NaCl	2.4 g
Na_2SO_4	0.4 g
$MgCl_2 \cdot 6H_2O$	1.1 g
$CaCl_2 \cdot 6H_2O$	0.2 g
KCl	0.07 g
$NaHCO_3$	0.02 g
KBr	0.01 g
$SrCl_2 \cdot 6H_2O$	0.004 g
H_3BO_3	0.003 g
$Na_2SiO_3 \cdot 9H_2O$	0.5 mg
NaF	0.3 mg
Demineralized water to	100 ml

From Aaronson, S., *Experimental Microbial Ecology,* Academic Press, London, 1970, 62. With permission.

Sea Water Medium for Halophilic Bacteria[342]

Peptone	5.0 g
Lab-Lemco®	2.0 g
KNO$_3$	0.5 g
Agar	15.0 g
Aged sea water	1 liter
pH 7.8	

Sterilize by autoclaving at 121°C for 15 min. Mix any precipitate that may form on heating.

6. Neisseriaceae
Media for N. gonorrhoeae[343]

Columbia agar base	90 ml
(sterile, cooled to 60°C)	
Defibrinated, sterile horse	10 ml
blood	
pH 7.4	

Add the blood to the molten agar, mix without causing bubbles, hold at 60°C until the heated blood has attained a light chocolate color.

The following antibiotics (µg/ml) may be added to the medium to make it more selective:[344]

Vancomycin	3
Colistin	7.5
Nystatin	12.5
Trimethoprim	5.0

Autoclaved hemoglobin may be added in place of whole blood.[345]

The following enrichment supplements may be added to improve the growth of demanding strains:

Supplement B (Difco Laboratories)
Isovitalex (Bioquest B-D)
GC Supplement (Oxoid Ltd.)

A transport medium called "Transgrow" was developed from the above media by increasing the agar content (from 1 to 2%) and the dextrose content (from 0.1 to 0.25%) and preparing the medium in a flat bottle with a CO_2 atmosphere.[344] Bicarbonate can be incorporated into the medium as replacement for a gaseous CO_2 atmosphere.[346]

Mueller Hinton Agar[262]

Beef infusion	300 ml
Casein hydrolysate	17.5 g
Starch	1.5 g
Agar	10 g
Distilled water to	1 liter

Emulsify the starch in a small amount of cold water, pour into the beef infusion, and add the casein hydrolysate and the agar. Make up the volume to 1 liter with distilled water. Dissolve the constituents by heating gently at 100°C with agitation. Filter if necessary. Adjust the pH to 7.4. Dispense in screw-capped bottles and sterilize by autoclaving at 121°C for 20 min. Pour plates.

The above media can also be used to isolate and cultivate *N. meningitidis.* The nonpathogenic *Neisseria* (*N. sicca, N. subflava, N. flavescens,* and *N. muessa*) are all less demanding in growth requirements and can be isolated on nutrient agar.

Media for the Growth of Moraxella spp.

Some species will grow on blood agar prepared with 5 to 10% v/v defibrinated horse or rabbit blood. All species will grow on coagulated serum and do not require hematin (x) or NAD (v) as growth factors.

Loeffler's Serum Medium

Sterile ox, sheep, or horse serum	300 ml
Nutrient broth	100 ml
Glucose	1 g

Dissolve the glucose in the broth and autoclave at 115°C for 20 min. Add the glucose broth to the serum using aseptic precautions and distribute in sterile test tubes or bottles. Inspissate by laying the tubes on a sloped tray and slowly raising the temperature to 80 to 85°C and holding at this temperature for 2 hr. The serum coagulates to a yellowish-white solid which is firm enough to inoculate as soon as it is cool.

Acinetobacter spp.

All strains are able to grow aerobically on ordinary nutrient media without the addition of serum.[347]

7. Affiliated Bacteria
Alcaligenes spp. Agar[348]

Peptone	5 g
Meat extract	3 g
Ammonium lactate	3 g
Ferric citrate	0.2 g
Agar powder	10 g
Distilled water	1 liter

Dissolve the ferric citrate in a small part of the water and add to complete the medium after both it and the remainder have been sterilized. Dissolve the peptone, meat extract, and lactate in the remainder of the water and adjust the pH to 7.0. Add the agar and steam the medium. Sterilize the two parts of the medium separately by steaming for 20 min on three successive days. Mix the parts and distribute aseptically.

Acetobacter spp. Agar[349]

Yeast extract	30 g
$CaCO_3$ (finely divided)	20 g
Agar	20 g
Distilled water	1 liter

Sterilize by autoclaving at 121°C for 15 min. When cool and molten, add 2 ml of filter-sterilized ethanol to every 13 ml of base agar. Disperse the chalk by shaking and pour plates.

A variation of this medium was proposed[350] in which bromcresol green indicator (1 ml of a 2.2% w/v solution per liter of base agar) replaced chalk as an indicator of acid production from alcohol.

Hoyer's Medium for Acetobacter spp.[350]

$(NH_4)_2SO_4$	1.0 g
K_2HPO_4	0.1 g
KH_2PO_4	0.9 g
$MgSO_4 \cdot 7H_2O$	0.25 g
$FeCl_3 \cdot 6H_2O$	0.02 g
Distilled water	800 ml

Dispense into 4-ml aliquots in test tubes and sterilize by autoclaving at 121°C for 15 min. Add 1 ml of a 15% v/v filter-sterilized solution of ethanol to each tube.

Brucella Media

Serum Dextrose Agar Base (g/l)[351]

Peptone	10.0
'Lab-Lemco' Powder	5.0
Dextrose	10.0
Sodium chloride	5.0
Agar	15.0
pH 7.5 (approximately)	

Suspend in 1 liter distilled water. Boil to dissolve completely. Sterilize by autoclaving at 121°C for 15 min. Cool to 50°C and add 5% of inactivated horse serum (i.e., serum held at 56°C for 30 min). Mix well before pouring.

The following antibiotics may be added to the above serum-dextrose agar:

Bacitracin	25,000 units/l
Polymyxin B	6,000 units/l
Actidione	100 mg/l

It is important to ensure thorough mixing in order to obtain uniform results.

Tween®-Dextrose Agar[352]

Nutrient agar (sterile, cooled to 60°C)	95 ml
Tween® 40	0.5 ml
Dextrose (20% w/v solution, filter sterilized)	5 ml

Glycerol-Dextrose Agar[352]

Nutrient agar (sterile, cooled to 60°C)	93 ml
Glycerol	2 ml
Dextrose (20% w/v solution, filter sterilized)	5 ml

Serum-Potato Infusion Agar[352]

Infuse 250 g thinly sliced potatoes in 1 liter distilled water overnight at 60°C. Filter.

Peptone	10.0 g
Meat extract	5.0 g
NaCl	5.0 g
Glycerol	20 ml
Agar	15.0 g
Potato infusion	1 liter
Final pH 6.8	

Sterilize by autoclaving at 121°C for 15 min, cool to 50°C, and add 10% v/v horse or ox serum (filter sterilized, free from *Brucella* agglutinins, and inactivated at 56°C for 30 min).

Albimi Agar[353]

Albimi Agar (Albimi Laboratories, Brooklyn, N.Y.)	1 liter
Ethyl violet	1.25 mg
Bacitracin	25,000 units
Polymyxin B	6,000 units
Cyclohexamide (Actidione)	100 mg

Sterilize the Albimi agar at 121°C for 15 min. Cool to 50°C and add the remaining ingredients as sterile solutions.

Starch-Glycerol-Blood Agar for Bordetella pertussis[354]

Glycerol starch agar

Peptone (Bengers or Evans)	10 g
Sodium chloride, NaCl	5 g
Glycerol	10 ml
Starch (Soluble, BDH)	2.5 g
Water	1 liter
Agar (Davis)	11 g

Dissolve the ingredients, except the agar, in the water and adjust to pH 7.8; add the agar; check the pH, which should be 7.5 to 7.6; distribute in 200-ml quantities and autoclave at 115°C for 10 min

Preparation of complete medium

Glycerol starch agar	200 ml
Penicillin (50 units/ml)	1.5 ml
M & B 938 (0.1% solution)	0.9 ml
Sterile defibrinated horse blood	100 ml

Melt the agar in the steamer at 100°C; cool to 40 to 45°C; warm the blood to 37°C and add all the ingredients; mix thoroughly and pour not more than 10 plates from this quantity of medium; store in the refrigerator; a short (10 to 15 min) period of drying may be necessary, but only when the plates are freshly poured; M & B 938 is 4:4 diamido-diphenylamine-hydrochloride (May and Baker)

From Cruickshank, R., Duguid, J. P., Marmion, B. P., and Swain, R. H. A., *Medical Microbiology,* Vol. 2, 12th ed., Churchill, Livingstone, Edinburgh, 1975, 132. With permission.

Blood-Glucose-Cystine Agar for Francisella tularensis[354]

Glucose cystine solution

Glucose	12.5 g
Cystine hydrochloride	0.5 g
Water	50 ml

Prepare the solution and sterilize it by Seitz filtration

Preparation of complete medium

Nutrient agar	85 ml
Glucose cystine solution	10 ml
Human blood, fresh	5 ml

Melt the agar, cool to 50°C, and add the remaining ingredients

From Cruickshank, R., Duguid, J. P., Marmion, B. P., and Swain, R. H. A., *Medical Microbiology,* Vol. 2, 12th ed., Churchill, Livingstone, Edinburgh, 1975, 132. With permission.

G. Facultatively Anaerobic Gram-negative Rods

1. Enterobacteriaceae

All members of this family grow readily on simple nutrient media. The many formulations of culture media listed are selective, indicating media designed to repress unwanted organisms and distinguish organisms that are described as pathogenic to their hosts. Enrichment media and media designed to demonstrate specific biochemical reactions are also published for this family.[355]

Enrichment Media[356]

GN Broth

Tryptose or polypeptone	20 g
Glucose	1 g
D-Mannitol	2 g
Sodium citrate	5 g
Sodium desoxycholate	0.5 g
Dipotassium phosphate	4 g
Monopotassium phosphate	1.5 g
Sodium chloride	5 g
Distilled water	1000 ml
(Final pH around 7.0)	

Dispense and sterilize at 116°C for 15 min.

Selenite F Broth[357]

Sodium acid selenite, NaHSeO$_3$	4 g
Peptone	5 g
Lactose	4 g
Disodium hydrogen phosphate, Na$_2$HPO$_4 \cdot$12H$_2$O	9.5 g
Sodium dihydrogen phosphate, NaH$_2$PO$_4 \cdot$2H$_2$O	0.5 g
Sterile water	1 liter

Dissolve the ingredients with sterile precautions and distribute the yellowish solution in 10-ml amounts in sterile screw-capped bottles. Steam at 100°C for 30 min. Excessive heat is detrimental to the medium and autoclaving must not be used to sterilize it. The amount of red precipitate should be very slight. The pH of the medium should be 7.1 and the phosphates may be varied slightly if necessary to attain this.

From Cruickshank, R., Duguid, J. P., Marmion, B. P., and Swain, R. H. A., *Medical Microbiology,* Vol. 2, 12th ed., Churchill, Livingstone, Edinburgh, 1975, 132. With permission.

Tetrathionate Broth[358]

Thiosulfate solution	
Sodium thiosulfate, Na$_2$S$_2$O$_3 \cdot$5H$_2$O	24.8 g
Sterile water to	100 ml

Mix the salt and water with sterile precautions and steam at 100°C for 30 min. It is a 1 *M* solution.

Iodine solution	
Potassium iodide, KI	20 g
Iodine, I	12.7 g
Sterile water to	100 ml

With sterile precautions, dissolve the potassium iodide in approximately 50 ml of warm water, add the iodine, and make up to a final volume of 100 ml. This gives a normal or 0.5 *M* solution.

Preparation of complete medium	
Calcium carbonate, CaCO$_3$	2.5 g
Nutrient broth	78 ml
Thiosulfate solution	15 ml
Iodine solution	4 ml
Phenol red, 0.02% in 20% ethanol	3 ml

Add the calcium carbonate to the broth and sterilize it by autoclaving at 121°C for 20 min. When cool, add the thiosulfate, iodine, and phenol red solutions with sterile precautions. Distribute in 10-ml amounts in sterile screw-capped bottles.

From Cruickshank, R., Duguid, J. P., Marmion, B. P., and Swain, R. H. A., *Medical Microbiology*, Vol. 2, 12th ed., Churchill, Livingstone, Edinburgh, 1975, 132. With permission.

Kauffmann-Müller Tetrathionate Broth[359]

Thiosulfate solution
Sodium thiosulfate, $Na_2S_2O_3 \cdot 5H_2O$	50 g
Sterile water	100 ml

Mix the salt and water with sterile precautions and steam at 100°C for 30 min.

Iodine solution
Potassium iodide, KI	25 g
Iodine, I	20 g
Sterile water	100 ml

With sterile precautions, dissolve the potassium iodide and add the iodine.

Ox bile solution
Desiccated ox bile	0.5 g
Water	5 ml

Dissolve with sterile precautions.

Preparation of complete medium
Nutrient broth, pH 7.4	90 ml
Calcium carbonate, $CaCO_3$	5 g
Brilliant green, 1 in 1000 aqueous	1 ml
Thiosulfate solution	10 ml
Iodine solution	2 ml
Ox bile solution	5 ml

Add the calcium carbonate to the broth and sterilize it by autoclaving at 121°C for 20 min. When cool, add the other solutions and distribute aseptically in approximately 10-ml amounts. Heat once in the steamer at 100°C for 10 min.

From Cruickshank, R., Duguid, J. P., Marmion, B. P., and Swain, R. H. A., *Medical Microbiology*, Vol. 2, 12th ed., Churchill, Livingstone, Edinburgh, 1975, 132. With permission.

Selective/Indicator Media
MacConkey Agar (Enterobacteriaceae)[360]

Peptone	20 g
Lactose	10 g
Sodium taurocholate	5 g
Neutral red	0.03 g
Agar	15 g
Distilled water	1 liter
pH 7.1	

Sterilize by autoclaving at 121°C for 15 min. This medium has been modified extensively in the following ways:

- By adding 5 g/l NaCl
- By substituting 1.5 g Bile Salts No. 3 for sodium taurocholate
- By adding 0.001 g/l crystal violet

Endo Agar (Enterobacteriaceae)[219]

Peptone	10.0 g
Lactose	10.0 g
K_2HPO_4	3.5 g
$NaHSO_3$	2.5 g
Agar	15.0 g
Distilled water	1 liter
pH 7.4	

Dissolve the ingredients in water. Add 4 ml of a 10% w/v alcoholic solution of basic fuchsin. Sterilize by autoclaving at 121°C for 15 min. Hold poured plates in the dark to prevent photooxidation.

Eosin-Methylene Blue Agar (Enterobacteriaceae)[362]

Peptone	10.0 g
Lactose	10.0 g
K_2HPO_4	2.0 g
Eosin (water soluble)	0.4 g
Methylene blue	0.07 g
Agar	15.0 g
Distilled water	1 liter
pH 6.8	

Sterilize by autoclaving at 121°C for 15 min. Cool to 60°C and mix the medium to suspend the precipitate before pouring.

Deoxycholate Agar (Enterobacteriaceae)[79]

Peptone	10.0 g
Lactose	10.0 g
NaCl	5.0 g
K_2HPO_4	2.0 g
Ferric citrate	1.0 g
Sodium deoxycholate	1.0 g
Sodium citrate	1.0 g
Neutral red	0.03 g
Agar	15.0 g
Distilled water	1 liter
pH 7.2	

Sterilize the peptone/lactose/agar solution separately by autoclaving at 121°C for 15 min. Add the other ingredients and boil the medium for 1 min. Mix and pour into dishes.

Bismuth-Sulfite Agar (*Salmonella* spp.)[363]

Bismuth sulfite glucose phosphate mixture

Bismuth ammoniocitrate, scales	30 g
Sodium sulfite, Na_2SO_3	100 g
Disodium hydrogen phosphate, $Na_2HPO_4 \cdot 12H_2O$	100 g
Glucose, commercial	50 g
Sterile water	1 liter

With sterile precautions dissolve the bismuth ammoniocitrate in 250 ml boiling water and the sodium sulfite in 500 ml boiling water. Mix the solutions and, while the mixture is boiling, add the sodium phosphate crystals. When this mixture is cool, add the glucose dissolved in 250 ml boiling water and cooled. This mixture will keep for months.

Iron citrate brilliant green mixture

Ferric citrate, brown scales	2 g
Brillian green	0.25 g
Sterile water	225 ml

With sterile precautions, mix solutions of the ferric citrate in 200 ml water and the brilliant green in 25 ml water. This mixture will keep for months.

Preparation of complete medium

Sterile 3% nutrient agar	100 ml
Bismuth sulfite glucose phosphate mixture	20 ml
Iron citrate-brilliant green mixture	4.5 ml

Melt the agar and cool to 60°C. Add the other ingredients with sterile precautions and pour plates.

From Cruickshank, R., Duguid, J. P., Marmion, B. P., and Swain, R. H. A., *Medical Microbiology,* Vol. 2, 12th ed., Churchill, Livingstone, Edinburgh, 1975, 132. With permission.

SS Agar (*Salmonella* and *Shigella* spp.)

Beef extract	5 g
Proteose peptone	5 g
Lactose	10 g
Bile Salts No. 3	8.5 g
Sodium citrate	8.5 g
Sodium thiosulfate	8.5 g
Ferric citrate	1 g
Agar	13.5 g
Brilliant green	0.00033 g
Neutral red	0.025 g
Distilled water	1000 ml
pH 7.0	

The medium is heated to a boiling temperature to dissolve the ingredients, cooled to 42 to 45°C, and poured into Petri dishes. Do not autoclave SS agar.

From Harris, A. H. and Coleman, M. R., Eds., *Diagnostic Procedures and Reagents,* 4th ed., American Public Health Association, Washington, D.C., 1963. With permission.

Deoxycholate-Citrate Agar (*Salmonella* and *Shigella* spp.)[7][9]

Meat infusion	10.0 g
Proteose peptone	10.0 g
Lactose	10.0 g
Sodium citrate	20.0 g
Ferric ammonium citrate	2.0 g
Sodium deoxycholate	5.0 g
Neutral red	0.02 g
Agar	14.0 g
Distilled water	1 liter
pH 7.5	

Heat to boiling to dissolve the ingredients, cool to 50°C, and pour into dishes. Do not autoclave.

Deoxycholate-Citrate Agar (Modified)[3][6][5]

Meat infusion	5.0 g
Peptone	5.0 g
Lactose	10.0 g
Sodium citrate	8.5 g
$Na_2S_2O_3 \cdot 5H_2O$	8.5 g
Ferric citrate	1.0 g
Sodium deoxycholate	5.0 g
Neutral red	0.02 g
Agar	15.0 g
Distilled water	1 liter
pH 7.3	

Heat to boiling to dissolve the ingredients, cool to 50°C, and pour into dishes. Do not autoclave.

Brilliant Green-Phenol Red Agar (*Salmonella* spp.)[359]

Meat extract	5.0 g
Peptone	10.0 g
NaCl	5.0 g
Lactose	15.0 g
Phenol red	0.08 g
Brilliant green	0.0125 g
Agar	1 liter
Distilled water	15.0 g
pH 6.9	

Sterilize by autoclaving at 121°C for 15 min.

Brilliant Green-Phenol Red Agar (Modified)[366]

Lab-Lemco® powder	5.0 g
Bacteriological peptone	10.0 g
Yeast extract powder	3.0 g
Disodium hydrogen phosphate	1.0 g
Sodium dihydrogen phosphate	0.6 g
Lactose	10.0 g
Sucrose	10.0 g
Phenol red	0.09 g
Brilliant green	0.0047 g
Agar	12.0 g
Distilled water	1 liter
pH 6.9 (approximately)	

Dissolve the ingredients with gentle heating. Boil very briefly, cool to 50°C, and pour into dishes. Do not autoclave.

XLD Medium (Enterobacteriaceae)[367-372]

Yeast extract	3.0 g
L-Lysine HCl	5.0 g
Xylose	3.75 g
Lactose	7.5 g
Sucrose	7.5 g
Sodium desoxycholate	1.0 g
Sodium chloride	5.0 g
Sodium thiosulfate	6.8 g
Ferric ammonium citrate	0.8 g
Phenol red	0.08 g
Agar	12.5 g
Distilled water	1 liter
pH 7.4 (approximately)	

Heat gently until boiling, cool to 50°C, mix thoroughly, and pour into dishes. Do not autoclave.

Hektoen Enteric Agar (*Salmonella* and *Shigella* spp.)[373]

Proteose peptone	12.0 g
Yeast extract	3.0 g
Lactose	12.0 g
Sucrose	12.0 g
Salicin	2.0 g
Bile Salts No. 3	9.0 g
Sodium chloride	5.0 g
Sodium thiosulfate	5.0 g
Ammonium ferric citrate	1.5 g
Acid fuchsin	0.1 g
Bromthymol blue	0.065 g
Agar	14.0 g
Distilled water	1 liter
pH 7.5 (approximately)	

Heat gently until boiling. Cool to 50°C, mix thoroughly, and pour into dishes. Do not autoclave.

China Blue Lactose Agar (Enterobacteriaceae)[374]

Meat extract	3.0 g
Peptone	5.0 g
Lactose	10.0 g
NaCl	5.0 g
China blue	0.375 g
Agar	12.0 g
Distilled water	1 liter
pH 7.0	

Sterilize by autoclaving at 121°C for 15 min.

Violet Red Bile Agar (Enterobacteriaceae)[375]

Yeast extract	3.0 g
Peptone	7.0 g
Sodium chloride	5.0 g
Bile Salts No. 3	1.5 g
Lactose	10.0 g
Neutral red	0.03 g
Crystal violet	0.002 g
Agar	12.0 g
Distilled water	1 liter
pH 7.4 (approximately)	

Heat gently until completely dissolved. Boil for 1 min. Cool to 60°C, mix thoroughly, and pour into dishes. Do not autoclave. This medium has been modified to include 10 g/l dextrose to improve the detection of anaerogenic coliform organisms.[376]

Broth Selective/Indicator Media (Enterobacteriaceae)

Violet Peptone Bile Lactose Broth[377]

Peptone	10.0 g
Lactose	10.0 g
Bile Salts	5.0 g
Gentian violet	0.04 g
Distilled water	1 liter

pH 7.6 (approximately)

Sterilize by autoclaving at 121°C for 15 min.

MacConkey Broth[378]

Peptone	20.0 g
Lactose	10.0 g
Bile salts	5.0 g
Sodium chloride	5.0 g
Neutral red	0.075 g
Distilled water	1 liter

pH 7.4 (approximately)

This medium can be prepared and sterilized at 121°C for 15 min at double strength. The medium can then be diluted with equal parts of water under test or by the most probable number (MPN) technique. A modified formulation has been described[379] in which bromcresol purple at 0.01 g/l is substituted for neutral red. This modification reduces the risk of toxicity of the neutral red and provides a more sensitive indicator to pH changes.

Lauryl Sulfate Broth[380]

Tryptose	20.0 g
Lactose	5.0 g
Sodium chloride	5.0 g
Dipotassium hydrogen phosphate	2.75 g
Potassium dihydrogen phosphate	2.75 g
Sodium lauryl sulfate	0.1 g
Distilled water	1 liter

pH 6.8 ± 0.2

Sterilize by autoclaving at 121°C for 15 min.

Brilliant Green Bile Lactose Broth[381]

Peptone	10.0 g
Lactose	10.0 g
Ox bile	20.0 g
Brilliant green	0.0133 g
Distilled water	1 liter

pH 7.4 (approximately)

Sterilize by autoclaving at 121°C for 15 min.

Lactose Ricinoleate Broth[382]

Peptone	5 g
Lactose	10 g
Sodium ricinoleate	1 g
Distilled water	1 liter

pH 7.6 (approximately)

Sterilize by autoclaving at 121°C for 15 min. This medium was proposed as an alternative to Brilliant Green Bile Lactose Broth, producing fewer false negative reactions.

Formate Lactose Glutamate Medium[378]

	Double strength (g)
Sodium glutamate	12.7
Lactose	20.0
Sodium formate	0.5
L-Cystine	0.04
L(−)-Aspartic acid	0.048
L(+)-Arginine	0.04
Thiamine	0.002
Nicotinic acid	0.002
Pantothenic acid	0.002
Magnesium sulfate (MgSO$_4$·7H$_2$O)	0.200
Ferric ammonium citrate	0.020
Calcium chloride (CaCl$_2$·2H$_2$O)	0.020
Dipotassium hydrogen phosphate	1.80
Bromcresol purple	0.020
Distilled water	1 liter

pH 6.7 (approximately)

Sterilize by autoclaving at 121°C for 15 min. The double strength medium is used in the multiple tube method to determine the MPN of test waters.

EE Broth[383]

Peptone	10.0 g
Dextrose	5.0 g
Disodium hydrogen phosphate, anhydrous	6.45 g
Potassium dihydrogen phosphate	2.0 g
Ox bile, purified	20.0 g
Brilliant green	0.0135 g
Distilled water	1 liter

pH 7.2 (approximately)

Heat at 100°C only for 30 min. Cool rapidly in running water. Do not autoclave. This medium has subsequently been modified by substituting 1 g/l sodium lauryl sulfate AR in place of the purified bile salt.[88]

Biochemical Reaction Media

The numerous media described for presumptive diagnosis of the Enterobacteriaceae are abstracted by Skerman[230] and described in full in the culture media manuals of Difco Laboratories, Bioquest, Merck, and Oxoid Limited.

Yersinia spp.

Yersinia species will grow on nutrient agar, nutrient broth, and MacConkey agar. *Y. pestis* will form small, white colonies on blood agar which enlarge to 3 to 4 mm diameter on prolonged incubation.

Y. pestis Medium[384]

Blood agar base medium containing

$CaCl_2 \cdot 6H_2O$	$0.1\ M\ Ca^{++}$
Glucose	1 mg/ml
Erythromycin	40 μg/ml
Ethyl violet	1 μg/ml
Cycloheximide	100 μg/ml
Nystatin	100 μg/ml
Sodium azide	5 μg/ml
Hemin	40 μg/ml
pH 6.9	

The selective agents are added after autoclaving the base medium and cooling to 45°C.

Y. pseudotuberculosis Medium[385]

Tryptose blood agar base (200 ml) sterilized and
    cooled to 50°C, to which is added

Peptic digest of sheep blood	10 ml
Novobiocin	4 mg
Erythromycin	1 mg
Mycostatin	40,000 units
Crystal violet	0.5 mg

Mix thoroughly and pour into dishes. A modification of this medium in which the antibiotics and crystal violet were replaced with 40 to 70 mg of potassium tellurite per 100 ml of base agar was claimed to give equally good results.[386]

Y. enterocolitica[387]

Most but not all strains grow on Deoxycholate-Citrate Agar.

Erwinia spp.

Yeast Extract-Glucose-Chalk Agar (YDC)[388]

Yeast extract	10.0 g
Glucose	20.0 g
$CaCO_3$ (finely divided)	20.0 g
Agar	15.0 g
Distilled water	1 liter

Sterilize by autoclaving at 121°C for 15 min. Mix thoroughly before pouring into dishes.

Yeast Extract Medium[389]

Yeast extract	1.0 g
K_2HPO_4	10.0 g
$MgSO_4 \cdot 7H_2O$	1.0 g
Tween® 80	1.0 g
Distilled water	1 liter

Sterilize the $MgSO_4$ solution separately and add to the autoclaved (121°C for 15 min) yeast extract/phosphate/Tween® solution.

Emerson's Medium[389]

Peptone	4.0 g
Meat extract	4.0 g
Dextrose	2.5 g
Yeast extract	1.0 g
Tryptone (casein digest)	5.0 g
Distilled water	1 liter

Adjust pH to 6.8 with phosphate buffer (K_2HPO_4, 6.9 g; KH_2PO_4, 1.4 g; water, 100 ml). Sterilize by autoclaving at 121°C for 15 min.

2. Vibrionaceae
Vibrio spp.
Bile Salt Gelatin Agar (*Vibrio cholerae*)[390]

Bile salt gelatin agar

Trypticase (Baltimore Biological Laboratories) or Bactotryptone	1 g
Sodium chloride, NaCl	1 g
Sodium taurocholate	0.5 g
Sodium carbonate, Na_2CO_3	0.1 g
Gelatin (Difco)	3 g
Agar	1.5 g
Water	100 ml

Adjust pH to 8.5 with NaOH and sterilize at 121°C for 15 min.

Potassium tellurite solution
Prepare a 0.5% solution in water and autoclave at 115°C for 20 min; this solution keeps indefinitely

Preparation of complete medium

| Bile salt gelatin agar | 100 ml |
| Potassium tellurite solution K_2TeO_3, 0.05% | 1 ml |

Make a 1 in 10 dilution of the stock potassium tellurite solution with sterile precautions and add it to the melted and cooled agar medium. The final pH of the medium must be 8.5 to 9.2. Pour plates.

From Cruickshank, R., Duguid, J. P., Marmion, B. P., and Swain, R. H. A., *Medical Microbiology,* Vol. 2, 12th ed., Churchill, Livingstone, Edinburgh, 1975, 132. With permission.

TCBS Medium (*Vibrio* spp.)[391]

Yeast extract powder	5 g
Bacteriological peptone	10 g
Sodium thiosulfate	10 g
Sodium citrate	10 g
Ox bile	8 g
Sucrose	20 g
Sodium chloride	10 g
Ferric citrate	1 g
Bromthymol blue	0.04 g
Thymol blue	0.04 g
Agar	14 g
Distilled water	1 liter

pH 8.6 (approximately)

Dissolve the ingredients with gentle heating. Adjust the pH to 8.6 with sodium carbonate. Boil for 1 min only. Do not autoclave.

Peptone-Water Enrichment[392]

Tryptone (casein digest)	10 g
Distilled water	1 liter

Adjust pH to 8.5 with NaOH. Sterilize by autoclaving at 121°C for 15 min. Sodium chloride may be added at 3, 7, or 10% w/v to test for halophilic *Vibrios*.

Salt-Colistin Broth (Halophilic *Vibrios*)[392]

Yeast extract	3.0 g
Peptone	10.0 g
NaCl	20.0 g
Colistin methane sulfonate	500 units/ml
Distilled water	1 liter

pH 7.4

Add the colistin salt to the previously sterilized medium (121°C for 15 min).

Glucose-Salt-Teepol Broth (Halophilic *Vibrios*)[393]

Peptone	10.0 g
Meat extract	3.0 g
NaCl	20.0 g
Teepol (Shell)	4.0 ml
Glucose	5.0 g
Methyl violet	0.002 g

pH 9.2

Sterilize by autoclaving at 121°C for 15 min.

Aeromonas spp.

These organisms will grow on MacConkey Agar and Desoxycholate Agar as well as simple nutrient media. They prefer media with a pH around 8.0 and will grow on the selective *Vibrio* media described above.

Yeast Extract Agar (*A. salmonicida*)[394]

Lab-Lemco®	5.0 g
Yeast extract	7.0 g
Peptone	9.5 g
NaCl	5.0 g
Agar	15.0 g
Distilled water	1 liter
pH 7.0	

Sterilize by autoclaving at 121°C for 15 min.

Yeast-Peptone Agar[395]

Yeast extract	3.0 g
Peptone	5.0 g
NaCl	5.0 g
Agar	15.0 g
Distilled water	1 liter
pH 7.4	

Sterilize by autoclaving at 121°C for 15 min.

Photobacterium spp. and Lucibacterium spp.

Sea Water-Yeast-Peptone Agar

Yeast extract	3.0 g
Peptone	5.0 g
Agar	15.0 g
Aged sea water diluted	1 liter
3:1 with distilled water	
pH 7.4	

Sterilize by autoclaving at 121°C for 15 min.

Yeast Peptone Broth

Yeast extract	3.0 g
Peptone	5.0 g
NaCl	5.0 g
Distilled water	1 liter
pH 7.4	

Sterilize by autoclaving at 121°C for 15 min.

Glycerol-Chalk Agar

Yeast extract	3.0 g
Peptone	5.0 g
NaCl	30.0 g
Glycerol	10.0 g
$CaCO_3$	5.0 g
Agar	15.0 g
Distilled water	1 liter

Sterilize by autoclaving at 121°C for 15 min. Mix thoroughly before pouring into dishes.

Sea Water Lemco®

Lab-Lemco®	10.0 g
Peptone	5.0 g
Aged sea water diluted	1 liter
3:1 with distilled water	
pH 7.4	

Sterilize by autoclaving at 121°C for 15 min.

Fish Medium[395]

Fish extract	200 g
(minced fresh plaice soaked in	
4 liters tap water for 2 hr,	
boiled for 1 hr, and filtered)	
Peptone	5.0 g
NaCl	30.0 g
Agar	20.0 g
Fish extract	1 liter
pH 7.3	

Sterilize by autoclaving at 121°C for 15 min.

3. Affiliated Bacteria
Zymomonas sp.

Zymomonas Media (*Z. anaerobia*)[396]

Peptone	10.0 g
Yeast extract	10.0 g
Glucose	20.0 g
Agar	15.0 g
Distilled water	1 liter
pH 6.8	

Sterilize by autoclaving at 121°C for 15 min.

Complex Medium (Z. mobilis)[397]

Yeast extract	2.0 g
Peptone	2.0 g
Glucose	1.0 g
Tris-maleate buffer	1 liter
(0.05 M, pH 6.8) or	
PO_4 buffer (0.67 M, pH 7.0)	

Sterilize by autoclaving at 121°C for 15 min.

Synthetic Medium[397]

Tris-maleate buffer	1 liter
$MgSO_4$	237.7 mg
$CaCl_2$	1.1 mg
$FeSO_4 \cdot 7H_2O$	5.0 mg
$ZnSO_4 \cdot 7H_2O$	7.2 mg
$MnSO_4 \cdot H_2O$	4.2 mg
$CuSO_4 \cdot H_2O$	1.4 mg
$CoSO_4 \cdot 5H_2O$	1.4 mg
KCl	50.0 mg
NaCl	50.0 mg
KH_2PO_4	10.0 mg
NH_4Cl	1000.0 mg
Dextrose	2000.0 mg
Calcium pantothenate	50.0 μg

Plus 60 mg of each of the following amino acids:
Alanine
Arginine
Aspartic acid
Cysteine
Glutamic acid
Glycine
Histidine
Isoleucine
Leucine
Lysine
Methionine
Ornithine
Oxyproline
Phenylalanine
Proline
Serine
Threonine
Tryptophan
Tyrosine
Valine

Sterilize the dextrose separately by filtration and add to the autoclaved (121°C for 15 min) medium. For some strains the mixture of amino acids can be replaced by vitamin-free casamino acids (Difco) at 4 g/l.

Flavobacterium spp.

Sodium Caseinate Agar[398]

Sodium caseinate	2 g
Yeast extract	0.5 g
Proteose peptone	0.5 g
K_2HPO_4	0.5 g
Agar	15 g
Distilled water	1 liter
pH 7.5	

Sterilize the medium by autoclaving at 121°C for 15 min.

Heparin Medium (*F. leparinum*)[399]

Trypticase (BBL)	3.5 g
Phytone (BBL)	0.6 g
Glucose	0.5 g
NaCl	1 g
K_2HPO_4	0.5 g
Agar	15 g
Distilled water	1 liter
pH 6.5	

Sterilize at 121°C for 15 min. Cool to 50°C and add a sterile solution of heparin to a final concentration of 0.002% w/v.

Haemophilus spp.

All species of the genus *Haemophilus* grow freely on heated blood agar, Levinthal agar, or Fildes agar. Specific requirements for X and V factors may be determined by impregnated paper discs.[400,401]

Charcoal Blood Agar[300]

Lab-Lemco®	10.0 g
Peptone	10.0 g
Starch	10.0 g
Bacteriological charcoal	4.0 g
Sodium chloride	5.0 g
Nicotinic acid	0.001 g
Agar	12.0 g
Distilled water	1 liter
pH 7.4 (approximately)	

Sterilize the medium by autoclaving at 121°C for 15 min. Cool to 50°C and add 10% v/v sterile horse blood.

Fildes Peptic Digest of Blood[1][2]

Sodium chloride, 0.85% solution	150 ml
Hydrochloric acid, concentrated	6 ml
Defibrinated blood	50 ml
Granulated pepsin	1 g
Sodium hydroxide, 20% solution	12 ml

Add salt solution, HCl, defibrinated blood, and pepsin to a sterile 250-ml glass-stoppered bottle and shake well. Place mixture in water bath at 55°C until it is digested (between 2 and 24 hr). Add 12 ml NaOH to the digested blood and adjust final reaction to pH 7.0 to 7.2. Add chloroform to 0.25% concentration and store in the refrigerator.

The culture medium is prepared by adding the blood digest at 2% v/v to sterile molten nutrient agar or to nutrient broth.

Levinthal's Agar[4][0][2]

Sterile nutrient agar	100 ml
Sterile rabbit or human blood	5 ml

Melt the agar, add the blood, and heat the mixture in boiling water. Allow the deposit to settle and distribute the clear supernatant.

Diphosphothiamine Medium (*H. piscium*)[4][0][3]

Proteose peptone	20.0 g
Glucose	10.0 g
NaCl	5.0 g
Tween® 40	0.05 g
Distilled water	1 liter
pH 7.3	

Sterilize by autoclaving at 121°C for 15 min. Cool to 50°C and add aseptically 1.0 μg diphosphothiamine or 30 μg adenosine triphosphate per milliliter.

Defined Medium for Growth of *Haemophilus influenzae*

Component	Concentration (μg/ml)	
	MI	MI-Cit
L-Arginine	300	
L-Aspartic acid	500	500
L-Cystine	200	200
L-Glutamic acid	1300	1300
Glycine	30	30
L-Leucine	300	300
L-Lysine	50	50
L-Methionine	100	100
L-Serine	100	100
L-Tyrosine	200	200
L-Histidine	10	10
Citrulline		150
Uracil	100	
Hypoxanthine	20	20
Inosine	2000	2000
Hemin	10	10
Nicotinamide adenine dinucleotide	4	4
Thiamine	4	4
Calcium pantothenate	4	4
Polyvinyl alcohol	20	20
Sodium lactate	800	800
Glycerol	3000	3000
Tween® 80	20	20
Ethylenediaminetetraacetate	4	4
Nitrilotriethanol	400	400
NaCl	5800	5800
K_2SO_4	1000	1000
$MgCl_2$	200	200
$CaCl_2$	22	22
KH_2PO_4	2700	2700
K_2HPO_4	3500	3500

From Herriott, R. M., Meyer, E. Y., Vogt, M., and Modan, M., *J. Bacteriol.*, 101, 513–516, 1970. With permission.

Pasteurella spp.

These organisms grow best on blood-containing media. Most strains fail to grow on bile-containing media.

YPC Medium (*P. multocida*)[405]

Yeast extract	5.0 g
Proteose peptone	15.0 g
L-Cystine	0.5 g
Glucose	2.0 g
Sucrose	2.5 g
Na_2SO_3	0.2 g
KH_2PO_4	4.0 g
Agar	15.0 g
Distilled water	1 liter
pH 7.2	

Sterilize by autoclaving at 121°C for 15 min.

Selective Media for *Pasteurella* spp.[406]

Difco Tryptose Agar containing 5% v/v peptic digest of blood with the following additions (μg/ml):

P. multocida	
Novobiocin	10
Potassium tellurite	5
Erythrocin	5
Actidione	100
P. haemolytica	
Neomycin	1.5
Novobiocin	2.0
Actidione	100

Actinobacillus spp.

Species of the genus *Actinobacillus* grow slowly on nutrient agar; for better growth, the addition of blood or glycerol is required.

A. *lignieresi* Medium [407]

Sterile nutrient agar	100 ml
Sterile horse blood	5 ml

Plus 1 μg/ml oleandomycin and 200 units/ml nystatin.

A. *actinomycetem-comitans*[408]

This organism grows on nutrient agar and MacConkey Agar, but requires a CO_2-enriched atmosphere. After subculture on blood agar, the organism grows well in nutrient broth without CO_2.

A. *actinoides* [409]

This organism cannot be grown on ordinary culture media, but can be cultivated under increased CO_2 pressure in the condensation fluid of a coagulated serum slope after incubation for several weeks.

Cardiobacterium spp.

C. hominis Medium[410]

This organism grows well on Tryptone Soya Agar, Tryptose Blood Agar, and Brain Heart Infusion Agar, with and without supplementation with blood.

Yeast extract	5.0 g
K_2HPO_4	7.0 g
KH_2PO_4	3.0 g
$MgSO_4 \cdot 7H_2O$	0.01 g
$(NH_4)_2SO_4$	0.1 g
Distilled water	1 liter
pH 7.0	

Sterilize by autoclaving at 121°C for 15 min.

Calymatobacterium spp.

C. granulomatis Medium

This organism will not grow on ordinary culture media, but will grow in the chick embryo yolk sac and, after adaption, in egg yolk in vitro.

Embryonic Egg Yolk Agar[411]

Add an equal quantity of aseptically withdrawn yolk from 4- to 5-day-old chick embryos to sterile meat infusion agar cooled to 60°C and mix gently. Distribute into sterile tubes and allow to set in a slanting position. Inoculate the condensation water.

H. Anaerobic Gram-negative Rods and Cocci

1. *Bacteroidaceae*

Organisms in this family are strict anaerobes and special techniques may be required to successfully transfer them to culture media and cultivate them.[412,413] For many strains, 10% v/v CO_2 is required, and the removal of oxygen must be rapid. Plates of media must be placed in an anaerobic container within 10 min of inoculation.

Different media have been developed to isolate organisms in the Bacteroidaceae family because the two major genera differ in their reactions to dyes and inhibitors. *Fusobacterium* spp. are resistant to dyes such as crystal violet and brilliant green, while *Bacteroides* spp. are inhibited. *Bacteroides* spp., however, are resistant to antibiotics such as neomycin and kanamycin.

Bacteroides and Fusobacterium spp.

To isolate *Bacteroides* spp. use VL Blood Agar plus 100 μg/ml kanamycin or 100 μg/ml neomycin plus 7.5 μg/ml vancomycin. For *B. melaninogenicus* use VL Medium containing 5% v/v laked blood, 0.5 μg/ml menadione and 100 μg/ml kanamycin.[414] To isolate *Fusobacterium* spp. use VL Blood Agar containing 0.05 mg/ml sodium azide and 0.05 mg/ml ethyl violet or 0.02 mg/ml brilliant green.[415]

VL Medium[415]

Tryptone	10.0 g
NaCl	5.0 g
Meat extract	2.0 g
Yeast extract	5.0 g
Cysteine HCl	0.3 g
Glucose	2.0 g
Agar	6.0 g
Distilled water	1 liter
pH 7.4	

Sterilize by autoclaving at 121°C for 15 min; 10% v/v beef bile may be added to promote the growth of many species of *Bacteroides*.

Fusobacterium **Medium**[416]

Casitone	15.0 g
Yeast extract	5.0 g
Glucose	5.0 g
NaCl	5.0 g
L-Cysteine	0.75 g
Crystal violet	0.01 g
Streptomycin	0.01 g
Agar	15.0 g
Distilled water	1 liter
pH 7.2	

Add the streptomycin and 5% v/v sterile serum to the medium after it has been sterilized by autoclaving at 121°C for 15 min.

Reinforced Clostridial Medium (RCM)[417]

Yeast extract	3.0 g
Lab-Lemco®	10.0 g
Peptone	10.0 g
Soluble starch	1.0 g
Dextrose	5.0 g
Cysteine hydrochloride	0.5 g
Sodium chloride	5.0 g
Sodium acetate	3.0 g
Agar	0.5 g
Distilled water	1 liter
pH 6.8 (approximately)	

Sterilize by autoclaving at 115°C for 20 min. To prepare a solid medium, add agar at 15 g/l. Ethyl violet (1/20,000) and sodium azide (1/20,000) may be added to RCM to select *Fusobacterium* spp.[418]

Medium to Isolate Anaerobes from Bovine Rumen

Media, Mixtures, and Solutions

Mineral I (% w/v in distilled water)
K_2HPO_4	0.6

Mineral II (% w/v in distilled water)
NaCl	1.2
$(NH_4)_2SO_4$	1.2
KH_2PO_4	0.6
$CaCl_2$, anhydrous	0.12
$MgSO_4 \cdot 7H_2O$	0.25

Volatile fatty acid mixture (VFA) (ml)
Acetic acid	17
Propionic acid	6
n-Butyric acid	4
Isobutyric acid	1
n-Valeric acid	1
Isovaleric acid	1
DL-α-methylbutyric acid	1

Cysteine-Na_2S reducing mixture[a] (% w/v in distilled water)
Cysteine HCl	1.5
$Na_2S \cdot 9H_2O$	1.5

Dissolve the cysteine in a little distilled water, adjust pH to 10.0 using 10 N NaOH, add sulfide, make up to volume, Seitz filter

Hemin

1% (w/v) solution in 50% (v/v) ethanol + 50% (v/v) 0.05 N NaOH

Resazurin

0.1% (w/v) in distilled water

Cysteine[a]

2.5% (w/v) in distilled water, Seitz filter

$NaHCO_3$ or Na_2CO_3[a]

8% (w/v) in distilled water, Seitz filter, or autoclave (121°/15 min).

Dilution solution (% v/v)
Mineral I	7.5
Mineral II	7.5
Resazurin solution	0.1

Make up to volume less 7% with distilled water, autoclave (120°C for 15 min), gas; add cysteine solution (2% v/v) and $NaHCO_3$ solution (5% v/v) while gassing; continue gassing until reduced and seal

Media for Roll Tubes

(a) Rumen fluid medium

Glucose	0.05 ⎫
Cellobiose	0.05 ⎬ % w/v
Soluble starch	0.05 ⎭
Mineral I	7.5 ⎫
Mineral II	7.5 ⎪
Rumen fluid	20.0 ⎬ % v/v
Resazurin solution	0.1 ⎭

Dissolve ingredients and adjust to pH 6.8; add 2.0% (w/v) agar; make up to volume less 7% with distilled water; autoclave (121°C for 15 min), gas; add cysteine-Na$_2$S (2% v/v) Na$_2$CO$_3$ (5% v/v) while gassing; continue gassing and seal

(b) Medium 10 (M10)

Glucose	0.05 ⎫
Cellobiose	0.05 ⎪
Starch	0.05 ⎬ % w/v
Yeast extract	0.05 ⎪
Trypticase	0.2 ⎭
Resazurin solution	0.1 ⎫
Haemin solution	1.0 ⎪
VFA mixture	0.31 ⎬ % v/v
Mineral I	7.5 ⎪
Mineral II	7.5 ⎭

Dissolve ingredients and adjust to pH 6.8; add 2.0% (w/v) agar; make up to volume less 7% with distilled water; autoclave (121°C for 15 min), gas; add cysteine-Na$_2$S (2% v/v) Na$_2$CO$_3$ (5% v/v) while gassing; continue gassing and seal

M10 medium for slopes — as for (b) above but with (%) w/v

Glucose	0.1
Cellobiose	0.1
Starch	0.1
Maltose	0.1
and Agar	1.5

[a] Make up fresh each time a medium is prepared. The other solutions can be stored at 4°C and used as required.

From Latham, M. J. and Sharpe, M. E., in *Isolation of Anaerobes,* Shapton, D. A. and Board, R. G., Eds., Academic Press, London, 1971, 134—147. With permission.

Leptotrichia

Leptotrichia spp. Medium[420]

Peptone	10.0 g
Meat extract	3.0 g
Yeast extract	5.0 g
Glucose	1.0 g
Sodium thioglycollate	0.5 g
Agar	20.0 g
Crystal violet	0.002 g
Distilled water	1 liter
pH 7.2	

Sterilize at 121°C for 15 min.

Thioglycollate Medium (Brewer)[420]

Meat extract	1.0 g
Yeast extract	2.0 g
Peptone	5.0 g
Dextrose	5.0 g
NaCl	5.0 g
Sodium thioglycollate	1.1 g
Methylene blue	0.002 g
Agar	1.0 g
Distilled water	1 liter
pH 7.2	

Sterilize by autoclaving at 121°C for 15 min.

Desulfovibrio

D. desulfuricans Medium (for pure cultures)[307]

Peptone	5.0 g
Beef extract	3.0 g
Yeast extract	0.2 g
$MgSO_4$	1.5 g
Na_2SO_4	1.5 g
$(NH_4)_2SO_4 \cdot FeSO_4 \cdot 6H_2O$	0.1 g
Glucose	5.0 g
Agar	15.0 g
Distilled water	1 liter
pH 7.4	

Sterilize the glucose and ferrous ammonium sulfate separately and add aseptically to the basal medium, sterilized at 121°C for 15 min.

Isolation Medium for D. desulfuricans[307]

NH_4Cl	1.0 g
$MgSO_4 \cdot 7H_2O$	2.0 g
Na_2SO_4	1.0 g
K_2HPO_4	0.5 g
$CaCl_2 \cdot 2H_2O$	0.1 g
Sodium lactate	5.0 g
(70% w/w solution)	
$(NH_4)_2SO_4 \cdot FeSO_4 \cdot 6H_2O$	0.5 g
Distilled water	1 liter
pH 7.6	

Sterilize the ferrous ammonium sulfate solution separately and add it to the bulk medium which has been sterilized at 121°C for 15 min.

Media for *Desulfovibrio* spp.[303]

Medium B (g)		
KH_2PO_4	0.5	Tap water, one liter; adjust
NH_4Cl	1	reaction to between pH
$CaSO_4$	1	7 and 7.5; this medium
$MgSO_4 \cdot 7H_2O$	2	always contains a pre-
Na lactate	3.5	cipitate; NaCl should
Yeast extract	1	be added for marine
Ascorbic acid	1	strains or sea water
Thioglycollic acid	1	used in place of tap
$FeSO_4 \cdot 7H_2O$	0.5	water
Medium C (g)		
KH_2PO_4	0.5	Distilled water, one liter; pH
NH_4Cl	1	7.5 ± 0.2; this medium
Na_2SO_4	4.5	may be cloudy after
$CaCl_2 \cdot 6H_2O$	0.06	autoclaving, but should
$MgSO_4 \cdot 7H_2O$	0.06	clear on cooling; add
Na lactate	6	extra NaCl for salt
Yeast extract	1	water strains
$FeSO_4 \cdot 7H_2O$	0.1	
Sodium citrate $\cdot 2H_2O$	0.3	
Medium D (g)		
KH_2PO_4	0.5	Distilled water, one liter; pH
NH_4Cl	1	7.5 ± 0.2; sterilize by
$CaCl_2 \cdot 2H_2O$	0.1	filtration; use extra NaCl
$MgCl_2 \cdot 6H_2O$	1.6	for salt water strains
Yeast extract	1	
$FeSO_4 \cdot 7H_2O$	0.1	
Na pyruvate	3.5	
or Choline chloride	1	
Medium E (g)		
KH_2PO_4	0.5	Tap water, one liter; adjust
NH_4Cl	1	to pH 7.6 with NaOH
Na_2SO_4	1	after boiling to dissolve
$CaCl_2 \cdot 6H_2O$	1	agar; autoclave and
$MgCl_2 \cdot 7H_2O$	2	use before it solidifies;
Na lactate	3.5	add extra NaCl for salt
Yeast extract	1	water strains
Ascorbic acid	1	
Thioglycollic acid	1	
$FeSO_4 \cdot 7H_2O$	0.5	
Agar	15	

Medium N
 Medium C with $(NH_4)_2SO_4$,
7 g, in place of
 Na_2SO_4, O_4

From Postgate, J. R., *Lab. Pract.*, 15, 1239—1244, 1966. With permission.

Isolation Medium for Sulfate-reducing Bacteria[421]

Nutrient agar	100 ml
Na_2SO_4	0.5 g
$FeSO_4$	0.005 g

Sterilize by autoclaving at 121°C for 15 min.

2. Veillonellaceae
Selective Medium for Veillonellaceae spp.[422]

Trypticase (BBL)	5.0 g
Yeast extract	3.0 g
Sodium thioglycollate	0.75 g
Tween® 80	1.0 g
Sodium lactate (50%)	25 ml
Distilled water	1 liter
pH 6.6	

Adjust pH with K_2CO_3. Agar at 0.1% w/v may be added to the medium. Sterilize by autoclaving at 121°C for 15 min. Streptomycin (5 μg/ml) or vancomycin (7.5 μg/ml) can be added after sterilization to make the medium selective.

3. Megasphaera
Isolation Medium[423]

K_2HPO_4	0.5 g
KH_2PO_4	0.5 g
NaCl	1.0 g
$(NH_4)_2SO_4$	0.5 g
$MgSO_4$	0.1 g
$CaCl_2$	0.1 g
Resazurin	0.001 g
Cysteine HCl	0.7 g
$NaHCO_3$	5.0 g
Distilled water	1 liter

Sterilize the cysteine and $NaHCO_3$ separately and add to the medium just before inoculation. The following additional nutrients may be added to the inorganic mix:

Soluble starch	5.0 g
Rumen fluid	30% v/v
Peptone	10.0 g
Yeast extract	5.0 g

Sterilize by autoclaving at 121°C for 15 min.

I. Chemolithotrophic Gram-negative Bacteria

1. Ammonia or Nitrate Oxidizing

Nitrobacter agilis Medium[424]

KNO_2	0.17 g
$CaCO_3$	10 g
$MgSO_4 \cdot 7H_2O$	0.14 g
$FeSO_4 \cdot 7H_2O$	0.03 g
$MnSO_4 \cdot 4H_2O$	0.01 g
NaCl	0.3 g
K_2HPO_4	0.14 g
Na_2CO_3	0.25 g
Distilled water	1 liter
Biotin	150 mg

Prepare the biotin separately as a filter-sterilized solution. Dissolve the Na_2CO_3 in a little of the water and sterilize by autoclaving at 121°C for 15 min. Dissolve the remaining salts in the rest of the water and sterilize by autoclaving at 121°C for 15 min. Mix the carbonate and salts solutions and add the biotin. Distribute the complete medium into tubes or flasks aseptically.

From Lapage, S. P., Shelton, J. E., and Mitchell, T. G., in *Methods in Microbiology*, Vol. 3A, Norris, J. R. and Ribbons, D. W., Eds., Academic Press, London, 1970, 1–228. With permission.

Nitrosomas europaea Medium

Medium A

Na_2HPO_4	13.4 g
KH_2PO_4	0.773 g
$NaHCO_3$	0.5 g
$(NH_4)_2SO_4$	2.5 g
Distilled water	1 liter

Medium B

$MgSO_4 \cdot 7H_2O$	8.5 mg
$CaCl_2 \cdot 2H_2O$	310 mg
Sodium ferric ethylenediamine di-o-hydroxyphenylacetate (Sequestrene® 138-Fe, Geigy Chemical Co.)	3.0 mg
Distilled water	100 ml

Dissolve the components of Medium A in the distilled water and check the pH (approximately 8.0). Sterilize the medium by filtration through a Millipore® filter. If it is desired to heat-sterilize Medium A, the bicarbonate should be prepared separately and the two parts autoclaved. Autoclave Medium B at 121°C for 15 min and add 6 ml aseptically to 1 liter of Medium A.

Nitrocystis oceanus

Prepare the medium in the same way as for *Nitrosomonas europaea*, except substitute aged filtered sea water for the distilled water.

From Lapage, S. P., Shelton, J. E., and Mitchell, T. G., in *Methods in Microbiology*, Vol. 3A, Norris, J. R. and Ribbons, D. W., Eds., Academic Press, London, 1970, 1–288. With permission.

2. Sulfur Metabolizing
Thiobacillus Media

Medium S

Sulfur	10 g	Distilled water, one liter;
(or $Na_2S_2O_3 \cdot 5H_2O$)	5 g	sterilize by steaming if
$(NH_4)_2SO_4$	2–4 g	sulfur is the substrate;
KH_2PO_4	2–4 g	otherwise, autoclave
$CaCl_2$	0.25 g	
$MgSO_4$	0.5 g	
$FeSO_4$	10 mg	

Medium R

NH_4Cl	0.5 g	Tap water, one liter; adjust
$MgCl_2 \cdot 6H_2O$	0.5 g	pH to about 7.0; iron,
KH_2PO_4	2 g	phosphate, and bicar-
$Na_2S_2O_3 \cdot 5H_2O$	5 g	bonate solutions steril-
KNO_3	2 g	ized independently;
$NaHCO_3$	1 g	used for *T. denitrifi-*
$FeSO_4 \cdot 7H_2O$	10 mg	*cans* only

Medium F

$(NH_4)_2SO_4$	0.15 g	Tap water, one liter; add 10
KCl	0.05 g	ml of a separately steril-
$MgSO_4 \cdot 7H_2O$	0.5 g	ized 10% $FeSO_4$
KH_2PO_4	0.05 g	$\cdot 7H_2O$ solution; check
$Ca(NO_3)_2$	0.01 g	that the pH value is
		around 3.5 after inocu-
		lation

Yields of most thiobacilli improve if a trace element solution is added to the culture medium. This contains ethylenediaminetetraacetate, 50 g; $ZnSO_4 \cdot 7H_2O$, 22 g; $CaCl_2$, 5.54 g, $MnCl_2 \cdot 4H_2O$, 5.06 g; $FeSO_4 \cdot 7H_2O$, 4.99 g; $(NH_4)_6Mo_7O_{24} \cdot 4H_2O$, 1.1 g; $CuSO_4 \cdot 5H_2O$, 1.57 g; $CoCl_2 \cdot 6H_2O$, 1.61 g; water, one liter; to pH 6.0 with KOH. Added at 1 part per 100, this supplement supplies inorganic iron which may, therefore, be omitted from the first two prescriptions above.

From Postgate, J. R., *Lab. Pract.*, 15, 1239—1244, 1966. With permission.

T. thioparus Medium[307]

$Na_2S_2O_3 \cdot 5H_2O$	5.0 g
$(NH_4)_2SO_4$	0.4 g
K_2HPO_4	4.0 g
$CaCl_2$	0.25 g
$MgSO_4 \cdot 7H_2O$	0.5 g
$FeSO_4$	0.01 g
Distilled water	1 liter
pH 7.0	

T. denitrificans Medium[307]

$Na_2S_2O_3 \cdot 5H_2O$	5.0 g
KNO_3	5.0 g
$NaHCO_3$	1.0 g
K_2HPO_4	0.2 g
$MgCl_2$	0.1 g
Distilled water	1 liter
pH 7.0	

T. thiooxidans Medium[307]

$(NH_4)_2SO_4$	0.2 g
$MgSO_4 \cdot 7H_2O$	0.1—0.5 g
$FeSO_4$	0.01 g
$CaCl_2$	0.25 g
KH_2PO_4	3.0—5.0 g
Powdered sulfur	10.0 g
Distilled water	1 liter
pH 7.0	

General Thiobacillus spp. Medium[307]

$Na_2S_2O_3$	10.0 g
K_2HPO_4	4.0 g
KH_2PO_4	4.0 g
$CaCl_2$	0.1 g
$MgSO_4 \cdot 7H_2O$	0.1 g
$(NH_4)_2SO_4$	0.1 g
$FeCl_3 \cdot 6H_2O$	0.02 g
$MnSO_4 \cdot 7H_2O$	0.02 g
Glucose (or asparagine)	1.5 g
Distilled water	1 liter
pH 7.0	

Do not autoclave these media. Steam at 100°C for 1 hr on three consecutive days.

T. thioparus and *T. thiooxidans* Medium[425]

$Na_2S_2O_3 \cdot 5H_2O$	1.0 g
NH_4Cl	0.1 g
KH_2PO_4	0.1 g
$MgCl_2 \cdot 6H_2O$	0.05 g
Water to	100 ml
pH 6.8	

Sterilize by autoclaving at 121°C for 15 min.

T. ferrooxidans Medium[426]

$FeSO_4 \cdot 7H_2O$	130.0 g
$MgSO_4 \cdot 7H_2O$	1.0 g
$(NH_4)_2SO_4$	0.5 g
Distilled water	1000 ml

Dissolve the ingredients and adjust the pH to between 2.0 and 2.5 with sulfuric acid. Autoclave at 121°C for 15 min and allow to stand. A voluminous precipitate of ferric hydroxide settles out. Remove the supernatant aseptically and distribute as required, preferably in layers not more than 1 cm deep in Erlenmeyer flasks. To prepare an agar medium, dissolve the ferrous sulfate in 300 ml of the water and sterilize separately. Dissolve the other ingredients plus 20 g of agar in the remaining water and sterilize. Mix the two solutions just before pouring plates.

Nonaciduric Species of *Thiobacillus*[427]

$(NH_4)_2SO_4$	0.1 g
K_2HPO_4	4.0 g
KH_2PO_4	4.0 g
$MgSO_4 \cdot 7H_2O$	0.1 g
$CaCl_2$	0.1 g
$FeCl_3 \cdot 6H_2O$	0.02 g
$MnSO_4 \cdot 4H_2O$	0.02 g
$Na_2S_2O_3 \cdot 5H_2O$	10 g
Distilled water	1000 ml

Adjust the pH to 6.6 and steam for 1 hr on three successive days.

T. ferrooxidans Medium[428]

Part 1	
KH_2PO_4	0.4 g
$MgSO_4 \cdot 7H_2O$	0.1 g
$(NH_4)_2SO_4$	0.1 g
$CaCl_2$	0.03 g
$MnSO_4 \cdot 4H_2O$	0.02 g
NaCl	1 g
$Al_2(SO_4)_3 \cdot 12H_2O$	1.4 g
Distilled water	1 liter
Part 2	
$FeSO_4 \cdot 7H_2O$	10 g
H_2SO_4 concentrated	0.09 ml
Distilled water	100 ml

Prepare the two parts separately by dissolving the respective constituents in the liquids. Distribute the basal medium (Part 1) in 90-ml amounts in 250-ml capacity conical flasks. Place the $FeSO_4$ solution in a bottle and sterilize it with the flasks of basal medium by autoclaving at 121°C for 15 min. Before use, add 10 ml of $FeSO_4$ solution to each flask aseptically.

From Lapage, S. P., Shelton, J. E., and Mitchell, T. G., in *Methods in Microbiology*, Vol. 3A, Norris, J. R. and Ribbons, D. W., Eds., Academic Press, London, 1970, 1—288. With permission.

J. Methane-producing Bacteria

Methanobacterium spp. Enrichment Medium[429]

K_2HPO_4	5.0 g
$MgSO_4 \cdot 7H_2O$	0.1 g
$(NH_4)_2SO_4$	0.3 g
$FeSO_4 \cdot 7H_2O$	0.02 g
Yeast autolysate	5.0 ml
$CaCO_3$	100 g
Na_2CO_3	0.5 g
$Na_2S \cdot 9H_2O$	0.1 g
Ethanol	10.0 ml
Distilled water	1 liter
pH 7.2	

Sterilize the Na_2CO_3, Na_2S, and ethanol separately and add to the basal medium after sterilization at 121°C for 15 min. A solid medium can be prepared by adding 15 to 20 g agar per liter.

Methanobacillus Medium for Mixed Cultures[430]

Ethanol	10 g
K_2HPO_4	6 g
KH_2PO_4	9 g
NH_4Cl	5 g
$MgCl_2$	1 g
$FeSO_4 \cdot 7H_2O$	0.01 g
$CaCl_2$	0.01 g
Tap water	1 liter
pH 7.4	

Separately sterilize the ethanol by filtrating. Sterilize the basal medium by autoclaving at 115°C for 20 min. Aseptically add the ethanol to the basal medium.

Methanol Salts Medium[431]

KH_2PO_4	1.36 g
Na_2HPO_4	2.13 g
$(NH_4)_2SO_4$	0.5 g
$MgSO_4 \cdot 7H_2O$	0.2 g
$CaCl_2 \cdot 2H_2O$	0.01 g
$FeSO_4 \cdot 7H_2O$	5 mg
$MnSO_4 \cdot 4H_2O$	2.5 mg
$Na_2MoO_4 \cdot 2H_2O$	2.5 mg
Distilled water	1 liter
Methanol	5 ml

Add all the ingredients except the methanol to the water and autoclave the mixture at 121°C for 15 min. Seitz-filter the methanol to sterilize it and add aseptically to the previously sterilized salts solution.

Methanosarcina barkeri Medium[432]

A.		
NH$_4$Cl		0.5 g
CaCl$_2$·2H$_2$O		0.01 g
MgCl$_2$·2H$_2$O		0.01 g
FeCl$_3$·6H$_2$O		2 mg
MnSO$_4$·4H$_2$O		1 mg
Na$_2$MoO$_4$·2H$_2$O		1 mg
K$_2$HPO$_4$		3.48 g
KH$_2$PO$_4$		2.72 g
Distilled water		1 liter

B. Na$_2$S·9H$_2$O, 10% w/v in distilled water
C. Na$_2$CO$_3$, 10% w/v in distilled water
D. Methanol

Prepare the salts solution (A) by dissolving the constituents in the water and sterilize by autoclaving at 121°C for 15 min. Sterilize solutions B and C by autoclaving at 121°C for 15 min. Seitz-filter the methanol. To 1 liter of Salts Medium A, add aseptically 2 ml of B, 20 ml of C, and 10 ml of D. After mixing, adjust the pH with 10% HCl to 6.8.

From Lapage, S. P., Shelton, J. E., and Mitchell, T. G., in *Methods in Microbiology,* Vol. 3A, Norris, J. R. and Ribbons, D. W., Eds., Academic Press, London, 1970, 1—288. With permission.

Methanococcus Medium[433]

Sodium formate	15.0 g
(NH$_4$)$_2$SO$_4$	1.0 g
CaCl$_2$·2H$_2$O	0.01 g
MgCl$_2$·2H$_2$O	0.01 g
FeCl$_3$·6H$_2$O	0.02 g
MnSO$_4$·4H$_2$O	0.01 g
Na$_2$MoO$_4$·2H$_2$O	0.001 g
K$_2$HPO$_4$	2.0 g
Sodium thioglycollate	0.5 g
Distilled water	1 liter
pH 8.0	

Sterilize by autoclaving at 121°C for 15 min. If the medium is to be stored, keep the thioglycollate as a separate solution and add just before use.

K. Aerobic and Anaerobic Gram-positive Cocci

1. Micrococcaceae

These organisms grow readily on nutrient agar or broth. The large number of media published reflects the necessity to isolate the organisms from mixed ecologies by using selective agents or the inclusion of enzyme substrates to give a tentative identification of genus or pathogenicity.

Milk Agar[434]

Sterile nutrient agar	200 ml
(containing 3% w/v agar)	
Fresh milk	100 ml

Heat the milk to 60°C, shake, and sterilize at 121°C for 15 min. Cool to 50°C and add to the molten sterile agar. Mix thoroughly.

Glycerol Monoacetate Agar[435]

Heart infusion broth (Difco)	100 ml
Glycerol monoacetate (BDH)	1.0 g
Agar	2.0 g

Dissolve the agar and glyceride in the broth with steaming and autoclave at 121°C for 15 min. Cool to 50°C and mix thoroughly to suspend the glyceride.

Milk-Salt Agar[436]

Peptone	5.0 g
Lab-Lemco®	3.0 g
NaCl	65.0 g
Agar	15.0 g
Distilled water	1 liter
pH 7.4	

Sterilize by autoclaving at 121°C for 15 min. Cool to 60°C and add 10% v/v sterile skim milk or homogenized milk. Mix thoroughly.

Mannitol-Salt Agar[437]

Lab-Lemco®	1.0 g
Peptone	10.0 g
Mannitol	10.0 g
Sodium chloride	75.0 g
Phenol red	0.025 g
Agar	15.0 g
Distilled water	1 liter
pH 7.5 (approximately)	

Sterilize by autoclaving at 121°C for 15 min.

Gelatin-Mannitol-Salt Agar[3]

Yeast extract	2.5 g
Tryptone	10.0 g
Lactose	2.0 g
Mannitol	10.0 g
Sodium chloride	75.0 g
Dipotassium hydrogen phosphate	5.0 g
Gelatin	30.0 g
Agar	15.0 g
Distilled water	1 liter
pH 7.1 (approximately)	

Sterilize by autoclaving at 121°C for 15 min. Acid production from mannitol is best demonstrated by adding a drop of 0.04% bromthymol blue indicator to the sites of the individual colonies: yellow indicates acid production. Gelatin hydrolysis may be demonstrated by adding a drop of a saturated aqueous solution of ammonium sulfate or, preferably, a 20% aqueous solution of sulfosalicylic acid to an individual colony ("Stone reaction"). A positive Stone reaction is denoted by the presence of a clear zone around gelatinase-producing colonies after contact with the reagent for 10 min.

Salt-Meat Broth[438]

Peptone	10 g
Lab-Lemco®	10 g
Neutral ox heart tissue	30 g
Sodium chloride	100 g
Distilled water	1 liter
pH 7.6 (approximately)	

Sterilize by autoclaving at 121°C for 15 min. Ensure even distribution of meat particles when distributing into containers.

Deoxyribonuclease Agar[439]

Tryptose	20 g
Deoxyribonucleic acid	2 g
Sodium chloride	5 g
Agar	12 g
Distilled water	1 liter
pH 7.3 (approximately)	

Sterilize by autoclaving at 121°C for 15 min. After inoculation and incubation, flood the surface of the plate with $N/1$ HCl and look for clear zones of DNase activity.

Lithium-Glycine-Mannitol-Tellurite Agar[440]

Tryptone	10.0 g
Yeast extract	5.0 g
Mannitol	10.0 g
K_2HPO_4	5.0 g
$LiCl_2$	5.0 g
Glycine	10.0 g
Phenol red	0.025 g
Potassium tellurite	0.2 g
Distilled water	1 liter
pH 7.2	

The tellurite must be added as a separately filter-sterilized solution after the rest of the medium has been sterilized at 121°C for 15 min.

Egg Yolk-Sodium Azide Agar[441]

Peptone	10.0 g
Lab-Lemco®	5.5 g
NaCl	3.0 g
$Na_2 HPO_4 \cdot 12H_2O$	0.2 g
Agar	15.0 g
Distilled water	1 liter
pH 7.0	

Sterilize the basal medium by autoclaving at 121°C for 15 min. Cool to 50°C and add a presterilized solution of sodium azide (final concentration in medium, 0.015% w/v) and 15% v/v of an aseptically mixed solution composed of equal parts of egg yolk and sterile saline.

Phenolphthalein Phosphate-Polymyxin Agar[442]

Sterile nutrient agar	100 ml
Phenolphthalein phosphate	1 ml
(1% w/v aq. solution)	
Polymyxin sulfate	125 units/ml medium

After inoculation and incubation the plates are exposed to ammonia vapor in order to detect phosphatase production.

Lithium-Glycine-Tellurite-Egg Yolk-Pyruvate Agar[279]

Tryptone	10 g
Lab-Lemco®	5 g
Yeast extract	1 g
Sodium pyruvate	10 g
Glycine	12 g
Lithium chloride	5 g
Agar	20 g
Distilled water	1 liter
pH 6.8 (approximately)	

Sterilize the basal medium by autoclaving at 121°C for 15 min. Cool to 50°C and add potassium tellurite (presterilized solution) to a concentration of 0.3% w/v and 5% v/v of an egg yolk/saline solution aseptically prepared using equal parts of egg yolk and normal saline. Sulfamethazine may be added at 50 μg/ml to inhibit swarming of *Proteus* spp.[443]

2. Streptococcaceae
S. pyogenes Medium

Sterile molten blood agar base	100 ml
Defibrinated sterile horse or	5—10 ml
sheep blood	

Cool the agar base to 50°C before adding the blood. Mix thoroughly before pouring plates. Crystal violet (0.0002 g % w/v) may be added as a selective agent for *S. pyogenes*.

Antigenic Streptococcal Hemolysin Broth[444]

Beef infusion	10.0 g
Tryptone	20.0 g
Dextrose	2.0 g
$NaHCO_3$	2.0 g
NaCl	2.0 g
Na_2HPO_4	0.4 g
Distilled water	1 liter
pH 7.8	

Sterilize by autoclaving at 121°C for 15 min. Check pH after sterilization.

Brain Heart Infusion Agar[445]

Calf brain infusion solids	12.5 g
Beef heart infusion solids	5.0 g
Proteose peptone	10.0 g
Sodium chloride	5.0 g
Dextrose	2.0 g
Disodium phosphate	2.5 g
Agar	10.0 g
Distilled water	1 liter
pH 7.4 (approximately)	

Sterilize by autoclaving at 121°C for 15 min. Many fastidious *Streptococci* will grow in this medium without the addition of blood.

Tryptose Phosphate Broth[446]

Tryptose	20.0 g
Dextrose	2.0 g
Sodium chloride	5.0 g
Disodium hydrogen phosphate	2.5 g
Distilled water	1 liter
pH 7.3 (approximately)	

Sterilize by autoclaving at 121°C for 15 min.

Dextrose Broth[447]

Lab-Lemco®	3 g
Tryptose	10 g
Dextrose	5 g
Sodium chloride	5 g
Distilled water	1 liter
pH 7.2 (approximately)	

Sterilize by autoclaving at 121°C for 15 min. The acid pH produced by organisms in this unbuffered broth may prevent survival and early subculture is advisable.

Esculin-Thallium Medium (Bovine Mastitis)[448]

Lab-Lemco®	10.0 g
Peptone	10.0 g
Esculin	1.0 g
Sodium chloride	5.0 g
Crystal violet	0.0013 g
Thallous sulfate	0.33 g
Agar	15.0 g
Distilled water	1 liter
pH 7.4 (approximately)	

Sterilize by autoclaving at 121°C for 15 min. Cool to 50°C and add 5 to 7% v/v sterile bovine or sheep blood. Mix well.

Azide-Crystal Violet Broth[449]

Beef heart infusion broth	100 ml
Tryptose	1.0 g
Glucose	0.02 g
pH 7.6	

Sterilize by autoclaving at 121°C for 15 min. When cool, add 5% v/v sterile defibrinated rabbit blood. On the day before use add the following aqueous solutions (% v/v) previously sterilized by autoclaving:

Sodium azide (0.1% w/v solution)	7.5% v/v
Crystal violet (0.004% w/v solution)	5 % v/v

Phenylethylalcohol Agar[450]

Tryptone	15.0 g
Soya peptone	5.0 g
NaCl	5.0 g
Phenylethylalcohol	2.5 g
Agar	15.0 g
Distilled water	1 liter
pH 7.3	

Sterilize the medium by autoclaving at 121°C for 15 min.

Tetrazolium-Thallium-Glucose Agar[451]

Peptone	10.0 g
Lab-Lemco®	10.0 g
Glucose	10.0 g
2-3-5-Triphenyltetrazolium chloride	0.1 g
Thallous acetate	1.0 g
Agar	14.0 g
Distilled water	1 liter

Separately sterilize stock solutions of glucose, tetrazolium, and thallous acetate. Add to the previously sterilized basal agar. The many modifications of this medium are described by Barnes.[452]

Tetrazolium-Azide-Glucose Agar[453]

Tryptose	20.0 g
Yeast extract	5.0 g
Dextrose	2.0 g
Disodium phosphate·$2H_2O$	4.0 g
Sodium azide	0.4 g
Tetrazolium chloride	0.1 g
Agar	10.0 g
Distilled water	1 liter

pH 7.2 (approximately)

Boil very briefly to dissolve the medium. Excessive heating must be avoided.

Esculin-Azide Broth[454]

Peptone	20.0 g
Yeast extract	5.0 g
Bile	10.0 g
Sodium citrate	1.0 g
Esculin	1.0 g
Ferric ammonium citrate	0.5 g
Sodium azide	0.25 g

pH 7.2

Sterilize by autoclaving at 121°C for 15 min.

Group D Streptococci Media

Hartman et al.[455] list 43 agar media and 33 broth media. The table of selective agents used and source references to the media are reproduced with permission in Table 16.

Leuconostoc spp. Media
Tomato Juice Broth[516]

Tryptone	10.0 g
Soya peptone	5.0 g
Yeast extract	5.0 g
Tomato juice	200 ml
NaCl	5.0 g
Tween® 80	0.5 ml
K_2HPO_4	1.5 g
Glucose	5.0 g
Bromcresol purple	0.016 g
Distilled water	800 ml

pH 7.0

Sterilize by autoclaving at 121°C for 15 min.

Table 16
SELECTIVE MEDIA FOR ISOLATION
OF FECAL STREPTOCOCCI

Agar Media

Medium	Selective and differential agent(s),[a] conditions, comments, and synonyms	Ref.
A-1	0.006% NaN_3, 0.0001% crystal violet	456
A-2	0.01% NaN_3	457
A-3	0.01% NaN_3, 2% sodium citrate, 0.001% tetrazolium blue (Citrate-Azide agar)	458
A-4	0.01% NaN_3, 0.0002% basic fuchsin, anaerobic	459
A-5	0.02% NaN_3 (Azide-Blood Agar base)	460, 461
A-6	0.02% NaN_3, 0.0002% crystal violet, 0.1% sodium citrate (Streptocel® agar)	462
A-7	0.02% NaN_3, 5% sucrose	463
A-8	0.029% NaN_3, 0.002% acridine orange, 0.93% sodium glutamate, 0.0015% TTC^b	464
A-9	0.03% NaN_3, 3.25 IU/ml penicillin	465
A-10	0.03% NaN_3, 6.5 IU/ml penicillin, 0.001% methylene blue	466
A-11	0.04% NaN_3	467
A-12	0.04% NaN_3, 0.001% methylene blue (after enrichment in Medium B-20)	468
A-13	0.04% NaN_3, 0.01% TTC (M-entero-coccus agar)	453
A-14	0.04% NaN_3, 0.0005% ethyl violet, 0.01% TTC (Ethyl Violet Azide or EVA agar)	469, 470
A-15	0.04% NaN_3, 0.05% Tween® 80, 0.2% sodium carbonate, 0.01% TTC (Tween-Carbonate agar)	471
A-16	0.04% NaN_3, 0.05% Tween® 80, 0.2% sodium bicarbonate, 0.01% TTC (TC agar)	472
A-17	0.04% NaN_3, 0.0015% bromcresol purple, 1.0% sodium glycerophos-phate, 0.0636% sodium carbonate, 0.01% TTC (KF agar)	473
A-18	0.04% NaN_3, 2% sodium citrate, 0.001% tetrazolium blue (Citrate-Azide agar)	474
A-19	0.04% NaN_3, 0.001% methylene blue (Enterococcus Confirmatory slant); 0.04% NaN_3, 0.001% methylene blue, 6.5% NaCl, 6.5 IU/ml peni-cillin, pH 8.0 (Enterococcus Con-firmatory broth)	466, 475
A-20	0.04% NaN_3, 4.5% NaCl, 0.003% water-soluble aniline blue, pH 8.3, 45°C (for confirmation only; see also Medium B-13)	476
A-21	0.05% NaN_3, 0.0002% crystal violet	477
A-22	0.05% NaN_3, 0.0002% crystal violet, 39.5°C	478
A-23	0.05% NaN_3, 0.000125% crystal violet, 0.008% bromthymol blue, 0.5% sorbitol	479

[a] Blood (usually 5%) was included in some of these media to serve as an indicator of the type of hemolysis and, in one medium, as a source of Fe^{+++}.

[b] TTC = 2,3,5-triphenyltetrazolium chloride.

Table 16 (continued)
SELECTIVE MEDIA FOR ISOLATION
OF FECAL STREPTOCOCCI

Agar Media

Medium	Selective and differential agent(s),[a] conditions, comments, and synonyms	Ref.
A-24	0.225% NaN_3, 0.5% glycine, 0.1% esculin, pH 9.0	480, 481
A-25	0.225% NaN_3, 0.5% glycine, 0.1% esculin, 1.0% lactose, pH 9.0 (ELA agar)	482
A-26	Hydrazoic acid vapor	483
A-27	0.0033% thallium sulfate, 0.000133% crystal violet, 0.1% esculin (TKT agar)	484
A-28	0.08% thallium acetate or thallium nitrate	485
A-29	0.1% thallium acetate, 0.01% TTC (TLTC agar)	486, 451
A-30	0.1% thallium acetate, 0.01% TTC, 45°C	487
A-31	0.1% thallium acetate, 0.01% TTC, 2% sodium citrate	472
A-32	0.1% thallium acetate, 0.00005% crystal violet	488
A-33	0.1% thallium acetate, 0.01% TTC, 0.5% tyrosine, 0.2% sorbitol, 37 and 45°C	489
A-34	0.024% sodium selenite, 0.071% aniline blue (see A-20)	490
A-35	1% sodium taurocholate, 0.05% analine blue, 0.1% esculin, pH 8.0	491
A-36	1% sodium taurochlorate, 0.035% potassium tellurite, 0.1% esculin, 2.5% NaCl, pH 8.0—9.0	492
A-37	0.001% potassium tellurite, 0.00025% crystal violet (S-1)	493
A-38	0.002% potassium tellurite, 0.00008% crystal violet, 0.0075% trypan blue *Mitis-Salivarius* agar)	456
A-39	0.0067% potassium tellurite	479
A-40	0.00005% neomycin sulfate, 0.0025% phenol red, 1% mannitol (MN Trypticase-Soy agar)	494, 462
A-41	0.0014% neomycin sulfate, 0.003% polymyxin B sulfate (MSDH agar)	495
A-42	0.25% phenylethyl alcohol	461
A-43	4% NaCl	496

Broth Media

B-1	0.006% NaN_3, 0.0001% crystal violet	497
B-2	0.006—0.007% NaN_3, 0.0002% crystal violet	498, 449
B-3	0.01 NaN_3, 0.0001% crystal violet	499
B-4	0.01% NaN_3, 0.0025% bromcresol purple	488
B-5	0.02% NaN_3	500
B-6	0.02% NaN_3 (Azide-Dextrose or AD broth)	501

Table 16 (continued)
SELECTIVE MEDIA FOR ISOLATION
OF FECAL STREPTOCOCCI

Broth Media

Medium	Selective and differential agent(s),[a] conditions, comments, and synonyms	Ref.
B-7	0.02% NaN_3 (AD broth, 45°C, presumptive test, see also B-33)	502
B-8	0.02% NaN_3 (presumptive test, see also B-17)	503
B-9	0.02% NaN_3, 0.0002% crystal violet, 0.1% sodium citrate (Streptocel® broth for enrichment)	462
B-10	0.02% NaN_3, 0.0005% crystal violet	504
B-11	0.02% NaN_3, 0.003% bromthymol blue	505
B-12	0.02% NaN_3, 1% sodium taurocholate, 3% NaCl (N-N broth)	484
B-13	0.025% NaN_3, 0.0032% bromcresol purple, 45°C	506
B-14	0.025% NaN_3, 0.0032% bromcresol purple, 37°C, followed by 45°C	379
B-15	0.025% NaN_3, 0.00025% water-soluble aniline blue, 3% NaCl, pH 8.3 (presumptive medium, see Medium A-20)	476
B-15a	0.04% NaN_3, 0.0032% bromthymol blue, pH 8.0, 45°C (Enterococcus Presumptive broth; see A-19)	475
B-16	0.04% NaN_3, 0.0032% bromthymol blue, 45°C	507
B-17	0.04% NaN_3, 0.0005% ethyl violet (Ethyl Violet-Azide or EVA broth)	503
B-18	0.04% NaN_3, 1% sucrose, 0.01% TTC, 0.000012% ethyl violet optional (for membrane filter)	508
B-19	0.04% NaN_3, 1% sodium glycerophosphate, 0.0015% bromcresol purple, 0.0636% sodium carbonate, 0.01% TTC (KF broth)	473
B-20	0.05% NaN_3, 0.0002% crystal violet	477
B-21	0.05% NaN_3, 0.0015% bromcresol purple, 0.5% glycerol (BAGG broth)	509
B-22	0.05% NaN_3, 0.0032% bromcresol purple, 45.5°C (SF medium)	510
B-23	0.05% NaN_3, 6.5% NaCl, 45°C	511
B-24	0.01% potassium tellurite, 0.00005% crystal violet	512
B-25	0.02% potassium tellurite	513
B-26	0.05% thallium acetate	513
B-27	0.1% thallium acetate	451
B-28	0.1% thallium acetate, 0.00005% crystal violet	488
B-29	Tetrathionate	513, 514
B-32	0.05% Tween® 80, 0.53% sodium carbonate, pH 10.0	515
B-33	6.5% NaCl, 45°C (confirmatory test; see B-7)	502

From Hartman, P. A., Reinbold, G. W., and Sarawat, D. S., in *Advances in Applied Microbiology*, Vol. 8, Umbreit, W. W., Ed., Academic Press, New York, 1966, 253–289. With permission.

Yeast-Glucose-Citrate Broth[517]

Peptone	20.0 g
Meat extract	10.0 g
Yeast extract	6.0 g
NaCl	5.0 g
Ammonium citrate	5.0 g
Glucose	10.0 g
Distilled water	1 liter
pH 6.5	

Sterilize by autoclaving at 121°C for 15 min.

Glucose-Tween®-Manganese Broth[518]

Peptone	5.0 g
Meat extract	5.0 g
Yeast extract	5.0 g
Glucose	5.0 g
$MnSO_4 \cdot 4H_2O$	0.1 g
Tween® 80	0.5 ml
Distilled water	1 liter
pH 6.5	

Sterilize by autoclaving at 121°C for 15 min.

***Leuconostoc oenos* Medium**[519]

Peptone	10.0 g
Yeast extract	5.0 g
Glucose	10.0 g
Tomato juice	250 ml
Cysteine HCl	0.5 g
$MnSO_4 \cdot 4H_2O$	0.1 g
Distilled water	750 ml
pH 4.8	

Sterilize the cysteine HCl (as a 1% w/v solution) by filtration and add it to the autoclaved basal medium (121°C for 15 min), at 0.5 to 10 ml of basal medium immediately before use.

***Pediococcus cereviseae* and *Aerococcus viridans* Medium**[520]

Tryptone	12.5 g
Yeast extract	7.5 g
Glucose	10.0 g
Sodium citrate	5.0 g
NaCl	5.0 g
K_2HPO_4	5.0 g
$MgSO_4$	0.8 g
$MnCl_2$	0.14 g
$FeSO_4$	0.04 g
Thiamine HCl	0.0001 g
Tween® 80	0.2 g
pH 6.7	

Sterilize by autoclaving at 121°C for 15 min. Avoid reheating the medium. Agar may be added at 15 g/l to provide a solid medium.

Pediococcus spp. Medium[521]

Tomato juice agar (Oxoid)	100 ml
Tween® 80	0.1 ml
pH 6.1	

Sterilize by autoclaving at 121°C for 15 min.

Tryptone Glycine Medium (Peptococcus glycinophilus)[522]

Tryptone	5.0 g
Yeast extract	5.0 g
Glycine	3.0 g
$MgSO_4 \cdot 7H_2O$	0.2 g
$FeSO_4 \cdot 7H_2O$	0.01 g
$MnSO_4 \cdot 4H_2O$	0.005 g
KH_2PO_4-Na_2HPO_4 buffer	5 ml
(0.1 M, pH 7.0)	
Distilled water	1 liter
pH 7.0	

Sterilize by autoclaving at 121°C for 15 min.

Peptone Yeast Glutamate Medium (Peptococcus aerogenes)[523]

Peptone	20.0 g
Yeast extract	10.0 g
Monosodium glutamate	4.0 g
Sodium thioglycollate	1.0 g
Distilled water	1 liter
pH 7.2	

Sterilize by autoclaving at 121°C for 15 min.

Sarcina ventriculi Medium[524]

Glucose	20.0 g
Yeast extract	20.0 g
Distilled water	1 liter
pH 6.0	

Sterilize by autoclaving at 121°C for 15 min. Boil and cool immediately before use.

Sarcina maxima Medium[524]

Peptone	10.0 g
Yeast extract	5.0 g
Glucose	10.0 g
Cysteine HCl	0.5 g
Distilled water	1 liter
pH 6.0	

Sterilize the cysteine HCl separately and add it to the sterilized basal medium (121°C for 15 min) immediately before use.

L. Endospore-forming Rods and Cocci

1. Bacillus Species

All the organisms in this genus grow on nutrient agar and in nutrient broth. The addition of carbohydrates improves the growth of some strains. Characteristic pigment production may require the addition of trace minerals and sporulation may only take place in the presence of specific metals. Some species grow in the presence of NH_4-nitrogen uric acid or urea; other more demanding strains require vitamins, peptides, and unknown constituents of complex media.

B. stearothermophilus Medium[524a]

Growth Medium

Tryptone	10.0 g
Yeast extract	5.0 g
K_2HPO_4	2.0 g
Distilled water	1 liter
pH 7.2	

Sterilize by autoclaving at 121°C for 15 min.

Sporulation Medium

Nutrient Broth (Difco B3)	8.0 g
Yeast extract	4.0 g
$MnCl_2 \cdot 4H_2O$	10 ppm
Agar	20.0 g
Distilled water	1 liter
pH 7.2	

Sterilize by autoclaving at 121°C for 15 min.

Defined Liquid Medium for B. stearothermophilus[524b]

L-Alanine	0.84 ml
L-Arginine HCl	0.64 ml
L-Asparagine H_2O	0.50 ml
L-Aspartate (monopotassium salt)	1.30 ml
L-Cystine	0.50 ml
L-Glutamate HCl	4.00 ml
L-Glutamine	0.50 ml
Glycine	0.50 ml
L-Histidine HCl H_2O	0.42 ml
L-Isoleucine	1.00 ml
L-Leucine	1.64 ml
L-Lysine HCl	1.40 ml
L-Methionine	0.52 ml
L-Phenylalanine	0.86 ml
L-Proline	1.00 ml
L-Serine	1.40 ml
L-Threonine	0.84 ml
L-Tryptophan	0.30 ml
L-Tyrosine	0.56 ml
L-Valine	1.26 ml
Biotin (10 mg/100 ml)	1.00 ml

Defined Liquid Medium for B. stearothermophilus[524b]
(continued)

Thiamine HCl (10 mg/100 ml)	1.00 ml
Nicotinic acid (10 mg/100 ml)	1.00 ml
Anhydrous $CaCl_2$ (5%)	0.01 ml
$FeCl_3 \cdot 6H_2O$ (0.05%)	0.01 ml
$ZnSO_4 \cdot 7H_2O$ (5%)	0.01 ml
$MnCl_2$ (10 mM)	0.01 ml
Glucose (20%)	1.00 ml
Mineral salts solution	10.00 ml
(10 g NH_4Cl, 10 g NaCl, and	
4 g $MgSO_4$ per liter)	
Potassium phosphate buffer	5.00 ml
(125 g K_2HPO_4 and	
30 g KH_2PO_4 per 500 ml)	
Total volume	100.00 ml

Add all ingredients in the order indicated above without stirring to prevent precipitation. All stock amino acid solutions are 1% (w/v). Adjust the final volume to 100 ml with deionized water and filter sterilize the solution. The pH is 7.3.

B. popillae Medium[524c]

Yeast extract	15.0 g
K_2HPO_4	3.0 g
Distilled water	1 liter

For the maintenance of stock cultures, add

Glucose	2.0 g
Tryptone	5.0 g
Agar	20.0 g
pH 7.2	

to the basal medium. Sterilize all components of the medium, with the exception of glucose, at 121°C for 15 min. The medium can be used with or without agar.

B. popillae Medium[524d]

Mueller-Hinton Broth	10.0 g
Yeast extract	10.0 g
K_2HPO_4	3.0 g
Trehalose	0.5–1.0 g
Distilled water	1 liter

Dissolve and filter-sterilize the ingredients. If a solid medium is required, add the sterilized broth to equal volumes of double-strength agar (40 g/l) in distilled water which has been autoclaved at 121°C for 15 min. Mix when the agar has cooled to 50°C.

B. cereus Medium[524e]

Peptone	5.0 g
Lab-Lemco®	5.0 g
Na_3 citrate	5.0 g
$LiCl_2$	5.0 g
Polymyxin sulfate	50 units/ml
Agar	15.0 g
Distilled water	1 liter
pH 7.0	

Dissolve the ingredients in the water and sterilize by autoclaving at 121°C for 15 min. Cool to 50°C and add egg yolk (1:1 with physiological saline) at 2.5% v/v.

B. coagulans Medium[524f]

Yeast extract	5.0 g
Proteose peptone	5.0 g
Dextrose	5.0 g
K_2HPO_4	4.0 g
$MnSO_4 \cdot H_2O$	50 ppm
$CaCl_2$	45 ppm
Agar	20.0 g
Distilled water	1 liter
pH 5.0	

Dissolve the ingredients in water and sterilize at 121°C for 15 min. The Mn^{++} and Ca^{++} may be added separately after autoclaving to prevent precipitation. Avoid overheating at the low pH value.

Bacillus spp. Medium[524g]

Peptone	6.0 g
Tryptone	3.0 g
Yeast extract	3.0 g
Lab-Lemco®	1.5 g
Mn^{++} (SO_4 or Cl)	1 ppm
Agar	25.0 g
Distilled water	1 liter
pH 7.0	

Sterilize the medium by autoclaving at 121°C for 15 min.

Bacillus fastidiosus Medium[525]

A mixture of defined and natural ingredients was described to grow a fastidious specie.

Uric acid	1.0 g
Yeast extract	2.5 g
$Na_2HPO_4 \cdot 12H_2O$	6.0 g
Agar	15.0 g
Distilled water	900 ml

Mineral solution (in 100 ml distilled water)

KH_2PO_4	100 mg
$CaCl_2$	10 mg
$MgSO_4 \cdot 7H_2O$	30 mg
NaCl	10 mg
$FeCl_3 \cdot 6H_2O$	1 mg
pH 7.0	

Dissolve the ingredients and sterilize at 121°C for 15 min.

2. Clostridium Species

Members of the genus *Clostridium* are anaerobic spore-forming bacteria. Anaerobic incubation in liquid medium gives effective enrichment. Isolation may be obtained by plating on solid media with or without selective agents. A wide variety of media is required to grow the full range of *Clostridia* from clinical material, fishery products, and marine and fresh-water sediments.

Enrichment Media

- Thioglycollate broth
- Cooked meat medium
- Maize extract
- Liver broth

The addition of neomycin sulfate (100 µg/ml) to enrichment media favors the growth of Gram-positive anaerobic bacilli.[526] Phenethyl alcohol (0.25% v/v) may be added to inhibit the growth of Gram-negative anaerobic organisms.[527] Fluid media for the growth of *Clostridium* species may contain

1. Reducing agents such as thioglycollic acid, cystine, cysteine, dithionite, sulfite.
2. Agents which help retain low eH values after heating, such as glucose, semisolid agar, chopped meat.

Many of the media previously described for anaerobic asporogenous organisms are also suitable for clostridia.

Dithionate-Thioglycollate (HS-T) Broth

I Pancreatic digest of casein (e.g., tryptone, Oxoid L12)	15.0 g
Yeast extract (e.g., Oxoid L21)	6.0 g
Enzymatic hydrolysate of soybean flour (e.g., soya peptone, Oxoid L44)	3.0 g
Agar (e.g., No. 1, Oxoid L11)	0.75 g
Sodium chloride	2.50 g
Dipotassium hydrogen phosphate	2.00 g
Sodium citrate	1.00 g
L-Cystine	0.50 g
L-Asparagine	1.25 g
Magnesium sulfate·$7H_2O$	0.40 g
Calcium chloride·$2H_2O$	0.004 g
Cobalt sulfate·$7H_2O$	0.001 g
Cupric sulfate·$5H_2O$	0.001 g
Zinc sulfate·$7H_2O$	0.001 g
Ferrous sulfate·$7H_2O$	0.001 g
Manganese chloride·$4H_2O$	0.002 g
II Lecithin	0.30 g
III Tween® 80 (polyethylene sorbitan monooleate)	3.0 g
Glycerol	5.0 g
IV Dextrose	6.0 g
Sodium dithionite	0.40 g
Sodium thioglycollate	0.50 g
Resazurin	0.001 g
V Distilled water up to	1000 ml

Mixture I is dissolved in V by heating, after which lecithin (II) is suspended in a heated mortar with a mixture of III and some of the hot solution of Mixture I in V. The suspension is transferred to the rest of the solution to which Mixture IV is then added. The pH value is adjusted before autoclaving to 7.2 to 7.3 with 1 *N* hydrochloric acid or 1 *N* sodium hydroxide. After autoclaving the pH value should be ca. 7.1 (7.0 to 7.2). The broth is run off into ca. 18 × 200-mm tubes, about 15 ml in each, and into containers (bottles), 100 ml in each, and autoclaved for 15 min at 121°C, after which it is allowed to cool to room temperature. The final product is a yellowish, almost clear liquid which turns pale pink under aerobic conditions. Before use, no more than the upper third of the broth at most should have turned pink. The medium may be stored at ca. -4°C for up to 1 month and can be regenerated once by heating to ca. 100°C. The medium is available from Oxoid Ltd., London, under the name of "Clausen Medium."

From Clausen, O. G., Aasgaard, N. B., and Solberg, O., *Ann. Microbiol.* (Paris), 124B, 205–216, 1973. With permission.

Solid media may have to be prereduced in order to enable very exacting strains to grow. Freshly prepared blood agar containing 0.05% w/v cysteine hydrochloride will grow demanding species such as *C. oedematiens* type D.[529] Carbon dioxide often improves the growth of *Clostridia,* either as 5% v/v CO_2 in the atmosphere or by including 0.1% w/v sodium bicarbonate in the culture medium.[529]

3. Nitrogen-fixing Clostridia

Mineral Compositions of Liquid Media used for Growing Nitrogen-fixing Clostridia

	A	B	C	D	E
K_2HPO_4	1.0	0.5	0.25	0.5	0.8
KH_2PO_4	—	0.5	0.75	0.5	0.2
$MgSO_4 \cdot 7H_2O$	0.5	0.1	0.125	0.5	0.2
NaCl	0.01—0.02	0.1	0.125	0.5	0.2
$FeSO_4 \cdot 7H_2O$	0.01—0.02	0.01	—	0.01	—
$Fe_2(SO_4)_3$	—	—	0.0025	—	0.01
$MnSO_4 \cdot 7H_2O$	0.01—0.02	0.01	0.0025	0.01	—
$CaCO_3$	Trace	4—5	—	—	—
$CaSO_4$	—	—	—	—	0.1

From Skinner, F. A., *Soc. Appl. Bacteriol. Tech. Ser.*, 5, 58, 1971. With permission.

Trace elements and growth factors may be added to these media:[54]

K_2HPO_4	0.8 g
KH_2PO_4	0.2 g
$MgSO_4 \cdot 7H_2O$	0.2 g
NaCl	0.2 g
$FeSO_4 \cdot 7H_2O$	0.01 g
$MnSO_4 \cdot 7H_2O$	0.01 g
$CaCl_2$	0.01 g
Glucose	10.0 g
Yeast extract	0.001—0.01 g
$Na_2MoO_4 \cdot 2H_2O$	0.025 mg
Trace element mixture (see below)	1.0 ml
Soil extract	100 ml
Sodium thioglycollate	1.0 g
Distilled water	1 liter
pH 7.2	

Trace element mixture:

$K_2MoO_4 \cdot 2H_2O$	0.05 g
$NaBO_4$	0.05 g
$CoNO_3$	0.05 g
$CdSO_4$	0.05 g
$CuSO_4$	0.05 g
$ZnSO_4$	0.05 g
$MnSO_4$	0.05 g
$FeCl_3$	Trace
Distilled water	1 liter

All clostridia require organic growth factors to multiply and sporulate. The mineral media given above require supplementation to grow small inocula:

- Yeast extract — 2×10^{-3} μg/ml of medium.[530]
- Soil extract — 10 ml/l of medium.[63]
- Potato extract — 0.2 to 0.5% v/v mineral medium.[531]

C. (Zymobacterium) oroticum Medium[300]

Tryptone	20.0 g
Glucose	5.0 g
Yeast extract	0.5 g
Orotic acid	2.0 g
Sodium thioglycollate	0.5 g
$MgSO_4 \cdot 7H_2O$	0.2 g
$FeSO_4 \cdot 7H_2O$	0.005 g
$MnSO_4 \cdot 4H_2O$	0.005 g
$Na_2MoO_4 \cdot 2H_2O$	0.005 g
Riboflavin	15 mg
KH_2PO_4-Na_2HPO_4	1 liter
(0.05 M, pH 7.4)	
Agar	15 g

Dissolve the ingredients (except the agar) in half the phosphate buffer solution and adjust to pH 7.4 with NaOH. Add the remainder of the phosphate solution and the agar, steam to dissolve the agar, and sterilize at 121°C for 15 min. If the medium is not to be used immediately, omit the thioglycollate and add it just before use.

C. acidiurici Medium[532]

Uric acid	2.0 g
KOH (10 M)	1.2 ml
Yeast extract	1.0 g
$K_2HPO_4 \cdot 3H_2O$	0.91 g
$MgSO_4 \cdot 7H_2O$	0.035 g
$FeSO_4 \cdot 7H_2O$	1.75 mg
$CaCl_2 \cdot 2H_2O$	4.2 mg
Thioglycollic acid	2.0 ml
Distilled water	1 liter
pH 7.0—7.2	

Dissolve the uric acid in the water to which the KOH has been added. Boil if necessary. Add the remaining ingredients and adjust the pH to 7.2. Sterilize the medium by autoclaving at 121°C for 15 min. Store the medium at 37°C to prevent precipitation of the uric acid.

C. aceticum Medium[533]

A. Fructose	10 g
Distilled water	200 ml
B. $NaHCO_3$	20 g
Distilled water	200 ml

C. Vitamin B_{12} (2 mg/100 ml distilled water)	1 ml
D. Peptone	10 g
K_2HPO_4	10 g
$Co(NO_3)_2$	19.7 mg
Distilled water	600 ml
E. Vitamin solution	10 ml
F. Heavy metal solution	25 ml
Vitamin solution	
Biotin	0.1 mg
Nicotinic acid	1 mg
Thiamin	0.5 mg
p-Aminobenzoic acid	0.5 mg
Pantothenic acid	0.25 mg
Pyridoxin	2.5 mg
Distilled water	100 ml
Heavy metal solution	
EDTA	1.5 g
$FeSO_4 \cdot 7H_2O$	0.2 g
$ZnSO_4 \cdot 7H_2O$	0.1 g
$MnCl_2 \cdot 4H_2O$	0.02 g
H_3BO_3	0.03 g
$CuCl_2 \cdot 2H_2O$	0.001 g
$NiCl_2 \cdot 6H_2O$	0.002 g
$Na_2MoO_4 \cdot 2H_2O$	0.003 g
Distilled water	1 liter

Dissolve the EDTA in the water first. Prepare the vitamin B_{12} solution (C), vitamin solution (E), and heavy metal solution (F) separately and sterilize by filtration. Dissolve the fructose in the appropriate amount of water and sterilize by autoclaving (A) at 121°C for 15 min. Dissolve the bicarbonate (B) in the water and sterilize the solution by filtration under positive pressure of CO_2. Prepare Medium D and sterilize by autoclaving at 121°C for 15 min. Mix the sterile solutions A, B, D, E, and F, add 1 ml of solution C, and mix again. Adjust the pH aseptically to 8.0 and distribute the medium into sterile tubes or bottles.

From Lapage, S. P., Shelton, J. E., and Mitchell, T. G., in *Methods in Microbiology*, Vol. 3A, Norris, J. R. and Ribbons, D. W., Eds., Academic Press, London, 1970, 1–288. With permission.

C. sticklandii Medium[534]

Yeast extract	2.0 g
L-Arginine HCl	1.5 g
L-Lysine HCl	1.5 g
NH_4Cl	0.5 g
$CaCl_2 \cdot 2H_2O$	0.01 g
$MgSO_4 \cdot 7H_2O$	0.2 g
$FeSO_4 \cdot 7H_2O$	0.01 g
$MnSO_4 \cdot 4H_2O$	0.001 g
$Na_2MoO_4 \cdot 2H_2O$	0.001 g
K_2HPO_4 -KH_2PO_4	1 liter
(0.04 M, pH 7.5)	

Dissolve the ingredients in the phosphate buffer solution and sterilize at 121°C for 15 min.

Part 1

Ethanol	20 ml
Potassium acetate	4.3 g
M Phosphate buffer, pH 7.1	3.5 ml
Acetic acid (glacial)	2.5 ml
$(NH_4)_2SO_4$	0.5 g
$MgSO_4 \cdot 7H_2O$	0.2 g
$CaSO_4 \cdot 2H_2O$	10 mg
$MnSO_4 \cdot 4H_2O$	2.5 mg
$Na_2MoO_4 \cdot 2H_2O$	2.5 mg
Biotin	10 μg
p-Aminobenzoic acid	1 mg
Distilled water	900 ml

Part 2

K_2CO_3	5.9 g
Distilled water	100 ml

Part 3

$Na_2S \cdot 9H_2O$	1 g
Distilled water	10 ml

Part 4

$FeSO_4 \cdot 7H_2O$	0.1 g
H_2SO_4, 0.1 N	10 ml

Dissolve the solid ingredients of each part of the medium in the respective liquids. Sterilize each part in separate containers by autoclaving at 121°C for 15 min. Add Part 2 to Part 1 aseptically, then add 2 ml of Part 3 and 0.5 ml of Part 4. Mix well and check the pH; adjust if necessary to approximately 7.0. Distribute aseptically.

From Lapage, S. P., Shelton, J. E., and Mitchell, T. G., in *Methods in Microbiology,* Vol. 3A, Norris, J. R. and Ribbons, D. W., Eds., Academic Press, London, 1970, 1—288. With permission.

Differential Reinforced Clostridial Medium (DRCM)[536]

Peptone	10.0 g
Lab-Lemco®	10.0 g
Sodium acetate (hydrated)	5.0 g
Yeast extract	1.5 g
Soluble starch	1.0 g
Glucose	1.0 g
L-Cysteine HCl	0.5 g
Distilled water	1 liter
pH 7.1—7.2	

Dissolve the peptone, Lab-Lemco®, sodium acetate, and yeast extract in 800 ml distilled water. Make a starch paste in the remaining 200 ml water and boil to dissolve. Mix together and dissolve the glucose and L-cysteine. Adjust to pH 7.2 with 10 N NaOH. Filter through paper pulp and distribute in 25-ml volumes in small bottles. Sterilize at 121°C for 15 min. Immediately before use add sodium sulfate (0.04% w/v) and ferric citrate (0.07% w/v). This can be done by preparing the sulfite solution at 4% w/v and the citrate solution at 7% w/v. Sterilize by filtration. When the DRCM is required, the two solutions are added together and 0.5 ml of the mixture is aseptically added to 25 ml of DRCM.

Lactose Egg-Yolk Agar for Clostridial Identification[529]

Nutrient broth	100 ml
Agar	1.2 g
Lactose	1.0 g
Neutral red (1% w/v solution)	0.3 ml

Sterilize at 121°C for 15 min. Cool to 50°C and add 4 ml of egg yolk emulsion (equal parts of egg yolk and sterile 0.9% w/v saline solution).

Clostridium chauvoei Medium[537]

Nutrient agar	75 ml
Glucose (50% w/v solution)	2 ml
Liver extract (Difco)	3 ml
Sterile sheep blood	5 ml

The nutrient agar should contain the equivalent of 1.8% w/v agar and can be sterilized with the liver extract at 121°C for 15 min. Cool to 50°C and add 2 ml of filter-sterilized glucose solution. Add 5 ml of sheep blood and mix thoroughly before pouring.

C. oedematiens Medium (Types B, C, and D)[538]

Peptone	10.0 g
Yeast extract	5.0 g
Liver digest	5.0 g
Glucose	10.0 g
Salts solution (see below)	5.0 ml
Agar	20.0 g
Distilled water	1 liter
pH 7.6–7.8	
Cysteine HCl	100 mg
Glutamine	50 mg
Dithiothreitol	100 mg
Horse blood	100 ml

Salts solution:

$MgSO_4 \cdot 7H_2O$	40.0 g
$MnSO_4 \cdot 4H_2O$	2.0 g
$FeCl_3$	0.4 g
10 N HCl	0.5 ml
Distilled water	1 liter

Sterilize the basal medium at 121°C for 15 min and add the separately filter-sterilized solutions of cysteine, glutamine and dithiothreitol. Add the sterile horse blood and mix thoroughly.

4. *Desulfotomaculum spp. Medium*[539]

Part 1

K_2HPO_4	0.5 g
NH_4Cl	1 g
Na_2SO_4	1 g
$CaCl_2 \cdot 2H_2O$	0.1 g
$MgSO_4 \cdot 7H_2O$	2 g
Sodium lactate (70%)	3.5 g
Yeast extract	1 g
Distilled water	980 ml

Part 2

$FeSO_4 \cdot 7H_2O$	0.5 g
Distilled water	10 ml

Part 3

Sodium thioglycollate	0.1 g
Ascorbic acid	0.1 g
Distilled water	10 ml

Dissolve the ingredients of each part in the appropriate quantities of water. Adjust the pH of Parts 1 and 3 to 7.4. Sterilize the three parts separately by autoclaving at $121°C$ for 15 min. Combine the three parts, mix, and distribute the medium aseptically as required. If the medium is to be stored, Part 1 should be boiled immediately before use. When the whole batch of medium is to be employed immediately following preparation, all of the constituents may be mixed before sterilization. This medium is also prepared, where indicated, with 2.5% NaCl.

From Lapage, S. P., Shelton, J. E., and Mitchell, T. G., in *Methods in Microbiology,* Vol. 3A, Norris, J. R. and Ribbons, D. W., Eds., Academic Press, London, 1970, 1—288. With permission.

5. *Sporosarcina ureae Medium*[2]

Peptone	5.0 g
Meat extract	5.0 g
NH_4Cl	5.0 g
Distilled water	1 liter
pH 8.5	

Sterilize the medium at $121°C$ for 15 min. Agar may be added at 1.5% w/v if required.

M. Asporogenous Gram-positive Rods
1. *Lactobacillaceae*
György and Rose Medium (Lactobacilli)[541]

K_2HPO_4	2.5 g
Lactose	35 g
Sodium acetate	25 g
N-Z Case® (enzymatic digest of casein; Sheffield Chemical Division)	25 g

György and Rose Medium (Lactobacilli)[541] (continued)

Adenine, guanine, uracil, and xanthine	0.01 g of each
Alanine, cystine, and tryptophan	0.2 g of each
Asparagine	0.1 g
Thiamine HCl	0.2 mg
Riboflavin	0.2 mg
Pyridoxine HCl	1.2 mg
Nicotinic acid	0.6 mg
p-Aminobenzoic acid	0.01 mg
Folic acid	0.01 mg
Biotin	0.004 mg
Salts solution (see below)	5 ml
Ascorbic acid	5 mg
Distilled water	1 liter

Salts solution	
$MgSO_4 \cdot 7H_2O$	10 g
$FeSO_4 \cdot 7H_2O$	0.5 g
NaCl	0.5 g
$MnSO_4 \cdot 4H_2O$	0.337 g
Distilled water	250 ml

Mix all the ingredients except the ascorbic acid with the water and adjust the pH to 6.8 with NaOH. Prepare the ascorbic acid as a 1% w/v solution in distilled water and sterilize in a well-filled screw-capped bottle by autoclaving at 121°C for 15 min. Autoclave the bulk of the medium under the same conditions and add the ascorbic acid (0.5 ml) afterwards. Distribute the complete medium aseptically in bottles or tubes.

From Lapage, S. P., Shelton, J. E., and Mitchell, T. G., in *Methods in Microbiology*, Vol. 3A, Norris, J. R. and Ribbons, D. W., Eds., Academic Press, London, 1970, 1–288. With permission.

MRS Medium[542]

Peptone	10.0 g
Meat extract	10.0 g
Yeast extract	5.0 g
K_2HPO_4	2.0 g
Ammonium citrate	2.0 g
Glucose	20.0 g
Tween® 80	1.0 ml
Sodium acetate	5.0 g
$MgSO_4 \cdot 7H_2O$	0.2 g
$MnSO_4 \cdot 4H_2O$	0.2 g
Distilled water	1 liter
pH 6.2–6.6	

Dissolve the ingredients in distilled water and sterilize by autoclaving at 121°C for 15 min.

APT Medium[543]

Tryptone	10.0 g
Yeast extract	5.0 g
K_2HPO_4	5.0 g
NaCl	5.0 g
Sodium citrate	5.0 g
Glucose	10.0 g
Tween® 80	1 ml
$MgSO_4 \cdot 7H_2O$	0.8 g
$MnCl_2 \cdot 4H_2O$	0.14 g
$FeSO_4 \cdot 7H_2O$	0.008 g
Distilled water	1 liter
pH 6.7–7.0	

Dissolve the ingredients in distilled water and sterilize at 121°C for 15 min.

Rogosa Medium[544]

Trypticase	10.0 g
Tryptose	3.0 g
Yeast extract	5.0 g
K_2HPO_4	3.0 g
KH_2PO_4	3.0 g
Ammonium citrate	2.0 g
Glucose	20.0 g
Tween® 80	1 ml
Sodium acetate	1.0 g
Cysteine	0.2 g
$MgSO_4 \cdot 7H_2O$	0.2 g
$MnSO_4 \cdot 4H_2O$	0.2 g
$FeSO_4 \cdot 7H_2O$	0.03 g
Distilled water	1 liter
pH 6.8	

Dissolve the ingredients in distilled water and sterilize at 121°C for 15 min.

2. Listeria spp.
Thiocyanate Broth[545]

Nutrient broth (Oxoid No. 2 CM 67)	25.0 g
Potassium thiocyanate	3.75 g
Tween® 80	1.0 g
Distilled water	1 liter
pH 7.4	

Sterilize by autoclaving at 121°C for 15 min.

Nalidixic Acid-Blood Medium[215]

Nutrient broth (Oxoid No. 2 CM 67)	25.0 g
Nalidixic acid	0.04 g
Distilled water	1 liter
pH 7.4	

Dissolve nalidixic acid in a small quantity of $N/1$ NaOH and add it to the broth after it has been sterilized at 121°C for 15 min; 5% v/v sterile horse blood may be added when the broth has cooled below 45°C.

3. Erysipelothrix rhusiopathiae
Crystal Violet-Esculin Agar[546]

Proteose peptone	5.0 g
Meat extract	3.0 g
Tryptone	5.0 g
NaCl	5.0 g
Glucose	5.0 g
Esculin	1.0 g
Crystal violet	0.002 g
Agar	15.0 g
Distilled water	1 liter
pH 7.5	

Sterilize by autoclaving at 121°C for 15 min. Cool to 50°C and add 10% v/v citrated blood. Mix well and pour plates.

Crystal Violet-Azide-Esculin Agar[547]

Proteose peptone	5.0 g
Meat extract	3.0 g
Tryptone	5.0 g
NaCl	5.0 g
Glucose	5.0 g
Esculin	1.0 g
Crystal violet	0.01 g
NaN_3	1.0 g
Agar	15.0 g
Distilled water	1 liter
pH 7.5	

Sterilize by autoclaving at 121°C for 15 min. Cool to 50°C and add 10% v/v citrated bovine blood. Mix well and pour plates.

N. Actinomycetales and Related Organisms
1. Coryneform Group

The organisms grow well on nutrient agar, especially if glucose is added. Coagulated serum (Loeffler's medium) is used as a cultivation medium when the coryneform morphology of the organism is being studied. Tellurite-blood agar is the most common selective/indicator medium used, and many variations are described.

Loeffler's Coagulated Serum

Sterile ox, sheep, or horse serum	2 vol
1% w/v glucose broth	1 vol

Add the mixture to stoppered, presterilized tubes which are sloped in an inspissator. Slowly raise the temperature to 80 to 85°C; maintain this for 4 hr, at which time the serum coagulates to a yellow-white solid. A further period of heating at 85°C (20 min) may be carried out on two successive days.

McLeod's Medium[548]

Meat infusion agar in which the infusion is not heated above 75°C and is sterilized by filtration	100 ml
Sterile, defibrinated rabbit blood	7—10 ml
Potassium tellurite (0.5% aqueous solution)	0.04 g

Mix together and heat at 75°C for 10 to 20 min before pouring into dishes.

Hoyle's Medium[549]

Lab-Lemco®	10.0 g
Peptone	10.0 g
NaCl	5.0 g
Agar	15.0 g
Distilled water	1 liter
pH 7.8	

Sterilize at 121°C for 15 min and cool to 50°C. To every 100 ml of molten agar add

- 5 ml of saponin-lysed horse blood.
- 1 ml of 3.5% w/v potassium tellurite solution.

Cystine-Tellurite-Blood Agar[550]

Blood agar base (presterilized)	100 ml
Sterile defibrinated blood	5 ml
Potassium tellurite	0.45 g
Cystine	0.05 g
pH 7.6	

Dry cystine may be added directly to the molten medium held at 50°C. Mix well and pour plates.

Glucose-Serum-Tellurite Agar[551]

Proteose No. 3 Agar (Difco)	45.0 g
Agar	5.0 g
Phosphate-buffered water (pH 7.2)	
Na_2HPO_4 ($M/15 = 9.47$ g/l)	72 ml
NaH_2PO_4 ($M/15 = 8.00$ g/l)	28 ml
Distilled water	900 ml

Sterilize at 121°C for 15 min. Cool to 50°C and to every 100 ml of molten sterile agar add

Sterile serum	5 ml
Glucose (10% w/v aqueous solution)	4 ml
Potassium tellurite	1 ml
(0.5% w/v aqueous solution)	

Mix thoroughly and pour into plates or tubes.

Tinsdale's Medium[552]

Peptone	20.0 g
Yeast extract	5.0 g
NaCl	5.0 g
L-Cystine	0.24 g
NaHSO$_3$	0.43 g
Agar	15.0 g
Distilled water	1 liter
pH 7.4	

Sterilize by autoclaving at 121°C for 15 min. Cool to 50°C and add 100 ml sterile bovine serum and 10 ml 3% w/v aqueous filter-sterilized potassium tellurite. Mix thoroughly and pour.

Enrichment Medium

Sterile nutrient broth	100 ml
Potassium tellurite	1.0 ml
(3.5% w/v aqueous solution)	
Copper sulfate	0.25 ml
(10% w/v aqueous solution)	
Lysed horse blood	5.0 ml

No further sterilization is required. Distribute in 2.5- to 5.0-ml amounts. Incubate the enrichment cultures for 6 hr only before subculturing to blood agar.

Arthrobacter spp. Medium[553]

Peptone	5.0 g
Lab-Lemco®	3.0 g
Agar	15.0 g
Soil extract (500 g soil	1 liter
heated in 2.4 liters tap	
water and filtered)	
pH 7.0	

Sterilize at 121°C for 15 min.

Cellulomonas spp. Medium[300]

Yeast extract	0.5 g
$NaNO_3$	0.5 g
K_2HPO_4	1.0 g
$MgSO_4 \cdot 7H_2O$	0.5 g
KCl	0.5 g
$FeSO_4 \cdot 7H_2O$	0.01 g
Distilled water	1 liter
pH 7.2	

Dissolve the extract and salts in water. Distribute into 10-ml volumes and add a strip of filter paper to each tube. Sterilize the medium at 121°C for 15 min. Agar may be added at 15 g/l and filter paper then placed aseptically on the surface of the solidified medium.

Propionibacterium spp.[554]

Trypticase	17.0 g
Soya peptone	3.0 g
Glucose	6.0 g
Sodium thioglycollate	0.5 g
NaCl	2.5 g
Na_2SO_3	0.1 g
L-Cystine	0.25 g
Distilled water	1 liter
pH 7.4	

Dissolve the ingredients in the order given and sterilize at 121°C for 15 min. Use freshly prepared medium or ensure that the medium is stored at room temperature in tightly capped containers.

Eubacterium spp. Medium[555]

Tryptone	20.0 g
Meat extract	15.0 g
Glucose	5.0 g
$Na_2HPO_4 \cdot 12H_2O$	4.0 g
Cysteine HCl	0.5 g
Agar	15.0 g
Distilled water	1 liter
pH 7.4	

Sterilize at 121°C for 15 min. Use freshly prepared medium.

2. Actinomycetaceae

Actinomyces spp. Medium

All species of the genus *Actinomyces* grow in thioglycollate broth (see Section X.L.2, *Clostridia*), although some may require the addition of 10% v/v serum or whole blood. To isolate colonies, the thioglycollate medium is usually subcultured to a highly nutritious medium containing glucose, e.g., Brain Heart Infusion Agar.

Brain Heart Infusion Agar[556]

Calf brain infusion solids	12.5 g
Beef heart infusion solids	5.0 g
Peptone	10.0 g
NaCl	5.0 g
Glucose	2.0 g
Na_2HPO_4	2.5 g
Agar	15.0 g
Distilled water	1 liter
pH 7.4	

Sterilize by autoclaving at 121°C for 15 min.

Arachnia

Arachnia propionica will grow on the same medium as *Propionibacterium* spp.

Bifidobacterium spp. Medium[557]

Soluble starch	10.0 g
$CaCO_3$	2.0 g
$(NH_4)_2SO_4$	2.0 g
K_2HPO_4	1.0 g
$MgSO_4 \cdot 7H_2O$	1.0 g
$FeSO_4 \cdot 7H_2O$	0.001 g
$MnCl_2 \cdot 4H_2O$	0.001 g
$ZnSO_4 \cdot 7H_2O$	0.001 g
Agar	20.0 g
Distilled water	1 liter
pH 7.2	

Suspend the starch in water by making a paste and diluting it with water. Dissolve the salts in the remaining water and add the chalk. Mix the starch, salts, and agar together. Sterilize at 121°C for 15 min and mix well before pouring to suspend the chalk. *Bifidobacterium* spp. will also grow in MRS Medium (see Section X.M.1, Lactobacillaceae) when incubated under anaerobic conditions.[558]

Bacterionema[559]

Brain heart infusion (prepared at half strength)	100 ml
Yeast extract (0.15 g % w/v)	100 ml
Hemin	40 μg

Sterilize at 121°C for 15 min. Cool rapidly and use freshly prepared.

Rothia

Rothia dentocariosa will grow on Brain Heart Infusion Agar or Tryptone Soya Agar.[560]

3. *Mycobacterium Species*

The organisms in this genus vary widely in their nutritional demands. Some are obligate parasites (*M. leprae* and *M. lepraemurium*) and have never been isolated in vitro. Some saprophytic strains ("rapid" growers) can be isolated on very simple substrates. The intermediate forms, which are often pathogenic, require complex media and long incubation periods.

The "medical" mycobacteria (*M. tuberculosis* and other species) were first cultivated on slices of potato saturated with glycerol solution or similar nutrients. Coagulated egg media became more effective and many formulas based on Lowenstein-Jensen's original formula have been described. Defined media (with and without agar) were later developed, but have not yet replaced coagulated egg media.

Potato Media

Select large potatoes, wash, and peel. Using a large borer, cut out a cylinder of potato and wash it in water to remove excess starch. Cut the cylinder obliquely in half and place each half into a 150 × 25-mm test tube. The thickest end should rest on a plug of cotton wool.

Plain Potato Medium — Fill the tubes with sterile water and steam for 30 min. Pour off the water, cover the tubes, and autoclave at 116°C for 20 min.

Alkaline Potato Medium — Prepare as above, but use 0.7% w/v sodium bicarbonate instead of plain water.

Glycerol Potato Medium — Prepare as above, but use 5% v/v aqueous glycerol solution in place of water.

Dorset's Egg Medium (M. bovis)[561]

Eggs (yolk and white)	750 ml
Nutrient broth (presterilized)	250 ml
Malachite green	12.5 ml
(2% w/v aqueous solution)	

Blend the ingredients together using aseptic precautions. Fill out into tubes or bottles. Slant and solidify by heating at 85°C for 1 hr.

Lowenstein-Jensen Medium[562]

A. KH_2PO_4	2.4 g
$MgSO_4 \cdot 7H_2O$	0.24 g
Mg citrate	0.6 g
Asparagine	3.6 g
Glycerol	12.0 ml
Potato starch	30.0 g
Distilled water	600 ml

Dissolve the salts and asparagine in half the distilled water. Make a paste of the starch and dilute it with water, then add the glycerol. Mix together and sterilize by autoclaving at 121°C for 15 min.

B. Eggs (yolks and whites)	1 liter

Wash the eggs carefully in soap and water, dry, and break them aseptically into a sterile flask. Homogenize the egg mixture thoroughly.

C. Malachite green (2% w/v	20 ml
aqueous solution)	

Blend Solutions A, B, and C together and fill out into sterile tubes or bottles. Slope and solidify by heating at 85°C for 50 min.

ATS Medium[563]

A. Egg yolk   500 ml

   Clean eggs and separate yolk from white.

B. Potato starch   20.0 g
   Glycerol   10.0 ml
   Water   500 ml

Make a paste from the starch, dilute with water, and add glycerol. Heat to boiling and cool to 50°C.

C. Malachite green (1% w/v in   20 ml
   50% alcohol)

Blend solutions A, B, and C together aseptically. Fill out into tubes or bottles. Slope and solidify by heating at 90°C for 1 hr.

Petragnani Medium[563]

A. Milk   225 ml
   Potato starch   9.0 g
   Peptone   1.5 g
   Diced potato   150.0 g

   Simmer at 100°C for 1 hr, stirring continuously.

B. Homogenized eggs   8 eggs
   Glycerol   18 ml
   Malachite green   15 ml
      (2% w/v aqueous solution)
Mix together.

C. Glucose   1.5 g
   Asparagine   1.5 g
   Distilled water   50 ml

Dissolve the glucose and asparagine in the water. Blend solutions A, B, and C together in a sterile flask. Pass through one layer of sterile gauze. Fill out into tubes or bottles. Slope and solidify by heating at 80°C for 1 hr.

Pyruvic Acid Media
   It has been shown that the replacement of glycerol by pyruvic acid encourages the growth of "dysgonic" strains.

Egg (yolk and white)	2 liter
Mineral salt solution	1 liter
KH_2PO_4	7.0 g
$Na_2HPO_4 \cdot 2H_2O$	4.0 g
Sodium pyruvate	12.5 g
Distilled water	1 liter
Malachite green (2% w/v aqueous solution)	40 ml

Sterilize the mineral salt solution by autoclaving at 121°C for 15 min. When cool, blend aseptically with the egg fluid and malachite green. Fill out into tubes and bottles. Slope and solidify by heating at 85°C for 1 hr.[564]

Malachite green may be toxic at the usual levels in the presence of pyruvic acid. Reducing the level by half (0.025% to 0.0125% w/v) is recommended.[565]

Kirchner's Enrichment Medium

There are many published modifications of the original formula.[566] The following formula is one of the modifications:

$Na_2HPO_4 \cdot 12H_2O$	19.0 g
KH_2PO_4	2.5 g
$MgSO_4$	0.6 g
Sodium citrate	2.5 g
Asparagine	5.0 g
Glycerol	20 ml
Phenol red (0.4% w/v aqueous solution)	3 ml
Distilled water	1 liter
pH 7.4–7.6	

Dissolve the ingredients, adding the glycerol last, and sterilize by autoclaving at 121°C for 15 min. Before use, add 10% v/v sterile serum and 100 units/ml penicillin.[567]

Vaccine Production (B.C.G.)

Medium I – Sauton's Medium

L-Asparagine	4.0 g
Ferric NH_4 citrate	0.05 g
Citric acid	2.0 g
Glycerol	40.0 ml
K_2HPO_4	0.5 g
$MgSO_4$	0.5 g
Triton® WR 1339	0.25 g
Distilled water	1 liter

Medium II – Glycerol-free Medium

L-Asparagine	4.0 g
Ferric NH_4 citrate	0.05 g
Monosodium glutamate	4.0 g
Casitone (Bacto)	1.0 g
L-Glutamine	4.0 g
KH_2PO_4	1.0 g
Na_2HPO_4	2.5 g
$CaCl_2$	0.001 g
$CuSO_4$	0.0005 g
$ZnSO_4$	0.0005 g
Triton® WR 1339	0.25 g
Distilled water	1 liter

Both media are sterilized at 121°C for 15 min.[568]

Tween® -Albumin Medium (Dubos)[569]

Sodium glutamate	0.5 g
KH_2PO_4	1.5 g
Na_2HPO_4	1.5 g
$(NH_4)_2SO_4$	0.5 g
Sodium citrate	0.4 g
Ferric NH_4 citrate	0.04 g
$MgSO_4 \cdot 7H_2O$	0.05 g
Pyridoxine HCl	0.001 g
Biotin	0.0005 g
Tween® 80	0.5 ml
Distilled water	1 liter
pH 6.6	

Sterilize by autoclaving at 121°C for 15 min. Cool to 45°C and add

Glucose (50% w/v sterile solution in 0.005 M citric acid)	10 ml
Albumin (Plasma fraction V, filter-sterilized 5% w/v in 0.85% NaCl solution);	100 ml
Catalase (0.1% w/v in 0.85% NaCl solution, filter sterilized)	3 ml

Heat glucose and albumin at 56°C for 30 min. Dispense into tubes or bottles aseptically.

Middlebrook 7H10 Agar[570]

This is a variation on the preceding Dubos formula in which 15 g agar and 0.00025 g/l malachite green are added. In place of Tween® 80, oleic acid is substituted at 0.006 ml per liter of medium.

M. johnei Medium[571]

Casamino acids (Difco)	2.5 g
L-Asparagine	0.3 g
Na_2HPO_4, anhydrous	2.5 g
KH_2PO_4	1.0 g
Sodium citrate	1.5 g
$MgSO_4 \cdot 7H_2O$	0.6 g
Glycerol	25 ml
Tween® 80 (1% solution)	50 ml
Standard Davis agar (New Zealand)	15 g
Distilled water	to 800 ml

Dissolve the constituents with minimum heat to give a total volume of 800 ml. Sterilize this basal medium at 116°C for 15 min and store at 4 to 6°C until required.

Crude mycobactin

> Grow *Mycobacterium phlei* for 3—4 weeks in a beef infusion broth containing 10% glycerol and 4% Difco Bactopeptone, pH 7.2; kill the growth by autoclaving, then collect by filtration; wash with large amounts of distilled water, drain, and dry over $CaCl_2$; extract 100 g of dried culture for 30 min with 3 successive 50-ml amounts of acetone in a 1-liter flask fitted with a reflux condenser

> Bulk the acetone extracts, evaporate them to dryness, and extract the resulting waxy residue in a Soxhlet for 18—20 hr with petroleum ether (40—60°C); dissolve the hard brick-red deposit (ca. 3 g) that is left in warm absolute alcohol and centrifuge the solution at 710 g for 30 min, thereby removing any bacilli that may have been carried over; evaporate the supernatant to dryness and grind the product to a fine powder of crude mycobactin free from acid-fast bacilli and debris

Final solid medium

Basal medium	800 ml
Crude mycobactin	0.16 g
Penicillin	100,000 units
Chloramphenicol	0.05 g
Pimaricine (Royal Netherlands Fermentation Industries Ltd., Delft, Holland)	0.05 g
Sterile bovine serum	200 ml

Melt the basal medium, cool to 56°C in a water bath, and add the remaining constituents, the bovine serum having been inactivated by heat at 56°C for 1 hr; thoroughly mix the medium and check the final pH value (7.2) before distributing aseptically in 5-ml amounts into sloped sterile screw-capped bottles.

From Lapage, S. P., Shelton, J. E., and Mitchell, T. G., in *Methods in Microbiology*, Vol. 3A, Norris, J. R. and Ribbons, D. W., Eds., Academic Press, London, 1970, 1—288. With permission.

4. *Actinoplanaceae*
Czapek Peptone Agar[572]

Peptone	5.0 g
Sucrose	30.0 g
K_2HPO_4	1.0 g
$MgSO_4 \cdot 7H_2O$	0.5 g
KCl	0.5 g
$FeSO_4 \cdot 7H_2O$	0.01 g
Agar	15.0 g
Distilled water	1 liter
pH 7.0—7.3	

Sterilize the medium by autoclaving at 121°C for 15 min.

Oatmeal Agar[557]
See Section X.N.7.

Glycerol-Glycine Agar[557]

Glycerol	20.0 g
Glycine	2.5 g
K_2HPO_4	1.0 g
NaCl	2.0 g
$MgSO_4 \cdot 7H_2O$	0.5 g
$FeSO_4 \cdot 7H_2O$	0.1 g
$CaCO_3$	0.2 g
Agar	18.0 g
Distilled water	1 liter
pH 7.2	

Sterilize by autoclaving at 121°C for 15 min.

Inorganic Salts-Starch Agar[557]

Solution 1

Soluble starch	10 g
Tap water	500 ml

Make a paste of the starch with a small amount of the water and dilute to volume.

Solution II

K_2HPO_4	1.0 g
$MgSO_4 \cdot 7H_2O$	1.0 g
NaCl	1.0 g
$(NH_4)_2SO_4$	2.0 g
$CaCO_3$	2.0 g
Tap water	500 ml

Mix Solutions I and II and add 20 g of agar. Heat to dissolve the agar and sterilize at 121°C for 15 min.

Glycerol-Asparagine Agar[557]

Glycerol	10.0 g
Asparagine	1.0 g
K_2HPO_4	1.0 g
Agar	20.0 g
Tap water	1 liter
pH 7.0	

Sterilize by autoclaving at 121°C for 15 min.

Peptone Agar[557]

Tryptone	1.0 g
Yeast extract	1.0 g
NaCl	8.5 g
Agar	17.0 g
Tap water	1 liter
pH 7.0	

Sterilize by autoclaving at 121°C for 15 min.

5. Dermatophilaceae
Dermatophilus sp.

These organisms grow well on blood agar (10% horse blood), Brain Heart Infusion Agar, and Loeffler Serum Agar.

Dermatophilus sp. Agar[573]

Yeast extract	5.0 g
Glycerol	50.0 g
CaCO₃	1.0 g
Agar	15.0 g
Distilled water	1 liter

Sterilize at 121°C for 15 min. Mix well to suspend the chalk before pouring plates.

Geodermatophilus spp.[573]

Yeast extract	5.0 g
N-Z amine, type A	5.0 g
Glucose	10.0 g
Soluble starch	20.0 g
CaCO₃	1.0 g
Agar	15.0 g

Sterilize by autoclaving at 121°C for 15 min. Suspend the chalk before pouring.

Geodermatophilus spp. Broth[573]

Yeast extract	1.0 g
Glucose	1.0 g
Soluble starch	1.0 g
CaCO₃	1.0 g

Sterilize by autoclaving at 121°C for 15 min. Suspend the chalk before pouring.

Geodermatophilus spp. Agar[573]

Yeast extract	5.0 g
Malt extract	15.0 g
Soluble starch	10.0 g
Sucrose	10.0 g
CaCO₃	2.0 g

Sterilize by autoclaving at 121°C for 15 min. Suspend the chalk before pouring.

6. Nocardiaceae
There is a considerable overlap of culture media among some of the families of organisms in the Actinomycetes group. This is especially true between Nocardiaceae and Streptomycetaceae. Therefore, all the isolation media will be grouped together under the following section.

7. Streptomycetaceae
Tryptone-Yeast Extract Broth[574]

Tryptone	5.0 g
Yeast extract	3.0 g
Distilled water	1 liter
pH 7.0	

Sterilize by autoclaving at 121°C for 15 min.

Glycerol-Glycine Agar[575]

Glycerol	20 ml
Glycine	2.5 g
NaCl	1.0 g
K_2HPO_4	1.0 g
$FeSO_4$	0.1 g
$MgSO_4$	0.1 g
$CaCO_3$	0.1 g
Agar	15.0 g
Distilled water	1 liter
pH 7.0	

Dissolve the salts, glycine, agar, and glycerol in that order. Sterilize by autoclaving at 121°C for 15 min.

Yeast Extract-Malt Extract Agar[576]

Yeast extract	4.0 g
Malt extract	10.0 g
Dextrose	4.0 g
Agar	20.0 g
Distilled water	1 liter
pH 7.3	

Sterilize by autoclaving at 121°C for 15 min.

Czapek's Agar[577]

Sucrose (or glycerol or glucose)	30.0 g
$NaNO_3$	2.0 g
K_2HPO_4	1.0 g
$MgSO_4 \cdot 7H_2O$	0.5 g
KCl	0.5 g
$FeSO_4$	0.01 g
Agar	15.0 g
Distilled water	1 liter
pH 7.0	

Sterilize by autoclaving at 121°C for 15 min.

Glucose-Asparagine Agar[577]

Glucose	10.0 g
Asparagine	0.5 g
K_2HPO_4	0.5 g
Agar	15.0 g
Distilled water	1 liter
pH 6.8	

Sterilize by autoclaving at 121°C for 15 min.

Glycerol-Asparagine Agar (Ushinsky)[577]

Glycerol	35.0 ml
NaCl	5.0 g
$CaCl_2$	0.1 g
$MgSO_4$	0.3 g
K_2HPO_4	2.5 g
Ammonium lactate	6.5 g
Na asparaginate	3.5 g
Agar	20.0 g
Distilled water	1 liter
pH 7.0	

Dissolve the salts, lactate, asparaginate, agar, and glycerol in that order. Sterilize by autoclaving at 121°C for 15 min.

Calcium-Malate Agar[577]

Ca malate	10.0 g
K_2HPO_4	0.5 g
NH_4Cl	0.5 g
Agar	15.0 g
Distilled water	1 liter
pH 7.0	

Glycerol at 10 ml/l is often added to this medium. Sterilize by autoclaving at 121°C for 15 min.

Starch-Casein Agar (Marine Actinomycetes)[578]

Soluble starch	10.0 g
Casein	1.0 g
Agar	15.0 g
Sea water	1 liter
pH 7.0—7.5	

Solubilize the casein in a little NaOH. Make a paste of the starch in some of the water and dilute to volume. Adjust pH if necessary. Sterilize by autoclaving at 121°C for 15 min.

Oatmeal Agar[557]

Oatmeal	20.0 g
Distilled water	1 liter

Steam for 20 min. Filter through muslin and restore volume to 1 liter.

Trace salts solution	1 ml
$FeSO_4 \cdot 7H_2O$	0.1 g
$MnCl_2 \cdot 4H_2O$	0.1 g
$ZnSO_4 \cdot 7H_2O$	0.1 g
Distilled water	100.0 ml
Agar	18.0 g
pH 7.2	

Sterilize by autoclaving at 121°C for 15 min. Mix thoroughly before pouring to ensure even distribution of the oatmeal.

Glycerol-Arginine-Antibiotic Agar[579]

Glycerol	20.0 g
L-Arginine	2.5 g
NaCl	1.0 g
$CaCO_3$	0.1 g
$FeSO_4 \cdot 7H_2O$	0.1 g
$MgSO_4 \cdot 7H_2O$	0.1 g
Agar	20.0 g
Distilled water	1 liter
pH 6.8	

Dissolve the ingredients in the water, adding the glycerol last. Sterilize by autoclaving at 121°C for 15 min and cool to 50°C. Add 50 μg/ml of medium of freshly prepared filter-sterilized solutions of the following antifungal agents:

- Pimaricin
- Cycloheximide
- Nystatin

Mix thoroughly and pour into dishes.

Glycerol-Asparagine Agar[580]

L-Asparagine	1.0 g
Glycerol	10.0 g
K_2HPO_4	1.0 g
Trace salts solution (see Oatmeal Agar)	1.0 ml
Agar	20.0 g
Distilled water	1 liter
pH 7.2	

Dissolve the salts, asparagine, agar, and glycerol in that order. Sterilize by autoclaving at 121°C for 15 min.

Tomato Paste-Oatmeal Agar[581]

Solution I	
Heinz Baby Oatmeal Food	20.0 g
Contadina Tomato Paste	20.0 g
Hot tap water	500 ml
Solution II	
Agar	15.0 g
Tap water	500 ml

Mix Solutions I and II together and autoclave at 121°C for 15 min. Final pH should be 6.8.

Amidex Agar[581]

Yeast extract	1.0 g
Beef extract	1.0 g
N-Z amine, type A	2.0 g
$CoCl_2 \cdot 6H_2O$	0.02 g
Agar	20.0 g
Distilled water	1 liter
pH 7.3	

Sterilize by autoclaving at 121°C for 15 min.

Glucose-Yeast Extract Agar[582]

Glucose	10.0 g
Yeast extract	10.0 g
Agar	25.0 g
Distilled water	1 liter
pH 6.8	

Sterilize the medium by autoclaving at 121°C for 15 min. The authors describe a selective technique for growing Actinomycetes and inhibiting other organisms as follows: Dry the plates of medium overnight and inoculate the soil suspension on the surface. Place the dish in a hot-air oven at 110°C for 10 min only. Remove and incubate at 30°C for 5 to 7 days.

Arginine-Glycerol-Salts Agar[583]

Arginine HCl	1.0 g
Glycerol	12.5 g
K_2HPO_4	1.0 g
NaCl	1.0 g
$MgSO_4 \cdot 7H_2O$	0.5 g
$Fe_2(SO_4)_3 \cdot 6H_2O$	0.01 g
$CuSO_4 \cdot 5H_2O$	0.001 g
$ZnSO_4 \cdot 7H_2O$	0.001 g
$MnSO_4 \cdot H_2O$	0.001 g
Agar	15 g
Distilled water	1 liter
pH 6.9—7.1	

Dissolve the ingredients in the water, adding the glycerol last. Sterilize by autoclaving at 121°C for 15 min.

Glycerol-Casein Agar[584]

Glycerol	10.0 g
Casein (vitamin-free)	0.3 g
KNO_3	2.0 g
NaCl	2.0 g
K_2HPO_4	2.0 g
$MgSO_4 \cdot 7H_2O$	0.05 g
$CaCO_3$	0.02 g
$FeSO_4 \cdot 7H_2O$	0.01 g
Agar	15.0 g
Distilled water	1 liter
pH 7.0	

Dissolve the casein in a little NaOH, add to the salts solution, dissolve the agar, and add the glycerol. Adjust pH if necessary. Sterilize by autoclaving at 121°C for 15 min.

Peptone-Glucose Agar[585]

Peptone	0.5 g
Glucose	1.5 g
KH_2PO_4	0.5 g
$MgSO_4 \cdot 7H_2O$	0.2 g
$Fe_2(SO_4)_3 \cdot 9H_2O$	Trace
Agar	17.0 g
Distilled water	1 liter
pH 6.8	

Sterilize by autoclaving at 121°C for 15 min. Cool to 50°C and add the antibiotics pimaricin and/or nystatin as freshly prepared, filter-sterilized solutions at 100 ppm w/v in the final medium.

Glycerol-Soil Extract Agar[586]

Peptone	5.0 g
Meat extract	3.0 g
Glycerol	70.0 ml
Agar	15.0 g
Soil extract	150.0 ml
Tap water	850.0 ml
pH 7.0	

Prepare the soil extract by adding 400 g of sieved, air-dried garden soil to 960 ml of tap water and autoclaving at 121°C for 1 hr. Allow the supernatant to clear by settling. Decant the extract and add it to the medium. Sterilize the medium by autoclaving at 121°C for 15 min.

O. Mycoplasmas

Many different media have been described by workers in the field of mycoplasma isolation. There is no precise knowledge with which to measure the significance of differences in formulation or methods of preparation. What is known suggests that mycoplasmas isolated from different hosts vary in their demands on culture media. Wild strains are probably easier to isolate than parasitic strains and pathogenic strains are probably more host adapted, therefore requiring special cultural conditions.

Fallon and Whittlestone[161] have published a table of mycoplasmas of medical or veterinary importance classified by hosts and sites of origin. The extensive list of references in the table will enable interested workers to refer to the numerous culture media described for the isolation of specific strains. Unfortunately, there is not yet a "mycoplasma" medium described which is capable of supporting the growth of all mycoplasmas, although most will grow in embryonated eggs and tissue culture preparations. Therefore, failure to obtain growth of mycoplasma from suspected material using the usual media does not mean that mycoplasmas are not present. In such cases, the use of tissue culture media or other viral techniques should be fully explored.

In general, media for the growth of mycoplasmas consist of three main ingredients:

1. A meat extract or heart infusion (which may contain added peptone).
2. A yeast extract to provide growth factors.
3. Serum or serum protein fractions.

A variety of selective agents may be added to the medium to inhibit the growth of bacteria during the long incubation periods.

Edward's Mycoplasma Medium[587]

Heart Infusion Broth	70 ml
5% yeast extract	10 ml
Horse serum	20 ml

Presterilize these ingredients and add together aseptically. Penicillin (50,000 units) and thallium acetate (1 ml of a 1/80 solution) can be added to 100 ml of complete medium to prevent bacterial growth.

Heart Infusion Broth — To 1 lb of minced horse or ox heart, add 1 liter of water. Keep in the cold overnight; steam or simmer for 2 hr. Adjust the pH to 8.4, filter, then lower the pH to 7.8. Make up to original volume with water and add 1% peptone and 0.5% NaCl. Sterilize by autoclaving at 121°C for 15 min.

Heart Infusion Agar — Add 20 g washed agar to 1 liter broth and sterilize as before.

Yeast extract — Dissolve 5 g dry yeast extract powder in 100 ml water. Filter sterilize and store in a refrigerator.

Horse serum — Collect horse blood aseptically into a sterilized 2-liter filter flask. Allow the blood to clot and the serum to separate. Decant off the serum through the side arm into sterilized bottles. Test for sterility; if necessary, filter sterilize. Store in a refrigerator.

Hayflick's Medium[588]

Mycoplasma or PPLO broth (Oxoid or Difco)	70 ml
Horse serum	20 ml
Yeast extract	10 ml
Thallous acetate (1% w/v solution)	2.5 ml
Penicillin	20,000 units
pH 7.8	

Presterilize each separate ingredient and mix together aseptically. If a solid medium is required, use Bacto PPLO Agar which has been sterilized and cooled to 50°C before mixing the other ingredients. The addition of 20 µg/ml of DNA (sodium salt of calf thymus deoxyribonucleic acid) is commonly used as a growth supplement and appears to be essential for the primary isolation of cattle mycoplasma.[589]

M. pneumoniae Biphasic Medium[590]

1. Small bottles or tubes containing 1 ml of Mycoplasma or PPLO Agar (Oxoid or Difco) are sterilized.

2. 2-ml volumes of broth medium are added on top of the agar, when required, for inoculation.

Mycoplasma or PPLO Broth	100 ml
Horse serum (unheated)	30 ml
Yeast extract (see below) (25% w/v solution)	15 ml
Thallous acetate (1% w/v solution)	4 ml
Penicillin (100,000 units/ml)	0.6 ml
Glucose (10% w/v solution)	15 ml
Phenol red (0.2% w/v solution)	3 ml
Methylene blue chloride (1/1000)	1.5 ml

Separately sterilize the ingredients and mix together aseptically.

Yeast extract solution — Suspend 250 g baker's yeast in 1 liter deionized water. Boil 30 min, cool, and clarify by centrifugation. Discard the sediment. Recentrifuge if necessary. Dispense in 20-ml volumes. Autoclave at 116°C for 10 min and store at -30°C.

Mycoplasma T-Strain Medium

Tryptone Soya Broth	1 liter
Agar	12.0 g

Adjust pH to 6.0 with HCl and sterilize by autoclaving at 121°C for 15 min. Before use add to 76-ml volumes of medium[591]

Horse serum (unheated)	20 ml
Yeast extract	5 ml
Penicillin	1,000 units/ml

A later modification of this medium includes urea and phenol red. To 95 ml of sterile Mycoplasma or PPLO broth at pH 6.0, add[592]

Horse serum	4 ml
Urea	0.05 g
Phenol red	0.001 g
Penicillin	100,000 units

M. gallisepticum Medium[593]

Ox liver	500 g
Ox muscle	400 g
Ox blood	100 g
Pig stomach	600 g
Tap water	5 liters

Cut stomach wall into small pieces after removing fat and adhering tissues. Remove fat and bile ducts from the muscle and liver. Mince the tissues together and add to the water. Add 50 ml of concentrated HCl. Digest at 48 to 52°C for 24 hr and heat the mixture to 80°C to stop digestion. Allow to settle overnight. Decant off supernatant and filter through paper. Adjust pH to 7.6 with 10% v/v NaOH. To each 825 ml of broth add the following:

Horse serum	150 ml
Glucose	2.0 g
Penicillin	500,000 units
Thallium acetate (1/80 solution)	10 ml

Sterilize by passing through a membrane filter.

M. mycoides Medium[594]

Tryptose	20.0 g
Dextrose	5.0 g
NaCl	5.0 g
$Na_2 HPO_4$	2.5 g
Glycerol	5.0 g
Yeast extract (Difco)	1.0 g
Pig serum (heated at 56°C for 30 min)	100 ml
Penicillin	100,000 units
Distilled water	1 liter

Dissolve the ingredients in the distilled water, adding the glycerol and serum last. Filter sterilize the complete medium.

M. suipneumoniae Medium[595]

Hartley's tryptic digest broth (sterile)	300 ml
Lactalbumin hydrolysate in Hanks balanced salt solution (50 g/l, sterile)	100 ml
Pig serum from pneumonia-free pigs (heated at 56°C for 30 min)	200 ml
Hanks balanced salt solution (autoclaved at 120°C for 10 min)	400 ml
Yeast extract (filter-sterilized)	5 ml

M. suipneumoniae Medium[595]
(continued)

Penicillin	200 units/ml
Thallium acetate	0.125 g/l
pH 7.6	

Mix the ingredients together aseptically and store at −25°C.

Avian Mycoplasma Medium[596]

Brain Heart Infusion Broth	37.0 g
Yeast autolysate (Albimi Laboratories)	5.0 g
Tris buffer	0.25 g
Turkey serum	100 ml
2,3,5-Triphenyltetrazolium chloride	0.05 g
Distilled water	1 liter
pH 8.0	

Sterilize the medium by filtration. Penicillin and thallous acetate may be added as bacterial inhibitors.

M. synovial Biphasic Medium[597]

Mycoplasma or PPLO Agar	1 liter
Soluble starch	5.0 g
Tryptone	5.0 g
Thallium acetate	0.25 g
Phenol red	0.025 g
Diphosphopyridine nucleotide (DPN)	0.1 g
Swine serum (inactivated)	100 ml
Penicillin	1,000,000 units

Add the tryptone, soluble starch, thallium acetate, and phenol red to the molten agar and sterilize by autoclaving at 121°C for 15 min. Cool to 50°C before adding other constituents. The swine serum, penicillin and DPN are mixed, dissolved, filter-sterilized, and stored frozen in 10-ml quantities. The broth overlay is similarly prepared:

Mycoplasma or PPLO broth	1 liter
Cysteine HCl	0.1 g
Phenol red	0.025 g
Thallium acetate	0.25 g
DPN	0.1 g
Swine serum (inactivated)	100 ml
Penicillin	1,000,000 units

Place 1- to 2-ml quantities of the agar medium in sterile tubes and allow it to set as a butt or slant. Pour the sterile broth on top at approximately twice the agar volume.

REFERENCES

1. **Dean, A. C. R.** and **Hinshelwood, C.**, *Growth, Function and Regulation in Bacterial Cells,* Oxford University Press, Oxford, 1966, 1.
2. **Buchanan, R. E.** and **Gibbons, N. E.,** Eds., *Bergey's Manual of Determinative Bacteriology,* 8th ed., Williams & Wilkins, Baltimore, 1974.
3. **Chapman, G. H.,** *J. Bacteriol.,* 51, 409–410, 1946.
4. **Baird-Parker, A. C.,** *J. Appl. Bacteriol.,* 25, 12–19, 1962.
5. **Holt, L. B.,** *J. Gen. Microbiol.,* 27, 327–330, 1962.
6. **Hayflick, L.,** *Tex. Rep. Biol. Med.,* 23, 185–190, 1965.
7. **Perez, A. G., Kim, J. H., Gelbard, A. S.,** and **Djordjevic, B.,** *Exp. Cell Res.,* 70, 307–309, 1972.
8. **Stokes, E. J.,** *Clinical Bacteriology,* E. Arnold, London, 1955, 24.
9. **Krumwiede, E.** and **Kuttner, A. G.,** *J. Exp. Med.,* 67, 429–441, 1938.
10. **Waterworth, P. M.,** *Br. J. Exp. Pathol.,* 36, 186–190, 1955.
11. **Harris, A. H.** and **Coleman, M. B.,** Eds., *Diagnostic Procedures and Reagents,* 4th ed., American Public Health Association, Washington, D.C., 1963, 133–134.
12. **Fildes, P.,** *Br. J. Exp. Pathol.,* 1, 129–131, 1920.
13. **Ley, H. L.** and **Mueller, J. H.,** *J. Bacteriol.,* 52, 453–456, 1946.
14. **Wilson, G. S.** and **Miles, A. A.,** in *Topley and Wilson's Principles of Bacteriology and Immunity,* 5th ed., E. Arnold, London, 1965, 697–700.
15. **Mackie, I. M.,** *Process Biochem.,* 9, 12–14, 1974.
16. **Wingerd, W. H.,** *J. Dairy Sci.,* 23, 544–546, 1971.
17. **Brewer, J. H.** and **McLaughlin, C. B.,** *Bacteriol. Proc.,* G107, 60, 1957.
18. **Ford, J. E., Perry, K. D.,** and **Briggs, C. A. E.,** *J. Gen. Microbiol.,* 18, 273–276, 1958.
19. **Ziska, P.,** *Arch. Hyg.,* 152, 73–76, 1968.
19a. **Stowe, H.** and **Mahoney, N. C.,** unpublished work, 1971.
20. **Bridson, E. Y.** and **Brecker, A.,** in *Methods in Microbiology,* Vol. 3A, Norris, J. R. and Ribbons, D. W., Eds., Academic Press, London, 1970, 229–295.
21. **Jones, J. G.** and **Mercier, P. L.,** *Process Biochem.,* 9, 21–24, 1974.
22. **Smith, E. L.** and **Kimmel, J. R.,** in *The Enzymes,* Vol. 4, 2nd ed., Boyer, P. D., Lardy, H., and Myrbäck, K., Eds., Academic Press, New York, 1960, 133–173.
23. **Nord, F. F.** and **Bier, M.,** *Biochim. Biophys. Acta,* 12, 56–57, 1953.
24. **McGinnis, J., Salman, A.,** and **Pubols, M.,** 25th Annu. Meet. Institute of Food Technology, Kansas City, May 1965, 21.
25. **Niemann, C.,** *Science,* 143, 1287–1296, 1964.
26. **El-Gharbawi, M.** and **Whitaker, J. R.,** *Biochemistry,* 2, 476–481, 1963.
27. **Whitaker, J. R.,** *Food Res.,* 22, 483–493, 1957.
28. **Bergkvist, R.,** *Acta Chem. Scand.,* 17, 1541–1551, 1963.
29. **Feder, J.,** *Biochemistry,* 6, 2088–2093, 1967.
30. **Keay, L.** and **Wildi, B. S.,** *Biotechnol. Bioeng.,* 179–249, 1970.
31. **Herbert, D., Phipps, P. J.,** and **Strange, R. E.,** in *Methods in Microbiology,* Vol. 5B, Norris, J. R. and Ribbons, D. W., Eds., Academic Press, London, 1971, 209–344.
32. **Ziska, P.,** *Arch. Hyg.,* 151, 370–376, 1967.
33. **Nekvasilova, K., Sidlo, J.,** and **Haza, J.,** *J. Gen Microbiol.,* 62, 3–16, 1970.
34. **Phillips, A. W.** and **Gibbs, P. A.,** *Biochem. J.,* 81, 551–556, 1961.
35. **Moss, C. W.** and **Speck, M. L.,** *J. Bacteriol.,* 91, 1098–1104, 1966.
36. **Katz, A., Dreyer, W. I.,** and **Anfinsen, C. B.,** *J. Biol. Chem.,* 234, 2897–2900, 1959.
37. **Kennedy, H. E., Speck, M. L.,** and **Airand, L. W.,** *J. Bacteriol.,* 70, 70–77, 1955.
38. **Henry, S. M., Jacobs, G.,** and **Achmeteli, A.,** *Appl. Microbiol.,* 15, 1489–1491, 1967.
39. **Weinberg, E. D.,** *Adv. Microb. Physiol.,* 4, 1–60, 1970.
40. **Sykes, J.,** Ed., *Constituents of Bacteriologic Culture Media,* Cambridge University Press, Cambridge, 1956, 6.
41. *U.S. Pharmacopeia,* 19th ed., Mack, Easton, Pa., 1975, 745.
42. **Gray, R. D.,** *J. Hyg.,* 57, 249–265, 1959.
43. **Kheshgi, S.** and **Saunders, R.,** *J. Biochem. Microbiol. Technol. Eng.,* 1, 115–127, 1959.
44. **Wood, T.** and **Bender, A. E.,** *Biochem. J.,* 67, 366–373, 1957.
45. **Bender, A. E., Wood, T.,** and **Palgrave, J. A.,** *J. Sci. Food Agric.,* 9, 812–817, 1958.
46. *U.S. Pharmacopeia,* 19th ed., Mack, Easton, Pa., 1975, 590.
47. **Huddleston, I. F., Hasley, D. E.,** and **Torrey, J. P.,** *J. Infect. Dis.,* 40, 325–329, 1927.
48. **Huddleston, I. F.,** *Brucelloses in Man and Animals,* Commonwealth Fund, New York, 1939, 13.
49. **Lodenkamper, H.** and **Ruhlcke, U.,** *Zentralbl. Bakteriol. Parasitenkd. Infektionskr. Hyg. Abt. 1 Orig.,* 232, 554–569, 1975.
50. **Diem, K.,** Ed., *Documenta Geigy Scientific Tables,* 6th ed., Macclesfield, Cheshire, 1962, 510.

51. Griffin, O. T., *Process Biochem.*, 7, 17–20, 1972.
52. Piendl, A. and Wagner, D., *Process Biochem.*, 7, 26–28, 1972.
53. MacWilliam, I. C., *J. Inst. Brew. London*, 74, 38–54, 1968.
54. Skinner, F. A., *Soc. Appl. Bacteriol. Tech. Ser.*, 5, 62, 1971.
55. Jensen, H. L. and Spencer, D., *Proc. Linn. Soc. N.S.W.*, 72, 73–76, 1947.
56. Meiklejohn, J., *Proc. 6th Int. Congr. Soil Science*, Paris, 1956, C243; as cited in Skinner, F. A., *Soc. Appl. Bacteriol. Tech. Ser.*, 5, 62, 1971.
57. *Standard Methods for the Examination of Dairy Products*, 13th ed., American Public Health Association, Washington, D.C., 1972, 62.
58. Lockhead, A. G. and Chase, F. E., *Soil Sci.*, 55, 185–195, 1943.
59. Lockhead, A. G. and Thexton, R. H., *J. Bacteriol.*, 63, 219–226, 1952.
60. Lockhead, A. G. and Burton, M. O., *Can. J. Bot.*, 31, 7–22, 1953.
61. James, N., *Can. J. Microbiol.*, 4, 363–370, 1958.
62. Augier, J., *Ann. Inst. Pasteur Paris*, 91, 759–761, 1956.
63. Augier, J., *Ann. Inst. Pasteur Paris*, 92, 817–821, 1957.
64. Hoem, W. M., *Kirk-Othmer Encyclopedia of Chemical Technology*, 2nd ed., Interscience, New York, 1964.
65. Yamamoto, A. and Raiser, G., *Biochim. Biol. Sper.*, 6, 135–140, 1967.
66. MacConkey, A. T., *J. Hyg.*, 8, 322–334, 1908.
67. Shimada, K., Bricknell, K. S., and Finegold, S. M., *J. Infect. Dis.*, 119, 273–281, 1969.
68. Aries, V., Crowther, J. S., Drasar, B. S., and Hill, M. J., *Gut*, 10 (7), 575–576, 1969.
69. Burman, W. P., *Proc. Soc. Wat. Treat. Exam.*, 4, 10–26, 1955.
70. Stokes, E. J., *Clinical Bacteriology*, E. Arnold, London, 1955, 88.
70a. Northolt, M. D., personal communication, 1971.
71. Northolt, M. D., *Antonie van Leeuwenhoek J. Microbiol. Serol.*, 38, 632, 1972.
72. Hofmann, A. F., *J. Lipid Res.*, 3, 127–129, 1962.
73. Eneroth, P., *J. Lipid Res.*, 4, 11–14, 1963.
74. Sjovall, J., *Methods Biochem. Anal.*, 12, 97–99, 1964.
75. Anthony, A., *J. Chromatogr.*, 13, 567–569, 1964.
76. Peric-Golia, L. and Socic, H., *Am. J. Pathol.*, 215, 1284–1286, 1968.
77. Turnberg, L. A. and Mote, A. A., *Clin. Chim. Acta*, 24, 253–257, 1969.
78. Panveliwalla, D., Lewis, B., Wooton, I. D. P., and Tabaqchali, S., *J. Clin. Pathol.*, 23, 309–311, 1970.
78a. Mossel, D. A. A., personal communication, 1971.
79. Leifson, E., *J. Pathol. Bacteriol.*, 40, 581–584, 1935.
80. Mallman, W. L. and Darby, C. W., *Am. J. Public Health*, 31, 127–134, 1941.
81. Jameson, J. E. and Emberley, W. W., *J. Gen. Microbiol.*, 15, 198–204, 1956.
82. Jameson, J. E. and Emberley, W. W., *J. Gen. Microbiol.*, 18, 238–242, 1958.
83. Pollard, A. L., *Science*, 103, 758–759, 1946.
84. Chapman, G. H., *J. Bacteriol.*, 53, 504, 1947.
85. Public Health Laboratory Service Standing Committee Report, *J. Hyg.*, 66, 641–647, 1968.
86. Gray, R. D., *J. Hyg.*, 57, 249–265, 1959.
87. Public Health Laboratory Service Standing Committee Report, *J. Hyg.*, 66, 67–82, 1968.
88. Mossel, D. A. A., Harrewijn, G. A., and Zadelhoff, C. F. M., *Health Lab. Science*, 11, 260–267, 1974.
89. Mossel, D. A. A. and Ratto, M. A., *Appl. Microbiol.*, 20, 273–275, 1970.
90. Araki, C. and Araki, K., *Bull. Chem. Soc. Jpn.*, 30, 287–290, 1957.
91. Araki, C. and Hirase, S., *Bull. Chem. Soc. Jpn.*, 33, 291–295, 1960.
92. Duckworth, M. and Yaphe, W., *Carbohydr. Res.*, 16, 189–197, 1971.
93. Rees, D. A., *Adv. Carbohydr. Chem. Biochem.*, 24, 267–290, 1969.
94. Chapman, V. J., *Seaweeds and Their Uses*, 2nd ed., Methuen & Co., London, 1970, 151–193.
95. Barth-Reller, L., Schoenknecht, F. D., Kenny, M. A., and Sherris, J. C., *J. Infect. Dis.*, 130, 454–463, 1974.
96. Onöz, E. and Hoffman, K., *Arztl. Lab.*, 15, 335–338, 1969.
97. von Kirschninck, H., *Pharmazie*, 7, 422–423, 1958.
98. Wilson, W. J., *J. Hyg.*, 38, 507–512, 1938.
99. Foster, J. W. and Wynne, E. S., *J. Bacteriol.*, 55, 495–497, 1948.
100. Murrell, W. G., Olsen, A. M., and Scott, W. J., *Aust. J. Sci. Res. Ser. B*, 3, 234–241, 1950.
101. Olsen, A. M. and Scott, W. J., *Aust. J. Sci. Res. Ser. B.*, 3, 219–223, 1950.
102. Hardwick, W. A., Guirard, B., and Foster, J. W., *J. Bacteriol.*, 61, 145–147, 1951.
103. Jacobs, S. E. and Harris, N. D., *Congr. Int. Microbiol.*, Rome, 1953, 1, 603.
104. Jacobs, S. E. and Harris, N. D., *J. Pharm. Pharmacol.*, 6, 877–879, 1954.

105. Jacobs, S. E. and Harris, N. D., *J. Appl. Bacteriol.*, 23, 294–317, 1960.
106. Jacobs, S. E. and Harris, N. D., *J. Appl. Bacteriol.*, 24, 172–181, 1961.
107. Cooper K. E., in *Analytical Microbiology*, Vol. 1., Kavanagh, F., Ed., Academic Press, New York, 1963, 1–45.
108. Bechtle, R. M. and Scherr, G. H., *Antibiot. Chemother.* (Basel), 8, 599–606, 1958.
109. Hjerten, S., *J. Chromatogr.*, 61, 73–80, 1971.
110. Brishammar, S., Hjerten, S., and Hofsten, B., *Biochim. Biophys. Acta*, 53, 578–583, 1961.
111. Wieme, R. J., *Proc. 7th Colloq. Protides of the Biological Fluids*, Peeters, H., Ed., Elsevier, New York, 1960, 18.
112. Hjerten, S., *Biochim. Biophys. Acta*, 53, 514–523, 1961.
113. Renn, D. W. and Mueller, M. P., Dried Agarose Gel: Method of Preparation Thereof and Production of Aqueous Agarose Gel, U.S. Patent 3527712, 1970.
114. Egorov, A. M., Vakhabov, A. K. H., and Chernyak, V. Y. A., *J. Chromatogr.*, 46, 143–148, 1969.
115. Hjerten, S., *Biochim. Biophys. Acta*, 79, 393–398, 1964.
116. Sluyterman, L. A. E. and Wijdenes, J., *Biochim. Biophys. Acta*, 200, 593–595, 1970.
117. Hjerten, S., *Arch. Biochem. Biophys.*, 99, 466–475, 1962.
118. Kohn, J., *J. Clin. Pathol.*, 6, 249, 1953.
119. Lautrop, H., *Acta Pathol. Microbiol. Scand.*, 39, 357–363, 1956.
120. Hartman, P. A., *Miniaturized Microbiological Methods*, Academic Press, New York, 1968, 21–22.
121. Tolle, A., Zeidler, H., and Heeschen, W., *Milchwissenschaft*, 23, 65–68, 1968.
122. Smith, D. B., Cook, W. H., and Neal, J. L., *Arch. Biochim. Biophys.*, 53, 192–198, 1954.
123. Chapman, V. J., *Seaweeds and Their Uses*, 2nd Ed., Methuen, London, 1970, 145–146.
124. Meynell, G. G. and Meynell, E., *Theory and Practice in Experimental Bacteriology*, Cambridge University Press, Cambridge, 1970, 54–55.
125. Temple, K. L., *J. Bacteriol.*, 57, 383, 1949.
126. Kingsbury, J. M. and Barghoorn, E. S., *Appl. Microbiol.*, 2, 5–8, 1954.
127. Ingelman, B. and Laurell, H., *J. Bacteriol.*, 53, 364–365, 1947.
128. Pramer, D., *Appl. Microbiol.*, 5, 392–395, 1957.
129. Skerman, V. B. D., *A Guide to the Identification of the Genera of Bacteria*, 2nd ed., Williams & Wilkins, Baltimore, 1968, 215–216.
130. Lorian, V. and Gray, N., *Appl. Microbiol.*, 14, 836, 1966.
131. Myrvik, A. L., Whitaker, J. M., and Cannon, R. E., *Can. J. Microbiol.*, 22, 1002–1006, 1976.
132. Rose, A. H., *Chemical Microbiology*, 2nd ed., Butterworths, London, 1968, 60.
133. Lichstein, H. C., *Annu. Rev. Microbiol.*, 14, 17–42, 1960.
134. Taylor, A. W. and Stewart, J., *J. Pathol. Bacteriol.*, 53, 87–91, 1941.
135. Boyd, M. J., Logan, M. A., and Tytell, A. A., *J. Biol. Chem.*, 174, 1013–1016, 1948.
136. Payne, J. W., *Peptide Transport in Bacteria and Mammalian Gut*, ASP, Amsterdam, 1971, 17–42.
137. Davis, B. D. and Mingioli, E. S., *J. Bacteriol.*, 60, 17–28, 1950.
138. Tatum, E. L. and Lederberg, J., *J. Bacteriol.*, 53, 673–684, 1947.
139. Wolin, M. J., Manning, G. B., and Nelson, W. O., *J. Bacteriol.*, 78, 147, 1959.
140. Kann, E. and Mills, R. C., *J. Bacteriol.*, 69, 659–664, 1955.
141. Jebb, W. H. H. and Tomlinson, A. H., *J. Gen. Microbiol.*, 5, 951–965, 1951.
142. Yee, R. B., Pan, S. F., and Gezon, H. M., *J. Bacteriol.*, 75, 51–55, 1958.
143. Aaronson, S., *Experimental Microbial Ecology*, Academic Press, London, 1970, 62.
144. Kihara, H. and Snell, E. E., *J. Biol. Chem.*, 197, 791–800, 1952.
145. Demain, A. L. and Hendlin, D., *J. Bacteriol.*, 75, 46–50, 1958.
146. Leach, F. R. and Snell, E. C., *Biochim. Biophys. Acta*, 34, 292–293, 1959.
147. Payne, J. W. and Gilvarg, C., *J. Biol. Chem.*, 243, 6291–6299, 1968.
148. Woolley, D. W. and Merrifield, R. B., *Science*, 128, 238–240, 1958.
149. Berkeley, R. C. W. and Hedges, A. J., *Lab. Pract.*, 23, 356–361, 1974.
150. Marshall, J. H. and Kelsey, J. C., *J. Hyg.*, 58, 367–372, 1960.
151. Williams-Smith, H. and Tucker, J. F., *J. Hyg.*, 76, 97–108, 1976.
152. Maskell, R., Okubadejo, O. A., and Payne, R. H., *Lancet*, 1, 834–835, 1976.
153. Simmovitch, L. and Graham, A. F., *Can. J. Microbiol.*, 1, 721–723, 1955.
154. Crawford, L. V., *Biochim. Biophys. Acta*, 30, 428–431, 1958.
155. Barton-Wright, E. C., *Practical Methods for the Microbiological Assay of the Vitamin B-Complex and Amino Acids*, United Trade Press, London, 1961.
156. Ten Ham, E. J., in *Effects of Sterilization on Components in Nutrient Media*, van Bragt, J., Mossel, D. A. A., Pierik, R. L. M., and Veldstra, H., Eds., H. Veenman & Zonen, Wageningen, Holland, 1971, 121–123.

157. **Willis, A. T., O'Connor, J. J., and Smith, J. A.,** *Nature,* 210, 654, 1966.
158. **Mitruka, B. M.,** *Gas Chromatographic Applications in Microbiology and Medicine,* John Wiley & Sons, New York, 1975.
159. **Goren, M. B.,** *Bacteriol. Rev.,* 36, 33–64, 1972.
160. **Adler, H. E. and Shifline, M.,** *Annu. Rev. Microbiol.,* 14, 141–160, 1960.
161. **Fallon, R. J. and Whittlestone, P.,** in *Methods in Microbiology,* Vol. 3B, Norris, J. R. and Ribbons, D. W., Eds., Academic Press, London, 1969, 211–267.
162. **Gray, T. R. G.,** *J. Gen. Microbiol.,* 31, 483–490, 1963.
163. **Guirard, B. M. and Snell, E. E.,** in *The Bacteria,* Vol. 4, Gunsalus, I. C. and Stanier, R. Y., Eds., Academic Press, London, 1962, 40.
164. **Kempner, E. S.,** *Appl. Microbiol.,* 15, 1525–1526, 1967.
165. **Bovallius, A. and Zacharias, B.,** *Appl. Microbiol.,* 22, 260–262, 1971.
166. **Marshall, J. H. and Kelsey, J. C.,** *J. Hyg.,* 58, 367–372, 1960.
167. **Shankar, K. and Bard, R. C.,** *J. Bacteriol.,* 63, 279–290, 1952.
168. **Rose, A. H.,** *Chemical Microbiology,* 2nd ed., Butterworths, London, 1968, 73.
169. **Weinberg, E. D.,** *Science,* 184, 952–956, 1974.
170. **Weinberg, E. D.,** *Adv. Microbiol. Physiol.,* 4, 1–60, 1970.
171. **Macham, L.,** *Process Biochem.,* 11, 12–17, 1976.
172. **Murrell, W. G.,** *Adv. Microbiol. Physiol.,* 1, 133–251, 1967.
173. **Barth Reller, L., Schoenknecht, F. D., Kenny, M. A., and Sherris, J. C.,** *J. Infect. Dis.,* 130, 454–463, 1974.
174. **Brock, T. D.,** *J. Bacteriol.,* 72, 320–323, 1956.
175. **Bulger, R. J.,** *Annu. Intern. Med.,* 67, 523–532, 1967.
176. **Davis, S. D., Iannetta, A., and Wedgewood, R. J.,** *J. Infect. Dis.,* 124, 610–612, 1971.
177. **Donovick, R., Bayan, A. P., Canales, P., and Pansey, F.,** *J. Bacteriol.,* 56, 125–137, 1948.
178. **Duncan, I. B. R.,** *Antimicrob. Agents Chemother.,* 5, 9–15, 1974.
179. **Garrod, L. P. and Waterworth, P. M.,** *J. Clin. Pathol.,* 22, 534–538, 1969.
180. **Gilbert, D. N., Kutscher, E., Ireland, P., Barnett, J. A., and Sandford, J. P.,** *J. Infect. Dis.,* 124(Suppl.), 537–545, 1971.
181. **Gourevitch, A., Rossomano, V. Z., Puglisi, T. A., Tynda, J. M., and Lein, J.,** *Ann. N.Y. Acad. Sci.,* 76, 31–41, 1958.
182. **Hameister, V. W. and Wahlig, H.,** *Arzneim. Forsch.,* 21, 1658–1660, 1971.
183. **Mederios, A. A., O'Brien, T. F., Wacker, W. E. C., and Yulung, N. F.,** *J. Infect. Dis.,* 124(Suppl.), 559–567, 1971.
184. **Newton, B. A.,** *Nature,* 172, 160–161, 1953.
185. **Raymond, E. A. and Traub, W. H.,** *Appl. Microbiol.,* 21, 192–194, 1971.
186. **Rubenis, M., Kosij, V. M., and Jackson, G. G.,** *Antimicrob. Agents Chemother.,* 1962, 153–156.
187. **Traub, W. H. and Raymond, E. A.,** *Appl. Microbiol.,* 23, 4–7, 1972.
188. **Weinberg, E. D.,** *Bacteriol. Rev.,* 21, 46–48, 1957.
189. **Young, L. S. and Hewitt, W. L.,** *Antimicrob. Agents Chemother.,* 4, 617–625, 1973.
190. **Zimelis, V. M. and Jackson, G. G.,** *J. Infect. Dis.,* 127, 663–669, 1973.
191. **Mueller, J. H. and Miller, P. A.,** *J. Bacteriol.,* 67, 271–277, 1954.
192. **Donald, C., Passey, B. I., and Swaby, R. J.,** *J. Gen. Microbiol.,* 7, 211–220, 1952.
193. **Bard, R. C. and Gunsalus, K.,** *J. Bacteriol.,* 59, 387–400, 1950.
194. **MacLeod, R. A. and Snell, E. E.,** *J. Biol. Chem.,* 170, 351–365, 1947.
195. **Webb, M.,** *J. Gen. Microbiol.,* 2, 275–287, 1948.
196. **Abelson, P. H. and Aldous, E.,** *J. Bacteriol.,* 60, 401–413, 1950.
197. **Shankar, K. and Bard, R. C.,** *J. Bacteriol.,* 63, 279–290, 1952.
198. **Munro, A. L. S.,** in *Methods in Microbiology,* Vol. 2, Norris, J. R. and Ribbons, D. W., Eds., Academic Press, London, 1970, 39–89.
199. **Hutner, S. H., Cury, A., and Baker, H.,** *Anal. Chem.,* 30, 849–867, 1958.
200. **Lewis, J. C.,** *Anal. Biochem.,* 14, 495–496, 1966.
201. **Good, N. E., Winget, G. D., Winter, W., Connolly, T. N., Izaura, S., and Singh, R. M. M.,** *Biochemistry,* 5, 467–477, 1966.
202. **Williamson, J. D. and Cox, P.,** *J. Gen. Virol.,* 2, 309–312, 1968.
203. **Mallette, M. F.,** *J. Bacteriol.,* 94, 283–290, 1967.
204. **Douglas, J.,** *Lab. Pract.,* 20, 414–416, 1971.
205. **Bates, R. G.,** in *Handbook of Biochemistry,* 2nd ed., Sober, H. A., Ed., Chemical Rubber Co., Cleveland, 1968, J-190–J-207.
206. **Albert, A.,** in *The Strategy of Chemotherapy,* Cowan, S. T. and Rowatt, E., Eds., Cambridge University Press, Cambridge, 1958.
207. **Fleming, A.,** *Br. J. Exp. Pathol.,* 10, 226–230, 1929.

208. **Goldberg, H. S.,** *Antibiotics — Their Chemistry and Non-medical Uses,* van Nostrand, Princeton, N.J., 1959, 546.
209. **Kavanagh, F.,** in *Analytical Microbiology,* Kavanagh, F., Ed., Academic Press, New York, 1963, 256—259.
210. **Finegold, S. M., Sugihara, P. T., and Sutter Vera, L.,** *Soc. Appl. Bacteriol. Tech. Ser.,* 5, 99—108, 1971.
211. **Thayer, J. D. and Martin, J. E.,** *Public Health Rep.,* 81, 559—592, 1966.
212. **Seth, A.,** *Br. J. Vener. Dis.,* 46, 201—202, 1970.
213. **Riddell, R. H. and Buck, A. C.,** *J. Clin. Pathol.,* 23, 481—483, 1970.
214. **Odegaard, K.,** *Acta Pathol. Microbiol. Scand. Sect. B,* 79, 545—548, 1971.
215. **Beerens, H. and Tahon-Castel, M. M.,** *Ann. Inst. Pasteur Paris,* 3, 90—96, 1966.
216. **Drigalski, V. and Conradi, H.,** *Z. Hyg. Infektionkr.,* 39, 283—300, 1902.
217. **Churchman, J. W.,** *J. Exp. Med.,* 16, 221—247, 1912.
218. **Krumwiede, C. and Pratt, J. S.,** *J. Exp. Med.,* 19, 20—27, 1914.
219. **Endo, S.,** *Zentralbl. Bakteriol. Parasitenkd. Ifektionskr. Hyg. Abt. 1 Orig.,* 35, 109—110, 1904.
220. **Hartman, P. A., Reinbold, G. W., and Saraswat, D. S.,** *Adv. Appl. Microbiol.,* 8, 253—289, 1966.
221. **Bryan, C. S.,** *Am. J. Public Health,* 22, 749—751, 1932.
222. **Kline, E. K.,** *Am. J. Public Health,* 24, 314—318, 1935.
223. **Quastel, J. H. and Wheatley, A. H. M.,** *Biochem. J.,* 25, 629—632, 1931.
224. **Petroff, S. A. and Gump, W. S.,** *J. Lab. Clin. Med.,* 20, 689—698, 1935.
225. **Conn, H. J.,** *Biological Stains,* Williams & Wilkins, Baltimore, 1961.
226. **Gurr, E.,** *Synthetic Dyes in Biology, Medicine and Chemistry,* Academic Press, London, 1971.
227. **Fung, D. Y. C. and Miller, R. D.,** *Appl. Microbiol.,* 25, 793—799, 1973.
228. **Cruickshank, J. C.,** *J. Pathol. Bacteriol.,* 60, 328—329, 1948.
229. **Alton, G. G. and Jones, L. M.,** *Laboratory Techniques in Brucellosis,* World Health Organization, Geneva, 1967, 27.
230. **Skerman, V. B. D.,** *Abstracts of Microbiological Methods,* Interscience, New York, 1969.
231. **Costin, D.,** *Zentralbl. Bakteriol. Parasitenkd. Infektionskr. Hyg. Abt. 1 Orig.,* 198(5), 385—463, 1965.
232. **Hartman, P. A.,** *Miniaturized Microbiological Methods,* Academic Press, New York, 1968.
233. Sub-committee on the Approval of Culture Media, *Am. J. Public Health,* 44, 935—945, 1954.
234. Co-ordinating Committee on Laboratory Methods, *Am. J. Public Health,* 53, 1305—1320, 1963.
235. *Standard Methods for the Examination of Dairy Products,* 13th ed., American Public Health Association, Washington, D.C., 1972, 64—70.
236. **Stouthamer, A. H.,** in *Methods in Microbiology,* Vol. 1, Norris, J. R. and Ribbons, D. W., Eds., Academic Press, London, 1969, 629—663.
237. **Jemmali, M.,** *J. Appl. Bacteriol.,* 32, 151—155, 1969.
238. **Ingram, M., Mossel, D. A. A., and deLange, P.,** *Chem. Ind.* (London), pp. 63—64, 1955.
239. **Finkelstein, R. A. and Lankford, C. E.,** *Appl. Microbiol.,* 5, 74—79, 1957.
240. **Lahn, D.,** *Bacteriol. Rev.,* 9, 1—48, 1945.
241. **Theil, C. C., Burton, H., and McClemont, J.,** *Proc. Soc. Appl. Bacteriol.,* 15, 53—64, 1952.
242. Medical Research Council Working Party, *Lancet,* 1, 425—435, 1959.
243. **Sykes, G.,** in *Methods in Microbiology,* Vol. 1, Norris, J. R. and Ribbons, D. W., Eds., Academic Press, London, 1969, 85.
244. **Elliott, E. C. and Georgala, D. L.,** in *Methods in Microbiology,* Vol. 1, Norris, J. R. and Ribbons, D. W., Eds., Academic Press, London, 1967, 1—20.
245. **Behagel, H. A.,** in *Effects of Sterilization on Components in Nutrient Media,* van Bragt, J., Mossel, D. A. A., Pierik, R. L. M., and Veldstra, H., Eds., H. Veenman & Zonen, Wageningen, Holland, 1971, 117—124.
246. **Mulvany, J. G.,** in *Methods in Microbiology,* Vol. 1, Norris, J. R. and Ribbons, D. W., Eds., Academic Press, London, 1969, 205—253.
247. **Grubert, G.,** in *Developments in Biological Standardization,* Vol. 23, Regamey, R. H., Gallardo, F. P., and Hennessen, W., Eds., S. Karger, Basel, 1973, 19—27.
248. **Palmer, E. D., Orris, L., and Nelson, N.,** *Am. Ind. Hyg. Assoc. J.,* 23, 257—264, 1962.
249. **Hoffman, R. K. and Warshowskey, B.,** *Appl. Microbiol.,* 6, 358—362, 1958.
250. **Toplin, I.,** *Biotechnol. Bioeng.,* 4, 331—340, 1962.
251. **Taquet, A., Tison, F., and Polspoel, B.,** *Ann. Inst. Pasteur Lille,* 14, 139—143, 1963.
252. **Himmelfarb, P.,** *Appl. Microbiol.,* 9, 534—537, 1961.
253. **Molin, N., Satmark, L., and Thorell, M.,** *Food Technol.* (London), 17, 797—801, 1963.
254. **Kayser, A. M.,** in *Effects of Sterilization on Components in Nutrient Media,* van Bragt, J., Mossel, D. A. A., Pierik, R. L. M., and Veldstra, H., Eds., H. Veenman & Zonen, Wageningen, Holland, 1971, 89—98.

255. **Ernst, R. R.,** in *Developments in Biological Standardization,* Vol. 2, Regamey, R. H., Gallardo, F. P., and Hennessen, W., Eds., S. Karger, Basel, 1973, 40–50.

256. **Becking, J. H.,** in *Effects of Sterilization on Components in Nutrient Media,* van Bragt, J., Mossel, D. A. A., Pierik, R. L. M., and Veldstra, H., Eds., H. Veenman & Zonen, Wageningen, Holland, 1971, 55–87.

257. **Mossel, D. A. A.,** in *Effects of Sterilization on Components of Culture Media,* van Bragt, J., Mossel, D. A. A., Pierik, R. L. M., and Veldstra, H., Eds., H. Veenman & Zonen, Wageningen, Holland, 1971, 27.

258. **Martin, R. J.,** *Lab. Pract.,* 20, 653–656, 1971.

259. **Postgate, J. R.,** *J. Gen. Microbiol.,* 30, 481–484, 1963.

260. **Waterworth, P. M.,** *J. Clin. Pathol.,* 22, 273–277, 1969.

261. **Vera, H. D.,** *Am. J. Public Health,* 40, 1267–1272, 1950.

262. **Mueller, J. H. and Hinton, J.,** *Proc. Soc. Exp. Biol. Med.,* 48, 330–333, 1941.

263. **Ellner, P. D., Stoessel, C. J., Drakeford, E., and Vasi, F.,** *Tech. Bull. Regist. Med. Technol.,* 3, 36, 1966; (reprinted in *Am. J. Clin. Pathol.,* 45, 502–504, 1966.)

264. **Gottlieb, D.,** *Appl. Microbiol.,* 9, 55–65, 1961.

265. **Hayward, N. J.,** *Br. Med. J.,* 1, 811–814 and 916, 1941.

266. **Hayward, N. J.,** *J. Pathol. Bacteriol.,* 55, 285, 1943.

267. **McClung, L. S. and Toabe, R.,** *J. Bacteriol.,* 53, 139–147, 1947.

268. **Willis, A. T.,** *J. Pathol. Bacteriol.,* 80, 379–390, 1960.

269. **Willis, A. T. and Gowland, G.,** *J. Pathol. Bacteriol.,* 83, 219–226, 1962.

270. **Willis, A. T. and Hobbs, G.,** *J. Pathol. Bacteriol.,* 75, 299–305, 1958.

271. **Willis, A. T. and Hobbs, G.,** *J. Pathol. Bacteriol.,* 77, 511–521, 1959.

272. **Willis, A. T.,** *Anaerobic Bacteriology in Clinical Medicine,* Butterworths, London, 1960.

273. **Chu, H. P.,** *J. Gen. Microbiol.,* 3, 255–273, 1949.

274. **Colmer, A. R.,** *J. Bacteriol.,* 55, 777–785, 1948.

275. **Knight, B. C. J. G. and Proom, H.,** *J. Gen. Microbiol.,* 4, 508–538, 1950.

276. **McGaughey, C. A. and Chu, H. P.,** *J. Gen. Microbiol.,* 2, 334–340, 1948.

277. **Stone, M. J. and Rowlands, A.,** *J. Dairy Res.,* 19, 51–62, 1952.

278. **Alder, V. G., Gillespie, W. A., and Herdan, G.,** *J. Pathol. Bacteriol.,* 66, 205–210, 1953.

279. **Baird-Parker, A. C.,** *J. Appl. Bacteriol.,* 25, 12–19, 1962.

280. **Baird-Parker, A. C.,** *J. Appl. Bacteriol.,* 25, 441–444, 1962.

281. **Baird-Parker, A. C.,** *J. Gen. Microbiol.,* 30, 409–427, 1963.

282. **Carantonis, L. M. and Spink, M. S.,** *J. Pathol. Bacteriol.,* 86, 217–220, 1963.

283. **Carter, C. H.,** *J. Bacteriol.,* 79, 753–754, 1960.

284. **Colbeck, J. C.,** *Can. Serv. Med. J.,* 12, 563–580, 1956.

285. **Gillespie, W. A. and Alder, V. G.,** *J. Pathol. Bacteriol.,* 64, 187–200, 1952.

286. **Graber, C. D., Latta, R., Fairchild, J. P., and Vogel, E. H., Jr.,** *Am. J. Clin. Pathol.,* 30, 314–317, 1958.

287. **Grogan, J. B. and Artz, C. P.,** *Clin. Res.,* 7, 159, 1959.

288. **Hopton, J.,** *J. Appl. Bacteriol.,* 24, 121–124, 1961.

289. **Innes, A. G.,** *J. Appl. Bacteriol.,* 23, 108–113, 1960.

290. **Jay, J. M.,** *J. Appl. Bacteriol.,* 26, 69–74, 1963.

291. **Shah, D. B., Russell, K. E., and Wilson, J. B.,** *J. Bacteriol.,* 85, 1181–1182, 1963.

292. **Willis, A. T. and Turner, G. C.,** *J. Pathol. Bacteriol.,* 84, 337–347, 1962.

293. **Shapton, D. A. and Hindes, W. R.,** *Chem. Ind.* (London), pp. 230–234, February 9, 1963.

294. **Brewer, J. H.,** *JAMA,* 115, 598–600, 1940.

295. *U.S. Pharmacopeia,* 19th ed., Mack, Easton, Pa., 1975, 592.

296. **Skerman, V. B. D.,** *A Guide to the Identification of the Genera of Bacteria,* 2nd ed., Williams & Wilkins, Baltimore, 1967, 213.

297. **Shepard, M. C.,** *Ann. N.Y. Acad. Sci.,* 143, 505–514, 1967.

298. **Hershey, A. D.,** *Virology,* 1, 108–112, 1955.

299. **Powell, E. O. and Errington, F. P.,** *J. Gen. Microbiol.,* 31, 315–327, 1963.

300. **Lapage, S. P., Shelton, J. E., and Mitchell, T. G.,** in *Methods in Microbiology,* Vol. 3A, Norris, J. R. and Ribbons, D. W., Eds., Academic Press, London, 1970, 1–228.

301. **Bose, S. K.,** in *Bacterial Photosynthesis,* Gest, H., San Pietro, A., and Vernon, L. P., Eds., Antioch Press, Yellow Springs, Ohio, 1963, 501–510.

302. **Murray, R. G. E. and Douglas, H. C.,** *J. Bacteriol.,* 59, 157–167, 1950.

303. **Postgate, J. R.,** *Lab. Pract.,* 15, 1239–1244, 1966.

304. **Van Niel, C. B.,** *Bacteriol. Rev.,* 8, 1–64, 1944.

305. **Larsen, H.,** *J. Bacteriol.,* 64, 187–196, 1952.

306. **Hutner, S. H.,** *J. Gen. Microbiol.,* 4, 286–294, 1950.

307. Collins, V. G., *Proc. Soc. Water Treat. Exam.*, 12, 40–73, 1963.
308. Carr, N. G., in *Methods in Microbiology*, Vol. 3B, Norris, J. R. and Ribbons, D. W., Eds., Academic Press, London, 1969, 53–77.
309. Burton, S. D. and Morita, R. Y., *J. Bacteriol.*, 88, 1755–1761, 1964.
310. Fox, D. L. and Lewin, R. A., *Can. J. Microbiol.*, 9, 753–768, 1963.
311. Jeffers, E. E., *Int. Bull. Bacteriol. Nomencl. Taxon.*, 14, 115–136, 1964.
312. Veldkamp, H., *J. Gen. Microbiol.*, 26, 331–342, 1961.
313. Dondero, N. C., Phillips, R. A., and Heukelekian, H., *Appl. Microbiol.*, 9, 219–227, 1961.
314. Rouf, M. A. and Stokes, J. L., *Arch. Mikrobiol.*, 49, 132–149, 1964.
315. Poindexter, J. S., *Bacteriol. Rev.*, 28, 231–295, 1964.
316. Kucera, S. and Wolfe, R. S., *J. Bacteriol.*, 74, 344–349, 1957.
317. Zavarine, G. A., *Z. Allg. Mikrobiol.*, 4, 390–395, 1964.
318. Breznak, J. A. and Canale-Parola, E., *J. Bacteriol.*, 97, 386–395, 1969.
319. Power, D. A. and Pelczar, M. J., *J. Bacteriol.*, 77, 789–793, 1959.
320. Meyer, P. E. and Hunter, E. F., *J. Bacteriol.*, 93, 784–789, 1967.
321. Shenberg, E., *J. Bacteriol.*, 93, 1598–1606, 1967.
322. Korthof, G., *Zentralbl. Bakteriol. Parasitenkd. Infektionskr. Hyg. Abt. 1 Orig.*, 125, 429–434, 1932.
323. Hylemon, P. B., Well, J. S., Krieg, N. R., and Jannasch, H. W., *Int. J. Syst. Bacteriol.*, 23, 340–380, 1973.
324. Canale-Parola, E., Rosenthal, S. L., and Kupfer, D., *Antonie van Leeuwenhoek J. Microbiol. Serol.*, 32, 113–124, 1966.
325. Kuzdas, C. D. and Morse, E. V., *J. Bacteriol.*, 71, 251–252, 1956.
326. Stolp, H. and Starr, M. P., *Antonie van Leeuwenhoek J. Microbiol. Serol.* 29, 217–248, 1963.
327. King, E. O., Ward, M. K., and Roney, D. E., *J. Lab. Clin. Med.*, 44, 301–307, 1954.
328. Brammer, W. J. and Clarke, P. H., *J. Gen. Microbiol.*, 37, 307–319, 1964.
329. Doudoroff, M., Kaplan, N., and Hassid, W. Z., *J. Biol. Chem.*, 148, 67–75, 1943.
330. Schubert, R. and Blum, U., *Zentralbl. Bakteriol. Parasitenkd. Infektionskr. Hyg. Abt. 1 Orig.*, 158, 583–587, 1974.
331. Brown, V. I. and Lowbury, E. J. L., *J. Clin. Pathol.*, 18, 752–756, 1965.
332. Mossel, D. A. A. and Indacochea, L., *J. Med. Microbiol.*, 4, 380–382, 1971.
333. Abdon, M. A. F., *Zentralbl. Bakteriol. Parasitenkd. Infektionskr. Hyg. Abt. 1 Orig.*, 221, 182–195, 1972.
334. Norris, J. R., *Lab. Pract.*, 8, 239–243, 1959.
335. Kleczkowska, J., Nutman, P. S., and Skinner, F. A., in *Identification Methods for Microbiologists*, Gibbs, B. M. and Shapton, D. A., Eds., Academic Press, London, 1968, 51–65.
336. Pittman, H. A., *J. Dept. Agric. W. Aust.*, 12(2nd ser.), 105, 1935; (as quoted by Skerman, V. B. D., *A Guide to the Identification of the Genera of Bacteria*, 2nd ed., Williams & Wilkins, Baltimore, 1967, 213.)
337. Eneroth, P., *J. Lipid Res.*, 4, 11–14, 1963.
338. Foster, J. W. and Davis, R. H., *J. Bacteriol.*, 91, 1924–1931, 1966.
339. Payne, J. I., Sehgal, S. N., and Gibbons, N. E., *Can. J. Microbiol.*, 6, 9–15, 1960.
340. Onishi, H., McCance, M. E., and Gibbons, N. E., *Can. J. Microbiol.*, 2, 365–373, 1965.
341. Gray, T. R. G., *J. Gen. Microbiol.*, 31, 483–490, 1963.
342. ZoBell, C. E. and Feltham, C. B., *Bull. Scripps Inst. Oceanogr.*, 3, 279–296, 1934.
343. Ellner, P. D., Stoessel, C. J., Drakeford, E., and Vasi, F., *Am. J. Clin. Pathol.*, 36 502–508, 1966.
344. Martin, J. E. and Lester, A., *Health Serv. Mental Health Admin. Health Rep.*, 86(1), 30–33, 1971.
345. Thayer, J. D. and Martin, J. E., *Public Health Rep.*, 81, 559–562, 1966.
346. Talley, R. S. and Baugh, C. L., *Appl. Microbiol.*, 29, 469–471, 1975.
347. Thornley, M. J., *J. Gen. Microbiol.*, 49, 211–257, 1967.
348. Abd-El-Malek, Y. and Gibson, T., *J. Dairy Res.*, 294–301, 1952.
349. Frateur, J., *Cellule*, 53, 287–392, 1950.
350. Carr, J. G., in *Identification Methods for Microbiologists*, Gibbs, B. M. and Shapton, D. A., Eds., Academic Press, London, 1968, 1–8.
351. Jones, L. M. and Morgan, W. J. B., *Bull. WHO*, 19, 200–203, 1958.
352. Alton, G. G. and Jones, L. M., *Laboratory Techniques in Brucellosis*, World Health Organization, Geneva, 1967, 10–11.
353. Joint Food and Agriculture Organization/World Health Organization Expert Committee on Brucellosis, *WHO Tech. Rep. Ser.*, 148, 50, 1958.
354. Cruickshank, R., Duguid, J. P., Marmion, B. P., and Swain, R. H. A., *Medical Microbiology*, Vol. 2, 12th ed., Churchill, Livingstone, Edinburgh, 1975, 132.

355. **Kauffmann, F.**, *The Bacteriology of Enterobacteriaceae*, Munksgaard, Copenhagen, 1966, 361–370.

356. **Hajna, A. A.**, *Public Health Lab.*, 13, 59–62, 1955.

357. **Leifson, E.**, *Am. J. Hyg.*, 24, 423–432, 1936.

358. **Knox, R., Gell, P. G. H., and Pollock, M. R.**, *J. Pathol. Bacteriol.*, 54, 469–483, 1942.

359. **Kauffman, F.**, *Z. Hyg.*, 117, 26–32, 1935.

360. **MacConkey, A. T.**, *J. Hyg.*, 5, 333–379, 1905.

360a. *U.S. Pharmacopeia*, 19th ed., Mack, Easton, Pa., 1975, 590.

361. **Levine, M.**, *Abstr. Bacteriol.*, 2, 13, 1918.

362. **Levine, M.**, *J. Infect. Dis.*, 23, 43–47, 1918.

363. **Wilson, W. J.**, *J. Hyg.*, 38, 507–512, 1938.

364. **Harris, A. H. and Coleman, M. B.**, Eds., *Diagnostic Procedures and Reagents*, 4th ed., American Public Health Association, Washington, D.C. 1963.

365. **Hynes M.**, *J. Pathol. Bacteriol.*, 54, 193–207, 1942.

366. **Edel, W. and Kampelmacher, E. H.**, *Bull. WHO*, 41, 297–306, 1969.

367. **Taylor, W. I.**, *Am. J. Clin. Pathol.*, 44, 471–475, 1965.

368. **Taylor, W. I. and Harris, B.**, *Am. J. Clin. Pathol.*, 44, 476–479, 1965.

369. **Taylor, W. I. and Harris, B.**, *Am. J. Clin. Pathol.*, 48, 350–355, 1967.

370. **Taylor, W. I. and Schelhart, D.**, *Am. J. Clin. Pathol.*, 48, 356–362, 1967.

371. **Taylor, W. I. and Schelhart, D.**, *Appl. Microbiol.*, 16, 1387–1392, 1968.

372. **Taylor, W. I. and Schelhart, D.**, *Appl. Microbiol.*, 18, 393–395, 1969.

373. **King, S. and Metzger, W. I.**, *Appl. Microbiol.*, 16, 577–578, 1968.

374. Methodenbuch Band VI, *Chemische, physikalische und bakteriologische Untersuchungsverfahren für Milch, Milcherzeugnisse und Molkereihilfstoffe*, Verband Deutscher Landwirtschaftlicher Untersuchungs- und Forschungsanstalten, Neumann Verlag, Radebeul and Berlin, 1970.

375. *Standard Methods for the Examination of Dairy Products*, 12th ed., American Public Health Association, Washington, D. C., 1967, 56, 58, and 241.

376. **Mossel, D. A. A., Mengerink, W. H. J., and Scholts, H. H.**, *J. Bacteriol.*, 84, 381, 1962.

377. Methodenbuch Band VI, *Untersuchung von Milch, Milcherzeugnissen und Molkereihilfstoffen*, Verband Deutscher Landwirtschaftlicher Untersuchungs- und Forschungsanstalten, Neumann Verlag, Radebeul and Berlin, 1950, 116.

378. Department of Health and Social Security, *The Bacteriological Examination of Water Supplies*, 4th ed., Her Majesty's Stationery Office, London, 1969.

379. **Childs, E. and Allen, L. A.**, *J. Hyg.*, 51, 468–477, 1953.

380. *Standard Methods for the Examination of Water and Waste Water*, 13th ed., American Public Health Association, Washington, D.C., 1971, 651, 664–665.

381. **Mackenzie, E. F. W., Windle Taylor, E., and Gilbert, W. E.**, *J. Gen. Microbiol*, 2, 197–204, 1948.

382. Public Health Laboratory Service Standing Committee on the Bacteriological Examination of Water Supplies, *J. Hyg.*, 67, 367–374, 1969.

383. **Mossel, D. A. A. and Ratto, M. A.**, *Appl. Microbiol.*, 20, 273–275, 1970.

384. **Knisely, R. F., Swaney, L. M., and Friedlander, H.**, *J. Bacteriol.*, 88, 491–496, 1964.

385. **Paterson, J. S. and Cook, R.**, *J. Pathol. Bacteriol.*, 85, 241–242, 1963.

386. **Brzin, B.**, *Zentralbl. Bakteriol. Parasitenkd. Infektionskr. Hyg. Abt. 1 Orig.*, 189, 543–545, 1963.

387. **Collins, C. H. and Lyne, P. M.**, *Microbiological Methods*, 4th ed., Butterworths, London, 1976, 391.

388. **White, J. N. and Starr, M. P.**, *J. Appl. Bacteriol.*, 34, 459–475, 1971.

389. **Reinhardt, J. F. and Powell, D.**, *Phytopathology*, 50, 685–686, 1960.

390. **Monsur, K. A.**, *Bull. WHO*, 28, 387–389, 1963.

391. **Kobayashi, T., Enomoto, S., Sakazaki, R., and Kuwahara, S.**, *Jpn. J. Bacteriol.*, 18, 10–11, 387–391, 1963.

392. **Sakazaki, R.**, in *Food-borne Infections and Intoxications*, Reimann, H., Ed., Academic Press, New York, 1969, 116–117.

393. **Akiyama, S., Takizawa, K., Ichinoe, H., Enomoto, S., Kobayashi, T., and Sakazaki, R.**, *Jpn. J. Bacteriol.*, 18, 255–262, 1963.

394. **Smith, I. W.**, *J. Gen. Microbiol.*, 33, 263–274, 1963.

395. **Hendrie, M. S., Hodgkiss, W., and Shewan, J. M.**, *J. Gen. Microbiol.*, 64, 151–169, 1970.

396. **Bexon, J. and Dawes, E. A.**, *J. Gen. Microbiol.*, 60, 421–425, 1970.

397. **Belaich, J. P. and Senez, J. C.**, *J. Bacteriol.*, 89, 1195–1200, 1965.

398. **Weeks, O. B. and Beck, S. M.**, *J. Gen. Microbiol.*, 23, 217–229, 1960.

399. **Payza, A. N. and Korn, E. D.**, *J. Biol. Chem.*, 223, 853–858, 1956.

400. Everall, P. H., *J. Med. Lab. Technol.*, 11, 181–184, 1953.
401. Evans, N. M., Bell, S. M., Smith, D. D., *J. Clin. Microbiol.*, 1, 89–95, 1975.
402. Levinthal, W. and Fernbach, H., *Z. Hyg.*, 96, 456–519, 1922.
403. Griffin, P. J., *Arch. Biochem.*, 30, 100, 1951.
404. Herriott, R. M., Meyer, E. Y., Vogt, M., and Modan, M., *J. Bacteriol.*, 101, 513–516, 1970.
405. Namioka, S. and Murata, M., *Cornell Vet.*, 51, 498–501, 1961.
406. Morris, E. J., *J. Gen. Microbiol.*, 19, 305–317, 1958.
407. Phillips, J. E., *J. Pathol. Bacteriol.*, 82, 205–209, 1961.
408. Mitchell, R. G. and Gillespie, W. A., *J. Clin. Pathol.*, 17, 511–512, 1964.
409. Smith, T., *J. Exp. Med.*, 33, 441–443, 1921.
410. Slotnick, I. J. and Dougherty, M., *Antonie van Leeuwenhoek J. Microbiol. Serol.*, 30, 261–272, 1964.
411. Anderson, K., DeMonbreun, W. A., and Goodpasture, E. W., *J. Exp. Med.*, 81, 25–29, 1945.
412. Hungate, R. E., in *Methods in Microbiology*, Vol. 3B, Norris, J. R. and Ribbons, D. W., Eds., Academic Press, London, 1969, 117–132.
413. Drasar, B. S. and Crowther, J. S., in *Isolation of Anaerobes*, Shapton, D. A. and Board, R. G., Eds., Academic Press, London, 1971, 93–97.
414. Finegold, S. M., Miller, A. B., and Posnide, D. J., *Ernaehrungsforschung*, 10, 517–521, 1965.
415. Beerens, H. and Fievez, L., in *Isolation of Anaerobes*, Shapton, D. A. and Board, R. G., Eds., Academic Press, London, 1971, 109–113.
416. Omata, R. R. and Ichinoe, H., *J. Bacteriol.*, 72, 677–680, 1966.
417. Hirsch, A. and Grinsted, E., *J. Dairy Res.*, 21, 101–110, 1954.
418. Barnes, E. M. and Goldberg, H., *J. Appl. Bacteriol.*, 25, 94–96, 1962.
419. Latham, M. J. and Sharpe, M. E., in *Isolation of Anaerobes*, Shapton, D. A. and Board, R. G., Eds., Academic Press, London, 1971, 134–147.
420. Jackins, H. C. and Barker, H. A., *J. Bacteriol.*, 61, 101–114, 1951.
421. Stickland, L. H., *Biochem. J.*, 25, 215–220, 1931.
422. Rogosa, M., *J. Bacteriol.*, 87, 162–170, 1964.
423. Hamlin, L. J. and Hungate, R. E., *J. Bacteriol.*, 72, 548–554, 1956.
424. Krulwich, T. A. and Funk, H. B., *J. Bacteriol.*, 90, 729–733, 1965.
425. London, J., *Arch. Mikrobiol.*, 46, 329–337, 1963.
426. Temple, K. L. and Colmer, A. R., *J. Bacteriol.*, 62, 605, 1951.
427. Parker, C. D. and Frisk, J., *J. Gen. Microbiol.*, 8, 344, 1953.
428. Hutchinson, M., Johnstone, K. I., and White, D., *J. Gen. Microbiol.*, 44, 373–381, 1966.
429. Barker, H. A. and Taha, S. M., *J. Bacteriol.*, 43, 347, 1942.
430. Johns, A. T. and Barker, H. A., *J. Bacteriol.*, 80, 837–841, 1960.
431. Peel, D. and Quayle, J. R., *Biochem. J.*, 81, 465–469, 1961.
432. Blaylock, B. A. and Stadtman, T. C., *Arch. Biochem. Biophys.*, 116, 138–152, 1966.
433. Stadtman, T. C. and Barker, H. A., *J. Bacteriol.*, 62, 269–280, 1951.
434. Christie, R. and Keogh, E. V., *J. Pathol. Bacteriol.*, 51, 189–192, 1940.
435. Jacobs, S. I., Willis, A. T., and Goodburn, G. M., *J. Pathol. Bacteriol.*, 87, 151–154, 1964.
436. Gilbert, R. J., and Kendall, M., and Hobbs, B. C., in *Isolation Methods for Microbiologists*, Shapton, D. A. and Gould, G. W., Eds., Academic Press, London, 1969, 9–15.
437. Chapman, G. H., *J. Bacteriol.*, 50, 201–203, 1945.
438. Fairbrother, R. W. and Southall, J. E., *Mon. Bull. Ministr. Health Public Health Lab. Serv.*, 9, 170–172, 1950.
439. DiSalvo, J. W., *Med. Tech. Bull.*, 9, 191–196, 1958.
440. Vogel, R. A. and Johnson, M., *Public Health Lab.*, 18, 131–134, 1960.
441. International Association of Microbiological Societies, in *Micro-organisms in Food*, Thatcher, F. S. and Clarke, D. S., Eds., University of Toronto Press, Toronto, 1968.
442. Gilbert, R. J., Kendall, M., and Hobbs, B. C., in *Isolation Methods for Microbiologists*, Shapton, D. A. and Gould, G. W., Eds., Academic Press, London, 1969, 120.
443. Smith, B. A. and Baird-Parker, A. C., *J. Appl. Bacteriol.*, 27, 78–82, 1964.
444. Todd, E. W. and Hewitt, L. F., *J. Pathol. Bacteriol.*, 35, 973–974, 1932.
445. Rosenow, L., *J. Dent. Res.*, 1, 205–249, 1919.
446. *Diagnostic Procedures and Reagents*, 4th ed., American Public Health Association, Washington, D.C., 1963, 141.
447. Stokes, E. J., *Clinical Bacteriology*, Edward Arnold, London, 1955, 27.
448. Edwards, S. J., *J. Comp. Pathol. Ther.*, 46, 211–217, 1933.
449. Pike, R. M., *Am. J. Hyg.*, 41, 211–220, 1945.
450. Lilley, B. D. and Brewer, J. H., *J. Am. Pharm. Assoc. Sci. Ed.*, 42, 6–8, 1953.
451. Barnes, E. M., *J. Appl. Bacteriol.*, 19, 193–203, 1956.

452. **Barnes, E. M.,** *Lab. Pract.,* 25, 145–147, 1976.

453. **Slanetz, L. W. and Bartley, C. H.,** *J. Bacteriol.,* 74, 591–595, 1957.

454. **Buck, J. D.,** *Am. J. Public Health,* 62, 419–421, 1972.

455. **Hartman, P. A., Reinbold, G. W., and Sarawat, D. S.,** in *Advances in Applied Microbiology,* Vol. 8, Umbreit, W. W., Ed., Academic Press, New York, 1966, 253–289.

456. **Chapman, G. H.,** *J. Bacteriol.,* 48, 113–114, 1944.

457. **Snyder, M. L. and Lichstein, H. C.,** *J. Infect. Dis.,* 67, 113–115, 1940.

458. **Reinbold, G. W., Swern, M., and Hussong, R. V.,** *J. Dairy Sci.,* 36, 1–6, 1953.

459. **Fitzgerald, R. J. and Keyes, P. H.,** *J. Am. Dent. Assoc.,* 61, 23–33, 1960.

460. **Anon.,** *Difco Manual,* 9th ed., Difco Laboratories, Detroit, 1953.

461. **Anon.,** *BBL Products,* 4th ed., Baltimore Biological Laboratory, Baltimore, 1956.

462. Baltimore Biological Laboratory, Baltimore, personal communication, 1963.

463. **Sherman, J. M., Niven, C. F., Jr., and Smiley, K. L.,** *J. Bacteriol.,* 45, 249–263, 1943.

464. **Raibaud, P., Caulet, M., Galpin, J. V., and Mocquot, G.,** *J. Appl. Bacteriol.,* 24, 285–306, 1961.

465. **White, J. C. and Sherman, J. M.,** *J. Bacteriol.,* 48, 262, 1944.

466. **Winter, C. E. and Sandholzer, L. A.,** *J. Bacteriol.,* 51, 588, 1946.

467. **Harrison, A. P., Jr. and Hansen, P. A.,** *J. Bacteriol.,* 59, 197–210, 1950.

468. **Fujiwara, K., Sekiya, K., and Bamba, K.,** *Jpn. J. Bacteriol.,* 11, 411–415, 1956.

469. **Mallmann, W. L. and Kereluk, K.,** *Bacteriol. Proc.,* p. 142, 1957.

470. **Kereluk, K.,** personal communication, 1960.

471. **Burkwall, M. K. and Hartman, P. A.,** *Appl. Microbiol.,* 12, 18–23, 1964.

472. **Lachica, V. F. and Hartman, P. A.,** *Bacteriol. Proc.,* pp. 2–3, 1965.

473. **Kenner, B. A., Clark, H. F., and Kabler, P. W.,** *Appl. Microbiol.,* 9, 15–20, 1961.

474. **Saraswat, D. S., Clark, W. S., Jr., and Reinbold, G. W.,** *J. Milk Food Technol.,* 26, 114–117, 1963.

475. **Winter, C. E. and Sandholzer, L. A.,** *U.S. Fish Wildl. Serv. Fish. Leafl.,* 201, 1–9, 1946 (rev. 1957).

476. **Guthof, O. and Dammann, G.,** *Arch. Hyg. Bakteriol.,* 142, 559–568, 1958.

477. **Packer, R. A.,** *J. Bacteriol.,* 46, 343–349, 1943.

478. **Mossel, D. A. A., van Diepen, H. M. J., and de Bruin, A. S.,** *J. Appl. Bacteriol.,* 20, 265–272, 1957.

479. **Kjellander, J.,** *Acta Pathol. Microbiol. Scand. Suppl.,* 48(136), 1–124, 1960.

480. **Colobert, L. and Morelis, P.,** *Ann. Inst. Pasteur Paris,* 94, 120–122, 1958.

481. **Morelis, P. and Colobert, L.,** *Ann. Inst. Pasteur Paris,* 95, 667–680, 1958.

482. **Horie, S. and Saheki, K.,** *Bull. Jpn. Soc. Sci. Fish.,* 26, 623–626, 1960.

483. **Gerencser, V. F. and Weaver, R. H.,** *Appl. Microbiol.,* 7, 113–115, 1959.

484. **Mieth, H.,** *Zentralbl. Bakteriol. Parasitenkd. Infektionskr. Hyg. Abt. 1 Orig.,* 179, 456–482, 1960.

485. **Rantasalo, I.,** *Ann. Med. Intern. Fenn.,* 36, 341–348, 1947.

486. **Barnes, E. M.,** *J. Gen. Microbiol.,* 14, v, 1956.

487. **Franklin, J. G. and Sharpe, M. E.,** *J. Dairy Res.,* 30, 87–99, 1963.

488. **McKenzie, D. A.,** *Vet. Rec.,* 53, 473–480, 1941.

489. **Mead, G. C.,** *Nature,* 197, 1323–1324, 1963.

490. **Guthof, O.,** *Zentralbl. Bakteriol. Parasitenkd. Infektionskr. Hyg. Abt. 1 Orig.,* 158, 87–95, 1952.

491. **Koch, F. E.,** *Zentralbl. Bakteriol. Parasitenkd. Infektionskr. Hyg. Abt. 1 Orig.,* 134, 348–367, 1935.

492. **Schafer, W.,** *Zentralbl. Bakteriol. Parasitenkd. Infektionskr. Hyg. Abt. 1 Orig.,* 160, 54–62, 1953.

493. **Williams, R. E. O. and Hirch, A.,** *J. Hyg.,* 48, 504–524, 1950.

494. **Greer, J. E. and Britt, E. M.,** *Bacteriol. Proc.,* 63, 1959.

495. **Vera, H. D.,** personal communication, 1963.

496. **Snyder, M. L.,** *J. Infect. Dis.,* 66, 1–16, 1940.

497. **Mastromatteo, L. and Pisu, I.,** *Boll. Ist. Sieroter. Milan.,* 38, 347–358, 1959.

498. **Pike, R. M.,** *Proc. Soc. Exp. Biol. Med.,* 57, 186–187, 1944.

499. **Edwards, S. J.,** *J. Comp. Pathol. Ther.,* 51, 250–263, 1938.

500. **Mallmann, W. L.,** *Sewage Works J.,* 12, 875–878, 1940.

501. **Rothe, W. C.,** cited in *Difco Manual,* 9th ed., Difco Laboratories, Detroit, 1953, 48.

502. **Splittstoesser, D. F., Wright, R., and Hucker, G. J.,** *Appl. Microbiol.,* 9, 303–308, 1961.

503. **Litsky, W., Mallmann, W. L., and Fifield, C. W.,** *Am. J. Public Health,* 43, 873–879, 1953.

504. **Ritter, C., Shull, I. F., and Quinley, R. L.,** *Am. J. Public Health,* 46, 612–618, 1956.

505. **Raj, H., Wiebe, W. J., and Liston, J.,** *Appl. Microbiol.,* 9, 295–303, 1961.

506. Hannay, C. L. and Norton, I. L., *Proc. Soc. Appl. Bacteriol.*, 10, 39–45, 1947.
507. Wang, W-L. L. and Dunlop, S. G., *Public Health Rep.*, 66, 1212–1218, 1951.
508. Slanetz, L. W., Bent, D. F., and Bartley, C. H., *Public Health Rep.*, 70, 67–72, 1955.
509. Hajna, A. A., *Public Health Lab.*, 9, 80–81, 1951.
510. Hajna, A. A. and Perry, C. A., *Am. J. Public Health*, 33, 550–556, 1943.
511. Ostrolenk, M. and Hunter, A. C., *J. Bacteriol.*, 51, 735–741, 1946.
512. Pike, R. M., *J. Bacteriol.*, 50, 297–300, 1945.
513. Cooper, K. E. and Ramadan, F. M., *J. Gen. Microbiol.*, 12, 180–190, 1955.
514. Cooper, K. E., Wood, N., Elliot, E., Caswell, M., and Small, W., *J. Pathol. Bacteriol.*, 54, 345–353, 1942.
515. Chesbro, W. R. and Evans, J. B., *J. Bacteriol.*, 78, 858–862, 1959.
516. Langston, C. W. and Bouma, C., *Appl. Microbiol.*, 8, 212–222, 1960.
517. Garvie, E. I., *J. Dairy Res.*, 27, 283–295, 1960.
518. Whittenbury, R., *J. Gen. Microbiol.*, 35, 13–26, 1964.
519. Garvie, E. I., *J. Gen. Microbiol.*, 48, 431–438, 1967.
520. Deibel, R. H., Niven, C. E., and Wilson, G. D., *Appl. Microbiol.*, 9, 156–161, 1961.
521. Coster, E. and White, H. R., *J. Gen. Microbiol.*, 37, 15–31, 1964.
522. Klein, S. M. and Sagers, R. D., *J. Biol. Chem.*, 241, 197–205, 1966.
523. Horler, D. F., Westlake, D. W. S., and McConnell, W. B., *Can. J. Microbiol.*, 12, 47–53, 1966.
524. Holt, S. C. and Canale-Parola, E., *J. Bacteriol.*, 93, 399–410, 1967.
524a. Heintz, M. T., Urban, S., Schiller, I., Gay, M., and Bühlman, X., *Pharm. Acta Helv.*, 51, 137–143, 1976.
524b. Rowe, J. J., Goldberg, I. D., and Amelunxen, R. E., *J. Bacteriol.*, 124, 279–284, 1975.
524c. Rhodes, R. A., Sharpe, E. S., Hall, H. H., and Jackson, R. W., *Appl. Microbiol.*, 14, 189–195, 1966.
524d. Sharpe, E. S., St. Julian, G., and Crowell, C., *Appl. Microbiol.*, 19, 681–688, 1970.
524e. Donovan, K., *J. Appl. Bacteriol.*, 21, 100–103, 1958.
524f. Put, H. M. C. and Wybinga, S. J., *J. Appl. Bacteriol.*, 26, 428–434, 1963.
524g. Williams, D. J. and Clegg, L. F. L., *J. Appl. Bacteriol.*, 20, 167–174, 1957.
525. Schefferle, H. E., *J. Appl. Bacteriol.*, 28, 412–420, 1965.
526. Lowbury, E. J. L. and Lilly, H. A., *J. Pathol. Bacteriol.*, 70, 105–109, 1955.
527. Dowell, V. R., Hill, E. O., and Altemeier, W. A., *J. Bacteriol.*, 88, 1811–1815, 1964.
528. Clausen, O. G., Aasgaard, N. B., and Solberg, O., *Ann. Microbiol.* (Paris), 124B, 205–216, 1973.
529. Hobbs, G., Williams, K., and Willis, A. T., in *Isolation of Anaerobes*, Shapton, D. A. and Board, R. G., Eds., Academic Press, London, 1971, 2.
530. Lampen, J. O. and Peterson, W. H., *Arch. Biochem.*, 2, 443–454, 1943.
531. Jensen, H. and Spencer, D., *Proc. Linn. Soc. N.S.W.*, 72, 73–84, 1947.
532. Rabinowitz, J. C., *Methods Enzymol.*, 6, 703–713, 1963.
533. El-Ghazzawi, E., *Arch. Microbiol.*, 57, 1–19, 1967.
534. Stadtman, T. C. and White, F. H., *J. Bacteriol.*, 67, 651–657, 1954.
535. Stadtman, E. R. and Barker, H. A., *J. Biol. Chem.*, 180, 1085–1093, 1949.
536. Gibbs, B. M. and Freame, B., *J. Appl. Bacteriol.*, 28, 95–106, 1965.
537. Batty, I. and Walker, P. D., *J. Appl. Bacteriol.*, 28, 112–118, 1965.
538. Moore, W. B., *J. Gen. Microbiol.*, 53, 415–424, 1968.
539. Postgate, J. R., *Appl. Microbiol.*, 11, 265–267, 1963.
540. James, N., *Can. J. Microbiol.*, 4, 363–370, 1958.
541. György, P. and Rose, C. S., *J. Bacteriol.*, 69, 483–490, 1955.
542. de Man, J. C., Rogosa, M., and Sharpe, M. E., *J. Appl. Bacteriol.*, 23, 130–135, 1960.
543. Deibel, R. H., Evans, J. B., and Niven, C. F., *J. Bacteriol.*, 74, 818–824, 1957.
544. Rogosa, M., Mitchell, J. A., and Wiseman, R. F., *J. Dent. Res.*, 30, 682–690, 1951.
545. Lehnert, C. H., *Arch. Exp. Vet. Med.*, 18, 981–984, 1964.
546. Byrne, J. L., Connell, R., Frank, J. F., and Moynihan, I. W., *Can. J. Comp. Med.*, 16, 129–148, 1952.
547. Connell, R. and Langford, E. V., *Can. J. Comp. Med.*, 17, 448–453, 1953.
548. Anderson, J., Happold, F., McLeod, J. W., and Thompson, J., *J. Pathol. Bacteriol.*, 34, 667–671, 1931.
549. Hoyle, L., *Lancet*, 1, 175–176, 1941.
550. Frobisher, M. and Parsons, E. I., *Am. J. Public Health*, 43, 1441–1442, 1953.
551. Whitley, O. R. and Damon, S. R., *Public Health Rep.*, 64, 201–212, 1949.
552. Moore, M. S. and Parsons, E. I., *J. Infect. Dis.*, 102, 88–96, 1958.
553. Gordon, R. E. and Smith, M. M., *J. Bacteriol.*, 66, 41–48, 1953.
554. Moss, C. W., Dowell, V. R., Farshtchi, D., Raines, L. J., and Cherry, W. B., *J. Bacteriol.*, 97, 561–570, 1969.

555. King, J. W. and Rettger, L. F., *J. Bacteriol.*, 44, 307–316, 1942.

556. Slack, J. M., Landfried, S., and Gerencser, M. A., *J. Bacteriol.*, 97, 873–884, 1969.

557. Küster, E., *Int. Bull. Bacteriol. Nomencl. Taxon.*, 9, 97–104, 1959.

558. Scardovi, V. and Trovatelli, L. D., *Zentralbl. Bakteriol. Parasitenkd. Infektionskr. Hyg. Abt. 2*, 123, 64–88, 1969.

559. Gilmour, M. N., Howell, A., and Bibby, B. G., *Bacteriol. Rev.*, 25, 131–141, 1961.

560. Georg, L. K. and Brown, J. M., *Int. J. Syst. Bacteriol.*, 17, 79–88, 1967.

561. Cruickshank, R., *Medical Microbiology*, E. & S. Livingstone, Edinburgh, 1965, 754.

562. Holm, J. and Lester, V., *Acta Tuberc. Scand.*, 16, 3–4, 1941; (abstracted in *Public Health Rep.*, 62, 847–854, 1947).

563. *Tuberculosis Laboratory Methods*, Department of Medicine and Surgery, Veterans Administration, Washington, D.C., July 1960, 31–33.

564. Stonebrink, B., *Proc. Tuberc. Res. Counc.*, 44, 67–81, 1957.

565. Marks, J., *Mon. Bull. Minist. Health Public Health Lab.*, 22, 150–152, 1963.

566. Kirchner, O., *Zentralbl. Bakteriol. Parasitenkd. Infektionskr. Hyg. Abt. 1 Orig.*, 124, 403–406, 1932.

567. Cruickshank, R., Ed., *Mackie and McCartney's Handbook of Bacteriology*, 10th ed., E. & S. Livingstone, Edinburgh, 1960, 215.

568. Muggleton, P. W., in *Proceedings of the Specialists Conference on Culture Collections, Ottawa*, Martin, S. M., Ed., University of Toronto Press, Toronto, 1962, 202–205.

569. Dubos, R. and Middlebrook, G., *Am. Rev. Tuberc.*, 56, 334–345, 1947.

570. Middlebrook, G., Cohn, M. L., Dye, W. E., Russell, W. F., and Levy, D., *Acta Tuberc. Scand.*, 38, 66–81, 1960.

571. Stuart, P., *Br. Vet. J.*, 121, 289–318, 1965.

572. Kane, W. D., *J. Elisha Mitchell Sci. Soc.*, 82, 220–230, 1966.

573. Luedemann, G. M., *J. Bacteriol.*, 96, 1848–1858, 1968.

574. Pridham, T. G. and Gottlieb, D., *J. Bacteriol.*, 56, 107–114, 1948.

575. Lindenbein, W., *Arch. Microbiol.*, 17, 361–383, 1952.

576. Pridham, T. G., Anderson, P., Foley, C., Lindenfelser, L. A., Hesseltine, C. W., and Benedict, R. G., *Antibiot. Annu.*, 947–953, 1956/57.

577. Waksman, S. A., *Bacteriol. Rev.*, 21, 1–29, 1957.

578. Grein, A. and Meyers, S. P., *J. Bacteriol.*, 76, 457–463, 1958.

579. Porter, J. N., Wilhelm, J. J., and Tresner, H. D., *Appl. Microbiol.*, 8, 174–178, 1960.

580. Pridham, T. G. and Lyons, A. J., *J. Bacteriol.*, 81, 431–441, 1961.

581. Gottlieb, D., *Appl. Microbiol.*, 9, 55–65, 1961.

582. Agate, A. D. and Bhat, J. V., *Antonie van Leeuwenhoek J. Microbiol. Serol.*, 29, 297–304, 1963.

583. El-Nakeeb, M. A. and Lechevalier, H. A., *Appl. Microbiol.*, 11, 75–77, 1963.

584. Küster, E. and Williams, S. T., *Nature*, 202, 928–929, 1964.

585. Tsao, P. H. and Thieleke, D. W., *Can. J. Microbiol.*, 12, 1091–1094, 1966.

586. Gordon, R. E., Barnett, D. A., Handerhan, J. E., Pang, C. H.-N., *Int. J. Syst. Bacteriol.*, 24, 54–63, 1974.

587. Edward, D. G. ff., *J. Gen. Microbiol.*, 1, 238–243, 1947.

588. Chanock, R. M., Hayflick, L., and Barile, M. F., *Proc. Natl. Acad. Sci. U.S.A.*, 48, 41–49, 1962.

589. Edwards, D. G. ff. and Fitzgerald, W. A., *Vet. Rec.*, 64, 395, 1952.

590. Smith, C. B., Chanock, R. M., Friedewald, W. T., and Alford, R. H., *Ann. N.Y. Acad. Sci.*, 143, 471–483, 1967.

591. Shepard, M. C., *Ann. N.Y. Acad. Sci.*, 143, 505–514, 1967.

592. Shepard, M. C. and Lunceford, C. D., *Appl. Microbiol.*, 20, 539–542, 1971.

593. Chu, H. P., personal communication.

594. Gourlay, R. N., *Res. Vet. Sci.*, 5, 473–482, 1964.

595. Goodwin, R. F. W., Pomeroy, A. P., and Whittlestone, P., *J. Hyg.*, 65, 85–96, 1967.

596. Dierks, R. E., Newman, J. A., and Pomeroy, B. S., *Ann. N.Y. Acad. Sci.*, 143, 170–189, 1967.

597. Chalquest, P. R., *Avian Dis.*, 6, 36–43, 1962.

CULTURE MEDIA (NATURAL AND SYNTHETIC): BLUE-GREEN ALGAE

N. G. Carr

The relationship of algal culture and physiology to both natural and laboratory environments has been reviewed with respect to phytoplankton in general, from fresh-water and marine environments, by Fogg.[1] Knowledge of the inorganic nutrition of blue-green algae[2] has been successfully applied to their laboratory cultivation, and the growth requirements of those species that have attracted physiological study have been described.[3-7]

It is central to a discussion of the growth of phototrophs to realize that the primary nutrient is light and that the most important carbon source is carbon dioxide. Blue-green algae, in contrast to the eukaryotic, oxygen-evolving, photosynthetic organisms, possess only chlorophyll a. This has an absorption maxima at 675 nm and is the major light-harvesting pigment for photosystem I. Phycobilin pigments such as phycocyanin (absorption maximum 620 nm), allophycocyanin (650 nm), or phycoerythrin (550 nm) act as the light-harvesting pigment for photosystem II. These molecules consist of linear tetrapyrolle chromophores attached to a globulin-type protein. Light sources that are employed for the growth of blue-green algae should possess a significant proportion of their emission in the 500- to 700-nm range. Although tungsten lamps may be used, their considerable heat output is often an undesirable factor. Fluorescent tube lighting, preferably of the warm-white category, provides a more appropriate ratio of useful-to-redundant emission. The amount of light flux within the photosynthetic range that is required depends upon the species of blue-green alga employed, the growth rate desired, and the final yield of organism that is to be obtained. Obviously, a culture which may grow from, say, 10^6 to 10^9 organisms ml^{-1} will change from having excess light to light limitation under most regimes of laboratory illumination. In view of the three variables listed above, absolute values of light to be supplied do not have great significance. Some species of blue-green algae will utilize quite high light intensities, around 10 Klux, before light saturation occurs. However, it is important to note that many species require much lower light intensities and are inhibited by light in excess of their growth requirement. It is necessary to establish for each organism used in defined conditions the light intensity that is optimal for that culture.

An adequate supply of carbon dioxide to populations growing rapidly to high density usually necessitates gassing the culture with air (or N_2) enriched with carbon dioxide to a value of 2 to 5%. This procedure will cause a reduction in pH value of the medium, which may be minimized by inclusion of $NaHCO_3$; 0.5 g/l^{-1} will stabilize the pH when 3% CO_2 in air is employed. Most species of blue-green algae grow best at pH values between 7.0 and 9.0, acidic conditions being a frequent cause of poor growth. Many species are sensitive to high concentrations of dissolved oxygen, and a $CO_2:N_2$ mix is often advantageous. Double-distilled water is advised for all media and special care should be employed in the cleaning of (and removal of the cleaning agent from) glassware.

The culture media employed in the growth of blue-green algae have evolved over many years, alterations frequently being made to enhance the growth of a particular species. Media that have been used for elective isolation of cultures of blue-green algae are usually simpler than those employed for axenic cultures. Some of the procedures that have been proven successful in isolating blue-green algae have been summarized.[4,6,7] Blue-green algae are widely distributed in nature, and sometimes comprise the major microbial species in fresh water, while in marine situations they are less dominant. The examples of analysis of water bodies given in Table 1 provide a basis for a comparison between the range of nutrients present in natural water and a recommended medium for elective culture.

Some compounds present in laboratory media may have a dual function: first, as a nutrient ion, and second, as a component in buffering the pH variation of the culture or maintaining some other nutrient in an accessible form (Table 2). These secondary roles often require a much greater concentration to be present, and some advantage may be obtained by diluting a given growth media for certain species. The criteria for a "good growth media" are more than merely that which will produce the maximum crop. The ease with which organisms can be disrupted and their failure to adhere to the surface of the growth vessel are often of equal importance. In some species of blue-green algae, both these features are altered by variation in the strength of a given medium and in some cases by reduction in phosphate content.

Table 1
CONSTITUENTS OF NATURAL WATER BODIES COMPARED WITH A MEDIUM FOR ELECTIVE CULTURE OF BLUE-GREEN ALGAE

Eutrophic lakes provide favorable conditions for the growth of blue-green algae; much lower concentrations of algal biomass are found in oligotrophic water or marine water from oceans. The enrichment culture contains a higher mineral composition than any of the typical natural water bodies.

Oligotrophic lake[8] (mg/l)		Eutrophic lake[9] (mg/l)		Ocean water (after Fogg[1]) (mg/l)		Enrichment medium for blue-green algae (after Allen[7]) (mg/l)	
Na	1.30	Na	17.2	Na	10,500	$NaNO_3$	1,500
K	0.64	K	2.4	K	380	K_2HPO_4	369
Ca	4.4	Ca	34.2	Ca	400	$MgSO_4 \cdot 7H_2O$	75
Mg	1.2	Mg	9.1	Mg	1,350	$CaCl_2 \cdot 2H_2O$	36
Cl	2.0	HCO_3	106.8	HCO_2	140	Na_2CO_3	20
SO_4	10.6	Cl	22.2	Cl	19,000	$Na_2SiO_3 \cdot 9H_2O$	58
NO_3	0.14	SO_4	17.1	SO_4	2,600	Ferric citrate	6
NH^4	0.14	N (total)	1.1	$NO_3 \cdot N$	0.001—0.6	Citric acid	6
SiO_2	2.9	P (total)	0.112	$PO_4 \cdot P$	0.07	EDTA	1
P	0.03	SiO_2	—	SiO_2	6.4		
Fe	0.24	Fe (total)	0.094	Fe	0.01		
				B	4.6		

0.8 mg/l of a trace element solution containing (g/l)

H_3BO_3	3.1
$MnSO_4 \cdot 4H_2O$	2.23
$ZnSO_4 \cdot 7H_2O$	0.287
$(NH_4)_6Mo_7O_{24} \cdot 4H_2O$	0.088
$Co(NO_3)_2 \cdot 4H_2O$	0.146
$Na_2WO_4 \cdot 2H_2O$	0.33
KBr	0.119
KI	0.083
$Cl(NO_3)_2 \cdot 4H_2O$	0.154
$NiSO_4(NH_4)SO_4 \cdot 6H_2O$	0.198
$VOSO_4 \cdot 2H_2O$	0.02
$Al_2(SO_4)_3 \cdot K_2SO_4 \cdot 24H_2O$	0.474

Additional ocean water values: Mn 0.002, Mo 0.01, Co 0.0005, Cu 0.003, Zn 0.01

Fe autoclaved separately
Final pH 9—10

Table 2
MEDIA EMPLOYED IN THE LABORATORY CULTURE OF BLUE-GREEN ALGAE[a]

The chemical constituents of the media below represent the nutritional requirements of various species of blue-green algae in laboratory culture. Workers intending to use a particular growth medium are advised to consult the original reference.

Table A

Allen[10] (g/l)

$NaNO_3$	1.50
K_2HPO_4	0.039
$MgSO_4 \cdot 7H_2O$	0.075
Na_2CO_3	0.020
$CaCl_2$	0.027
$NaSiO_3 \cdot 9H_2O$	0.058
EDTA	0.001
Citric acid	0.006
Ferrous citrate	0.006

1 ml/l of medium of a solution containing (g/l)

H_3BO_3	2.86
$MnCl_2 \cdot 4H_2O$	1.81
$ZnSO_4 \cdot 7H_2O$	0.22
$Na_2MoO_4 \cdot 2H_2O$	0.39
$CuSO_4 \cdot 5H_2O$	0.079
$Co(NO_3)_2 \cdot 6H_2O$	0.049

Table B

Allen and Arnon[14] (g/l)

	N_2 fixing	Non-N fixing
$MgSO_4 \cdot 7H_2O$	0.246	0.246
$CaCl_2 \cdot 2H_2O$	0.074	0.074
NaCl	0.232	0.232
K_2HPO_4	0.348	0.348
NH_4Cl	0	0.054

Either medium contains 1 ml/l of a solution of (g/l)

Sodium citrate	10.0
$FeSO_4 \cdot 7H_2O$	19.9
$MnSO_4 \cdot 4H_2O$	1.8
MoO_3	0.15
$ZnSO_4 \cdot 7H_2O$	0.22
$CuSO_4 \cdot 5H_2O$	0.079
H_3BO_3	2.86
NH_4VO_3	0.03
$Co(NO_3)_2 \cdot 6H_2O$	0.05
$NiSO_4 \cdot 7H_2O$	0.05
$CrK(SO_4)_2 \cdot 12H_2O$	0.096
$Na_2WO_4 \cdot 2H_2O$	0.018
$K_2TiO(C_2O_4)_2 \cdot 2H_2O$	0.074

Table C

Castenholtz[11] (g/l)

Nitriloacetic acid	0.10
$CaSO_4 \cdot 2H_2O$	0.06
$MgSO_4 \cdot 7H_2O$	0.10
NaCl	0.008
KNO_3	0.103
$NaNO_3$	0.689
$NaHPO_4$	0.111
$FeCl_3$	0.003

0.5 ml/l of medium of a solution containing (g/l)

$MnSO_4 \cdot H_2O$	2.28
$ZnSO_4 \cdot 7H_2O$	0.50
H_3BO_3	0.50
$CuSO_4 \cdot 5H_2O$	0.025
$Na_2MoO_4 \cdot 2H_2O$	0.025
$CoCl_2 \cdot 6H_2O$	0.045
Conc. H_2SO_4	0.5 ml

Table D

Gerloff, Fitzgerald, and Skoog[12] (g/l)

$Ca(NO_3)_2 \cdot 4H_2O$	0.04
K_2HPO_4	0.01
$MgSO_4 \cdot 7H_2O$	0.025
$NaCO_3$	0.02
$Na_2SiO_3 \cdot 9H_2O$	0.025
Ferric citrate	0.003
Citric acid	0.003

Table E

Kratz and Myers,[13] Medium C (g/l)

$MgSO_4 \cdot 7H_2O$	0.25
K_2HPO_4	1.00
$Ca(NO_3)_2 \cdot 4H_2O$	0.025
KNO_3	1.00
Sodium citrate $\cdot 2H_2O$	0.165
$Fe(SO_4)_2 \cdot 6H_2O$	0.004

1 ml/l of medium of a solution containing (g/l)

H_3BO_3	2.86
$MnCl_2 \cdot 4H_2O$	1.81
$ZnSO_4 \cdot 7H_2O$	0.222
MoO_3 (85%)	0.018
$CuSO_4 \cdot 5H_2O$	0.079

Table 2 (continued)
MEDIA EMPLOYED IN THE LABORATORY CULTURE OF BLUE-GREEN ALGAE[a]

Table F		Table G	
Provasoli, McLaughlin, and Droop[15] (g/l)		Stanier et al.,[5] Medium BG-11 (g/l)	
K_2HPO_4	0.005	$NaNO_3$	1.500
$MgSO_4 \cdot 7H_2O$	5.0	K_2HPO_4	0.040
NaCl	18.0	$MgSO_4 \cdot 7H_2O$	0.075
$Na_2SiO_3 \cdot 9H_2O$	0.15	$CaCl_2 \cdot 2H_2O$	0.036
$NaNO_3$	0.05	Citric acid	0.006
KCl	0.60	Ferric ammonium citrate	0.006
$CuCl_2 \cdot 6H_2O$	0.30		
TRIS	1.00	**1 ml/l of medium of a solution containing (g/l)**	
EDTA	0.03		
Vitamin B_{12}	2 μg		
$MnCl_2 \cdot 4H_2O$	1.2 mg/l	H_2BO_3	2.86
$ZnCl_2$	0.150 mg/l	$MnCl_2 \cdot 4H_2O$	1.81
$CuCl_2$	0.012 mg/l	$ZnSO_4 \cdot 7H_2O$	0.222
H_3BO_3	6.0 mg/l	$Na_2MoO_4 \cdot 2H_2O$	0.39
$CoCl_2$	0.03 mg/l	$CuSO_4 \cdot 5H_2O$	0.079
$FeCl_3$	0.80 mg/l	$Co(NO_3)_2 \cdot 6H_2O$	0.0494

Note: Adjusted to pH 7.1 after autoclaving.

[a] The medium of Allen is of particular value in the growth of blue-green algae on solid agar. Agar (1.5%) is autoclaved at double final strength and is added to the sterile double strength salt medium after cooling to 50°C. The medium BG-11 is very similar to that used by Allen, both these often being employed in the culture of unicellular blue-gree algae. The Castenholtz recipe has been used to maintain many species including thermophilic blue-green algae. Like the Gerloff, Fitzgerald, and Skoog formulation, this is a relatively ionically weak solution with low phosphate concentrations. When dense cultures of species are required, media with greater nutrient levels and better buffering capacity are employed, such as those of Kratz and Myers and Allen and Arnon. The latter is frequently used to compare properties of organisms grown in the presence or absence of fixed nitrogen. The medium of Provasoli, McLaughlin, and Droop is designed for marine species of blue-green algae which frequently require vitamin B_{12}

REFERENCES

1.   Fogg, G. E., *Algal Culture and Phytoplankton Ecology*, University of Wisconsin Press, Madison, 1975, 175.
2.   Healey, F. P., *CRC Crit. Rev. Microbiol.*, 3, 69—113, 1973.
3.   Allen, M. B., *Arch. Mikrobiol.*, 17, 34—53, 1952.
4.   Carr, N. G., in *Methods in Microbiology*, Vol. 3B, Norris, J. R. and Ribbans, D. W., Eds., Academic Press, London, 1969, 53—77.
5.   Stanier, R. Y., Kunisawa, R., Mandel, M., and Cohen-Bazire, G., *Bacteriol. Rev.,* 35, 171—205, 1971.
6.   Carr, N. G., Komárek, J., and Whitton, B. A., in *The Biology of Blue-Green Algae*, Carr, N. G. and Whitton, B. A., Eds., Blackwell Scientific Publications, Oxford, 1973, 525—530.
7.   Allen, M. M., in *Handbook of Phycological Methods*, Stein, J. R., Ed., Cambridge University Press, Cambridge, Engl., 1973, 127-138.
8.   Christie, A. E., *Phytoplankton Populations in Several Ice Covered Lakes in South Ontario*, Publ. 2025, Division of Research, Ontario Water Research Commission, Ontario, Can., 1969.
9.   Smith, R. V., Freshwater Biological Investigation Unit, Mackamore, Antrim, Northern Ireland, personal communication.
10.   Allen, M. M., *J. Phycol.*, 4, 1—4, 1968.
11.   Castenholtz, R. W., *Schweiz. Z. Hydrol.*, 32, 538—551, 1970.
12.   Gerloff, G. C., Fitzgerald, G. P., and Skoog, F., *Am. J. Bot.*, 37, 216—218, 1950.
13.   Kratz, W. A. and Myers, J., *Am. J. Bot.*, 42, 282—287, 1955.
14.   Allen, M. B. and Arnon, D. I., *Plant Physiol.*, 30, 366—372, 1955.
15.   Provasoli, L., McLaughlin, J. J. A., and Droop, M. R., *Arch. Mikrobiol.*, 25, 392—428, 1957.

CULTURE MEDIA FOR ALGAE

M. F. Turner and M. R. Droop

In the eight tables following this introduction, the term "algae" has been accorded its traditional sense, to include also the prokaryotic cyanophytes together with those unpigmented unicells formerly claimed by zoologists that are structurally similar to the pigmented members of the classes dealt with here.

The tables could have been extended almost indefinitely, since there are almost as many variations on media as there are authors. Some media, however, occur again and again, and have proved successful in different hands, often for a wide range of algae from different classes. These versatile media are of greatest use in the maintenance of culture collections and as a first recourse in the isolation of a new organism whose requirements are unknown. More fastidious organisms have required careful tailoring of media to individual and sometimes fairly specific requirements, and their use is essential for successful cultivation in these instances. Other media have been designed with the object of studying some particular physiological feature.

The tables fall naturally into two groups: the first four deal with freshwater media, and Tables 5 to 8 with marine and brackish media. Within these two categories are defined and undefined media. A defined medium is one in which the nature of all the constituents is known in precise chemical terms. An undefined medium, on the other hand, utilizes materials of unknown or only partially known constitution, e.g., natural waters, agar, soil extract, casamino acids. The media have again been subdivided, according to whether they are "organic" or "mineral." Here the criterion for a medium to be included as a "mineral" medium has not been that it is entirely inorganic in a strict chemical sense, but rather that it is suitable for the growth of cultures that are not bacteria-free (not axenic), with little danger of bacteria or other organisms overgrowing the algae.

With few exceptions, the media included are post 1940 to date. This covers a period of considerable innovation in the design of media; references to earlier media are well documented in the following: 170 (1929), 171 (1942), and 25 (1946). Other general references to algal cultivation include 172 (1951), 26 (1951), 27 (1952), 62 (1957), 61 (1958), 142 (1961), 112 (1969), 173 (1973), and 169 (1973). It is hoped that most media of general application that have been widely used and devised since 1940 have been included, and that most of the major variants in terms of general approach to their design have been exemplified.

Each table has been arranged so that, in general, those media having the widest application appear at the beginning. However, the organisms listed with respect to use are, almost without exception, the organisms for which the medium was first designed, even though in some cases the use has subsequently been considerably extended.

In many instances the method of preparation of certain media is as important as the composition, and reference should be made to the original publication whenever possible.

The layout employed for the constituents is for the most part self-explanatory, although occasionally some compounds, e.g., citrate and glycylglycine, have been taken out of their chemical context because they have special functions.

The quantities entered in the tables, *which, unless otherwise stated, are in mg per liter,* refer to the chemical species indicated at the start of the particular row in which they occur. In calculating the amounts to be entered, note has been taken of water of crystallization. The form in which the species occurs is indicated in the footnotes, which are placed preceding Table 1.

Generally, the names of acids end in "-ate," the cation being included elsewhere in the table, except for H^+, which has been omitted, as has the anion Cl^-.

Errors are inevitable; the authors have used their judgment to correct obvious errors when they occurred in the sources, but hope that they have avoided making too many themselves.

Footnotes

a As $CO_3^=$.
b As HCO_3^-.
c As NO_3^-.
d As NH_4^+.
e As K_2HPO_4.
f As KH_2PO_4.
g As $SO_4^=$.
h As Cl^-.
i As Na_2SiO_3.
j As Na^+.
k As Mg^{++}.
l As K^+.
m As Ca^{++}.
n As H_3BO_3.
o As $(NH_4)_2HPO_4$.
p As $KAl(SO_4)_2$.
q As Na_2HPO_4.
r As NaH_2PO_4.
s As VCl_2.
t As $VOSO_4$.
u As $V_2O_4(SO_4)_3$.
v As NH_4VO_3.
w As K_2CrO_4.
x As $K[Cr(SO_4)_2]$.
y As $Cr(NO_3)_3$.
z As $MnCl_2$.
aa As $MnSO_4$.
ab As $FeCl_3$.
ac As $FeSO_4$.
ad As $(NH_4)_2SO_4 \cdot FeSO_4$.
ae As $FeCl_2$.
af As $Fe_2(SO_4)_3$.
ag As $FeC_6H_5O_7$.
ah As ferric ammonium citrate.
ai As $Fe_2(C_4H_4O_6)_3$.
aj As $CoCl_2$.
ak As $Co(NO_3)_2$.
al As $CoSO_4$.
am As $NiSO_4$.
an As $(NH_4)_2SO_4 \cdot NiSO_4$.
ao As $CuCl_2$.
ap As $CuSO_4$.
aq As As_2O_3.
ar As H_2SeO_3.
as As KBr.
at As $RbSO_4$.
au As MoO_2.
av As MoO_3.
aw As H_2MoO_4.
ax As Na_2MoO_4.
ay As K_2MoO_4.
az As $(NH_4)_2MoO_4$.
ba As $(NH_4)_6Mo_7O_{24}$.

bb $SnCl_2$.
bc As KI.
bd As H_2WO_4.
be As Na_2WO_4.
bf As $HgCl_2$.
bg As $PbCl_2$.
bh As $Bi(NO_3)_2$.
bi Triple-distilled water.
bj As $CaCO_3 + HCl$.
bk As $Fe + HCl + HNO_3$.
bl Redistilled water.
bm As NH_4MgPO_4.
bn As citrate.
bo 1.5 drops of a saturated solution of $FeSO_4$.
bp As $TiO(C_2O_4)_x \cdot yH_2O$.
bq As FeEDTA complex.
br As Fe-Versene-diol complex.
bs As $[CH_2 \cdot N(CH_2COO)_2]_2FeNa$.
bt As $[CH_2 \cdot N(CH_2COOH) \cdot CH_2COONa]_2 \cdot 2H_2O$.
bu As NaFe-diethylene-triamine-pentaacetate.
bv As $[CH_2 \cdot N(CH_2COO)_2]_2MgNa_2$.
bw As $RbCl$.
bx As hydrochloride.
by As $K_2C_3H_5(OH)_2PO_4$.
bz As NaF.
ca As K_3PO_4.
cb As $NiCl_2$.
cc As $ZrOCl_2$.
cd As $RuCl_3$.
ce As $RhCl_3$.
cf As $(NH_4)_2Ce(NO_3)_6$.
cg As Zn^{++}.
ch As $Na_2C_3H_5(OH)_2PO_4$.
ci As $(NH_4)_2SO_4 \cdot FeSO_4$ dissolved in sulfosalicylic acid.
cj As $(C_2H_5O)PO(OH)_2$.
ck TRIS-HCl, pH 7.6.
cl As $(C_2H_5O)_4Si$.
cm As gluconate.
cn As acetate.
co As $NO_3^- + Cl^-$.
cp As $NaBr$.
cq As DL-.
cr As L-.
cs Neutralized with NaOH.
ct As fumarate.
cu As lactate.

cv   pH 7.5.

cw   Filtered.

cx   Aged.

cy   Concentrated.

cz   Or D_2O, 1,000 ml.

da   As Na_2HASO_4.

db   Adjusted with KOH after autoclaving.

dc   Autoclaved apart.

dd   Salinity 25 to 27°/$_{oo}$

de   Salinity 34 to 37°/$_{oo}$

df   Usually 80% sea water.

dg   As SiO_2.

dh   200 ml of 41.3 mM TRIS, pH 6.8 to 6.9 before autoclaving, added to 800 ml of sea water.

di   Open ocean, salinity 33°/$_{oo}$

dj   Treated with activated charcoal, 2 g/l, agitated for 1 hr, then filtered.

dk   Salinity 30 to 34°/$_{oo}$.

dl   Added as 50 ml of stock solution, adjusted to pH 8.2 by addition of HCl.

dm   "Difco."

dn   "Oxo" L25.

do   "Bacto."

dp   "Vitalia" algal flour, manufactured by Th. Myklestad, Oslo, Norway.

dq   Extract of lemon rind.

dr   "Sigma."

ds   Added only for diatom cultures.

dt   Addition of Na_2SO_4 (0.06%); allows heavy growth.

du   Vitamin-free.

dv   A limiting quantity.

dw   "Oxoid."

dx   "Rila Marine Mix" (Utility Chemical Co., Paterson, N.J.)

dy   As K_2SiO_3.

dz   pH 7.1 to 7.3 before autoclaving.

ea   Geigy Industrial Chemicals, New York.

eb   pH adjusted to 7.6—7.8 with aqueous HCl or NaOH before incorporation with the rest of the medium.

ec   Baltimore Biological Laboratories.

ed   "Chelated" (Alrose Chemical Co., Providence, R.I.)

ee   Nutritional Biochemicals Corporation, Cleveland, Oh.

ef   As Na_4 EDTA.

eg   From yeast.

eh   Wilson.

ei   As dextrose.

ej   As $(NH_4)_2HC_6H_5O_7$.

ek   Allen and Hanbury's Ltd.

el   Dehydrated.

em   "Basic carbonate," assumed to be 3 $MgCO_3 \cdot Mg(OH)_2 \cdot 3H_2O$.

en   As Na_3VO_4.

eo   "Triturate," i.e., 60 mg of triturate containing 1 g of thiamine HCl + 99 g of sucrose.

ep   Succinate.

eq   Monohydrate.

er   As NaI.

es   Or 0.02M sodium acetate.

et $FeC_6H_6O_7 \cdot H_2O$.

eu   Enough to adjust to pH 6.0—6.2.

ev   "Ionagar No. 2" (Colab Laboratories, Chicago Heights, Ill.

ew   pH to 3.8 with HCl.

ex   Sterilized by filtration.

ey   Added from a 40 mg/ml solution adjusted to pH 7.0 with KOH before addition.

ez   As $COOH \cdot C_6H_4 \cdot COOK$.

fa   As $Na_2B_4O_7$.

fb   As $Na_2Si_2O_5$.

fc   Reached on aeration with the gas mixture shown, after initial adjustment to pH 6.2—6.5.

fd   20°/$_{oo}$

fe   This replaces extract of lemon rind (LF) in the original formulation of the medium.

ff   As $[CH_2N(CH_2COO)_2]_2$ FeNa.

fg   As $TiCl_3$.

fh   As H_2SeO_4.

fi   2-(N-Morpholino)ethane sulfonic acid.

fj   Tris-(hydroxymethyl) aminomethane.

fk   N-[Tris-(hydroxymethyl)-methyl]glycine.

fl   Ethylenediaminetetraacetic acid.

fm   Nitrilotriacetic acid.

fn   Liver, 1:20.

fo   TRIS-HCl, pH 7.4.

fp   As $TiO(COO:COOK)_2$.

fq   As TiO_2.

fr   As citric acid.

fs   As KF.

ft   N'-(2-Hydroxyethyl)-ethylenediamine-N,N,N'-triacetic acid.

fu   Distilled.

fv   Glass-distilled.

fw   Cation-free.

fx   Deionized.

fy   The author gives 2.9 mg/l H_3PO_3; this has been interpreted as H_3PO_4, and the calculation for $PO_4^{\equiv}$ has been made accordingly.

fz Na_2SiO_3 solution, mixed with aqueous HCl of appropriate strength (normally 0.19N) to adjust to pH 7.4—7.6.

ga   Optional.

gb   As 10 drops of a 1% solution of ammonium ferric citrate, added after autoclaving.

gc   3 to 5 cm of sandy garden loam in a test tube, overlying wheat grains.

gd   As "a little," i.e., a small pinch of $CaCO_3$ per test tube.

ge   A grain or two, underlying sandy garden loam.

gf   As sequestrine AA.

gg   Tubes containing ingredients are steamed twice (on consecutive days).

gh   300 g of garden soil in 1 liter of distilled water.

gi   40 g of sphagnum in 1 liter of distilled water, boiled for 20 min, decanted, then filtered.

gj   1 kg of calcareous garden loam or potting soil plus 1 liter of glass-distilled water, steamed for 1 hr, decanted, filtered, and sterilized after 2 to 3 days.

gk   Sea water, filtered, then pasteurized at 70°C; the sterile ingredients are added to the pasteurized sea water.

gl   Basic $MgCO_3$ here is 4 $MgCO_3 \cdot Mg(OH)_2 \cdot 4H_2O$ (Mallinckrodt).

gm   About ten times the amount of soil.

gn   As guanylate.

go   As "Liver L".

gp   25 g of dried goat manure is added to 1 liter of distilled water and sterilized in the autoclave at 15 lb/in.² for 20 min; the extract is cooled, filtered, and added (1 part) to BBM (9 parts).

gq   As lactic acid.

gr   Various additions can be made, e.g., starch or cheese, or they may be omitted, depending on the organism being cultivated.

gs   As yeast autolysate.

gt   Diethylenetriaminepentaacetic acid.

gu   N,N-Bis(2'-hydroxyethyl)-glycine.

gv   As sodium glutamate.

gw   As sodium hydrogen glutamate.

gx   As glutamic acid.

gy   As acetic acid.

gz   Analysis excludes components of undefined additions, e.g. Na from natural sea water or nutrient extracts.

ha   Of a dried clay-free alkali extract.

Table 1
UNDEFINED FRESHWATER MEDIA (MINERAL)

	25 (1946) 36 (1946)	61 (1958)	61 (1958)	141 (1971)
Reference				
Author's designation	Biphasic or soil-water[gg]	4	77	e + s Agar
			Use	
Constituents	*Chilomonas Polytoma Astasia*	*Cyanophora* Unicellular prasinophytes and chlorophytes *Trachelomonas Vacuolaria*	*Chromulina* sp. *Carteria inversa Pandorina morum*	A wide range of algae
(1) Carbonate	Present[gd]			
(2) Nitrate		73.0[j]	137[l,m]	123[l]
(3) Ammonium				
(4) Phosphate		10.9[e]	21.6[e]	10.9[e]
(5) Sulfate		7.79[k]	11.6[k]	7.79[k]
(6) Silicon				
(7) Na		27.1[c]		
(8) Mg		1.97[g]	2.93[g]	1.97[g]
(9) K		8.98[e]	94.4[c,e]	86.3[c,e]
(10) Ca	Present[gd]		5.04[c]	
(11) Fe				
(12) B			5.77[ab]	
(13) Al				
(14) V				
(15) Mn				
(16) Co				
(17) Ni				
(18) Cu				
(19) Zn				
(20) Br				
(21) Mo				
(22) Cd				
(23) I				
(24) W				
(25) EDTA[fl]				
(26) Citrate				
(27) Biotin				
(28) Cyanocobalamin				
(29) Thiamine hydrochloride				
(30) Garden loam	Present[gc]			
(31) Peat extract				
(32) Soil extract		100 ml[gh]	9.90 ml[gh]	100 ml
(33) Sphagnum extract			4.95 ml[gi]	
(34) Wheat or barley[gr]	Present[ge]			
(35) Agar				10,000[ga]
(36) Water	Present[gm]	To 1,000 ml[fu]	To 1,000 ml[fu]	To 1,000 ml[fu]
(37) pH before autoclaving		5.0–8.0	4.2	
(38) Final pH				
(39) Aeration				

Table 1 (continued)
UNDEFINED FRESHWATER MEDIA (MINERAL)

	38 (1958)	132 (1963)	137 (1965)	156 (1972)
Reference				
Author's designation	F		Dm	
			Use	
Constituents	*Coelosphaerium kuetzingianum Gloeotrichia echinulata*	*Nostoc* (symbiotic)	Cyanophytes (unicellular coccoid forms)	*Gomphonema parvulum*
(1) Carbonate	11.3^j			37.2^m
(2) Nitrate	90.5^j	61.3^l	$743^{j,m}$	
(3) Ammonium			33.7^h	
(4) Phosphate	5.41^e	68.2^e	545^e	2.73^e
(5) Sulfate	9.74^k	107^k	97.4^k	9.59^k
(6) Silicon	5.73^i			210^{fb}
(7) Na	$51.6^{a,c,i}$	19.7^h	$275^{c,bt}$	$173^{fb,bt}$
(8) Mg	2.47^g	27.1^g	24.7^g	2.43^g
(9) K	4.45^e	$94.8^{e,c}$	449^e	2.25^e
(10) Ca	9.81^h	18.1^h	4.24^c	12.0^c
(11) Fe	0.5^{ag}	0.804^{ac}	0.880^{af}	0.0559^{ac}
(12) B	0.0217^n	0.5^n		
(13) Al	0.0011^p			
(14) V	0.0002^t			0.005^z
(15) Mn	0.0220^{aa}	0.502^z	0.240^z	
(16) Co	0.0012^{ak}		0.0595^{ak}	
(17) Ni	0.0012^{an}			
(18) Cu		0.0201^{ap}	0.0240^{ap}	
(19) Zn	0.0026^g	0.0505^g	1.20^g	
(20) Br	0.0032^{as}			
(21) Mo	0.0019^{ba}	0.01^{av}	0.2839^{av}	
(22) Cd	0.0022^c			
(23) I	0.0025^{bc}			
(24) W	0.0007^{be}			
(25) EDTA^{fl}			24.3^{bt}	3.93^{bt}
(26) Citrate	$2.95^{fr} + 1.69^{ag}$			
(27) Biotin				0.001
(28) Cyanocobalamin				
(29) Thiamine hydrochloride				
(30) Garden loam				
(31) Peat extract				15 ml
(32) Soil extract	40 ml			
(33) Sphagnum extract				
(34) Wheat or barley^{gr}				
(35) Agar		20,000	10,000	$20,000^{ga}$
(36) Water	To 1,000 ml^{fx}	To 1,000 ml	To 1,000 ml^{fv}	To 1,000 ml^{fu}
(37) pH before autoclaving		7.2—7.4		
(38) Final pH				
(39) Aeration				

Table 1 (continued)
UNDEFINED FRESHWATER MEDIA (MINERAL)

Reference	3 (1957)	3 (1957)	3 (1957)	38 (1958)
Author's designation	II	IV	VI	K
		Use		
Constituents	*Chlamydomonas Haematococcus*	*Chlamydomonas Haematoccus*	*Chlamydomonas Haematococcus*	Chlorophyceae
(1) Carbonate				
(2) Nitrate	153[l]	153[l]	155[d]	229[l,m]
(3) Ammonium			45.1[c]	
(4) Phosphate	54.5[e]	54.5[e]	54.5[e]	27.9[e]
(5) Sulfate	39.0[k]	39.0[k]	39.0[k]	12.9[k]
(6) Silicon		23.0[i]		
(7) Na		37.7[i]		
(8) Mg	9.87[g]	9.87[g]	9.87[g]	3.26[g]
(9) K	142[c,e]	142[c,e]	44.9[e]	152[c,e]
(10) Ca				8.15[c]
(11) Fe	5.0[ed]	5.0[ed]	5.0[ed]	0.333[ag]
(12) B				0.0271[n]
(13) Al				0.0013[p]
(14) V				0.00026[t]
(15) Mn				0.0275[aa]
(16) Co				0.0015[ak]
(17) Ni				0.0015[an]
(18) Cu				
(19) Zn				0.0033[g]
(20) Br				0.0040[as]
(21) Mo				0.0024[ba]
(22) Cd				0.0028[c]
(23) I				0.0032[bc]
(24) W				0.0009[be]
(25) EDTA[fl]				
(26) Citrate				1.97[fr] + 1.13[ag]
(27) Biotin				
(28) Cyanocobalamin				
(29) Thiamine hydrochloride				
(30) Garden loam				
(31) Peat extract				
(32) Soil extract	25 ml	25 ml	25 ml	3.0 ml
(33) Sphagnum extract				
(34) Wheat or barley[gr]				
(35) Agar				
(36) Water	To 1,000 ml	To 1,000 ml	To 1,000 ml	To 1,000 ml[fx]
(37) pH before autoclaving				
(38) Final pH				
(39) Aeration				

Table 1 (continued)
UNDEFINED FRESHWATER MEDIA (MINERAL)

Reference	157 (1973)	157 (1973)
Author's designation	GSP	
	Use	
Constituents	*Vacuolaria virescens* (unialgal)	*Gonyostomum semen* (unialgal)
(1) Carbonate		0.120[m]
(2) Nitrate	61.3[l]	1.38[l]
(3) Ammonium		0.169[h]
(4) Phosphate	5.45[e]	0.396[e]
(5) Sulfate	3.90[k]	1.13[k,aa,ac,cg]
(6) Silicon		
(7) Na		Trace[ax]
(8) Mg	0.987[g]	0.220[g]
(9) K	43.2[c,e]	1.20[c,e]
(10) Ca		0.080[a]
(11) Fe	0.167[ag]	0.020[ac]
(12) B		0.0044[n]
(13) Al		
(14) V		
(15) Mn		0.0369[aa]
(16) Co		0.0105[al]
(17) Ni		
(18) Cu		0.000636[ap]
(19) Zn		0.114[g]
(20) Br		
(21) Mo		0.0198[ax]
(22) Cd		
(23) I		
(24) W		
(25) EDTA[fl]		1.25
(26) Citrate	1.0[fr] + 0.564[ag]	
(27) Biotin		0.025
(28) Cyanocobalamin		0.00025
(29) Thiamine hydrochloride		0.025
(30) Garden loam		33,300[dc]
(31) Peat extract	100 ml	
(32) Soil extract	100 ml	
(33) Sphagnum extract		
(34) Wheat or barley[gr]		
(35) Agar		
(36) Water	To 1,000 ml[fu]	To 1,000 ml[fu]
(37) pH before autoclaving		5.5–5.8
(38) Final pH	6.0–6.2[fc]	
(39) Aeration	4% CO_2	4% CO_2

Compiled by M. F. Turner and M. R. Droop.

Table 2
UNDEFINED FRESHWATER MEDIA (ORGANIC)

Reference	2 (1953)	2 (1953)	3 (1957)	61 (1958)
Author's designation	A	B	I	21
			Use	
Constituents	*Cyanophora* Chlorophyceae	*Ochromonas* *Peridinium*	*Anacystis* *Navicula* Unicellular chlorophytes	*Brachiomonas* *Chlorogonium* Euglenophytes
(1) Carbonate				
(2) Bicarbonate				
(3) Nitrate		52.5[m]	92.0[l]	
(4) Ammonium				
(5) Phosphate	10.9[e]	10.9[e]	54.5[e]	
(6) Sulfate	7.79[k]	7.79[k]	39.0[k]	
(7) Silicon				
(8) Na	4.69[bn]			
(9) Mg	1.97[g]			280[cn]
(10) K	8.98[e]	1.97[g]	9.87[g]	
(11) Ca		8.98[e]	103[c,e]	
(12) Fe	2.0[g]	17.0[c]	5.0[ed]	3.61[h]
(13) B				
(14) V				
(15) Mn				
(16) Co				
(17) Cu				
(18) Zn				
(19) Mo				
(20) I				
(21) EDTA[fl]				
(22) Citrate	12.9[j]			
(23) Tartaric acid				
(24) Acetate				720[j]
(25) Dextrin				
(26) Glucose		500	10,000[ei]	
(27) Glycerol				
(28) DL-lactate				
(29) Lecithin				
(30) Starch				
(31) Sucrose				
(32) DL-Alanine				
(33) L-Arginine hydrochloride				
(34) DL-Aspartic acid				
(35) L-Cystine				
(36) Glutamate				
(37) Glycine				
(38) L-Histidine HCl·H_2O				
(39) L-Hydroxyproline				
(40) DL-Isoleucine				
(41) DL-Leucine				
(42) DL-Lysine hydrochloride				
(43) DL-Methionine				

Table 2 (continued)
UNDEFINED FRESHWATER MEDIA (ORGANIC)

Reference	2 (1953)	2 (1953)	3 (1957)	61 (1958)
Author's designation	A	B	I	21
			Use	
Constituents	*Cyanophora* Chlorophyceae	*Ochromonas* Peridinium	*Anacystis Navicula* Unicellular chlorophytes	*Brachiomonas Chlorogonium* Euglenophytes
(44) DL-Phenylalanine				
(45) L-Proline				
(46) DL-Serine				
(47) DL-Threonine				
(48) DL-Tryptophan				
(49) L-Tyrosine				
(50) DL-Valine				
(51) Guanylate				
(52) Uracil				
(53) Urea			50	
(54) Cholesterol				
(55) *p*-Aminobenzoic acid				
(56) Biotin				
(57) Choline chloride				
(58) Cyanocobalamin				
(59) *m*-Inositol				
(60) Lipoic acid				
(61) Riboflavine				
(62) Thiamine hydrochloride				
(63) Apple juice				1,000[dm]
(64) Beef extract				
(65) Casein (acid-hydrolyzed)				
(66) Cream				
(67) Goat-manure extract[gp]				
(68) Liver extract				
(69) Malt extract				
(70) Potato extract			5,000	
(71) "Proteose peptone"				
(72) "Thiopeptone"	600[eh]			
(73) "Trypticase"	160[ec]	500[ec]		
(74) Tryptone				2,000[dm]
(75) "Tween 80"				
(76) Yeast extract	50[dm]			2,000[dm]
(77) Soil extract			25 ml	
(78) Agar			20,000	
(79) Cyanide				
(80) Potassium hydroxide				
(81) Sulfuric acid				
(82) Water	To 1,000 ml	To 1,000 ml	To 1,000 ml	To 1,000 ml[fu]
(83) pH adjusted with				
(84) pH before autoclaving	6.5	6.5		6.0
(85) Final pH				

Table 2 (continued)
UNDEFINED FRESHWATER MEDIA (ORGANIC)

Reference	95 (1963)	61 (1958)	10 (1958)	144 (1962)
Author's designation	OBBM	11		
			Use	
Constituents	Cyanophytes, diatoms and chlorophytes from soil	*Cyathomonas truncata* (bacteria-free)	*Ochromonas malhamensis* (B_{12} assay)	*Ochromonas danica* (maintenance medium)
(1) Carbonate				
(2) Bicarbonate				
(3) Nitrate	163[j]			
(4) Ammonium			128[ej]	
(5) Phosphate	36.7[e] + 109[f]		210[f]	
(6) Sulfate	30.4[k,ac,cg]		77.9[k]	39.0[k]
(7) Silicon				
(8) Na	69.4[c,h]		Trace[ax]	84.5[cn]
(9) Mg	6.64[g]		19.7[g]	9.87[g]
(10) K	75.3[e,f]		86.2[f]	90.4[bn]
(11) Ca	6.11[h]		54.2[h]	
(12) Fe	0.897[ac]		2.00[ac]	
(13) B	1.8[n]		0.105[n]	
(14) V				
(15) Mn	0.358[z]		20.0[aa]	
(16) Co	0.089[ak]		0.629[al]	
(17) Cu	0.358[ap]		0.102[ap]	
(18) Zn	1.8[g]		25.0[g]	
(19) Mo	0.424[av]		19.8[ax]	
(20) I			0.0076[bc]	
(21) EDTA[fl]	44.8		50	
(22) Citrate			669[ej]	146[l]
(23) Tartaric acid				
(24) Acetate				217[j]
(25) Dextrin				
(26) Glucose		2,000	10,000	
(27) Glycerol				1,000
(28) DL-lactate				400[ey]
(29) Lecithin				
(30) Starch				4,000
(31) Sucrose				2,000
(32) DL-Alanine				
(33) L-Arginine hydrochloride				
(34) DL-Aspartic acid				
(35) L-Cystine			100	
(36) Glutamate				
(37) Glycine				
(38) L-Histidine HCl·H_2O				
(39) L-Hydroxyproline				
(40) DL-Isoleucine				
(41) DL-Leucine				
(42) DL-Lysine hydrochloride				
(43) DL-Methionine			200	

Table 2 (continued)
UNDEFINED FRESHWATER MEDIA (ORGANIC)

Reference	95 (1963)	61 (1958)	10 (1958)	144 (1962)
Author's designation	OBBM	11		
			Use	
Constituents	Cyanophytes, diatoms and chlorophytes from soil	*Cyathomonas truncata* (bacteria-free)	*Ochromonas malhamensis* (B_{12} assay)	*Ochromonas danica* (maintenance medium)
(44) DL-Phenylalanine				
(45) L-Proline				
(46) DL-Serine				
(47) DL-Threonine				
(48) DL-Tryptophan			100	
(49) L-Tyrosine				
(50) DL-Valine				
(51) Guanylate				
(52) Uracil				
(53) Urea				
(54) Cholesterol				
(55) *p*-Aminobenzoic acid			1.0	
(56) Biotin			0.01	
(57) Choline chloride			2.0	
(58) Cyanocobalamin				
(59) *m*-Inositol			10	
(60) Lipoic acid				
(61) Riboflavine				
(62) Thiamine hydrochloride			2.0	
(63) Apple juice				
(64) Beef extract		1,000[dm]		
(65) Casein (acid-hydrolyzed)			5,000[du,ek]	
(66) Cream				
(67) Goat-manure extract[gp]	100 ml			
(68) Liver extract				200[ee,go]
(69) Malt extract				
(70) Potato extract				
(71) "Proteose peptone"				
(72) "Thiopeptone"				
(73) "Trypticase"				6,000[ec]
(74) Tryptone				
(75) "Tween 80"			1.0 ml	
(76) Yeast extract				2,000[gs]
(77) Soil extract				
(78) Agar				
(79) Cyanide			1.06[j]	
(80) Potassium hydroxide	27.8			
(81) Sulfuric acid	0.00165[cy]			
(82) Water	To 1,000 ml[fu]	To 1,000 ml[fu]	To 1,000 ml	To 1,000 ml
(83) pH adjusted with				KOH or H_2SO_4
(84) pH before autoclaving		6.5–7.0	5.5	6.7–7.2
(85) Final pH				

Table 2 (continued)
UNDEFINED FRESHWATER MEDIA (ORGANIC)

Reference	151 (1962)	141 (1971)	14 (1966)	4 (1950)
Author's designation		(Ochr.)	II	Soln. 42
Use				
Constituents	*Poteriochromonas stipitata* (maintenance medium)	*Ochromonas*	*Navicula pelliculosa* (synchronous cells)	*Stigeoclonium* and green epiphytes
(1) Carbonate				
(2) Bicarbonate	432[d,j]		11.3[j]	6.00[m]
(3) Nitrate				
(4) Ammonium	91.3[b]		52.5[m]	6.20[l]
(5) Phosphate	210[f]			
(6) Sulfate	419[k,aa,ad,ap,cg]		5.45[e]	4.75[f]
(7) Silicon			9.74[k]	19.2[j,k]
			7.90[i,dv]	
(8) Na	46.3[b]		21.6[a,i]	4.60[g]
(9) Mg	59.2[g]		2.47[g]	2.43[g]
(10) K	86.2[f]		4.49[e]	5.86[c,f]
(11) Ca	59.9[h]		17.0[c]	4.00[a]
(12) Fe	2.00[ad,ex]		0.5	
(13) B				
(14) V			0.1	
(15) Mn	60.0[aa,ex]		0.1	
(16) Co	0.801[al,ex]		0.1	
(17) Cu	9.93[ap,ex]		0.1	
(18) Zn	39.8[g,ex]		0.3	
(19) Mo	3.52[ba,ex]		0.1	
(20) I				
(21) EDTA[fl]	640			
(22) Citrate				
(23) Tartaric acid				
(24) Acetate				
(25) Dextrin				
(26) Glucose	10,000[ex]	1,000		
(27) Glycerol				
(28) DL-lactate				
(29) Lecithin				
(30) Starch				
(31) Sucrose				
(32) DL-Alanine				
(33) L-Arginine hydrochloride				
(34) DL-Aspartic acid				
(35) L-Cystine				
(36) Glutamate				
(37) Glycine				
(38) L-Histidine HCl·H_2O				
(39) L-Hydroxyproline				
(40) DL-Isoleucine				
(41) DL-Leucine				
(42) DL-Lysine hydrochloride				
(43) DL-Methionine				

Table 2 (continued)
UNDEFINED FRESHWATER MEDIA (ORGANIC)

Reference	151 (1962)	141 (1971)	14 (1966)	4 (1950)
Author's designation		(Ochr.)	II	Soln. 42
			Use	
Constituents	*Poteriochromonas stipitata* (maintenance medium)	*Ochromonas*	*Navicula pelliculosa* (synchronous cells)	*Stigeoclonium* and green epiphytes
(44) DL-Phenylalanine				
(45) L-Proline				
(46) DL-Serine				
(47) DL-Threonine				
(48) DL-Tryptophan				
(49) L-Tyrosine				
(50) DL-Valine				
(51) Guanylate				
(52) Uracil				
(53) Urea				
(54) Cholesterol				
(55) *p*-Aminobenzoic acid				
(56) Biotin	0.005[ex]			
(57) Choline chloride				
(58) Cyanocobalamin	0.0005[ex]			
(59) *m*-Inositol				
(60) Lipoic acid				
(61) Riboflavine				
(62) Thiamine hydrochloride	2.0[ex]			
(63) Apple juice				
(64) Beef extract				
(65) Casein (acid-hydrolyzed)	2,000			
(66) Cream				
(67) Goat-manure extract[gp]				
(68) Liver extract		1,000[dw,el]		
(69) Malt extract				
(70) Potato extract				
(71) "Proteose peptone"				
(72) "Thiopeptone"				
(73) "Trypticase"				
(74) Tryptone		1,000[do]	1,000[dm]	
(75) "Tween 80"				100[dm]
(76) Yeast extract				
(77) Soil extract				
(78) Agar				
(79) Cyanide				
(80) Potassium hydroxide				
(81) Sulfuric acid				
(82) Water	To 1,000 ml	To 1,000 ml[fu]	To 1,000 ml	To 1,000 ml
(83) pH adjusted with			HCl	
(84) pH before autoclaving	5 or 7		7.0	
(85) Final pH				

Table 2 (continued)
UNDEFINED FRESHWATER MEDIA (ORGANIC)

Reference	5 (1952)	5 (1952)	3 (1957)	3 (1957)
Author's designation	BAT	BATY	III	V
			Use	
Constituents	*Chlamydomonas moewusii*	*Chalmydomonas moewusii*	*Chlamydomonas Haematococcus*	*Chlamydomonas Haematococcus*
(1) Carbonate				
(2) Bicarbonate				
(3) Nitrate	387^d	387^d	153^l	
(4) Ammonium	113^c	113^c		
(5) Phosphate	109^e	109^e	54.5^e	54.5^e
(6) Sulfate	77.9^k	77.9^k	39.0^k	39.0^k
(7) Silicon				
(8) Na	28.0^{cn}	28.0^{cn}	169^{cn}	
(9) Mg	19.7^g	19.7^g	9.87^g	9.87^g
(10) K	89.8^e	89.8^e	$142^{c,e}$	44.9^e
(11) Ca	27.3^h	27.3^h		
(12) Fe	0.5^{ai}	0.5^{ai}	5.0^{ed}	5.0^{ed}
(13) B	0.1^n	0.1^n		
(14) V				
(15) Mn	0.1^h	0.1^h		
(16) Co				
(17) Cu	0.1^h	0.1^h		
(18) Zn	0.3^h	0.3^h		
(19) Mo	0.1^{ax}	0.1^{ax}		
(20) I				
(21) EDTA^{fl}				
(22) Citrate				
(23) Tartaric acid				
(24) Acetate	72.0^j	72.0^j	434^j	
(25) Dextrin				
(26) Glucose				
(27) Glycerol				
(28) DL-lactate				
(29) Lecithin				
(30) Starch				
(31) Sucrose				
(32) DL-Alanine				
(33) L-Arginine hydrochloride				
(34) DL-Aspartic acid				
(35) L-Cystine				
(36) Glutamate				
(37) Glycine				
(38) L-Histidine HCl·H$_2$O				
(39) L-Hydroxyproline				
(40) DL-Isoleucine				
(41) DL-Leucine				
(42) DL-Lysine hydrochloride				
(43) DL-Methionine				

Table 2 (continued)
UNDEFINED FRESHWATER MEDIA (ORGANIC)

Reference	5 (1952)	5 (1952)	3 (1957)	3 (1957)
Author's designation	BAT	BATY	III	V
		Use		
Constituents	Chlamydomonas moewusii	Chalmydomonas moewusii	Chlamydomonas Haematococcus	Chlamydomonas Haematococcus
(44) DL-Phenylalanine				
(45) L-Proline				
(46) DL-Serine				
(47) DL-Threonine				
(48) DL-Tryptophan				
(49) L-Tyrosine				
(50) DL-Valine				
(51) Guanylate				
(52) Uracil				150
(53) Urea				
(54) Cholesterol				
(55) p-Aminobenzoic acid				
(56) Biotin				
(57) Choline chloride				
(58) Cyanocobalamin				
(59) m-Inositol				
(60) Lipoic acid				
(61) Riboflavine				
(62) Thiamine hydrochloride				
(63) Apple juice				
(64) Beef extract				
(65) Casein (acid-hydrolyzed)				
(66) Cream				
(67) Goat-manure extract[gp]				
(68) Liver extract				
(69) Malt extract				
(70) Potato extract				
(71) "Proteose peptone"			5,000	
(72) "Thiopeptone"				
(73) "Trypticase"				
(74) Tryptone	400[do,dm]	400[do,dm]		
(75) "Tween 80"				
(76) Yeast extract		100[dm]		
(77) Soil extract			25 ml	25 ml
(78) Agar				
(79) Cyanide				
(80) Potassium hydroxide				
(81) Sulfuric acid				
(82) Water	To 1,000 ml	To 1,000 ml	To 1,000 ml	To 1,000 ml
(83) pH adjusted with				
(84) pH before autoclaving				
(85) Final pH				

Table 2 (continued)
UNDEFINED FRESHWATER MEDIA (ORGANIC)

Reference	61 (1958)	11 (1959)	15 (1960)	12 (1962)
Author's designation	9			C
			Use	
Constituents	Carteria Eudorina Gonium Chlamydomonas etc.	Polytoma uvella	Chlorella vulgaris	Chlorella
(1) Carbonate				
(2) Bicarbonate				
(3) Nitrate	123[l]		123[l]	1240[l]
(4) Ammonium				
(5) Phosphate	10.9[e]		14.0[f]	190[e,dc]
(6) Sulfate	7.79[k]		7.79[k]	96.1[k]
(7) Silicon				
(8) Na		280[cn]		
(9) Mg	1.97[g]			92.0[h]
(10) K	86.3[c,e]		1.97[g]	24.3[g]
(11) Ca			83.1[c,f]	938[c,e]
(12) Fe				20.0[h]
				4.0[o,bs or bu]
(13) B				
(14) V				0.5[n]
(15) Mn				0.01[v]
(16) Co				0.5[aa]
(17) Cu				
(18) Zn				0.02[ap]
(19) Mo				0.05[g]
(20) I				0.01[av]
(21) EDTA[fl]				
(22) Citrate				Present[bs or bu]
(23) Tartaric acid				
(24) Acetate		720[j]		
(25) Dextrin				
(26) Glucose				
(27) Glycerol				10,000
(28) DL-lactate				
(29) Lecithin				
(30) Starch				
(31) Sucrose				
(32) DL-Alanine				
(33) L-Arginine hydrochloride				
(34) DL-Aspartic acid				
(35) L-Cystine				
(36) Glutamate				
(37) Glycine				
(38) L-Histidine HCl·H_2O				
(39) L-Hydroxyproline				
(40) DL-Isoleucine				
(41) DL-Leucine				
(42) DL-Lysine hydrochloride				
(43) DL-Methionine				

Table 2 (continued)
UNDEFINED FRESHWATER MEDIA (ORGANIC)

Reference	61 (1958)	11 (1959)	15 (1960)	12 (1962)
Author's designation	9			C
		Use		
Constituents	*Carteria* *Eudorina* *Gonium* *Chlamydomonas* etc.	*Polytoma* *uvella*	*Chlorella* *vulgaris*	*Chlorella*
(44) DL-Phenylalanine				
(45) L-Proline				
(46) DL-Serine				
(47) DL-Threonine				
(48) DL-Tryptophan				
(49) L-Tyrosine				
(50) DL-Valine				
(51) Guanylate				
(52) Uracil				
(53) Urea				
(54) Cholesterol				
(55) *p*-Aminobenzoic acid				
(56) Biotin				
(57) Choline chloride				
(58) Cyanocobalamin				
(59) *m*-Inositol				
(60) Lipoic acid				
(61) Riboflavine				
(62) Thiamine hydrochloride				
(63) Apple juice				
(64) Beef extract				
(65) Casein (acid-hydrolyzed)				
(66) Cream				
(67) Goat-manure extract[gp]				
(68) Liver extract				
(69) Malt extract				
(70) Potato extract				
(71) "Proteose peptone"	1,000[dm]		1,000[dm]	
(72) "Thiopeptone"				
(73) "Trypticase"		4,000[ec]		
(74) Tryptone				
(75) "Tween 80"				
(76) Yeast extract		2,000[do,dm]		500
(77) Soil extract				
(78) Agar	10,000	ga	10,000	ga
(79) Cyanide				
(80) Potassium hydroxide				
(81) Sulfuric acid				
(82) Water	To 1,000 ml[fu]	To 1,000 ml	To 1,000 ml	To 1,000 ml
(83) pH adjusted with				
(84) pH before autoclaving	6.0—7.0			
(85) Final pH		8.3		

Table 2 (continued)
UNDEFINED FRESHWATER MEDIA (ORGANIC)

Reference	12 (1962)	13 (1964)	164 (1964)	141 (1971)
Author's designation	BAD	Apple Juice Medium	Volvocacean agar	(Polyt.)
Use				
Constituents	*Chlorella*	*Chlamydomonas acidophila* *Carteria acidicola*	Volvocales	*Polytoma*
(1) Carbonate				
(2) Bicarbonate				
(3) Nitrate	248^l		48.8^l	
(4) Ammonium			4.35^o	
(5) Phosphate	94.9^e		11.5^o	
(6) Sulfate	96.1^k		$34.3^{k,m}$	
(7) Silicon				
(8) Na			56.1^{cn}	338^{cn}
(9) Mg	24.3^g		1.57^g	
(10) K	$235^{c,e}$		30.8^c	
(11) Ca			$12.4^{g,h}$	
(12) Fe	4.00^{bs} *or* bu		0.796^{ac}	
(13) B	0.5^n			
(14) V	0.01^v			
(15) Mn	0.5^{aa}			
(16) Co				
(17) Cu	0.02^{ap}			
(18) Zn	0.05^g			
(19) Mo	0.01^{av}			
(20) I				
(21) EDTA[fl]	Present[bs] *or* bu			
(22) Citrate			4.16^{gf}	
(23) Tartaric acid				
(24) Acetate			144^j	868^j
(25) Dextrin		10,000		
(26) Glucose	2,000	20,000		
(27) Glycerol				
(28) DL-lactate				
(29) Lecithin				
(30) Starch				
(31) Sucrose				
(32) DL-Alanine				
(33) L-Arginine hydrochloride				
(34) DL-Aspartic acid				
(35) L-Cystine				
(36) Glutamate				
(37) Glycine				
(38) L-Histidine $HCl \cdot H_2O$				
(39) L-Hydroxyproline				
(40) DL-Isoleucine				
(41) DL-Leucine				
(42) DL-Lysine hydrochloride				
(43) DL-Methionine				

Table 2 (continued)
UNDEFINED FRESHWATER MEDIA (ORGANIC)

Reference	12 (1962)	13 (1964)	164 (1964)	141 (1971)
Author's designation	BAD	Apple Juice Medium	Volvocacean agar	(Polyt.)
		Use		
Constituents	*Chlorella*	*Chlamydomonas acidophila Carteria. acidicola*	Volvocales	*Polytoma*
(44) DL-Phenylalanine				
(45) L-Proline				
(46) DL-Serine				
(47) DL-Threonine				
(48) DL-Tryptophan				
(49) L-Tyrosine				
(50) DL-Valine				
(51) Guanylate				
(52) Uracil				
(53) Urea				
(54) Cholesterol				
(55) *p*-Aminobenzoic acid				
(56) Biotin				
(57) Choline chloride				
(58) Cyanocobalamin				
(59) *m*-Inositol				
(60) Lipoic acid				
(61) Riboflavine				
(62) Thiamine hydrochloride				
(63) Apple juice		250 ml		
(64) Beef extract	1,000		200	
(65) Casein (acid-hydrolyzed)				
(66) Cream				
(67) Goat-manure extract[gp]				
(68) Liver extract				
(69) Malt extract		10,000		
(70) Potato extract		10,000[el]		
(71) "Proteose peptone"				
(72) "Thiopeptone"				
(73) "Trypticase"			400	1,000[do]
(74) Tryptone				
(75) "Tween 80"			400	1,000
(76) Yeast extract				
(77) Soil extract	ga			
(78) Agar		25,000	10,000	10,000[ga]
(79) Cyanide			2.41	
(80) Potassium hydroxide				
(81) Sulfuric acid				
(82) Water	To 1,000 ml	To 1,000 ml	To 1,000 ml	To 1,000 ml[fu]
(83) pH adjusted with			Not adjusted	
(84) pH before autoclaving		6.0		
(85) Final pH				

Table 2 (continued)
UNDEFINED FRESHWATER MEDIA (ORGANIC)

Reference	141 (1971)	6 (1953)	7 (1953)	8 (1955)
Author's designation				E
		Use		
Constituents	Desmids	*Peranema trichophorum*	*Euglena gracilis*	*Euglena*
(1) Carbonate		30.0[m]		
(2) Bicarbonate				
(3) Nitrate	61.3[l]		387[d]	372[d]
(4) Ammonium	Trace[gb]	128[ej]	113[c]	108[c]
(5) Phosphate	5.45[e]	105[f]	273[e]	285[e]
(6) Sulfate	3.90[k]	156[k]	195[k]	192[k]
(7) Silicon				
(8) Na		178[cn,gn,gw]	845[cn]	845[cn]
(9) Mg	0.987[g]	39.5[g]	49.4[g]	48.7[g]
(10) K	43.2[c,e]	43.1[f]	225[e]	235[e]
(11) Ca		20.0[a]		
(12) Fe	Present[gb]	0.5[ad]	1.0[ac]	50.3[ac,bq]
(13) B		0.5		
(14) V				
(15) Mn		15[aa]		
(16) Co		0.2[al]		
(17) Cu		1.25[ap]		
(18) Zn		20.0[g]		
(19) Mo		5.0[az]		
(20) I				
(21) EDTA[fl]		12.5		Present[bq]
(22) Citrate	Present[gb]	669[ej]		
(23) Tartaric acid			2.0	
(24) Acetate		174[j]	2,170[j]	2,170[j]
(25) Dextrin				
(26) Glucose				
(27) Glycerol				
(28) DL-lactate				
(29) Lecithin		60		
(30) Starch				
(31) Sucrose				
(32) DL-Alanine		400		
(33) L-Arginine hydrochloride		900		
(34) DL-Aspartic acid		800		
(35) L-Cystine		100		
(36) Glutamate		686[gw] + 1,000[cr,gx]		
(37) Glycine		400		
(38) L-Histidine HCl·H$_2$O		660		
(39) L-Hydroxyproline		4.8		
(40) DL-Isoleucine		40		
(41) DL-Leucine		40		
(42) DL-Lysine hydrochloride		400		
(43) DL-Methionine		298		

Table 2 (continued)
UNDEFINED FRESHWATER MEDIA (ORGANIC)

Reference	141 (1971)	6 (1953)	7 (1953)	8 (1955)
Author's designation				E
		Use		
Constituents	Desmids	*Peranema trichophorum*	*Euglena gracilis*	*Euglena*
(44) DL-Phenylalanine		32		
(45) L-Proline		32		
(46) DL-Serine		80		
(47) DL-Threonine		80		
(48) DL-Tryptophan		140		
(49) L-Tyrosine		32		
(50) DL-Valine		40		
(51) Guanylate		28.1[j]		
(52) Uracil		20		
(53) Urea				
(54) Cholesterol		2.0		
(55) *p*-Aminobenzoic acid				
(56) Biotin				
(57) Choline chloride				
(58) Cyanocobalamin		0.0005		
(59) *m*-Inositol				
(60) Lipoic acid		2,500 units		
(61) Riboflavine		0.1		
(62) Thiamine hydrochloride		2.0		
(63) Apple juice				
(64) Beef extract				
(65) Casein (acid-hydrolyzed)				
(66) Cream		0.3—0.5 ml		
(67) Goat-manure extract[gp]				
(68) Liver extract				
(69) Malt extract				
(70) Potato extract				
(71) "Proteose peptone"				
(72) "Thiopeptone"				
(73) "Trypticase"				
(74) Tryptone			5,000[dm]	5,000[dm]
(75) "Tween 80"				
(76) Yeast extract				
(77) Soil extract				
(78) Agar	10,000			
(79) Cyanide				
(80) Potassium hydroxide				
(81) Sulfuric acid				
(82) Water	To 1,000 ml[fu]	To 1,000 ml	To 1,000 ml	To 1,000 ml
(83) pH adjusted with			NaOH	
(84) pH before autoclaving		5.8—6.0	7.0	7.0
(85) Final pH				

Table 2 (continued)
UNDEFINED FRESHWATER MEDIA (ORGANIC)

	Reference	141 (1971)
	Author's designation	(E.g.)

		Use
		Euglena
	Constituents	*gracilis*
(1)	Carbonate	
(2)	Bicarbonate	
(3)	Nitrate	
(4)	Ammonium	
(5)	Phosphate	
(6)	Sulfate	
(7)	Silicon	
(8)	Na	169[cn]
(9)	Mg	
(10)	K	
(11)	Ca	3.61[h]
(12)	Fe	
(13)	B	
(14)	V	
(15)	Mn	
(16)	Co	
(17)	Cu	
(18)	Zn	
(19)	Mo	
(20)	I	
(21)	EDTA[fl]	
(22)	Citrate	
(23)	Tartaric acid	
(24)	Acetate	434[j]
(25)	Dextrin	
(26)	Glucose	
(27)	Glycerol	
(28)	DL-lactate	
(29)	Lecithin	
(30)	Starch	
(31)	Sucrose	
(32)	DL-Alanine	
(33)	L-Arginine hydrochloride	
(34)	DL-Aspartic acid	
(35)	L-Cystine	
(36)	Glutamate	
(37)	Glycine	
(38)	L-Histidine HCl·H_2O	
(39)	L-Hydroxyproline	
(40)	DL-Isoleucine	
(41)	DL-Leucine	
(42)	DL-Lysine hydrochloride	
(43)	DL-Methionine	

Table 2 (continued)
UNDEFINED FRESHWATER MEDIA (ORGANIC)

Reference   141 (1971)

Author's designation   (E.g.)

Constituents	Use *Euglena gracilis*
(44) DL-Phenylalanine	
(45) L-Proline	
(46) DL-Serine	
(47) DL-Threonine	
(48) DL-Tryptophan	
(49) L-Tyrosine	
(50) DL-Valine	
(51) Guanylate	
(52) Uracil	
(53) Urea	
(54) Cholesterol	
(55) *p*-Aminobenzoic acid	
(56) Biotin	
(57) Choline chloride	
(58) Cyanocobalamin	
(59) *m*-Inositol	
(60) Lipoic acid	
(61) Riboflavine	
(62) Thiamine hydrochloride	
(63) Apple juice	
(64) Beef extract	1,000
(65) Casein (acid-hydrolyzed)	
(66) Cream	
(67) Goat-manure extract[gp]	
(68) Liver extract	
(69) Malt extract	
(70) Potato extract	
(71) "Proteose peptone"	
(72) "Thiopeptone"	
(73) "Trypticase"	
(74) Tryptone	2,000[do]
(75) "Tween 80"	
(76) Yeast extract	2,000
(77) Soil extract	
(78) Agar	10,000[ga]
(79) Cyanide	
(80) Potassium hydroxide	
(81) Sulfuric acid	
(82) Water	To 1,000 ml[fu]
(83) pH adjusted with	
(84) pH before autoclaving	
(85) Final pH	

Compiled by M. F. Turner and M. R. Droop.

Table 3
DEFINED FRESHWATER MEDIA (MINERAL)

	Reference	1 (1942)	1 (1942)	20 (1948)	22 (1951)
	Author's designation	10	12	VIII	
				Use	
	Constituents	*Oscillatoria* *Botryococcus* *Asterionella* *Nitzschia* *Tabellaria*	*Staurastrum* *Botryococcus* *Asterionella* *Nitzschia* *Tabellaria*	A wide range of planktonic algae	Base for phyto-flagellates
(1)	Carbonate	11.3[j]	11.3[j]		
(2)	Bicarbonate				
(3)	Nitrate	30.2[m]	22.7[m]	45.3[m]	
(4)	Ammonium				67.4[h]
(5)	Phosphate	5.45 or 2.73[e]	2.73[e]	2.73[e,dc]	109[e]
(6)	Sulfate	9.74[k]	29.2[k]	3.99[k]	312[k]
(7)	Silicon	5.75[i]	4.55[dy]	4.60[i]	
(8)	Na	18.1[a,i]	8.68[a]	7.53[i]	Trace[ax]
(9)	Mg	2.47[g]	7.40[g]	1.01[g]	79.0[g]
(10)	K	4.49 or 2.25[e]	17.5[e,h,dy]	2.25[e,dc]	89.8[e]
(11)	Ca	9.77[c]	7.33[c]	14.7[c]	20.0
(12)	Fe	0.275[ab]	0.172[ab]	0.167[ag,dc]	8.0[ac]
(13)	Li				
(14)	B				20.0[n]
(15)	F				
(16)	Al				
(17)	Ti				
(18)	V				
(19)	Cr				
(20)	Mn			0.011[aa]	20.0[aa]
(21)	Co				4.00[al]
(22)	Ni				
(23)	Cu				4.00[ap]
(24)	Zn				50.0
(25)	As				
(26)	Se				
(27)	Br				
(28)	Rb				
(29)	Sr				
(30)	Mo				6.00[ax]
(31)	Cd				
(32)	Sn				
(33)	I				
(34)	Ba				

Table 3 (continued)
DEFINED FRESHWATER MEDIA (MINERAL)

Reference	1 (1942)	1 (1942)	20 (1948)	22 (1951)
Author's designation	10	12	VIII	
			Use	
Constituents	*Oscillatoria* *Botryococcus* *Asterionella* *Nitzschia* *Tabellaria*	*Staurastrum* *Botryococcus* *Asterionella* *Nitzschia* *Tabellaria*	A wide range of planktonic algae	Base for phyto- flagellates
(35) W				
(36) Hg				
(37) Pb				
(38) Bi				
(39) EDTA[fl]				500
(40) NTA[fm]				
(41) Citrate			$0.984^{dc,fr} + 0.564^{ag,dc}$	
(42) TRIS[fj]				
(43) TRICINE[fk]				
(44) *p*-Aminobenzoic acid				
(45) Biotin				
(46) Choline hydrogen citrate				
(47) Cyanocobalamin				
(48) N^5-Formyltetrahydro- pteroylglutamic acid, Ca salt				
(49) *p*-Hydroxybenzoic acid				
(50) Inositol				
(51) Lipoic acid				
(52) Nicotinic acid				
(53) Pantothenic acid, Ca salt				
(54) Pteroylglutamic acid				
(55) Putrescine dihydro- chloride				
(56) Pyridoxal hydrochlor- ide				
(57) Pyridoxamine dihydro- chloride				
(58) Riboflavine, Na, PO_4				
(59) Thiamine hydrochlor- ide				
(60) Hydrochloric acid				
(61) Sulfuric acid				
(62) Potassium hydroxide				
(63) Water	To 1,000 ml[fv]	To 1,000 ml[fv]	To 1,000 ml	To 1,000 ml
(64) pH adjusted with				KOH
(65) pH before autoclaving				6.5–6.9
(66) Final pH				
(67) Aeration				
(68) Tartrate				

Table 3 (continued)
DEFINED FRESHWATER MEDIA (MINERAL)

Reference	8 (1955)	38 (1958)	95 (1963)	16 (1937)
Author's designation	V	A	BBM	
Use				
Constituents	*Phormidium Vaucheria Chlorococcum Chlorella Scenedesmus*	Cyanophytes *Chlorella pyrenoidosa*	Chlorophytes, cyanophytes, and diatoms from soil	*Nostoc muscorum*
(1) Carbonate				$600-3,000^m$
(2) Bicarbonate				
(3) Nitrate	$313^{l,m}$	146^j	182^j	
(4) Ammonium		3.37^h		
(5) Phosphate	$47.4^e + 47.5^f$	27.0^e	$40.8^e + 121^f$	273^e
(6) Sulfate	192^k	19.5^k	$33.7^{k,ac,cg}$	$134^{k,m}$
(7) Silicon				
(8) Na		54.1^c	$77.2^{c,h}$	78.7^h
(9) Mg	48.7^g	4.94^g	7.37^g	19.7^g
(10) K	$254^{c,e,f}$	22.3^e	$83.5^{e,f}$	225^e
(11) Ca	1.00^c	3.54^h	6.79^h	$424-2,030^{a,g}$
(12) Fe	50.2^{ac}	0.5^{ag}	0.996^{ac}	1.03^{ab}
(13) Li				
(14) B	0.25^n	0.0217^n	2.00^n	
(15) F				
(16) Al		0.0011^p		
(17) Ti				
(18) V		0.0002^t		
(19) Cr				
(20) Mn	0.341^{aa}	0.0220^{aa}	0.398^z	
(21) Co		0.0012^{ak}	0.100^{ak}	
(22) Ni		0.0012^{an}		
(23) Cu	0.01^{ap}		0.398^{ap}	
(24) Zn	0.024^h	0.0026^g	2.00^g	
(25) As				
(26) Se				
(27) Br		0.0032^{as}		
(28) Rb				
(29) Sr				
(30) Mo	0.0053^{aw}	0.0019^{ba}	0.471^{av}	
(31) Cd		0.0022^c		
(32) Sn				
(33) I		0.0025^{bc}		
(34) Ba				

Table 3 (continued)
DEFINED FRESHWATER MEDIA (MINERAL)

Reference	8 (1955)	38 (1958)	95 (1963)	16 (1937)
Author's designation	V	A	BBM	
			Use	
Constituents	*Phormidium* *Vaucheria* *Chlorococcum* *Chlorella* *Scenedesmus*	Cyanophytes *Chlorella* *pyrenoidosa*	Chlorophytes, cyanophytes, and diatoms from soil	*Nostoc* *muscorum*
(35) W		0.0007[be]		
(36) Hg				
(37) Pb				
(38) Bi				
(39) EDTA[fl]	Present		49.8	
(40) NTA[fm]				
(41) Citrate		2.95[fr] + 1.69[ag]		
(42) TRIS[fj]				
(43) TRICINE[fk]				
(44) *p*-Aminobenzoic acid				
(45) Biotin				
(46) Choline hydrogen citrate				
(47) Cyanocobalamin				
(48) N^5-Formyltetrahydro-pteroylglutamic acid, Ca salt				
(49) *p*-Hydroxybenzoic acid				
(50) Inositol				
(51) Lipoic acid				
(52) Nicotinic acid				
(53) Pantothenic acid, Ca salt				
(54) Pteroylglutamic acid				
(55) Putrescine dihydro-chloride				
(56) Pyridoxal hydrochlor-ide				
(57) Pyridoxamine dihydro-chloride				
(58) Riboflavine, Na, PO_4				
(59) Thiamine hydrochlor-ide				
(60) Hydrochloric acid				
(61) Sulfuric acid			0.00183	
(62) Potassium hydroxide			30.9	
(63) Water	To 1,000 ml[fw]	To 1,000 ml[fx]	To 1,000 ml[fu]	To 1,000 ml
(64) pH adjusted with				
(65) pH before autoclaving	ca. 6.8			
(66) Final pH				ca. 7 with aeration
(67) Aeration				1% CO_2
(68) Tartrate				

Table 3 (continued)
DEFINED FRESHWATER MEDIA (MINERAL)

Reference	17 (1942)	18 (1942)	168 (1949)	23 (1950)
Author's designation				Modified Chu No. 10
		Use		
Constituents	*Anabaena cylindrica*	*Chroococcus*	*Anabaena cylindrica*	Cyanophytes (22 species)
(1) Carbonate		849[j]		11.3[j]
(2) Bicarbonate				
(3) Nitrate		632[l,m]		30.2[m]
(4) Ammonium		16.9[h]		
(5) Phosphate	109[e]	698[f]	109[e]	5.45[e]
(6) Sulfate	77.9[k]	100[k,af]	77.9[k]	9.74[k]
(7) Silicon				5.75[i]
(8) Na		651[a]	Trace[ax]	18.1[a,i]
(9) Mg	19.7[g]	24.7[g]	19.7[g]	2.47[g]
(10) K	89.8[e]	674[c,f]	89.8[e]	4.49[e]
(11) Ca	36.6[h]	6.11[c]	36.1[h]	9.77[c]
(12) Fe	0.413[ab]	1.12[af]	0.4[h]	0.5[ag]
(13) Li				
(14) B		1.0	0.1[n]	
(15) F				
(16) Al				
(17) Ti				
(18) V				
(19) Cr				
(20) Mn		2.0	0.1[h]	
(21) Co				
(22) Ni				
(23) Cu		0.01	0.01[g]	
(24) Zn		0.05	0.01[g]	
(25) As				
(26) Se				
(27) Br				
(28) Rb				
(29) Sr				
(30) Mo	0.101[ay]	0.01	0.1[ax]	
(31) Cd				
(32) Sn				
(33) I				
(34) Ba				

Table 3 (continued)
DEFINED FRESHWATER MEDIA (MINERAL)

Reference	17 (1942)	18 (1942)	168 (1949)	23 (1950)
Author's designation				Modified Chu No. 10
			Use	
Constituents	*Anabaena cyclindrica*	*Chroococcus*	*Anabaena cylindrica*	Cyanophytes (22 species)
(35) W				
(36) Hg				
(37) Pb				
(38) Bi				
(39) EDTA[fl]				
(40) NTA[fm]				
(41) Citrate				$2.95^{fr} + 1.69^{ag}$
(42) TRIS[fj]				
(43) TRICINE[fk]				
(44) *p*-Aminobenzoic acid				
(45) Biotin				
(46) Choline hydrogen citrate				
(47) Cyanocobalamin				
(48) N^5-Formyltetrahydro-pteroylglutamic acid, Ca salt				
(49) *p*-Hydroxybenzoic acid				
(50) Inositol				
(51) Lipoic acid				
(52) Nicotinic acid				
(53) Pantothenic acid, Ca salt				
(54) Pteroylglutamic acid				
(55) Putrescine dihydro-chloride				
(56) Pyridoxal hydrochlor-ide				
(57) Pyridoxamine dihydro-chloride				
(58) Riboflavine, Na, PO₄				
(59) Thiamine hydrochlor-ide				
(60) Hydrochloric acid				
(61) Sulfuric acid				
(62) Potassium hydroxide				
(63) Water	To 1,000 ml[fv]	To 1,000 ml[fv]	To 1,000 ml[fv]	To 1,000 ml
(64) pH adjusted with				
(65) pH before autoclaving			7.4	
(66) Final pH	7.3			
(67) Aeration				
(68) Tartrate				

Table 3 (continued)
DEFINED FRESHWATER MEDIA (MINERAL)

Reference	24 (1950)	27 (1952)	8 (1955)	33 (1955)
Author's designation		No. 3	PG	
			Use	
Constituents	*Coccochloris peniocystis*	Cyanophytes	*Nostoc muscorum*	*Anabaena cylindrica*
(1) Carbonate	11.3[j]			
(2) Bicarbonate				
(3) Nitrate	30.1[j]	730[j]	734[l,m]	
(4) Ammonium		16.9[h]	18.0[h]	
(5) Phosphate	5.48[q]	137[e]	541[e]	190[e]
(6) Sulfate	9.87[j]	200[k]	96.1[k]	96.1[k]
(7) Silicon	5.75[i]			
(8) Na	36.7[a,c,g,i,q]	271[c]		92.0[h]
(9) Mg	2.50[h]	50.5[g]	24.3[g]	24.3[g]
(10) K	4.51[h]	112[e]	903[c,e]	156[e]
(11) Ca	9.79[h]	18.1[h]	4.01[c]	20.0[h]
(12) Fe	0.5[ag]	0.2[ad] + trace[ac]	50.2[ac]	4.0[bq]
(13) Li	0.000182[h]			
(14) B	0.00428[n]	0.02[n]	0.25[n]	0.5[n]
(15) F	0.000073[fs]			
(16) Al	0.000350[g]			
(17) Ti	0.00133[fq]			0.01[bp]
(18) V	0.000093[s]			0.01[v]
(19) Cr	0.000298[w]			0.01[x]
(20) Mn	0.00432[z]	0.02[aa]	0.341[aa]	0.5[aa]
(21) Co	0.000450[ak]			0.01[ak]
(22) Ni	0.000496[am]			0.01[am]
(23) Cu	0.000566[ap]	0.1[ap]	0.01[ap]	0.02[ap]
(24) Zn	0.0009[g]	0.15[g]	0.024[h]	0.05[g]
(25) As	0.000168[aq]			
(26) Se	0.000121[fh]			
(27) Br	0.000746[as]			
(28) Rb	0.000142[at]			
(29) Sr	0.000530[g]			
(30) Mo	0.000708[au]	0.02[az]	0.0053[aw]	0.1[av]
(31) Cd	0.000136[h]			
(32) Sn	0.000584[bb]			
(33) I	0.000849[bc]			
(34) Ba	0.000733[h]			

Table 3 (continued)
DEFINED FRESHWATER MEDIA (MINERAL)

Reference	24 (1950)	27 (1952)	8 (1955)	33 (1955)
Author's designation		No. 3	PG	
		Use		
Constituents	*Coccochloris peniocystis*	Cyanophytes	*Nostoc muscorum*	*Anabaena cylindrica*
(35) W	0.000164[bd]			0.01[be]
(36) Hg	0.000164[bf]			
(37) Pb	0.000166[bg]			
(38) Bi	0.000096[bh]			
(39) EDTA[fl]			Present	Present
(40) NTA[fm]				
(41) Citrate	2.95[fr] + 1.69[ag]			
(42) TRIS[fj]				
(43) TRICINE[fk]				
(44) *p*-Aminobenzoic acid				
(45) Biotin				
(46) Choline hydrogen citrate				
(47) Cyanocobalamin				
(48) N[5]-Formyltetrahydro-pteroylglutamic acid, Ca salt				
(49) *p*-Hydroxybenzoic acid				
(50) Inositol				
(51) Lipoic acid				
(52) Nicotinic acid				
(53) Pantothenic acid, Ca salt				
(54) Pteroylglutamic acid				
(55) Putrescine dihydro-chloride				
(56) Pyridoxal hydrochlor-ide				
(57) Pyridoxamine dihydro-chloride				
(58) Riboflavine, Na, PO_4				
(59) Thiamine hydrochlor-ide				
(60) Hydrochloric acid				
(61) Sulfuric acid				
(62) Potassium hydroxide				
(63) Water	To 1,000 ml[fv]	To 1,000 ml[fu]	To 1,000 ml[fw]	To 1,000 ml[fv]
(64) pH adjusted with				
(65) pH before autoclaving			ca. 8.5	
(66) Final pH				
(67) Aeration			Air	5% CO_2
(68) Tartrate				

Table 3 (continued)
DEFINED FRESHWATER MEDIA (MINERAL)

Reference	34 (1955)	34 (1955)	37 (1958)	43 (1960)
Author's designation	A	D	No. 11	No. 32
	Use			
Constituents	*Anabaena variabilis* *A. cylindrica* *Anacystis* *Nostoc muscorum*	Cyanophytes	*Microcystis aeruginosa* (unialgal)	*Microcystis aeruginosa*
(1) Carbonate	849[j]		11.3[j]	11.3[j]
(2) Bicarbonate				
(3) Nitrate	665[d,l,m]	735[j,m]	362[j]	620[j]
(4) Ammonium	11.3[c]			
(5) Phosphate	698[f]	545[e]	21.3[e]	57.0[e]
(6) Sulfate	97.4[k]	58.5[k]	29.2[k]	48.7[k]
(7) Silicon			5.73[i]	11.5[i]
(8) Na	651[a]	271[c]	152[a,c,i]	257[a,c,i]
(9) Mg	24.7[g]	14.8[g]	7.40[g]	12.3[g]
(10) K	674[c,f]	449[e]	17.5[e]	46.9[e]
(11) Ca	4.24[c]	1.70[c]	9.81[h]	11.7[h]
(12) Fe	1.0[ag]	0.880[af]	1.0[ag]	0.750[br]
(13) Li				
(14) B			0.0434[n]	0.0434[n]
(15) F				
(16) Al			0.0022[p]	0.0022[p]
(17) Ti				
(18) V			0.0004[t]	0.0004[u]
(19) Cr				0.0004[y]
(20) Mn		0.400[z]	0.0439[aa]	0.0439[aa]
(21) Co		0.1[ak]	0.0024[ak]	0.0024[ak]
(22) Ni			0.0024[an]	0.0024[an]
(23) Cu		0.400[ap]		0.0052[ap]
(24) Zn		2.01[g]	0.0052[g]	0.0052[g]
(25) As				
(26) Se				
(27) Br			0.0064[as]	0.0064[as]
(28) Rb				
(29) Sr				
(30) Mo		0.473[av]	0.0038[ba]	0.0038[ba]
(31) Cd			0.0045[c]	0.0045[c]
(32) Sn				
(33) I			0.0051[bc]	0.0051[bc]
(34) Ba			0.0015[be]	0.0015[be]
(35) W				
(36) Hg				
(37) Pb				
(38) Bi				

Table 3 (continued)
DEFINED FRESHWATER MEDIA (MINERAL)

Reference	34 (1955)	34 (1955)	37 (1958)	43 (1960)
Author's designation	A	D	No. 11	No. 32
		Use		
Constituents	*Anabaena variabilis* *A. cylindrica* *Anacystis* *Nostoc muscorum*	Cyanophytes	*Microcystis aeruginosa* (unialgal)	*Microcystis aeruginosa*
(39) EDTA[fl]		50	1.0	4.67[br]
(40) NTA[fm]				
(41) Citrate	3.39[ag]		5.91[fr] + 3.39[ag]	
(42) TRIS[fj]				
(43) TRICINE[fk]				
(44) *p*-Aminobenzoic acid				
(45) Biotin				
(46) Choline hydrogen citrate				
(47) Cyanocobalamin				
(48) N^5-Formyltetrahydro-pteroylglutamic acid, Ca salt				
(49) *p*-Hydroxybenzoic acid				
(50) Inositol				
(51) Lipoic acid				
(52) Nicotinic acid				
(53) Pantothenic acid, Ca salt				
(54) Pteroylglutamic acid				
(55) Putrescine dihydro-chloride				
(56) Pyridoxal hydrochlor-ide				
(57) Pyridoxamine dihydro-chloride				
(58) Riboflavine, Na, PO_4				
(59) Thiamine hydrochlor-ide				
(60) Hydrochloric acid				
(61) Sulfuric acid				
(62) Potassium hydroxide				
(63) Water	To 1,000 ml[fv]	To 1,000 ml[fv]	To 1,000 ml[fx]	To 1,000 ml
(64) pH adjusted with				
(65) pH before autoclaving				
(66) Final pH			8.0—8.5	
(67) Aeration				
(68) Tartrate				

Table 3 (continued)
DEFINED FRESHWATER MEDIA (MINERAL)

Reference	44 (1960)	46 (1961)	48 (1961)	48 (1961)
Author' designation		ASM	Medium 1	Medium 2
Use				
Constituents	*Tolypothrix tenuis*	*Microcystis aeruginosa*	*Nostoc muscorum*	*Nostoc muscorum*
(1) Carbonate				
(2) Bicarbonate				
(3) Nitrate		62.0[j]		
(4) Ammonium				
(5) Phosphate	137[e]	9.50[e]	82.0[e]	82.0[e]
(6) Sulfate	195[k]	19.2[k]	77.9[k]	77.9[k]
(7) Silicon				5.75[i]
(8) Na		24.0[c,bt]	Trace[ax]	15.6[i,bt]
(9) Mg	49.4[g]	9.73[g,h]	19.7[g]	19.7[g]
(10) K	112[e]	7.82[e]	67.4[e]	67.4[e]
(11) Ca	5.45[h]	4.01[h]	6.82[h]	6.82[h]
(12) Fe	4.02[ac]	0.112[ab]	0.413[ab]	1.61[ac]
(13) Li				
(14) B	0.500[n]	0.108[n]	0.105[n]	0.175[n]
(15) F				
(16) Al				
(17) Ti				
(18) V				
(19) Cr				
(20) Mn	0.503[z]	0.385[z]	0.111[z]	1.97[aa]
(21) Co		0.00118[aj]		0.248[aj]
(22) Ni				
(23) Cu	0.0201[ap]	0.000013[ao]	0.0102[ap]	0.373[ao]
(24) Zn	0.0505[g]	0.0523[h]	0.0091[g]	4.80[h]
(25) As				
(26) Se				
(27) Br				
(28) Rb				
(29) Sr				
(30) Mo	0.0100[av]		0.186[ax]	1.40[ax]
(31) Cd				
(32) Sn				
(33) I				
(34) Ba				

Table 3 (continued)
DEFINED FRESHWATER MEDIA (MINERAL)

Reference	44 (1960)	46 (1961)	48 (1961)	48 (1961)
Author's designation		ASM	Medium 1	Medium 2
		Use		
Constituents	*Tolypothrix tenuis*	*Microcystis aeruginosa*	*Nostoc muscorum*	*Nostoc muscorum*
(35) W				
(36) Hg				
(37) Pb				
(38) Bi				
(39) EDTA[fl]		5.85[bt]		39.3[bt]
(40) NTA[fm]				
(41) Citrate				
(42) TRIS[fj]				
(43) TRICINE[fk]				
(44) *p*-Aminobenzoic acid				
(45) Biotin				
(46) Choline hydrogen citrate				
(47) Cyanocobalamin				
(48) N^5-Formyltetrahydro-pteroylglutamic acid, Ca salt				
(49) *p*-Hydroxybenzoic acid				
(50) Inositol				
(51) Lipoic acid				
(52) Nicotinic acid				
(53) Pantothenic acid, Ca salt				
(54) Pteroylglutamic acid				
(55) Putrescine dihydrochloride				
(56) Pyridoxal hydrochloride				
(57) Pyridoxamine dihydrochloride				
(58) Riboflavine, Na, PO_4				
(59) Thiamine hydrochloride				
(60) Hydrochloric acid				
(61) Sulfuric acid				
(62) Potassium hydroxide				
(63) Water	To 1,000 ml	To 1,000 ml[fu,fx]	To 1,000 ml[fu]	To 1,000 ml[fu]
(64) pH adjusted with				
(65) pH before autoclaving		7.4–7.6	7.2	7.2
(66) Final pH		ca. 6.8		
(67) Aeration				
(68) Tartrate				

Table 3 (continued)
DEFINED FRESHWATER MEDIA (MINERAL)

Reference	53 (1963)	136 (1964)	133 (1968)	135 (1970)
Author's designation		ASM-1		
Use				
Constituents	*Aphanocapsa thermalis*	Planktonic cyanophytes	Cyanophytes (35 species, including *Gloeocapsa alpicola*)	Thermophilic cyanophytes at 45–70°C
(1) Carbonate	Present[l]			
(2) Bicarbonate			11.3[j]	
(3) Nitrate	1,040[j,l,m]	124[j]	1,090[j]	566[j,l]
(4) Ammonium				
(5) Phosphate	273[e]	19.0[e,q]	21.3[e]	74.2[q]
(6) Sulfate	108[k,af]	19.2[k]	29.2[k]	72.4[k,m]
(7) Silicon			5.73[i]	
(8) Na	143[c,bt]	51.5[c,e,bt]	424[a,c,i]	225[c,h,q]
(9) Mg	24.7[g]	9.73[g,h]	7.40[g]	9.87[g]
(10) K	611[c,e]	7.82[e]	17.5[e]	39.8[c]
(11) Ca	21.2[c]	8.02[h]	9.75[h]	14.0[g]
(12) Fe	4.19[af]	0.223[ab]	1.00[ag]	0.100[ab]
(13) Li				
(14) B	0.500[n]	0.433[n]	0.500[n]	
(15) F				0.0438[n]
(16) Al				
(17) Ti	0.01[fp]			
(18) V	0.01[v]			
(19) Cr	0.01[x]			
(20) Mn	0.503[z]	0.385[z]	0.503[z]	0.370[aa]
(21) Co	0.01[ak]	0.0047[aj]	0.01[ak]	0.0056[aj]
(22) Ni	0.01[am]			
(23) Cu	0.0201[ap]	0.000051[ao]	0.0201[ap]	0.0032[ap]
(24) Zn	0.0505[g]	0.209[h]	0.0505[g]	0.0569[g]
(25) As				
(26) Se				
(27) Br				
(28) Rb				
(29) Sr				
(30) Mo	0.01[av]		0.155[ax]	0.0050[ax]
(31) Cd				
(32) Sn				
(33) I				
(34) Ba				

Table 3 (continued)
DEFINED FRESHWATER MEDIA (MINERAL)

	53 (1963)	136 (1964)	133 (1968)	135 (1970)
Reference				
Author's designation		ASM-1		
			Use	
Constituents	*Aphanocapsa thermalis*	Planktonic cyanophytes	Cyanophytes (35 species, including *Gloeocapsa alpicola)*	Thermophilic cyanophytes at 45—70°C
(35) W	0.01[be]			
(36) Hg				
(37) Pb				
(38) Bi				
(39) EDTA[fl]	50.2[bt]	5.85[bt]	1.0	100
(40) NTA[fm]			5.91[fr] + 3.39[ag]	
(41) Citrate				
(42) TRIS[fj]				
(43) TRICINE[fk]				
(44) *p*-Aminobenzoic acid				
(45) Biotin				
(46) Choline hydrogen citrate				
(47) Cyanocobalamin				
(48) N^5-Formyltetrahydro-pteroylglutamic acid, Ca salt				
(49) *p*-Hydroxybenzoic acid				
(50) Inositol				
(51) Lipoic acid				
(52) Nicotinic acid				
(53) Pantothenic acid, Ca salt				
(54) Pteroylglutamic acid				
(55) Putrescine dihydro-chloride				
(56) Pyridoxal hydrochloride				
(57) Pyridoxamine dihydro-chloride				
(58) Riboflavine, Na, PO_4				
(59) Thiamine hydrochloride				
(60) Hydrochloric acid				0.00046[cy]
(61) Sulfuric acid				
(62) Potassium hydroxide				
(63) Water	To 1,000 ml	To 1,000 ml	To 1,000 ml	To 1,000 ml[fu]
(64) pH adjusted with	0.5M K_2CO_3			1M NaOH
(65) pH before autoclaving	8.0		7.8	8.2
(66) Final pH				7.5—7.6
(67) Aeration	2.5% CO_2			
(68) Tartrate				

Table 3 (continued)
DEFINED FRESHWATER MEDIA (MINERAL)

Reference	57 (1971)	159 (1972)	153 (1973)	35 (1957)
Author's designation	BG-11	MCY11	Modified Chu solution	
			Use	
Constituents	Unicellular cyanophytes	*Porphyridium aerugineum* (axenic)	*Woloszynskia apiculata*	*Monodus subterraneus*
(1) Carbonate	11.3[j]			
(2) Bicarbonate				
(3) Nitrate	1,090[j]	322[j]	24.4[j] 37.2[m]	
(4) Ammonium	Trace[ah]			1,240[j]
(5) Phosphate	21.8[e]	27 1[ch]	4.75[e]	109[e]
(6) Sulfate	29.2[k]	39.0[k]	9.59[k]	76.8[k]
(7) Silicon				
(8) Na	414[a,c]	133[c,ch]	9.51[b,h,bt]	460[c]
(9) Mg	7.47[g,bv]	9.87[g]	2.43[g]	19.5[g]
(10) K	18.0[e]	15.7[h]	3.91[e]	89.9[e]
(11) Ca	9.81[h]	9.98[h]	12.0[c]	25.2[h]
(12) Fe	ca. 0.6[ah]	0.493[ab]	0.069[ac]	1.12[ag]
(13) Li				
(14) B	0.500[n]	2.00[n]	0.0108[fa]	0.2[n]
(15) F				
(16) Al				
(17) Ti				
(18) V				
(19) Cr				
(20) Mn	0.503[z]	0.4[h]	0.0055[z]	0.2[z]
(21) Co	0.01[ak]	0.01[h]	0.000063[al]	0.02[aj]
(22) Ni				
(23) Cu	0.0201[ap]		0.000064[ap]	0.02[ap]
(24) Zn	0.0505[g]	0.05[h]	0.000066[g]	0.02[g]
(25) As				
(26) Se				
(27) Br				
(28) Rb				
(29) Sr				
(30) Mo	0.155[ax]		0.000095[ax]	0.2[av]
(31) Cd				
(32) Sn				
(33) I				
(34) Ba				

Table 3 (continued)
DEFINED FRESHWATER MEDIA (MINERAL)

Reference	57 (1971)	159 (1972)	153 (1973)	35 (1957)
Author's designation	BG-11	MCY11	Modified Chu solution	
			Use	
Constituents	Unicellular cyanophytes	*Porphyridium aerugineum* (axenic)	*Woloszynskia apiculata*	*Monodus subterraneus*
(35) W				
(36) Hg				
(37) Pb				
(38) Bi				
(39) EDTA[fl]	0.82[bv]	10.0	0.58[bt]	
(40) NTA[fm]				Present[fr] + 3.78[ag]
(41) Citrate	5.91[fr] + Present[ah]			
(42) TRIS[fj]				
(43) TRICINE[fk]		986		
(44) *p*-Aminobenzoic acid				
(45) Biotin				
(46) Choline hydrogen citrate				
(47) Cyanocobalamin		0.0035	0.0007	
(48) N^5-Formyltetrahydro-pteroylglutamic acid, Ca salt				
(49) *p*-Hydroxybenzoic acid				
(50) Inositol				
(51) Lipoic acid				
(52) Nicotinic acid				
(53) Pantothenic acid, Ca salt				
(54) Pteroylglutamic acid				
(55) Putrescine dihydro-chloride				
(56) Pyridoxal hydrochlor-ide				
(57) Pyridoxamine dihydro-chloride				
(58) Riboflavine, Na, PO$_4$				
(59) Thiamine hydrochlor-ide				
(60) Hydrochloric acid				
(61) Sulfuric acid				
(62) Potassium hydroxide				
(63) Water	To 1,000 ml	To 1,000 ml[fu]	To 1,000 ml	To 1,000 ml
(64) pH adjusted with		NaOH		
(65) pH before autoclaving		7.6		
(66) Final pH	7.1			
(67) Aeration				
(68) Tartrate				

Table 3 (continued)
DEFINED FRESHWATER MEDIA (MINERAL)

Reference	1 (1942)	1 (1942)	1 (1942)	19 (1943)
Author's designation	9	16	11	
	Use			
Constituents	*Asterionella* *Nitzschia* *Tabellaria*	*Fragilaria* *Nitzschia*	*Pediastrum* *Staurastrum*	*Chlorella* *vulgaris*
(1) Carbonate	6.00[m]	11.3[j]	21.7[l]	
(2) Bicarbonate				
(3) Nitrate	15.1—30.2[m]	37.8[m]	24.5[l]	1,550[l]
(4) Ammonium				
(5) Phosphate	1.09[e]	0.545[e]	0.545[e]	1,710[f]
(6) Sulfate	3.90[k]	15.6[k]	9.74[k]	1,920[k]
(7) Silicon	4.55[dy]	4.55[dy]	1.82[dy]	
(8) Na		8.68[a]		
(9) Mg	0.987[g]	3.95[g]	2.47[g]	487[g]
(10) K	13.6[e,dy]	13.1[e,dy]	49.3[a,c,e,dy]	1,680[c,f]
(11) Ca	8.88—13.8[a,c]	19.5[c,h]	0.183[h]	
(12) Fe	0.344[ab]	0.344[ab]	0.344[ab]	2.79[ac]
(13) Li				
(14) B				
(15) F				
(16) Al				
(17) Ti				
(18) V				
(19) Cr				
(20) Mn				
(21) Co				
(22) Ni				
(23) Cu				
(24) Zn				
(25) As				
(26) Se				
(27) Br				
(28) Rb				
(29) Sr				
(30) Mo				
(31) Cd				
(32) Sn				
(33) I				
(34) Ba				

Table 3 (continued)
DEFINED FRESHWATER MEDIA (MINERAL)

Reference	1 (1942)	1 (1942)	1 (1942)	19 (1943)
Author's designation	9	16	11	
	Use			
Constituents	*Asterionella Nitzschia Tabellaria*	*Fragilaria Nitzschia*	*Pediastrum Staurastrum*	*Chlorella vulgaris*
(35) W				
(36) Hg				
(37) Pb				
(38) Bi				
(39) EDTA[fl]				
(40) NTA[fm]				9.45[l]
(41) Citrate				
(42) TRIS[fj]				
(43) TRICINE[fk]				
(44) *p*-Aminobenzoic acid				
(45) Biotin				
(46) Choline hydrogen citrate				
(47) Cyanocobalamin				
(48) N^5-Formyltetrahydropteroylglutamic acid, Ca salt				
(49) *p*-Hydroxybenzoic acid				
(50) Inositol				
(51) Lipoic acid				
(52) Nicotinic acid				
(53) Pantothenic acid, Ca salt				
(54) Pteroylglutamic acid				
(55) Putrescine dihydrochloride				
(56) Pyridoxal hydrochloride				
(57) Pyridoxamine dihydrochloride				
(58) Riboflavine, Na, PO_4				
(59) Thiamine hydrochloride				
(60) Hydrochloric acid				
(61) Sulfuric acid				
(62) Potassium hydroxide				
(63) Water	To 1,000 ml[fv]	To 1,000 ml[fv]	To 1,000 ml[fv]	To 1,000 ml
(64) pH adjusted with				
(65) pH before autoclaving				
(66) Final pH				
(67) Aeration				5% CO_2
(68) Tartrate				

Table 3 (continued)
DEFINED FRESHWATER MEDIA (MINERAL)

Reference	20 (1948)	21 (1949)	21 (1949)	21 (1949)
Author's designation	V	A	B	C
		Use		
Constituents	Ankistrodesmus falcatus	Scenedesmus	Scenedesums	Scenedesmus
(1) Carbonate				1,360[j,l]
(2) Bicarbonate	12.0[m]			
(3) Nitrate		230[j,m]	234[j,m]	234[j,m]
(4) Ammonium	6.74[h]			
(5) Phosphate	2.73[e]	2.73[e]	6.11[e]	6.11[e]
(6) Sulfate	20.0[k]	20.0[k]	20.0[k]	20.0[k]
(7) Silicon	4.60[i]	4.60[i]	5.06[i]	
(8) Na	7.53[i]	88.7[c,i]	89.4[c,i]	604[a,c]
(9) Mg	5.05[g]	5.05[g]	5.05[g]	5.05[g]
(10) K	2.25[e]	2.25[e]	5.03[e]	893[a,e]
(11) Ca	3.96[b]	3.66[c]	4.88[c]	4.88[c]
(12) Fe	0.167[ag,dc]	0.167[ag,dc]	0.450[ag,dc]	0.450[ag]
(13) Li				
(14) B				
(15) F				
(16) Al				
(17) Ti				
(18) V				
(19) Cr				
(20) Mn	0.011[aa]	0.011[aa]	0.011[aa]	0.011[aa]
(21) Co				
(22) Ni				
(23) Cu				
(24) Zn				
(25) As				
(26) Se				
(27) Br				
(28) Rb				
(29) Sr				
(30) Mo				
(31) Cd				
(32) Sn				
(33) I				
(34) Ba				

Table 3 (continued)
DEFINED FRESHWATER MEDIA (MINERAL)

Reference	20 (1948)	21 (1949)	21 (1949)	21 (1949)
Author's designation	V	A	B	C
		Use		
Constituents	*Ankistrodesmus falcatus*	*Scenedesmus*	*Scenedesmus*	*Scenedesmus*
(35) W				
(36) Hg				
(37) Pb				
(38) Bi				
(39) EDTA[fl]				
(40) NTA[fm]				
(41) Citrate	0.984[dc,fr] + 0.564[ag,dc]	0.984[dc,fr] + 0.564[ag,dc]	2.66[dc,fr] + 1.52[ag,dc]	2.66[fr] + 1.52[ag]
(42) TRIS[fj]				
(43) TRICINE[fk]				
(44) *p*-Aminobenzoic acid				
(45) Biotin				
(46) Choline hydrogen citrate				
(47) Cyanocobalamin				
(48) N⁵-Formyltetrahydro-pteroylglutamic acid, Ca salt				
(49) *p*-Hydroxybenzoic acid				
(50) Inositol				
(51) Lipoic acid				
(52) Nicotinic acid				
(53) Pantothenic acid, Ca salt				
(54) Pteroylglutamic acid				
(55) Putrescine dihydrochloride				
(56) Pyridoxal hydrochloride				
(57) Pyridoxamine dihydrochloride				
(58) Riboflavine, Na, PO₄				
(59) Thiamine hydrochloride				
(60) Hydrochloric acid				1,000
(61) Sulfuric acid				
(62) Potassium hydroxide				
(63) Water	To 1,000 ml	To 1,000 ml[fv]	To 1,000 ml[fv]	To 1,000 ml[fv]
(64) pH adjusted with				
(65) pH before autoclaving				
(66) Final pH				
(67) Aeration				
(68) Tartrate				

Table 3 (continued)
DEFINED FRESHWATER MEDIA (MINERAL)

Reference	26 (1951)	5 (1952)	9 (1953)	30 (1953)
Author's designation	Medium A	BPM	JM	
Use				
Constituents	*Chlorella vulgaris*	*Chlamydomonas moewusii*	*Chalamydomonas moewusii*	*Ankistrodesmus braunii*
(1) Carbonate	0.60[m]		10.0[l]	
(2) Bicarbonate				
(3) Nitrate		387[d]	45.6[l,m]	497[l]
(4) Ammonium	73.5[bm]	113[c]		
(5) Phosphate	387[bm] + 2.81[fy]	328[e]	9.27[e]	96.1[q] + 286[r]
(6) Sulfate	1,200[l,af]	77.9[k]	6.62[k]	97.4[k]
(7) Silicon				
(8) Na		Trace[ax]	Trace[ax]	300[h,q,r]
(9) Mg	99.1[bm]	19.7[g]	1.68[g]	24.7[g]
(10) K	974[g]	269[e]	27.2[a,c,e]	313[c]
(11) Ca	0.40[a]	27.3[h]	11.4[c]	3.66[h]
(12) Fe	2.79[af]	0.367[ai]	0.367[ai]	2.01[ac]
(13) Li				
(14) B		0.1[n]	0.1[n]	
(15) F				
(16) Al				
(17) Ti				
(18) V				
(19) Cr				
(20) Mn	0.5[z]	0.137[z]	0.137[z]	0.056[z]
(21) Co				
(22) Ni				
(23) Cu		0.1[ao]	0.1[ao]	
(24) Zn		0.3[h]	0.3[h]	0.227[g]
(25) As				
(26) Se				
(27) Br				
(28) Rb				
(29) Sr				
(30) Mo		0.1[ax]	0.1[ax]	
(31) Cd				
(32) Sn				
(33) I				
(34) Ba				

Table 3 (continued)
DEFINED FRESHWATER MEDIA (MINERAL)

Reference	26 (1951)	5 (1952)	9 (1953)	30 (1953)
Author's designation	Medium A	BPM	JM	
			Use	
Constituents	*Chlorella vulgaris*	*Chlamydomonas moewusii*	*Chlamydomonas moewusii*	*Ankistrodesmus braunii*
(35) W				
(36) Hg				
(37) Pb				
(38) Bi				
(39) EDTA[fl]				
(40) NTA[fm]				
(41) Citrate				
(42) TRIS[fj]				
(43) TRICINE[fk]				
(44) *p*-Aminobenzoic acid				
(45) Biotin				
(46) Choline hydrogen citrate				
(47) Cyanocobalamin				
(48) N^5-Formyltetrahydro-pteroylglutamic acid, Ca salt				
(49) *p*-Hydroxybenzoic acid				
(50) Inositol				
(51) Lipoic acid				
(52) Nicotinic acid				
(53) Pantothenic acid, Ca salt				
(54) Pteroylglutamic acid				
(55) Putrescine dihydro-chloride				
(56) Pyridoxal hydrochlor-ide				
(57) Pyridoxamine dihydro-chloride				
(58) Riboflavine, Na, PO_4				
(59) Thiamine hydrochloride				
(60) Hydrochloric acid				
(61) Sulfuric acid				
(62) Potassium hydroxide				
(63) Water	To 1,000 ml	To 1,000 ml	To 1,000 ml	To 1,000 ml[fu]
(64) pH adjusted with				
(65) pH before autoclaving				6.0
(66) Final pH				
(67) Aeration				
(68) Tartrate		1.46	1.46	

Table 3 (continued)
DEFINED FRESHWATER MEDIA (MINERAL)

Reference	31 (1953)	38 (1958)	41 (1959)	40 (1960)
Author's designation	"Nitrate-A"	J		
			Use	
Constituents	Chlorella ellipsoidea	Haematococcus	Polytomella caeca (base)	Chlorella vulgaris C. pyrenoidosa
(1) Carbonate				
(2) Bicarbonate				
(3) Nitrate	3,070[l]	77.5[d]		767[l]
(4) Ammonium		22.5[c]	67.4[h]	
(5) Phosphate	872[f]	41.7[e]	349[f]	872[f]
(6) Sulfate	974[k]	39.0[k]	39.0[k]	99.8[k]
(7) Silicon				
(8) Na				
(9) Mg	247[g]	9.87[g]	9.87[g]	25.3[g]
(10) K	2,290[c,f]	34.3[e]	144[f]	843[c,f]
(11) Ca		2.73[h]		
(12) Fe	0.603[ac]	0.834[ag]	0.413[ab]	1.08[ac]
(13) Li				
(14) B	0.5[n]	0.109[n]		
(15) F				0.5
(16) Al		0.0054[p]		
(17) Ti	0.01[fp]			
(18) V	0.01[v]	0.0010[t]		
(19) Cr	0.01[x]			
(20) Mn	0.5[z]	0.110[aa]		0.25
(21) Co	0.01[ak]	0.0059[ak]		
(22) Ni	0.01[am]	0.0059[an]		
(23) Cu	0.02[ap]			
(24) Zn	0.05[g]	0.0131[g]		0.02
(25) As				0.05
(26) Se				
(27) Br		0.0160[as]		
(28) Rb				
(29) Sr				
(30) Mo	0.01[av]	0.0096[ba]		0.05
(31) Cd		0.0112[c]		
(32) Sn				
(33) I		0.0127[bc]		
(34) Ba				

Table 3 (continued)
DEFINED FRESHWATER MEDIA (MINERAL)

Reference	31 (1953)	38 (1958)	41 (1959)	40 (1960)
Author's designation	"Nitrate-A"	J		
			Use	
Constituents	*Chlorella ellipsoidea*	*Haematococcus*	*Polytomella caeca* (base)	*Chlorella vulgaris C. pyrenoidosa*
(35) W	0.01[be]	0.0037[be]		
(36) Hg				
(37) Pb				
(38) Bi				
(39) EDTA[fl]				
(40) NTA[fm]		4.92[fr] + 2.82[ag]		
(41) Citrate				
(42) TRIS[fj]				
(43) TRICINE[fk]				
(44) *p*-Aminobenzoic acid				
(45) Biotin				
(46) Choline hydrogen citrate				
(47) Cyanocobalamin				
(48) N⁵-Formyltetrahydro-pteroylglutamic acid, Ca salt				
(49) *p*-Hydroxybenzoic acid				
(50) Inositol				
(51) Lipoic acid				
(52) Nicotinic acid				
(53) Pantothenic acid, Ca salt				
(54) Pteroylglutamic acid				
(55) Putrescine dihydro-chloride				
(56) Pyridoxal hydrochlor-ide				
(57) Pyridoxamine dihydro-chloride				
(58) Riboflavine, Na, PO₄				
(59) Thiamine hydrochlor-ide			1.0	
(60) Hydrochloric acid				
(61) Sulfuric acid				
(62) Potassium hydroxide				
(63) Water	To 1,000 ml	To 1,000 ml[fx]	To 1,000 ml[fv]	To 1,000 ml[cz]
(64) pH adjusted with			1N KOH, 1N HCl	
(65) pH before autoclaving				
(66) Final pH	5.3–5.4			
(67) Aeration	5% CO_2			5% CO_2 + 95% N_2
(68) Tartrate				

Table 3 (continued)
DEFINED FRESHWATER MEDIA (MINERAL)

Reference	40 (1960)	42 (1960)	42 (1960)	45 (1960)
Author's designation		N2	N2b	
			Use	
Constituents	*Scenedesmus obliquus*	*Chlorella pyrenoidosa*	*Chlorella pyrenoidosa*	*Chlorella vulgaris*
(1) Carbonate				
(2) Bicarbonate	122[l]			
(3) Nitrate	541[d,m]	123[l]	123[l]	155[d]
(4) Ammonium	90.2[c]			45.1[c]
(5) Phosphate	134[q]	14.0[f]	1.38[q] + 10.3[f]	1,270[e] + 5,080[f]
(6) Sulfate	383[k]	3.90[k]	3.90[k]	480[k]
(7) Silicon				
(8) Na	72.6[h,q]		0.668[q]	0.230[bs]
(9) Mg	97.0[g]	0.987[g]	0.987[g]	122[g]
(10) K	78.1[b]	83.1[c,f]	81.6[c,f]	3,140[e,f]
(11) Ca	74.7[c]	0.092[h]	0.092[h]	
(12) Fe	5.63[ac]	0.010[ab]	0.010[ab]	0.559[bs]
(13) Li				
(14) B	1.0			
(15) F				
(16) Al		Trace	Trace	
(17) Ti				
(18) V				
(19) Cr				
(20) Mn	0.5			0.440
(21) Co				
(22) Ni				
(23) Cu	0.04			0.003
(24) Zn	0.1			0.070
(25) As				
(26) Se				
(27) Br		Trace	Trace	
(28) Rb				
(29) Sr				
(30) Mo	0.1			0.020
(31) Cd				
(32) Sn				
(33) I				
(34) Ba				

Table 3 (continued)
DEFINED FRESHWATER MEDIA (MINERAL)

Reference	40 (1960)	52 (1960)	42 (1960)	45 (1960)
Author's designation		N2	N2b	
		Use		
Constituents	*Scenedesmus obliquus*	*Chlorella pyrenoidosa*	*Chlorella pyrenoidosa*	*Chlorella vulgaris*
(35) W				
(36) Hg		Trace	Trace	
(37) Pb				
(38) Bi				
(39) EDTA[fl]				2.92[bs]
(40) NTA[fm]				
(41) Citrate				
(42) TRIS[fj]				
(43) TRICINE[fk]				
(44) *p*-Aminobenzoic acid				
(45) Biotin				
(46) Choline hydrogen citrate				
(47) Cyanocobalamin				
(48) N[5]-Formyltetrahydro-pteroylglutamic acid, Ca salt				
(49) *p*-Hydroxybenzoic acid				
(50) Inositol				
(51) Lipoic acid				
(52) Nicotinic acid				
(53) Pantothenic acid, Ca salt				
(54) Pteroylglutamic acid				
(55) Putrescine dihydro-chloride				
(56) Pyridoxal hydrochlor-ide				
(57) Pyridoxamine dihydro-chloride				
(58) Riboflavine, Na, PO_4				
(59) Thiamine hydrochlor-ide				
(60) Hydrochloric acid				
(61) Sulfuric acid				
(62) Potassium hydroxide				
(63) Water	To 1,000 mℓ[cz]	To 1,000 mℓ	To 1,000 mℓ	To 1,000 mℓ
(64) pH adjusted with				
(65) pH before autoclaving				6.15
(66) Final pH				
(67) Aeration	5% CO_2 + 95% N_2			5% CO_2
(68) Tartrate				

Table 3 (continued)
DEFINED FRESHWATER MEDIA (MINERAL)

Reference	47 (1961)	49 (1961)	12 (1962)	51 (1963)
Author's designation			BGM	
			Use	
Constituents	*Chlorella pyrenoidosa*	*Chlorella ellipsoidea*	*Chlorella*	*Chlamydomonas mundana*
(1) Carbonate				
(2) Bicarbonate				
(3) Nitrate	767[l]	190[l]	1,240[l]	
(4) Ammonium				
(5) Phosphate	872[f]	216[f]	190[e]	169[h]
(6) Sulfate	195[k]	246[k]	96.1[k]	218[e] + 419[f]
(7) Silicon				97.4[k]
(8) Na	1.65[bs]		92.0[h]	42.4[h,bt]
(9) Mg	49.4[g]	62.2[g]	24.3[g]	24.7[g]
(10) K	842[c,f]	209[c,f]	938[c,e]	352[e,f]
(11) Ca	0.502[h]		20.0[h]	1.36[h]
(12) Fe	4.00[bs]	0.603[ac]	4.0[bs] *or* bu	2.01[ac]
(13) Li				
(14) B	0.500[n]	0.500[n]	0.5[n]	
(15) F				
(16) Al				
(17) Ti				
(18) V	0.0100[v]		0.01[v]	
(19) Cr				
(20) Mn	0.503[z]	0.503[z]	0.5[aa]	0.813[aa]
(21) Co	0.01[ak]			
(22) Ni				
(23) Cu	0.0201[ap]	0.0201[ap]	0.02[ap]	
(24) Zn	0.0505[g]	0.0505[g]	0.05[g]	
(25) As				
(26) Se				
(27) Br				
(28) Rb				
(29) Sr				
(30) Mo	0.0956[ba]	0.01[av]	0.01[av]	
(31) Cd				
(32) Sn				
(33) I				
(34) Ba				

Table 3 (continued)
DEFINED FRESHWATER MEDIA (MINERAL)

Reference	47 (1961)	49 (1961)	12 (1962)	51 (1963)
Author's designation			BGM	
		Use		
Constituents	*Chlorella pyrenoidosa*	*Chlorella ellipsoidea*	*Chlorella*	*Chlamydomonas mundana*
(35) W				
(36) Hg				
(37) Pb				
(38) Bi				
(39) EDTA[fl]	21.0[bs]		21.0[bs]	19.6[bt]
(40) NTA[fm]				
(41) Citrate				
(42) TRIS[fj]				
(43) TRICINE[fk]				
(44) *p*-Aminobenzoic acid				
(45) Biotin				
(46) Choline hydrogen citrate				0.0025
(47) Cyanocobalamin				
(48) N^5-Formyltetrahydro-pteroylglutamic acid, Ca salt				
(49) *p*-Hydroxybenzoic acid				
(50) Inositol				
(51) Lipoic acid				
(52) Nicotinic acid				
(53) Pantothenic acid, Ca salt				
(54) Pteroylglutamic acid				
(55) Putrescine dihydrochloride				
(56) Pyridoxal hydrochloride				
(57) Pyridoxamine dihydrochloride				
(58) Riboflavine, Na, PO_4				
(59) Thiamine hydrochloride				
(60) Hydrochloric acid				
(61) Sulfuric acid				
(62) Potassium hydroxide				
(63) Water	To 1,000 ml	To 1,000 ml	To 1,000 ml	To 1,000 ml
(64) pH adjusted with	KOH[db]			
(65) pH before autoclaving				
(66) Final pH	6.8[db]			
(67) Aeration				1–2% CO_2
(68) Tartrate				

Table 3 (continued)
DEFINED FRESHWATER MEDIA (MINERAL)

Reference	52 (1963)	55 (1963)	160 (1964)	165 (1968)
Author's designation			TB1M	Medium M
Use				
Constituents	*Chlorella pyrenoidosa C. vulgaris* (base for study of P)	*Chlorella vulgaris*	Chlorophytes	*Golenkinia*
(1) Carbonate				
(2) Bicarbonate				
(3) Nitrate	613[l]	581[d]	124[l]	620[l]
(4) Ammonium		169[c]		
(5) Phosphate		127[e] + 542[f]	95.0[q]	237[f]
(6) Sulfate	97.4[k]	160[k]	31.7[k,cg]	48.0[k]
(7) Silicon				
(8) Na			46.0[q]	
(9) Mg	24.7[g]	40.4[g]	7.30[g]	12.2[g]
(10) K	387[c]	327[e,f]	78.2[c]	489[c,f]
(11) Ca	1.0[bq]		4.01[h]	20.0[h]
(12) Fe	5.0[bq]	10[bq]	0.994[ac]	0.559[ac]
(13) Li				
(14) B		0.500[n]	2.00[n]	4.33[fa]
(15) F				
(16) Al				
(17) Ti				
(18) V				
(19) Cr				
(20) Mn	1.0[bq]	0.503[z]	0.401[z]	0.110[z]
(21) Co	1.0[bq]		0.100[ak]	0.118[aj]
(22) Ni				
(23) Cu	1.0[bq]	0.0201[ap]	0.400[ap]	0.127[ap]
(24) Zn	1.0[bq]	0.0505[g]	1.96[g]	0.981[g]
(25) As				
(26) Se				
(27) Br				
(28) Rb				
(29) Sr				
(30) Mo			0.470[av]	1.34[ba]
(31) Cd				
(32) Sn				
(33) I				
(34) Ba				

Table 3 (continued)
DEFINED FRESHWATER MEDIA (MINERAL)

Reference	52 (1963)	55 (1963)	160 (1964)	165 (1968)
Author's designation			TB1M	Medium M
			Use	
Constituents	*Chlorella pyrenoidosa C. vulgaris* (base for study of P)	*Chlorella vulgaris*	Chlorophytes	*Golenkinia*
(35) W				
(36) Hg				
(37) Pb				
(38) Bi				
(39) EDTA[fl]	Present[bq]	Present[bq]	49.7	
(40) NTA[fm]				6.31
(41) Citrate				
(42) TRIS[fj]			606	
(43) TRICINE[fk]				
(44) *p*-Aminobenzoic acid				0.0343
(45) Biotin				0.0017
(46) Choline hydrogen citrate				5.14
(47) Cyanocobalamin				0.000343
(48) N^5-Formyltetrahydro-pteroylglutamic acid, Ca salt				0.000685
(49) *p*-Hydroxybenzoic acid				0.0343
(50) Inositol				3.43
(51) Lipoic acid				0.0137
(52) Nicotinic acid				0.343
(53) Pantothenic acid, Ca salt				0.343
(54) Pteroylglutamic acid				0.0137
(55) Putrescine dihydro-chloride				0.137
(56) Pyridoxal hydrochlor-ide				0.0685
(57) Pyridoxamine dihydro-chloride				0.0685
(58) Riboflavine, Na, PO_4				0.0343
(59) Thiamine hydrochlor-ide				0.343
(60) Hydrochloric acid				
(61) Sulfuric acid			3.68[cy]	
(62) Potassium hydroxide			26.4	
(63) Water	To 1,000 ml	To 1,000 ml[fv]	To 1,000 ml[fx] *or* [fv]	To 1,000 ml
(64) pH adjusted with	HCl or NaOH		1N HCl	KOH
(65) pH before autoclaving	7.0		8.8-7.0	6.5
(66) Final pH		6.1		
(67) Aeration	1% CO_2			
(68) Tartrate				

Table 3 (continued)
DEFINED FRESHWATER MEDIA (MINERAL)

Reference	56 (1965)	149 (1965)	39 (1959)	157 (1973)
Author's designation				Chloromonad medium
Use				
Constituents	*Chara globularis C. zeylanica*	*Chara Nitella* (base)	*Cyanidium caldarium*	*Vacuolaria virescens* (unialgal)
(1) Carbonate	22.6[j]			4.80[m]
(2) Bicarbonate				
(3) Nitrate	60.5[m]	0.341[l,m]		55.2[l]
(4) Ammonium			361[g]	6.74[h]
(5) Phosphate	0.307−3.07[e]	19.0[f]	190[f]	15.8[e]
(6) Sulfate	39.0[k]	0.096[k]	1,060[d,k]	45.3[k,aa,ac,cg]
(7) Silicon	5.75[i]			
(8) Na	26.8[a,i]			Trace[ax]
(9) Mg	9.87[g]	0.024[g]	24.3[g]	8.78[g]
(10) K	0.252−2.52[e]	7.84[c,f]	78.2[f]	47.8[c,e]
(11) Ca	19.5[c]	0.200[c,h]	20.0[h]	3.20[a]
(12) Fe	0.2[h]		4.0	0.804[ac]
(13) Li				
(14) B	0.2[n]			
(15) F			0.5	0.175[n]
(16) Al				
(17) Ti				
(18) V			0.01	
(19) Cr				
(20) Mn	0.001[h]		0.5	1.48[aa]
(21) Co				0.419[al]
(22) Ni				
(23) Cu	0.002[h]		0.02	0.0255[ap]
(24) Zn	0.05[h]		0.05	4.55[g]
(25) As				
(26) Se				
(27) Br				
(28) Rb				
(29) Sr				
(30) Mo	0.05[ax]		0.01	0.793[ax]
(31) Cd				
(32) Sn				
(33) I				
(34) Ba				
(35) W				
(36) Hg				
(37) Pb				
(38) Bi				
(39) EDTA[fl]				
(40) NTA[fm]				50
(41) Citrate	10			
(42) TRIS[fj]				
(43) TRICINE[fk]	500			

Table 3 (continued)
DEFINED FRESHWATER MEDIA (MINERAL)

Reference	56 (1965)	149 (1965)	39 (1959)	157 (1973)
Author's designation		Use		Chloromonad medium
Constituents	*Chara globularis* C. *zeylanica*	*Chara Nitella* (base)	*Cyanidium caldarium*	*Vacuolaria virescens* (unialgal)
(44) p-Aminobenzoic acid				
(45) Biotin				1.0
(46) Choline hydrogen citrate				
(47) Cyanocobalamin				0.01
(48) N^5-Formyltetrahydro-pteroylglutamic acid, Ca salt				
(49) p-Hydroxybenzoic acid				
(50) Inositol				
(51) Lipoic acid				
(52) Nicotinic acid				
(53) Pantothenic acid, Ca salt				
(54) Pteroylglutamic acid				
(55) Putrescine dihydro-chloride				
(56) Pyridoxal hydrochloride				
(57) Pyridoxamine dihydro-chloride				
(58) Riboflavine, Na, PO_4				
(59) Thiamine hydrochloride				1.0
(60) Hydrochloric acid				
(61) Sulfuric acid			98.1	
(62) Potassium hydroxide				
(63) Water	To 1,000 ml[fx]	To 1,000 ml[fx]	To 1,000 ml	To 1,000 ml[fu]
(64) pH adjusted with	HCl			
(65) pH before autoclaving	7.0			6.3–6.5
(66) Final pH			2.0	
(67) Aeration			5% CO_2	4% CO_2
(68) Tartrate				

Compiled by M. F. Turner and M. R. Droop.

Table 4
DEFINED FRESHWATER MEDIA (ORGANIC)

Reference	27 (1952)	2 (1953)	34 (1955)	34 (1955)
Author's designation	No. 4		B	C
		Use		
Constituents	Cyanophytes	*Cyanophora paradoxa* (vitamin requirements)	*Anabaena cylindrica A. variabilis Nostoc muscorum Anacystis*	Cyanophytes
(1) Carbonate		60.0[m]	396[j,dc]	
(2) Bicarbonate				
(3) Nitrate	730[j]	23.2[d]	626[l,m]	626[l,m]
(4) Ammonium	16.9[h]	6.76[c]		
(5) Phosphate	137[e]	69.8[f]	698[f]	545[e]
(6) Sulfate	200[k]	19.5[k]	97.4[k]	97.4[k]
(7) Silicon				
(8) Na	512[c,gv]	37.3[cn,gw]	342[a,bn]	38.7[bn]
(9) Mg	50.5[g]	4.94[g]	24.7[g]	24.7[g]
(10) K	112[e]	28.7[f]	674[c,f]	836[c,e]
(11) Ca	18.1[h]	40.0[a]	4.24[c]	4.24[c]
(12) Fe	0.2[ad] + trace[ac]	3.00[bk]	0.880[af]	0.880[af]
(13) B	0.02[n]		0.5[n]	0.5[n]
(14) V				
(15) Mn	0.02[aa]	0.15[h]	0.5[z]	0.5[z]
(16) Co		0.004[h]		
(17) Cu	0.1[ap]	0.001[h]	0.02[ap]	0.02[ap]
(18) Zn	0.15[g]	0.1[h]	0.05[g]	0.05[g]
(19) Mo	0.02[az]	0.1[ax]	0.01[av]	0.01[av]
(20) I				
(21) DTPA[gt]				
(22) EDTA[fl]				
(23) HEDTA[ft]				
(24) NTA[fm]				
(25) Citrate		270[fr]	106[j]	106[j]
(26) TRIS[fj]				
(27) BICINE[gu]				
(28) Glycylglycine				
(29) Phthalate				
(30) Pyromellitic acid				
(31) Acetate		43.4[j]		
(32) *n*-Butanol				
(33) Ethanol				
(34) Glucose				
(35) Glyceryl				
(36) Lactate				
(37) Malic acid				
(38) Succinate				
(39) Sucrose				

Table 4 (continued)
DEFINED FRESHWATER MEDIA (ORGANIC)

Reference	27 (1952)	2 (1953)	34 (1955)	34 (1955)
Author's resignation	No. 4		B	C
		Use		
Constituents	Cyanophytes	Cyanophora paradoxa (vitamin requirements)	Anabaena cylindrica A. variabilis Nostoc muscorum Anacystis	Cyanophytes
(40) DL-Alanine				
(41) L-Arginine				
(42) Asparagine				
(43) DL-Aspartic acid				
(44) L-Cysteine hydrochloride				
(45) L-Cystine				
(46) Glutamate	759[gv]	129[gw]		
(47) Glutamine				
(48) Glycine				
(49) Histidine				
(50) L-Isoleucine				
(51) L-Leucine				
(52) L-Lysine hydrochloride				
(53) Methionine				
(54) Phenylalanine				
(55) L-Tryptophan				
(56) L-Valine				
(57) Adenine sulfate				
(58) Urea				
(59) Adenosine				
(60) *p*-Aminobenzoic acid				
(61) Biotin				
(62) Cyanocobalamin				
(63) N^5-Formyltetrahydro-pteroylglutamic acid				
(64) Nicotinamide				
(65) Pantothenic acid, Ca salt				
(66) Pteroylglutamic acid				
(67) Pyridoxine hydrochloride				
(68) Riboflavine				
(69) Thiamine hydrochloride				
(70) Thymidine				
(71) Thymine				
(72) HCl (0.1N)				
(73) Fresh water	To 1,000 ml	To 1,000 ml	To 1,000 ml[fv]	To 1,000 ml[fv]
(74) pH adjusted with				
(75) pH before autoclaving		6.3		
(76) Final pH				
(77) Aeration				

Table 4 (continued)
DEFINED FRESHWATER MEDIA (ORGANIC)

	54 (1964)	134 (1971)	76 (1957)	2 (1953)
Reference				
Author's designation				
		Use		
Constituents	*Anabaena variabilis Anacystis nidulans*	*Plectonema boryanum*	*Chilomonas paramecium*	*Peridinium* sp. (vitamin requirements)
(1) Carbonate		11.3[j]		60.0[m]
(2) Bicarbonate				
(3) Nitrate	316[l,m]	244[j,m]		30.7[l]
(4) Ammonium			64.1[h]	
(5) Phosphate	273[e]	5.45[e]	405[f]	14.0[f]
(6) Sulfate	48.7[k]	9.74[k]	3.90[k]	39.0[k]
(7) Silicon		4.35[i]		
(8) Na	23.5[bn]	86.1[a,c,i]		136[gw]
(9) Mg	12.3[g]	2.47[g]		9.87[g]
(10) K	418[c,e]	4.49[e]	0.987[g]	25.1[c,f]
(11) Ca	3.05[c]	17.5[c]	167[f]	80.0[a,h]
(12) Fe	0.041[ab]	58.3[ag]	1.0[h]	2.00[bk]
			1.0[h]	
(13) B	0.25[n]	42.0[n]		
(14) V				
(15) Mn	0.251[z]	3.89[z]		0.8[h]
(16) Co		0.5[aj]		0.01[h]
(17) Cu	0.01[ap]	3.3730[ao]		0.001[h]
(18) Zn	0.0252[g]	19.20[h]		0.8[h]
(19) Mo	0.005[av]			0.5[ax]
(20) I				
(21) DTPA[gt]				
(22) EDTA[fl]				300
(23) HEDTA[ft]				
(24) NTA[fm]				
(25) Citrate	64.3[j]	345[fr] + 198[ag]		
(26) TRIS[fj]				
(27) BICINE[gu]				
(28) Glycylglycine				
(29) Phthalate				
(30) Pyromellitic acid				
(31) Acetate				
(32) *n*-Butanol				
(33) Ethanol				
(34) Glucose				500
(35) Glyceryl				
(36) Lactate			2,500[gq]	
(37) Malic acid				
(38) Succinate				
(39) Sucrose				

Table 4 (continued)
DEFINED FRESHWATER MEDIA (ORGANIC)

Reference	54 (1964)	134 (1971)	76 (1957)	2 (1953)
Author's designation				
			Use	
Constituents	*Anabaena variabilis* *Anacystis nidulans*	*Plectonema boryanum*	*Chilomonas paramecium*	*Peridinium* sp. (vitamin requirements)
(40) DL-Alanine				700
(41) L-Arginine				
(42) Asparagine				
(43) DL-Aspartic acid				
(44) L-Cysteine hydrochloride				
(45) L-Cystine				858[gw]
(46) Glutamate				
(47) Glutamine				700
(48) Glycine				
(49) Histidine				
(50) L-Isoleucine				
(51) L-Leucine				
(52) L-Lysine hydrochloride				
(53) Methionine				
(54) Phenylalanine				
(55) L-Tryptophan				
(56) L-Valine				
(57) Adenine sulfate				
(58) Urea				
(59) Adenosine				
(60) *p*-Aminobenzoic acid				
(61) Biotin				
(62) Cyanocobalamin				
(63) N⁵-Formyltetrahydro-pteroylglutamic acid				
(64) Nicotinamide				
(65) Pantothenic acid, Ca salt				
(66) Pteroylglutamic acid				
(67) Pyridoxine hydrochloride				
(68) Riboflavine				
(69) Thiamine hydrochloride			2.0	
(70) Thymidine				
(71) Thymine				
(72) HCl (0.1N)				
(73) Fresh water	To 1,000 ml[fv]	To 1,000 ml[fu]	To 1,000 ml	To 1,000 ml
(74) pH adjusted with	NaOH		KOH	
(75) pH before autoclaving	8.0			6.2
(76) Final pH			4.2	
(77) Aeration		5% CO_2		

Table 4 (continued)
DEFINED FRESHWATER MEDIA (ORGANIC)

	2 (1953)	81 (1953)	81 (1953)	89 (1957)
Reference				
Author's designation		"EDTA base"		
Use				
Constituents	Synura sp. (vitamin requirements)	Ochromonas Poteriochromonas (study of substrates)	Ochromonas Poteriochromonas	Ochromonas malhamensis above 35°C
(1) Carbonate		89.9[m]	336[m,em]	603[m,gl]
(2) Bicarbonate		309[d]	309[d]	309[d]
(3) Nitrate	14.6[j]			
(4) Ammonium		98.8[b,az]	291[b,az,ej]	91.3[b]
(5) Phosphate	9.77[f]	210[f]	210[f]	174[f]
(6) Sulfate	7.79[k]	376[k,aa,ad,ap,cg]	137[aa,ad,ap,cg]	234[k]
(7) Silicon				
(8) Na	25.8[c,cn,gw]			
(9) Mg	1.97[g]	59.2[g]	133[em]	319[g,gl]
(10) K	9.27[f,h]	86.2[f]	86.2[f]	71.8[f]
(11) Ca	13.0[h]	60.1[a]	60.1[a]	60.1[a] + 10
(12) Fe	0.7[bk]	1.2[ad]	2.0[ad]	1.2[ad]
(13) B		1.2		0.01
(14) V				0.001
(15) Mn	2.0[h]	36[aa]	40[aa]	0.205
(16) Co	0.03[h]	0.48[al]	0.5[al]	0.01
(17) Cu	0.005[h]	3.0[ap]	1.0[ap]	0.008
(18) Zn	10.0[h]	48[g]	40[g]	1.1
(19) Mo	0.01[ax]	20[az]	20[az]	0.005
(20) I				
(21) DTPA[gt]				
(22) EDTA[fl]	50	530		
(23) HEDTA[ft]				200
(24) NTA[fm]				
(25) Citrate			1,000[ej] + 27.0[fr]	
(26) TRIS[fj]				
(27) BICINE[gu]				200
(28) Glycylglycine				
(29) Phthalate				
(30) Pyromellitic acid				1,000
(31) Acetate	17.4[j]			
(32) n-Butanol				
(33) Ethanol				
(34) Glucose				
(35) Glyceryl				
(36) Lactate				
(37) Malic acid				
(38) Succinate				
(39) Sucrose			12,000	10,000

Table 4 (continued)
DEFINED FRESHWATER MEDIA (ORGANIC)

	2 (1953)	81 (1953)	81 (1953)	89 (1957)
Reference				
Author's designation		"EDTA base"		
Use				
Constituents	*Synura* sp. (vitamin requirements)	*Ochromonas Poteriochromonas* (study of substrates)	*Ochromonas Poteriochromonas*	*Ochromonas malhamensis* above 35°C
(40) DL-Alanine				
(41) L-Arginine			331[bx]	413[bx]
(42) Asparagine				
(43) DL-Aspartic acid				
(44) L-Cysteine hydrochloride				
(45) L-Cystine				20
(46) Glutamate	85.8[gw]		3,000[cr,gx]	2,000[cr,gx]
(47) Glutamine				
(48) Glycine				
(49) Histidine			370[bx,cr,eq]	740[bx,cr,eq]
(50) L-Isoleucine				40
(51) L-Leucine				
(52) L-Lysine hydrochloride				400
(53) Methionine		400[cq]	600[cq]	500[cq]
(54) Phenylalanine				80[cq]
(55) L-Tryptophan				40
(56) L-Valine				50
(57) Adenine sulfate				
(58) Urea				
(59) Adenosine				10
(60) *p*-Aminobenzoic acid				
(61) Biotin		0.005	0.004	0.05
(62) Cyanocobalamin		0.004	0.005	0.02
(63) N^5-Formyltetrahydro-pteroylglutamic acid				0.1
(64) Nicotinamide				
(65) Pantothenic acid, Ca salt				
(66) Pteroylglutamic acid				0.1
(67) Pyridoxine hydrochloride				
(68) Riboflavine				
(69) Thiamine hydrochloride		2.0	2.0	20
(70) Thymidine				10
(71) Thymine				20
(72) HCl (0.1N)				
(73) Fresh water	To 1,000 ml	To 1,000 ml	To 1,000 ml	To 1,000 ml
(74) pH adjusted with				
(75) pH before autoclaving	5.5	4.9−5.3	4.9−5.3	4.8−5.2
(76) Final pH				
(77) Aeration				

Table 4 (continued)
DEFINED FRESHWATER MEDIA (ORGANIC)

Reference	61 (1958)	122 (1960)	122 (1960)	147 (1962)
Author's designation	14			
		Use		
Constituents	*Synura petersenii S. caroliniana* (axenic)	*Ochromonas malhamensis*	*Ochromonas danica*	*Poteriochromonas stipitata*
(1) Carbonate		$336^{m,em}$	$227^{m,em}$	
(2) Bicarbonate		309^{d}		
(3) Nitrate	7.75^{d}			
(4) Ammonium	2.25^{c}	$283^{b,ej}$	169^{h}	337^{h}
(5) Phosphate	1.38^{r}	210^{f}	210^{f}	509^{f}
(6) Sulfate			390^{k}	$216^{k,aa,ad,ap,cg}$
(7) Silicon	2.96^{i}			
(8) Na	$9.88^{i,r,bn}$			
(9) Mg	0.478^{h}	133^{em}	$205^{g,em}$	5.92^{g}
(10) K	2.10^{h}	86.2^{f}	86.2^{f}	210^{f}
(11) Ca	4.09^{h}	60.1^{a}	20.0^{a}	28.9^{h}
(12) Fe	0.5^{h}	1.0^{ad}	1.0^{ad}	4.00^{ad}
(13) B		0.05^{n}	0.05^{n}	
(14) V		0.005^{en}	0.005^{en}	
(15) Mn	0.01^{h}	0.25^{aa}	0.25^{aa}	60.0^{aa}
(16) Co		0.05^{al}	0.05^{al}	0.398^{al}
(17) Cu		0.04^{ap}	0.04^{ap}	9.93^{ap}
(18) Zn		0.5^{g}	0.5^{g}	39.8^{g}
(19) Mo		0.025^{ba}	0.025^{ba}	0.761^{ba}
(20) I				
(21) DTPA^{gt}				
(22) EDTA^{fl}				640
(23) HEDTA^{ft}				
(24) NTA^{fm}			200	
(25) Citrate	12.9^{j}	$1,000^{ej}$		
(26) TRIS^{fj}				
(27) BICINE^{gu}				
(28) Glycylglycine				
(29) Phthalate				
(30) Pyromellitic acid				
(31) Acetate				
(32) *n*-Butanol				
(33) Ethanol				
(34) Glucose			10,000	20,000
(35) Glyceryl				
(36) Lactate				
(37) Malic acid				
(38) Succinate				
(39) Sucrose		16,000		

Table 4 (continued)
DEFINED FRESHWATER MEDIA (ORGANIC)

	Reference	61 (1958)	122 (1960)	122 (1960)	147 (1962)
	Author's designation	14			
				Use	
	Constituents	*Synura petersenii* *S. caroliniana* (axenic)	*Ochromonas malhamensis*	*Ochromonas danica*	*Poteriochromonas stipitata*
(40)	DL-Alanine				
(41)	L-Arginine		413[bx]	331[bx]	100
(42)	Asparagine				
(43)	DL-Aspartic acid				
(44)	L-Cysteine hydrochloride				
(45)	L-Cystine				
(46)	Glutamate		3,000[cr,gx]	3,000[cr,gx]	600[cr,gx]
(47)	Glutamine				
(48)	Glycine				
(49)	Histidine	150	407[bx,cr]	324[bx,cr]	100[cr]
(50)	L-Isoleucine				500
(51)	L-Leucine				500
(52)	L-Lysine hydrochloride				
(53)	Methionine		600[cq]		
(54)	Phenylalanine				500[cr]
(55)	L-Tryptophan				
(56)	L-Valine				500
(57)	Adenine sulfate				500
(58)	Urea				
(59)	Adenosine				
(60)	*p*-Aminobenzoic acid				
(61)	Biotin		0.004	0.01	0.02
(62)	Cyanocobalamin	0.0001	0.01		0.005
(63)	N⁵-Formyltetrahydro-pteroylglutamic acid				
(64)	Nicotinamide				
(65)	Pantothenic acid, Ca salt				
(66)	Pteroylglutamic acid				
(67)	Pyridoxine hydrochloride				
(68)	Riboflavine				
(69)	Thiamine hydrochloride		2.0	1.0	2.0
(70)	Thymidine				
(71)	Thymine				
(72)	HCl (0.1N)				
(73)	Fresh water	To 1,000 ml[fu]	To 1,000 ml	To 1,000 ml	To 1,000 ml
(74)	pH adjusted with				
(75)	pH before autoclaving	6.0	5.0	5.0	7.0 (light) *or* 5.0 (dark)
(76)	Final pH				
(77)	Aeration				

Table 4 (continued)
DEFINED FRESHWATER MEDIA (ORGANIC)

Reference	50 (1968)	80 (1950)	127 (1951)	5 (1952)
Author's designation				BA
			Use	
Constituents	Navicula pelliculosa	Chlamydomonas	Polytomella agilis	Chlamydomonas moewusii
(1) Carbonate	11.3[j]			
(2) Bicarbonate				
(3) Nitrate	52.5[m]			387[d]
(4) Ammonium		234[cn]		
(5) Phosphate	5.45[e]	137[e]		113[c]
(6) Sulfate	9.74[k]	195[k]	334[q] + 349[f]	109[e]
(7) Silicon	7.90[i,dv]		1.46[k]	77.9[k]
(8) Na	124[a,i,cu]			
(9) Mg	2.47[g]	49.4[g]	162[q]	28.0[cn]
(10) K	4.49[e]	112[e]	2.76[g,h]	19.7[g]
(11) Ca	17.0[c]	50	144[f]	89.8[e]
(12) Fe	0.5	10	5.11[h]	27.3[h]
(13) B	0.1	20	0.062[ab]	0.5[ai]
(14) V				0.1[n]
(15) Mn	0.1	4.0	0.0028[z]	
(16) Co	0.1	0.5		0.1[z]
(17) Cu	0.1	2		
(18) Zn	0.3	20		0.1[ao]
(19) Mo	0.1	4.0	0.0048[h]	0.3[h]
(20) I				0.1[ax]
(21) DTPA[gt]				
(22) EDTA[fl]		200		
(23) HEDTA[ft]				
(24) NTA[fm]				
(25) Citrate				
(26) TRIS[fj]				
(27) BICINE[gu]				
(28) Glycylglycine	661			
(29) Phthalate				
(30) Pyromellitic acid				
(31) Acetate		766[d]		72.0[j]
(32) n-Butanol				
(33) Ethanol				
(34) Glucose			2,000	
(35) Glyceryl				
(36) Lactate	397[j]			
(37) Malic acid				
(38) Succinate				
(39) Sucrose				

Table 4 (continued)
DEFINED FRESHWATER MEDIA (ORGANIC)

Reference	50 (1968)	80 (1950)	127 (1951)	5 (1952)
Author's designation				BA
			Use	
Constituents	*Navicula pelliculosa*	*Chlamydomonas*	*Polytomella agilis*	*Chlamydomonas moewusii*
(40) DL-Alanine				
(41) L-Arginine				
(42) Asparagine				
(43) DL-Aspartic acid				
(44) L-Cysteine hydrochloride				
(45) L-Cystine				
(46) Glutamate			100	
(47) Glutamine				
(48) Glycine		2,000		
(49) Histidine				
(50) L-Isoleucine				
(51) L-Leucine				
(52) L-Lysine hydrochloride				
(53) Methionine				
(54) Phenylalanine				
(55) L-Tryptophan				
(56) L-Valine				
(57) Adenine sulfate				
(58) Urea				
(59) Adenosine				
(60) *p*-Aminobenzoic acid				
(61) Biotin				
(62) Cyanocobalamin			0.006	
(63) N^5-Formyltetrahydro-pteroylglutamic acid				
(64) Nicotinamide				
(65) Pantothenic acid, Ca salt				
(66) Pteroylglutamic acid				
(67) Pyridoxine hydrochloride				
(68) Riboflavine				
(69) Thiamine hydrochloride			1.0	
(70) Thymidine				
(71) Thymine				
(72) HCl (0.1N)				
(73) Fresh water	To 1,000 ml	To 1,000 ml	To 1,000 ml	To 1,000 ml
(74) pH adjusted with		Ammonia vapor or KOH		
(75) pH before autoclaving				
(76) Final pH				
(77) Aeration	Air			

Table 4 (continued)
DEFINED FRESHWATER MEDIA (ORGANIC)

Reference	28 (1953)	28 (1953)	29 (1953)	31 (1953)
Author's designation	Medium I	Medium II		'Urea-EH'
		Use		
Constituents	*Chlamydomonas reinhardii*	*Chlamydomonas reinhardii* (grown in the dark)	*Chlorella vulgaris*	*Chlorella ellipsoidea* (mass culture)
(1) Carbonate				
(2) Bicarbonate				
(3) Nitrate	232[d]	232[d]	581[d]	
(4) Ammonium	67.6[c]	67.6[c]	169[c]	
(5) Phosphate	54.5[e] + 69.8[f]	627[e] + 2,530[r]	1,270[e] + 5,070[f]	872[f]
(6) Sulfate	117[k]	117[k]	160[k]	974[k]
(7) Silicon				
(8) Na	117[bn]	1,070[r,bn,cn]	46.9[bn]	4.57[bt]
(9) Mg	29.6[g]	29.6[g]	40.4[g]	247[g]
(10) K	73.6[e,f]	516[e]	3,130[e,f]	359[f]
(11) Ca	14.4[h]	14.4[h]		30.0
(12) Fe	2.07[ab]	2.07[ab]	Present[bo]	11.2
(13) B	0.175[n]	0.175[n]	0.5[n]	20.0
(14) V				
(15) Mn	0.0985[aa]	0.0985[aa]	0.5030[z]	4.0
(16) Co	0.0495[aj]	0.0495[aj]		1.0
(17) Cu	0.0159[ap]	0.0159[ap]	0.0201[ap]	4.0
(18) Zn	0.227[g]	0.227[g]	0.0505[g]	20.0
(19) Mo	0.0793[ax]	0.0793[ax]		4.0
(20) I				
(21) DTPA[gt]				
(22) EDTA[fl]				29.0[bt]
(23) HEDTA[ft]				
(24) NTA[fm]				
(25) Citrate	322[j]	322[j]	129[j]	
(26) TRIS[fj]				
(27) BICINE[gu]				
(28) Glycylglycine				
(29) Phthalate				
(30) Pyromellitic acid				
(31) Acetate		868[j]		
(32) *n*-Butanol				
(33) Ethanol				
(34) Glucose				
(35) Glyceryl				
(36) Lactate				
(37) Malic acid				
(38) Succinate				
(39) Sucrose				

Table 4 (continued)
DEFINED FRESHWATER MEDIA (ORGANIC)

	Reference	28 (1953)	28 (1953)	29 (1953)	31 (1953)
	Author's designation	Medium I	Medium II		'Urea-EH'
			Use		
	Constituents	*Chlamydomonas reinhardii*	*Chlamydomonas reinhardii* (grown in the dark)	*Chlorella vulgaris*	*Chlorella ellipsoidea* (mass culture)
(40)	DL-Alanine				
(41)	L-Arginine				
(42)	Asparagine				
(43)	DL-Aspartic acid				
(44)	L-Cysteine hydrochloride				
(45)	L-Cystine				
(46)	Glutamate				
(47)	Glutamine				
(48)	Glycine				
(49)	Histidine				
(50)	L-Isoleucine				
(51)	L-Leucine				
(52)	L-Lysine hydrochloride				
(53)	Methionine				
(54)	Phenylalanine				
(55)	L-Tryptophan				
(56)	L-Valine				
(57)	Adenine sulfate				
(58)	Urea				3,000
(59)	Adenosine				
(60)	*p*-Aminobenzoic acid				
(61)	Biotin				
(62)	Cyanocobalamin				
(63)	N^5-Formyltetrahydro-pteroylglutamic acid				
(64)	Nicotinamide				
(65)	Pantothenic acid, Ca salt				
(66)	Pteroylglutamic acid				
(67)	Pyridoxine hydrochloride				
(68)	Riboflavine				
(69)	Thiamine hydrochloride				
(70)	Thymidine				
(71)	Thymine				
(72)	HCl (0.1N)				
(73)	Fresh water	To 1,000 ml	To 1,000 ml	To 1,000 ml[fv]	To 1,000 ml
(74)	pH adjusted with				
(75)	pH before autoclaving				5.2
(76)	Final pH	6.8	6.2	6.0–6.1	
(77)	Aeration			0.5% CO_2	

Table 4 (continued)
DEFINED FRESHWATER MEDIA (ORGANIC)

Reference	129 (1961)	142 (1961)	146 (1961)	12 (1962)
Author's designation	S66			M
		Use		
Constituents	*Haematococcus Balticola Stephanosphaera*	*Polytoma uvella*	*Pyrobotrys (= Chlamydo- botrys)*	*Chlorella*
(1) Carbonate		113[j]		
(2) Bicarbonate				
(3) Nitrate	61.3[l]			1,240[l]
(4) Ammonium			27.1[h]	
(5) Phosphate	5.45[e]	109[e]	28.5[e]	190[e,dc]
(6) Sulfate	5.58[m]	79.8[k]	33.1[k,m,ac,cg]	96.1[k]
(7) Silicon				
(8) Na	118[h]	647[a,cn]	91.9[cn]	92.0[h]
(9) Mg	4.78[h]	20.2[g]	0.973[g]	24.3[g]
(10) K	47.4[c,e,h]	89.8[e]	23.5[e]	938[c,e]
(11) Ca	2.33[g]		2.81[g]	20.0[h]
(12) Fe	0.5		11.2[ac]	4.00[bs] *or* bu
(13) B			2.16[n]	0.5[n]
(14) V				0.01[v]
(15) Mn	0.05		0.549[aa]	0.5[aa]
(16) Co	0.0005		0.206[ak]	
(17) Cu	0.005		0.0003[ap]	0.02[ap]
(18) Zn	0.005		2.29[g]	0.05[g]
(19) Mo	0.0005		0.384[ax]	0.01[av]
(20) I				
(21) DTPA[gt]				
(22) EDTA[fl]			200	Present[bs]
(23) HEDTA[ft]				
(24) NTA[fm]				
(25) Citrate	39.3[fr]			
(26) TRIS[fj]				
(27) BICINE[gu]				
(28) Glycylglycine	500			
(29) Phthalate				
(30) Pyromellitic acid				
(31) Acetate		1,440[j]	236[j]	
(32) *n*-Butanol				
(33) Ethanol				
(34) Glucose				10,000
(35) Glyceryl				
(36) Lactate				
(37) Malic acid				
(38) Succinate				
(39) Sucrose				

Table 4 (continued)
DEFINED FRESHWATER MEDIA (ORGANIC)

Reference	129 (1961)	142 (1961)	146 (1961)	12 (1962)
Author's designation	S66			M
			Use	
Constituents	*Haematococcus Balticola Stephanosphaera*	*Polytoma uvella*	*Pyrobotrys (= Chlamydobotrys)*	*Chlorella*
(40) DL-Alanine				
(41) L-Arginine				
(42) Asparagine		2,000	1,500	
(43) DL-Aspartic acid				
(44) L-Cysteine hydrochloride				
(45) L-Cystine				
(46) Glutamate				
(47) Glutamine				
(48) Glycine	250			
(49) Histidine				
(50) L-Isoleucine				
(51) L-Leucine				
(52) L-Lysine hydrochloride				
(53) Methionine				
(54) Phenylalanine				
(55) L-Tryptophan				
(56) L-Valine				
(57) Adenine sulfate				
(58) Urea				
(59) Adenosine				
(60) *p*-Aminobenzoic acid				
(61) Biotin				
(62) Cyanocobalamin	0.0001		0.02	
(63) N^5-Formyltetrahydro-pteroylglutamic acid				
(64) Nicotinamide				
(65) Pantothenic acid, Ca salt				
(66) Pteroylglutamic acid				
(67) Pyridoxine hydrochloride				
(68) Riboflavine				
(69) Thiamine hydrochloride	0.1		0.2	
(70) Thymidine				
(71) Thymine				
(72) HCl (0.1N)				
(73) Fresh water	To 1,000 ml	To 1,000 ml	To 1,000 ml	To 1,000 ml
(74) pH adjusted with			NaOH or HCl	
(75) pH before autoclaving	8.0	7.5	7.0	
(76) Final pH				
(77) Aeration				

Table 4 (continued)
DEFINED FRESHWATER MEDIA (ORGANIC)

Reference	131 (1962)	139 (1962)	13 (1964)	148 (1964)
Author's designation		BM2-A		
			Use	
Constituents	Chlorella pyrenoidosa	Chlamydomonas mundana	Chlamydomonas acidophila Carteria turfosa C. acidicola	Prototheca zopfii
(1) Carbonate				
(2) Bicarbonate				
(3) Nitrate		123^l		
(4) Ammonium		33.7^h	95.7^{ej}	273^g
(5) Phosphate	915^f	136^e	210^f	$1,400^f$
(6) Sulfate	195^k	$83.1^{k,ac}$	195^k	$883^{d,k}$
(7) Silicon				
(8) Na		$961^{cn,bt}$		460^{cn}
(9) Mg	49.4^g	19.7^g	49.4^g	$39.5^{g,dc}$
(10) K	376^f	$190^{c,e}$	$195^{f,bn}$	575^f
(11) Ca	5.00	6.82^h	15.0^h	
(12) Fe	2.00	3.01^{ac}	2.0	
(13) B	0.5	0.105^n	0.02	
(14) V	0.01		0.01	
(15) Mn	0.5	4.37^z	2.0	
(16) Co	0.01		0.1	
(17) Cu	0.04	0.0127^{ap}	0.2	
(18) Zn	0.5	0.227^g	1.0	
(19) Mo	0.02	0.0082^{aw}	0.08	
(20) I				
(21) DTPA^{gt}				
(22) EDTA^{fl}	Present	53.4^{bt}		
(23) HEDTA^{ft}				
(24) NTA^{fm}				
(25) Citrate			$676^{l,ej} + 492^{fr}$	
(26) TRIS^{fj}		1,210		
(27) BICINE^{gu}				
(28) Glycylglycine				
(29) Phthalate				
(30) Pyromellitic acid				
(31) Acetate		$2,450^j$		$1,180^j$
(32) n-Butanol				
(33) Ethanol				
(34) Glucose			10,000	
(35) Glyceryl				
(36) Lactate				
(37) Malic acid				
(38) Succinate				
(39) Sucrose				

Table 4 (continued)
DEFINED FRESHWATER MEDIA (ORGANIC)

Reference	131 (1962)	139 (1962)	13 (1964)	148 (1964)
Author's designation		BM2-A		
			Use	
Constituents	*Chlorella pyrenoidosa*	*Chlamydomonas mundana*	*Chlamydomonas acidophila Carteria turfosa C. acidicola*	*Prototheca zopfii*
(40) DL-Alanine				
(41) L-Arginine				
(42) Asparagine				
(43) DL-Aspartic acid				
(44) L-Cysteine hydrochloride				
(45) L-Cystine				
(46) Glutamate			$1,000^{cr,gx}$	
(47) Glutamine				
(48) Glycine				
(49) Histidine				
(50) L-Isoleucine				
(51) L-Leucine				
(52) L-Lysine hydrochloride				
(53) Methionine				
(54) Phenylalanine				
(55) L-Tryptophan				
(56) L-Valine				
(57) Adenine sulfate				
(58) Urea	440			
(59) Adenosine				
(60) *p*-Aminobenzoic acid		0.080		
(61) Biotin		0.0025		
(62) Cyanocobalamin		0.0004	0.005	
(63) N^5-Formyltetrahydro-pteroylglutamic acid				
(64) Nicotinamide		0.1		
(65) Pantothenic acid, Ca salt		0.4		
(66) Pteroylglutamic acid		0.004		
(67) Pyridoxine hydrochloride		0.25		
(68) Riboflavine		0.05		
(69) Thiamine hydrochloride		0.05	0.6	0.1
(70) Thymidine				
(71) Thymine				
(72) HCl (0.1N)				
(73) Fresh water	To 1,000 ml	To 1,000 ml	To 1,000 ml	To 1,000 ml
(74) pH adjusted with				NaOH
(75) pH before autoclaving	6.0	7.3	3.5	7.2
(76) Final pH				
(77) Aeration				

Table 4 (continued)
DEFINED FRESHWATER MEDIA (ORGANIC)

Reference	130 (1965)	166 (1971)	158 (1974)	32 (1956)
Author's designation			M3	
			Use	
Constituents	*Chlamydomonas mundana* (base)	*Scenedesmus* (synchronous cultures)	*Pandorina unicocca* (axenic)	*Euglena gracilis*
(1) Carbonate				48.0[m]
(2) Bicarbonate				
(3) Nitrate		496[l]	62.0[l]	
(4) Ammonium	90.2[h]			142[ep]
(5) Phosphate	332[f]	95.0[q] + 285[r]	47.5[ch]	210[f]
(6) Sulfate	28.8[k,ac]	98.2[k,af]	19.2[k]	156[k]
(7) Silicon				
(8) Na		337[h,q,r,bn]	23.9[bt,ch]	Trace[en]
(9) Mg	4.87[g]	24.3[g]	4.87[g]	39.5[g]
(10) K	137[f]	313[c]	39.1[c]	86.2[f]
(11) Ca	8.02[h]	4.01[h]	4.01[h]	32.2[a]
(12) Fe	5.59[ac]	0.838[af]	0.201[ab]	2.0[ad]
(13) B		0.988[n]		0.1[n]
(14) V				0.01[en]
(15) Mn	0.275[aa]	0.942[z]	0.115[z]	0.5[aa]
(16) Co			0.005[aj]	0.1[al]
(17) Cu		0.0392[ap]		0.08[ap]
(18) Zn		0.0963[g]	0.0242[h]	1.0[g]
(19) Mo		0.0196[av]	0.0163[ax]	0.350[ba]
(20) I				
(21) DTPA[gt]	39.3			
(22) EDTA[fl]			5.85[bt]	
(23) HEDTA[ft]				
(24) NTA[fm]				
(25) Citrate		104[j]		
(26) TRIS[fj]	3,630			
(27) BICINE[gu]				
(28) Glycylglycine				
(29) Phthalate				
(30) Pyromellitic acid				
(31) Acetate	1,180[gy]			
(32) *n*-Butanol				
(33) Ethanol				
(34) Glucose				
(35) Glyceryl			37.5[ch]	
(36) Lactate				
(37) Malic acid				1,000[cq]
(38) Succinate				458[d]
(39) Sucrose				15,000

Table 4 (continued)
DEFINED FRESHWATER MEDIA (ORGANIC)

Reference	130 (1965)	166 (1971)	158 (1974)	32 (1956)
Author's designation			M3	
			Use	
Constituents	*Chlamydomonas mundana* (base)	*Scenedesmus* (synchronous cultures)	*Pandorina unicocca* (axenic)	*Euglena gracilis*
(40) DL-Alanine				
(41) L-Arginine				
(42) Asparagine				
(43) DL-Aspartic acid				2,000
(44) L-Cysteine hydrochloride				
(45) L-Cystine				
(46) Glutamate				3,000[cr,gx]
(47) Glutamine				
(48) Glycine				2,500
(49) Histidine				
(50) L-Isoleucine				
(51) L-Leucine				
(52) L-Lysine hydrochloride				
(53) Methionine				
(54) Phenylalanine				
(55) L-Tryptophan				
(56) L-Valine				
(57) Adenine sulfate				
(58) Urea				
(59) Adenosine				
(60) *p*-Aminobenzoic acid				
(61) Biotin	0.002			
(62) Cyanocobalamin	0.0025		0.0001	
(63) N^5-Formyltetrahydro-pteroylglutamic acid				
(64) Nicotinamide				
(65) Pantothenic acid, Ca salt				
(66) Pteroylglutamic acid				
(67) Pyridoxine hydrochloride				
(68) Riboflavine				
(69) Thiamine hydrochloride	0.005			0.6[eo]
(70) Thymidine				
(71) Thymine				
(72) HCl (0.1N)				
(73) Fresh water	To 1,000 ml	To 1,000 ml	To 1,000 ml[fv]	To 1,000 ml
(74) pH adjusted with	HCl			NaOH or H_2SO_4
(75) pH before autoclaving	7.2			3.6
(76) Final pH				
(77) Aeration		2—3% CO_2		

Table 4 (continued)
DEFINED FRESHWATER MEDIA (ORGANIC)

Reference	105 (1959)	105 (1959)	122 (1960)	143 (1960)
Author's designation	A	B		
		Use		
Constituents	*Euglena gracilis* (low pH)	*Euglena gracilis*	*Euglena gracilis*	*Astasia longa* (synchronous cultures)
(1) Carbonate	300^m			
(2) Bicarbonate				
(3) Nitrate				
(4) Ammonium	54.6^o	135^h	95.7^{ej}	273°
(5) Phosphate	$143^o + 279^f$	54.5^e	210^f	$719^o + 698^f$
(6) Sulfate	$260^{k,aa,cg}$	$227^{k,aa,cg}$	195^k	77.9^k
(7) Silicon				
(8) Na	3.83^{ax}	169^{cn}		$1,520^{bn,cn}$
(9) Mg	49.4^g	49.4^g	49.4^g	19.7^g
(10) K	115^f	44.9^e	$195^{f,bn}$	287^f
(11) Ca	200^a	50.0^h	15.0^h	7.22^h
(12) Fe	5.00^{ab}	2.00^{ab}	0.4^{ad}	0.660^{af}
(13) B	0.1^n	0.05^n	0.02^n	
(14) V			0.002^{en}	
(15) Mn	20.1^{aa}	10.0^{aa}	0.1^{aa}	0.5^z
(16) Co	0.4^{aj}	0.2^{aj}	0.02^{al}	0.315^{al}
(17) Cu	0.199^{ap}	0.1^{ap}	0.016^{ap}	0.0051^{ap}
(18) Zn	20.0^g	10.0^g	0.2^g	0.0910^g
(19) Mo	8.0^{ax}	4.0^{ax}	0.01^{ba}	0.0793^{ax}
(20) I	0.02^{er}	0.01^{er}		
(21) DTPA^{gt}				
(22) EDTA^{fl}		500		
(23) HEDTA^{ft}				
(24) NTA^{fm}				
(25) Citrate			$676^{l,ej} + 492^{fr}$	332^j
(26) TRIS^{fj}				
(27) BICINE^{gu}				
(28) Glycylglycine				
(29) Phthalate				
(30) Pyromellitic acid				
(31) Acetate		434^j		$3,600^j$
(32) *n*-Butanol		3,000		
(33) Ethanol				
(34) Glucose			10,000	
(35) Glyceryl				
(36) Lactate				
(37) Malic acid	$2,000^{cq}$			
(38) Succinate				
(39) Sucrose				

Table 4 (continued)
DEFINED FRESHWATER MEDIA (ORGANIC)

Reference	105 (1959)	105 (1959)	122 (1960)	143 (1960)
Author's designation	A	B		
Use				
Constituents	*Euglena gracilis* (low pH)	*Euglena gracilis*	*Euglena gracilis*	*Astasia longa* (synchronous cultures)
(40) DL-Alanine				
(41) L-Arginine				
(42) Asparagine				
(43) DL-Aspartic acid				
(44) L-Cysteine hydrochloride				100[ex]
(45) L-Cystine				
(46) Glutamate	5,000[cr,gx]		1,000[cr,gx]	
(47) Glutamine				
(48) Glycine				
(49) Histidine				
(50) L-Isoleucine				
(51) L-Leucine				
(52) L-Lysine hydrochloride				
(53) Methionine				1.75[cr,ex]
(54) Phenylalanine				
(55) L-Tryptophan				
(56) L-Valine				
(57) Adenine sulfate				
(58) Urea				
(59) Adenosine				
(60) *p*-Aminobenzoic acid				
(61) Biotin				
(62) Cyanocobalamin	0.0002	0.0004	0.005	0.0005
(63) N^5-Formyltetrahydro-pteroylglutamic acid				
(64) Nicotinamide				
(65) Pantothenic acid, Ca salt				
(66) Pteroylglutamic acid				
(67) Pyridoxine hydrochloride				
(68) Riboflavine				
(69) Thiamine hydrochloride	1.0	1.0	0.6	0.01
(70) Thymidine				
(71) Thymine				
(72) HCl (0.1N)				
(73) Fresh water	To 1,000 ml	To 1,000 ml	To 1,000 ml	To 1,000 ml
(74) pH adjusted with				
(75) pH before autoclaving	3.3	8.0	3.5	6.8
(76) Final pH				
(77) Aeration				

Table 4 (continued)
DEFINED FRESHWATER MEDIA (ORGANIC)

Reference	152 (1962)	152 (1962)	128 (1963)
Author's designation	Medium 1	Medium 2	
		Use	
Constituents	*Euglena gracilis*	*Euglena gracilis*	*Astasia longa*
(1) Carbonate			
(2) Bicarbonate			
(3) Nitrate			
(4) Ammonium	674^h	546°	273°
(5) Phosphate	140^f	$1{,}440^o + 1{,}400^f$	$719^o + 698^f$
(6) Sulfate	77.9^k	77.9^k	77.9^k
(7) Silicon			
(8) Na		12.7^{bn}	151^{bn}
(9) Mg	19.7^g	19.7^g	19.7^g
(10) K	$440^{f,ez}$	575^f	287^f
(11) Ca	7.22^h		72.2^h
(12) Fe	0.660^{af}		0.802^{af}
(13) B			
(14) V			
(15) Mn	0.5^z	0.5^z	0.5^z
(16) Co	0.263^{ak}	0.263^{ak}	0.322^{aj}
(17) Cu	0.0025^{ap}	0.0025^{ap}	0.0051^{ap}
(18) Zn	0.0910^g	0.0910^g	0.091^g
(19) Mo	0.119^{aw}	0.119^{aw}	0.0793^{ax}
(20) I			
(21) DTPA^{gt}			
(22) EDTA^{fl}			
(23) HEDTA^{ft}			
(24) NTA^{fm}			
(25) Citrate		34.7^j	415^j
(26) TRIS^{fj}			
(27) BICINE^{gu}			
(28) Glycylglycine			
(29) Phthalate	$1{,}620^{ez}$		
(30) Pyromellitic acid			
(31) Acetate			
(32) *n*-Butanol			
(33) Ethanol			$9{,}220^{es}$
(34) Glucose	$10{,}000^{ex}$	$10{,}000^{ex}$	
(35) Glyceryl			
(36) Lactate			
(37) Malic acid			
(38) Succinate			
(39) Sucrose			

Table 4 (continued)
DEFINED FRESHWATER MEDIA (ORGANIC)

Reference	152 (1962)	152 (1962)	128 (1963)
Author's designation	Medium 1	Medium 2	
		Use	
Constituents	*Euglena gracilis*	*Euglena gracilis*	*Astasia longa*
(40) DL-Alanine			
(41) L-Arginine			
(42) Asparagine			
(43) DL-Aspartic acid			
(44) L-Cysteine hydrochloride			
(45) L-Cystine			
(46) Glutamate			
(47) Glutamine			
(48) Glycine			
(49) Histidine			
(50) L-Isoleucine			
(51) L-Leucine			
(52) L-Lysine hydrochloride			
(53) Methionine			
(54) Phenylalanine			
(55) L-Tryptophan			
(56) L-Valine			
(57) Adenine sulfate			
(58) Urea			
(59) Adenosine			
(60) *p*-Aminobenzoic acid			
(61) Biotin			
(62) Cyanocobalamin	0.00005[ex]	0.00005[ex]	0.01
(63) N^5-Formyltetrahydro-pteroylglutamic acid			
(64) Nicotinamide			
(65) Pantothenic acid, Ca salt			
(66) Pteroylglutamic acid			
(67) Pyridoxine hydrochloride			
(68) Riboflavine			
(69) Thiamine hydrochloride	1.0[ex]	1.0[ex]	0.02
(70) Thymidine			
(71) Thymine			
(72) HCl (0.1N)	10 ml	10 ml	
(73) Fresh water	To 1,000 ml	To 1,000 ml	To 1,000 ml
(74) pH adjusted with			
(75) pH before autoclaving	4.5	7.0	6.7
(76) Final pH			
(77) Aeration			

Compiled by M. F. Turner and M. R. Droop.

Table 5
UNDEFINED MARINE MEDIA (MINERAL)[gz]

Reference	97 (1927)	97 (1927)	98 (1934)	62 (1957)
Author's designation			"Erdschreiber"	E3
		Use		
Constituents	Diatoms and flagellates	Diatoms and flagellates	A wide range of algae	Diatoms and flagellates
(1) Bicarbonate				
(2) Nitrate	73.0[j]	123[l]	73.0[j]	61.3[l]
(3) Ammonium				
(4) Phosphate	13.4[q]	54.5[e]	13.4[q]	5.45[e]
(5) Sulfate				
(6) Sulfide				
(7) Silicon		1.82[dy]		
(8) Na	33.5[c,q]		33.5[c,q]	
(9) Mg				
(10) K		127[c,e,dy]		43.2[c,e]
(11) Ca				
(12) Fe		1.40[af]		
(13) B				
(14) Al				
(15) V				
(16) Mn				
(17) Co				
(18) Ni				
(19) Cu				
(20) Zn				
(21) As				
(22) Br				
(23) Mo				
(24) Cd				
(25) I				
(26) W				

Table 5 (continued)
UNDEFINED MARINE MEDIA (MINERAL)[gz]

Reference	97 (1927)	97 (1927)	98 (1934)	62 (1957)
Author's designation			"Erdschreiber"	E3
Use				
Constituents	Diatoms and flagellates	Diatoms and flagellates	A wide range of algae	Diatoms and flagellates
(27) EDTA[fl]				
(28) Citrate				
(29) TRIS[fj]				
(30) Glyceryl				
(31) *p*-Aminobenzoic acid				
(32) Biotin				
(33) Choline dihydrogen citrate				
(34) Cyanocobalamin				
(35) N[5]-Formyltetrahydro-pteroylglutamic acid				
(36) Inositol				
(37) Nicotinic acid				
(38) Orotic acid				
(39) Pantothenic acid, Ca salt				
(40) Pteroylglutamic acid				
(41) Putrescine dihydro-chloride				
(42) Pyridoxamine dihydro-chloride				
(43) Pyridoxine hydrochloride				
(44) Riboflavine				
(45) Thiamine hydrochloride				
(46) Thymine				
(47) Soil extract			50 ml	25[ha]
(48) HCl				
(49) Sea salt				
(50) Sea water	To 1,000 ml[cw]	To 1,000 ml[cw]	To 1,000 ml	500 ml[dc]
(51) Fresh water	50 ml	50 ml		To 1,000 ml[fu]
(52) pH before autoclaving				Not adjusted
(53) Final pH				

Table 5 (continued)
UNDEFINED MARINE MEDIA (MINERAL)^{gz}

Reference	62 (1957)	109 (1958)	61 (1958)	61 (1958)
Author's designation	E13		22	1
		Use		
Constituents	Diatoms and flagellates	Flagellates (N-ethyl carbazole reaction)	Cryptophytes Chrysophytes Chlorophytes	Cryptophytes Dinoflagellates Haptophytes Chrysophytes Chlorophytes
(1) Bicarbonate				
(2) Nitrate	30.7^l		66.3^j	66.3—69.5^j
(3) Ammonium		16.9^h		
(4) Phosphate	2.73^e	15.9^r	9.91^e	9.91—10.4^e
(5) Sulfate				
(6) Sulfide				
(7) Silicon				
(8) Na		3.83^r	10,800^{c,h}	24.6—25.8^c
(9) Mg				
(10) K	21.6^{c,e}		8.16^e	8.16—8.55^e
(11) Ca				
(12) Fe		1.3^{bs}		
(13) B				
(14) Al				
(15) V				
(16) Mn		0.1		
(17) Co		0.005		
(18) Ni				
(19) Cu		0.005		
(20) Zn		0.01		
(21) As				
(22) Br				
(23) Mo		0.005		
(24) Cd				
(25) I				
(26) W				

Table 5 (continued)
UNDEFINED MARINE MEDIA (MINERAL)[gz]

Reference	62 (1957)	109 (1958)	61 (1958)	61 (1958)
Author's designation	E13		22	1
		Use		
Constituents	Diatoms and flagellates	Flagellates (N-ethyl carbazole reaction)	Cryptophytes Chrysophytes Chlorophytes	Cryptophytes Dinoflagellates Haptophytes Chrysophytes Chlorophytes
(27) EDTA[fl]		Present[bs]		
(28) Citrate				
(29) TRIS[fj]				
(30) Glyceryl				
(31) *p*-Aminobenzoic acid				
(32) Biotin		0.001		
(33) Choline dihydrogen citrate				
(34) Cyanocobalamin		0.001		
(35) N^5-Formyltetrahydro-pteroylglutamic acid				
(36) Inositol				
(37) Nicotinic acid				
(38) Orotic acid				
(39) Pantothenic acid, Ca salt				
(40) Pteroylglutamic acid				
(41) Putrescine dihydro-chloride				
(42) Pyridoxamine dihydro-chloride				
(43) Pyridoxine hydrochloride				
(44) Riboflavine				
(45) Thiamine hydrochloride		0.2		
(46) Thymine				
(47) Soil extract	12–5[h]		91 ml[gj]	47.6–90.9 ml[gj]
(48) HCl				
(49) Sea salt				
(50) Sea water	750 ml[dc]	1,000 ml[dd]	To 1,000 ml[gk]	To 1,000 ml[gk]
(51) Fresh water	To 1,000 ml[fu]			
(52) pH before autoclaving	Not adjusted		8.0–8.2	8.0–8.2
(53) Final pH				

Table 5 (continued)
UNDEFINED MARINE MEDIA (MINERAL)[gz]

Reference	161 (1967)	118 (1968)	79 (1969)	79 (1969)
Author's designation	IMR/2	ES	FCRG Type I	FCRG Type II
		Use		
Constituents	Planktonic diatoms and flagellates	Seaweeds and other algae	Planktonic diatoms	A wide range of phytoplankton
(1) Bicarbonate				
(2) Nitrate	15.5[l]	51.1[j]	30.7[l]	30.7[l]
(3) Ammonium				
(4) Phosphate	2.37[e]	3.01[ch]	4.75[f]	4.75[f]
(5) Sulfate				
(6) Sulfide				
(7) Silicon	7.02[i]		6.92[i]	
(8) Na	11.9[i,bt]	20.7[c,bt,ch]	2.24[i,bt]	Trace[bt]
(9) Mg				
(10) K	11.7[c,e]		21.0[c,f]	21.0[c,f]
(11) Ca				
(12) Fe	0.103[ab]	0.5[bq] + 0.025[ab]	0.0103[ab]	0.0103[ab]
(13) B		0.5[n]		
(14) Al				
(15) V				
(16) Mn	0.101[aa]	0.1[aa]	0.065[aa]	0.065[aa]
(17) Co	0.0005[aj]	0.0025[al]	0.000063[al]	0.000063[al]
(18) Ni				
(19) Cu	0.0005[ap]		0.000064[ap]	0.000064[ap]
(20) Zn	0.0284[g]	0.0125[g]	0.000068[g]	0.000068[g]
(21) As				
(22) Br				
(23) Mo	0.0258[ax]		0.0001[ax]	0.0001[ax]
(24) Cd				
(25) I				
(26) W				

Table 5 (continued)
UNDEFINED MARINE MEDIA (MINERAL)[gz]

Reference	161 (1967)	118 (1968)	79 (1969)	79 (1969)
Author's designation	IMR/2	ES	FCRG Type I	FCRG Type II
		Use		
Constituents	Planktonic diatoms and flagellates	Seaweeds and other algae	Planktonic diatoms	A wide range of phytoplankton
(27) EDTA[fl]	2.36[bt]	3.4[bq] + 1.96[bt]	3.93[bt]	3.93[bt]
(28) Citrate				
(29) TRIS[fj]		100[dr]		200
(30) Glyceryl		2.38[ch]		
(31) *p*-Aminobenzoic acid				
(32) Biotin	0.0005	0.001	0.001	0.001
(33) Choline dihydrogen citrate				
(34) Cyanocobalamin	0.005	0.002	0.001	0.001
(35) N^5-Formyltetrahydro-pteroylglutamic acid				
(36) Inositol				
(37) Nicotinic acid				
(38) Orotic acid				
(39) Pantothenic acid, Ca salt				
(40) Pteroylglutamic acid				
(41) Putrescine dihydro-chloride				
(42) Pyridoxamine dihydro-chloride				
(43) Pyridoxine hydrochloride				
(44) Riboflavine				
(45) Thiamine hydrochloride	0.1	0.1	0.01	0.01
(46) Thymine				
(47) Soil extract			10 ml[ga]	10 ml[ga]
(48) HCl			5.0 ml	
(49) Sea salt				
(50) Sea water	900 ml	980 ml[dj]	964 ml[cw,dc,dk]	968 ml[cw,dc,dk]
(51) Fresh water	To 1,000 ml[fu]	To 1,000 ml	To 1,000 ml[fu]	To 1,000 ml[fu]
(52) pH before autoclaving				
(53) Final pH				

Table 5 (continued)
UNDEFINED MARINE MEDIA(MINERAL)[gz]

Reference	112 (1969)	116 (1970)	8 (1955)	110 (1961)
Author's designation	MA[v]		S and B	SWI
			Use	
Constituents	Phytoplankton (bacterized)	Flagellates	*Porphyridium*	*Porphyra*
(1) Bicarbonate				
(2) Nitrate	61.3[l]	30.7[l]	310[l]	44.3[l]
(3) Ammonium				
(4) Phosphate	5.45[e]	4.75[r]	57.0[e]	6.14[f]
(5) Sulfate			19.2[k]	
(6) Sulfide				
(7) Silicon		8.30[i,eb]		
(8) Na	2.47[bt]	15.7[i,r,bt]		
(9) Mg			4.86[g]	
(10) K	43.2[c,e]	19.3[e]	242[c,e]	30.4[c,f]
(11) Ca				
(12) Fe	0.502[ac]	0.558[ab,eb]	50.3[bq]	0.5[bq]
(13) B				
(14) Al				
(15) V				
(16) Mn	0.0813[aa]	0.277[aa,eb]		
(17) Co		0.0029[al,eb]		
(18) Ni				
(19) Cu		0.0064[ap,eb]		
(20) Zn		0.131[g,eb]		
(21) As				
(22) Br				
(23) Mo		0.0963[ax,eb]		
(24) Cd				
(25) I				
(26) W				

Table 5 (continued)
UNDEFINED MARINE MEDIA (MINERAL)[gz]

Reference	112 (1969)	116 (1970)	8 (1955)	110 (1961)
Author's designation	MA[v]		S and B	SWI
		Use		
Constituents	Phytoplankton (bacterized)	Flagellates	*Porphyridium*	*Porphyra*
(27) EDTA[fl]	15.7[bt]	6.36[bt,eb]	Present[bq]	Present[bq]
(28) Citrate				
(29) TRIS[fj]		769[bx,dh]		500[dr]
(30) Glyceryl				
(31) *p*-Aminobenzoic acid				
(32) Biotin		0.001		
(33) Choline dihydrogen citrate				
(34) Cyanocobalamin	0.00005	0.002		
(35) N^5-Formyltetrahydro-pteroylglutamic acid				
(36) Inositol				
(37) Nicotinic acid				
(38) Orotic acid				
(39) Pantothenic acid, Ca salt				
(40) Pteroylglutamic acid				
(41) Putrescine dihydro-chloride				
(42) Pyridoxamine dihydro-chloride				
(43) Pyridoxine hydrochloride				
(44) Riboflavine				
(45) Thiamine hydrochloride	0.006	0.5		
(46) Thymine				
(47) Soil extract				
(48) HCl				
(49) Sea salt				
(50) Sea water	To 1,000 ml	To 1,000 ml[di]	500 ml	1,000 ml[cw]
(51) Fresh water			To 1,000 ml[fx]	
(52) pH before autoclaving			ca. 8.5	8.0—8.2
(53) Final pH		7.6—7.8		

Table 5 (continued)
UNDEFINED MARINE MEDIA (MINERAL)[gz]

Reference	110 (1961)	70 (1963)	113 (1963)	162 (1970)
Author's designation	SWII	ASM		
		Use		
Constituents	Porphyra	Spiridia filamentosa Ceramium diaphanum	Asparagopsis (unialgal)	Dasya pedicellata
(1) Bicarbonate				
(2) Nitrate	44.3[l]	73.0[j]	31.0[j]	73.0[j]
(3) Ammonium				
(4) Phosphate	3.16[ch] + 3.14[f]	5.45[e]	2.85[r]	5.30[q]
(5) Sulfate		1,950[k]		
(6) Sulfide				
(7) Silicon				
(8) Na	1.53[ch]	9,470[c,h,bt]	12.7[c,r,bt]	29.6[c,q]
(9) Mg		924[g,h]		
(10) K	29.2[c,f]	319[e,h]		
(11) Ca		300[h]		
(12) Fe	0.5[bq]	1.1[h]	0.056[ac]	0.083[ag]
(13) B		2[n]		
(14) Al				
(15) V				
(16) Mn		0.4[h]	0.0055[z]	0.123[aa]
(17) Co		0.01[h]		
(18) Ni				
(19) Cu				
(20) Zn		0.05[h]		
(21) As				
(22) Br				
(23) Mo				
(24) Cd				
(25) I				
(26) W				

Table 5 (continued)
UNDEFINED MARINE MEDIA (MINERAL)[gz]

	Reference	110 (1961)	70 (1963)	113 (1963)	162 (1970)
	Author's designation	SWII	ASM		
			Use		
	Constituents	*Porphyra*	*Spiridia filamentosa Ceramium diaphanum*	*Asparagopsis* (unialgal)	*Dasya pedicellata*
(27)	EDTA[fl]	Present[bq]	7.85[bt]	2.92[bt]	
(28)	Citrate				0.282[ag]
(29)	TRIS[fj]	500[dr]	1,000		
(30)	Glyceryl	2.50[ch]			
(31)	*p*-Aminobenzoic acid		0.01		
(32)	Biotin		0.0005		
(33)	Choline dihydrogen citrate		0.5		
(34)	Cyanocobalamin		0.00055		
(35)	N^5-Formyltetrahydro-pteroylglutamic acid		0.0002		
(36)	Inositol		1.0		
(37)	Nicotinic acid		0.1		
(38)	Orotic acid		0.26		
(39)	Pantothenic acid, Ca salt		0.1		
(40)	Pteroylglutamic acid		0.0025		
(41)	Putrescine dihydro-chloride		0.04		
(42)	Pyridoxamine dihydro-chloride		0.02		
(43)	Pyridoxine hydrochloride		0.04		
(44)	Riboflavine		0.005		
(45)	Thiamine hydrochloride		0.2		
(46)	Thymine		0.8		
(47)	Soil extract		30 ml		
(48)	HCl				
(49)	Sea salt				
(50)	Sea water	1,000 ml[cw]		1,000 ml	1,000 ml[fd]
(51)	Fresh water		To 1,000 ml		Present
(52)	pH before autoclaving	8.0–8.2	7.6		
(53)	Final pH				

Table 5 (continued)
UNDEFINED MARINE MEDIA (MINERAL)[gz]

Reference	154 (1972)	99 (1935)	101 (1951)	103 (1954)
Author's designation				
			Use	
Constituents	*Porphyropsis coccinea* (unialgal)	**Dinoflagellates** (some axenic)	*Gymnodinium*	*Gymnodinium splendens* (axenic)
(1) Bicarbonate				
(2) Nitrate	31.0[j]	60.1[l]	119[l]	124[l]
(3) Ammonium				
(4) Phosphate	2.85[r]	5.34[e]	18.3[e]	19.0[e]
(5) Sulfate				
(6) Sulfide				
(7) Silicon				
(8) Na	12.7[c,r,bt]			1.24[bt]
(9) Mg				
(10) K		42.3[c,e]	90.2[c,e]	93.8[c,e]
(11) Ca				
(12) Fe	0.056[ac]	0.0338[ab]	0.322[ab]	0.335[ab]
(13) B	0.0543[n]			
(14) Al	0.0027[p]			
(15) V	0.0005[t]			
(16) Mn	0.0605[z,aa]		0.0317[z]	0.0329[z]
(17) Co	0.0030[ak]			
(18) Ni	0.0029[an]			
(19) Cu				
(20) Zn	0.0065[g]			
(21) As				
(22) Br	0.008[as]			
(23) Mo	0.0048[ba]			
(24) Cd	0.0056[c]			
(25) I	0.0063[bc]			
(26) W	0.0018[be]			

Table 5 (continued)
UNDEFINED MARINE MEDIA (MINERAL)[gz]

	Reference	154 (1972)	99 (1935)	101 (1951)	103 (1954)
	Author's designation				
				Use	
	Constituents	*Porphyropsis coccinea* (unialgal)	Dinoflagellates (some axenic)	*Gymnodinium*	*Gymnodinium splendens* (axenic)
(27)	EDTA[fl]	2.90[bt]			7.85[bt]
(28)	Citrate				
(29)	TRIS[fj]				
(30)	Glyceryl				
(31)	*p*-Aminobenzoic acid				
(32)	Biotin	0.001			
(33)	Choline dihydrogen citrate				
(34)	Cyanocobalamin	0.001			0.00001
(35)	N^5-Formyltetrahydro-pteroylglutamic acid				
(36)	Inositol				
(37)	Nicotinic acid				
(38)	Orotic acid				
(39)	Pantothenic acid, Ca salt				
(40)	Pteroylglutamic acid				
(41)	Putrescine dihydrochloride				
(42)	Pyridoxamine dihydrochloride				
(43)	Pyridoxine hydrochloride				
(44)	Riboflavine				
(45)	Thiamine hydrochloride	0.2			
(46)	Thymine				
(47)	Soil extract		19.6 ml	38.5 ml	
(48)	HCl				
(49)	Sea salt				
(50)	Sea water	1,000 ml[cx,dk,ex]	980.4 ml[cx]	To 1,000 ml[cx]	750 ml[cx]
(51)	Fresh water				To 1,000 ml[fv]
(52)	pH before autoclaving				
(53)	Final pH				

Table 5 (continued)
UNDEFINED MARINE MEDIA(MINERAL)[gz]

Reference	104 (1955)	107 (1955)	145 (1975)	96 (1910)
Author's designation			GPM	"Miquel Sea-water"
			Use	
Constituents	*Gonyaulax polyedra* (unialgal)	*Gymnodinium brevis*	Dinoflagellates	Pelagic diatoms
(1) Bicarbonate		0.669[j]		
(2) Nitrate	122[l]		123[l]	248[l]
(3) Ammonium		0.311[h]		
(4) Phosphate	18.6[e]	0.322[f]	19.1[e]	12.7[q]
(5) Sulfate				
(6) Sulfide		0.123[j]		
(7) Silicon				
(8) Na		17.3[b,bt]	4.71[bt]	6.11[q]
(9) Mg		0.022[h]		
(10) K	92.0[c,e]	0.132[f]	93.1[c,e]	156[c]
(11) Ca				8.71[h]
(12) Fe	0.329[ab]		0.3[ab]	ca. 24[ab]
(13) B			6.0[n]	
(14) Al				
(15) V				
(16) Mn	0.0323[z]		1.2[z]	
(17) Co			0.03[aj]	
(18) Ni				
(19) Cu				
(20) Zn			0.15[h]	
(21) As				
(22) Br				
(23) Mo				
(24) Cd				
(25) I				
(26) W				

Table 5 (continued)
UNDEFINED MARINE MEDIA (MINERAL)[gz]

Reference	104 (1955)	107 (1955)	145 (1975)	96 (1910)
Author's designation			GPM	"Miquel Sea-water"
		Use		
Constituents	*Gonyaulax polyedra* (unialgal)	*Gymnodinium brevis*	Dinoflagellates	Pelagic diatoms
(27) EDTA[fl]	9.80	109[bt]	29.9[bt]	
(28) Citrate				
(29) TRIS[fj]				
(30) Glyceryl				
(31) *p*-Aminobenzoic acid				
(32) Biotin		0.00046	0.002	
(33) Choline dihydrogen citrate				
(34) Cyanocobalamin		0.0009	0.001	
(35) N^5-Formyltetrahydro-pteroylglutamic acid				
(36) Inositol				
(37) Nicotinic acid				
(38) Orotic acid				
(39) Pantothenic acid, Ca salt				
(40) Pteroylglutamic acid				
(41) Putrescine dihydrochloride				
(42) Pyridoxamine dihydrochloride				
(43) Pyridoxine hydrochloride				
(44) Riboflavine				
(45) Thiamine hydrochloride		9.22	1.0	
(46) Thymine				
(47) Soil extract	19.6 ml	18.4 ml	15 ml	
(48) HCl				0.024 ml[cy]
(49) Sea salt				
(50) Sea water	735 ml[cx]	876 ml[cx]	750 ml	1,000 ml
(51) Fresh water	To 1,000 ml[fu]	To 1,000 ml[fu]	To 1,000 ml[fv]	
(52) pH before autoclaving			Not adjusted	
(53) Final pH				

Table 5 (continued)
UNDEFINED MARINE MEDIA(MINERAL)[gz]

		106 (1914)	100 (1938)	102 (1951)	63 (1958)
Reference					
Author's designation					AK
				Use	
Constituents		*Thalassiosira gravida*	*Phaeodactylum*	*Asterionella japonica* and other diatoms	*Asterionella japonica*
(1)	Bicarbonate	152[j]			
(2)	Nitrtate	238[l]	68.1[l]	124[l]	124[l]
(3)	Ammonium				
(4)	Phosphate	12.1[q]	5.30[q]	Various[e]	19.0[e]
(5)	Sulfate	2,680[k]	16.0[k]		
(6)	Sulfide				
(7)	Silicon				0.562[i]
(8)	Na	10,700[b,h,q]	2.57[q]		0.920[i]
(9)	Mg	1,300[g,h]	4.04[g]		
(10)	K	537[c,h]	43.0[c]	78.2[c] + various[e]	93.8[c,e]
(11)	Ca	424[h]	7.22[h]		
(12)	Fe	ca. 24[ab]	ca. 10[ab]	0.559[ag]	0.559[ab]
(13)	B				
(14)	Al				
(15)	V				
(16)	Mn			0.0549[z]	0.0549[z]
(17)	Co			0.0589[aj]	
(18)	Ni				
(19)	Cu				
(20)	Zn				
(21)	As			0.0749[da]	
(22)	Br				
(23)	Mo				
(24)	Cd				
(25)	I				
(26)	W				

Table 5 (continued)
UNDEFINED MARINE MEDIA (MINERAL)[gz]

Reference	106 (1914)	100 (1938)	102 (1951)	63 (1958)
Author's designation				AK
			Use	
Constituents	*Thalassiosira gravida*	*Phaeodactylum*	*Asterionella japonica* and other diatoms	*Asterionella japonica*
(27) EDTA[fl]				
(28) Citrate			1.89[ag]	
(29) TRIS[fj]				
(30) Glyceryl				
(31) *p*-Aminobenzoic acid				
(32) Biotin				
(33) Choline dihydrogen citrate				
(34) Cyanocobalamin				
(35) N^5-Formyltetrahydro-pteroylglutamic acid				
(36) Inositol				
(37) Nicotinic acid				
(38) Orotic acid				
(39) Pantothenic acid, Ca salt				
(40) Pteroylglutamic acid				
(41) Putrescine dihydro-chloride				
(42) Pyridoxamine dihydro-chloride				
(43) Pyridoxine hydrochloride				
(44) Riboflavine				
(45) Thiamine hydrochloride				
(46) Thymine				
(47) Soil extract				20 ml
(48) HCl	0.023 ml[cy]	0.01 ml[cy]		
(49) Sea salt				
(50) Sea water	38.4 ml	1,000 ml	1,000 ml[cw]	980 ml
(51) Fresh water	To 1,000 ml[fu]			
(52) pH before autoclaving				
(53) Final pH				

Table 5 (continued)
UNDEFINED MARINE MEDIA (MINERAL)[gz]

Reference	63 (1958)	111 (1962)	111 (1962)	114 (1964)
Author's designation	AR_N	f	f-1	
		Use		
Constituents	*Asterionella japonica*	*Cyclotella nana* and other diatoms	*Cyclotella Detonula*	*Stephanopyxis turris*
(1) Bicarbonate				
(2) Nitrate	124[l]	109[j]	109[j]	31.0[j]
(3) Ammonium				
(4) Phosphate	19.0[e]	6.88[r]	6.88[r]	2.85[q]
(5) Sulfate				
(6) Sulfide				
(7) Silicon	2.81[i]	2.96–5.93[i]	2.96–5.93[i]	5.61[dg]
(8) Na	4.60[i]	47.1–51.9[c,i,r]	47.1–51.9[c,i,r]	13.4[c,q,bt]
(9) Mg				
(10) K	93.8[c,e]			
(11) Ca				
(12) Fe	0.307[ab]	1.3[bs,ea]	1.3[bs,ea]	0.0558[ac]
(13) B	5.95[n]			
(14) Al				
(15) V				
(16) Mn	1.10[aa]	0.1[z]	0.1[z]	0.0055[z]
(17) Co	0.0029[al]	0.005[aj]	0.005[aj]	
(18) Ni				
(19) Cu	0.0013[ap]	0.005[ap]	0.005[ap]	
(20) Zn	0.164[g]	0.01[g]	0.01[g]	
(21) As				
(22) Br				
(23) Mo		0.005[ax]	0.005[ax]	
(24) Cd				
(25) I				
(26) W				

Table 5 (continued)
UNDEFINED MARINE MEDIA (MINERAL)[gz]

	Reference	63 (1958)	111 (1962)	111 (1962)	114 (1964)
	Author's designation	AR_N	f	f-1	
			Use		
	Constituents	*Asterionella japonica*	*Cyclotella nana* and other diatoms	*Cyclotella Detonula*	*Stephanopyxis turris*
(27)	EDTA[fl]	100	Present[bs,ea]	Present[bs,ea]	2.92[bt]
(28)	Citrate				
(29)	TRIS[fj]	1,000		500[dz]	
(30)	Glyceryl				
(31)	*p*-Aminobenzoic acid				
(32)	Biotin		0.001	0.001	
(33)	Choline dihydrogen citrate				
(34)	Cyanocobalamin		0.001	0.001	0.0007
(35)	N^5-Formyltetrahydro-pteroylglutamic acid				
(36)	Inositol				
(37)	Nicotinic acid				
(38)	Orotic acid				
(39)	Pantothenic acid, Ca salt				
(40)	Pteroylglutamic acid				
(41)	Putrescine dihydro-chloride				
(42)	Pyridoxamine dihydro-chloride				
(43)	Pyridoxine hydrochloride				
(44)	Riboflavine				
(45)	Thiamine hydrochloride		0.2	0.2	
(46)	Thymine				
(47)	Soil extract				
(48)	HCl				
(49)	Sea salt				
(50)	Sea water	1,000 ml	To 1,000 ml[de]	To 1,000 ml[de]	1,000 ml
(51)	Fresh water				
(52)	pH before autoclaving				
(53)	Final pH				

Table 5 (continued)
UNDEFINED MARINE MEDIA (MINERAL)[gz]

	Reference	108 (1958)	150 (1969)
	Author's designation	ASW8	Medium C
		Use	
	Constituents	*Ulva*	*Cladophora glomerata* (unialgal)
(1)	Bicarbonate		14.5[j]
(2)	Nitrate	219[j]	26.3[m]
(3)	Ammonium		
(4)	Phosphate	9.04[ch]	2.73[e]
(5)	Sulfate		7.79[k]
(6)	Sulfide		
(7)	Silicon		
(8)	Na	85.5[c,ch]	5.47[b]
(9)	Mg		1.97[g]
(10)	K		7.49[e,h]
(11)	Ca		8.49[c]
(12)	Fe	0.8[h]	0.115[ff]
(13)	B	6.0[n]	
(14)	Al		
(15)	V		
(16)	Mn	1.2[h]	
(17)	Co	0.03[h]	
(18)	Ni		
(19)	Cu		
(20)	Zn	0.15[h]	
(21)	As		
(22)	Br		
(23)	Mo		
(24)	Cd		
(25)	I		
(26)	W		

Table 5 (continued)
UNDEFINED MARINE MEDIA (MINERAL)[gz]

Reference	108 (1958)	150 (1969)
Author's designation	ASW8	Medium C
	Use	
Constituents	*Ulva*	*Cladophora glomerata* (unialgal)
(27) EDTA[fl]	30	0.591[ff]
(28) Citrate		
(29) TRIS[fj]	1,000[dr]	
(30) Glyceryl	7.15[ch]	
(31) *p*-Aminobenzoic acid	0.005	
(32) Biotin	0.0005	
(33) Choline dihydrogen citrate		
(34) Cyanocobalamin	0.0001	
(35) N[5]-Formyltetrahydro- pteroylglutamic acid		
(36) Inositol	2.5	
(37) Nicotinic acid	0.05	
(38) Orotic acid		
(39) Pantothenic acid, Ca salt	0.05	
(40) Pteroylglutamic acid	0.001	
(41) Putrescine dihydro- chloride		
(42) Pyridoxamine dihydro- chloride		
(43) Pyridoxine hydrochloride		
(44) Riboflavine		
(45) Thiamine hydrochloride	0.25	
(46) Thymine	1.5	
(47) Soil extract		
(48) HCl		
(49) Sea salt		
(50) Sea water	800 ml	500 ml
(51) Fresh water	To 1,000 ml	To 1,000 ml[fu]
(52) pH before autoclaving	8.0	
(53) Final pH		

Compiled by M. F. Turner and M. R. Droop.

Table 6
UNDEFINED MARINE MEDIA (ORGANIC)[gz]

Reference	62 (1957)	62 (1957)	69 (1964)	119 (1949)
Author's designation	E6	STP	SWM-1	
			Use	
Constituents	Diatoms, flagellates, and sterility testing	Diatoms and flagellates	10 species of unicellular algae from 10 classes	*Porphyridium cruentum*
(1) Carbonate				
(2) Nitrate	30.7[l]	123[l]	62.0[j]	387[d]
(3) Ammonium				113[c]
(4) Phosphate	2.73[e]	5.45[e]	9.50[r]	54.5[e]
(5) Sulfate				19.5[k]
(6) Sulfide				
(7) Silicon			5.62[i]	
(8) Na		68.0[gw]	35.9[c,i,r,bt]	
(9) Mg				4.94[g]
(10) K	21.6[c,e]	81.8[c,e]		44.9[e]
(11) Ca				
(12) Fe			0.112[ab]	0.172[ab]
(13) B			2.16[n]	
(14) V				
(15) Cr				
(16) Mn			0.385[z]	
(17) Co			0.0012[aj]	
(18) Ni				
(19) Cu			0.000013[ao]	
(20) Zn			0.0523[h]	
(21) Mo				
(22) EDTA[fl]			8.77[bt]	
(23) NTA[fm]				
(24) Citrate				
(25) MES[fi]				
(26) TRIS[fj]				
(27) Glycylglycine			661	
(28) Acetate				
(29) Glucose	250			
(30) Glyceryl				
(31) Sucrose		1,000		
(32) 5-Sulfosalycilic acid				
(33) DL-Alanine		100		
(34) Betaine hydrochloride				
(35) Glutamate		429[gw]		
(36) Glycine		100		
(37) Histidine hydrochloride				
(38) Valine				
(39) Adenylic acid				

Table 6 (continued)
UNDEFINED MARINE MEDIA (ORGANIC)[gz]

Reference	62 (1957)	62 (1957)	69 (1964)	119 (1949)
Author's designation	E6	STP	SWM-1	
		Use		
Constituents	Diatoms, flagellates, and sterility testing	Diatoms and flagellates	10 species of unicellular algae from 10 classes	*Porphyridium cruentum*
(40) Cytidylic acid				
(41) Urea				
(42) *p*-Aminobenzoic acid		0.01	0.01	
(43) Biotin		0.0005	0.001	
(44) Choline dihydrogen citrate		0.5		
(45) Cyanocobalamin		0.00005	0.001	
(46) N^5-Formyltetrahydro-pteroylglutamic acid		0.0002		
(47) Inositol		1.0	5.0	
(48) Lipoic acid				
(49) Nicotinic acid		0.1	0.1	
(50) Orotic acid		0.26		
(51) Pantothenic acid, Ca salt		0.1	0.1	
(52) Pteroylglutamic acid		0.0025	0.002	
(53) Putrescine dihydrochloride		0.04		
(54) Pyridoxamine dihydro-chloride		0.02		
(55) Pyridoxine hydrochloride		0.04		
(56) Riboflavine		0.005		
(57) Thiamine hydrochloride		0.2	0.5	
(58) Thymine		0.8	0.003	
(59) Algal flour extract				
(60) Beef extract				
(61) "Casamino acids"				
(62) "Casitone"				
(63) Lemon factor[dq]				
(64) Liver extract	250[dn]			
(65) Peptone				
(66) "Phytone"[ec]				
(67) "Trypticase"		200		
(68) "Tryptone"	250[do]			2,000[dm]
(69) Yeast autolysate		200		
(70) Agar				
(71) Soil extract	2.5 ml	12–5[ha]		
(72) Sodium hydroxide				
(73) Fresh water	To 1,000 ml[fu]	To 1,000 ml[fu]		
(74) Sea salt				
(75) Sea water	500 ml[dc]	800 ml	To 1,000 ml	To 1,000 ml
(76) pH adjusted with	Not adjusted			
(77) pH before autoclaving		7.5–7.6	7.5	

Table 6 (continued)
UNDEFINED MARINE MEDIA (ORGANIC)[gz]

Reference	78 (1969)	125 (1965)	120 (1957)	85 (1959)
Author's designation	SWM-3			OX7
		Use		
Constituents	*Bonnemaisonia hamifera* (unialgal)	*Hemiselmis virescens*	Dinoflagellates	*Oxyrrhis marina* (axenic)
(1) Carbonate				
(2) Nitrate	124[j]	29.6[l]	73.0[j]	
(3) Ammonium				
(4) Phosphate	9.50[r]	4.58[r]	13.4[q]	5.45[e]
(5) Sulfate				
(6) Sulfide				
(7) Silicon	5.62[i]	8.01[i,fz]		
(8) Na	58.9[c,i,r,bt]	15.2[i,r,bt]	33.5[c,q]	561[cn]
(9) Mg				
(10) K		18.7[c]		4.49[e]
(11) Ca				
(12) Fe	0.112[ab]	0.538[ab]		
(13) B	10.8[n]			
(14) V				
(15) Cr				
(16) Mn	1.92[z]	0.267[aa]		
(17) Co	0.0059[aj]	0.0028[al]		
(18) Ni				
(19) Cu	0.000064[ao]	0.0061[ap]		
(20) Zn	0.262[h]	0.126[g]		
(21) Mo		0.93[ax]		
(22) EDTA[fl]	8.77[bt]	6.14[bt]		
(23) NTA[fm]				
(24) Citrate				
(25) MES[fi]				
(26) TRIS[fj]	500	965[bx,dh]		
(27) Glycylglycine				
(28) Acetate				1,440[j]
(29) Glucose				
(30) Glyceryl				
(31) Sucrose				
(32) 5-Sulfosalycilic acid				
(33) DL-Alanine				
(34) Betaine hydrochloride				
(35) Glutamate				
(36) Glycine		300[dc]		
(37) Histidine hydrochloride				
(38) Valine				250
(39) Adenylic acid				

Table 6 (continued)
UNDEFINED MARINE MEDIA (ORGANIC)[gz]

	78 (1969)	125 (1965)	120 (1957)	85 (1959)
Reference				
Author's designation	SWM-3			OX7
			Use	
Constituents	*Bonnemaisonia hamifera* (unialgal)	*Hemiselmis virescens*	Dinoflagellates	*Oxyrrhis marina* (axenic)
(40) Cytidylic acid				
(41) Urea				
(42) *p*-Aminobenzoic acid	0.01			
(43) Biotin	0.001	0.00096		0.01
(44) Choline dihydrogen citrate				
(45) Cyanocobalamin	0.001	0.0019		0.001
(46) N[5]-Formyltetrahydro-pteroylglutamic acid				0.004[ga]
(47) Inositol	5.0			
(48) Lipoic acid				0.01[ga]
(49) Nicotinic acid	0.1			
(50) Orotic acid				
(51) Pantothenic acid, Ca salt	0.1			
(52) Pteroylglutamic acid	0.002			0.02[ga]
(53) Putrescine dihydrochloride				
(54) Pyridoxamine dihydro-chloride				
(55) Pyridoxine hydrochloride				
(56) Riboflavine				
(57) Thiamine hydrochloride	0.5	0.482		4.0
(58) Thymine	0.003			
(59) Algal flour extract			1–8 ml[dp]	
(60) Beef extract				
(61) "Casamino acids"				
(62) "Casitone"				125[do]
(63) Lemon factor[dq]				3.0 ml
(64) Liver extract	10.0			
(65) Peptone				
(66) "Phytone"[ec]				
(67) "Trypticase"				
(68) "Tryptone"				
(69) Yeast autolysate				
(70) Agar				
(71) Soil extract	50 ml			30[ha]
(72) Sodium hydroxide				
(73) Fresh water		To 1,000 ml	50 ml[fv]	To 1,000 ml[fv]
(74) Sea salt				
(75) Sea water	To 1,000 ml	772 ml[cw,di]	To 1,000 ml	500 ml
(76) pH adjusted with		HCl[fz]		
(77) pH before autoclaving	7.5	7.6–7.8		

Table 6 (continued)
UNDEFINED MARINE MEDIA (ORGANIC)[gz]

Reference	86 (1959)	86 (1959)	86 (1959)	86 (1959)
Author's designation	M4	M5	M6	M8
Use				
Constituents	Zooxanthellae	Zooxanthellae	Zooxanthellae	Zooxanthellae
(1) Carbonate		24.2[j]	24.2[j]	24.2[j]
(2) Nitrate	613[l]	21.9[j]	22.0[d,j]	22.3[d,j]
(3) Ammonium	0.027[g]		0.0225[c]	0.113[c]
(4) Phosphate	16.3[e]	3.01[ch]	3.01[ch]	3.01[ch]
(5) Sulfate	2,340[k]	2,340[k]	2,340[k]	
(6) Sulfide		0.0589[j]		
(7) Silicon	1.98[i]			
(8) Na	9,450[h,i,bt]	10,300[a,c,h,bt,ch]	10,300[a,c,h,bt,ch]	1,600[a,c,h,ch]
(9) Mg	831[g,h]	592[g]	592[g]	
(10) K	715[c,e,h]	315[h]		
(11) Ca	100[h]	200[h]	100[h]	
(12) Fe	0.1	0.2	0.2	0.1[h]
(13) B	2.0	2.0	2.0	
(14) V				
(15) Cr				
(16) Mn	0.4	0.4	0.4	
(17) Co	0.01	0.01	0.01	
(18) Ni				
(19) Cu				
(20) Zn	0.05	0.05	0.05	
(21) Mo				
(22) EDTA[fl]	7.85[bt]	7.85[bt]	7.85[bt]	
(23) NTA[fm]		100	100	50
(24) Citrate				
(25) MES[fi]				
(26) TRIS[fj]	1,000	1,000	1,000	1,000
(27) Glycylglycine				
(28) Acetate				
(29) Glucose				
(30) Glyceryl		2.38[ch]	2.38[ch]	2.38[ch]
(31) Sucrose				
(32) 5-Sulfosalicylic acid				
(33) DL-Alanine				
(34) Betaine hydrochloride				
(35) Glutamate				
(36) Glycine				
(37) Histidine hydrochloride				
(38) Valine				
(39) Adenylic acid		10	10	

Table 6 (continued)
UNDEFINED MARINE MEDIA (ORGANIC)[gz]

Reference	86 (1959)	86 (1959)	86 (1959)	86 (1959)
Author's designation	M4	M5	M6	M8
	Use			
Constituents	Zooxanthellae	Zooxanthellae	Zooxanthellae	Zooxanthellae
(40) Cytidylic acid				10
(41) Urea		0.1		0.3
(42) *p*-Aminobenzoic acid	0.01	0.005	0.005	0.005
(43) Biotin	0.0015	0.00025	0.00025	0.00035
(44) Choline dihydrogen citrate	0.5	0.25	0.25	0.25
(45) Cyanocobalamin	0.00105	0.000125	0.000125	0.000225
(46) N^5-Formyltetrahydro-pteroylglutamic acid	0.0002	0.0001	0.0001	0.0001
(47) Inositol	1.0	0.5	0.5	0.5
(48) Lipoic acid				
(49) Nicotinic acid	0.1	0.05	0.05	0.05
(50) Orotic acid	0.02	0.01	0.01	0.01
(51) Pantothenic acid, Ca salt	0.1	0.05	0.05	0.05
(52) Pteroylglutamic acid	0.0025	0.00125	0.00125	0.00125
(53) Putrescine dihydrochloride	0.04	0.02	0.02	0.02
(54) Pyridoxamine dihydro-chloride	0.02	0.01	0.01	0.01
(55) Pyridoxine hydrochloride	0.04	0.02	0.02	0.02
(56) Riboflavine	0.005	0.0025	0.0025	0.0025
(57) Thiamine hydrochloride	0.3	0.1	0.1	0.2
(58) Thymine	0.8	0.4	0.4	0.4
(59) Algal flour extract				
(60) Beef extract				
(61) "Casamino acids"				
(62) "Casitone"				
(63) Lemon factor[dq]				
(64) Liver extract			10[ee,fn]	10[ee,fn]
(65) Peptone				
(66) "Phytone"[ec]		10		10
(67) "Trypticase"	10			10
(68) "Tryptone"				
(69) Yeast autolysate			10	10
(70) Agar				
(71) Soil extract				
(72) Sodium hydroxide				
(73) Fresh water	To 1,000 ml	To 1,000 ml	To 1,000 ml	To 1,000 ml
(74) Sea salt				
(75) Sea water				800 ml
(76) pH adjusted with				
(77) pH before autoclaving	7.6—7.8	7.8—8.0	7.8—8.0	7.8

Table 6 (continued)
UNDEFINED MARINE MEDIA (ORGANIC)[gz]

Reference	123 (1960)	138 (1968)	140 (1975)	61 (1958)
Author's designation			MLH	6
			Use	
Constituents	*Katodinium dorsalisulcum* (axenic)	*Crypthecodinium cohnii* (axenic)	*Crypthecodinium cohnii* (axenic)	Chrysophytes Haptophytes
(1) Carbonate	0.0283[j]	300[m]		
(2) Nitrate	21.9[j]			
(3) Ammonium		27.0[h]	54.1[g]	57.3–59.0[l]
(4) Phosphate	7.54[cj]	69.8[f]	75.0[ch]	
(5) Sulfate	2,340[k]	2,790[k]	2,830[k]	5.10–5.24[e]
(6) Sulfide	0.0411[j]			
(7) Silicon				
(8) Na	10,300[c,h,bt]	11,800[h]	8,250[h,ch,cn]	
(9) Mg	592[g]	707[g]	681[g]	
(10) K	315[h]	810[f,h,bn]	352[h]	40.3–41.5[c,e]
(11) Ca	200[h]	200[a]	301[h]	
(12) Fe	0.6	2.5[ad]	0.1[ad]	
(13) B	2.0	0.04[n]		
(14) V		0.08[v]		
(15) Cr		0.04[x]		
(16) Mn	0.4	2.00[aa]		
(17) Co	0.01	0.04[al]		
(18) Ni		0.04[am]		
(19) Cu		0.16[ap]		
(20) Zn	0.05	2.00[g]		
(21) Mo		0.8[ba]		
(22) EDTA[fl]	7.85[bt]			
(23) NTA[fm]	100	20	11.5	
(24) Citrate		583[l]		
(25) MES[fi]			1,560	
(26) TRIS[fj]	1,000	Present[eu]		
(27) Glycylglycine				
(28) Acetate			886[j]	
(29) Glucose		5,000	3,960	
(30) Glyceryl			59.3[ch]	
(31) Sucrose		1,000		
(32) 5-Sulfosalycilic acid			0.175	
(33) DL-Alanine		100		
(34) Betaine hydrochloride		100	1,490	
(35) Glutamate		1,000[gx]		
(36) Glycine				
(37) Histidine hydrochloride			153	
(38) Valine				
(39) Adenylic acid	10[eg]			

Table 6 (continued)
UNDEFINED MARINE MEDIA (ORGANIC)[gz]

Reference	123 (1960)	138 (1968)	140 (1975)	61 (1958)
Author's designation			MLH	6
			Use	
Constituents	*Katodinium dorsalisulcum* (axenic)	*Crypthecodinium cohnii* (axenic)	*Crypthecodinium cohnii* (axenic)	Chrysophytes Haptophytes
(40) Cytidylic acid				
(41) Urea				
(42) *p*-Aminobenzoic acid	0.005			
(43) Biotin	0.00025	0.002	0.002	
(44) Choline dihydrogen citrate	0.25			
(45) Cyanocobalamin	0.000125		0.001	
(46) N⁵-Formyltetrahydro-pteroylglutamic acid	0.001			
(47) Inositol	0.5			
(48) Lipoic acid				
(49) Nicotinic acid	0.05			
(50) Orotic acid	0.01			
(51) Pantothenic acid, Ca salt	0.05			
(52) Pteroylglutamic acid	0.00125			
(53) Putrescine dihydrochloride	0.02			
(54) Pyridoxamine dihydro-chloride	0.01			
(55) Pyridoxine hydrochloride	0.02			
(56) Riboflavine	0.0025			
(57) Thiamine hydrochloride	0.1	0.5	1.0	
(58) Thymine	0.4			
(59) Algal flour extract				
(60) Beef extract				
(61) "Casamino acids"				
(62) "Casitone"				
(63) Lemon factor[dq]				93.5—96.2[dn]
(64) Liver extract				
(65) Peptone				
(66) "Phytone"[ec]				
(67) "Trypticase"				93.5—96.2[dm]
(68) "Tryptone"				
(69) Yeast autolysate	0.5			
(70) Agar		15,000[ev]	10,000	
(71) Soil extract				3.85—6.54 ml[gh]
(72) Sodium hydroxide			12.0[ew]	
(73) Fresh water	To 1,000 ml	To 1,000 ml	To 1,000 ml	To 1,000 ml
(74) Sea salt				234—641 ml
(75) Sea water				
(76) pH adjusted with			9*M* NaOH	
(77) pH before autoclaving	7.4—8.0	6.0—6.2	6.6	7.2—7.6

Table 6 (continued)
UNDEFINED MARINE MEDIA (ORGANIC)[gz]

	124 (1963)	61 (1958)	126 (1954)	121 (1960)
Reference				
Author's designation		8		
	Use			
Constituents	*Prymnesium parvum*	*Pleurochrysis scherffellii* (axenic) *Stichochrysis immobilis* (axenic)	*Phaeodactylum tricornutum*	Littoral diatoms
(1) Carbonate				
(2) Nitrate	146[j]		53.1[j]	52.5[m]
(3) Ammonium				
(4) Phosphate	33.4[q]		9.20[q,dc]	10.9[e]
(5) Sulfate	1,170[k]			
(6) Sulfide				
(7) Silicon				4.94[i]
(8) Na	4,000[c,e,h]		24.1[c,e]	8.09[i]
(9) Mg	296[g]			
(10) K	420[h]			8.98[e]
(11) Ca	27.3[h]			17.0[c]
(12) Fe	0.207[ab]		0.1[ag,dc]	0.5
(13) B	1.75[n]			
(14) V				0.1
(15) Cr				
(16) Mn	1.39[z]		0.02[z]	0.1
(17) Co	0.00074[aj]			0.1
(18) Ni				
(19) Cu				
(20) Zn	0.0341[g]			0.1
(21) Mo	0.397[ax]			0.3
				0.1
(22) EDTA[fl]				
(23) NTA[fm]				
(24) Citrate			0.339[ag,dc]	
(25) MES[fi]				
(26) TRIS[fj]	1,000			
(27) Glycylglycine				
(28) Acetate				
(29) Glucose				
(30) Glyceryl				
(31) Sucrose				
(32) 5-Sulfosalycilic acid				
(33) DL-Alanine				
(34) Betaine hydrochloride				
(35) Glutamate				
(36) Glycine				
(37) Histidine hydrochloride				
(38) Valine				
(39) Adenylic acid				

Table 6 (continued)
UNDEFINED MARINE MEDIA (ORGANIC)[gz]

Reference	124 (1963)	61 (1958)	126 (1954)	121 (1960)
Author's designation		8		
Use				
Constituents	*Prymnesium parvum*	*Pleurochrysis scherffellii* (axenic) *Stichochrysis immobilis* (axenic)	*Phaeodactylum tricornutum*	Littoral diatoms
(40) Cytidylic acid				
(41) Urea				
(42) *p*-Aminobenzoic acid				
(43) Biotin				
(44) Choline dihydrogen citrate				0.001
(45) Cyanocobalamin	Required			
(46) N[5]-Formyltetrahydro-pteroylglutamic acid				
(47) Inositol				
(48) Lipoic acid				
(49) Nicotinic acid				
(50) Orotic acid				
(51) Pantothenic acid, Ca salt				
(52) Pteroylglutamic acid				
(53) Putrescine dihydrochloride				
(54) Pyridoxamine dihydro-chloride				
(55) Pyridoxine hydrochloride				
(56) Riboflavine				
(57) Thiamine hydrochloride	0.01			
(58) Thymine				
(59) Algal flour extract				
(60) Beef extract		909[dm]		
(61) "Casamino acids"	1,000[dm,du]			
(62) "Casitone"				
(63) Lemon factor[dq]				
(64) Liver extract			5,000[dw]	
(65) Peptone				
(66) "Phytone"[ec]				
(67) "Trypticase"				1,000[dm]
(68) "Tryptone"				
(69) Yeast autolysate				
(70) Agar		9,090	15,000	10,000[ga]
(71) Soil extract		90.9 ml[gh]		
(72) Sodium hydroxide				
(73) Fresh water	To 1,000 ml[fv]	To 1,000 ml	To 1,000 ml[fv]	
(74) Sea salt				
(75) Sea water		455 ml	750 ml[cw,cx,dd]	To 1,000 ml[cw]
(76) pH adjusted with				
(77) pH before autoclaving	8.2	7.0		

Table 6 (continued)
UNDEFINED MARINE MEDIA (ORGANIC)[gz]

	Reference	14 (1966)	108 (1958)
	Author's designation	I	ASW III
		Use	
	Constituents	*Cylindrotheca fusiformis*	*Ulva*
(1)	Carbonate		
(2)	Nitrate	117[j]	123[l]
(3)	Ammonium		
(4)	Phosphate		10.9[e]
(5)	Sulfate		
(6)	Sulfide		
(7)	Silicon	9.88[i]	
(8)	Na	61.0[c,i,bt]	68.0[gw]
(9)	Mg		
(10)	K		86.3[c,e]
(11)	Ca		
(12)	Fe	0.5	0.1[h]
(13)	B	0.1	
(14)	V		
(15)	Cr		
(16)	Mn	0.1	0.4[h]
(17)	Co		
(18)	Ni		
(19)	Cu	0.1	
(20)	Zn	0.3	
(21)	Mo	0.1	
(22)	EDTA[fl]	9.42[bt]	
(23)	NTA[fm]		
(24)	Citrate		
(25)	MES[fi]		
(26)	TRIS[fj]		
(27)	Glycylglycine		1,000[dr]
(28)	Acetate		
(29)	Glucose		
(30)	Glyceryl		
(31)	Sucrose		
(32)	5-Sulfosalycilic acid		
(33)	DL-Alanine		
(34)	Betaine hydrochloride		
(35)	Glutamate		429[gw]
(36)	Glycine		500
(37)	Histidine hydrochloride		
(38)	Valine		
(39)	Adenylic acid		

Table 6 (continued)
UNDEFINED MARINE MEDIA (ORGANIC)[gz]

	Reference	14 (1966)	108 (1958)
	Author's designation	I	ASW III
		Cylindrotheca fusiformis	*Ulva*
	Constituents		
(40)	Cytydylic acid		
(41)	Urea		
(42)	*p*-Aminobenzoic acid		0.01
(43)	Biotin		0.0005
(44)	Choline dihydrogen citrate		0.5
(45)	Cyanocobalamin		0.00005
(46)	N^5-Formyltetrahydro-pteroylglutamic acid		0.0002
(47)	Inositol		1.0
(48)	Lipoic acid		
(49)	Nicotinic acid		0.1
(50)	Orotic acid		0.26
(51)	Pantothenic acid, Ca salt		0.1
(52)	Pteroylglutamic acid		0.0025
(53)	Putrescine dihydrochloride		0.04
(54)	Pyridoxamine dihydrochloride		0.02
(55)	Pyridoxine hydrochloride		0.04
(56)	Riboflavine		0.005
(57)	Thiamine hydrochloride	0.5	0.2
(58)	Thymine		0.8
(59)	Algal flour extract		
(60)	Beef extract		
(61)	"Casamino acids"		
(62)	"Casitone"		
(63)	Lemon factor[dq]		
(64)	Liver extract		10[ee,fn]
(65)	Peptone		
(66)	"Phytone"[ec]		
(67)	"Trypticase"		
(68)	"Tryptone"	1,000[dm]	
(69)	Yeast autolysate		
(70)	Agar		
(71)	Soil extract		40 ml
(72)	Sodium hydroxide		
(73)	Fresh water	To 1,000 ml	
(74)	Sea salt	40,000[dx]	
(75)	Sea water		To 1,000 ml
(76)	pH adjusted with	NaOH	
(77)	pH before autoclaving	8.0	7.5

Compiled by M. F. Turner and M. R. Droop.

Table 7
DEFINED MARINE MEDIA (MINERAL)

Reference	62 (1957)	62 (1957)	62 (1957)	93 (1964)
Author's designation	ASP	ASP 2	ASP 6	CF-1
	Use			
Constituents	Diatoms and flagellates	Diatoms and flagellates	Diatoms and flagellates	Diatoms and flagellates
(1) Carbonate				
(2) Bicarbonate				$145^{j,dc}$
(3) Nitrite				
(4) Nitrate	61.3^l	36.5^j	219^j	124^j
(5) Ammonium				
(6) Phosphate	10.9^e	2.73^e	38.3^{by}	9.54^e
(7) Sulfate	$2,340^k$	$1,950^k$	$3,120^k$	$1,950^k$
(8) Sulfide				
(9) Silicon	2.47^i	14.8^i	6.92^i	$2.77^{i,ds}$
(10) Na	$9,450^{h,i,bt}$	$7,120^{c,h,i,bt}$	$9,530^{c,h,i}$	$7,190^{b,c,h,i}$
(11) Mg	$1,130^{g,h}$	494^g	790^g	494^g
(12) K	$362^{c,e,h}$	$317^{e,h}$	$399^{h,by}$	$323^{e,h}$
(13) Ca	400^h	100^h	150^h	27.3^h
(14) Fe	0.1^h	0.8^h	2^h	0.289^{ab}
(15) Li				
(16) B	2.0^n	6.0^n	2.0^n	
(17) F				
(18) Al				
(19) Ti				
(20) V				
(21) Cr				
(22) Mn	0.4^h	1.2^h	1.0^h	0.555^z
(23) Co	0.001^h	0.003^h	0.01^h	0.0297^{aj}
(24) Ni				
(25) Cu	0.0004^h	0.0012^h	0.02^h	
(26) Zn	0.05^h	0.15^h	0.5^h	0.106^h
(27) Se				
(28) Br				
(29) Rb				
(30) Sr				
(31) Zr				
(32) Mo			0.5^{ax}	
(33) Ru				

Table 7 (continued)
DEFINED MARINE MEDIA (MINERAL)

Reference	62 (1957)	62 (1957)	62 (1957)	93 (1964)
Author's designation	ASP	ASP 2	ASP 6	CF-1
	Use			
Constituents	Diatoms and flagellates	Diatoms and flagellates	Diatoms and flagellates	Diatoms and flagellates
(34) Rh				
(35) Cd				
(36) Sn				
(37) I				
(38) Cs				
(39) Ba				
(40) Ce				
(41) EDTA[fl]	7.85[bt]	23.6[bt]		
(42) Versenol,[ft] trisodium salt			30	
(43) NTA[fm]				
(44) Citrate				
(45) TRIS[fj]		1,000	1,000	
(46) Glyceryl			30.3[by]	
(47) *p*-Aminobenzoic acid	0.005	0.01	0.01	
(48) Biotin	0.00025	0.001	0.0005	0.005
(49) Choline hydrogen citrate	0.250		0.5	
(50) Cyanocobalamin	0.000225	0.002	0.00055	0.0015
(51) N^5-Formyltetrahydro-pteroylglutamic acid	0.0001		0.0002	
(52) Ionsitol	0.5	5.0	1.0	
(53) Nicotinic acid	0.05	0.1	0.1	
(54) Orotic acid	0.130		0.26	
(55) Pantothenic acid, Ca salt	0.05	0.1	0.1	
(56) Pteroylglutamic acid	0.00125	0.002	0.0025	
(57) Putrescine dihydro-chloride	0.02		0.04	
(58) Pyridoxamine dihydro-chloride	0.01		0.02	
(59) Pyridoxine hydro-chloride	0.02		0.04	
(60) Riboflavine	0.0025		0.005	
(61) Thiamine hydrochlor-ide	0.1	0.5	0.2	5.0
(62) Thymine	0.4	3.0	0.8	
(63) Fresh water	To 1,000 ml	To 1,000 ml	To 1,000 ml	To 1,000 ml
(64) pH adjusted with				HCl
(65) pH before autoclaving	7.6	7.6−7.8	7.4−7.6	7.5−7.8

Table 7 (continued)
DEFINED MARINE MEDIA (MINERAL)

Reference	67 (1962)	68 (1962)	70 (1963)	70 (1963)
Author's designation			ASP 1	ASP 7
Use				
Constituents	*Calothrix Nostoc*	*Porphyridium cruentum*	*Dasya pedicillata Spyridia filamentosa*	Rhodophytes, e.g., *Spermothamnion turneri*
(1) Carbonate				
(2) Bicarbonate		29.1[j]		
(3) Nitrite				
(4) Nitrate		613[l]	73.0[j]	36.5[j]
(5) Ammonium				
(6) Phosphate	137[e]	48.9[f]	10.9[e]	6.03[ch]
(7) Sulfate	71.7[l]	2,570[k]	2,340[k]	3,510[k]
(8) Sulfide				
(9) Silicon			2.47[i]	6.92[i]
(10) Na	1,970[h]	10,600[b,h,bt]	9,470[c,h,i,bt]	9,870[c,h,i,bt,ch]
(11) Mg	89.7[h]	1,320[g,h]	1,130[g,h]	888[g]
(12) K	171[e,g]	407[c,f]	324[e,h]	367[h]
(13) Ca	24[h]	409[h]	400[h]	300[h]
(14) Fe	2.07[h,ag]	0.496[ab]	0.1[h]	0.3[h]
(15) Li				
(16) B	0.1[n]	0.105[n]	2.0[n]	6.0[n]
(17) F				
(18) Al				
(19) Ti				
(20) V				
(21) Cr				
(22) Mn	0.1[h]	0.111[z]	0.4[h]	1.2[h]
(23) Co		0.0037[aj]	0.01[h]	0.03[h]
(24) Ni				
(25) Cu	0.01[g]	0.0149[ao]		
(26) Zn	0.01[g]	0.0192[h]	0.05[h]	0.15[h]
(27) Se				
(28) Br				
(29) Rb				
(30) Sr				
(31) Zr				
(32) Mo	0.1[ax]	0.201[ba]		
(33) Ru				

Table 7 (continued)
DEFINED MARINE MEDIA (MINERAL)

Reference	67 (1962)	68 (1962)	70 (1963)	70 (1963)
Author's designation			ASP 1	ASP 7
Use				
Constituents	*Calothrix* *Nostoc*	*Porphyridium* *cruentum*	*Dasya* *pedicillata* *Spyridia* *filamentosa*	Rhodophytes, e.g., *Spermothamnion* *turneri*
(34) Rh				
(35) Cd				
(36) Sn				
(37) I				
(38) Cs				
(39) Ba				
(40) Ce				
(41) EDTA[fl]		14.6[bt]	7.85[bt]	23.6[bt]
(42) Versenol,[ft] trisodium salt				70
(43) NTA[fm]				
(44) Citrate	9.84[fr] + 5.64[ag]		1,000	1,000
(45) TRIS[fj]		2,420[bx,cv]		
(46) Glyceryl				4.77[ch]
(47) *p*-Aminobenzoic acid			0.005	0.01
(48) Biotin			0.00025	0.001
(49) Choline hydrogen citrate			0.25	
(50) Cyanocobalamin			0.000225	0.001
(51) N^5-Formyltetrahydro-pteroylglutamic acid			0.0001	
(52) Inositol			0.5	5
(53) Nicotinic acid			0.05	0.1
(54) Orotic acid			0.13	
(55) Pantothenic acid, Ca salt			0.05	0.1
(56) Pteroylglutamic acid			0.00125	0.002
(57) Putrescine dihydro-chloride			0.02	
(58) Pyridoxamine dihydro-chloride			0.01	
(59) Pyridoxine hydrochlor-ide			0.02	
(60) Riboflavine			0.0025	
(61) Thiamine hydrochlor-ide			0.1	0.5
(62) Thymine			0.4	3
(63) Fresh water	To 1,000 ml[fu]	To 1,000 ml	To 1,000 ml	To 1,000 ml
(64) pH adjusted with				
(65) pH before autoclaving			7.6	7.8—8.0

Table 7 (continued)
DEFINED MARINE MEDIA (MINERAL)

Reference	70 (1963)	65 (1960)	71 (1963)	71 (1963)
Author's designation	ASP 12		AC	MGC
Use				
Constituents	Rhodophytes, e.g., *Bangia fuscopurpurea Porphyra tenera* (Conchocelis)	*Gymnodinium breve* (axenic)	*Amphidinium carteri*	*Gymnodinium splendens*
(1) Carbonate			17.0^j	
(2) Bicarbonate		0.726^j		
(3) Nitrite		0.541^l		
(4) Nitrate	73.0^j	0.613^l	73.0^j	730^j
(5) Ammonium		0.337^h		
(6) Phosphate	$7.49^{ca,ch}$	$0.545^e + 0.349^f$	5.45^e	37.7^{cj}
(7) Sulfate	$2,730^k$	$2,340^k$	$1,950^k$	$2,340^k$
(8) Sulfide		0.507^j		
(9) Silicon	14.8^i			
(10) Na	$11,100^{c,h,i,bt}$	$11,409^h$	$7,120^{a,c,h}$	$9,710^{c,h}$
(11) Mg	$1,169^{g,h}$	$1,130^{g,h}$	494^g	592^g
(12) K	$373^{h,ca}$	315^h	$424^{e,h}$	315^h
(13) Ca	400^h	253^h	100^h	100^h
(14) Fe	0.1^h	0.02^{ae}	0.8^h	$2^{ci} + 0.3^h$
(15) Li	0.2^h			
(16) B	2.0^n	0.1^n	6.0	6.0
(17) F				
(18) Al		0.1^h		
(19) Ti		0.1^{fq}		
(20) V		0.02^v		
(21) Cr		0.1		
(22) Mn	0.4^h	0.2^z	1.2^h	1.2^h
(23) Co	0.01^h	0.1^{aj}	0.03^h	0.03^h
(24) Ni		0.02^{cb}		
(25) Cu		0.02^{ao}	0.012^h	
(26) Zn	0.05^h	0.1^h	0.150^h	0.150^h
(27) Se		0.1^{ar}		
(28) Br	10^j			
(29) Rb	0.2^h	0.2^{bw}		
(30) Sr	2.0^h	0.1^h		
(31) Zr		0.1^{cc}		
(32) Mo	0.5^{ax}	0.1^{ax}		
(33) Ru		0.02^{cd}		

Table 7 (continued)
DEFINED MARINE MEDIA (MINERAL)

Reference	70 (1963)	65 (1960)	71 (1963)	71 (1963)
Author's designation	ASP 12		AC	MGC
Use				
Constituents	Rhodophytes, e.g., *Bangia fuscopurpurea Porphyra tenera* (Conchocelis)	*Gymnodinium breve* (axenic)	*Amphidinium carteri*	*Gymnodinium splendens*
(34) Rh		0.02[ce]		
(35) Cd		0.02[h]		
(36) Sn		0.02[bb]		
(37) I	0.01[l]			
(38) Cs		0.1[h]		
(39) Ba		0.02[h]		
(40) Ce		0.02[cf]		
(41) EDTA[fl]	7.85[bt]	2.36[bt]	30	30
(42) Versenol,[ft] trisodium salt				
(43) NTA[fm]	100			100
(44) Citrate				
(45) TRIS[fj]	1,000	20[dl]	1,000	1,000
(46) Glyceryl	2.38[ch]			
(47) *p*-Aminobenzoic acid				
(48) Biotin	0.001	0.0005	0.0005	
(49) Choline hydrogen citrate				
(50) Cyanocobalamin	0.0002	0.001	0.00003	0.0001
(51) N^5-Formyltetrahydro-pteroylglutamic acid				
(52) Inositol				
(53) Nicotinic acid				
(54) Orotic acid				
(55) Pantothenic acid, Ca salt				
(56) Pteroylglutamic acid				
(57) Putrescine dihydro-chloride				
(58) Pyridoxamine dihydro-chloride				
(59) Pyridoxine hydrochlor-ide				
(60) Riboflavine				
(61) Thiamine hydrochlor-ide	0.1	1.0	0.5	
(62) Thymine				
(63) Fresh water	To 1,000 ml	To 1,000 ml[bi]	To 1,000 ml	To 1,000 ml
(64) pH adjusted with				
(65) pH before autoclaving	7.8—8.0		7.8	7.8—8.0

Table 7 (continued)
DEFINED MARINE MEDIA (MINERAL)

	115 (1969)	145 (1975)	64 (1958)	72 (1964)
Reference				
Author's designation	NH 15	NS		
			Use	
Constituents	*Gymnodinium breve* and related forms	*Cachonina niei* (axenic)	*Isochrysis Monochrysis Prymnesium*	*Coccolithus huxieyi* (unialgal)
(1) Carbonate				
(2) Bicarbonate	0.726[j]	61.0[j]		119[j]
(3) Nitrite				
(4) Nitrate	6.13[l]	219[j]	365[j]	73.0[j]
(5) Ammonium	0.337[h]			
(6) Phosphate	5.73[e]	30.1[ch]	13.4[ca]	5.30[q]
(7) Sulfate	2,340[k]	2,400[k]	1,950[k]	558[m]
(8) Sulfide	0.100[j]			
(9) Silicon	0.250[i]		19.8[i]	
(10) Na	9,440[h,bt]	7,680[b,c,h,bt,ch]	7,250[c,h,i,bt]	11,900[a,c,h,q,bt]
(11) Mg	1,130[g,h]	606[g]	494[g]	598[h]
(12) K	323[c,e,h]	393[h]	331[h,ca]	431[h,as]
(13) Ca	253[h]	199[h]	100[h]	233[g]
(14) Fe	0.25[ab]	0.3[ab]	0.4	0.1[bq]
(15) Li				
(16) B	0.25[n]	6.0[n]	6.0	0.006[h]
(17) F				
(18) Al				
(19) Ti	0.25[fg]			0.028[h]
(20) V	0.025[v]			
(21) Cr	0.01[w]			
(22) Mn	0.05[z]	1.2[z]	1.2	0.65[g]
(23) Co		0.03[aj]	0.003	0.0063[g]
(24) Ni				
(25) Cu			0.0012	0.0013[g]
(26) Zn		0.15[h]	0.150	2.3[g]
(27) Se				
(28) Br				22[as]
(29) Rb				0.071[h]
(30) Sr				3.8[h]
(31) Zr	0.1[cc]			
(32) Mo				0.2[ax]
(33) Ru				

Table 7 (continued)
DEFINED MARINE MEDIA (MINERAL)

	Reference	115 (1969)	145 (1975)	64 (1958)	72 (1964)
	Author's designation	NH 15 NS			
				Use	
Constituents		*Gymnodinium breve* and related forms	*Cachonina niei* (axenic)	*Isochrysis Monochrysis Prymnesium*	*Coccolithus huxleyi* (unialgal)
(34)	Rh				
(35)	Cd				
(36)	Sn				0.02[bc]
(37)	I				
(38)	Cs				
(39)	Ba	0.05[h]			
(40)	Ce				
(41)	EDTA[fl]	7.85[bt]	29.9[bt]	23.6[bt]	15.7[bt]
(42)	Versenol,[ft] trisodium salt				
(43)	NTA[fm]				
(44)	Citrate				
(45)	TRIS[fj]	400	610	1,000	250
(46)	Glyceryl		23.8[ch]		
(47)	*p*-Aminobenzoic acid				
(48)	Biotin	0.0005		0.0001	0.001
(49)	Choline hydrogen citrate				
(50)	Cyanocobalamin	0.001	0.001	Required	0.0002
(51)	N[5]-Formyltetrahydro-pteroylglutamic acid				
(52)	Inositol				
(53)	Nicotinic acid				
(54)	Orotic acid				
(55)	Pantothenic acid, Ca salt				
(56)	Pteroylglutamic acid				
(57)	Putrescine dihydro-chloride				
(58)	Pyridoxamine dihydro-chloride				
(59)	Pyridoxine hydrochlor-ide				
(60)	Riboflavine				
(61)	Thiamine hydrochlor-ide	10.0	1.0	0.5	0.2
(62)	Thymine				
(63)	Fresh water	To 1,000 ml[bi]	To 1,000 ml[fv]	To 1,000 ml	To 1,000 ml
(64)	pH adjusted with				HCl
(65)	pH before autoclaving		7.9	7.5–8.0	8.0

Table 7 (continued)
DEFINED MARINE MEDIA (MINERAL)

Reference	73 (1965)	60 (1954)	117 (1955)	63 (1958)
Author's designation			S36	AR_A
			Use	
Constituents	*Prymnesium parvum* (base)	*Nitzschia*	*Skeletonema costatum*	*Asterionella japonica* (unialgal)
(1) Carbonate				
(2) Bicarbonate		144[j]		139[j]
(3) Nitrite				
(4) Nitrate	146[j]	61.3[l]	61.3[l]	124[l]
(5) Ammonium				
(6) Phosphate	33.4[q]	5.30[q]	5.45[e]	19.0[e]
(7) Sulfate	1,170[k]	1,270[k]	353[m]	2,650[j]
(8) Sulfide				
(9) Silicon		1.98[i]	9.88[i]	2.81[i]
(10) Na	4,000[c,h,q]	10,600[b,h,i,q]	5,920[h,i,bt]	10,600[b,g,h,i]
(11) Mg	296[g]	591[g,h]	299[h]	1,220[h]
(12) K	420[h]	441[c,h,as]	245[c,e,h,as]	474[c,e,h,as]
(13) Ca	27.3[h]	416[h]	147[g]	437[h]
(14) Fe	0.207[ab]	0.167[ag]	0.1[bq]	0.307[ab]
(15) Li			0.0057[h]	
(16) B	1.75[n]	10.2[n]		10.5[n]
(17) F				1.36[bz]
(18) Al			0.039[g]	
(19) Ti				
(20) V				
(21) Cr				
(22) Mn	1.39[z]		0.728[aa]	1.10[z]
(23) Co	0.00074[aj]		0.0065[al]	0.0029[al]
(24) Ni				
(25) Cu			0.0012[ap]	0.0013[ap]
(26) Zn	0.0341[g]		2.27[g]	0.164[g]
(27) Se				
(28) Br		38.9[as]	10.1[as]	64.4[as]
(29) Rb			0.035[bw]	
(30) Sr			2.76[h]	13.1[h]
(31) Zr				
(32) Mo	0.397[ax]		0.186[ax]	
(33) Ru				

Table 7 (continued)
DEFINED MARINE MEDIA (MINERAL)

Reference	73 (1965)	60 (1954)	117 (1955)	63 (1958)
Author's designation			S36	AR_A
		Use		
Constituents	Prymnesium parvum (base)	Nitzschia	Skeletonema costatum	Asterionella japonica (unialgal)
(34) Rh				
(35) Cd				
(36) Sn				
(37) I				
(38) Cs				
(39) Ba				
(40) Ce				
(41) EDTA[fl]			15.7[bt]	100
(42) Versenol,[ft] trisodium salt				
(43) NTA[fm]				
(44) Citrate		0.564[ag]		
(45) TRIS[fj]	1,000		500	1,000
(46) Glyceryl				
(47) p-Aminobenzoic acid				
(48) Biotin				
(49) Choline hydrogen citrate				
(50) Cyanocobalamin	0.0001		0.0001	
(51) N^5-Formyltetrahydro-pteroylglutamic acid				
(52) Inositol				
(53) Nicotinic acid				
(54) Orotic acid				
(55) Pantothenic acid, Ca salt				
(56) Pteroylglutamic acid				
(57) Putrescine dihydrochloride				
(58) Pyridoxamine dihydrochloride				
(59) Pyridoxine hydrochloride				
(60) Riboflavine				
(61) Thiamine hydrochloride	0.01		1.0	
(62) Thymine				
(63) Fresh water	To 1,000 ml	To 1,000 ml	To 1,000 ml[fv]	To 1,000 ml[fu]
(64) pH adjusted with				
(65) pH before autoclaving	8.2—8.4			

Table 7 (continued)
DEFINED MARINE MEDIA (MINERAL)

Reference	77 (1958)	71 (1963)	74 (1965)	66 (1961)
Author's designation		DV		F$_2$
			Use	
Constituents	Phaeodactylum tricornutum	Amphora perpusilla	Cyclindrotheca fusiformis	Ectocarpus
(1) Carbonate				
(2) Bicarbonate				
(3) Nitrite				
(4) Nitrate	263[m]	365[j]	182[j]	124[l]
(5) Ammonium				
(6) Phosphate	54.5[e]	16.3[e]	11.3[e]	19.0[e]
(7) Sulfate	39.0[k]	1,950[k]	2,400[k]	558[m]
(8) Sulfide				
(9) Silicon	5.00[i]		14.8[i]	
(10) Na	7,880[h,i]	7,220[c,h]	5,800[c,h,i]	11,800[h,bt]
(11) Mg	9.87[g]	494[g]	607[g]	598[h]
(12) K	44.9[e]	328[e,h]	203[e,h]	498[c,e,h,as]
(13) Ca	84.9[c]	100[h]	204[h]	233[g]
(14) Fe	0.5	0.4[h]	0.5	0.502[ac]
(15) Li				0.0057[h]
(16) B	0.1	6.0	0.5	
(17) F				
(18) Al				0.0168[h]
(19) Ti				
(20) V				
(21) Cr				
(22) Mn	0.1	1.2[h]	0.1	0.0504[aa]
(23) Co	0.1	0.03[h]	0.1	0.0004[al]
(24) Ni				
(25) Cu	0.1	0.012[h]	0.1	0.0048[ap]
(26) Zn	0.3	0.150[h]	0.1	0.0050[g]
(27) Se				
(28) Br				21.8[as]
(29) Rb				0.0707[bw]
(30) Sr				2.14[h]
(31) Zr				
(32) Mo	0.1		0.1	0.0004[ax]
(33) Ru				

Table 7 (continued)
DEFINED MARINE MEDIA (MINERAL)

Reference	77 (1958)	71 (1963)	74 (1965)	66 (1961)
Author's designation		DV		F_2
		Use		
Constituents	*Phaeodactylum tricornutum*	*Amphora perpusilla*	*Cylindrotheca fusiformis*	*Ectocarpus*
(34) Rh				
(35) Cd				
(36) Sn				
(37) I				0.0191[bc]
(38) Cs				
(39) Ba				
(40) Ce				
(41) EDTA[fl]		30		39.3[bt]
(42) Versenol,[ft] trisodium salt				
(43) NTA[fm]				
(44) Citrate				
(45) TRIS[fj]		1,000	1,200[dr]	
(46) Glyceryl				
(47) *p*-Aminobenzoic acid		0.01		
(48) Biotin		0.0005		
(49) Choline hydrogen citrate		0.5		
(50) Cyanocobalamin		0.00305		
(51) N^5-Formyltetrahydro-pteroylglutamic acid		0.0002		
(52) Inositol		1.0		
(53) Nicotinic acid		0.1		
(54) Orotic acid		0.26		
(55) Pantothenic acid, Ca salt		0.1		
(56) Pteroylglutamic acid		0.0025		
(57) Putrescine dihydro-chloride		0.04		
(58) Pyridoxamine dihydro-chloride		0.02		
(59) Pyridoxine hydrochlor-ide		0.04		
(60) Riboflavine		0.005		
(61) Thiamine hydrochlor-ide		0.2	0.5	
(62) Thymine		0.8		
(63) Fresh water	To 1,000 ml	To 1,000 ml	To 1,000 ml[bl]	To 1,000 ml[fv]
(64) pH adjusted with				
(65) pH before autoclaving		7.5−7.8	8.0−8.1	

Table 7 (continued)
DEFINED MARINE MEDIA (MINERAL)

Reference	75 (1970)	59 (1954)	60 (1954)	155 (1975)
Author's designation		SSM		
		Use		
Constituents	*Platymonas Tetraselmis Prasinocladus*	*Stichococcus*	*Nannochloris Stichococcus*	*Dunaliella parva*
(1) Carbonate				
(2) Bicarbonate		145[j]	144[j]	
(3) Nitrite				
(4) Nitrate	146[j]	52.5[m]		310[j,l]
(5) Ammonium			17.9[h]	
(6) Phosphate	14.0[f]	10.9[e]	5.30[q]	9.50[e]
(7) Sulfate	1,950[k]	2,700[j]	1,270[k]	2,300[k]
(8) Sulfide				
(9) Silicon			1.98[i]	
(10) Na	10,700[c,h,ef]	10,400[b,g,h]	10,600[b,h,i,q]	34,600[c,h]
(11) Mg	494[g]	1,280[h]	591[g,h]	1,070[g,h]
(12) K	478[f,h]	8.98[e]	402[h,as]	46.9[c,e]
(13) Ca	274[h]	17.0[c]	416[h]	401[h]
(14) Fe	0.207[ab]	0.367[ai]	0.167[ag]	0.0838[ab]
(15) Li				
(16) B		0.1[n]	10.2[n]	2.0[n]
(17) F				
(18) Al				
(19) Ti				
(20) V				
(21) Cr				
(22) Mn	0.0555[z]	0.138[z]		0.385[z]
(23) Co	0.0005[aj]			0.0012[aj]
(24) Ni				
(25) Cu	0.0005[ap]	0.1[ao]		0.000013[ao]
(26) Zn	0.00046[g]	0.3[h]		0.0523[h]
(27) Se				
(28) Br			38.9[as]	
(29) Rb				
(30) Sr				
(31) Zr				
(32) Mo	0.0004[ax]	0.1[ax]		
(33) Ru				

Table 7 (continued)
DEFINED MARINE MEDIA (MINERAL)

Reference	75 (1970)	59 (1954)	60 (1954)	155 (1975)
Author's designation		SSM		
	Use			
Constituents	*Platymonas Tetraselmis Prasinocladus*	*Stichococcus*	*Nannochloris Stichococcus*	*Dunaliella parva*
(34) Rh				
(35) Cd				
(36) Sn				
(37) I				
(38) Cs				
(39) Ba				
(40) Ce				
(41) EDTA[fl]	14.6[ef]			8.77
(42) Versenol,[ft] trisodium salt				
(43) NTA[fm]				
(44) Citrate			0.564[ag]	
(45) TRIS[fj]	769[ck]			2,420[fo]
(46) Glyceryl				
(47) *p*-Aminobenzoic acid				
(48) Biotin	0.001			
(49) Choline hydrogen citrate				
(50) Cyanocobalamin	0.001	Required		
(51) N^5-Formyltetrahydro-pteroylglutamic acid				
(52) Inositol				
(53) Nicotinic acid				
(54) Orotic acid				
(55) Pantothenic acid, Ca salt				
(56) Pteroylglutamic acid				
(57) Putrescine dihydro-chloride				
(58) Pyridoxamine dihydro-chloride				
(59) Pyridoxine hydrochlor-ide				
(60) Riboflavine				
(61) Thiamine hydrochlor-ide	1.0			
(62) Thymine				
(63) Fresh water	To 1,000 ml	To 1,000 ml	To 1,000 ml	To 1,000 ml
(64) pH adjusted with				
(65) pH before autoclaving				

Table 8
DEFINED MARINE MEDIA (ORGANIC)

Reference	62 (1957)	62 (1957)	87 (1960)	163 (1970)
Author's designation	DC	RC	MKD	S
Use	Diatoms and flagellates	Diatoms and flagellates	Dinoflagellates Chrysophytes *Melosira* (all axenic)	Algae of salt marsh communities (axenic)
Constituents				
(1) Carbonate				
(2) Bicarbonate				
(3) Nitrate	365[j]	61.3[l]	36.5[j]	72.6[j]
(4) Ammonium			0.273[g]	194[d]
(5) Phosphate	153[by]	5.45[e]	5.45[e]	56.4[c]
(6) Sulfate	1,950[k]	894[j]	3,500[k]	30.1[ch], 3,510[k]
(7) Silicon	19.8[i]		0.988[i]	6.92[i]
(8) Na	7,500[c,h,i,bt,cn,cu,gw]	8,790[g,h,bt,cn,gw]	9,470[c,h,ct]	9,890[b,h,i,ct,ch]
(9) Mg	494[g]	598[h]	888[g]	888[g]
(10) K	441[h,by]	358[c,e,h]	372[e,h]	367[h]
(11) Ca	100[h]	70[h]	300[h]	300[h]
(12) Fe	0.8[h]	0.1[h]	0.03	0.3[h]
(13) Li		0.1		
(14) B	6.0[n]	2.0[n]	0.6	6.0[n]
(15) F				
(16) Al		0.5		
(17) Mn	1.2[h]	0.4[h]		
(18) Co	0.003[h]	0.001[h]	0.12	1.2[h]
(19) Cu	0.0012[h]	0.0004[h]	0.003	0.03[h]
(20) Zn	0.15[h]	0.05[h]		0.15[h]
(21) Se			0.015	
(22) Br		65		
(23) Rb		0.2		
(24) Sr		13		
(25) Mo				
(26) I		0.05		
(27) EDTA[fl]	23.6[bt]	7.85[bt]	2.36[bt]	23.6[bt]
(28) NTA[fm]				
(29) Citrate			200	7.00
(30) TRIS[fj]	5,000	5,000	1,000	1,000
(31) Glycylglycine				
(32) Acetate	217[j]	86.8[j]		
(33) Fumarate			35.6[j]	
(34) Gluconate				
(35) Glucose				
(36) Glyceryl	121[by]			90
(37) Glycerol				23.8[ch]
(38) Lactate	397[j]			
(39) D-Ribose				
(40) Sorbitol			50	
(41) Succinic acid				
(42) Sucrose	500	700		
(43) Alanine			50[cq]	44

Table 8 (continued)
DEFINED MARINE MEDIA (ORGANIC)

	62 (1957)	62 (1957)	87 (1960)	163 (1970)
Reference				
Author's designation	DC	RC	MKD	S
			Use	
Constituents	Diatoms and flagellates	Diatoms and flagellates	Dinoflagellates Chrysophytes *Melosira* (all axenic)	Algae of salt marsh communities (axenic)
(44) β-Alanine				
(45) Asparagine				660
(46) Aspartic acid				
(47) Cystine	429[gw]	429[gw]		73[gx]
(48) Glutamate	500[cq]			37
(49) Glycine				
(50) Histidine				
(51) Leucine				
(52) DL-Lysine				
(53) Methionine				
(54) L-Proline				
(55) Serine				
(56) Tryptophan				
(57) L-Valine				
(58) Adenine				
(59) Cytidylic acid				
(60) Thymine	0.8	0.16		
(61) Uracil				
(62) Xanthine			1.0	
(63) Urea				
(64) *p*-Aminobenzoic acid	0.01	0.002		
(65) L-Ascorbic acid				
(66) Biotin	0.0005	0.0006	0.01	
(67) Choline				
(68) Choline hydrogen citrate	0.5	0.1		
(69) Cyanocobalamin	0.00305	0.001	0.01	0.01
(70) N[5]-Formyltetrahydro-pteroylglutamic acid	0.0002	0.00004		
(71) Inositol	1.0	0.2		
(72) Nicotinic acid	0.1	0.02		
(73) Orotic acid	0.26	0.052		
(74) Pantothenic acid, Ca salt	0.1	0.02		
(75) Pteroylglutamic acid	0.0025	0.0005	0.02	
(76) Putrescine dihydrochloride	0.04	0.008		
(77) Pyridoxamine dihydro-chloride	0.02	0.004		
(78) Pyridoxine hydrochloride	0.04	0.008	0.02	
(79) Riboflavine	0.005	0.001		
(80) Thiamine hydrochloride	0.2	1.04	0.1	
(81) Ergosterol				
(82) Ubiquinone				
(83) Taurocholic acid, Na salt		3.0		
(84) Fresh water	To 1,000 ml	To 1,000 ml	To 1,000 ml	To 1,000 ml
(85) pH adjusted with				
(86) pH before autoclaving	7.6—8.0	7.2—7.4	7.6—7.8	

Table 8 (continued)
DEFINED MARINE MEDIA (ORGANIC)

Reference	84 (1958)	62 (1957)	86 (1959)	86 (1959)
Author's designation		S46	M9	CSl
		Use		
Constituents	*Phormidium persicinum* (axenic)	*Hemiselmis virescens*	*Gymnodinium adriaticum Exuviaella* sp. (both axenic)	*Gymnodinium adriaticum Exuviaella* sp. (both axenic)
(1) Carbonate			0.145[j]	0.145[j]
(2) Bicarbonate				
(3) Nitrate	73.0[j]	61.3[l]	36.5[j]	7.30[j]
(4) Ammonium			0.273[g]	
(5) Phosphate	22.4[ca]	5.45[e]	6.03[ch]	0.904[ch] + 5.45[e]
(6) Sulfate	1,950[k]	288[m]	3,510[k]	2,340[k]
(7) Silicon			9.88[i]	1.98[i]
(8) Na	9,860[c,h]	5,990[h,bt,cn]	10,300[c,h,i,bt,ch]	11,000[h,bt,gw]
(9) Mg	494[g]	299[h]	888[g]	1,070[g,h]
(10) K	237[h,ca]	264[c,e,h,as]	315[h]	319[e,h]
(11) Ca	80.0[h]	120[g]	300[h]	400[h]
(12) Fe	4.01[h]	0.1[bq]	0.4[h]	0.2[h]
(13) Li		0.006[h]		
(14) B			6.0	2.0
(15) F				
(16) Al		0.028[h]		
(17) Mn	0.16[h]	0.65[g]	1.2	0.4
(18) Co	0.002[h]	0.0063[g]	0.03	0.01
(19) Cu	0.004[h]	0.0013[g]		
(20) Zn	0.16[h]	2.3[g]	0.15	0.05
(21) Se				
(22) Br		22.0[as]		
(23) Rb		0.061[h]		
(24) Sr		3.8[h]		
(25) Mo		0.2[ax]		
(26) I		0.02[bc]		
(27) EDTA[fl]	200	15.7[bt]	23.6[bt]	7.85[bt]
(28) NTA[fm]			200	
(29) Citrate	0.118[fr]			
(30) TRIS[fj]	1,000	500	500	1,000
(31) Glycylglycine				
(32) Acetate		216[j]		
(33) Fumarate				
(34) Gluconate				
(35) Glucose		300		
(36) Glyceryl			4.77[ch]	0.715[ch]
(37) Glycerol				
(38) Lactate				30[cs,gq]
(39) D-Ribose				
(40) Sorbitol				
(41) Succinic acid				
(42) Sucrose				
(43) Alanine				

Table 8 (continued)
DEFINED MARINE MEDIA (ORGANIC)

Reference	84 (1958)	62 (1957)	86 (1959)	86 (1959)
Author's designation		S46	M9	CSI
	Use			
Constituents	*Phormidium persicinum* (axenic)	*Hemiselmis virescens*	*Gymnodinium adriaticum* *Exuviaella* sp. (both axenic)	*Gymnodinium adriaticum* *Exuviaella* sp. (both axenic)
(44) β-Alanine				
(45) Asparagine	200[cq]	300		
(46) Aspartic acid				
(47) Cystine				
(48) Glutamate		300[gx]		25.7[gw]
(49) Glycine		300		
(50) Histidine				
(51) Leucine				
(52) DL-Lysine				
(53) Methionine				
(54) L-Proline				
(55) Serine				
(56) Tryptophan				
(57) L-Valine				
(58) Adenine				
(59) Cytidylic acid				0.3
(60) Thymine			0.08	0.08
(61) Uracil				
(62) Xanthine				
(63) Urea			1.0	5.0
(64) *p*-Aminobenzoic acid			0.001	0.001
(65) L-Ascorbic acid				
(66) Biotin			0.00105	0.00015
(67) Choline				
(68) Choline hydrogen citrate			0.05	0.05
(69) Cyanocobalamin	0.002	0.0001	0.001	0.000105
(70) N^5-Formyltetrahydro-pteroylglutamic acid			0.00002	0.00002
(71) Inositol			0.1	0.1
(72) Nicotinic acid			0.01	0.01
(73) Orotic acid			0.002	0.002
(74) Pantothenic acid, Ca salt			0.01	0.01
(75) Pteroylglutamic acid			0.00025	0.00025
(76) Putrescine dihydrochloride			0.004	0.004
(77) Pyridoxamine dihydro-chloride			0.002	0.002
(78) Pyridoxine hydrochloride			0.004	0.004
(79) Riboflavine			0.0005	0.0005
(80) Thiamine hydrochloride		1.0	0.52	0.12
(81) Ergosterol				
(82) Ubiquinone				
(83) Taurocholic acid, Na salt				
(84) Fresh water	To 1,000 ml[fu]	To 1,000 ml[fv]	To 1,000 ml	To 1,000 ml
(85) pH adjusted with		HCl or NaOH		
(86) pH before autoclaving	7.0	8.0	7.8–8.0	7.8–8.0

Table 8 (continued)
DEFINED MARINE MEDIA (ORGANIC)

	87 (1960)	87 (1960)	71 (1963)	85 (1959) 167 (1971)
Reference				
Author's designation	MDV	MMK		S68 (modified)
			Use	
Constituents	Dinoflagellates, e.g., *Katodinium dorsalisulcum*	Mass culture of *Katodinium dorsalisulcum* (axenic)	*Gyrodinium californicum*	*Oxyrrhis marina*
(1) Carbonate				
(2) Bicarbonate				
(3) Nitrate			61.3^l	
(4) Ammonium	1.69^h			
(5) Phosphate	16.3^e	5.45^e	14.0^f	5.45^e
(6) Sulfate	$1,950^k$	$3,510^k$	117^k	279^m
(7) Silicon	19.8^i	0.988^i		
(8) Na	$7,120^{h,i,bt}$	$9,850^{h,bt,ct}$	$9,450^{h,cp,gw}$	$6,460^{h,cn}$
(9) Mg	494^g	888^g	$388^{g,h}$	299^h
(10) K	$328^{e,h}$	$372^{e,h}$	$202^{c,f,h}$	$225^{e,h,as}$
(11) Ca	100^h	300^h	50^h	116^g
(12) Fe	0.3	0.3	3.0^h	0.502^{ac}
(13) Li			0.1^h	
(14) B	6.0	6.0	2.0	
(15) F				
(16) Al			0.25^h	0.028^h
(17) Mn	1.2	1.2	10^h	0.0325^{aa}
(18) Co	0.03	0.03	0.03^h	0.000524^{al}
(19) Cu			0.003^h	0.0051^{ap}
(20) Zn	0.15	0.15	4.0^h	0.0227^g
(21) Se				
(22) Br			32.5^{cp}	22.2^{as}
(23) Rb			0.1^h	0.0707^h
(24) Sr			6.5^h	2.14^h
(25) Mo			0.5^{ax}	0.000475^{ax}
(26) I			0.025^{bc}	0.0382^{bc}
(27) EDTA[fl]	23.6^{bt}	23.6^{bt}	200	6.0
(28) NTA[fm]	0.003	200		
(29) Citrate				
(30) TRIS[fj]	100	1,000		
(31) Glycylglycine				500
(32) Acetate				
(33) Fumarate		35.6^j		$1,440^j$
(34) Gluconate				
(35) Glucose				
(36) Glyceryl				
(37) Glycerol				
(38) Lactate				
(39) D-Ribose		50		
(40) Sorbitol		50		
(41) Succinic acid				
(42) Sucrose				
(43) Alanine	10^{cq}	500^{cq}		

Table 8 (continued)
DEFINED MARINE MEDIA (ORGANIC)

Reference	87 (1960)	87 (1960)	71 (1963)	85 (1959) 167 (1971)
Author's designation	MDV	MMK		S68 (modified)
Use				
Constituents	Dinoflagellates, e.g., *Katodinium dorsalisulcum*	Mass culture of *Katodinium dorsalisulcum* (axenic)	*Gyrodinium californicum*	*Oxyrrhis marina*
(44) β-Alanine				
(45) Asparagine				
(46) Aspartic acid				
(47) Cystine			17.2^{gw}	
(48) Glutamate				
(49) Glycine				
(50) Histidine			2.0^{cq}	
(51) Leucine			10	
(52) DL-Lysine				
(53) Methionine				40
(54) L-Proline				
(55) Serine				
(56) Tryptophan				250
(57) L-Valine				
(58) Adenine				
(59) Cytidylic acid				
(60) Thymine	0.8			
(61) Uracil				
(62) Xanthine				
(63) Urea				
(64) *p*-Aminobenzoic acid	0.01			
(65) L-Ascorbic acid				0.05
(66) Biotin	0.0005	0.001		
(67) Choline				
(68) Choline hydrogen citrate	0.5		0.0001	0.0002
(69) Cyanocobalamin	0.00105			
(70) N^5-Formyltetrahydro-pteroylglutamic acid	0.0002			
(71) Inositol	1.0			
(72) Nicotinic acid	0.1			
(73) Orotic acid	0.02			
(74) Pantothenic acid, Ca salt	0.1			
(75) Pteroylglutamic acid	0.0025			
(76) Putrescine dihydrochloride	0.04			
(77) Pyridoxamine dihydro-chloride	0.02			
(78) Pyridoxine hydrochloride	0.04	0.01		
(79) Riboflavine	0.005			
(80) Thiamine hydrochloride	0.2	0.1		0.1
(81) Ergosterol				0.25^{fe}
(82) Ubiquinone				0.25^{fe}
(83) Taurocholic acid, Na salt				3.0
(84) Fresh water	To 1,000 ml	To 1,000 ml	To 1,000 ml	To 1,000 ml^{fv}
(85) pH adjusted with				8.0
(86) pH before autoclaving	7.8—7.9	7.4	7.5	

Table 8 (continued)
DEFINED MARINE MEDIA (ORGANIC)

Reference	85 (1959) 167 (1971)	64 (1958)	83 (1958)	92 (1968)
Author's designation	S69 (modified)		S50	S88
		Use		
Constituents	*Oxyrrhis marina*	Mass culture of *Prymnesium, Monochrysis,* and *Isochrysis*	Littoral chrysophytes and haptophytes	*Monochrysis lutheri* (dry mixture; use 20 g/l)
(1) Carbonate				
(2) Bicarbonate				
(3) Nitrate		73.0[j]	61.3[l]	61.3[l]
(4) Ammonium		27.3[g]		
(5) Phosphate	5.45[e]	13.4[ca]	5.45[e]	5.45[e]
(6) Sulfate	279[m]	2,410[d,k]	279[m]	1,250[k,m]
(7) Silicon				
(8) Na	6,460[h,cn]	7,110[c,h]	5,910[h,bt]	6,300[h,bt]
(9) Mg	299[h]	592[g]	299[h]	247[g]
(10) K	225[e,h,as]	331[h,ca]	253[c,e,h]	264[c,e,h,as]
(11) Ca	116[g]	50[h]	116[g]	116[g]
(12) Fe	0.502[ac]	2.0	0.5	0.5
(13) Li			0.006	0.006
(14) B				
(15) F				
(16) Al	0.028[h]		0.028	0.028
(17) Mn	0.0325[aa]	4.0	0.05	0.05
(18) Co	0.000524[al]	0.1	0.0005	0.0005
(19) Cu	0.0051[ap]	0.01	0.005	0.005
(20) Zn	0.0227[g]	0.5	0.005	0.005
(21) Se				
(22) Br	22.2[as]		22.0	22.0[as]
(23) Rb	0.0707[h]		0.061	0.061
(24) Sr	2.14[h]		3.8	3.8
(25) Mo	0.000475[ax]	0.5	0.0005	0.0005
(26) I	0.0382[bc]		0.02	0.02
(27) EDTA[fl]		50	39.3[bt]	39.3[bt]
(28) NTA[fm]				
(29) Citrate	39.3[fr]			
(30) TRIS[fj]				
(31) Glycylglycine	500		500	500
(32) Acetate	1,440[j]			
(33) Fumarate				
(34) Gluconate				
(35) Glucose				
(36) Glyceryl				
(37) Glycerol				
(38) Lactate				
(39) D-Ribose				
(40) Sorbitol				
(41) Succinic acid		300		
(42) Sucrose				
(43) Alanine				

Table 8 (continued)
DEFINED MARINE MEDIA (ORGANIC)

	85 (1959) 167 (1971)	64 (1958)	83 (1958)	92 (1968)
Reference				
Author's designation	S69 (modified)		S50	S88
Use				
Constituents	*Oxyrrhis marina*	Mass culture of *Prymnesium, Monochrysis,* and *Isochrysis*	Littoral chrysophytes and haptophytes	*Monochrysis lutheri* (dry mixture; use 20 g/l)
(44) β-Alanine				
(45) Asparagine		5,000[cq]		
(46) Aspartic acid				
(47) Cystine				
(48) Glutamate				
(49) Glycine			250	250
(50) Histidine	200[cr]			
(51) Leucine				
(52) DL-Lysine				
(53) Methionine				
(54) L-Proline	40			
(55) Serine				
(56) Tryptophan				
(57) L-Valine	250			
(58) Adenine				
(59) Cytidylic acid				
(60) Thymine				
(61) Uracil				
(62) Xanthine				
(63) Urea				
(64) *p*-Aminobenzoic acid				
(65) L-Ascorbic acid				
(66) Biotin	0.05	0.001		
(67) Choline				
(68) Choline hydrogen citrate				
(69) Cyanocobalamin	0.0002	0.001	0.0001	As required
(70) N^5-Formyltetrahydro-pteroylglutamic acid				
(71) Inositol				
(72) Nicotinic acid				
(73) Orotic acid				
(74) Pantothenic acid, Ca salt				
(75) Pteroylglutamic acid				
(76) Putrescine dihydrochloride				
(77) Pyridoxamine dihydrochloride				
(78) Pyridoxine hydrochloride				
(79) Riboflavine				
(80) Thiamine hydrochloride	0.1	0.5	1.0	As required
(81) Ergosterol	0.25[fe]			
(82) Ubiquinone	0.25[fe]			
(83) Taurocholic acid, Na salt	3.0			
(84) Fresh water	To 1,000 ml[fv]	To 1,000 ml	To 1,000 ml[fv]	To 1,000 ml[fv]
(85) pH adjusted with				
(86) pH before autoclaving	8.0	6.0	8.0	8.0

Table 8 (continued)
DEFINED MARINE MEDIA (ORGANIC)

Reference	94 (1968)	58 (1948)	58 (1948)	88 (1962)
Author's designation				S76
		Use		
Constituents	Chrysochromulina kappa C. brevifilum C. strobilis Dicrateria inornata	Phaeodactylum	Phaeodactylum tricornutum (medium for determination of sodium requirement)	Skeletonema costatum
(1) Carbonate				
(2) Bicarbonate				
(3) Nitrate	73.0^j	387^d	310^d	61.3^l
(4) Ammonium		113^c	$172^{c,o}$	
(5) Phosphate	30.1^{ch}	218^e	216°	5.45^e
(6) Sulfate	$2,730^k$	974^k	$1,170^k$	279^m
(7) Silicon	9.88^i	4.94^i	5.39^{cl}	9.88^i
(8) Na	$11,100^{c,h,i,bt,ch}$	$1,030^{h,i,bn}$	Required^{dt}	$5,920^{h,i,bt}$
(9) Mg	691^g	241^g	296^g	299^h
(10) K	315^h	180^e	542^{bn}	$265^{c,e,h,as}$
(11) Ca	300^h	35^{bj}	30^{cm}	116^g
(12) Fe	0.1^h	5.0^{bk}	5.0^{ac}	0.5
(13) Li				0.006
(14) B	2.0^n	0.5^n	1.0^n	
(15) F				
(16) Al				0.028
(17) Mn	0.4^h	0.0005	0.01^{aa}	0.05
(18) Co	0.01^h		0.1^{al}	0.0005
(19) Cu		0.05		0.005
(20) Zn	0.05^h	0.05	0.01^g	0.005
(21) Se	0.01^{ar}			
(22) Br				25as
(23) Rb				0.06
(24) Sr				4.0
(25) Mo		0.05	0.01^{ba}	0.0005
(26) I				0.02
(27) EDTA^{fl}	7.85^{bt}			39.3^{bt}
(28) NTA^{fm}				
(29) Citrate		643^j	874^l	
(30) TRIS^{fj}				
(31) Glycylglycine	500			500
(32) Acetate				
(33) Fumarate				
(34) Gluconate			292^m	
(35) Glucose				
(36) Glyceryl	23.8^{ch}			
(37) Glycerol	1,000			
(38) Lactate				
(39) D-Ribose				
(40) Sorbitol				
(41) Succinic acid				
(42) Sucrose				
(43) Alanine				

Table 8 (continued)
DEFINED MARINE MEDIA (ORGANIC)

		Reference	94 (1968)	58 (1948)	58 (1948)	88 (1962)
	Author's designation					S76
				Use		
	Constituents		*Chrysochromulina kappa* *C. brevifilum* *C. strobilis* *Dicrateria inornata*	*Phaeodactylum*	*Phaeodactylum tricornutum* (medium for determination of sodium requirement)	*Skeletonema costatum*
(44)	β-Alanine					
(45)	Asparagine					
(46)	Aspartic acid					
(47)	Cystine					
(48)	Glutamate					
(49)	Glycine					250
(50)	Histidine					
(51)	Leucine					
(52)	DL-Lysine					
(53)	Methionine					
(54)	L-Proline					
(55)	Serine					
(56)	Tryptophan					
(57)	L-Valine					
(58)	Adenine					
(59)	Cytidylic acid					
(60)	Thymine					
(61)	Uracil					
(62)	Xanthine					
(63)	Urea					
(64)	*p*-Aminobenzoic acid					
(65)	L-Ascorbic acid					
(66)	Biotin		0.001			
(67)	Choline					
(68)	Choline hydrogen citrate					
(69)	Cyanocobalamin		0.001			0.0001
(70)	N^5-Formyltetrahydro-pteroylglutamic acid					
(71)	Inositol					
(72)	Nicotinic acid					
(73)	Orotic acid					
(74)	Pantothenic acid, Ca salt					
(75)	Pteroylglutamic acid					
(76)	Putrescine dihydrochloride					
(77)	Pyridoxamine dihydro-chloride					
(78)	Pyridoxine hydrochloride					
(79)	Riboflavine					
(80)	Thiamine hydrochloride		0.1			
(81)	Ergosterol					
(82)	Ubiquinone					
(83)	Taurocholic acid, Na salt					
(84)	Fresh water		To 1,000 ml	To 1,000 ml	To 1,000 ml	To 1,000 ml[fv]
(85)	pH adjusted with					
(86)	pH before autoclaving		7.8–8.0	7.2–7.5	7.4	8.0

Table 8 (continued)
DEFINED MARINE MEDIA (ORGANIC)

Reference	90 (1963)	91 (1967)	80 (1950)	82 (1955)
Author's designation			III	
Use				
Constituents	*Asterionella japonica* *Chaetoceros dydimus* *Coscinodiscus* sp. (all unialgal)	*Nitzschia alba* *N. putrida* *N. leucosigma*	*Chlamydomonas*	*Prasiola stipitata* (axenic)
(1) Carbonate				
(2) Bicarbonate	29.1^j			145^j
(3) Nitrate	123^l	613^l	77.4^m	525^m
(4) Ammonium				
(5) Phosphate	$19.1^e + 0.383^{by}$	30.1^{ch}	109^e	109^e
(6) Sulfate	$2,370^j$	$1,950^k$	974^k	$2,340^k$
(7) Silicon	4.60^i	9.88^i		
(8) Na	$9,020^{b,g,h,i}$	$9,950^{h,i,ch,cu}$	$984-15,700^h$	$10,300^{b,h}$
(9) Mg	$1,150^h$	494^g	247^g	592^g
(10) K	$427^{c,e,h,as}$	$911^{c,h}$	$887^{e,cn}$	89.8^e
(11) Ca	361^h	273^h	25^c	170^c
(12) Fe	0.689^{ab}	0.5	6.0	0.367^{ai}
(13) Li	0.201^c			
(14) B	1.05^n	0.5	40	0.1
(15) F	0.453^{bz}			
(16) Al				
(17) Mn	0.437^z	0.5	10	0.137^z
(18) Co	0.045^{aj}	0.01		
(19) Cu	0.02^{ap}	0.01	5.0	0.1^{ao}
(20) Zn	0.203^g	0.01	30	0.3^h
(21) Se				
(22) Br	40.3^{as}			
(23) Rb				
(24) Sr				
(25) Mo	0.0466^{ax}	0.01	10	0.1^{ax}
(26) I	0.382^{bc}			
(27) EDTA^{fl}	10		500	100
(28) NTA^{fm}				
(29) Citrate	0.197^{fr}			
(30) TRIS^{fj}	500	$1,000^{dr}$		
(31) Glycylglycine				
(32) Acetate			$1,200^l$	
(33) Fumarate				
(34) Gluconate				
(35) Glucose				
(36) Glyceryl	0.303^{by}	23.8^{ch}		
(37) Glycerol				
(38) Lactate		$7,950^j$		
(39) D-Ribose				
(40) Sorbitol				
(41) Succinic acid				
(42) Sucrose				
(43) Alanine				

Table 8 (continued)
DEFINED MARINE MEDIA (ORGANIC)

	Reference	90 (1963)	91 (1967)	80 (1950)	82 (1955)
	Author's designation			III	
				Use	
	Constituents	*Asterionella japonica* *Chaetoceros dydimus* *Coscinodiscus* sp. (all unialgal)	*Nitzschia alba* *N. putrida* *N. leucosigma*	*Chlamydomonas*	*Prasiola stipitata* (axenic)
(44)	β-Alanine	0.15			
(45)	Asparagine				1,000
(46)	Aspartic acid	0.25			
(47)	Cystine	0.05			
(48)	Glutamate	0.15[gx]			
(49)	Glycine	0.15		2,500	
(50)	Histidine	0.1			
(51)	Leucine	0.1[cr]			
(52)	DL-Lysine				
(53)	Methionine	0.03			
(54)	L-Proline				
(55)	Serine	0.1			
(56)	Tryptophan	0.05			
(57)	L-Valine				
(58)	Adenine	1.0			
(59)	Cytidylic acid				
(60)	Thymine				
(61)	Uracil	1.0			
(62)	Xanthine	1.0			
(63)	Urea				
(64)	p-Aminobenzoic acid	0.02			
(65)	L-Ascorbic acid	2.0			
(66)	Biotin	0.01			
(67)	Choline	0.1			
(68)	Choline hydrogen citrate				
(69)	Cyanocobalamin	0.001	0.001		
(70)	N^5-Formyltetrahydro-pteroylglutamic acid				
(71)	Inositol	5.0			
(72)	Nicotinic acid	0.05			
(73)	Orotic acid				
(74)	Pantothenic acid, Ca salt	3.0			
(75)	Pteroylglutamic acid	0.001			
(76)	Putrescine dihydrochloride				
(77)	Pyridoxamine dihydro-chloride				
(78)	Pyridoxine hydrochloride	0.01			
(79)	Riboflavine	0.01			
(80)	Thiamine hydrochloride	0.3	1.0		
(81)	Ergosterol				
(82)	Ubiquinone				
(83)	Taurocholic acid, Na salt				
(84)	Fresh water	To 1,000 ml[fu]	To 1,000 ml[fu]	To 1,000 ml	To 1,000 ml
(85)	pH adjusted with			KOH	
(86)	pH before autoclaving			7.0	

Compiled by M. F. Turner and M. R. Droop.

REFERENCES

1. **Chu, S. P.,** *J. Ecol.,* 30, 284–325, 1942.
2. **Provasoli, L. and Pintner, I. J.,** *Ann. N. Y. Acad. Sci.,* 56, 839–851, 1953.
3. **Proctor, V. W.,** *Limnol. Oceanogr.,* 2, 125–139, 1957.
4. **Reynolds, N.,** *New Phytol.,* 49, 155–162, 1950.
5. **Lewin, R. A.,** *J. Gen. Microbiol.,* 6, 233–248, 1952.
6. **Storm, J. and Hutner, S. H.,** *Ann. N. Y. Acad. Sci.,* 56, 901–909, 1953.
7. **Lynch, V. H. and Calvin, M.,** *Ann. N. Y. Acad. Sci.,* 56, 890–900, 1953.
8. **Norris, L., Norris, R. E., and Calvin, M.,** *J. Exp. Bot.,* 6, 64–74, 1955.
9. **Lewin, R. A.,** *J. Genet.,* 51, 543–560, 1953.
10. **Ford, J. E.,** *J. Gen. Microbiol.,* 19, 161–172, 1958.
11. **Moewus, F. and Moewus, L.,** *Trans. Am. Microsc. Soc.,* 78, 163–172, 1959.
12. **Bendix, S. and Allen, M. B.,** *Arch. Mikrobiol.,* 41, 115–141, 1962.
13. **Fott, B. and McCarthy, A. J.,** *J. Protozool.,* 11, 116–120, 1964.
14. **Lewin, J. C., Reimann, B. E., Busby, W. F., and Volcani, B. E.,** in *Cell Synchrony,* Cameron, I. L. and Padilla, G. M., Eds., Academic Press, New York, 1966, pp. 169–188.
15. **Griffiths, D. J., Thresher, C. L., and Street, H. E.,** *Ann. Bot.,* 24, 1–11, 1960.
16. **Allison, F. E., Hoover, S. R., and Morris, H. J.,** *Bot. Gaz.,* 98, 433–463, 1937.
17. **Fogg, G. E.,** *J. Exp. Biol.,* 19, 78–87, 1942.
18. **Emerson, R. and Lewis, C. M.,** *J. Gen. Physiol.,* 25, 579–595, 1942.
19. **Pratt, R.,** *Am. J. Bot.,* 30, 404–408, 1943.
20. **Rodhe, W.,** *Symb. Bot. Ups.,* 10(1), 1–149, 1948.
21. **Österlind, S.,** *Symb. Bot. Ups.,* 10(3), 1–141, 1949.
22. **Hutner, S. H. and Provasoli, L.,** in *Biochemistry and Physiology of Protozoa,* Vol. 1, Lwoff, A., Ed., Academic Press, New York, 1951, pp. 27–128.
23. **Gerloff, G. C., Fitzgerald, G. P., and Skoog, F.,** *Am. J. Bot.,* 37, 216–218, 1950.
24. **Gerloff, G. C., Fitzgerald, G. P., and Skoog, F.,** *Am. J. Bot.,* 37, 835–840, 1950.
25. **Pringsheim, E. G.,** *Pure Cultures of Algae, Their Preparation and Maintenance,* The University Press, Cambridge, 1946, pp. xii and 119.
26. **Neish, A. C.,** *Can. J. Bot.,* 29, 68–78, 1951.
27. **Allen, M. B.,** *Arch. Mikrobiol.,* 17, 34–53, 1952.
28. **Sager, R. and Granick, S.,** *Ann. N. Y. Acad. Sci.,* 56, 831–838, 1953.
29. **Syrett, P. J.,** *Ann. Bot.,* 17, 1–19, 1953.
30. **Kessler, E.,** *Flora, Jena,* 140, 1–38, 1953.
31. **Tamiya, H., Hase, E., Shibata, K., Mituya, A., Iwamura, T., Nihei, T., and Sasa, T.,** in *Algal Culture from Laboratory to Pilot Plant,* Publication 600, Burlew, J. S., Ed., Carnegie Washington, D.C., 1953, pp. 204–232.
32. **Hutner, S. H., Bach, M. K., and Ross, G. I. M.,** *J. Protozool.,* 3, 101–112, 1956.
33. **Allen, M. B. and Arnon, D. I.,** *Plant Physiol.* (Lancaster), 30, 366–372, 1955.
34. **Kratz, W. A. and Myers, J.,** *Am. J. Bot.,* 42, 282–287, 1955.
35. **Müler, J. D. A. and Fogg, G. E.,** *Arch. Mikrobiol.,* 28, 1–17, 1957.
36. **Pringsheim, E. G.,** *J. Ecol.,* 33, 193–204, 1946.
37. **Hughes, E. O., Gorham, P. R., and Zehnder, A.,** *Can. J. Microbiol.,* 4, 225–236, 1958.
38. **Zehnder, A. and Hughes, E. O.,** *Can. J. Microbiol.,* 4, 399–408, 1958.
39. **Allen, M. B.,** *Arch. Mikrobiol.,* 32, 270–277, 1959.
40. **Chorney, W., Scully, N. J., Crespi, H. L., and Katz, J. J.,** *Biochim. Biophys. Acta,* 37, 280–287, 1960.
41. **Wise, D. L.,** *J. Protozool.,* 6, 19–23, 1959.
42. **Soeder, C. J.,** *Flora, Jena,* 148, 489–516, 1960.
43. **Zehnder, A. and Gorham, P. R.,** *Can. J. Microbiol.,* 6, 645–660, 1960.
44. **Kiyohara, T, Fujita, Y., Hattori, A., and Watanabe, A.,** *J. Gen. Appl. Microbiol.* (Tokyo), 6, 176–182, 1960.
45. **Reisner, G. S., Gering, R. K., and Thompson, J. F.,** *Plant Physiol.* (Lancaster), 35, 48–52, 1960.
46. **McLachlan, J. and Gorham, P. R.,** *Can. J. Microbiol.,* 7, 869–882, 1961.
47. **Schmidt, R. R.,** *Exp. Cell Res.,* 23, 209–217, 1961.
48. **Lazaroff, N. and Vishniac, W.,** *J. Gen. Microbiol.,* 25, 365–374, 1961.
49. **Miyachi, S. and Tamiya, H.,** *Biochim. Biophys. Acta,* 46, 200–202, 1961.
50. **Coombs, J., Lauritis, J. A., Darley, W. M., and Volcani, B. E.,** *Z. Pflanzenphysiol.,* 59, 124–152, 1968.
51. **Eppley, R. W., Gee, R., and Saltman, P.,** *Physiol. Plant.,* 16, 777–792, 1963.

52. **Galloway, R. A. and Krauss, R. W.**, in *Microalgae and Photosynthetic Bacteria,* Japanese Society of Plant Physiologists, Eds., University Press, Tokyo, 1963, pp. 569–575.

53. **Moyse, A. and Guyon, D.**, in *Microalgae and Photosynthetic Bacteria,* Japanese Society of Plant Physiologists, Eds., University Press, Tokyo, 1963, pp. 253–270.

54. **Kumar, H. D.**, *J. Exp. Bot.*, 15, 232–250, 1964.

55. **Syrett, P. J., Merrett, M. J., and Bocks, S. M.**, *J. Exp. Bot.*, 14, 249–264, 1963.

56. **Forsberg, C.**, *Life Sci.*, 4, 225–226, 1965.

57. **Stanier, R. Y., Kunisawa, R., Mandel, M., and Cohen-Bazire, G.**, *Bacteriol. Rev.*, 35, 171–205, 1971.

58. **Hutner, S. H.**, *Trans. N.Y. Acad. Sci.*, 10, 136–141, 1948.

59. **Lewin, R. A.**, *J. Gen. Microbiol.*, 10, 93–96, 1954.

60. **Ryther, J. H.**, *Biol. Bull. Mar. Biol. Lab. Woods Hole*, 106, 198–209, 1954.

61. **Committee on Cultures, Society of Protozoologists**, *J. Protozool.*, 5, 1–38, 1958.

62. **Provasoli, L., McLaughlin, J. J. A., and Droop, M. R.**, *Arch. Mikrobiol.*, 25, 392–428, 1957.

63. **Kain, J. M. and Fogg, G. E.**, *J. Mar. Biol. Assoc. U.K.*, 37, 397–413, 1958.

64. **McLaughlin, J. J. A.**, *J. Protozool.*, 5, 75–81, 1958.

65. **Aldrich, D. V. and Wilson, W. B.**, *Biol. Bull. Mar. Biol. Lab. Woods Hole*, 119, 57–64, 1960.

66. **Boalch, G. T.**, *J. Mar. Biol. Assoc. U.K.*, 41, 287–304, 1961.

67. **Stewart, W. D. P.**, *Ann. Bot.*, 26, 439–445, 1962.

68. **Jones, R. F.**, *J. Cell. Comp. Physiol.*, 60, 61–64, 1962.

69. **McLachlan, J.**, *Can. J. Microbiol.*, 10, 769–782, 1964.

70. **Provasoli, L.**, in *Proceedings, 4th International Seaweed Symposium,* Davy de Virville, A. and Feldmann, J., Eds., Pergamon Press, Oxford, 1964, pp. 9–17.

71. **Provasoli, L. and McLaughlin, J. J. A.**, in *Symposium on Marine Microbiology,* Oppenheimer, C. H., Ed., Charles C. Thomas, Springfield, Ill., 1963, pp. 105–113.

72. **Paasche, E.**, *Physiol. Plant. Suppl.*, 3, 1–82, 1964.

73. **Rahat, M. and Jahn, T. L.**, *J. Protozool.*, 12, 246–250, 1965.

74. **Lewin, J. C.**, *Phycologia*, 4, 141–144, 1965.

75. **Gooday, G. W.**, *J. Mar. Biol. Assoc. U.K.*, 50, 199–208, 1970.

76. **Hall, R. P.**, *J. Protozool.*, 4, 42–48, 1957.

77. **Lewin, J. C., Lewin, R. A., and Philpott, D. E.**, *J. Gen. Microbiol.*, 18, 418–426, 1958.

78. **Chen, L. C.-M., Edelstein, T., and McLachlan, J.**, *J. Phycol.*, 5, 211–220, 1969.

79. **Jordan, J. B.**, Research on the Marine Food Chain, Progress Report, Part 1, University of California, San Diego, Institute of Marine Resources, La Jolla, Calif., 1969, pp. 25–40.

80. **Hutner, S. H., Provasoli, L., Shatz, A., and Haskins, C. P.**, *Proc. Am. Philos. Soc.*, 94, 152–170, 1950.

81. **Hutner, S. H., Provasoli, L., and Filfus, J.**, *Ann. N.Y. Acad. Sci.*, 56, 852–862, 1953.

82. **Lewin, R. A.**, *Can. J. Bot.*, 33, 5–10, 1955.

83. **Droop, M. R.**, *J. Mar. Biol. Assoc. U.K.*, 37, 323–329, 1958.

84. **Pintner, I. J. and Provasoli, L.**, *J. Gen. Microbiol.*, 18, 190–197, 1958.

85. **Droop, M. R.**, *J. Mar. Biol. Assoc. U.K.*, 38, 605–620, 1959.

86. **McLaughlin, J. J. A. and Zahl, P. A.**, *Ann. N.Y. Acad. Sci.*, 77, 55–72, 1959.

87. **McLaughlin, J. J. A., Zahl, P. A., Nowak, A., Marchisotto, J., and Prager, J.**, *Ann. N.Y. Acad. Sci.*, 90, 856–865, 1960.

88. **Droop, M. R.**, *Vortr. Gesamtgeb. Bot.*, N.F. 1, 77–82, 1962.

89. **Hutner, S. H., Baker, H., Aaronson, S., Nathan, H. A., Rodriguez, E., Lockwood, S., Sanders, M., and Petersen, R. A.**, *J. Protozool.*, 4, 259–269, 1957.

90. **Soli, G.**, in *Symposium on Marine Microbiology,* Oppenheimer, C. H., Ed., Charles C Thomas, Springfield, Ill., 1963, pp. 122–126.

91. **Lewin, J. and Lewin, R. A.**, *J. Gen. Microbiol.*, 46, 361–367, 1967.

92. **Droop, M. R.**, *J. Mar. Biol. Assoc. U.K.*, 48, 689–733, 1968.

93. **Taylor, W. R.**, *Occas. Publ. Narragansett Mar. Lab.*, 2, 17–24, 1964.

94. **Pintner, I. J. and Provasoli, L.**, *Bull. Misaki Mar. Biol. Inst. Kyoto Univ.*, No. 12, 25–31, 1968.

95. **Bischoff, H. W. and Bold, H. C.**, *Univ. Tex. Publ.*, No. 6318, 1963.

96. **Allen, E. J. and Nelson, E. W.**, *J. Mar. Biol. Assoc. U.K.*, 8, 421–474, 1910.

97. **Schreiber, E.**, *Wiss. Meeresunters.*, N.F., Abt. Helgoland, 16(10), 1–34, 1927.

98. **Fφyn, B.**, *Arch. Protistenkd.*, 83, 1–56, 1934.

99. **Barker, H. A.**, *Arch. Mikrobiol.*, 6, 157–181, 1935.

100. **Ketchum, B. H. and Redfield, A. C.**, *Biol. Bull. Mar. Biol. Lab. Woods Hole*, 75, 165–169, 1938.

101. **Sweeney, B. M.**, *Am. J. Bot.*, 38, 669–677, 1951.

102. **Goldberg, E. D., Walker, T. J., and Whisenand, A.**, *Biol. Bull. Mar. Biol. Lab. Woods Hole*, 101, 274–284, 1951.

103. Sweeney, B. M., *Am. J. Bot.,* 41, 821–824, 1954.
104. Haxo, F. T. and Sweeney, B. M., in *The Luminescence of Biological Systems,* Johnson, F. H., Ed., American Association for the Advancement of Science, Washington, D.C., 1955, pp. 415–420.
105. Greenblatt, C. L. and Schiff, J. A., *J. Protozool.,* 6, 23–28, 1959.
106. Allen, E. J., *J. Mar. Biol. Assoc. U.K.,* 10, 417–439, 1914.
107. Wilson, W. B. and Collier, A., *Science, New York,* 121, 394–395, 1955.
108. Provasoli, L., *Biol. Bull. Mar. Biol. Lab. Woods Hole,* 114, 375–384, 1958.
109. Guillard, R. R. L. and Wangersky, P. J., *Limnol. Oceanogr.,* 3, 449–454, 1958.
110. Iwasaki, H., *Biol. Bull. Mar. Biol. Lab. Woods Hole,* 121, 173–187, 1961.
111. Guillard, R. R. L. and Ryther, J. H., *Can. J. Microbiol.,* 8, 229–239, 1962.
112. Droop, M. R., in *Methods in Microbiology,* Vol. 3B, Norris, J. R. and Ribbons, D. W., Eds., Academic Press, London and New York, 1969, pp. 269–313.
113. von Stosch, H. A., in *Proceedings, 4th International Seaweed Symposium,* Davy de Virville, A. and Feldmann, J., Eds., Pergamon Press, Oxford, 1964, pp. 9–17.
114. von Stosch, H. A. and Drebes, G., *Helgoländer Wiss. Meeresunters.,* 11, 209–257, 1964.
115. Collier, A., Wilson, W. B., and Borkowski, M., *J. Phycol.,* 5, 168–172, 1969.
116. Cheng, J. Y. and Antia, N. J., *J. Fish. Res. Board Can.,* 27, 335–346, 1970.
117. Droop, M. R., *J. Mar. Biol. Assoc. U.K.,* 34, 229–231, 1955.
118. Provasoli, L., in *Cultures and Collections of Algae,* Watanabe, A. and Hattori, A., Eds., Japanese Society of Plant Physiologists, 1968, pp. 63–75.
119. Pringsheim, E. G. and Pringsheim, O., *J. Ecol.,* 37, 57–64, 1949.
120. Nordli, E., *Nytt Mag. Bot.,* 5, 13–16, 1957.
121. Lewin, J. C. and Lewin, R. A., *Can. J. Microbiol.,* 6, 127–134, 1960.
122. Aaronson, S. and Scher, S., *J. Protozool.,* 7, 156–158, 1960.
123. Hulburt, E. M., McLaughlin, J. J. A., and Zahl, P. A., *J. Protozool.,* 7, 323–326, 1960.
124. Rahat, M. and Reich, K., *J. Gen. Microbiol.,* 31, 195–202, 1963.
125. Antia, N. J. and Kalmakoff, J., *Manuscr. Rep. Ser. (Oceanogr. Limnol.). Fish. Res. Board Can.,* No. 203, 1–24, 1965.
126. Spencer, C. P., *J. Mar. Biol. Assoc. U.K.,* 33, 265–290, 1954.
127. Little, P. A., Oleson, J. J., and Williams, J. H., *Proc. Soc. Exp. Biol. Med.,* 78, 510–513, 1951.
128. Buetow, D. E. and Padilla, G. M., *J. Protozool.,* 10, 121–123, 1963.
129. Droop, M. R., *Rev. Algol., N.S.,* 4, 247–259, 1961.
130. Macias, F. M., *J. Protozool.,* 12, 500–504, 1965.
131. Sorokin, C. and Kruass, R. W., *Plant Physiol. Lancaster,* 37, 37–42, 1962.
132. Watanabe, A. and Kiyohara, T., in *Microalgae and Photosynthetic Bacteria,* Japanese Society of Plant Physiologists, Eds., University Press, Tokyo, 1963, pp. 189–196.
133. Allen, M. M., *J. Phycol.,* 4, 1–4, 1968.
134. Cannon, R. E., Shane, M. S., and Bush, V. N., *Virology,* 45, 149–153, 1971.
135. Castenholz, R. W., *Schweiz. Z. Hydrol.,* 32, 538–551, 1970.
136. Gorham, P. R., McLachlan, J., Hammer, U. T., and Kim, W. K., *Verh. Int. Vere. Theor. Angew. Limmol.,* 15, 796–804, 1964.
137. Van Baalen, C., *J. Phycol.,* 1, 19–22, 1965.
138. Keller, S. E., Hutner, S. H., and Keller, D. E., *J. Protozool.,* 15, 792–795, 1968.
139. Eppley, R. W. and Maciasr, F. M., *Physiol. Plant.,* 15, 72–79, 1962.
140. Tuttle, R. C. and Loeblich, A. R., III, *Phycologia,* 14, 1–8, 1975.
141. N.E.R.C. Culture Centre of Algae and Protozoa, List of Strains, Cambridge, England, 1971, pp. 1–73.
142. Links, J., Verloop, A., and Havinga, E., *Antonie van Leeuwenhoek J. Microbiol. Serol.,* 27, 76–80, 1961.
143. Padilla, G. M. and James, T. W., *Exp. Cell Res.,* 20, 401–415, 1960.
144. Baker, H., Frank, O., Matovitch, V. B., Pasher, I., Aaronson, S., Hutner, S. H., and Sobotka, H., *Anal. Biochem.,* 3, 31–39, 1962.
145. Loeblich, A. R., III, *J. Phycol.,* 11, 80–86, 1975.
146. Pringsheim, E. G. and Wiessner, W., *Arch. Mikrobiol.,* 40, 231–246, 1961.
147. Isenberg, H. D., *J. Gen. Microbiol.,* 29, 373–388, 1962.
148. Callely, A. G. and Lloyd, D., *Biochem. J.,* 90, 483–489, 1964.
149. Imahori, K. and Iwasa, K., *Phycologia,* 4, 127–134, 1965.
150. Zuraw, E. A., *J. Phycol.,* 5, 83–85, 1969.
151. Isenberg, H. D., Berkman, J. I., and Sundheim, L. H., *J. Protozool.,* 9, 40–44, 1962.
152. Hurlbert, R. E. and Rittenberg, S. C., *J. Protozool.,* 9, 170–182, 1962.
153. von Stosch, H. A., *Br. Phycol. J.,* 8, 105–134, 1973.

154. **Murray, S. N., Dixon, P. S., and Scott, J. L.,** *Br. Phycol. J.,* 7, 323–333, 1972.
155. **Ben-Amotz, A.,** *J. Phycol.,* 11, 50–54, 1975.
156. **Dawson, P. A.,** *Br. Phycol. J.,* 7, 255–271, 1972.
157. **Heywood, P.,** *J. Phycol.,* 9, 156–159, 1973.
158. **Rayburn, W. R. and Starr, R. C.,** *J. Phycol.,* 10, 42–49, 1974.
159. **Ramus, J.,** *J. Phycol.,* 8, 97–111, 1972.
160. **Smith, R. L. and Wiedeman, V. E.,** *Can. J. Bot.,* 42, 1582–1586, 1964.
161. **Eppley, R. W., Holmes, R. W., and Strickland, J. D. H.,** *J. Exp. Mar. Biol. Ecol.,* 1, 191–208, 1967.
162. **Nygren, S.,** *Helgoländer Wiss. Meeresunters.,* 20, 126–129, 1970.
163. **Lee, J. J., Tietjen, J. H., Stone, R. J., Müller, W. A., Rullman J., and McEnery, M.,** *Helgoländer Wiss. Meeresunters.,* 20, 136–156, 1970.
164. **Starr, R. C.,** *Am. J. Bot.,* 51, 1013–1044, 1964.
165. **Ellis, R. J. and Machlis, L.,** *Am. J. Bot.,* 55, 590–599, 1968.
166. **Bishop, N. I. and Senger, H.,** in *Methods in Enzymology,* Vol. 23, San Pietro, A., Ed., Academic Press, New York and London, 1971, pp. 53–66.
167. **Droop, M. R. and Pennock, J. F.,** *J. Mar. Biol. Assoc. U.K.,* 51, 455–470, 1971.
168. **Fogg, G. E.,** *Ann. Bot.,* 13, 241–259, 1949.
169. **Carr, N. G., Komárek, J., and Whitton, B. A.,** in *The Biology of Blue-green Algae,* Carr, N. G. and Whitton, B. A., Eds., Blackwell Scientific Publications, Oxford, 1973, pp. 525–530.
170. **Kuffareth, H.,** *Rev. Algol.,* 4, 127–346, 1929.
171. **Bold, H. C.,** *Rev. Bot.,* 8(2), 69–137, 1942.
172. **Pringsheim, E. G.,** in *Manual of Phycology,* Smith, G. M., Ed., Chronica Botanica Co., Waltham, Mass., 1951, pp. 347–357.
173. **Stein, J. R., Ed.,** *Handbook of Phycological Methods: Culture Methods and Growth Measurements,* University Press, Cambridge, U.K., 1973, pp. xii and 448.

NATURAL AND SYNTHETIC CULTURE MEDIA: FUNGI*

John Tuite

Media for the cultivation of fungi appear to be almost as numerous and diverse as fungi themselves, and the numbers of media are increasing. For example, through the use of new microbial inhibitors, there has been a considerable increase in media for the isolation of a specific fungus, even when it occurs in low numbers and in environments where other microorganisms are numerous. Probably about a dozen culture media would suffice for the routine isolation, cultivation, and sporulation of many fungi. This arbitrary list includes: alphacel, cornmeal, Czapek solution, Fries, glucose asparagine, hemp seed, malt, oatmeal, potato dextrose, PDTC, Sabouraud, water agar, and V-8® juice agar. Increased research needs, however, often require more sophisticated media, which justifies many but not all of the new media. Often, the small variations in known media appear to have little value.

In an attempt to assist in the selection of appropriate media from the possibly bewildering list that is compiled here, I have selected 92 fungi that are diverse and of particular interest to mycologists, microbiologists, and plant pathologists. In Table 1 media are tabulated that are recommended for the isolation, growth, and reproduction of these fungi. Common media, unless their use is not obvious, are not listed in the table. Recognizing that the environmental conditions under which these fungi are cultivated are often very important, appropriate references are given where these details may be found. Hopefully, the many organisms not on the list may be associated with those that are by virtue of their taxonomy, and the literature and media may also be appropriate for them.

Recommendations of media and procedures are sometimes unsatisfactory, as witnessed in part by the several media listed for some fungi. Synthetic media seem to be more variable than natural media in their effects on different isolates. Particularly in regard to sporulation, isolates need to be properly maintained so that they are capable of reproduction. Once a fungus becomes mycelial, it usually cannot be induced to sporulate.

Some recipes are old and were designed when only agar of low gel strength was available. It may be, if the pH of medium is not below 6.0 such as malt agars are, that as little as 12 or 13 g of agar may be used. Economy can also be obtained by substituting cerelose for glucose when a highly defined medium is not required.

DEDICATION

This chapter is dedicated to the memory of Dorothy Fennell.

ACKNOWLEDGEMENTS

I wish to gratefully acknowledge the help of Carol Felkel, Betty Rice, Marilyn Coffey, Charles W. Bacon, E. S. Beneke, E. E. Butler, Lucas Calpouzos, R. W. Curtis, Dexter R. Douglas, Larry D. Dunkle, D. C. Erwin, F. I. Frosheiser, M. O. Garraway, T. T. Hebert, G. Hartmann, C. W. Hesseltine, R. S. Hunt, O. C. Huisman, E. G. Kuhlman, P. Knox-Davies, E. Levetin, J. W. Lorbeer, M. Lortie, J. Lovett, W. D. McClellan, S. M. Mertz, Jr., John Mircetich, J. E. Mitchell, E. Moore-Landecker, G. A. Neish, Paul Pecknold, O. K. Ribeiro, Dave Sonneborn, J. R. Stavely, G. E. Templeton, L. S. Watrud, H. E. Wheeler, Roy D. Wilcoxson, and W. Wynn.

* Purdue University Agricultural Experiment Station Journal Paper No. 55016.

Table 1

FUNGI AND SPECIFIC MEDIA RECOMMENDED FOR THEIR ISOLATION, GROWTH, AND REPRODUCTION, INCLUDING APPROPRIATE REFERENCES

Fungus		Isolation	Growth	Reproduction
Achyla ambisexualis	Media	109	35, 154	154, 222, 223
	Ref.	1	1—3	1—4
Allescheria boydii	Media	241	35, 76, 263	35, 241
	Ref.	5, 6	5, 6	5, 6
Allomyces spp.	Media	109, 320	131, 320	320
	Ref.	3, 7, 8	3, 7—10	3, 7, 8, 11
Alternaria alternata	Media	—	233, 234	290
	Ref.	—	12, 13	14, 15
A. solani	Media	—	48,[a] 128	48,[b] 122, 128, 181, 290
	Ref.	—	16—18	19—24
Aphanomyces euteiches	Media	32, 35,[c] 218	4, 143, 196, 315	4, 32, 143, 196, 226, 315
	Ref.	25, 26	27—30	28, 30—33
Ascobolus magnificus	Media	—	65, 317	65, 317
	Ref.	—	3, 34	3, 34
Ascochyta pisi	Media	—	51, 136A, 178A	48, 136A agar, 290
	Ref.	35	36	21, 36, 37
A. rabei	Media	—	170, 178A, 231[d]	28, 78, 170, 231[d]
	Ref.	—	38	38, 39
Aspergillus flavus	Media	1, 15, 139, 190	51, 269, 306	178A, 269, 306
	Ref.	40—43, 45, 46	47, 48	47—49
A. fumigatus	Media	190, 242	51, 133, 136A	133, 136A
	Ref.	6	47	47
A. glaucus group	Media	51, 139, 281	51, 139	51, 139
	Ref.	40, 42, 47	40, 42, 47	40, 42, 47
A. nidulans	Media	51, 139	51, 133, 136A, 211	51, 136A agar, 211
	Ref.	46, 50	47, 52	47, 52
A. niger	Media	127, 139, 190	44, 264	178A, 264
	Ref.	51	47, 53, 54	47, 55
Bipolaris (see Helminthosporium carbonum and H. maydis)		—	—	—
Blastocladiella emersonii	Media	—	200, 258	56, 200, 258
	Ref.	3, 7, 8	3, 7, 56, 57	3, 5, 58

Note: Media indicated according to recipe number. Refer to recipes beginning on page 437.

[a] +10% sucrose.
[b] +10% Marmite®.
[c] +300 ppm streptomycin.
[d] +15% sucrose.

Table 1 (continued)
FUNGI AND SPECIFIC MEDIA RECOMMENDED FOR THEIR ISOLATION, GROWTH, AND REPRODUCTION, INCLUDING APPROPRIATE REFERENCES

Fungus		Isolation	Growth	Reproduction
Botrytis cinerea	Media	—	16	136A agar, 178A, 303
	Ref.	—	17, 59–61	61–64
Calonectria crotolariae (see *Cylindrocladium crotolariae*)	Media	—	—	—
Ceratocystis fagacearum	Media	97, 166	97, 166, 167	10, 97, 136A, 166, 167, 178A, 178A no agar
	Ref.	65, 66	65, 67, 68	68–70
C. ulmi	Media	58	86, 99	2, 60, 178A, 279
	Ref.	71	72–74	75–78
Cercospora beticola	Media	—	267	267, 274, 275
	Ref.	—	79	80–82
C. nicotianae	Media	—	267	267, 290[e]
	Ref.	—	79, 83	79, 83
C. zebrina	Media	—	231[f]	34, 35, 290
	Ref.	—	84	84
Chaetomium globosum	Media	—	81, 319	84, 214, 290
	Ref.	—	17, 85–87	3, 85, 88
Choanephora cucurbitarum	Media	—	83	83, 141
	Ref.	—	17, 89	3, 89–91
Cladosporium cucumerinum	Media	—	—	178A
	Ref.	—	92	92–94
C. fulvum	Media	—	14	290
	Ref.	—	95, 96	96, 97
Cochliobolus (see *Helminthosporium carbonum, H. maydis*)		—	—	—
Colletotrichum coccodes	Media	63	25	25, 136A agar, 178A, 293
	Ref.	98, 99	100, 101	99, 101–103
C. dematium	Media	—	—	170
	Ref.	104	105	104, 105
C. graminicola	Media	—	—	170, 216
	Ref.	—	106	107–109
Coprinus lagopus	Media	100	31	77, 100
	Ref.	3	3, 110, 111	3, 112, 113
Cyathus stercoreus	Media	—	142	142
	Ref.	115	3, 114, 115	3, 114, 115

[e]   Adjust pH to 4.5.
[f]   +30% sucrose; 15% yeast extract.

Table 1 (continued)
FUNGI AND SPECIFIC MEDIA RECOMMENDED FOR THEIR ISOLATION, GROWTH, AND REPRODUCTION, INCLUDING APPROPRIATE REFERENCES

Fungus		Isolation	Growth	Reproduction
Cylindrocladium crotolariae	Media	45, 178B	136A broth	136A broth, 178a
	Ref.	116–118	116	116, 117, 119
C. scoparium	Media	53, 136A agar, 136A[g]	51	51[h] 178A, 231
	Ref.	120, 121	120, 121	119, 120
Dendrophoma obscurans	Media	—	88	88
	Ref.	—	17, 122	17, 122
Dictyostelium discoideum	Media	108	195, 197, 198	195, 197, 198
	Ref.	123–125	124–128	124–128
Drechslera (see Helminthosporium teres, H. turcicum)	Media	—	—	—
Endothia parasitica	Media	—	61, 62	61
	Ref.	—	17, 129	17, 129
Exserohilum (see Helminthosporium turcicum)	Media	—	—	—
Fomes annosus	Media	66, 67, 135	165[i]	136A agar
	Ref.	130–134	135, 136	133
Fusarium oxysporum	Media	118, 159, 177	6, 26, 117	6, 26, 117, 178A, 216, 232, 280
	Ref.	137–140	17	138, 141–143
F. roseum (see Gibberella zeae)	Media	—	—	—
Fusarium solani	Media	159, 177	91	91, 117, 178A, 216, 282, 291
	Ref.	137, 139	144	144–149
Gaeumannomyces graminis	Media	215[j]	82	82
	Ref.	150	151, 152	150, 151, 153
Geotrichum candidum	Media	190	8	8, 23, 78, 102, 217
	Ref.	—	154	154–158
Gibberella zeae	Media	159, 177, 190, 255	30, 262	21, 29, 30, 216, 255
	Ref.	159, 160	160–162	163–167, 168a, 168b
Glomerella cingulata	Media	—	81	24, 170
	Ref.	—	17, 87	169–172
Gnomonia fructicola	Media	—	173	35, 121, 178A
	Ref.	—		173–176
Helminthosporium maydis	Media	177	70, 71, 283	2, 72, 75, 120, 178A, 260
	Ref.		178, 179	177, 179–185

g +300 ppm streptomycin.
h +20% sucrose.
i   Replace $NH_4H_2PO_4$ with asparagine.
j   +30 ppm chlortetracycline.

Table 1 (continued)
FUNGI AND SPECIFIC MEDIA RECOMMENDED FOR THEIR ISOLATION, GROWTH, AND REPRODUCTION, INCLUDING APPROPRIATE REFERENCES

Fungus		Isolation	Growth	Reproduction
H. sorokinianum	Media	—		
	Ref.		14	245B, 266, 296
H. teres	Media	—		
	Ref.		17, 186, 187	187–189
H. turcicum	Media	—	190	190–192
	Ref.		268	229, 245A
H. victoriae	Media	—	120	120, 245C, 290
	Ref.		193	21, 193–196
Histoplasma capsulatum	Media	17, 239, 241	70, 71	2, 71, 245C
	Ref.		197, 198	76, 199, 200
Hypomyces solani (see Fusarium solani)	Media	6, 201	103, 239, 240, 247	178A, 247
	Ref.		6, 202, 203	6, 202, 204
Kabatiella caulivora	Media	136A agar, 188[k]	—	178A broth
	Ref.	205, 206	207	207, 208
Leptosphaerulina briosiana	Media	—		
	Ref.		231	170, 290
Macrophomina phaseolina	Media	185, 186		
	Ref.		209	209, 210
Monilinia fructicola	Media	211, 212	194	121, 178A, 192, 261, 309
	Ref.		213, 214	133, 178A, 179, 290
Myrothecium verrucaria	Media	153	54[a,c,g]	25, 54[b,h] 175
	Ref.		81, 86	221–224
			87, 222	
Nectria galligena	Media	220	226	225–227
	Ref.	225		146[l] 169
Neurospora spp.	Media	25	—	228–230
	Ref.		228	43, 163
Penicillium spp.	Media	47, 190	161, 162	3, 232
	Ref.		3, 231	46, 51, 51[m] 133
Periconia circinata	Media	233–235	51, 51[m] 133	233, 235, 237
	Ref.		233, 236	178A and 308
			71[n]	
Phoma betae	Media	238	239	238
	Ref.	19	17, 241	3, 212
		240		76, 242

k   +0.2% yeast extract.
l   pH 5.2, no vitamins.
m   +1% corn steep liquor.
n   +0.1% yeast extract.

Table 1 (continued)
FUNGI AND SPECIFIC MEDIA RECOMMENDED FOR THEIR ISOLATION, GROWTH, AND REPRODUCTION, INCLUDING APPROPRIATE REFERENCES

Fungus		Isolation	Growth	Reproduction
Phycomyces blakesleeanus	Media	—	171	101, 136, 241, potato slices
	Ref.		17, 243	3, 244
Phyllosticta maydis	Media	—	71	178A, 290, oat kernels
	Ref.		245, 246	246—248
Physarium polycephalum	Media	—	38, 203	35, 37, 204
	Ref.	168	3, 249, 251	3, 249—251
Phytophthora cactorum	Media	3, 249	209	125, 209, 290, 295
	Ref.	39?, 40?, 187, 271	253	252—255
P. capsici	Media	252	208, 209	110, 170, 209, 221
	Ref.		253, 258	253, 254, 258—260
P. cinnamomi	Media	39, 187	207, 209	109 and 246, 112, 125, 209, 294
	Ref.	256, 257	253, 261	254, 255, 261—265
P. infestans	Media	39, 187, 301	80, 126, 236, 237	124, 236, 237, 321
	Ref.	257, 261	266—269	266—268, 270
P. megasperma var. *sojae*	Media	35°	35, 170, 209	35, 122, 170, 209
	Ref.	271	253, 273	253, 254, 272, 273
P. parasitica	Media	39, 187	209	125, 209, 290, 294
	Ref.	257, 274	253	253, 255, 264, 275, 276
Pilobolus keinii	Media	277	57, 210	57, 210
	Ref.		277—279	277—279
Piricularia oryzae	Media		272	230
	Ref.		280, 281	165, 281—283
Piricularia grisea	Media		272	170, 290
	Ref.		—	21, 284, 285
Pyricularia (see *Piricularia*)	Media			
	Ref.			
Puccinia coronata f. sp. *avenae*	Media	313	313	313
	Ref.	286	286	286
P. graminis f. sp. *tritici*	Media	20, 313	20, 69, 313	20, 69, 313
	Ref.	287, 288	287, 288	287, 288
P. recondita	Media	20	20	20
	Ref.	286, 289	286, 289	286
Pyronema domesticum	Media		178A	35, 173, 212, 219
	Ref.	3	3	3, 290, 291
Pythium aphanidermatum	Media	152	143	35, 297
	Ref.	292—294	296, 297	165, 298, 299

° +100 ppm pimarcin.

Table 1 (continued)
FUNGI AND SPECIFIC MEDIA RECOMMENDED FOR THEIR ISOLATION, GROWTH, AND REPRODUCTION, INCLUDING APPROPRIATE REFERENCES

Fungus		Isolation	Growth	Reproduction
Rhizoctonia solani	Media	228, 308	178A, 316	183, 184, 188, 215, 316
	Ref.	300–302	17, 295, 303, 304	305–308
Rhizopus stolonifer	Media	—	68, 276	92, 178A
	Ref.		309, 310	311
Rhynchosporium secalis	Media	—	—	123, 178A
	Ref.		312, 313	313–316
Saprolegnia spp.	Media	89, 98, 286	35, 89, 109	89, 109
	Ref.	1, 3, 8, 317–319	1, 3, 317, 318	1, 3, 317, 318
Schizophyllum commune	Media	133	101, 248	101, 248, 290
	Ref.		17	3, 320–322
Septoria nodorum	Media		51, 265	50, 170, 178A, 290, 311
	Ref.		323	324–328
S. tritici	Media		70	50, 136A broth, 290
	Ref.		329	325, 328, 330
Sordaria fimicola	Media	33	18, 43, 259	18, 41, 65, 306
	Ref.	331	317, 331, 332	3, 331–333
Stemphylium botryosum	Media		—	136A agar, 178A, 290
	Ref.		334	21, 335–337
Thielaviopsis basicola	Media	278, 302, 304	17, 340, 341	52, 290
	Ref.	338, 339	284, 285	339, 342
Tilletia caries	Media	285	343, 345	284, 285
	Ref.	343, 344	—	343, 345
Trichoderma lignorum (see *T. viride*)	Media			—
	Ref.			
T. viride	Media	96, 190	165	136A agar, 178A
	Ref.	346	346, 347	347–350
Trichophyton mentagrophytes	Media	55	85	35, 164, 178A, 287
	Ref.	6, 351	6, 352, 353	6, 352, 354
Ustilago maydis	Media	178B	178A, 288, 289	178A, 288, 289
	Ref.	355	3, 356, 357	3, 356, 357
Venturia inaequalis	Media		14	5, 136B broth
	Ref.	358	359, 360	165, 361, 362
Verticillium albo-atrum	Media	48,[p] 256	—	48, 51, 178A
	Ref.	363–366	367	365, 368–370

[p]   +50 ppm streptomycin and chlortetracycline.

Table 2
SELECTED LIST OF MEDIA USED TO CULTURE FUNGI

Recipe number	Medium	Recipe number	Medium
1	ADM	52	Czapek solution yeast extract agar
2	Alphacel medium	53	Czapek tergitol® medium
3	Alphacel medium (modified)	54	Darby and Mandels Myrothecium media
4	Aphanomyces synthetic medium	55	Dermatophyte test medium
5	Apple potato infusion agar	56	Dilute inorganic salts solution
6	Armstrong Fusarium medium	57	Dung oatmeal agar
7	Asparagine glucose agar	58	Dutch elm medium
8	Asparagine yeast extract medium	59	Elliott agar
9	Asthana and Hawker medium A	60	Elm wood
10	Barnett maltose casamino medium	61	Endothia complete medium
11	Bean juice agar	62	Endothia minimal medium
12	Bean pod agar	63	Farley Colletotrichum medium
13	Bonner-Addicott medium	64	Fernandos medium
14	Boone minimal medium	65	Filter paper yeast agar (M-61)
15	Botran® isolation medium	66	Fomes annosus isolation medium no. 1
16	Botrytis species separation agar	67	Fomes annosus isolation medium no. 2
17	Brain heart infusion agar	68	Fothergill sucrose ammonium sulfate medium
18	Bretzloff medium	69	Foudin and Wynn medium
19	Bugbee phoma isolation medium	70	Fries medium (modified)
20	Bushnell Puccinia medium	71	Fries medium modified no. 3
21	Carnation leaf agar	72	Fungal sporulation medium
22	Carrot decoction agar	73	Gallic or tannic acid medium
23	Carrot glucose agar	74	Gallic oxalate medium
24	Carrot juice agar	75	Garraway medium
25	Cellulose asparagine medium	76	Gelatin liquefaction test medium
26	Cerelose ammonium nitrate medium	77	Glucose alanine agar
27	Chen and Zentmyer salt solution	78	Glucose asparagine agar[155]
28	Chick pea seed meal agar	79	Glucose asparagine agar[397]
29	CMC medium	80	Glucose asparagine ascorbic acid medium
30	Coons synthetic medium	81	Glucose asparagine manganese medium
31	Coprinus synthetic	82	Glucose asparagine medium
32	Corn extract broth or agar	83	Glucose asparagine thiamine agar
33	Corn meal acetate agar	84	Glucose calcium medium
34	Corn meal agar	85	Glucose casamino acid agar
35	Corn meal agar (Difco)	86	Glucose casamino acid medium
36	Corn meal glucose agar	87	Glucose casamino thiamine agar
37	Corn meal half-strength agar (CM/2) M-8A	88	Glucose casein hydrolysate medium
38	Corn meal oat flake agar	89	Glucose glutamate medium
39	Corn meal Phytophthora isolation medium no. 1 (P10PV)	90	Glucose minimal medium
40	Corn meal Phytophthora isolation medium no. 1 (P5VP)	91	Glucose nitrate yeast extract agar
		92	Glucose peptone agar[311]
41	Corn meal sugar yeast agar	93	Glucose peptone agar[400]
42	Corn meal tween 80® agar	94	Glucose peptone bengal medium
43	Corn meal yeast extract glucose agar	95	Glucose peptone yeast agar (SM/5)
44	Corn steep dextrose (cerelose)	96	Glucose peptone yeast antibiotic agar
45	Cylindrocladium isolation medium	97	Glucose phenylalanine agar
46	Czapek corn steep botran® medium	98	Glucose starch tellurite medium
47	Czapek corn steep tergitol	99	Glucose tryptone medium
48	Czapek-Dox agar (Difco® Czapek solution)	100	Glucose yeast extract agar (M-63A)
49	Czapek-Dox agar (modified)	101	Glucose yeast extract agar (M-64)
50	Czapek-Dox V-8® medium	102	Glucose yeast extract broth
51	Czapek solution agar (Difco® and BBL® Czapek-Dox)	103	Glycerine glucose broth
		104	Green leaf agar
		105	Guaiacol agar

Note: See recipes beginning on page 437.

Table 2 (continued)
SELECTED LIST OF MEDIA USED TO CULTURE FUNGI

Recipe number	Medium	Recipe number	Medium
106	Harrold medium	161	Neurospora "complete" medium
107	Hay extract agar	162	Neurospora "minimal" medium
108	Hay infusion agar	163	Neurospora sex synthetic
109	Hemp seed	164	Neutral wort agar
110	Hemp seed agar	165	Norkrans medium
111	Hemp seed agar modified	166	Oak wilt agar
112	Hemp seed decoction and agar	167	Oak wilt agar (modified)
113	Hoagland solution (1950)	168	Oat flake agar
114	Honey peptone medium	169	Oat wheat agar
115	Hyphomyces synthetic	170	Oatmeal agar
116	Joham medium	171	Odegard medium
117	Kerrs medium	172	Ohio medium (OAES)
118	Kerrs medium (modified)	173	Olive medium
119	Knop agar	174	Pablum agar
120	Lactose casein hydrolysate medium	175	Pablum agar (modified)
121	Leonian agar	176	Park medium
122	Lima bean agar (Difco®)	177	PCNB medium (modified)
123	Lima bean agar (Difco® modified)	178A	PDA
124	Lima bean agar (frozen)	178B	PDA (acidified)
125	Lima bean agar (frozen and cleared)	179	PDA[4 3 9]
126	Lima bean dextrose agar	180	PDA (modified)[8 6]
127	Littman oxgall agar	181	PDA (modified)[2 4]
128	Lukens and Sisler medium	182	PDA-BDP
129	Lutz medium (modified)	183	PDA casamino acid
130	Machlis medium a (modification of Ingraham medium)	184	PDA charcoal medium
		185	PDA-DOPCNB
131	Machlis medium b	186	PDA-DORB
132	Macrophomina isolation medium (CMR)	187	PDA Phytophthora isolation medium
133	Malt agar (Blakeslee formula)	188	PDA yeast extract
134	Malt agar (modified by Kaufmann)	189	PDTA
135	Malt benomyl agar	190	PDTCS (PDA-NPX)
136A	Malt extract broth or agar	191	Pea seed
136B	Malt extract broth	192	Peanut meal agar
137	Malt extract filter paper	193	Penicillin medium
138	Malt glucose yeast extract agar	194	Peptone glucose agar[2 1 4]
139	Malt sugar agar	195	Peptone glucose agar[1 2 4],[1 2 6]
140	Malt yeast extract agar[2 4 4],[4 2 1]	196	Peptone glucose medium
141	Malt yeast extract agar[3 9 8]	197	Peptone glucose yeast agar (SM-agar)
142	Maltose glycerine medium	198	Peptone glucose yeast broth
143	Maltose peptone broth	199	Peptone glucose yeast broth (medium A)
144	Maltose tartrate medium	200	Peptone yeast glucose agar (Cantino-PYG) (M-51)
145	Martin media		
146	Matsushima PHB agar	201	Peptone yeast extract maltose agar (M-53)
147	Melin-Norkrans medium (modified)	202	Petri medium
148	Microelement solution[3 9 8]	203	Physarium medium
149	Microelement solution[3]	204	Physarium sporulation medium
150	Milk agar	205	Phytone dextrose agar
151	Minimal agar medium	206	Phytophthora isolation medium
152	Mircetich isolation medium (MPVM)	207	Phytophthora synthetic no. 1
153	Monilinia selective medium	208	Phytophthora synthetic no. 2
154	Mullins and Barksdale synthetic	209	Phytophthora synthetic no. 3
155	Mycobiotic agar (Difco®)	210	Pilobolus hemin medium
156	Mycological agar	211	Pontecorvo minimal medium
157	Mycophil agar (BBL®)	212	Potato carrot agar
158	Mycosel agar (BBL®)	213	Potato carrot glucose broth
159	Nash and Snyder PCNB medium	214	Potato malt agar
160	Neopeptone glucose agar	215	Potato marmite agar

Table 2 (continued)
SELECTED LIST OF MEDIA USED TO CULTURE FUNGI

Recipe number	Medium	Recipe number	Medium
216	Potato sucrose agar	267	Sucrose leucine medium
217	Potato tomato agar	268	Sucrose proline agar (SPA)
218	Prune agar	269	Sucrose salts yeast medium (SMKY)
219	Pyronema medium (modification of Claussen's agar)	270	Sucrose synthetic medium
		271	Sucrose threonine medium
220	Rabbit food agar	272	Sucrose urea medium (medium B)
221	Rape seed agar	273	Sucrose yeast extract medium
222	Raper Achyla medium no. 1	274	Sugar beet leaf agar
223	Raper Achyla medium no. 2	275	Sugar beet molasses medium
224	Raulin-Thom medium	276	Synthetic mucor
225	Raulin-Thom medium	277	Talboy medium
226	Replacement medium	278	TBM-C
227	Reynold medium	279	Tchernoff medium
228	Rhizoctonia isolation medium	280	Tochinai solution
229	Rice cereal agar	281	Tomato salt agar
230	Rice polish agar	282	Toussoun medium
231	Richard solution	283	Toxin production medium
232	Richard solution (modified)[142,463]	284	Trione Tilletia medium
233	Richard solution (modified)[13]	285	Trione Tilletia synthetic medium
234	Richard V-8® medium	286	Tryptone yeast extract broth
235	Russell basidiomycete medium	287	Urease test medium
236	Rye dextrose agar	288	Ustilago genetics complete medium
237	Rye grain agar A	289	Ustilago genetics minimal medium
238	Rye grain agar B	290	V-8® agar
239	Sabhi blood medium	291	V-8® agar (modified)
240	Sabouraud agar (Emmons modification)	292	V-8® agar (cleared)[482]
241	Sabouraud dextrose agar (Difco®)	293	V-8® agar (cleared)[103]
242	Sabouraud dextrose agar (modified)	294	V-8® agar (cleared)[258]
243	Sachs agar	295	V-8® agar sterol (cleared)
244	Sachs agar (modified)	296	V-8® agar (modified)
245A	Sachs agar and sorghum kernels	297	V-8® agar sterol broth (cleared)
245B	Sachs agar and corn kernels	298	V-8® agar sterol broth (dilute)
245C	Sachs agar and corn leaves	299	V-8® isolation agar (VDYA)
245D	Sachs agar and barley straw	300	V-8® isolation agar (dilute)
246	Salt solution	301	V-8® isolation cleared agar
247	Salvin medium	302	V-8® isolation cleared agar (TBM-V8)
248	Schizophyllum fruiting medium (M-80)	303	V-8® glucose agar
249	Schmitthenner medium	304	V-8® isolation agar (VDYA-PCNB)
250	Sea water glucose agar	305	V-8® juice simulated medium
251	Sea water glucose peptone agar	306	Vogel medium N
252	Sea water isolation medium	307	Warcup medium
253	Soil extract agar[470]	308	Water agar
254	Soil extract agar[471]	309	Water agar leaf medium
255	Soil extract corn meal agar	310	Wheat leaf extract agar
256	Soil extract pga medium	311	Wheat meal agar
257	Soil horsehair chicken feather medium	312	Wilbrink agar
258	Sonneborn defined medium	313	Williams Puccinia medium
259	Sordaria synthetic medium	314	Wort agar
260	Sorghum grain	315	Yang and Schoulties medium
261	Soybean seed extract broth	316	Yeast dextrose agar
262	Starch glutamic acid medium	317	Yeast extract filter paper medium
263	Starch hydrolysis test medium	318	Yeast glucose (YpG)
264	Steinberg dibasal medium	319	Yeast malt extract agar
265	Sucrose asparagine medium	320	Yeast starch agar
266	Sucrose asparagine yeast agar (Say)	321	Yellow peas

RECIPES OF CULTURE MEDIA

(Ingredients on a liter basis)

Recipe 1
ADM[45]

Tryptone	15 g	30 ppm of tetracycline[a] can be added to inhibit bac-
Yeast extract	10.0 g	teria
Ferric citrate	0.5 g	
Agar	15 g	

[a] Indicates material should be added after medium is autoclaved; usually if material is an antibiotic and handled aseptically can be added without sterilization by filtration.

Recipe 2
ALPHACEL MEDIUM[76]

Alphacel (obtained from NBC, Cleveland, Ohio, a form of cellulose)	20 g	Filter coconut milk through several layers of cheesecloth; if desired autoclave for 15 min at 121°C and
$MgSO_4$	1 g	store at 6°C; adjust pH of medium
KH_2PO_4	1.5 g	to 5.6 and sterilize at 20 lb/20 min;
$NaNO_3$	1 g	I have not found the high tempera-
Coconut milk	50 ml	ture necessary
Agar	12 g	

Recipe 3
ALPHACEL MEDIUM MODIFIED[76]

Alphacel medium (Recipe 2)	
Tomato paste, Hunts®	10 g
Oatmeal, Beechnut® baby	10 g

Recipe 4
APHANOMYCES SYNTHETIC MEDIUM[32]

D-glucose	5.0 g	Adjust pH to 5.5 after autoclaving
L-asparagine	0.75 g	
Monobasic potassium phosphate	2.0 g	
Magnesium chloride	0.05 g	
Manganese chloride	0.005 g	
Zinc chloride	0.005 g	
Ferric chloride	0.005 g	
L-methionine or L-cystine	5 to 20 ppm	

Recipe 5
APPLE POTATO INFUSION AGAR[361]

Glucose	5 g	Steam leaves, potatoes and agar
Potatoes	40 g	each separately in water for 30
Apple leaves (dead and dry)	25 g	min, combine, and add glucose
Agar	17 g	

Recipe 6
ARMSTRONG FUSARIUM MEDIUM[143]

Glucose or sucrose	2%
KCl	0.003M
KH$_2$PO$_4$	0.008M
Ca(NO$_3$)$_2$	0.0356M
FeCl$_3$	0.2 ppm of each
MnSO$_4$	cation
ZnSO$_4$	

Recipe 7
ASPARAGINE GLUCOSE AGAR[157]

L-Asparagine	2.0 g
Glucose	8.0 g
KH$_2$PO$_4$	1.0 g
MgSO$_4$·7H$_2$O	0.5 g
Agar	1.5 g

Recipe 8
ASPARAGINE YEAST EXTRACT MEDIUM[158]

Asparagine	2.25 g
MgSO$_4$·7H$_2$O	0.123 g
Yeast extract	3.0 g
Potassium phosphate buffer	0.02 M, pH 7.0

Recipe 9
ASTHANA AND HAWKER MEDIUM A[371]

Glucose	5.0 g
KH$_2$PO$_4$	3.5 g
MgSO$_4$	0.75 g
Agar	15 g

Recipe 10
BARNETT MALTOSE CASAMINO MEDIUM[69]

Maltose	5 g	Adjust pH to 6.0; autoclave at 15 lb/15
Difco® casamino acids (technical grade)	1.0 g	min
KH$_2$PO$_4$	1.0 g	
MgSO$_4$·7H$_2$O	0.5 g	
Zn	0.2 mg	
Fe (as sulfates)	0.2 mg	See micronutrient solution No. 148
Mn	0.1 mg	
Biotin (0.5 g yeast extract may be substituted for some isolates)	5 µg	
Agar	20 g	

Recipe 11
BEAN JUICE AGAR[372]

Bean juice from canned green beans	215 ml
Agar	10 g
Water	285 ml

Recipe 12
BEAN POD AGAR[373]

Bean pod	200 g	Make a distilled-water extract of bean pods;
Sucrose	5 g	sterilize sucrose separately
Malt extract	5 g	
Agar	17 g	

Recipe 13
BONNER-ADDICOTT MEDIUM[3][74]

Glucose	20 g
KNO_3	81 mg
$MgSO_4 \cdot 7H_2O$	36 mg
$Ca(NO_3)_2 \cdot 4H_2O$	236 mg
KH_2PO_4	12 mg
KCl	65 mg
Ferric tartrate	1 mg
or $Fe_2(SO_4)_3$	
Agar	25 g

Recipe 14
BOONE MINIMAL MEDIUM[3][59]

Glucose	5 g	
KNO_3	3.12 g	
K_2HPO_4	0.75 g	
KH_2PO_4	0.75 g	
$MgSO_4 \cdot 7H_2O$	0.5 g	
NaCl	0.1 g	
$CaCl_2 \cdot H_2O$	0.1 g	
Trace elements	1 ml	
$ZnSO_4 \cdot 7H_2O$	58.4 mg	
$CuSO_4 \cdot 5H_2O$	31.6 mg	
$MnSO_4 \cdot 4H_2O$	16.2 mg	
H_3BO_3	11.4 mg	In distilled water to 400 ml
MoO_3	7.0 mg	
$FeC_6H_5O_7 \cdot 3H_2O$	214.2 mg	

Recipe 15
BOTRAN® ISOLATION MEDIUM[41]

K_2HPO_4	0.5 g	Adjust pH to 5.5 with 1 N HCl or
$MgSO_4 \cdot 7H_2O$	0.5 g	NaOH before autoclaving
Peptone	0.5 g	
Yeast extract	0.5 g	
Sucrose	20 g	
Rose bengal	25 mg	
Streptomycin sulfate[a]	50 mg	
Botran® (Upjohn Co.)[a]	10 ppm	If technical grade, dissolve in acetone
(2,6-dichloro-4-nitroaniline)	of active ingredient	

[a] See footnote, Recipe 1.

Recipe 16
BOTRYTIS SPECIES SEPARATION AGAR[59]

Glycerol	5 g	*Botrytis cinerea* will grow equally well with and
Sorbose	2.5 g	without sorbose; *B. alli* is restricted by the
Casein hydrolysate	5 g	latter
Yeast extract	3 g	
KCl	1 g	
KH_2PO_4	.15 g	
$NaNO_3$	3.0 g	
$MgSO_4$	0.5 g	
Agar	20 g	

Recipe 17
BRAIN HEART INFUSION AGAR[6]

Brain heart infusion	37 g	0.5 ml of whole blood may be added to
Cycloheximide[a] (dissolved in acetone)	0.5 g	the surface of the slant, according to
Chloramphenicol[a] (dissolved in 95% Etoh)	0.05 g	E. S. Beneke[493]
Agar	20 g	

[a] See footnote, Recipe 1.

Recipe 18
BRETZLOFF MEDIUM[331]

Sucrose	20 g	Very similar to Recipe 163;
Potassium nitrate	1.0 g	for more growth of *Sordaria*
Potassium dihydrogen phosphate	1.0 g	*fimicola* use glucose instead
$MgSO_4 \cdot 7H_2O$	0.5 g	of sucrose
$CaCl_2$	0.1 g	
Zn	2.0 mg	
Fe	0.2 mg	
Cu	0.1 mg	
Mn	0.02 mg	
Mo	0.02 mg	
B	0.01 mg	
Biotin	4 μg	

Recipe 19
BUGBEE PHOMA ISOLATION MEDIUM[240]

K_2HPO_4	4.0 g	Adjust to pH 7.0 with HCl before auto-
KH_2PO_4	1.5 g	claving; soil extract — suspend 1 kg of
Soil extract	25 ml	soil in 1 l of water, steam at 104—110°C
Boric acid	200 mg	for 30 min
Streptomycin sulfate[a]	100 mg	
Chlortetracycline[a]	100 mg	
Benomyl[a]	100 mg	
Sucrose[a]	10 g	
Agar	17 g	

[a] See footnote, Recipe 1.

Recipe 20
BUSHNELL PUCCINIA MEDIUM[288]

Glucose	3.0 g	Adjust pH to 6.4 with HCl be-
Peptone (Evans Medical Ltd. Speke, Liverpool)	4 g	fore agar is added; see also Recipe 313, Williams *Puccinia*
K_2HPO_4	1.0 g	medium
KCl	0.5 g	
$MgSO_4 \cdot 7H_2O$	0.5 g	
$FeSO_4$	0.01 g	
Casamino acids (Difco®)	4—8 g	
Agar	10 g for seeding of urediospores 15 g for subculture	

Recipe 21
CARNATION LEAF AGAR[168]

Carnation leaf	5 mm discs	Discs cut from fresh leaves are soaked in 60% EtOH
Water agar	2.0%	for 5–10 min, rinsed thoroughly, and placed five
		to a plate on water agar

Recipe 22
CARROT DECOCTION AGAR[375]

Carrot leaves	300 g	Leaves are finely ground in food chopper; steam
Agar	12 g	without pressure in 500 ml water for 1 hr, strain,
		and add 500 ml of melted agar

Recipe 23
CARROT GLUCOSE AGAR[155]

Carrot roots	250 g
Glucose	10 g
Agar	15 g

Recipe 24
CARROT JUICE AGAR[171]

Carrot juice (12 oz can)	355 ml
Agar	15 g

Recipe 25
CELLULOSE ASPARAGINE MEDIUM[225]

Cellulose (Whatman's standard grade chromatographic powder ball milled for 72 hr)	10 g	Cellulose is added as a 4% solution to the steamed medium and the medium is auto- claved for 20 min at 10 psi; final pH is 6.2.
L-asparagine	0.5 g	
Ammonium sulfate	0.5 g	
Potassium dihydrogen phosphate	1.0 g	
Potassium chloride	0.5 g	
Magnesium sulfate	0.2 g	
Calcium chloride	0.1 g	
Yeast extract (Difco®)	0.5 g	
Agar	20 g	

Recipe 26
CERELOSE AMMONIUM NITRATE MEDIUM[141]

Cerelose (crude form of glucose)	50 g	Recipe 232 is the same but designated as Richard
NH_4NO_3	10 g	solution
KH_2PO_4	5 g	
$MgSO_4 \cdot 7H_2O$	2.5 g	
$FeCl_3 \cdot 6H_2O$	0.02 g	

Recipe 27
CHEN AND ZENTMYER SALT SOLUTION[376]

EDTA	13.05 g	Sterilize by filtration through a 0.22-μm Milli-
KOH	7.5 g	pore® filter
$FeSO_4 \cdot 7H_2O$	24.9 g	
Deionized water		

Recipe 28
CHICK PEA SEED MEAL AGAR[39]

Chick pea seed (freshly crushed, white seeded)	2 g	Incorporate seed in melted agar and autoclave
Water agar	20 g	

Recipe 29
CMC MEDIUM[164]

Carboxymethylcellulose CMC7MP—Hercules Powder Co.; (dissolve by mixing in a blender in warm water)	15.0 g
NH_4NO_3	1.0 g
KH_2PO_4	1.0 g
$MgSO_4 \cdot 7H_2O$	0.5 g
Yeast extract	1.0 g

Recipe 30
COONS SYNTHETIC MEDIUM[377]

Sucrose	7.2 g
Glucose	3.6 g
$MgSO_4$	1.23 g
KH_2PO_4	2.72 g
KNO_3	2.02 g
Agar	15 g

Recipe 31
COPRINUS SYNTHETIC[3,110]

Solution A

Glucose	20 g	Autoclave solutions A and B separately, add 1
Asparagine	2 g	ml of B to 100 ml of A
NH_4 tartrate	0.5 g	
KH_2PO_4	1.0 g	
Na_2HPO_4	2.25 g	
Na_2SO_4	0.28 g	
Thiamine	100 μg	
Distilled water	1 liter	

Solution B

$CaCl_2$	0.1 g	Adjust pH 6.8 to obtain clear solution
$MgCl_2$	0.41 g	
Fe-citrate	5.31 mg	
Citric acid	5.31 mg	
$MnSO_4 \cdot 4H_2O$	4.43 mg	
$ZnSO_4 \cdot 7H_2O$	4.05 mg	
Distilled water	40 ml	

Amendments to Make a Complete Medium

Yeast extract	0.75 g
Hydrolyzed casein	0.75 g
Malt extract	0.60 g
Hydrolyzed nucleic acid	1.25 ml

(Add 1 g yeast nucleic acid and 1 g thymus nucleic acid in 15 ml 1N sodium hydroxide and 1 g of each in 15 ml 1N HCl, autoclave at 15 psi for 10 min, mix the hydrolysates, adjust to pH 6.0, filter hot, and adjust to 40 ml. Store at −20 or at 4°C with 1 ml chloroform[3]

Recipe 32
CORN EXTRACT BROTH OR AGAR[378,379]

Sweet corn seed	45 g	Broth prepared in a liter of distilled water; add 17 g of agar if solid medium desired to isolate *Aphanomyces euteiches*[379]

Recipe 33
CORN MEAL ACETATE AGAR[331]

Corn meal agar		Add 1 liter of corn meal agar (commercial versions vary in amounts of agar to start with), amending to bring it to 3—4%
Sodium acetate	7g	
Agar to bring final concentration to 3—4%		

Recipe 34
CORN MEAL AGAR

Corn meal	60 g	Steam corn meal in a fine cloth bag in 500 ml H_2O; use extract
Agar	15 g	

Recipe 35
CORN MEAL AGAR DIFCO[®][380]

Corn meal	50 g
Agar	15 g

Recipe 36
CORN MEAL GLUCOSE AGAR[381]

Corn meal	50 g	Acidify to pH 4.8 for isolation of some fungi; use lactic acid
Glucose	2 g	
Agar	15 g	

Recipe 37
CORN MEAL HALF-STRENGTH AGAR CM/2—M-8A[3,250]

Corn meal, agar, Difco[®] (without glucose)		Mix equal quantities of CMA and 2% agar; Collins and Tang[250] suggest buffering medium between pH 5 and 6
Agar	2%	

Recipe 38
CORN MEAL OAT FLAKE AGAR[3]

Corn meal	25 g	After growth of *Physarium* is able to be seen without a lens, sprinkle dry, sterilized oat flakes on surface of agar medium
Agar	20.0 g	
Pulverized oat flakes		

Recipe 39
CORN MEAL PHYTOPHTHORA ISOLATION MEDIUM NO. 1 (P10VP)[382]

Corn meal agar (Difco®)	17 g
Pimaricin[a] (dissolved in DMSO if insoluble parent compound used)	10 mg
Vancomycin[a]	200 mg
PCNB	100 mg

[a] See footnote, Recipe 1.

Recipe 40
CORN MEAL PHYTOPHTHORA ISOLATION MEDIUM NO. 2 (P5VP)[383]

Corn meal agar (Difco®)	17 g
Pimaricin[a]	5 mg
Vancomycin[a]	300 mg
PCNB	25 mg

[a] See footnote, Recipe 1.

Recipe 41
CORN MEAL SUGAR YEAST AGAR[332]

Corn meal agar (Difco®)	
Sucrose	10.0 g
Glucose	7.0 g
Yeast extract	1.0 g
KH_2PO_4	0.1 g

Recipe 42
CORN MEAL TWEEN 80 AGAR[6]

Corn meal agar	
Tween 80®	1%

Recipe 43
CORN MEAL YEAST EXTRACT GLUCOSE AGAR[3]

Difco® corn meal glucose (2 g), agar made as directed by manufacturer	
Yeast extract	1 g

Recipe 44
CORN STEEP DEXTROSE CERELOSE[54]

Dextrose (cerelose)	40 g	Add 2 drops of Dow Corning Antifoam AF® prior
Corn steep liquor (Staley®)	40 g	to autoclaving to avoid frothing in fermentation
$NaNO_3$	3 g	tanks
K_2HPO_4	0.5 g	
$MgSO_4 \cdot 7H_2O$	0.25 g	
$CaCO_3$	3.5 g	

Recipe 45
CYLINDROCLADIUM ISOLATION MEDIUM[118]

Glucose	15 g
KNO_3	0.5 g
Yeast extract	0.5 g
KH_2PO_4	1.0 g
$MgSO_4 \cdot 7H_2O$	0.5 g
Tergitol NPX® (Union Carbide)[a]	1 ml
Thiabendazole[a]	1 mg
Chloramphenicol[a] (dissolved in 95% EtOH)	100 mg
Chlortetracycline[a] (dissolved in 50% EtOH)	40 mg
Agar	20 g

[a] See footnote, Recipe 1.

Recipe 46
CZAPEK CORN STEEP BOTRAN® MEDIUM[235,237]

Czapek solution agar
Corn steep (concentrated)   1%
Botran® 75 W (dissolved in acetone)[a]   7—11 mg[235]
                                     or 20 mg[237]

Botran enhances fasicle production of
certain species of *Penicillium*; adjust
pH to 7.0 with 1 N NaOH

[a] See footnote, Recipe 1.

Recipe 47
CZAPEK CORN STEEP TERGITOL®[235]

Czapek solution agar no. 51
Corn steep (concentrated liquor)   10 g
Tergito® NPX   100 ppm
Agar   20 g

Adjust pH to 7.0 with 1 N KOH prior to autoclaving

Recipe 48
CZAPEK-DOX AGAR (DIFCO® CZAPEK SOLUTION)[384,385]

Same as Recipe 51 except for 2.0 g of $NaNO_3$

According to Thom and Church,[384] this difference has little effect on the aspergilli; strictly adhering to the original version of Dox,[385]
KH_2PO_4 should be used instead of K_2HPO_4;
Thom and Church used the latter because
the final reaction is neutral or slightly acidic;
both Difco and BBL use K_2HPO_4 as do most
versions

Recipe 49
CZAPEK-DOX AGAR MODIFIED[386]

Sodium nitrate   2.0 g
Potassium chloride   0.5 g
Magnesium glycerophosphate 0.5 g
Ferrous sulfate   0.01 g
Potassium sulfate   0.35 g
Sucrose   30.0 g
Oxoid® agar no. 3   12.0 g

The addition of magnesium glycerophosphate prevents the precipitation of magnesium phosphate
that may occur in Czapek-Dox and Czapek solution agar

Recipe 50
CZAPEK-DOX V-8® MEDIUM[325]

Czapek-Dox (Oxoid®)   45.4 g
V-8® juice   200 ml
Calcium carbonate   3.0 g
Agar, Oxoid® No. 3   10 g
Deionized water   800 ml

Recipe 51
CZAPEK SOLUTION AGAR (DIFCO® AND BBL® CAZPEK-DOX)[47],[387]

NaNO$_3$	3.0 g
K$_2$HPO$_4$	1.0 g
MgSO$_4 \cdot$7H$_2$O	0.5 g
KCl	0.5 g
FeSO$_4 \cdot$7H$_2$O	0.01 g
Sucrose	30.0 g
Agar	15.0 g

This medium may be modified by adding 1% concentration corn steep liquor; adjust pH to 7.0 with 1 N NaOH as the corn steep is acid; if medium is unadjusted, it may not solidify; xerophytic aspergilli such as members of the *Aspergillus glaucus* group require 200 g of sucrose

Recipe 52
CZAPEK SOLUTION YEAST EXTRACT AGAR[339],[342]

Czapek solution agar (Recipe 51) plus 5 g or 2.5 g of yeast extract

Recipe 53
CZAPEK TERGITOL® MEDIUM[121]

Czapek solution (Recipe 51?)	
Tergitol® NPX	1000 ppm
Agar	25 g

Adjust pH to 3.5 with lactic acid

Recipe 54
DARBY AND MANDELS MYROTHECIUM MEDIA[226]

Medium number[a]	Purpose	Ingredients (g/l)				Carbon source[b]
		NH$_4$NO$_3$	KH$_2$PO$_4$	K$_2$HPO$_4$	MgSO$_4 \cdot$7H$_2$O	
54A	Dry weight/liquid	1.2	21.8	27.9	0.9	Sucrose
54B	Mycelium + spores—agar	0.3	2.2	2.8	0.2	Sucrose
54C	Mycelium on agar	3.0	2.2	2.8	0.2	Sucrose
54D	Dry weight loss	6.0	2.2	2.8	0.2	Ground cloth
54E	Tensile strength loss	6.0	2.2	2.8	0.2	Cloth strips
54F	Tensile strength loss	0.3	0.7	0.9	0.05	Cloth strips
54G	Cellulose enzyme	1.0	1.36	—	0.03	Ground cellulose
54H	Spores	3.0	2.7	2.1	2.2	Filter paper

[a] All include 1 g of yeast extract (Difco®) and CP-grade chemicals.
[b] 2% by weight.

Recipe 55
DERMATOPHYTE TEST MEDIUM[351,398]

Phytone	10.0 g	The final pH of medium is 5.5 ± 0.1 and
Dextrose	10.0 g	should be yellow; store in refrigerator
Phenol red solution, Difco®	40 ml	Phenol red 0.5 g, 0.1N NaOH 15 ml,
HCl (0.8M)	6 ml	glass distilled H_2O made up to 100 ml
Cycloheximide[a]	0.5 g	Dissolve in acetone
Gentamicin sulfate[a] (active)	100 mg	Dissolve in sterile glass-distilled water
Chlortetracycline HCl[a]	100 mg	Autoclave medium at 12 psi for 10 min
Agar	20.0 g	

[a] See Footnote, Recipe 1.

Recipe 56
DILUTE INORGANIC SALTS SOLUTION[58]

$MgSO_4$	$2.5 \times 10^{-5}M$	pH of solution – 7.0
$CaCl_2$ at pH 7.0	$2.5 \times 10^{-5}M$	
NH_4NO_3	$2.5 \times 10^{-4}M$	
K_2HPO_4	$2.5 \times 10^{-4}M$	
KH_2PO_4	$2.5 \times 10^{-4}M$	

Recipe 57
DUNG OATMEAL AGAR[277]

Bovung®	125 g	Boil dung in 1 liter of tap H_2O for 2–3 hr,
(a commercial, dried cow manure)		filter through cheesecloth and filter paper
Oatmeal	2%	Boil oatmeal for 5 min, combine 500 ml of
		each, and add agar

Recipe 58
DUTCH ELM MEDIUM[71]

Potato dextrose agar	
Cycloheximide (actidione)[a]	200 ppm
Streptomycin sulphate[a]	10 ppm

[a] See footnote, Recipe 1.

Recipe 60
ELM WOOD[75,77]

Debark split twigs in a test tube or use slices of wood in petri dish; add a small amount of water and autoclave; for the production of perithecia of *Ceratocystis ulmi* when inoculated with two compatable strains

Recipe 59
ELLIOTT AGAR[389]

Glucose	5.0 g
Asparagine	1.0 g
Potassium monobasic phosphate	1.36 g
Sodium carbonate	1.06 g
Magnesium sulfate	0.5 g
Agar	15 g

Recipe 61
ENDOTHIA COMPLETE MEDIUM[129]

Glucose	10 g
Yeast extract (Difco®)	2.5 g
Malt extract (Difco®)	7.5 g
Salt solution complete medium	62.5 ml
from *Ustilago* genetics medium	
(Recipe 288)	
Thiamine	2 mg
Agar	20 g

Recipe 62
ENDOTHIA MINIMAL MEDIUM[129]

Glucose	10 g
Salt solution (from *Ustilago* genetics complete medium)	62.5 ml
Thiamine	2 mg

Recipe 63
FARLEY COLLETOTRICHUM MEDIUM[98]

Polygalacturonic acid (in 500 ml H_2O, adjust to pH 5)	10.0 g	Amendments do not need to be sterilized but made up aseptically
KH_2PO_4	1.5 g	
K_2HPO_4	4.0 g	
Soil extract	25 ml	
Agar	17.0 g	
PCNB (in 1 ml acetone)[a]	0.1 g	
Benomyl[a]	0.1 g	
Streptomycin sulfate[a]	0.1 g	
Tetracycline chloride[a]	0.1 g	
Chloramphenicol[a]	0.1 g	

[a] See footnote, Recipe 1. These ingredients are nonsterile.

Recipe 64
FERNANDOS MEDIUM[390]

Glucose	15 g
Asparagine	5.0 g
KH_2PO_4	6.8 g
$MgSO_4 \cdot 7H_2O$	5.0 g

Recipe 65
FILTER PAPER YEAST AGAR (M-61)[3,391]

Filter paper	12 g	Blend filter paper in water
Yeast extract (Difco®)	4 g	
Agar	24 g	

Recipe 66
FOMES ANNOSUS ISOLATION MEDIUM NO. 1[130,133]

Bacto®-peptone	5.0 g	Shake before pouring to resuspend
$MgSO_4$	0.25 g	PCNB; Edmonds and Driver[133]
KH_2PO_4	0.5 g	added rose bengal at 1:30,000
Pentachloronitrobenzene	190 ppm	
Streptomycin	100 ppm	
Agar	20 g	
Lactic acid (50%)[a]	2 ml	
Ethyl alcohol (95%)[a]	20 ml	

[a] See footnote, Recipe 1.

Recipe 67
FOMES ANNOSUS ISOLATION MEDIUM
NO. 2[131]

Malt extract	20.0 g
Agar	17.0 g
Streptomycin sulfate (1% solution)[a]	10 ml
Ortho-phenyl-phenol[a] (0.48 g in 20 ml of 95% EtOH)	2.5 ml
Lactic acid (50%)[a]	1.0 ml

[a] See footnote, Recipe 1.

Recipe 68
FOTHERGILL SUCROSE AMMONIUM SULFATE MEDIUM[309]

Sucrose	4% w/v	Add sterile phosphate solution after autoclaving to avoid possible precipitation
K_2HPO_4	0.048 M	
$MgSO_4 \cdot 7H_2O$	0.012 M	
$(NH_4)_2SO_4$	0.025 M	
Zn	2 ppm	
Fe	2 ppm	
Mn	2 ppm	

Recipe 69
FOUDIN AND WYNN MEDIUM[392]

		Burkholder and Nickell Trace Elements (mg/l)	
Czapek mineral solution (Recipe 51 with ferric EDTA instead of ferrous sulfate	14.4 mg	H_3BO_3	570
		$MnCl_2 \cdot H_2O$	360
Glucose	30 g	$ZnCl_2$	625
Burkholder and Nickell trace elements less ferric tartrate	2.0 ml	$Na_2MoO_4 \cdot 2H_2O$	252
		$CuCl_2 \cdot 2H_2O$	268
Agar	10 g		

Amino Acids

Alanine	114 mg	Lysine	287 mg
Arginine	159 mg	Methionine	110 mg
Aspartic acid	281 mg	Phenylalanine	125 mg
Glutamic acid	863 mg	Proline	376 mg
Glycine	71 mg	Serine	207 mg
Histidine	122 mg	Threonine	163 mg
Isoleucine	201 mg	Tyrosine	110 mg
Leucine	292 mg	Valine	242 mg

Recipe 70
FRIES MEDIUM MODIFIED[197]

$(NH_4)_2C_4H_4O_6$	5.0 g
NH_4NO_3	1.0 g
$MgSO_4 \cdot 7H_2O$	0.5 g
$CaCl_2$	0.13 g
K_2HPO_4	1.0 g
NaCl	0.1 g
Sucrose	30.0 g

Recipe 71
FRIES MEDIUM MODIFIED NO. 3[198,239,246]

Sucrose	30.0 g	In another version, Pringle and Scheffer[239]
Ammonium tartrate	5.0 g	excluded five trace elements, (Mn to Fe)
$NH_4 NO_3$	1.0 g	and included yeast extract
$MgSO_4 \cdot 7H_2O$	0.5 g	
$KH_2 PO_4$	1.0 g	
NaCl	0.1 g	
$CaCl_2 \cdot 2H_2O$	0.13 g	
$MnSO_4 \cdot 4H_2O$	0.1 mg	
Boric acid	1.0 mg	
$CuSO_4$	0.1 mg	
$ZnSO_4 \cdot 7H_2O$	0.1 mg	
$FeSO_4 \cdot XH_2O$[a]	20 mg	
Yeast extract	1.0 g	Added by Yoder[246]

[a] X = variable.

Recipe 72
FUNGAL SPORULATION MEDIUM

Glucose	20.0 g
KNO_3	1.0 g
$KH_2 PO_4$	1.0 g
$MgSO_4 \cdot 7H_2O$	0.5 g
NaCl	0.1 g
$CaCl_2 \cdot 2H_2O$	0.13 g
Trace elements solution	0.5 ml
Agar	15.0 g

Trace Elements

$FeCl_3$	116.8 mg	Make up to 100 ml, store in brown
$ZnSO_4 \cdot 7H_2O$	172.5 mg	bottle
$MnSO_4 \cdot H_2O$	67.6 mg	
$CuSO_4 \cdot 5H_2O$	49.9 mg	
Concentrated H_2SO_4	2 drops	

Recipe 73
GALLIC OR TANNIC ACID MEDIUM[394,395]

Difco® powdered malt	15.0 g
Gallic or tannic acid	5.0 g
Agar	20.0 g

Recipe 74
GALLIC OXALATE MEDIUM[393]

KH$_2$PO$_4$	1.0 g	Filter-sterilize, adjust to pH 4.2 with
MgSO$_4$·7H$_2$O	0.5 g	HCl; autoclave agar solution at
KNO$_3$	2.0 g	121°C for 15 min; cool to 60°C and
Thiamine·HCl	1.0 mg	combine with filter-sterilized por-
Minor element solution	10 ml	tion and pour plates
Gallic acid	160.0 mg	
Potassium oxalate	10.0 g	
Distilled water to 250 ml		
Agar	20 g in 750 ml H$_2$O	

Minor Element Solution

Fe$_2$SO$_4$·7H$_2$O	1.0 g
ZnSO$_4$·7H$_2$O	1.0 g
MnSO$_4$·H$_2$O	0.6 g
Distilled water to 1 liter	

Recipe 75
GARRAWAY MEDIUM[182,396]

D-Glucose	10.0 g	Garraway[182] suggests sterilizing a
D-(+)-xylose (J. D. Baker or Matheson Coleman & Bell)	2.0 g	solution of each compound separate- ly, except trace elements are com-
L-asparagine	4.0 g	bined; KH$_2$PO$_4$ and MgSO$_4$ are also
KH$_2$PO$_4$	1.5 g	mixed together; he finds xylose in-
MgSO$_4$·7H$_2$O	0.75 g	creases sporulation of race T iso-
CuSO$_4$	1.0 mg	lates of *Helminthosporium may-*
FeSO$_4$	1.0 mg	*dis.*[396]
MnSO$_4$	1.0 mg	
ZnSO$_4$	1.0 mg	
Agar	20 g	

Recipe 76
GELATIN LIQUEFACTION TEST MEDIUM[6]

Heart infusion broth	25 g	Adjust pH to 7.2—7.4, dispense 5 ml/
Gelatin	120 g	tube

Recipe 77
GLUCOSE ALANINE AGAR[112]

Glucose	10.0 g
DL-alanine	1.0 g
Thiamine·HCl	500 µg
K$_2$HPO$_4$	2.0 g
MgSO$_4$·7H$_2$O	0.2 g
Agar	20 g

Recipe 78
GLUCOSE ASPARAGINE AGAR[155]

Glucose	8.0 g	Used for the growth and sporulation of *Geothri-*
L-asparagine	2.0 g	*chum candidum*
KH_2PO_4	1.0 g	
$MgSO_4 \cdot 7H_2O$	0.5 g	
Pyridoxine	.2 mg	
Agar	15 g	

Recipe 79
GLUCOSE ASPARAGINE AGAR[397]

Glucose	2.0 g
Asparagine	2.0 g
K_3PO_4	1.25 g
$MgSO_4 \cdot 7H_2O$	0.75 g
Agar	15.0 g

Recipe 80
GLUCOSE ASPARAGINE ASCORBIC ACID MEDIUM[269]

D-glucose	5.0 g	Ascorbic acid used to keep iron in
L-asparagine (anhydrous)	0.5 g	ferrous state
KH_2PO_4	1.0 g	
$MgSO_4 \cdot 7H_2O$	0.5 g	
Zn^{+2}	0.2 mg	
Ca^{+2}	10.0 mg	
Cu^{+2}	0.02 mg	
Mn^{+2}	0.05 mg	
Mo^{+6}	0.01 mg	
Fe^{+2}	0.1 ppm	
β-sitosterol	10.0 mg	
Thiamine	100 μg	
L-ascorbic acid	200 mg	

Recipe 81
GLUCOSE ASPARAGINE MANGANESE MEDIUM[87]

Glucose	5 g	Adjust pH to 6.0 with KOH; many fungi respond
Asparagine	1.0 g	to the increased amount of manganese, see
KH_2PO_4	1.0 g	Recipe 149
$MgSO_4 \cdot 7H_2O$	0.5 g	
Fe	0.1 mg	
Zn	0.1 mg	
Ca	10.0 mg	
Mn	0.5 mg	
Thiamine	100 μg	
Biotin	5 μg	

Recipe 82
GLUCOSE ASPARAGINE MEDIUM[398]

Glucose	30.0 g
Asparagine	1.0 g
$MgSO_4 \cdot 7H_2O$	0.5 g
KH_2PO_4	1.5 g

Recipe 83
GLUCOSE ASPARAGINE THIAMINE AGAR[89],[90]

Glucose	2.0 g	For conidia and sporangia of *Choaenephora*
Asparagine	1.0 g	*cucurbitarum*, Barnett and Lilly[90]
KH_2PO_4	1.0 g	used this medium at pH 6.0; for zygospore
$MgSO_4 \cdot 7H_2O$	0.5 g	production they used 5 g glucose and 100 μg
$Fe(NO_3)_3 \cdot 9H_2O$[a]	724 mg)	of thiamin, adjusting pH to 6.0[89]
$ZnSO_4 \cdot 7H_2O$[a]	440 mg }	Recipe 148 dissolved in liter of water with sul-
$MnSO_4 \cdot 4H_2O$[a]	203 mg)	furic acid added to give a clear solution, 2
Thiamine	25 μg	ml/l
Agar	20 g	

[a] Ingredients of Recipe 148.

Recipe 84
GLUCOSE CALCIUM MEDIUM[85]

Glucose	2.0 g	
$CaCl_2$	10 ppm of Ca	pH 7.4
$NaNO_2$	2.0 g	
K_2HPO_4	1.0 g	
KCl	0.5 g	
$MgSO_4 \cdot 7H_2O$	0.5 g	
Agar	15 g	

Recipe 85
GLUCOSE CASAMINO ACID AGAR[353]

Glucose	40.0 g
Casamino acid (Difco®), vitamin free	2.5 g
$MgSO_4$	0.1 g
KH_2PO_4	1.8 g
Agar	20 g

Recipe 86
GLUCOSE CASAMINO ACID MEDIUM[72]

Glucose	15.0 g
Casamino acids	1.5 g
Yeast extract	1.0 g
KH_2PO_4	1.5 g
$MgSO_4$	1.0 g
Trace elements A–Z from Hoagland solution	10.0 ml

Recipe 87
GLUCOSE CASAMINO THIAMINE AGAR[91]

Glucose	3.0 g	
Casamino acids (Difco®)	2.0 g	
KH_2PO_4	1.0 g	Adjust pH to 6.0
$MgSO_4 \cdot 7H_2O$	0.5 g	
Microelement solution (no. 148 or 149)	2.0 ml	
Thiamine	25 μg	
Agar	20 g	

Recipe 88
GLUCOSE CASEIN HYDROLYSATE MEDIUM[399]

D-glucose	10.0 g	Adjust pH to 6 prior to addition of agar
Casein hydrolysate (NBC)	2.0 g	
KH_2PO_4	1.0 g	
$MgSO_4 \cdot 7H_2O$	0.5 g	
Thiamine·HCl	100 μg	
Biotin	5.0 mg	
Microelements	1.0 ml	
$Fe_2(SO_4)_3 \cdot (NH_4)_2SO_4 \cdot H_2O$	0.826 g	Add microelements to a liter of H_2O con-
$ZnSO_4 \cdot 7H_2O$	0.88 g	taining 10 ml of concentrated HCl
$MnSO_4 \cdot H_2O$	0.154 g	
$Ca(NO_3)_2 \cdot 4H_2O$	58.9 g	
$CuSO_4 \cdot 5H_2O$	0.078 g	
Agar	20.0 g	

Recipe 89
GLUCOSE GLUTAMATE MEDIUM[318]

EDTA	200 mg	If used to isolate saprolegnia, de-
K_2HPO_4	87 mg	crease bacteria contamination
KH_2PO_4	68 mg	with penicillin G and strepto-
$MgCl_2 \cdot 6H_2O$	160 mg	mycin, 0.5 g each, or potassium
$CaCl_2$	66 mg	tellurite (0.1 g); adjust pH to 6.5
$MnCl_2 \cdot 4H_2O$	75 mg	with KOH
$ZnCl_2$	40 mg	
$FeCl_3 \cdot 6H_2O$	1.3 mg	
DL-methionine	50 mg	
Sodium glutamate (mono)	500 mg	
D-glucose	3.0 g	
Water (predistilled, ion exchanged)		
Agar (Difco® purified)	1.5 g for culture	
	15 g for isolation	

Recipe 90
GLUCOSE MINIMAL MEDIUM[186]

Glucose	5.0 g
KNO_3	3.12 g
K_2HPO_4	0.75 g
KH_2PO_4	0.75 g
$MgSO_4 \cdot 7H_2O$	5.0 g
NaCl	1.0 g
$CaCl_2 \cdot 2H_2O$	0.1 g
$ZnSO_4 \cdot 7H_2O$	0.396 mg
$CuSO_4 \cdot 5H_2O$	0.079 mg
$MnSO_4 \cdot 4H_2O$	0.0405 mg
MoO_3	0.0175 mg
$FeC_6H_5O_7 \cdot 3H_2O$	0.5355 mg

Recipe 91
GLUCOSE NITRATE YEAST EXTRACT AGAR[144]

Glucose	0.055 M
KNO_3	0.0075 M
$MgSO_4$	0.001 M
Potassium phosphate buffer (pH 6.5)	0.0033 M
Yeast extract (Difco®)	0.5 g
Fe	0.2 mg
B	0.01 mg
Mo	0.02 mg
Zn	0.66 mg
Mn	0.02 mg
Cu	0.1 mg
Agar	20 g

Recipe 92
GLUCOSE PEPTONE AGAR[311]

Glucose	20 g
Peptone	10 g
Agar	20 g

Recipe 93
GLUCOSE PEPTONE AGAR[400]

Gluose	10.0 g
Bacto-peptone	2.0 g
KH_2PO_4	0.5 g
$MgSO_4 \cdot 7H_2O$	0.5 g
Agar	15 g

Recipe 94
GLUCOSE PEPTONE BENGAL MEDIUM[401]

Glucose	10 g
Neopeptone (Difco®) or Polypeptone (BBL®)	5.0 g
Rose bengal	0.035 g
Chlortetracycline[a]	35 mg
Agar	20 g

[a] See footnote, Recipe 1.

Recipe 95
GLUCOSE PEPTONE YEAST AGAR (SM/5)[402]

Peptone, Bacto®	2.0 g	
Glucose	2.0 g	pH 6.0–6.3
Yeast extract	0.2 g	
K_2HPO_4	0.2 g	
KH_2PO_4	0.3 g	
$MgSO_4 \cdot 7H_2O$	0.2 g	
Agar	20 g	

Recipe 96
GLUCOSE PEPTONE YEAST ANTIBIOTIC AGAR[403]

Glucose	5.0 g	Sterilize at 11 psi for 15 min
Peptone	1.0 g	
Yeast extract	2.0 g	
NH_4NO_3	1.0 g	
K_2HPO_4	1.0 g	
$MgSO_4 \cdot 7H_2O$	0.5 g	
$FeCl_3 \cdot 6H_2O$	Trace	
Oxgall	5.0 g	
Sodium propionate	1.0 g	
Chlortetracycline[a]	30 mg	
Streptomycin[a]	30 mg	
Agar	20 g	

[a] See footnote, Recipe 1.

Recipe 97
GLUCOSE PHENYLALANINE AGAR[65]

D-glucose	3.0 g	Adjust pH to 5.5–6.0 before autoclaving
DL-phenylalanine	0.5 g	
KH_2PO_4	1.0 g	
$MgSO_4 \cdot 7H_2O$	0.5 g	
Microelement solution (Recipe 148)	2.0 ml	
Agar	20 g	

Recipe 98
GLUCOSE STARCH TELLURITE MEDIUM[3 1 7]

Glucose	10.0 g	Used to isolate and purify cultures of sapro-
Soluble starch	5.0 g	legniaceae
Yeast extract	2.0 g	
Na_2HPO_4	0.597 g	
KH_2PO_4	2.043 g	
K_2TeO_3	0.1 g	
Agar, Oxoid® #3	12.0 g	

Recipe 99
GLUCOSE TRYPTONE MEDIUM[7 3]

Glucose	10.0 g	Adjust to pH 5.3
Tryptone	5.0 g	
Asparagine	1.0 g	
$MgSO_4 \cdot 7H_2O$	1.0 g	
KH_2PO_4	5.0 g	
$FeCl_3$	.01 g	

Recipe 100
GLUCOSE YEAST EXTRACT AGAR M-63A[3 ,1 1 2]

Glucose	10.0 g	Suggest separate sterilization of glucose and K_2HPO_4;
Yeast	3.0 g	yeast extract was replaced by 500 μg thiamine and
K_2HPO_4	2.0 g	0.1% DL-α-alanine for *Coprinus lagopus*
$MgSO_4 \cdot 7H_2O$	0.2 g	
Agar	20 g	

Recipe 101
GLUCOSE YEAST EXTRACT AGAR (M-64)[3]

Glucose	5.0 g
Yeast extract	1.0 g
Microelement solution (Recipe 149)	2.0 ml
Agar	20 g

Recipe 102
GLUCOSE YEAST EXTRACT BROTH [1 5 6]

Glucose	10.0 g
Yeast extract	5.0 g

Recipe 103
GLYCERINE GLUCOSE BROTH[6]

Glycerine	25.0 g	Dissolve asparagine in 500 ml water at 50°C,
Glucose	10.0 g	dissolve salts separately in 25-ml amounts,
L-asparagine	7.0 g	add salts in order to asparagine solution, and
K_2HPO_4	1.31 g	sterilize at 115°C for 25 min; for the produc-
Ammonium chloride	7.0 g	tion of coccidoidin, histoplasmin
$Na_3C_6H_5O_7 \cdot 5\frac{1}{2}H_2O$	0.90 g	
$MgSO_4 \cdot 7H_2O$	1.50 g	
Ferric citrate	0.30 g	

Recipe 104
GREEN LEAF AGAR[404]

Leaf blades (10- to 30-day-old wheat seedlings)	100 g	Bring water suspension of leaves to medium boil and maintain for 10 min, pour without squeezing through two layers of cheesecloth; may be stored at −10°C for 6 months
$CuSO_4 \cdot 5H_2O$	0.8–2.0 g	
Pentachloronitrobenzene	1.0–10.0 mg	
Agar	20 g	

Recipe 105
GUAIACOL AGAR[405]

Guaiacol	0.125 g
Streptomycin sulfate	500 mg
Agar	5.0 g

Recipe 106
HARROLD MEDIUM[406]

Sucrose	400 g
Malt extract	20 g
Yeast extract	5 g
Agar	20 g

Recipe 107
HAY EXTRACT AGAR[3],[407]

Partially decomposed hay (or weathered grass)	35 g	Hay infused in a liter of tap water for 0.5 hr at 110°C (5–10 psi), the infusion is filtered and made up to a liter, and other ingredients are added; adjust to pH 6.2 and sterilize at 121°C for 20 min
K_2HPO_4	2.0 g	
Agar	15 g	

Recipe 108
HAY INFUSION AGAR[123]

Hay, mostly *Poa* spp., weathered	10 g	Autoclave hay in a liter of distilled water for 20 min at 120°C; filter through cheesecloth and cotton
KH_2PO_4	1.5 g	
$Na_2HPO_4 \cdot 7H_2O$	0.62 g	
Agar	15 g	

Recipe 109
HEMP SEED[1]

Usually two or more seeds in a petri dish with one third filtered pond water and two thirds glass-distilled water	Emerson[1] suggests viable seeds work best that are boiled for 10–20 min, removed, and gently pinched to reveal the white hypocotyl within

Recipe 110
HEMP SEED AGAR[259]

Hemp seed	100 g
Agar	20 g

Recipe 111
HEMP SEED AGAR (MODIFIED)[408]

Hemp seed	10 g	Blend hemp seed in 100 ml water, autoclave for 10 min, filter through cheesecloth, and add 100 ml of filtrate to 900 ml of melted agar
Agar	20 g	

Recipe 112
HEMP SEED DECOCTION AND AGAR[255],[262]

Hemp seed	50.0 g	Autoclave for 30 min, filter through cheesecloth;
Distilled water		Savage et al.[255] strains through a cotton cloth and adds 17 g agar

Recipe 113
HOAGLAND SOLUTION 1950[409]

Solution 1

Ingredients	g/l	ml in a liter of solution	Comments
$1M$ KH_2PO_4	136	1	Add 1 ml of
$1M$ KNO_3	101.1	5	solution A to
$1M$ $Ca(NO_3)_2$	164.1	5	either solution
$1M$ $MgSO_4$	120.3	2	1 or 2 and bring
			to a liter; add
			solution B or its
			alternate, adjust
			pH to 6 with
			$0.1N$ H_2SO_4

Solution 2

$1M$ $NH_4H_2PO_4$	132.07	1
$1M$ KNO_3	101.1	6
$1M$ $Ca(NO_3)_2$	164.1	4
$1M$ $MgSO_4$	120.3	2

Solution A

Ingredients	g/l		ppm
H_3BO_3	2.86		0.5
$MnCl_2 \cdot 4H_2O$	1.81		0.5
$ZnSO_4 \cdot 7H_2O$	0.22	} 1 ml	0.05
$CuSO_4 \cdot 5H_2O$	0.08		0.02
$H_2MoO_4 \cdot H_2O$	0.02		0.01

Solution B

	g/l	ml in 1 liter of solution
Iron tartrate (necessary to add at regular intervals)	5.0	1

Alternate Solution B — Dissolve 26.1 g ethylene-diamine tetra-acetic acid in 268 ml of $1N$ KOH. Then add 24.9 g $FeSO_4 \cdot 7H_2O$ and dilute to 1 liter. After aerating overnight to produce the stable ferric complex, the pH should be about 5.5. A milliliter provides 5 ppm to 1 liter of solution — **one addition is enough.**

Recipe 114
HONEY PEPTONE MEDIUM[410],[411]

Honey	60 g
Difco® Bacto-peptone	10 g
Agar	20 g

Recipe 115
HYPHOMYCES SYNTHETIC[145]

D-glucose	10 g	Modification of Asthana and Hawker medium A
DL-isoleucine	5 g	
KNO_3	3.5 g	
KH_2PO_4	1.75 g	
$MgSO_4 \cdot 7H_2O$	0.75 g	
Agar	20 g	

Recipe 116
JOHAM MEDIUM[412],[413]

Glucose	40 g
K_2HPO_4	0.004 M
$MgSO_4$	0.0015 M
KCl	0.002 M
NH_4NO_3	0.0125 M
Thiamine chloride	0.1 ppm
Iron	2 ppm
Zinc	2 ppm
Manganese	2 ppm

Recipe 117
KERRS MEDIUM[138]

Sucrose	30.0 g
$NaNO_3$	2.0 g
KH_2PO_4	1.0 g
KCl	0.5 g
$MgSO_4 \cdot 7H_2O$	0.5 g
$FeSO_4$	0.01 g
Yeast extract (Difco®)	0.5 g
Agar	15 g

Recipe 118
KERRS MEDIUM MODIFIED[138]

Kerrs medium (Recipe 117) and:	
Streptomycin sulfate[a]	50 ppm
PCNB[a]	100 ppm
Rose bengal[a]	60 ppm

[a]See footnote, Recipe 1.

Recipe 119
KNOP AGAR[414]

$Ca(NO_3)_2 \cdot 4H_2O$	0.8 g
KNO_3	0.2 g
KH_2PO_4	0.2 g
$MgSO_4 \cdot 7H_2O$	0.2 g
$FeSO_4$	Trace
Agar	20 g

Recipe 120
LACTOSE CASEIN HYDROLYSATE MEDIUM[193]

Lactose	37.5 g	See microelement solution in Recipe 148
Casein hydrolysate	3.0 g	Adjust pH to 6.0
KH_2PO_4	1.0 g	
$MgSO_4$	0.5 g	
Microelements	2.0 ml	
Agar	15 g	

Recipe 121
LEONIAN AGAR[401]

Peptone	0.625 g
Maltose	6.25 g
Malt extract	6.25 g
KH_2PO_4	1.25 g
$MgSO_4 \cdot 7H_2O$	0.625 g
Agar	20.0 g

Recipe 122
LIMA BEAN AGAR (DIFCO®)[380]

Dried lima bean	62.5 g	Usually 23 g of the dehydrated product is recon-
Agar	15 g	stituted in a liter of water

Recipe 123
LIMA BEAN AGAR (DIFCO® MODIFIED)[315]

Dried lima bean infusion and agar (Difco®)	18 g
Agar	5 g

Recipe 124
LIMA BEAN AGAR (FROZEN)[270]

Frozen lima beans	150–250 g	Blend beans in water, add melted agar; store
Agar	15–20 g	medium in dark

Recipe 125
LIMA BEAN AGAR (FROZEN AND CLEARED)[255]

Frozen lima beans	285 g	Autoclave unblended beans in 1 liter dis-
Agar	17 g	tilled water for 15 min at 121°C, screen
		through a cotton cloth, add agar, and auto-
		clave

Recipe 126
LIMA BEAN DEXTROSE AGAR[266]

Very fine-ground, dry lima beans	15 g	Bean powder suspended in 800 ml water 45
Dextrose	10 g	min at 121°C is then mixed together and auto-
Agar	10 g	claved in small amounts for 30 min
Difco® yeast extract	2 g	

Recipe 127
LITTMAN OXGALL AGAR[46,415]

Peptone	10 g	Dissolve 1.25 g of crystal violet in 25 ml of 95% EtOH
Dextrose	10 g	and keep in a tightly stoppered bottle; add 0.2 ml/
Oxgall	15 g	liter of medium; Mitchell and Stauber[46] suggest the
Agar	20 g	addition of 30 ppm streptomycin sulfate prior to
Crystal violet	0.01 g	pouring

Recipe 128
LUKENS AND SISLER MEDIUM[18,19]

Glucose	20 g	Adjust pH to 6.0 with KOH
Ammonium sulfate	3.0 g	
Monobasic potassium phosphate	3.0 g	
Glycine	1.0 g	
Thiamine·HCl	20 μg	
Niacin	20 μg	
Biotin	1 μg	
i-Inositol	200 μg	
Pyridoxine	10 μg	
Folic acid	10 μg	
Boron	0.01 ppm	
Mn	0.01 ppm	
Zn	0.07 ppm	
Cu	0.01 ppm	
Mo	0.01 ppm	
Fe	0.05 ppm	

Recipe 129
LUTZ MEDIUM MODIFIED[416]

Vitrums malt extract[a]	10.0 g
Ammonium nitrate	1.0 g
Ammonium phosphate	1.0 g
Magnesium sulfate	0.1 g
Ferric sulfate	0.1 g
Manganese sulfate	0.025 g
Agar	25.0 g

[a] Obtained from Apoteksvaruncentral, Stockholm, Sweden.

Recipe 130
MACHLIS MEDIUM A (MODIFICATION OF INGRAHAM MEDIUM)[9]

Thiamine·HCl	0.15 mg	Adjust pH to 7.0 with KOH; autoclave glucose separately
DL-methionine	0.1 g	
$MgCl_2$	0.001 M	
KH_2PO_4	0.01 M	
$(NH_4)_2HPO_4$	0.005 M	
$CaCl_2$	0.0002 M	
$MnCl_2$	0.5 ppm Mn	
$ZnSO_4$	0.1 ppm Zn	
H_3BO_3	0.5 ppm B	
$CuSO_4$	0.1 ppm Cu	
$FeCl_3$	0.5 ppm Fe	
$(NH_4)_6Mo_7O_{24}$	0.2 ppm Mo	
$CoCl_3$	0.2 ppm Co	
Glucose	5.0 g	

Recipe 131
MACHLIS MEDIUM B[9,10]

Thiamine·HCl	0.15 mg	Autoclave glucose separately; solution of
DL-methionine	0.1 g	$CaCl_2$ and $MgCl_2$ separately autoclaved
Glucose	5.0 g	and mixed with sterile glucose; addition of
$MgCl_2$	0.0005 *M*	500 μM of L-glutamic acid decreases log
K_2HPO_4	0.005 *M*	phase of *Allomyces macrogynus*[10]
KH_2PO_4	0.005 *M*	
$(NH_4)_2HPO_4$	0.005 *M*	
$CaCl_2$	0.0005 *M*	
$MnCl_2$	0.5 ppm Mn	
$ZnSO_4$	0.1 ppm Zn	
$CuSO_4$	0.1 ppm Cu	
$FeCl_3$	1.0 ppm Fe	
$(NH_4)_6Mo_7O_{24}$	0.2 ppm Mo	
H_3BO_3	0.5 ppm B	
$CoCl_2$	0.2 ppm Co	

Recipe 132
MACROPHOMINA ISOLATION MEDIUM CMR[211]

Rice agar medium (decoction of 10 g polished rice, boil rice for 5 min, 20 g of agar)		Media must be freshly prepared and well mixed; incubate in the dark
Chloroneb[a] (Demosan® 65, wettable)	300 mg (active ingredient)	
Mercuric chloride	7 mg	
Streptomycin sulfate[a]	40 mg	
Potassium penicillin[a]	60 mg	
Rose bengal	90 mg	
Lactic acid[a] to bring pH to 6.0		

[a] See footnote, Recipe 1.

Recipe 133
MALT AGAR BLAKESLEE FORMULA[417]

Malt extract	20 g
Peptone	1.0 g
Dextrose	20 g
Agar	20 g

Recipe 134
MALT AGAR MODIFIED BY KAUFFMAN[916]

Malt extract	10.0 g
Yeast extract	5.0 g
Peptone	1.5 g
Maltose	5.0 g
Magnesium sulfate	0.5 g
Calcium nitrate	0.5 g
Monopotassium phosphate	0.25 g
Agar	30.0 g

Recipe 135
MALT BENOMYL AGAR[134]

Malt extract	20.0 g	Benomyl is added prior to autoclaving
Benomyl (Benlate®)	1.0 ppm (active)	
Agar	20.0 g	

Recipe 136a
MALT EXTRACT BROTH OR AGAR

Malt extract	20 g	Add 20 g of agar; do not oversterilize as agar will hydrolyize

Recipe 136b
MALT EXTRACT BROTH[165]

Malt extract	40 g

Recipe 137
MALT EXTRACT FILTER PAPER[418]

Malt extract	20 g	Moistened filter paper in a petri dish with 2% malt extract

Recipe 138
MALT GLUCOSE YEAST EXTRACT AGAR[419]

Malt extract	3 g
Glucose	10 g
Yeast extract	3 g
Peptone	5 g
Agar	20 g

Recipe 139
MALT SALT AGAR[40,420]

NaCl	75 g	Various concentrations from 7.5 to 20% of sodium
Malt extract	20 g	chloride have been used; do not autoclave at 121°C
Agar	15 g	for more than 15 min

Recipe 140
MALT YEAST EXTRACT AGAR[244,421]

Malt extract	10 g	Shigo[421] used 2 g of yeast extract
Yeast extract	1 g	
Agar	20 g	

Recipe 141
MALT YEAST EXTRACT AGAR[398]

Malt extract	20.0 g
Yeast extract	2.0 g
Agar	20 g

Recipe 142
MALTOSE GLYCERINE MEDIUM[114]

Maltose	5.0 g	Add strips of filter paper to surface of
Glucose	2.0 g	hardened agar
Glycerine	2.0 g	
Yeast extract	2.0 g	
Peptone	0.2 g	
Asparagine	0.2 g	
Magnesium sulfate	0.5 g	
Calcium nitrate	0.5 g	
Dihydrogen phosphate	0.5 g	
Ferrous sulfate	Trace	
Agar	20.0 g	

Recipe 143
MALTOSE PEPTONE BROTH[27]

Maltose	3 g
Peptone	1 g

Recipe 144
MALTOSE TARTRATE MEDIUM[422]

Maltose	20.0 g	The pH of the culture medium was adjusted to
Ammonium tartrate	2.8 g	5.0 before autoclaving
KH_2PO_4	1.0 g	
$MgSO_4 \cdot 7H_2O$	0.5 g	
Thiamine	50.0 μg	
Zinc	0.2 mg	
Iron	0.2 mg	See Recipe 148
Manganese	0.1 mg	
Agar	20.0 g	

Recipe 145
MARTIN MEDIA[423-425]

Ingredient	Martin[423] medium (145A) original	Martin[424] medium (145B) modification	Tsao[424] medium (145C) modification	Singh and Mitchell[425] (medium (145D) modification
Glucose	10 g	10 g	10 g	10 g
Peptone	5 g	5 g	0.5 g	5 g
KH_2PO_4	1 g	0.5 g	0.5 g	1 g
K_2HPO_4	–	0.5 g	0.5 g	–
$MgSO_4 \cdot 7H_2O$	0.5 g	0.5 g	0.5 g	0.5 g
Yeast extract	–	0.5 g	0.5 g	–
Rose bengal	33 mg	50 mg	50 mg	33 mg
Streptomycin[a]	30 mg	30 mg	30 mg	30 mg
Agar	20 g	18 g	17 g	20 g
Terraclor[b]	–	–	–	0.01–0.05%
or				
Pimaricin[a,b]	–	–	–	.001–.002%

[a] See footnote, Recipe 1.
[b] If used together, most fungi except *Pythium* are eliminated.

Recipe 146
MATSUSHIMA PHB AGAR[426]

p-Hydroxybenzoic acid	1.0 g	Adjust pH to 6 before auto-
NH_4NO_3	0.5 g	claving
K_2HPO_4	0.5 g	
$MgSO_4 \cdot 7H_2O$	0.25 g	
$Fe(NO_3)_3 \cdot 9H_2O$	1.5 mg	
$ZnSO_4 \cdot 7H_2O$	1.0 mg	
$MnSO_4 \cdot 4H_2O$	0.5 mg	
Thiamine hydrochloride	1.0 mg	
Pyridoxine hydrochloride	1.0 mg	
Biotin	50 μg	
Agar	15 g	

Recipe 147
MELIN-NORKRANS MEDIUM MODIFIED[427]

$CaCl_2$	0.05 g	After autoclaving, pH at 5.5 to 5.7; a com-
NaCl	0.025 g	panion nutrient solution excluded agar
KH_2PO_4	0.5 g	and malt extract and reduced sucrose to
$(NH_4)_2HPO_4$	0.25 g	2.5 g and thiamine to 25 μg
$MgSO_4 \cdot 7H_2O$	0.15 g	
$FeCl_3$ (1%)	1.2 ml	
Thiamine·HCl	100 μg	
Malt extract (paste)	3.0 g	
Sucrose	10.0 g	
Agar	15.0 g	

Recipe 148
MICROELEMENT SOLUTION[398]

$Fe(NO_3)_3 \cdot 9H_2O$	723.5 mg	Dissolved in 600 ml of distilled water, add suffi-
$ZnSO_4 \cdot 7H_2O$	439.8 mg	cient sulfuric acid to give a clear solution and
$MnSO_4 \cdot 4H_2P$	203.0 mg	bring up to volume
Water to 1.0 liter		Two ml per liter of medium is sufficient for
		most fungi and gives 0.2 mg of Fe and Zn and
		0.1 mg of Mn; see also Recipe 149

Recipe 149
MICROELEMENT SOLUTION[3]

$Fe(NO_3)_3 \cdot 9H_2O$	724 mg	Same preparation as Recipe 148; manganese in-
$ZnSO_4 \cdot 7H_2O$	440 mg	creased to stimulate growth of some fungi
$MnSO_4 \cdot 4H_2O$	406 mg	Add 2 mg/l to other media

Recipe 150
MILK AGAR[428]

Powdered milk	2.0 g	Used to isolate fungi that trap nematodes by
Agar	20.0 g	sprinkling 1 g or less of soil on its surface

Recipe 151
MINIMAL AGAR MEDIUM[186,429]

KNO_3	3.12 g
K_2HPO_4	0.75 g
KH_2PO_4	0.75 g
$MgSO_4 \cdot 7H_2O$	0.5 g
NaCl	0.1 g
$CaCl_2 \cdot 2H_2O$	0.1 g
Glucose	5.0 g
Agar	20.0 g
$ZnSO_4 \cdot 7H_2O$	0.396 mg
$CuSO_4 \cdot 5H_2O$	0.079 mg
$MnSO_4 \cdot 4H_2O$	0.0405 mg
MoO_3	0.0175 mg
$FeC_6H_5O_7 \cdot 3H_2O$	0.5355 mg

Recipe 152
MIRCETICH ISOLATION MEDIUM (MPVM)[292]

Sucrose	20 g	Make up stock solutions fresh and
Corn meal agar (Difco®)	17 g	store medium in the dark at
$ZnCl_2$	1 mg	35°F for up to a week
$CuSO_4 \cdot 5H_2O$	.02 mg	
MoO_3	.02 mg	
$MnCl_2$	.02 mg	
$FeSO_4 \cdot 7H_2O$	.02 mg	
$MgSO_4 \cdot 7H_2O$	10 mg	
$CaCl_2$ [a]	10 mg	
Thiamine hydrochloride	100 μg	
Pimaricin (mycoprozine)[a]	5 mg	
Vancomycin hydrochloride[a]	300 mg	
Pentachloronitrobenzene (technical grade)	100 mg	
Rose bengal	10 mg	
Agar	23 g	

[a] See footnote, Recipe 1.

Recipe 153
MONILINIA SELECTIVE MEDIUM[220]

PCNB	1.0 g	Autoclave agar and water for 5 min at 121°C, cool
Gerber® strained peaches	40.0 g	to 80°C and add other components, mix, and
Neomycin (water solution)[a]	20 mg	pour
Streptomycin (water solution)[a]	1.0 g	
Agar	20.0 g	

Recipe 154
MULLINS AND BARKSDALE SYNTHETIC[2]

Glucose	2.8 g	Adjust pH to 6.9
Tris(hydroxymethyl) aminomethane	1.2 g	
Monosodium L-glutamate	0.4 g	
L-methionine (15 mg/ml in 1 N HCl)	1 ml	
KCl (2 M)	1 ml	
$MgSO_4 \cdot 7H_2O$ (0.5 M)	1 ml	
$CaCl_2$ (0.5 M)	1 ml	
HEDTA (10 mg/ml)	2 ml	
KH_2PO_4 (1 M)	1.5 ml	
Metal mix #4	10 ml	Grind together 2
$Fe(NH_4)_2(SO_4)_2 \cdot 6H_2O$	28.9 g	mg/ml
$ZnSO_4 \cdot 7H_2O$	8.8 g	
$MnSO_4 \cdot H_2O$	3.1 g	

Recipe 155
MYCOBIOTIC AGAR DIFCO®[430]

Soytone	10 g	Sterilized for 10 min at 121°C; should not be
Dextrose	10 g	reheated
Agar	15 g	
Cycloheximide	0.4 g	
Chloromycetin	0.05 g	

Recipe 156
MYCOLOGICAL AGAR[430]

Bacto® soytone	10.0 g
Dextrose	10.0 g
Agar	15.0 g

Recipe 157
MYCOPHIL AGAR BBL®[431]

Phytone peptone	10 g
Glucose	10 g
Agar	16 g

Recipe 158
MYCOSEL AGAR BBL®[432]

Phytone	10 g	Autoclave at 118°C for 15 min
Glucose	10 g	
Cycloheximide (Actidione®)	0.4 g	
Chloromycetin	0.05 g	
Agar	15.5 g	

Recipe 159
NASH AND SNYDER PCNB MEDIUM[137,140]

Difco® peptone	15 g	See also PCNB Recipe 177; to avoid hydrolysis
KH_2PO_4	1.0 g	of peptone and agar, autoclave water and agar;
$MgSO_4 \cdot 7H_2O$	0.5 g	add other ingredients at 90°C but hold poured
Streptomycin[a]	300 mg	plates 4—5 days prior to use to check for con-
Agar	20 g	tamination
Pentachloronitrobenzene[a] (75%	1.0 g	
wettable powder, Terraclor®)		

[a] See footnote, Recipe 1.

Recipe 160
NEOPEPTONE GLUCOSE AGAR[433]

Neopeptone	2.0 g
Glucose	2.8 g
$MgSO_4 \cdot 7H_2O$	1.23 g
KH_2PO_4	2.72 g
Agar	20 g

Recipe 161
NEUROSPORA "COMPLETE" MEDIUM[231]

Glucose	5.0 g
Sucrose	5.0 g
Hydrolized casein	5.0 ml
Difco® yeast extract	2.5 g
Spray-dried malt syrup	5.0 g
Agar	15 g
Vitamin solution containing (mg/l):	10 ml
Thiamin	100
Riboflavin	50
Pyridoxin	50
Pantothenic acid	200
p-Aminobenzoic acid	50
Nicotinamide	200
Choline	200
Inositol	400
Alkali hydrolyzed yeast nucleic acid	500
Folic acid	4 μg pure substance

Recipe 162
NEUROSPORA "MINIMAL" MEDIUM[231]

Ammonium tartrate	5.0 g
NH_4NO_3	1.0 g
KH_2PO_4	1.0 g
$MgSO_4 \cdot 7H_2O$	0.5 g
NaCl	0.1 g
$CaCl_2$	0.1 g
Sucrose	15.0 g
Biotin	5×10^{-6} g
Bo	0.01 mg
Cu	0.1 mg
Fe	0.2 mg
Mn	0.02 mg
Mo	0.02 mg
Zn	2.0 mg

Note: pH of medium is 5.6.

Recipe 163
NEUROSPORA SEX SYNTHETIC[232]

KNO_3	0.1%	May be supplemented with 0.25% yeast extract,
KH_2PO_4	0.1%	0.5% malt syrup, 0.5% acid-hydrolyzed casein,
$MgSO_4$	0.05%	and 0.1% vitamin mixtures; trace elements[231] in
$CaCl_2$	0.01%	(mg/l): Bo 0.01, Cu 0.1, Fe 0.2, Mn 0.02, Zn 2.0
NaCl	0.01%	Adjust pH to 6.5 before autoclaving
Biotin	5 μg/l	
Trace elements	1.0 ml	
Sucrose	2.0%	
Agar	1.5%	

Recipe 164
NEUTRAL WORT AGAR[6]

Maltose	12.75 g	Adjust pH to 6.8 or 7 and autoclave at 120°C for
Malt extract	15.0 g	10 min
Dextrin	2.75 g	
Glycerol	2.35 g	
K_2HPO_4	1.00 g	
NH_4Cl	1.00 g	
Peptone	0.78 g	
Agar	15.0 g	

Recipe 165
NORKRANS MEDIUM[434]

Cellulose #3 filter paper, ground	5.0 g
$NH_4 H_2 PO_4$	2.0 g
$KH_2 PO_4$	0.6 g
$K_2 HPO_4$	0.4 g
$MgSO_4 \cdot 7H_2 O$	0.5 g
Ferric citrate (dissolve in citric acid)	10 mg
$ZnSO_4 \cdot 7H_2 O$	4.4 mg
$MnSO_4 \cdot 7H_2 O$	5.0 mg
$CaCl_2$	100 mg
$CoCl_2$	1.0 mg

Note: pH of medium is approximately 5.8.

Recipe 166
OAK WILT AGAR[66],[69]

Maltose	5.0 g	Adjust the pH to about 6 before autoclaving
Difco Casamino acids (tech. grade)	1.0 g	at 15 lb per 15 min
$KH_2 PO_4$	1.0 g	
$MgSO_4 \cdot 7H_2 O$	0.5 g	
Zinc sulfate	0.2 mg	
Ferrous sulfate	0.2 mg	For amount of cation see Recipe 148
Manganese sulfate	0.1 mg	Add 250 mg of streptomycin sulfate to in-
Biotin (0.5 g yeast extract may be substituted for some isolates)	5 μg	hibit bacteria[66]
Agar	20 g	

Recipe 167
OAK WILT AGAR MODIFIED[68]

Modify oak wilt agar (Recipe 166) as follows:
Add 1 mg thiamine
Decrease $KH_2 PO_4$ to 0.1 g
Adjust pH to 5.2

Recipe 168
OAT FLAKE AGAR[249]

Oat flakes, rolled (preferably slow-cooking kind)		5-mm thickness in 100 × 20 mm petri dish; pour enough melted agar to cover oat flakes and auto-
Agar	20 g	clave

Recipe 169
OAT WHEAT AGAR[229]

Oat kernels	35 g	The grains are left intact and are suspended in
Wheat kernels	35 g	agar
Agar	10 g	

Recipe 170
OATMEAL AGAR[435]

Rolled oats	60 g	Blend oats in 600 ml of H_2O in Waring® Blendor for 5 min, add 400 ml of melted agar; put in prescription bottles, tighten caps; do not attempt to cool these bottles rapidly as they will crack; autoclaving of 300-ml amounts for 90 min recommended by Goody and Lucas[435] seems unnecessary; 30 min for 200-ml amounts is adequate
Agar	12 g	

Recipe 171
ODEGARD MEDIUM[243]

Glucose	30 g
Asparagine	3.5 g
KH_2PO_4	1.0 g
$MgSO_4 \cdot 7H_2O$	0.5 g
Ca^{++}	80 mg
Fe^{+++}	0.32 mg
Zn^{++}	0.44 mg
Thiamine	50 mg
Redistilled water	

Recipe 172
OHIO MEDIUM[436]

Glucose	5 g	Autoclave at 11 psi for 15 min
Yeast extract	2 g	
$MgSO_4 \cdot 7H_2O$	1 g	
KH_2PO_4	1 g	
Streptomycin sulfate	50 mg	
Chloromycetin	50 mg	
Oxgall	1 g	
Sodium propionate	1 g	
Agar	20 g	

Recipe 173
OLIVE MEDIUM[437]

Glucose	2.0 g
Filter paper—macerated sheet, 45 mm diameter	
K_3PO_4	1.0 g
NH_4NO_3	1.0 g
$MgSO_4$	1.0 g
Agar	25 g

Recipe 174
PABLUM AGAR[419]

Pablum (Mead Johnson, Co.)	50 g	Cook for 10 min in 700 ml of water, strain through porous cloth, add to melted agar, and bring to a liter
Agar	15 g	

Recipe 175
PABLUM AGAR MODIFIED[408]

Pablum mixed cereal or pablum high-protein cereal	12.5 g	Cook cereal for 10 min in 175 ml of H_2O, filter through cheesecloth
Agar	15 g	

Recipe 176
PARK MEDIUM[438]

Glucose	0.7 g
$MgSO_4 \cdot 7H_2O$	0.5 g
KH_2PO_4	0.2 g
NH_4NO_3	0.1 g
Agar	15 g

Recipe 177
PCNB MEDIUM MODIFIED[139]

Peptone (Difco)	15 g	It does not need to be sterilized if made up fresh; adjust pH to 5.2.
KH_2PO_4	1.0 g	
$MgSO_4 \cdot 7H_2O$	0.5 g	
PCNB (Terraclor®, a commercial product is 75% active)	0.5 g	Active ingredient
Oxgall	1.0 g	
Chlortetracycline HCl[a]	50 mg	
Streptomycin sulfate[a]	100 mg	
Agar	20 g	

[a] See footnote, Recipe 1.

Recipe 178A   **Recipe 178B**
PDA   **PDA ACIDIFIED**

Potatoes, peeled and sliced	200 g	As 178A but acidified with 3—5 drops of 25% lactic acid per 100 ml
Dextrose	10 g	
Agar	12 g	

Recipe 179   **Recipe 180**
PDA[439]   **PDA MODIFIED**[86]

Potatoes, peeled	500 g	Potatoes	200 g
Dextrose	10 g	Dextrose	20.0 g
NaCl	1 g	$CaCO_3$	0.2 g
Agar	20 g	$MgSO_4 \cdot 7H_2O$	0.2 g
		Agar	15.0 g

Recipe 181
PDA MODIFIED[24]

Potatoes, instant mashed	10.0 g
Dextrose	10.0 g
Flake agar	15.3 g

Recipe 182
PDA-BDP[132,140]

PDA		
Benomyl (active ingredient)	8 ppm	Inhibits *Penicillium* but also a few impor-
Dichloran, Botran® (active ingredient)	8 ppm	tant wood rotters[132,140]
Phenol (in 50% EtOH)[a]	50 ppm	

[a] See footnote, Recipe 1.

Recipe 183
PDA CASAMINO ACID[308]

PDA	
Casamino acids	5 g

Recipe 184
PDA-CHARCOAL MEDIUM[306]

PDA plus 1% charcoal (Grade G-60, Atlas Powder Co.)

Recipe 185
PDA-DOPCNB[212]

PDA		
Dexon®[a] (*p*-dimethylamino benzenediazo sodium sulfonate)	50 mg[b]	Keep in dark until used
Oxgall[a]	2 g[b]	
PCNB[a]	100 mg[b]	
Chlortetracycline[a]	25 mg	
Streptomycin sulfate[a]	100 mg	

[a] See footnote, Recipe 1.
[b] Active.

Recipe 186
PDA-DORB

PDA		
Dexon®[a]	50 mg	Keep in dark prior to use
Oxgall[a]	1.5 g	
Rose bengal[a]	150 mg	
Chlortetracycline[a]	25 mg	
Streptomycin sulfate[a]	100 mg	

[a] See footnote, Recipe 1.

Recipe 187
PDA PHYTOPHTHORA ISOLATION MEDIUM[257]

PDA (1% agar) mg/l

	Active ingredients (mg)
Benomyl®[a]	10
Mycostatin®[a]	25
PCNB[a]	25
Rifampin[a]	10
Ampicillin[a]	500
Tachigaren®[a] (3 hydroxyl-5-methyl-isoxazole)	25 or 50

[a] Dissolve or suspend in 80% EtOH; also, see footnote with Recipe 1.

Recipe 188
PDA YEAST EXTRACT[307,441]

PDA (from fresh potatoes)		Some dehydrated brands of PDA, because they are low in
Yeast extract	5 g	thiamine, are routinely supplemented by yeast extract[441]

Recipe 189
PDTA[442,443]

Potato dextrose agar		Adjust pH to 4.0–4.2 with 50% lactic acid
Tergitol® NPX (Union Carbide®)	500 ppm	after autoclaving

Recipe 190
PDTCS PDA-NPX[443,444]

Difco® potato dextrose agar	1 liter	Frequently, streptomycin can be omitted if
Tergitol® NPX (Union Carbide®)	100 ppm	isolating from plant tissue that is not bad-
Streptomycin sulfate[a]	35 ppm	ly deteriorated
Chlortetracycline[a]	35 ppm	

[a] See footnote, Recipe 1.

Recipe 191
PEA SEED[266]

1. Soak dried yellow peas overnight
2. Place a 1-in. layer of soaked pea in 250-ml flasks
3. Add enough water to cover the peas

Recipe 192
PEANUT MEAL AGAR[445]

Add autoclaved peanut meal to 2% water agar about to solidify; amount of meal not critical

Recipe 193
PENICILLIN MEDIUM[446]

Corn steep liquor	40 ml	Add sterile $CaSO_4$ to medium after autoclaving;
Lactose	27.5 g	can be sterilized dry or in solution
Glucose	3.0 g	
$NaNO_3$	3.0 g	
$MgSO_4 \cdot 7H_2O$	0.25 g	
KH_2PO_4	0.5 g	
$ZnSO_4 \cdot 7H_2O$	0.044 g	
$MnSO_4 \cdot 4H_2O$	0.02 g	
$CaSO_4$	8.0 g	

Recipe 194
PEPTONE GLUCOSE AGAR[214]

Peptone	40 g	Asparagine (30 g) can be substituted for pep-
Glucose	5.0 g	tone; for production of pycnidia of *Macro-*
$MgSO_4 \cdot 7H_2O$	0.5 g	*phomina phaseoli*, dip filter paper in 20% pea-
KH_2PO_4	1.0 g	nut oil or other vegetable oil (ether solution),
Agar	20 g	autoclave, and place on agar surface

Recipe 195
PEPTONE GLUCOSE AGAR[124],[126]

Peptone	10.0 g
Glucose	10.0 g
$Na_2HPO_4 \cdot 12H_2O$	0.96 g
K_2HPO_4	1.45 g
Agar	20 g

Recipe 196
PEPTONE GLUCOSE MEDIUM[29]

Peptone	20 g
Glucose	5 g

Recipe 197
PEPTONE GLUCOSE YEAST AGAR
(SM-AGAR)[127]

Bactopeptone	10 g
Glucose	10 g
Yeast extract	1.0 g
K_2HPO_4	1.0 g
KH_2PO_4	1.5 g
$MgSO_4$	0.5 g
Agar	20 g

Recipe 198
PEPTONE GLUCOSE YEAST BROTH[128]

Peptone (Oxoid® Bacteriological)	14.3 g	Final pH 6.7; if a solid substrate is desired,
Yeast extract (Oxoid®)	7.15 g	agar at 0.5% is added to cultivate axenic
D-glucose	15.4 g	strains of *Dicytostelium discoideum*
$Na_2HPO_4 \cdot 12H_2O$	1.28 g	
KH_2PO_4	0.486 g	

Recipe 199
PEPTONE GLUCOSE YEAST BROTH MEDIUM A[127]

Bactopeptone	5 g	Adjust pH to 6.0—6.3
Glucose	5 g	
Bacto-yeast extract	0.5 g	
KH_2PO_4	2.25 g	
$K_2HPO_4 \cdot 12H_2O$	1.5 g	
$MgSO_4 \cdot 7H_2O$	0.5 g	

Recipe 200
PEPTONE YEAST GLUCOSE AGAR (CANTINO PYG — M-51)[7],[447]

Peptone (Difco® bacto)	1.25 g	$1 \times 10^{-2} - 5 \times 10^{-3}$ M $NaHCO_3$ added to
Yeast extract	1.25 g	broth induces resistant sporangia of
D-glucose	3.0 g	*Blastocladiella emersonii*[447]
Agar	20.0 g	

Recipe 201
PEPTONE YEAST EXTRACT MALTOSE AGAR
(M-53)[3]

Maltose	5.0 g
Glycerine	2.0 g
Peptone	0.2 g
Asparagine	0.2 g
Yeast extract	2.0 g
$MgSO_4 \cdot 7H_2O$	0.5 g
$Ca(NO_3)_2$	0.5 g
KH_2PO_4	0.5 g
$FeSO_4$	Trace
Agar	20.0 g

Recipe 202
PETRI MEDIUM[448,449]

Calcium nitrate	0.4 g
Magnesium sulfate	0.15 g
Potassium acid phosphate	0.15 g
Potassium chloride	0.05 g

Recipe 203
PHYSARIUM MEDIUM[450,451]

Number 1 at pH 3.5

Citric acid·H_2O	2.2 g	Combine 1 and 2, equals basic medium;
$CaCl_2 \cdot 2H_2O$	0.6 g	heat-sterilize it and dextrose solution
$MgSO_4 \cdot 7H_2O$	0.6 g	separately: filter-sterilize hemin; mix
$FeSO_4 \cdot 7H_2O$	0.084 g	together, final pH 5.0;
$MnCl_2 \cdot 4H_2O$	0.084 g	
$ZnSO_4 \cdot 7H_2O$	0.034 g	

Number 2

Tryptone (Difco®)	10.0 g
Yeast extract (Difco®)	1.5 g
KH_2PO_4	2.0 g

Dextrose	10.0 g	Dissolve 0.0063 g of hemin in 20 ml
Hemin (Nutritional Biochemicals® Corp.)	0.005 g	1% NaOH, bring to pH 8.0 with H_3PO_4;
		dilute to 25 ml, filter sterilize, and add
		20 ml to 1 liter of medium

Recipe 204
PHYSARIUM SPORULATION MEDIUM[251]

HCl	40 mg
$FeCl_2 \cdot 4H_2O$	59 mg
$MnCl_2 \cdot 4H_2O$	82 mg
$ZnSO_4 \cdot 7H_2O$	33 mg
Citric acid·H_2O	1200 mg
$CaCl_2 \cdot 2H_2O$	590 mg
$MgSO_4 \cdot 7H_2O$	590 mg
$CuCl_2 \cdot 2H_2O$	23 mg
KH_2PO_4	390 mg
$CaCO_3$	1100 mg
Niacin	98 mg
Niacinamide	98 mg

Recipe 205
PHYTONE DEXTROSE AGAR[452]

Phytone (BBL®)	15 g
Dextrose	15 g
Yeast extract	1 g
Agar	17 g

Recipe 206
PHYTOPHTHORA ISOLATION MEDIUM[252]

KH_2PO_4	1.0 g	
$MgSO_4 \cdot 7H_2O$	0.1 g	
$CaSO_4 \cdot 2H_2O$	0.1 g	
Thiamin·HCl	0.02 g	
DL-threonine	1.0 g	
Sucrose	5.0 g	
Corn oil	0.1 ml	
Tergitol® NPX	0.1 ml	
Vancomycin[a]	200 mg	
Pimaricin[a]	10 mg	
Polymyxin B[a]	175,000 units	Polymyxin used when bacteria are a problem

[a] See footnote, Recipe 1.

Recipe 207
PHYTOPHTHORA SYNTHETIC NO. 1[261]

Sucrose	2.4 g
Asparagine	0.27 g
KH_2PO_4	0.15 g
K_2HPO_4	0.15 g
$MgSO_4 \cdot 7H_2O$	0.10 g
Cholesterol	10 mg
Ascorbic acid	10 mg
Thiamine·HCl	2 mg
$ZnSO_4 \cdot 7H_2O$	4.4 mg
$FeSO_4 \cdot 7H_2O$	1.0 mg
$MnCl_2 \cdot 4H_2O$	0.07 mg
Agar	20 g

Recipe 208
PHYTOPHTHORA SYNTHETIC NO. 2 (MINIMAL MEDIUM)[258]

Glucose	20 g	Adjust pH to 6.0 with 1 N KOH; auto-
$MgSO_4 \cdot 7H_2O$	0.5 g	clave glucose solution separately
KH_2PO_4	2.0 g	
$NaNO_3$	0.5 g	
$CaCl_3$	3.4 mg	
$FeSO_4 \cdot 7H_2O$	1.0 mg	
Citric acid	1.4 mg	
$ZnSO_4 \cdot 7H_2O$	1.8 mg	
$MnSO_4 \cdot H_2O$	0.3 mg	
$CuSO_4 \cdot 5H_2O$	0.4 mg	
$(NH_4)_6 MO_7O_{24} \cdot 4H_2O$	0.3 mg	
Thiamine·HCl	1.0 mg	
Difco® purified agar	15 g	

Recipe 209
PHYTOPHTHORA SYNTHETIC NO. 3[253]

Glucose	4.5 g	1 ml of microelements and 1 ml of iron
L-asparagine	0.1 g	solution added to medium; adjust pH to
KNO_3	0.15 g	6.2 with N KOH and autoclave
KH_2PO_4	1.0 g	
$MgSO_4 \cdot 7H_2O$	0.5 g	
$CaCl_2$	0.1 g	
Thiamine hydrochloride (10 mg/10 ml) after autoclaving	1.0 mg	Millipore®-filtered 1 ml per liter of medium, according to Ribeiro,[494] can also
β-sitosterol (dissolved in 30 ml dichloromethane)	30 mg	be autoclaved with medium
Noble agar	14 g	
Microelements	1.0 ml	
$Na_2MoO_4 \cdot 2H_2O$	41.1 mg	Dissolve microelements in 100 ml H_2O
$ZnSO_4 \cdot 7H_2O$	87.8 mg	
$CuSO_4 \cdot 5H_2O$	7.85 mg	
$MnSO_4 \cdot H_2O$	15.4 mg	
$Na_2B_4O_7$	0.5 mg	
$FeCl_3 \cdot 6H_2O$	44.44 mg	Best to make iron solution separately
EDTA	2.6 mg	
KOH	1.5 mg	

Recipe 210
PILOBOLUS HEMIN MEDIUM[278,279]

Hemin (dissolved in 0.1N NaOH)	10 mg
Sodium acetate ($CH_3COONa \cdot 3H_2O$)	10 g
$(NH_4)_2SO_4$	0.66 g
K_2HPO_4	1.0 g
$MgSO_4 \cdot 7H_2O$	0.5 g
Thiamine	10 mg
Agar	15 g

Recipe 211
PONTECORVO MINIMAL MEDIUM[52]

Glucose	10.0 g	Adjust pH to 6.5 with NaOH; to obtain perithecia of
$NaNO_3$	6.0 g	*Asperigillus nidulans* use 1 g $NaNO_3$ and 20 g of glu-
KCl	0.52 g	cose
KH_2PO_4	1.52 g	
$MgSO_4 \cdot 7H_2O$	0.52 g	
Iron	Trace	
Zinc	Trace	

Recipe 212
POTATO CARROT AGAR[453]

Potatoes	20 g	Steep-sliced tissue in water for 1 hr; then boil for 5
Carrot roots	20 g	min
Agar	20 g	

Recipe 213
POTATO CARROT GLUCOSE BROTH[14]

Potatoes	20 g	Make a hot water extract (80°C for 10 min) and add
Carrots	20 g	glucose; 5 ml added to Whatman No. 17 filter paper
Glucose	1 g	in a petri dish

Recipe 214
POTATO MALT AGAR[88]

Potatoes	60 g	Add sterile strips of filter paper to hardened
Dry malt (Fleischman®)	10 g	agar
Agar	15 g	

Recipe 215
POTATO MARMITE AGAR[150,305]

Glucose	20 g	Steep potatoes 1 hr at 60°C
Potatoes	250 g	May add 30 ppm chlortetracycline after
Marmite® yeast extract (Marmite Ltd., London)	1 g	autoclaving to inhibit bacteria when used to isolate fungi[150]
Agar	20 g	

Recipe 216
POTATO SUCROSE AGAR[163]

Potatoes	200 g
Sucrose	20 g
Agar	20 g

Recipe 217
POTATO TOMATO AGAR[157]

Potatoes	200 g
Tomato juice	150 ml
Glucose	10 g
Agar	15 g

Recipe 218
PRUNE AGAR[454]

Prunes	40 g
Agar	20 g

Recipe 219
PYRONEMA MEDIUM (MODIFICATION OF CLAUSSEN'S AGAR)[290,455]

KH_2PO_4	0.5 g	Make up medium with and without inulin; portion
NH_4NO_3	0.5 g	of a divided petri dish should have no inulin, the
$MgSO_4 \cdot 7H_2O$	0.2 g	other inulin; inoculate *Pyronema* on inulin side;
$FePO_4$	0.01 g	apothecia also produced without inulin or divided-
Agar	20 g	dish technique and in greater numbers on corn
Water	950 ml	meal agar[455]
Inulin	20 g	

Recipe 220
RABBIT FOOD AGAR[456]

Rabbit food (commercial pellets)	2.5 g
Agar	15.0 g

Recipe 221
RAPE SEED AGAR[260]

Rape seed	100 g	Wash seed thoroughly in water, boil for 30 min in
Agar	20 g	1 liter of distilled water; filter extract through cotton

Recipe 222
RAPER ACHYLA MEDIUM NO. 1[457]

Soluble starch	3 g
Peptone	1 g
Lentil (hot water extract)	10 g
KH_2PO_4	Trace
$MgSO_4$	Trace
$CaCl_2$	Trace
$FeCl_3$	Trace
$ZnSO_4$	Trace
Agar	20 g

Recipe 223
RAPER ACHYLA MEDIUM NO. 2[3]

Soluble starch	3 g
Inositol	1 g
Peptone	1 g
Agar	20 g

Recipe 224
RAULIN-THOM MEDIUM[458]

Glucose	50 g
Tartaric acid	2.6 g
Diammonium tartrate	2.6 g
$(NH_4)_2HPO_4$	0.4 g
K_2CO_3	0.4 g
$MgCO_3$	0.25 g
$(NH_4)_2SO_4$	0.16 g
$ZnSO_4 \cdot 7H_2O$	0.6 g
$FeSO_4 \cdot 7H_2O$	0.06 g

Recipe 225
RAULIN-THOM MEDIUM[459]

Glucose	70 g	4.0 g of diammonium tartrate is added by Bentley and
Tartaric acid	4.0 g	Kiel[459] for production of penicillic acid
$(NH_4)_2HPO_4$	0.6 g	
K_2CO_3	0.6 g	
$MgCO_3$	0.4 g	
$(NH_4)_2SO_4$	0.25 g	
$ZnSO_4 \cdot 7H_2O$	0.07 g	
$FeSO_4 \cdot 7H_2O$	0.07 g	
Distilled water to 1500 ml — note volume		

Recipe 226
REPLACEMENT MEDIUM[29]

$CaCl_2$	$1.75 \times 10^{-3}\,M$	Adjust pH to 6.5
KCl	$10^{-3}M$	
$MgSO_4$	$10^{-3}M$	

Recipe 227
REYNOLD MEDIUM[460]

K_2HPO_4	0.7 g
KH_2PO_4	0.3 g
$MgSO_4 \cdot 5H_2O$	0.5 g
$FeSO_4 \cdot 7H_2O$	0.01 g
$ZnSO_4$	0.001 g
Chitin (powdered)	0.25% w/v

Recipe 228
RHIZOCTONIA ISOLATION MEDIUM[301]

K_2HPO_4	1.0 g	Should be stored in dark
$MgSO_4 \cdot 7H_2O$	0.5 g	
KCl	0.5 g	
$FeSO_4 \cdot 7H_2O$	10 mg	
$NaNO_2$	0.2 g	
Gallic acid[a]	0.4 g	
Dexon® (Chemagro®)[a] wettable powder	90 mg	
Chloramphenicol[a]	50 mg	
Streptomycin[a]	50 mg	
Agar	20 g	

Recipe 229
RICE CEREAL AGAR[192]

Rice cereal (Gerbers® baby)	20 g
Agar	20 g

Recipe 230
RICE POLISH AGAR[461]

Rice polish	20 g	Mix the rice polish with half the water in a large flask; steam for 15 min and blend the steep in a Waring® blendor for several min; mix steep and melted agar thoroughly; swirl the flask during cooling to prevent the rice polish from settling
Agar	17 g	

Recipe 231
RICHARD SOLUTION[462]

Potassium nitrate	10 g
Potassium monobasic phosphate	5 g
Magnesium sulfate	2.5 g
Ferric chloride	0.02 g
Sucrose	50 g

Recipe 232
RICHARD SOLUTION MODIFIED[142,463]

Same medium as Recipe 26 designated as cerelose ammonium nitrate medium

Recipe 233
RICHARD SOLUTION MODIFIED[13]

Recipe 231 (Fahmy's version) with the addition of 100 ml of V-8® juice

Recipe 234
RICHARD V-8® MEDIUM[12,15]

NH_4NO_3	10.0 g
KH_2PO_4	5.0 g
$MgSO_4 \cdot 7H_2O$	2.5 g
$FeCl_3 \cdot 6H_2O$	0.02 g
Glucose	50 g
V-8® juice	20 ml
Distilled water	1.0 liter

Recipe 235
RUSSELL BASIDIOMYCETE MEDIUM[464]

Oxoid® desiccated malt extract	30 g	Autoclaved at 10 lb for 10 min; dissolved 1 g of o-phenylphenol in 50 ml industrial alcohol, water added to bring to 100 ml
Oxoid® mycological peptone	5.0 g	
o-phenylphenol	0.06 g	
Agar	25 g	Also see Recipes 66 and 67

Recipe 236
RYE DEXTROSE AGAR[267]

Rye seed	50 g	Soak seed overnight in 1 liter H_2O, comminute 3 min, add
Dextrose	10 g	agar and dextrose, and steam
Agar	15 g	

Recipe 237
RYE GRAIN AGAR A[268]

Rye grain	60 g	Soak rye in distilled H_2O 36 hr; pour off and retain super-
Sucrose	20 g	natant; macerate and extract grain for 3 hr at 50°C in
Agar	15 g	distilled H_2O; filter through several layers of muslin and
		combine with original supernatant

Recipe 238
RYE GRAIN AGAR B[268]

Prepare as above but do not macerate grains and gently
boil for 1 hr; gives a clear medium

Recipe 239
SABHI BLOOD MEDIUM[201]

Sabouraud dextrose agar (Difco®) (Equal parts)		Can be made without whole blood but fewer pathogenic fungi are isolated; plates are
Brain-heart infusion agar (Difco®)		poured and after 3 days examined for con-
Chloromycetin	100 mg	taminants; do not store more than 2 weeks
Whole blood	10%	

Recipe 240
SABOURAUD AGAR (EMMONS' MODIFICATION)[465]

Dextrose	20 g	Adjust pH to 6.8−7.0; the BBL® version uses polypep-
Neopeptone	10 g	tone instead of neopeptone
Agar	20 g	

Recipe 241
SABOURAUD DEXTROSE AGAR[380]

Neopeptone, Difco®	10 g	For other variations consult Difco® and BBL® man-
Dextrose	40 g	uals and Recipes 240 and 242
Agar	15 g	

Recipe 242
SABOURAUD DEXTROSE AGAR MODIFIED[466]

Neopeptone (Bacto®)	10 g	
Dextrose	20 g	
Agar	20 g	
Cycloheximide	100 mg	
Chloramphenicol[a]	50 mg	Preferably add after autoclaving if pH of water above 6

[a] See footnote, Recipe 1.

Recipe 243
SACHS AGAR[188,194]

Calcium nitrate	1.0 g
Dipotassium hydrogen phosphate	0.25 g
$MgSO_4$	0.25 g
Ferric chloride	Trace
$CaCO_3$	4.0 g
Agar	20.0 g

Recipe 244
SACHS AGAR MODIFIED[284]

$Ca(NO_3)_2 \cdot 4H_2O$	1.0 g
$K_2HPO_4 \cdot 3H_2O$	0.25 g
KCl	0.25 g
$MgSO_4 \cdot 7H_2O$	0.25 g
$CaCO_3$	0.85 g
$FeCl_3$	Trace
Agar	20 g

Recipe 245A
SACHS AGAR AND PLANT MATERIALS[191]

Sachs agar
Sorghum seed, ¼ teaspoon

Immerse sorghum seed in a mixture of 1% NaClO and 0.1% $HgCl_2$ for 5 min, rinse twice, and boil for 2 min; put seed in a 2-in. glass square in a petri dish and add Sachs agar to half submerge the seed[191]

Recipe 245B
SACHS AGAR AND CORN KERNELS[188]

Immerse corn kernels in a mixture of 0.5% NaClO and 0.1% $HgCl_2$ for 5 min, boil kernels 1 min in sterile, distilled H_2O, rinse twice with sterile distilled water

Recipe 245C
SACHS AGAR AND CORN LEAVES[180]

Corn leaves 1½ X ½ in. are autoclaved for 1 hr and placed on the agar surface

Recipe 245D
SACHS AGAR AND BARLEY STRAW[194]

Treat 2-in. pieces of dry barley culms (straw) with leaf sheath for 24 hr with propylene oxide (1 ml/l capacity of container); 24 hr later Sachs agar is added to half-submerged culms[194]

Recipe 246
SALT SOLUTION[262]

Solution A

KH_2PO_4	75 mg	Mix two solutions after autoclaving
K_2HPO_4	75 mg	
$MgSO_4 \cdot 7H_2O$	50 mg	
$ZnSO_4 \cdot 7H_2O$	2.2 mg	
$FeSO_4 \cdot 7H_2O$	0.5 mg	
$MnCl_2 \cdot 4H_2O$	0.035 mg	
Distilled water	950 ml	

Solution B

$Ca(NO_3)_2 \cdot 2H_2O$	0.1 g
Distilled water	50 ml

Recipe 247
SALVIN MEDIUM[202]

Proteose peptone	10.0 g	Adjust pH to 6.5 to 7.5
Neopeptone	3.25 g	
Tryptone	3.25 g	
Glucose	2.0 g	
NaCl	5.0 g	
Disodium phosphate	2.5 g	
Agar	1.75 g	

Recipe 248
SCHIZOPHYLLUM FRUITING MEDIUM (M-80)[3,320]

Glucose	20 g	Peptone can be substituted for asparagine
L-asparagine	2.0 g	and is designated as a complete medium;
Thiamine hydrochloride	100 μg	M-80 has 120 μg of thiamine
KH_2PO_4	0.46 g	
K_2HPO_4	1.0 g	
$MgSO_4 \cdot 7H_2O$	0.5 g	
Noble agar	20 g	

Recipe 249
SCHMITTHENNER MEDIUM[467]

Sucrose	2.4 g
Asparagine	0.27 g
KH_2PO_4	0.15 g
K_2HPO_4	0.15 g
$MgSO_4 \cdot 7H_2O$	0.10 g
$ZnSO_4 \cdot 7H_2O$	4.4 mg
$FeSO_4 \cdot 7H_2O$	1.0 mg
$MnCl_2 \cdot 4H_2O$	0.07 mg
Thiamine	2.0 mg
Ascorbic acid	10.0 mg
Cholesterol	10.0 mg
Calcium chloride	10—20 ppm

Recipe 250
SEA WATER GLUCOSE AGAR[468]

Glucose	1.0 g	Filter freshly collected sea water through
Yeast extract (Difco®)	0.1 g	medium-porosity sintered glass (or through
Aged sea water	1 liter	cotton pads); store in the dark in a cotton-
Agar	18 g	stoppered container for 3—6 weeks;
		A gram of phytone may substitute for yeast
		extract

Recipe 251
SEA WATER GLUCOSE PEPTONE AGAR[468]

Peptone	0.1 g
Glucose	1.0 g
Dibasic potassium phosphate	0.05 g
Ferric citrate	0.01 g
Agar	18.0 g
Aged sea water (see Recipe 250)	

Recipe 252
SEA WATER ISOLATION MEDIUM[469]

Glucose	1.0 g	
Gelatin hydrolysate	1.0 g	
Liver extract	.01 g	
Yeast extract	0.1 g	
Streptomycin sulfate[a]	0.5 g	For culture studies antibiotics can be deleted
Penicillin G[a]	0.5 g	
Agar	12.0 g	
Sea water		

[a] See footnote, Recipe 1.

Recipe 253
SOIL EXTRACT AGAR[470]

Soil, air dried	1.0 kg	Steam soil-water mixture for 30 min and leave
Distilled water	1.0 liter	standing overnight; pour through filter paper
Dextrose	1.0 g	pulp, add rest of chemicals and melted agar,
Yeast extract	0.1 g	make up to 1 liter and adjust to pH 7.0
KH_2PO_4	0.2 g	
Agar	20 g	

Recipe 254
SOIL EXTRACT AGAR[471]

Autoclave 1 kg of soil plus 1 liter of water at 15 psi for 30
min; filter supernatant and add water to make a liter,
then add 15 g of agar

Recipe 255
SOIL EXTRACT CORN MEAL[160]

Soil extract (1 kg of potting soil autoclaved 30 min in 1500 ml H_2O, filter with Whatman® #4)		Adjust pH to 6.0—6.5 with NaOH or HCl Chloramphenicol[a] (50 mg) and streptomycin sulfate[a] (30 mg) are added to aid in
Corn meal (Difco®)	17 g	the isolation of *Gibberella zeae* from corn kernels

[a] See footnote, Recipe 1.

Recipe 256
SOIL EXTRACT PGA MEDIUM[363,364]

Soil extract (1 kg garden soil in 1 liter tap water, steam for 30 min, then decant and filter)	25 ml	Make PGA solution separately and adjust to pH 7 before adding to medium
KH_2PO_4	1.5 g	
K_2HPO_4	4.0 g	
Polygalacturonic acid (PGA)	2.0 g	
Agar	15 g	
Streptomycin[a]	50 mg	
Chlortetracycline[a]	50 mg	
Chloramphenicol[a]	50 mg	

[a] See footnote, Recipe 1.

Recipe 257
SOIL HORSEHAIR CHICKEN FEATHERS MEDIUM[472]

Soil in petri dishes		Place soil in petri dishes and autoclave
Horsehair, 1 cm long	20 pieces	2 hr for 3 successive days or over-
Chicken feathers	4–6 small pieces	night with ethylene oxide; autoclave

hair and feathers separately for 15 min at 120°C; place hair or feathers on soil surface; moisten soil thoroughly with sterile water

Recipe 258
SONNEBORN DEFINED MEDIUM[473]

Amino acid mix (see below)	20 ml	The medium is sterilized by Millipore®
Glutamic acid	400 mg	filtration; concentrated stock solutions
Methionine	50 mg	of medium components can be stored
Glucose	3 g	frozen
$MgSO_4$	10 mM	
$CaCl_2$	1 mM	
NH_4NO_3	5 mM	
NaCl	10 mM	
Na_3PO_4	1 mM or 0.1 mM	The former for spinner flask cultures; the latter for dish cultures
Trace metals and thiamine	1 ml	Stock solution, per ml: 0.6 mg $FeSO_4$, 0.1 mg $CuSO_4$, 0.2 mg $ZnSO_4$, 0.2 mg $MnSO_4$, and 40 μg thiamine
Tris-maleate, pH 6.8	1 mM	
KCl (+KOH)	25 mM	The medium is adjusted to pH 6.8 with KOH; KCl is added to bring total K^+ concentration to 25 mM

Amino Acid Mix (All L-Amino Acids):

Amino acid	g
Arginine·HCl	4.21
Glycine	1.5
Histidine·HCl	2.1
Isoleucine	5.25
Leucine	5.25
Lysine·HCl	7.32
Methionine	1.49
Phenylalanine	3.3
Serine	2.0
Threonine	4.75
Tryptophan	0.82
Tyrosine	3.62
Valine	4.68

Add 350 ml 1N HCl to dissolve other amino acid mixture and bring to 2 l with distilled water

Recipe 259
SORDARIA SYNTHETIC MEDIUM[332]

Glucose	20 g	pH 6.5
KNO_3	1.0 g	To make a complete medium add 2.5
K_2HPO_4	0.7 g	g casein hydrolysate and 1 g of yeast
KH_2PO_4	0.5 g	extract
$MgSO_4 \cdot 7H_2O$	1.0 g	
NaCl	0.1 g	
$CaCl_2$	0.1 g	
Trace element solution[487]	0.1 ml	
citric acid$\cdot H_2O$	5.0 g	
$Fe(NH_4)_2(SO_4)_2 \cdot 6H_2O$	1.0 g	
$CuSO_4 \cdot 5H_2O$	0.25 g	
$MnSO_4 \cdot H_2O$	0.05 g	
H_3BO_3 (anhydrous)	0.05 g	
$Na_2MoO_4 \cdot 2H_2O$	0.05 g	
Distilled water	95 ml	
D-biotin	0.01 g	

Recipe 260
SORGHUM GRAIN[183]

Steep sorghum grain in tap water for 24 hr with several washings; drain on cheesecloth; fill flasks one third full with grain

Recipe 261
SOYBEAN SEED EXTRACT BROTH[218]

Dry soybean seed	100 g	Boil seed in water for 10—15 min, filter extract
Sucrose	20 g	through cheesecloth, add sucrose, and auto-
		clave

Recipe 262
STARCH GLUTAMIC ACID MEDIUM[160]

Starch (Fischer®, used for idometry)	20.0 g
L-glutamic acid$\cdot$HCl	10.0 g
Yeast extract (Difco®)	1.0 g
$K_2HPO_4 \cdot 3H_2O$	19.0 g
Sodium citrate$\cdot 2H_2O$	12.1 g

Recipe 263
STARCH HYDROLYSIS TEST MEDIUM[6]

Peptone	5 g	Suspend starch in 40 ml cold, distilled water; add to
Beef extract	3 g	dissolved medium and autoclave
Potato starch	10 g	
Agar	15 g	

Recipe 264
STEINBERG DIBASAL MEDIUM[53]

Sucrose	50 g
NH_4NO_3	1.90 g
K_2HPO_4	0.35 g
$MgSO_4 \cdot 7H_2O$	0.25 g
Chlorides (in mg)	
Fe	0.4
Zn	0.4
Cu	0.1
Mn	0.1
Mo	0.2
Ga	.02

There are several versions of Steinberg's unit-optimal dibasal media

Recipe 265
SUCROSE ASPARAGINE MEDIUM[323]

Sucrose	5 g
Asparagine	3 g
KH_2PO_4	0.5 g
$MgSO_4 \cdot 5H_2O$	0.2 g
NaCl	0.1 g

Recipe 266
SUCROSE ASPARAGINE YEAST AGAR (SAY)[188]

Sucrose	8.0 g
L-asparagine	1.2 g
Dipotassium hydrogen phosphate	0.6 g
Yeast extract (Difco®)	1.0 g
Agar	20 g

Recipe 267
SUCROSE LEUCINE MEDIUM[79,83]

Sucrose	5.0 g
DL-leucine	1.636 g
Yeast extract	3.6 g
Citrate buffer, pH 6.5	0.1 M
Chlorides	ppm
Iron	0.8
Zinc	4.8
Copper	0.8
Manganese	1.6
Molybdenum	0.32
Calcium	6.0
Agar	18 g

Buffer and constituents of the media are combined after autoclaving

Recipe 268
SUCROSE PROLINE AGAR (SPA)[474]

Sucrose	6.0 g
Proline	2.7 g
Dipotassium hydrogen phosphate	1.3 g
Monopotassium dihydrogen phosphate	1.0 g
Potassium chloride	0.5 g
Magnesium sulfate	0.5 g
Ferrous sulfate	10 mg
Zinc sulfate	2 mg
Manganese chloride	1.6 mg
Agar	20 g

Recipe 269
SUCROSE SALTS YEAST MEDIUM (SMKY)[475]

Sucrose (technical grade)	200 g
$MgSO_4 \cdot 7H_2O$	0.5 g
KNO_3	3.0 g
Yeast extract	7.0 g

Recipe 270
SUCROSE SYNTHETIC MEDIUM[170]

Sucrose	20.0 g
KH_2PO_4	3.4 g
$MgSO_4 \cdot 7H_2O$	1.1 g
$NaNO_3$	1.3 g
$CaCl_3$	0.03 g
Trace element solution	1.0 ml[a]

[a] See Recipe 162.

Recipe 271
SUCROSE THREONINE MEDIUM[252]

Sucrose	5 g	If bacteria are a problem add Polymyxin
Corn oil	0.1 ml	
dl-Threonine	1.0 g	
KH_2PO_4	1.0 g	
$MgSO_4 \cdot 7H_2O$	0.1 g	
$CaSO_4 \cdot 2H_2O$	0.1 g	
Thiamine·HCl	0.02 g	
Tergitol® NPX	0.1 ml	
Vancomycin[a]	200 mg	
Pimaricin[a]	10 mg	
Polymyxin B[a]	175,000 units	
Agar	20 g	

[a] See footnote, Recipe 1.

Recipe 272
SUCROSE UREA MEDIUM
MEDIUM B[280]

Sucrose	30.0 g
KNO_3	3.0 g
Urea	0.5 g
KH_2PO_4	1.0 g
K_2HPO_4	1.0 g
$MgSO_4 \cdot 7H_2O$	0.5 g
$CaCl_2 \cdot 6H_2O$	0.1 g
$FeSO_4 \cdot 7H_2O$	0.75 mg
$MnSO_4 \cdot 5H_2O$	0.22 mg
$CuSO_4 \cdot 5H_2O$	0.6 mg
$ZnCl_2$	7.5 mg
H_3BO_3	0.09 mg
$(NH_4)_6Mo_7O_{24} \cdot 4H_2O$	0.06 mg
Biotin	5.0 μg
Thiamine	1.0 mg

Recipe 273
SUCROSE YEAST EXTRACT MEDIUM[476]

Sucrose	40 g	Used for the production of ochratoxin by *Aspergillus* spp.
Yeast extract	20 g	

Recipe 274
SUGAR BEET LEAF AGAR[81]

Fresh sugar beet leaves	250 g	Boil leaves in 800 ml distilled water for 15
Agar	15 g	min

Recipe 275
SUGAR BEET MOLASSES MEDIUM[80]

Sugar beet molasses	150 g
Agar	15 g

Recipe 276
SYNTHETIC MUCOR[477]

Dextrose	40 g
Asparagine	2 g
KH_2PO_4	0.5 g
$MgSO_4$	0.25 g
Thiamine chloride	0.5 mg
Agar	15.0 g

Recipe 277
TALBOY MEDIUM[478]

Prune extract	100 ml	Simmer 50 g chopped prunes in a liter of water
Lactose	5 g	until soft, strain through muslin, filter and
Difco® yeast extract	1 g	make up to a liter; can be stored sterile and
Agar	30 g	cold until needed

Recipe 278
TBM-C[339]

Carrot juice (200 g of peeled carrot roots, autoclaved)	970 ml	Adjust pH to 5.2; the microbial agents are shaken for 1 hr in 30 ml, brought to 50°C,
Yeast extract	2 g	and added to autoclaved medium at the
PCNB (technical or 75% wettable)[a]	1 g	same temperature; see also Recipe 302
Nystatin[a]	50 mg	
Chloramphenicol[a]	250 mg	
K penicillin G[a]	60 mg	
Agar	20 g	

[a] See footnote, Recipe 1.

Recipe 279
TCHERNOFF MEDIUM[77]

Glucose	20 g
L-asparagine	2.0 g
KH_2PO_4	1.5 g
$MgSO_4 \cdot 7H_2O$	1 g
$ZnSO_4$	20 mg
$FeCl_3$	10 mg
Pyridoxine	1 mg
Thiamine	1 mg

Recipe 280
TOCHINAI SOLUTION[479]

Peptone	10.0 g
KH_2PO_4	0.5 g
$MgSO_4 \cdot 7H_2O$	.25 g
Maltose	20.0 g

Recipe 281
TOMATO SALT AGAR[42]

Tomato juice agar (Difco®)	25 g	Do not oversterilize as agar will be hydro-
NaCl	60 g	lyzed
Agar	15 g	

Recipe 282
TOUSSOUN MEDIUM[145]

D-glucose	10 g
KH_2PO_4	1.75 g
$MgSO_4 \cdot 7H_2O$	0.75 g
DL-Isoleucine	5.0 g
Agar	20.0 g

Recipe 283
TOXIN PRODUCTION MEDIUM[179]

Same medium as Recipe 72 but use 1.0 g of NH_4NO_3, instead of KNO_3 and delete agar

Recipe 284
TRIONE TILLETIA MEDIUM[345]

Wheat germ	10 g
Gluten	5 g
L-α-lecithin	5 g
Edamin® (amino acid mixture enzymatic digested lactalbumin from Sheffield Chemical Co., Norwich, N.Y.)	1 g
Sucrose	40 g
Thiamine	1 mg
Inorganic salts from Trione Dwarf Bunt synthetic medium (Recipe 285)	
Agar	15 g

Recipe 285
TRIONE TILLETIA SYNTHETIC MEDIUM[343]

Sucrose	20 g	Adjust pH to 6.0
L-asparagine	3 g	
KH_2PO_4	613 mg	
$MgSO_4 \cdot 7H_2O$	246 mg	
$K_2HPO_4 \cdot 3H_2O$	114 mg	
$CaCl_2$	55.5 mg	
Sodium ferric diethylenetriamine pentaacetate	20 mg	
$ZnSO_4 \cdot 7H_2O$	3.52 mg	
$CuSO_4 \cdot 5H_2O$	0.38 mg	
$MnSO_4 \cdot H_2O$	0.031 mg	
$Na_2MoO_4 \cdot 2H_2O$	0.025 mg	
Thiamine·HCl	5 mg	

Recipe 286
TRYPTONE YEAST EXTRACT BROTH[319]

Tryptone	10 g	Adjust pH to 7.3 with KOH prior to autoclaving; filter-
Yeast extract	5 g	sterilize carbenicillin if not in sterile container
NaCl	5 g	Lower dosage of carbenicillin may also be suitable in
Carbenicillin	2—4 g	purifying cultures of *Saprolegnia*

Recipe 287
UREASE TEST MEDIUM[6,480]

Peptone	1 g	Autoclave at 115°C for 15 min
NaCl	5 g	
KH_2PO_4	2 g	
Glucose	5 g	
Phenol red (0.2% solution in 50% ethanol)	5 ml	
Urea[a] (20% aqueous solution; sterilize by filtration)	100 ml	
Agar	20 g	

[a] See footnote, Recipe 1.

Recipe 288
USTILAGO GENETICS COMPLETE MEDIUM[3,356]

Glucose	10 g	Hydrolyzed casein and
Hydrolyzed casein	2.5 g	nucleic acid can be replaced
Hydrolyzed nucleic acid (see below)	5.0 g	by 10 g of yeast extract
Vitamin solution (see below)	10 ml	
Yeast extract	1.0 g	
NH_4NO_3	1.5 g	
Salt solution (see below)	62.5 ml	

Vitamin Solution

Thiamine	100 mg
Riboflavin	50 mg
Pyridoxin	50 mg
Ca pantothenate	200 mg
Nicotinic acid	200 mg
Choline chloride	200 mg
Inositol	400 mg
Distilled water to a liter	

Salt Solution

KH_2PO_4	16 g
Na_2SO_4	4 g
KCl	8 g
$MgSO_4 \cdot 7H_2O$	2 g
$CaCl_2$	1 g
Trace elements (see below)	8 ml
Distilled water to 1 liter	

Trace Element Solution

H_3BO_3	30 mg
$MnCl_2 \cdot 4H_2O$	70 mg
$ZnCl_2$	200 mg
$Na_2MoO_4 \cdot 2H_2O$	20 mg
$FeCl_3 \cdot 6H_2O$	50 mg
$CuSO_4 \cdot 5H_2O$	200 mg
Distilled water	500 ml

Hydrolyzed Nucleic Acid

Heat 1 g yeast nucleic acid and 1 thymus nucleic acid in 15 ml N NaOH at 15 psi for 10 min; repeat for 1 g of each nucleic acid in 15 ml 1N HCl; mix the hydrolysates, adjust to pH 6.0, filter hot, and bring to 40 ml; store frozen

Recipe 289
USTILAGO GENETICS MINIMAL MEDIUM[3,357]

Glucose	10 g	1% activated charcoal may be added for the production of dikaryons of *U. maydis*
KNO_3	3 g	
Salt solution (see *Ustilago* genetics complete medium)	62.5 ml	

Recipes 290—304
V-8® JUICE MEDIA

Recipe number	V-8® (ml)	CaCO₃	Glucose	Sucrose	Yeast extract	Sterols	Agar	Bacterial antibiotics	Fungal inhibitors	Preparation	Ref.
290	200	3.0	—	—	—	—	15	—	—	—	481
291	300	—	—	—	—	—	20	—	—	Adjust to pH 5.5	145
292	200	4.5	—	—	—	—	15	—	—	Centrifuge 300 ml juice at 3000 rpm with CaCO₃	482
293	300	—	—	—	—	—	20	—	—	Filter V-8® with Celite® (medium is ~ pH 4.5 and soft)	103
294	200	2.9	—	—	—	—	15	—	—	Centrifuge 1,000 g 15 min with CaCO₃	258
295	200	3.0	—	—	—	30 mg β-sitosterol	15	—	—	Centrifuge	483
296	200	—	—	—	—	—	30	—	—		189
297	200	2.5	—	—	—	30 mg cholesterol in 9% EtOH	0	—	—	Centrifuge 13,200 g 30 min; add cholesterol after centrifuging	299
298	50	2.0	—	—	—	—	15	—	—	—	256
299	200	3.0	5.0	—	2.0	—	20	30 mg Chlortetracycline[a] 30 mg Streptomycin[a]	5.0 g Oxgall® 1.0 g Sodium propionate	Bring V-8® juice and CaCO₃ to boil in 800 ml H₂O and filter through cotton and cheesecloth, 11 psi 15 min	403
300	10	—	—	—	—	—	40	100 mg Vancomycin[a]	50 mg Nystatin,[a] 10 mg PCNB	—	484
301	40	0.6	—	1.0	0.2	—	20	10 mg Chloramphenicol,[a] 10 mg Endomycin,[a] and 100 mg Neomycin[a]	—	Steam V-8 juice with CaCO₃ and filter through Celite®	261
302	200	—	—	—	29	—	20	250 mg Chloramphenicol,[a] 60 mg Potassium Penicillin G[a]	50 mg Nystatin,[a] 10 mg PCNB	Adjust pH to 5.2; shake antibiotic in water 1 hr	339
303	300	20	—	—	—	—	15	—	—	Clarified by centrifugation at 8,000—10,000 rpm	485
304	200	1.0	2.0	—	—	—	20	100 mg Streptomycin,[a] 2 mg Chlortetracycline[a]	0.5 g PCNB, 1.0 g Oxgall®, 50 mg Nystatin[a]	Adjust pH to 5.2	338

Note: Expressed in g/l except where noted.
[a] Add after autoclaving.

Recipe 305
V-8® JUICE SIMULATED MEDIUM[486]

Sucrose	3.1 g	KH_2PO_4	8.5
Asparagine	6.2 g	Na_2HPO_4	8.5
Arginine	1.13	$MgSO_4 \cdot 7H_2O$	34.8
Glutamic acid	2.44	$CaCl_2 \cdot 2H_2O$	66.0
NH_2-butyric	4.78	Na_4 EDTA	0.5
Histidine·HCl hydrate	0.75	$FeSO_4 \cdot 7H_2O$	0.69
Lysine·HCl	0.74	$CuSO_4 \cdot H_2O$	0.21
Tyrosine	1.30	$MnSO_4 \cdot H_2O$	0.3
Thiamine	5.0 μg	Na_2MoO_4	0.013
Riboflavin	3.0 μg	H_3BO_3	0.9
Niacin	80.0 μg		
Ca pantothenate	28.2 μg	Ingredients represent	
Pyridoxine HCl	18.0 μg	10% vegetable juice	
Folic acid	0.53 μg		
B_{12}	0.003 μg		
Biotin	0.18 μg		
Vitamin C	180 μg		

Note: Amounts expressed in milligrams unless otherwise indicated.

Recipe 306
VOGEL MEDIUM N[487]

Na_3 citrate·2½H_2O	3 g
KH_2PO_4 (anhydrous)	0.5 g
NH_4NO_4 (anhydrous)	2 g
$MgSO_4 \cdot 7H_2O$	0.2 g
$CaCl_2 \cdot 2H_2O$	0.1 g
Trace element solution (see below)	5 ml
Biotin solution	2.5 ml
Distilled water	750 ml
Sucrose	20 g/l
Agar, if desired	15 g/l

Trace Element Solution

Citric acid·H_2O	5.0 g	Dissolve successively
$ZnSO_4 \cdot 7H_2O$	5.0 g	
$Fe(NH_4)_2(SO_4)_2 \cdot 6H_2O$	1.0 g	
$CuSO_4 \cdot 5H_2O$	0.25 g	
$MnSO_4 \cdot H_2O$	0.05 g	
H_3BO_3 (anhydrous)	0.05 g	
$Na_2MoO_4 \cdot 2H_2O$	0.05 g	
Distilled water	95 ml	
Biotin (Merck®) solution	5.0 mg	
Distilled water	50 ml	

Note: To prepare a "complete medium", add 0.5% yeast extract and 0.5% casein hydrolysate.

Recipe 307
WARCUP MEDIUM[488]

Czapek-Dox	1 liter	Prior to pouring acidify with phos-
Yeast extract	5 g	phoric acid to pH 4.0.

Recipe 308
WATER AGAR

Agar	15 g	Amount of agar depends on the brand, purity, and intended use of medium; agar concentration generally ranges from 12—20 g but 40 g is used occasionally in isolation of fungi when a firm agar is sought or to discourage bacteria to some degree

Recipe 309
WATER AGAR LEAF MEDIUM[217,489]

Agar	15 g	Cut small pieces of leaves from growing seedlings or one of the grasses; place in water and autoclave at 120 psi 15 min; blot dry and place one or two pieces per plate on poured water agar
Leaf pieces (barley, wheat, oats, pearl millet, or sorghum)		

Recipe 310
WHEAT LEAF EXTRACT AGAR[490]

Wheat leaves	100 g	Heat leaves at 100°C in water for 30 min, add sucrose and melted agar, and combine
Sucrose	20 g	
Agar	20 g	

Recipe 311
WHEAT MEAL AGAR[327]

Wheat meal	20 g
Agar	20 g

Recipe 312
WILBRINK AGAR[491]

Sucrose	10 g
Peptone	5 g
K_2HPO_4	0.5 g
$MgSO_4 \cdot 7H_2O$	0.25 g
Agar	18 g

Recipe 313
WILLIAMS PUCCINIA MEDIUM[287]

D-glucose	30 g	See also Bushnell *Puccinia* medium (Recipe 20)
Peptone (Evans Medical Ltd., Speke, Liverpool)	5.0 g	
$NaNO_3$	1.0 g	
K_2HPO_4	0.5 g	
KCl	0.25 g	
$MgSO_4 \cdot 7H_2O$	0.25 g	
$FeSO_4 \cdot 7H_2O$	5.0 mg	
Trisodium citrate	1.5 g	
Agar (Bacto®)	15 g	

Recipe 314
WORT AGAR[380]

See neutral wort agar (Recipe 164) for ingredients, however pH is not adjusted and, according to Difco, is 4.8

Recipe 315
YANG AND SCHOULTIES MEDIUM[30]

D-glucose	5.5 g
DL-asparagine	4.0 g
Glutathione	0.1 g
$CaCl_2 \cdot 2H_2O$	0.02 g
$MgCl \cdot 6H_2O$	0.4 g
KH_2PO_4	0.7 g
K_2HPO_4	0.4 g

Recipe 316
YEAST DEXTROSE AGAR[307]

Yeast extract	7.5 g
Dextrose	20 g
Agar	15 g

Recipe 317
YEAST EXTRACT FILTER PAPER MEDIUM[34]

Yeast extract	4 g	Comminute filter paper
Filter paper	7 g	
Agar	25 g	

Recipe 318
YEAST GLUCOSE (YpG)[1]

Yeast extract (Difco®)	4.0 g
Glucose	20 g
K_2HPO_4	1.0 g
$MgSO_4 \cdot 7H_2O$	0.5 g

Recipe 319
YEAST MALT EXTRACT AGAR[86]

Yeast extract	4.0 g
Malt extract	10.0 g
Dextrose	4.0 g
Agar	15.0 g

Recipe 320
YEAST STARCH AGAR (Yp Ss)[492]

Yeast extract (Difco®)	4.0 g
Starch, soluble	15.0 g
K_2HPO_4	1.0 g
$MgSO_4 \cdot 7H_2O$	0.5 g
Agar	20 g

Recipe 321
YELLOW PEAS[266]

Dried yellow peas

Soak peas overnight in water and place a layer of soaked peas 1" thick in 250 ml erlenmeyer flasks; barely cover the peas with water and autoclave for 40 min

REFERENCES

1. Emerson, R., *Mycologia,* 50, 589–621, 1958.
2. Mullins, J. T. and Barksdale, A. W., *Mycologia,* 57, 352–359, 1965.
3. Stevens, R. B., Ed., *Mycology Guidebook,* University of Washington Press, Seattle and London, 1974.
4. Mullins, J. T. and Warren, C. O., *Am. J. Bot.,* 62, 770–774, 1975.
5. El-Ani, A. S., *Mycologia,* 66, 661–668, 1974.
6. Beneke, E. S. and Rogers, A. L., *Medical Mycology Manual,* Burgess, Minneapolis, 1970.
7. Lovett, J. S., in *Methods in Developmental Biology,* Wilt, F. H. and Wessells, N. K., Eds., Thomas Y. Crowell, New York, 1967, 341–358.
8. Sparrow, F. K., *Aquatic Phycomycetes,* University of Michigan Press, Ann Arbor, 1960.
9. Machlis, L., *Am. J. Bot.,* 40, 450–460, 1953.
10. Machlis, L., *Am. J. Bot.,* 44, 113–119, 1957.
11. Carlie, J. J. and Machlis, L., *Am. J. Bot.,* 52, 478–483, 1965.
12. Fulton, N. D., Bollenbacker, K., and Templeton, G. E., *Phytopathology,* 55, 49–51, 1965.
13. Templeton, G. E., in *Microbial Toxins,* Vol. VIII, Kadis, S., Ciegler, A., and Ajl, S. J., Eds., Academic Press, New York, 1972, 169–173.
14. Lloyd, H. L., *Mycopathol. Mycol. Appl.,* 38, 33–39, 1969.
15. Pearson, R. C. and Hall, D. H., *Phytopathology,* 65, 1352–1359, 1975.
16. Brian, P. W., Curtis, P. J., Hemming, H. G., Unwin, C. H., and Wright, J. M., *Nature,* 164, 534–535, 1949.
17. Lilly, V. G. and Barnett, H. L., *W. V. Agric. Exp. Stn. Bull.,* 362T, 1953.
18. Lukens, R. J. and Sisler, H. D., *Phytopathology,* 48, 235–244, 1958.
19. Lukens, R. J., *Phytopathology,* 50, 867–868, 1960.
20. Charlton, K. M., *Trans. Br. Mycol. Soc.,* 36, 349–355, 1953.
21. Aragaki, M., *Phytopathology,* 54, 565–569, 1964.
22. Barksdale, T. H., *Phytopathology,* 59, 443–446, 1969.
23. Douglas, D. R. and Pavek, J. J., *Phytopathology,* 61, 239, 1971.
24. Douglas, D. R., *Can. J. Bot.,* 50, 629–634, 1972.
25. Jones, F. R., Dreschsler, C., *J. Agric. Res.,* 30, 293–323, 1925.
26. Sundheim, L. and Wiggin, K., *Meld. Nor. Landbrukshoegsk.,* 51(35), 1–17, 1972.
27. Lockwood, J. L. and Ballard, J. C., *Phytopathology,* 49, 406–410, 1959.
28. Papavizas, G. C. and Ayers, W. A., *Mycologia,* 56, 816–830, 1964.
29. Mitchell, J. E. and Yang, C. Y., *Phytopathology,* 56, 917–922, 1966.
30. Yang, C. Y. and Schoulties, C. L., *Mycopathol. Mycol. Appl.,* 46, 5–15, 1972.
31. Llanos, C. M. and Lockwood, J. L., *Phytopathology,* 50, 826–830, 1960.
32. Haglund, W. A. and King, T. H., *Phytopathology,* 52, 315–317, 1962.
33. Cunningham, J. L. and Hagedorn, D. J., *Phytopathology,* 52, 616–621, 1962.
34. Yu, C. C., *Am. J. Bot.,* 41, 21–30, 1954.
35. Wallen, V. R., Wong, S. I., and Jeun, J., *Can. J. Bot.,* 45, 2243–2247, 1967.
36. Leach, C. M. and Trione, E. J., *Plant Physiol.,* 40, 808–812, 1965.
37. Leach, C. M., *Can. J. Bot.,* 40, 1577–1602, 1962.
38. Hafiz, A., *Trans. Br. Mycol. Soc.,* 34, 259–269, 1951.
39. Kaiser, W. J., *Mycologia,* 65, 444–457, 1973.
40. Christensen, C. M., *Cereal Chem.,* 23, 322–329, 1946.
41. Bell, D. K. and Crawford, J. L., *Phytopathology,* 57, 939–941, 1967.
42. Christensen, C. M., *Phytopathology,* 59, 145–148, 1969.
43. Dhingra, O. D., Nicholson, J. F., and Sinclair, J. B., *Plant Dis. Rep.,* 57, 195–197, 1973.
44. Rambo, G., Tuite, J., and Zachariah, G. L., *Cereal Chem.,* 52, 757–764, 1975.
45. Bothast, R. J., and Fennell, D. I., *Mycologia,* 66, 365–369, 1974.
46. Mitchell, T. G. and Stauber, P. C., in *Microbial Aspects Of The Deterioration Of Materials,* Gilbert, R. J. and Lovelock, D. W., Eds., Academic Press, New York, 1975, 206.
47. Raper, K. B. and Fennell, D. I., *The Genus Aspergillus,* Williams and Wilkins, Baltimore, 1965.
48. Olutiola, P. O., *Trans. Br. Mycol. Soc.,* 66, 131–136, 1976.
49. Moore, P. M. and Peberdy, J. F., *Trans. Br. Mycol. Soc.,* 66, 421–425, 1976.
50. Ward, J. E., Jr. and Cowley, G. T., *Mycologia,* 64, 200–205, 1972.
51. Tuite, J., *Aspergillus,* in *Mycotoxic Fungi – Mycotoxins – Mycotoxicoses: An Encyclopedia Handbook,* Marcel Dekker, New York, 1977.

52. Pontecorvo, G., Roper, J. A., Hemmons, L. M., MacDonald, K. D., and Bufton, A. W. J., *Adv. Genet.,* 5, 141–238, 1958.
53. Steinberg, R. A., *Am. J. Bot.,* 33, 210–214, 1946.
54. Curtis, R. W., *Plant Physiol.,* 32, 56–59, 1957.
55. Steinberg, R. A., *Am. J. Bot.,* 23, 227–231, 1936.
56. Soll, D. R., Bromberg, R., and Sonneborn, D. R., *Dev. Biol.,* 20, 183–217, 1969.
57. Goldstein, A. and Cantino, E. C., *J. Gen. Microbiol.,* 28, 689–699, 1962.
58. Murphy, M. N. and Lovett, J. S., *Dev. Biol.,* 14, 68–95, 1966.
59. Netzler, D. and Dishon, I., *Phytopathology,* 57, 795–796, 1967.
60. Clark, C. A. and Lorbeer, J. W., *Phytopathology,* 65, 338–341, 1975.
61. Clark, C. A. and Lorbeer, J. W., *Phytopathology,* 67, 96–100, 1977.
62. Leach, C. M., *Can. J. Bot.,* 40, 151–161, 1962.
63. Nelson, K. E., Kosuge, T., and Nightingale, A., *Am. J. Enol. Vitic.,* 14, 118–128, 1963.
64. Tan, K. K. and Epton, H. A. S., *Trans. Br. Mycol. Soc.,* 61, 145–157, 1973.
65. True, R. P., Barnett, H. L., Dorsey, C. K., and Leach, J. G., *W. V. Agric. Exp. Stn. Bull.,* 448T, 1960.
66. Skelly, J. M. and Wood, F. A., *Phytopathology,* 64, 1483–1485, 1974.
67. Beckman, C. H., Kuntz, J. E., and Riker, A. J., *Phytopathology,* 43, 441–447, 1953.
68. Bell, W. R. and Fergus, C. L., *Can. J. Bot.,* 45, 1235–1242, 1967.
69. Barnett, H. L., *Mycologia,* 45, 450–457, 1953.
70. Fenn, P., Durbin, R. D., and Kuntz, J. E., *Phytopathology,* 65, 1381–1389, 1975.
71. Schneider, I. R., *Plant Dis. Rep.,* 40, 816–821, 1956.
72. Dimond, A. E., Plumb, G. H., Stoddard, E. M., and Horsfall, J. G., *Conn. Agric. Exp. Stn. New Haven Bull.,* 531, 1949.
73. Beckman, C. H., *Phytopathology,* 46, 605–609, 1956.
74. Gagnon, C., *Can. J. Bot.,* 39, 1087–1093, 1961.
75. Shafer, T. and Liming, O. N., *Phytopathology,* 40, 1035–1042, 1950.
76. Sloan, B. J., Routien, J. B., and Miller, V. P., *Mycologia,* 52, 47–63, 1960.
77. Holmes, F. W., *Neth. J. Plant Pathol.,* 71, 97–112, 1965.
78. MacHardy, W. E. and Beckman, C. H., *Phytopathology,* 63, 98–103, 1973.
79. Stavely, J. R. and Nimmo, J. A., *Phytopathology,* 58, 1372–1376, 1968.
80. Calpouzos, L. and Stallknecht, G. F., *Phytopathology,* 55, 1370–1371, 1965.
81. Calpouzos, L. and Stallknecht, G. F., *Phytopathology,* 56, 702–704, 1966.
82. Ruppel, E. G., *Phytopathology,* 62, 134–136, 1972.
83. Stavely, J. R. and Nimmo, J. A., *Phytopathology,* 59, 496–498, 1969.
84. Berger, R. D. and Hanson, E. W., *Phytopathology,* 53, 286–294, 1963.
85. Basu, S. N., *J. Gen. Microbiol.,* 5, 231–238, 1951.
86. Haynes, W. C., Wickerham, L. J., and Hesseltine, C. W., *Appl. Microbiol.,* 3, 361–368, 1955.
87. Barnett, H. L. and Lilly, V. G., *Mycologia,* 58, 585–591, 1966.
88. Ames, L. M., *A Monograph of the Chaetomiaceae,* U.S. Army Research and Development Series No. 2, 1961.
89. Barnett, H. L. and Lilly, V. G., *Mycologia,* 48, 617–627, 1956.
90. Barnett, H. L. and Lilly, V. G., *Phytopathology,* 40, 80–89, 1950.
91. Barnett, H. L. and Lilly, V. G., *Mycologia,* 47, 26–29, 1955.
92. Strider, D. L. and Winstead, N. N., *Phytopathology,* 50, 583–587, 1960.
93. Crossan, D. F., Morehart, A. L., and Biehn, W., *Phytopathology,* 54, 1038–1039, 1964.
94. Ematty, D. A., *Phytopathology,* 64, 565–567, 1974.
95. Barr, R. and Tomes, M. L., *Am. J. Bot.,* 48, 512–515, 1961.
96. Lowther, R. L., *Can. J. Bot.,* 42, 1365–1386, 1964.
97. Patrick, Z. A., Kerr, E. A., and Bailey, D. L., *Can. J. Bot.,* 49, 189–193, 1971.
98. Farley, J. D., *Phytopathology,* 62, 1288–1293, 1972.
99. Stevenson, W. R., Green, R. J., and Bergeson, G. B., *Plant Dis. Rep.,* 60, 248–251, 1976.
100. Loprieno, N., *Caryologia,* 14, 219–229, 1961.
101. Chesters, C. G. C. and Hornby, D., *Trans. Br. Mycol. Soc.,* 48, 573–581, 1965.
102. Chesters, C. G. C. and Hornby, D., *Trans. Br. Mycol. Soc.,* 48, 583–594, 1965.
103. Barksdale, T. H., *Phytopathology,* 57, 1173–1175, 1967.
104. Brehara, L. and Wright, W. R., *Plant Dis. Rep.,* 57, 445–448, 1973.
105. Ghosh, A. K., *Indian Phytopathol.,* 19, 245–250, 1966.
106. Nicholson, R. L., Turpin, C. A., and Warren, H. L., *Phytopath. Z.,* 324–326, 1974.
107. Wheeler, H., Politis, D. J., and Poneleit, C. G., *Phytopathology,* 64, 293–296, 1974.
108. Politis, D. J., *Mycologia,* 67, 56–62, 1975.
109. Lapp, M. S. and Skoropad, W. P., *Can. J. Bot.,* 54, 2239–2242, 1976.

110. Fries, L., *Physiol. Plant.,* 6, 551–563, 1953.
111. Lewis, D., *Genet. Res.,* 2, 141–155, 1961.
112. Madelin, M. F., *Ann. Bot.,* 20, 307–330, 1956.
113. Casselton, L. A. and Casselton, P. J., *Trans. Br. Mycol. Soc.,* 49, 579–581, 1966.
114. Brodie, H. J., *Mycologia,* 41, 652–659, 1949.
115. Brodie, H. J., *The Bird's Nest Fungi,* University of Toronto Press, Toronto, 1975.
116. Rowe, R. C., Beute, M. K., and Wells, J. C., *Plant Dis. Rep.,* 57, 387–389, 1973.
117. Linderman, R. G., *Phytopathology,* 64, 567–569, 1974.
118. Phipps, P. M., Beute, M. K., and Barker, K. R., *Phytopathology,* 66, 1255–1259, 1976.
119. Sobers, E. K. and Littrell, R. H., *Plant Dis. Rep.,* 58, 1017–1019, 1974.
120. Bugbee, W. M. and Anderson, N. A., *Phytopathology,* 53, 1267–1271, 1963.
121. Thies, W. G. and Patton, R. F., *Phytopathology,* 60, 599–601, 1970.
122. Binder, F. L. and Lilly, V. G., *Can. J. Bot.,* 54, 566–571, 1976.
123. Cavender, J. C. and Raper, K. B., *Am. J. Bot.,* 52, 294–296, 1965.
124. Gregg, J. H., in *Methods in Developmental Biology,* Wilt, F. H. and Wessells, N. K., Eds., Thomas Y. Crowell, New York, 1967, 359–376.
125. Olive, L. S., *The Mycetozoans,* Academic Press, New York, 1975.
126. Bonner, J. T., *J. Exp. Zool.,* 106, 1–26, 1947.
127. Sussman, M., *J. Gen. Microbiol.,* 25, 375–378, 1961.
128. Watts, D. J. and Ashworth, J. M., *Biochem. J.,* 119, 171–174, 1970.
129. Puhalla, J. E. and Anagnostakis, S. L., *Phytopathology,* 61, 169–173, 1971.
130. Kuhlman, E. G. and Hendrix, F. F., Jr., *Phytopathology,* 52, 1310–1312, 1962.
131. Artman, J. D., Frazier, D. H., and Morris, C. L., *Plant Dis. Rep.,* 53, 108–111, 1969.
132. Hunt, R. S. and Cobb, F. W., Jr., *Can. J. Bot.,* 49, 2064–2065, 1971.
133. Edmonds, R. L. and Driver, C. H., *Phytopathology,* 64, 1313–1321, 1974.
134. Maloy, O. C., *Plant Dis. Rep.,* 58, 902–904, 1974.
135. Jennison, M. W., Newcomb, M. D., and Henderson, R., *Mycologia,* 47, 275–304, 1955.
136. Johansson, M. and Hagerby, E., *Physiol. Plant.,* 32, 23–32, 1974.
137. Nash, S. M. and Snyder, W. C., *Phytopathology,* 52, 567–572, 1962.
138. Kerr, A., *Aust. J. Biol. Sci.,* 16, 55–69, 1963.
139. Papavizas, G. C., *Phytopathology,* 57, 848–852, 1967.
140. Smith, S. N. and Snyder, W. C., *Phytopathology,* 65, 190–196, 1975.
141. Scheffer, R. P. and Walker, J. C., *Phytopathology,* 43, 116–125, 1953.
142. Collins, W. W. and Nielsen, L. W., *Phytopathology,* 66, 489–493, 1976.
143. Armstrong, G. M. and Armstrong, J. K., *Phytopathology,* 66, 542–545, 1976.
144. Cochrane, V. W. and Cochrane, J. C., *Plant Physiol.,* 41, 810–814, 1966.
145. Toussoun, T. A., *Phytopathology,* 52, 1141–1144, 1962.
146. Curtis, C. R., *Phytopathology,* 54, 1141–1145, 1964.
147. Kraft, J. M. and Berry, J. W., Jr., *Plant Dis. Rep.,* 56, 398–400, 1972.
148. Meyers, J. A. and Cook, R. J., *Phytopathology,* 62, 1148–1153, 1972.
149. Matuo, T. and Snyder, W. C., *Phytopathology,* 63, 562–565, 1973.
150. MacNish, G. C., *Aust. J. Biol. Sci.,* 29, 163–174, 1976.
151. Weste, G. and Thrower, L. B., *Phytopathology,* 53, 354, 1963.
152. Sivasithamparam, K., Stukely, M., and Parker, C. A., *Can. J. Microbiol.,* 21, 293–300, 1975.
153. Asher, M. J. C., *Ann. Appl. Biol.,* 70, 215–223, 1972.
154. Robinson, P. M. and Smith, J. M., *Trans. Br. Mycol. Soc.,* 66, 413–420, 1976.
155. Butler, E. E., *Phytopathology,* 50, 665–672, 1960.
156. Barash, I., *Phytopathology,* 58, 1364–1371, 1968.
157. Butler, E. E. and Petersen, L. J., *Mycologia,* 64, 365–374, 1972.
158. Wells, J. M. and Spalding, D. H., *Phytopathology,* 65, 1299–1302, 1975.
159. Tuite, J., Shaner, G., Rambo, G., Foster, J., and Caldwell, R. W., *Cereal Sci. Today,* 19, 238–241, 1974.
160. Bacon, C. W., Robbins, J. D., and Porter, J. K., *Appl. Environ. Microbiol.,* 33, 445–449, 1977.
161. Phillips, D. J., *Phytopathology,* 55, 328–329, 1965.
162. Bonn, W. G. and Capellini, R. A., *Can. J. Bot.,* 48, 1335–1337, 1970.
163. Gordon, W. L., *Can. J. Bot.,* 30, 209–251, 1952.
164. Cappellini, R. A. and Peterson, J. L., *Mycologia,* 57, 962–966, 1965.
165. Tuite, J., *Methods in Plant Pathology,* Burgess, Minneapolis, 1969.
166. Nyall, R. F., *Phytopathology,* 60, 1175–1177, 1970.
167. Wolf, J. C. and Mirocha, C. J., *Can. J. Bot.,* 19, 725–743, 1973.
168a. Tschanz, A. T., Horst, R. K., and Nelson, P. E., *Mycologia,* 67, 1101–1108, 1975.
168b. Tschanz, A. T., Horst, R. K., and Nelson, P. E., *Mycologia,* 68, 327–340, 1976.

169. Wheeler, H. E., *Mycologia*, 48, 349–353, 1956.
170. Wheeler, H. E., Driver, C. H., and Campa, C., *Am. J. Bot.*, 46, 361–365, 1959.
171. Baxter, L. W., Jr. and Fagan, S. G., *Plant Dis. Rep.*, 58, 300–303, 1974.
172. Baxter, L. W., Jr. and Fagan, S. G., *Plant Dis. Rep.*, 60, 945–947, 1976.
173. McOnie, K. C. and Snyder, W. C., *Phytopathology*, 56, 197–202, 1966.
174. McOnie, K. C., *Phytopathology*, 54, 490–491, 1964.
175. McOnie, K. C. and Snyder, W. C., *Can. J. Bot.*, 44, 149–154, 1966.
176. Gourley, C. D., *Can. J. Bot.*, 50, 49–51, 1972.
177. Leonard, K. J., *Plant Dis. Rep.*, 56, 834–836, 1972.
178. Bhullar, B. and Daly, J. M., *Can. J. Bot.*, 53, 1796–1800, 1975.
179. Mertz, S. M., Jr. and Arntzen, C. J., *Plant Physiol.*, 1977, in press.
180. Nelson, R. R., *Phytopathology*, 47, 191–192, 1957.
181. Fukuki, K. A. and Aragaki, M., *Mycologia*, 65, 705–709, 1973.
182. Garraway, M. O., *Phytopathology*, 63, 900–902, 1973.
183. Lim, S. M., *Phytopathology*, 65, 1117–1120, 1975.
184. Ayers, J. E., Nelson, R. R., Castor, L. L., and Blanco, M. H., *Plant Dis. Rep.*, 60, 331–335, 1976.
185. Warren, H., *Phytopathology*, 65, 623–626, 1975.
186. Tinline, R. D., Stauffer, J. F., and Dickson, J. G., *Can. J. Bot.*, 38, 275–282, 1960.
187. Clark, R. V., *Can. J. Bot.*, 49, 2175–2186, 1971.
188. Shoemaker, R. A., *Can. J. Bot.*, 33, 562–576, 1955.
189. Hodges, C. F. and Watschke, G. A., *Phytopathology*, 65, 398–400, 1975.
190. Piening, L. J., *Can. J. Microbiol.*, 9, 479–490, 1963.
191. McDonald, W. C., *Phytopathology*, 53, 771–773, 1963.
192. Keeling, B. L. and Banttari, E. E., *Phytopathology*, 65, 464–467, 1975.
193. Malca, I. and Ullstrup, A. J., *Bull. Torrey Bot. Club*, 89, 240–249, 1962.
194. Luttrell, E. S., *Phytopathology*, 48, 281–287, 1958.
195. Rodriguez, A. E. and Ullstrup, A. J., *Phytopathology*, 52, 599–601, 1962.
196. Hamid, A. H. and Aragaki, M., *Phytopathology*, 65, 280–283, 1975.
197. Luke, H. H. and Wheeler, H. E., *Phytopathology*, 45, 453–458, 1955.
198. Pringle, R. B. and Braun, A. C., *Phytopathology*, 47, 369–371, 1957.
199. Nelson, R. R., *Phytopathology*, 50, 774–775, 1960.
200. Yoder, O. C. and Scheffer, R. P., *Phytopathology*, 59, 1954–1959, 1969.
201. Gorman, J. W., *Am. J. Med. Technol.*, 33, 151–157, 1967.
202. Salvin, S. B., *J. Bacteriol.*, 54, 655–660, 1947.
203. McVeigh, I. and Morton, K., *Mycopath. Mycol. Appl.*, 25, 294–308, 1965.
204. Anderson, K. L. and Marcus, S., *Mycopath. Mycol. Appl.*, 36, 179–187, 1968.
205. Cole, H., Jr. and Couch, H. B., *Am. J. Bot.*, 46, 12–16, 1959.
206. Leach, C. M., *Phytopathology*, 52, 1184–1190, 1962.
207. Colotelo, N. and Grinchenko, A. H. H., *Can. J. Bot.*, 40, 439–446, 1962.
208. Helms, K., *Phytopathology*, 65, 197–201, 1975.
209. Pandey, M. C. and Wilcoxon, R. D., *Am. J. Bot.*, 54, 1170–1175, 1967.
210. Leath, K. T., *Phytopathology*, 61, 70–72, 1971.
211. Meyer, W. A., Sinclair, J. B., and Khare, M. N., *Phytopathology*, 63, 613–620, 1973.
212. Papavizas, G. C. and Klag, N. G., *Phytopathology*, 65, 182–187, 1975.
213. Norton, D. C., *Phytopathology*, 43, 633–636, 1953.
214. Knox-Davies, P. S., *S. Afr. J. Agric. Sci.*, 9, 595–600, 1966.
215. Goth, R. W. and Ostazeski, S. A., *Phytopathology*, 55, 1156, 1965.
216. Oosthuizen, M. M. J. and Potgieter, D. J. J., *Phytochemistry*, 13, 1027–1029, 1974.
217. Chidambaram, P. and Mathur, S. B., *Trans. Br. Mycol. Soc.*, 64, 165–168, 1975.
218. Ilyas, M. B., Ellis, M. A., and Sinclair, J. B., *Phytopathology*, 66, 355–359, 1976.
219. Ayanru, D. K. G. and Green, R. J., Jr., *Phytopathology*, 64, 595–601, 1974.
220. Phillips, D. J. and Harvey, J. M., *Phytopathology*, 65, 1233–1236, 1975.
221. Hall, R., *Cytologia*, 28, 181–193, 1963.
222. Van Den Ende, G. and Cornelis, J. J., *Neth. J. Plant Pathol.*, 76, 183–191, 1970.
223. Baxter, L. W., Jr., Zehr, E. I., and Epps, W. M., *Plant Dis. Rep.*, 58, 844–845, 1974.
224. Jones, D. R., Graham, W. G., and Ward, E. W. B., *Phytopathology*, 65, 1409–1417, 1975.
225. Eggins, H. O. W. and Pugh, G. J. F., *Nature*, 193, 94–95, 1962.
226. Darby, R. T. and Mandels, G. R., *Mycologia*, 46, 276–288, 1954.
227. Tulloch, M., Mycological Papers No. 130, Commonwealth Mycological Institute, Bul., 1, Kew, Surrey, England, 1972.
228. Dehorter, B., *Can. J. Bot.*, 54, 600–604, 1976.

229. **Lortie, M.,** *Can. J. Bot.,* 42, 123–124, 1964.
230. **Dubin, H. J. and English, H.,** *Phytopathology,* 64, 1201–1203, 1974.
231. **Beadle, G. W. and Tatum, E. L.,** *Am. J. Bot.,* 32, 678–686, 1945.
232. **Westergaard, M. and Mitchell, H. K.,** *Am. J. Bot.,* 34, 573–577, 1947.
233. **Raper, K. B. and Thom, C.,** *A Manual of the Penicillia,* Williams & Wilkins, Baltimore, 1949.
234. **Mislivec, P. B. and Tuite, J.,** *Mycologia,* 62, 67–74, 1970.
235. **Caldwell, R. W.,** The Occurrence, Pathogenicity and Toxicity of *Penicillium* Species Isolated From Dent Corn Kernels, Ph.D. thesis, Purdue University, West Lafayette, Ind., 1976.
236. **Pitt, J. I.,** *Mycologia,* 65, 1135–1157, 1973.
237. **Mislivec, P. B.,** *Mycologia,* 67, 194–198, 1975.
238. **Dunkle, L. D., Odvody, G. N., and Jones, B. A.,** *Phytopathology,* 65, 1321–1322, 1975.
239. **Pringle, R. B. and Scheffer, R. P.,** *Phytopathology,* 53, 785–787, 1963.
240. **Bugbee, W. M.,** *Phytopathology,* 64, 706–708, 1974.
241. **Bugbee, W. M.,** *Can. J. Bot.,* 50, 1705–1709, 1972.
242. **Warren, R. C.,** *Ann. Appl. Biol.,* 71, 193–200, 1972.
243. **Odegard, K.,** *Physiol. Plant,* 5, 583–609, 1952.
244. **Hocking, D.,** *Trans. Br. Mycol. Soc.,* 50, 207–220, 1976.
245. **Comstock, J. C., Martinson, C. A., and Gengenbach, B. G.,** *Phytopathology,* 63, 1357–1361, 1973.
246. **Yoder, O. C.,** *Phytopathology,* 63, 1361–1366, 1973.
247. **Arny, D. C. and Nelson, R. R.,** *Phytopathology,* 61, 1170–1172, 1971.
248. **Mukunya, D. M. and Boothroyd, C. W.,** *Phytopathology,* 63, 529–532, 1973.
249. **Gray, W. D. and Alexopoulos, C. J.,** *Biology of the Mycomycetes,* Ronald Press, New York, 1968.
250. **Collins, O. R. and Tang, H. C.,** *Mycologia,* 65, 232–236, 1973.
251. **Daniel, J. W. and Baldwin, H. H.,** in *Methods in Cell Physiology,* Vol. 1, Prescott, D. M., Ed., Academic Press, New York, 1964, 9–41.
252. **Banihashemi, Z. and Mitchell, J. E.,** *Phytopathology,* 65, 1424–1430, 1975.
253. **Ribeiro, O. K., Erwin, D. C., and Zentmyer, G. A.,** *Mycologia,* 67, 1012–1019, 1975.
254. **Banihashemi, Z. and Mitchell, J. E.,** *Phytopathology,* 66, 443–448, 1976.
255. **Savage, E. J., Clayton, C. W., Hunter, J. H., Brenneman, J. A., Laviola, C., and Gallegly, M. E.,** *Phytopathology,* 58, 1004–1021, 1968.
256. **Kunimoto, R. K., Aragaki, M., Hunter, J. E., and Ko, W. H.,** *Phytopathology,* 66, 546–548, 1976.
257. **Masago, H., Yoshikawa, M., Fukada, M., and Nakanishi, N.,** *Phytopathology,* 67, 425–428, 1977.
258. **Timmer, L. W., Castro, J., Erwin, D. C., Belser, W. L., and Zentmyer, G. A.,** *Am. J. Bot.,* 57, 1211–1218, 1970.
259. **Satour, M. M. and Butler, E. E.,** *Phytopathology,* 57, 510–515, 1967.
260. **Satour, M. M. and Butler, E. E.,** *Phytopathology,* 58, 183–192, 1968.
261. **Hoitink, H. A. J. and Schmitthenner, A. F.,** *Phytopathology,* 59, 708–709, 1969.
262. **Hoitink, H. A. J. and Schmitthenner, A. F.,** *Phytopathology,* 64, 1371–1374, 1974.
263. **Hwang, S. C., Ko, W. H., and Aragaki, M.,** *Mycologia,* 67, 1233–1234, 1975.
264. **Boccas, B. and Zentmyer, G. A.,** *Phytopathology,* 66, 477–484, 1976.
265. **Zentmyer, G. A. and Ribeiro, O. K.,** *Phytopathology,* 67, 91–95, 1977.
266. **Thurston, H. D.,** *Phytopathology,* 47, 186, 1957.
267. **Hodgson, W. A. and Grainger, P. N.,** *Can. J. Plant Sci.,* 44, 583, 1964.
268. **Caten, C. E. and Jinks, J. L.,** *Can. J. Bot.,* 46, 329–348, 1968.
269. **Cuppett, V. M. and Lilly, V. G.,** *Mycologia,* 65, 67–77, 1973.
270. **Smoot, J. J., Gough, F. J., Lamey, H. A., Eichenmuller, J. J., and Gallegly, M. E.,** *Phytopathology,* 48, 165–171, 1958.
271. **Haas, J. H. and Buzzell, R. I.,** *Phytopathology,* 66, 1361–1362, 1976.
272. **Morgan, F. L. and Hartwig, E. E.,** *Phytopathology,* 55, 1277–1279, 1965.
273. **Laviolette, F. A. and Athow, K. L.,** *Phytopathology,* 67, 267–268, 1977.
274. **Tsao, P. H. and Ocana, G.,** *Nature,* 223, 636–638, 1969.
275. **Menyonga, J. M. and Tsao, P. H.,** *Phytopathology,* 56, 354–360, 1966.
276. **Tsao, P. H.,** *Phytopathology,* 61, 1412–1413, 1971.
277. **Hesseltine, C. W., Whitehill, A. R., Pidacks, C., Hagen, M. T., Bohonas, N., Hutchings, B. L., and Williams, J. H.,** *Mycologia,* 45, 7–19, 1953.
278. **Page, R. M.,** *Mycologia,* 52, 480–490, 1960.
279. **Levetin, E. and Caroselli, N. E.,** *Mycologia,* 68, 1254–1258, 1976.
280. **Tanaka, S.,** Nutrition of *Piricularia oryzae* in vitro, in *Proc. Symp. The Rice Blast Organism,* John Hopkins Press, Baltimore, 1963, 23–30.

281. Kato, H. and Dimond, A. E., *Phytopathology,* 56, 864–865, 1966.
282. Chakrabarti, N. K. and Wilcoxson, R. D., *Phytopathology,* 60, 171–172, 1970.
283. Kato, H., Yamaguchi, T., and Nishihara, N., *Ann. Phytopath. Soc. Jap.,* 42, 507–510, 1976.
284. Hebert, T. T., *Phytopathology,* 61, 83–87, 1971.
285. Yaegashi, H. and Hebert, T. T., *Ann. Phytopath. Soc. Jap.,* 42, 556–562, 1976.
286. Green, G. J., *Can. J. Bot.,* 54, 1198–1205, 1976.
287. Williams, P. G., *Phytopathology,* 61, 994–1002, 1971.
288. Bushnell, W. R., *Can. J. Bot.,* 54, 1490–1498, 1976.
289. Raymundo, S. A. and Young, H. C., *Phytopathology,* 64, 262–263, 1974.
290. Moore-Landecker, E. J., *Am. J. Bot.,* 50, 37–44, 1963.
291. Moore-Landecker, E. J., *Mycologia,* 67, 1119–1127, 1975.
292. Mircetich, S. M., *Phytopathology,* 61, 357–360, 1971.
293. Mircetich, S. M. and Kraft, J. M., *Mycopath. Mycol. Appl.,* 50, 151–161, 1973.
294. Lumsden, R. D., Ayers, W. A., and Dow, R. L., *Can. J. Microbiol.,* 21, 606–612, 1975.
295. Parmeter, J. R., Ed., *Rhizoctonia Solani,* University of California Press, Berkeley, 1970.
296. Hendrix, J. W., *Phytopathology,* 55, 790–797, 1965.
297. Kraft, J. M. and Erwin, D. C., *Phytopathology,* 57, 374–376, 1967.
298. Adams, P. B., *Phytopathology,* 61, 1149–1150, 1971.
299. Ayers, W. A. and Lumsden, R. D., *Phytopathology,* 65, 1094–1100, 1975.
300. Papavizas, G. C. and Davey, C. B., *Plant Dis. Rep.,* 43, 404–410, 1959.
301. Ko, W. and Hora, F. K., *Phytopathology,* 61, 707–710, 1971.
302. Weinhold, A. R., *Phytopathology,* 67, 566–569, 1977.
303. Papavizas, G. C. and Ayers, W. A., *Phytopathology,* 55, 111–116, 1965.
304. Vest, G. and Anderson, N. A., *Phytopathology,* 58, 802–807, 1968.
305. Whitney, H. S. and Parmeter, J. R., Jr., *Can. J. Bot.,* 41, 879–886, 1963.
306. Butler, E. E. and Bolkan, H., *Phytopathology,* 63, 542–543, 1973.
307. Tu, C. C. and Kimbrough, J. W., *Phytopathology,* 65, 730–731, 1975.
308. Orellana, R. G., Sloger, C., and Miller, V. L., *Phytopathology,* 66, 464–467, 1976.
309. Fothergill, P. G. and Yeoman, M. M., *J. Gen. Microbiol.,* 17, 631–639, 1957.
310. Webster, R. K., Ogawa, J. M., and Moore, C. J., *Phytopathology,* 58, 997–1003, 1968.
311. Gauger, W. L., *Am. J. Bot.,* 48, 427–429, 1961.
312. Jackson, L. F. and Webster, R. K., *Phytopathology,* 66, 719–725, 1976.
313. Caldwell, R. M., *J. Agric. Res.,* 55, 175–198, 1937.
314. Jackson, L. F. and Webster, R. K., *Phytopathology,* 66, 726–728, 1976.
315. Schein, R. D. and Kerelo, J. W., *Plant Dis. Rep.,* 40, 814–815, 1956.
316. Habgood, R. M., *Trans. Br. Mycol. Soc.,* 66, 201–204, 1976.
317. Dick, M. W., *Mycologia,* 57, 828–831, 1965.
318. Seymour, R. L., *Nova Hedwiga Z. Kryptogamenkd.,* 19, 1–124, 1970.
319. Neish, G. A., *Mycologia,* 67, 1192–1197, 1975.
320. Raper, J. R. and Krongelb, G. S., *Mycologia,* 50, 707–740, 1958.
321. Niederpruem, D. J., *J. Bacteriol.,* 85, 1300–1308, 1963.
322. Latham, A. J., *Phytopathology,* 60, 596–598, 1970.
323. Kent, S. S. and Strobel, G. A., *Trans. Br. Mycol. Soc.,* 67, 354–358, 1977.
324. Richards, G. S., *Phytopathology,* 41, 571–578, 1951.
325. Cooke, B. M. and Jones, D. G., *Trans. Br. Mycol. Soc.,* 54, 221–226, 1970.
326. Calpouzos, L. and Lapis, D. B., *Phytopathology,* 60, 791–794, 1970.
327. Shearer, B. L. and Zadoks, J. C., *Neth. J. Plant Pathol.,* 78, 153–159, 1972.
328. Lee, N. P. and Jones, D. G., *Trans. Br. Mycol. Soc.,* 62, 208–211, 1974.
329. Narvaez, I. M., Studies on *Septoria* leaf blotch of wheat, Ph.D. thesis, Purdue University, West Lafayette, Ind., 1957.
330. Narvaez, I. and Caldwell, R. M., *Phytopathology,* 47, 529, 1957.
331. Bretzloff, C. W., Jr., *Am. J. Bot.,* 41, 58–67, 1954.
332. Jha, K. K. and Olive, L. S., *Mycologia,* 67, 45–55, 1975.
333. Hall, R., *Can. J. Microbiol.,* 17, 132–134, 1971.
334. Barash, I., Karr, A. L., Jr., and Strobel, G. A., *Plant Physiol.,* 55, 646–651, 1975.
335. Smith, O. F., *J. Agric. Res.,* 61, 831–846, 1940.
336. Leach, C. M., *Trans. Br. Mycol. Soc.,* 57, 295–315, 1971.
337. Borges, O. L., Stanford, E. H., and Webster, R. K., *Phytopathology,* 66, 715–716, 1976.
338. Papavizas, G. C., *Phytopathology,* 54, 1475–1481, 1964.
339. Maduewesi, J. N. C., Sneh, B., and Lockwood, J. L., *Phytopathology,* 66, 526–530, 1976.
340. Steinberg, R. A., *Am. J. Bot.,* 37, 711–714, 1950.
341. Lucas, G. B., *Mycologia,* 47, 793–798, 1955.

342.   Papavizas, G. C. and Adams, P. B., *Phytopathology*, 59, 371–378, 1969.
343.   Trione, E. J., *Phytopathology*, 54, 592–596, 1964.
344.   Trione, E. J., *Phytopathology*, 62, 1096–1097, 1972.
345.   Trione, E. J., *Am. J. Bot.*, 61, 914–919, 1974.
346.   Danielson, R. M. and Davey, C. B., *Soil Biol. Biochem.*, 5, 485–494, 1973.
347.   Aube, C. and Gagnon, C., *Can. J. Microbiol.*, 15, 703–706, 1969.
348.   Danielson, R. M. and Davey, C. B., *Soil Biol. Biochem.*, 5, 505–515, 1973.
349.   Miller, J. J. and Reid, J., *Can. J. Bot.*, 39, 259–262, 1961.
350.   Rifai, M. A., Mycological Papers No. 116. Commonwealth Mycological Institute, Kew, Surrey, England, 1969.
351.   Rebell, G. and Taplin, D., *Dermatophytes — Their Recognition and Identification*, University of Miami Press, Coral Gables, Fla., 1970.
352.   Robbins, W. J. and Ma, R., *Am. J. Bot.*, 32, 509–523, 1945.
353.   Georg, L. K. and Camp, L. B., *J. Bacteriol.*, 74, 113–121, 1957.
354.   Kane, J. and Fischer, J. B., *Mycopath. Mycol. Appl.*, 50, 127–143, 1973.
355.   Christensen, J. J., Corn Smut Caused by *Ustilago Maydis*, Monogr. No. 2, American Phytopathological Society, St. Paul, Minn., 1963.
356.   Holiday, R., *Genet. Res.*, 2, 204–230, 1961.
357.   Day, P. R. and Anagnostakis, S. L., *Nature New Biol.*, 231, 19–20, 1971.
358.   Keitt, G. W. and Langford, M. H., *Am. J. Bot.*, 28, 805–820, 1941.
359.   Boone, D. M., Stauffer, J. F., Stahmann, M. A., and Keitt, G. W., *Am. J. Bot.*, 43, 199–204, 1956.
360.   Ross, R. G., *Can. J. Bot.*, 46, 1555–1560, 1968.
361.   Boone, D. M. and Keitt, G. W., *Am. J. Bot.*, 43, 226–233, 1956.
362.   Ross, R. G. and Hamlin, S. A., *Can. J. Bot.*, 43, 959–965, 1965.
363.   Menzies, J. D. and Griebel, G. E., *Phytopathology*, 57, 703–709, 1967.
364.   Green, R. J., Jr. and Papavizas, G. C., *Phytopathology*, 58, 567–570, 1968.
365.   Green, R. J., Jr., *Phytopathology*, 59, 874–876, 1969.
366.   McKeen, C. D. and Thorpe, H. J., *Can. J. Microbiol.*, 17, 1139–1141, 1971.
367.   Kaiser, W. J., *Phytopathology*, 54, 481–487, 1964.
368.   Brandt, W. H., *Can. J. Bot.*, 42, 1017–1023, 1964.
369.   Conroy, J. J., Green, R. J., Jr., and Ferris, J. M., *Phytopathology*, 62, 362–366, 1972.
370.   Jones, J. P., Crill, P., and Volin, R. B., *Phytopathology*, 65, 647–648, 1975.
371.   Asthana, R. P. and Hawker, L. E., *Ann. Bot.*, 50, 325–343, 1936.
372.   Romanowski, R. D., Kuc, J., and Quakenbush, F. W., *Phytopathology*, 52, 1259–1263, 1962.
373.   Dhanvantari, B. N., *Can. J. Bot.*, 45, 1525–1543, 1967.
374.   Bonner, J. and Addicott, F., *Bot. Gaz.* (Chicago), 99, 144–170, 1937.
375.   Kilpatrick, R. A. and Johnson, H. W., *Phytopathology*, 46, 180–181, 1956.
376.   Chen, D. W. and Zentmyer, G. A., *Mycologia*, 62, 397–402, 1970.
377.   Coons, G. H. and Strong, M. C., *Mich. Agric. Exp. Stn. Bull.*, 115, 1931.
378.   Bhalla, H. S. and Mitchell, J. E., *Phytopathology*, 60, 1010–1012, 1970.
379.   Mitchell, J. E., personal communication, 1977.
380.   Anon., *Difco Manual*, Difco Laboratories, Detroit, 1953.
381.   Bruehl, G. W. and Lai, P., *Phytopathology*, 56, 766–768, 1966.
382.   Tsao, P. H., Tummakate, A., and Bhavakul, K., *Trans. Br. Mycol. Soc.*, 66, 557–558, 1976.
383.   Mircetich, S. M. and Matherton, M. E., *Phytopathology*, 66, 549–558, 1976.
384.   Thom, C. and Church, M. B., *The Aspergilli*, Williams & Wilkens, Baltimore, 1926.
385.   Dox, A. W., *U.S. Dep. Agric. Bur. Anim. Ind. Bull.*, 120, 1910.
386.   Anon., *The Oxoid Manual of Culture Media, Ingredients and other Laboratory Services*, Oxoid Ltd., London, 1971.
387.   Thom, C., *U.S. Dep. Agric. Bur. Anim. Ind. Bull.*, 118, 1910.
388.   Taplin, D., Zaias, N., Rebell, G., and Blank, H., *Arch. Dermatol.*, 99, 203–209, 1969.
389.   Elliott, J. A., *Am. J. Bot.*, 4, 439–476, 1917.
390.   Srivastava, D. N., Echandi, E., and Walker, J. C., *Phytopathology*, 49, 145–148, 1959.
391.   Ingold, C. T. and Dring, V. J., *Ann. Bot.*, 21, 465–477, 1957.
392.   Foudin, A. S. and Wynn, W. K., *Phytopathology*, 62, 1032–1040, 1972.
393.   Backman, P. A. and Rodriguez-Kabana, R. R., *Phytopathology*, 66, 234–236, 1976.
394.   Davidson, R. W., Campbell, W. A., and Blaisdell, D. J., *J. Agric. Res.*, 57, 683–695, 1938.
395.   Noble, M. K., *Can. J. Bot.*, 43, 1097–1139, 1965.
396.   Garraway, M. O., personal communication, 1977.
397.   Brown, W., *Ann. Bot.*, 39, 373–408, 1925.
398.   Lilly, V. G. and Barnett, H. L., *Physiology of the Fungi*, McGraw-Hill, New York, 1951.

399. Binder, F. L. and Lilly, V. G., *Mycologia*, 67, 1025–1031, 1975.
400. Goos, R. D. and Tschirsch, M., *Mycologia*, 54, 353–367, 1962.
401. Cooke, W. B., *A Laboratory Guide to Fungi in Polluted Waters, Sewage, and Sewage Treatment Systems*, Publ. No. 999-Wp-1, U.S. Public Health Service, Cincinnati, 1963.
402. Kerr, N. S. and Sussman, M. J., *Gen. Microbiol.*, 19, 173–177, 1958.
403. Papavizas, G. C. and Davey, C. B., *Soil Sci.*, 88, 112–117, 1959.
404. Wiese, M. V. and Ravenscroft, A. V., *Phytopathology*, 63, 1198–1201, 1973.
405. Kulik, M. M., *Phytopathology*, 65, 1325–1326, 1975.
406. Harrold, E. E., *Ann. Bot.*, 14, 127–147, 1950.
407. Raper, K. B., *J. Agric. Res.*, 55, 289–316, 1937.
408. Kuhlman, E. G., *Mycologia*, 64, 325–341, 1972.
409. Hoagland, D. R. and Aron, D. I., *Calif. Agric. Exp. Stn. Circ.*, 347, 1950.
410. Backus, M. P. and Stauffer, J. F., *Mycologia*, 47, 429–463, 1955.
411. Grosklags, J. H. and Swift, M. E., *Mycologia*, 49, 305–317, 1957.
412. Johnson, S. P. and Joham, H. E., *Plant Dis. Rep.*, 38, 602–606, 1954.
413. Watkins, G. M., *Phytopathology*, 51, 110–113, 1961.
414. Alexopoulos, C. J., *Am. J. Bot.*, 47, 37–43, 1960.
415. Littman, M. L., *Science*, 106, 109–111, 1947.
416. Singer, R., *The Agaricales*, Hafner, New York, 1962, 17.
417. Blakeslee, A. F., *Phytopathology*, 5, 68–69, 1915.
418. Groves, J. W. and Skolko, A. J., *Can. J. Res.*, 22C, 190–199, 1944.
419. Benjamin, R. K., *Aliso*, 4, 321–433, 1959.
420. Kotheimer, J. B. and Christensen, C. M., *Wallerstein Lab. Commun.*, 24, 21–27, 1961.
421. Shigo, A. L., *Plant Dis. Rep.*, 47, 820–823, 1963.
422. Timnick, M. B., Barnett, H. L., and Lilly, V. G., *Mycologia*, 44, 141–149, 1952.
423. Martin, J. P., *Soil Sci.*, 69, 215–232, 1950.
424. Tsao, P. H., *Phytopathology*, 54, 549–555, 1964.
425. Singh, R. S. and Mitchell, J. E., *Phytopathology*, 51, 440–444, 1961.
426. Matsushima, T., *Trans. Mycol. Soc. Jpn.*, 2, 118–120, 1961.
427. Marx, D. H., *Phytopathology*, 59, 153–163, 1969.
428. Tolmsoff, W. J., *Phytopathology*, 49, 113–114(Abstr.), 1959.
429. Huang, H. C. and Tinline, R. D., *Can. J. Bot.*, 54, 1344–1354, 1976.
430. Anon., *Difco Supplementary Literature*, Difco Laboratories, Detroit, 1968.
431. Anon., *BBL Manual of Products and Laboratory Procedure*, Baltimore Biological Laboratory, Cockeysville, Md., 1973.
432. Georg, L. K., Agello, L., and Gordon, M. A., *Science*, 114, 387–389, 1951.
433. Mathur, R. S., Barnett, H. L., and Lilly, V. G., *Phytopathology*, 40, 104–114, 1960.
434. Norkrans, B., *Physiol. Plant.*, 16, 11–19, 1963.
435. Gooding, G. V. and Lucas, G. B., *Phytopathology*, 49, 277–281, 1959.
436. Williams, L. E. and Schmitthenner, A. C., *Ohio Agric. Exp. Stn. Res. Circ.*, 39, 1956.
437. Olive, L. S., *Am. J. Bot.*, 37, 757–763, 1950.
438. Park, D., *Trans. Br. Mycol. Soc.*, 44, 377–390, 1961.
439. McCallan, S. E. A., Wellman, R. H., and Wilcoxon, F., *Boyce Thompson Inst. Plant Res. Contrib.*, 12, 49–77, 1941.
440. Hunt, R. S., personal communication, 1977.
441. Stokes, J. L., Gunness, M., and Foster, J. W., *J. Bacteriol.*, 47, 293–299, 1944.
442. Banihashemi, Z. and de Zeeuw, D. J., *Plant Dis. Rep.*, 53, 589–591, 1969.
443. Steiner, G. W. and Watson, R. D., *Phytopathology*, 55, 728–730, 1965.
444. Hornby, D. and Ullstrup, A. J., *Phytopathology*, 57, 76–82, 1967.
445. Knox-Davies, P., personal communication, 1977.
446. Moyer, A. J. and Coghill, R. D., *J. Bacteriol.*, 51, 79–93, 1946.
447. Cantino, E. C., *Mycologia*, 48, 225–240, 1961.
448. Tucker, C. M., *Mo. Univ. Res. Bull.*, 153, 1931.
449. Waterhouse, G., Mycological Paper, No. 47, Commonwealth Mycological Institute, Kew, Surrey, England, 1954.
450. Daniel, J. W. and Rusch, H. P., *J. Gen. Microbiol.*, 25, 47–59, 1961.
451. Chin, B. and Bernstein, I. A., *J. Bacteriol.*, 96, 330–337, 1968.
452. Sobers, K. K. and Seymour, C. P., *Phytopathology*, 53, 1443–1446, 1963.
453. Langeron, M., *Precis de Mycologie*, Chronica Botanica, Waltham, Mass., 1945, 456–457.
454. Riker, A. J. and Riker, R. S., *Introduction to Research on Plant Diseases*, John S. Swift, St. Louis, 1936.
455. Moore-Landecker, E. J., personal communication, 1977.

456.   **Alexander, M. T.,** Media, in *The American Type Culture Collection, Catalogue of Strains I,* 12th ed., Hall, H. D., Ed., American Type Culture Collector, Rockville, Md., 1976, 394.

457.   **Raper, J. R.,** *Am. J. Bot.,* 26, 639–650, 1939.

458.   **Ciegler, A.,** *Can. J. Microbiol.,* 18, 631–636, 1972.

459.   **Bentley, R. and Keil, J. G.,** *J. Biol. Chem.,* 237, 867–873, 1962.

460.   **Reynolds, D. M.,** *J. Gen. Microbiol.,* 11, 150–159, 1954.

461.   **Latterell, F. M.,** personal communication, 1966.

462.   **Fahmy, T.,** *Phytopathology,* 13, 543–550, 1923.

463.   **Winstead, N. N. and Walker, J. C.,** *Phytopathology,* 44, 153–158, 1954.

464.   **Russell, P.,** *Nature,* 177, 1038–1039, 1956.

465.   **Emmons, C. W., Binford, C. H., and Utz, J. P.,** *Medical Mycology,* Lea & Febiger, Philadelphia, 1963, 347.

466.   **Rosenthal, S. A. and Furnari, D. J.,** *Invest. Dermatol.,* 28, 367–371, 1957.

467.   **Schmitthenner, A. F.,** *Phytopathology,* 62, 788, 1972.

468.   **Johnson, T. W., Jr. and Sparrow, F. K., Jr.,** *Fungi in Oceans and Estuaries,* Hafner, New York, 1961, 26–27.

469.   **Fuller, M. S., Fowles, B. E., and McLaughlin, D. J.,** *Mycologia,* 56, 745–756, 1964.

470.   **Flentje, N. T.,** *Trans. Br. Mycol. Soc.,* 39, 343–356, 1956.

471.   **Miller, J. J., Peers, D. J., and Neal, R. W.,** *Can. J. Bot.,* 29, 26–31, 1951.

472.   **Ajello, L. and Cheng, S.,** *Mycologia,* 59, 689–697, 1967.

473.   **Sonneborn, D.,** personal communication, 1977.

474.   **Shoemaker, R. A.,** *Can. J. Bot.,* 40, 809–836, 1962.

475.   **Diener, U. L. and Davis, N. D.,** *Phytopathology,* 56, 1390–1393, 1966.

476.   **Davis, N. D., Sansing, G. A., Ellenbury, T. V., and Diener, U. L.,** *Appl. Microbiol.,* 23, 433–435, 1972.

477.   **Hesseltine, C. W.,** *Mycologia,* 46, 358–366, 1954.

478.   **Talboy, P. W.,** *Plant Pathol.,* 9, 57–58, 1960.

479.   **Tochinai, Y.,** *J. Coll. Agric. Hokkaido Imp. Univ.,* 14, 171–236, 1916.

480.   **Philpot, C.,** *Sabouraudia,* 5, 189–193, 1967.

481.   **Miller, P. M.,** *Phytopathology,* 45, 461–462, 1955.

482.   **Romero, S. and Gallegly, M. E.,** *Phytopathology,* 53, 899–903, 1963.

483.   **Erwin, D. C. and McCormick, W. H.,** *Mycologia,* 63, 972–977, 1971.

484.   **McCain, A. H., Holtzmann, O. V., and Trujillo, E. E.,** *Phytopathology,* 57, 1134–1135, 1967.

485.   **McClellan, W. D.,** personal communication, 1977.

486.   **Hylin, J. W., Hylin, V., and Aragaki, M.,** *Mycologia,* 62, 200–202, 1970.

487.   **Vogel, H. J.,** *Microbiol. Genet. Bull.,* 13, 42–43, 1956.

488.   **Warcup, J. H.,** *Nature,* 166, 117–118, 1950.

489.   **Spinivasan, M. C., Chidambaram, P., Mathur, S. B., and Neergaard, P.,** *Trans. Br. Mycol. Soc.,* 56, 31–35, 1971.

490.   **Sharp, E. L., Sally, B. K., and McNeal, F. H.,** *Plant Dis. Rep.,* 60, 135–138, 1976.

491.   **Koike, H.,** *Phytopathology,* 55, 317–319, 1965.

492.   **Emerson, R.,** *Lloydia,* 4, 77–144, 1941.

493.   **Beneke, E. S.,** personal communication, 1977.

494.   **Ribeiro, O. K.,** personal communication, 1977.

CULTURE MEDIA (NATURAL AND SYNTHETIC): LICHENS

Vernon Ahmadjian

It has not been possible to grow lichens on the synthetic media that are used to culture plants because the association of fungus and alga that makes up a lichen is delicately balanced, and if one varies the normal environmental conditions to which lichens are adapted, the association breaks down. For example, under sustained conditions of moisture which the lichen would encounter on a culture medium, or on media that provide mineral or organic nutrients, the symbionts outgrow each other and eventually the disorganized thallus becomes contaminated with mold. Lichen thalli have been cultured on a variety of natural or natural-like media under conditions that simulate their normal environment. The fungal and algal symbionts will grow on synthetic media in axenic culture, but their growth rates are much slower than those of most free-living algae and fungi.

Constituent	Concentration (mg/liter)
Lichen Thalli	
Agar (purified)[1]	20,000
Filter paper (over sand)[8]	—
Sand[7]	—
Silica gel[6]	—
Soil[4,5,9]	—
Stone[3]	—
Wood[2]	—
Lichen Fungi[a3]	
Glucose	10,000
Asparagine	2,000
KH_2PO_4	1,000
$Mg \cdot SO_4 \cdot 7H_2O$	500
$Fe(NO_3)_3 \cdot 9H_2O$	0.2
$ZnSO_4 \cdot 7H_2O$	0.2
$MnSO_4 \cdot 4H_2O$	0.1
Thiamine	100 μg
Biotin	5 μg
Lichen Fungi[3]	
Malt extract	20,000
Yeast extract	2,000
Agar	20,000

Constituent	Concentration (mg/liter)
Lichen Algae[b][3]	
$NaNO_3$	250
$CaCl_2$	25
$MgSO_4 \cdot 7H_2O$	75
K_2HPO_4	75
KH_2PO_4	175
$NaCl$	25
H_3BO_3	11.4
$FeSO_4 \cdot 7H_2O$	5
$ZnSO_4 \cdot 7H_2O$	8.8
$MnCl_2 \cdot 4H_2O$	1.4
MoO_3	0.7
$CuSO_4 \cdot 5H_2O$	1.6
$Co(NO_3)_2 \cdot 6H_2O$	0.5
EDTA[c]	50
Peptone[d]	10,000
Glucose	20,000

[a] Lilly and Barnett's medium; for solid medium add 20,000 mg agar/liter.
[b] Bold's mineral medium plus glucose and N source.
[c] EDTA dissolved and neutralized with KOH.
[d] Other good N sources are asparagine and casamino acids.

REFERENCES

1. **Ahmadjian, V.,** *Am. J. Bot.,* 49, 277–283, 1962.
2. **Ahmadjian, V.,** *Science,* 151, 199–201, 1966.
3. **Ahmadjian, V.,** *The Lichen Symbiosis,* Blaisdell/Ginn, Lexington, Mass., 1967.
4. **Ahmadjian, V. and Heikkila, H.,** *Lichenologist,* 4, 259–267, 1970.
5. **Dibben, M. J.,** *Lichenologist,* 5, 1–10, 1971.
6. **Galun, M., Marton, K., and Behr, L.,** *Arch. Mikrobiol.,* 83, 189–192, 1972.
7. **Kershaw, K. A. and Millbank, J. W.,** *Lichenologist,* 4, 83–87, 1969.
8. **Kershaw, K. A. and Millbank, J. W.,** *Lichenologist,* 4, 214–217, 1970.
9. **Wetmore, C. M.,** *Ann. Mo. Bot. Gard.,* 57, 158–209, 1971.

DIETS AND CULTURE MEDIA FOR PROTOZOA

D. McLaughlin

INTRODUCTION

The diets of free-living protozoa range in complexity from inorganic salts dissolved in water to living organisms such as bacteria, other protozoa, fungi, and complex organic nutrients. Parasitic species depend on the body fluids or tissue cells of the host, intestinal microorganisms, or nonliving species ingested by the host.

Two general methods have been followed in the study of food requirements. One procedure involved establishing the protist in an axenic crude medium. In this case, growth is often minimal and the organisms are only maintained in successive transfers. In the other general procedure, the crude medium is either refined or replaced by a semi- or chemically defined medium which will support maximum growth. In this instance, any constituent which cannot be omitted is considered essential. Both procedures require the sterilization of the organism, and the particular axenizing technic selected requires rigorous tests for sterility.

Recent advances in protistan nutrition and axenic methods are correlated. Kidder[1] showed that knowledge of protozoan diets accelerated after "pure-mixed" methods of culture paved the way to "sterilization" of hymenostomes. Axenic cultivation of *Tetrahymena* and parasitic flagellates became routine after Kidder and Dewey[2] reviewed the nutritional requirements of ciliates in pure culture, and Lwoff[3] showed that certain leishmanias or trypanosomes could be maintained indefinitely in a peptone medium containing agar, horse serum, hematin, and ascorbic acid. Imbalances in culture media could be detected effectively after Hutner[4] explained the "accordian" design. Other papers (not exclusively nutritional) which support the point are the following: Hutner and Provasoli,[5] von Brand,[6] Seaman,[7] Soldo and van Wagtendonk,[8] and Loefer and Scherbaum.[9] These investigations were concerned with details ranging from the nitrogen requirements of phytoflagellates and ciliates to the metabolism of various macromolecules.

As there are more than 50,000 known species of protozoa, the subject encompassed by the title of this article would swell several soon-to-be outdated volumes. To keep matters manageable, this review will be restricted to selected aspects of investigations on in vitro axenic cultivation of the four major groups: the amebas or sarcodinids, the flagellates (both pigmented and colorless), the ciliates, and the parasitic sporozoa. No attempt will be made to delineate the specific inorganic or organic nutritional requirements of any special group or species. As a consequence, the tables listed subsequently should only serve as a guide to the periodic literature. Therefore, it will be necessary to refer to the original work to confirm the validity of each study or to duplicate the experimental conditions.

DIETS OF SARCODINIDS

The results of early attempts to establish free-living and parasitic ameba in axenic culture have been reviewed.[10,59] Growth on solid surfaces proved distinctly superior to that of fluids, but either free-living or heat-killed bacteria were necessary for survival.[59,61,64] The first significant breakthrough in the axenic cultivation of free-living sarcodinids was provided by Reich's (1933) successful establishment of *Mayorella palestinensis* in axenic culture. A fluid medium containing 1% peptone plus 1% glucose supported growth, but he was unable to replace peptone with amino acids, ammonium acetate, or ammonium lactate. In 1953, Storm and Hutner developed a chemically defined medium for *Hartmannella rhysodes* and *Acanthamoeba castellanii.*[59] As a result of this bold scholarly achievement and the contributions of other researchers,[59] requirements for inorganic ions, nitrogen, carbohydrates, vitamins, and analogous factors have been established for sarcodinids as well as other free-living protozoa.[60]

The nutritional requirements of parasitic amebas appear to be quite different. They are complicated initially by the fact that only a limited number have been isolated in pure culture,[11,13-15,17,18,20] and most of these required serum for growth.[62,65] Culture media for these parasitic and coprozoic organisms ranged from simple salt solutions[19] to the TYP + *E. coli*[18] and the serum-enriched PYG media.[21] Table 1 contains a partial list of the various test media used for the cultivation of sarcodinids.

DIETS OF FLAGELLATES

The obligatory or preferential nutritional requirements which characterize the various categories of Mastigophora have been the subject of numerous studies, including those of Hutner and Provasoli,[5,66] Lwoff,[3] von Brand,[6] Shorb,[67] Guttman and Wallace,[68] and Hall.[60] The observations of Hutner and Provasoli concern the food requirements of free-living, parasitic, and phytoflagellates; those of Lwoff are on the nutrition of parasitic flagellates (Trypanosomidae, Trichomonadinae); those of Gutman and Wallace and von Brand are on Trypanosomidae and Bodinidae; and those of Hall are on flagellates, amebas, and several ciliates.

Although a copious quantity of literature is available on the diets of flagellates, many protozoologists have been disinclined to include all members of this group in the phylum Protozoa. For example, the phytoflagellates occupy a systematic position at intersections of plant and animal lines of descent. Consequently, much attention has been directed to the composition and synthesis of starch by flagellates,[5,66,69] organic carbon utilization in light and dark,[70,73] carbohydrate-free media,[71] nitrogen distribution, nutrition and metabolism,[72] growth factors,[60,74] and inorganic nutrition.[75,76]

This review contends that unusual attention to animal or plant-like characteristics of a species courts delay in the study of protozoan nutrition. Therefore, information on chlorophyll- and nonchlorophyll-bearing species is included in Table 2.

DIETS OF CILIATES

For a number of reasons, investigations into the nutritional requirements of ciliates are confined to a relatively small number of species. Probably the most important obstacle has been the unavailability of pure cultures.[1,218] The first decisive step to alleviate this problem was taken when Lwoff (1923) isolated *Tetrahymena (Glaucoma) pyriformis* in a peptone medium and Glaser and Coria (1933) cultivated *Paramecium caudatum* in a suspension of heat-killed yeast cells, liver extract, and pieces of rabbit kidney.[74] These observations furnished a timely stimulus; consequently, within a few years, strains of *Tetrahymena, Paramecium, Glaucoma, Colpoda, Colpidium* and *Stylonychia* were isolated by other workers. With a few exceptions, culture media for these protozoa included solutions of commercial peptones, yeast extract or yeast autolysates supplemented with organic salts or other crudes as water extract of baked lettuce, cerophyl wheat, and orange juice plus heat-killed or live bacteria.[2,180,236]

The next progressive step was obtained when Kline developed a chemically defined culture medium for *Colpidium striatum.*[219] However, Kline's strain was lost and several other strains failed to grow in this medium. *Paramecium bursaria* had been maintained in a peptone medium, but only *P. aurelia*[220] and *P. multimicronucleatum*[221,222] were cultivable in a medium that was chemically defined.

More revealing results were obtained with *T. geleii W.*[2,223,224] The simplest medium developed for this protist contained 11 amino acids, glucose, 11 vitamins, several inorganic salts, and plant and tissue extracts. These substrates were generally combined with certain unidentified growth factors containing vitamins and several minerals of natural origin. Subsequently, the development of defined media for the cultivation of other species of *Tetrahymena, Paramecium,* and *Glaucoma* supervened.[2,182,225]

Since 1955 more refined axenic media and improved axenic techniques have disclosed additional areas of inexplicable ciliate eccentricities. As a result, the attention of investigators in the area of diets of free-living ciliates has been maintained, and many problems of nutrition have been illuminated.[60,181-186,230,265,292,297]

A second difficulty has been a lack of adequate means of separation and characterization of the parasitic and free-living members of this group. Ciliates of the genera *Anophrys, Colpidium, Colpoda, Metopus,* and *Uronema* contain mostly free-living species, but some members are intestinal parasites of sea urchins. Even such genera as *Balantidium* and *Nyctotherus,* which include parasites only, cannot be distinguished from free-living species by morphological criteria. Obviously, the initial stages in the development of endoparasitism in ciliates do not demand appreciable changes in structure.[298]

For these reasons, the information listed in Table 3 will be limited to the six genera mentioned initially, i.e., the Peritrichs and a few other free-living ciliates and endo- and ectoparasites that have been cultivated in vitro.

DIETS OF SPOROZOA

Parasitism is ubiquitous among Protozoa. All of the four major groups contain representatives that have adopted this mode of life, but the classes Sporozoa and Cnidosporida are composed of parasites exclusively. Because the nutrition of amoeboid, flagellated, and ciliated parasites has already been considered, this section will deal solely with the nonmotile, spore-producing parasites.

Prior to 1950, much of the research on sporozoa was devoted to the antimetabolite approach to chemotherapy of malaria.[301-303] When the problem of determining the relationship between diet and in vitro cultivation was undertaken, it was through animal nutrition studies.[301] In 1950, Trager reported the results of efforts to cultivate *Plasmodium lophurae* extracellularly. A complex medium containing duck erythrocyte extract, yeast adenylic acid, cozymase, ATP, and sodium pyruvate supported survival and development for 2 days and the parasites remained normal for 3 days. He later (1952) demonstrated the need for L-malate and concentrates rich in coenzyme A to maintain 90 to 95% of the parasites.[302]

Much information on the nutrition of the malarial parasites has been accumulated during the last 25 years. The results of many of the investigations are discussed at considerable length in several review papers.[302-304,342] Table 4 contains information on selected aspects of malarial nutrition and the nutrition of other parasites included in sporozoa.

Table 1

SUMMARY OF GROWTH OF SARCODINIDS ON VARIOUS MEDIA

Composition	Species	Assessment	Ref.
Microorganisms as Food Source			
Agar + heat-killed *A. aerogenes*	*Acanthamoeba polyphaga, A. castellanii, Hartmannella vermiformis*	Growth of 248 strains; *A. polyphaga* dominant	23
Peptone-yeast extract (PYE) + GLC agar + *A. aerogenes*	*Naegleria gruberi*	PYE liver concentrate + chicken embryo extract or fetal calf serum required for axenic growth[a]	26
E. coli + agar + WB saline + trace elements	*H. vermiformis*	Serum-enriched (PYGS) medium replaced bacteria	12
Proteose peptone (PP) + *Tetrahymena, Colpidium,* or *Chilomonas* + wheat grains	*Amoeba proteus, A. discoides, Pelomyxa carolinensis*	Rapid growth increase with *Tetrahymena,* etc. as food source; useful for stock cultures	48, 62
Simple Media			
1.54×10^{-1} *M* NaCl + 3×10^{-2} *M* KCl + 2×10^{-3} *M* CaCl$_2$ + 2×10^{-2} *M* MgCl$_2$ + 1.1×10^{-2} *M* NaHCO$_3$, pH 5.6—7.7	*Acanthamoeba culbertsoni,*[b] *A. castellanii,*[c] *A. polyphaga,*[d] *Naegleria* sp.,[b]	0.89% NaCl supplement required for *Acanthamoeba*	19
1.0 m*M* Na acetate-2-[14]C + 3 m*M* Na acetate or 3 m*M* acetaldehyde	*Polytomella coeca*	Acetaldehyde	51
Riboflavin + L-glu + L-trp + L-ile + L-ser + L-pro	*Schizopyrenus russelli*	Encystment varied with riboflavin and amino acid concentration; riboflavin required for growth	31
4% Neopeptone® (Difco)	*Acanthamoeba* sp.	Doubling time, 100 hr[e]	57

a 21°C specified.
b 37°C specified.
c 20°C specified.
d No change in growth at 20 and 37°C.
e Aerated bottles specified for cultivation at 22°C.

Table 1 (continued)
SUMMARY OF GROWTH OF SARCODINIDS ON VARIOUS MEDIA

Composition	Species	Assessment	Ref.
Simple Media (continued)			
0.75% PP + 0.75% yeast extract (YE) + 1.5% glc (PGY)	*A. castellanii*	Population maximum 72 hr; encystment in PGY + 50 mM MgCl$_2$	24
2 × Trypticase soy broth	*H. culbertsoni*	4.5 gen after 7–8 days	37
2% Mycological peptone + 0.9% mal	*A. castellanii*	O$_2$ uptake with succinate, L-pro, DL-α-glycerophosphate, and glu	32
1.0% PP + 1.0% tryptone + 0.5% NaCl, pH 6.8	*H. culbertsoni*	Agar + MgCl + taurine required for trophozoite transformation	33
1.5% PP + 1.5% glc	*A. castellanii* (Neff)	Growth 5–7 days	46
Agar + 0.5% NaCl	*N. aerobia, N. gruberi, Didascalus thorntoni, S. russelli, S. atopus, H. rhysodes, H. culbertsoni, H. castellanii, H. astromyxis, H. palestinensis, H. glebae*	Variations in growth with species	50
Agar + 0.015 M MgCl$_2$ + 0.02 M taurine	*H. culbertsoni*	Encystment inhibited by 1 × 10^{-5} M mitomycin, 1 × 10^{-7} M cycloheximide, or 1 × 10^{-6} actinomycin D	52
1.5% Agar + 0.01% mal + 0.01% YE + distilled water and sea water	*A. griffini*	*Aerobacter aerogenes* required for growth	34
2.0% PP + 0.2% YE + 0.1 M glc + 4 mM MgSO$_4$·7H$_2$O + 0.4 mM CaCl$_2$ + 3.4 mM C$_6$H$_5$Na$_3$O$_7$·2H$_2$O + 50 μM Fe(NH$_4$)$_2$(SO$_4$)$_2$·6H$_2$O + 2.5 mM KH$_2$PO$_4$ + 2.5 mM Na$_2$HPO$_4$·7H$_2$O, pH 6.5 (PYG)	*Acanthamoeba* spp.	Growth 5–7 days	12

Table 1 (continued)

SUMMARY OF GROWTH OF SARCODINIDS ON VARIOUS MEDIA

Composition	Species	Assessment	Ref.
Simple Media (continued)			
PPG medium Replacement media (RM) RM 1: 0.1 M KCl + 0.2 M Tris buffer + 8 mM MgCl$_2$ + 0.4 mM CaCl$_2$ + 1 mM NaHCO$_3$ RM 2: 0.25 M NaCl + 65 mg% MgCl$_2$ + 4 mg% CaCl$_2$ RM 3: As RM-2, 0.1 M NaCl instead of 0.25 M RM 4: RM-2 + 0.2 M PO$_4$ buffer, pH 7.0 RM 5: RM-2 + 0.2 M PO$_4$ buffer, pH 8.0 RM 6: RM-2 + 0.2 M Tris buffer, pH 8.0	*Mayorella palestinensis* *H. castellanii*	Clumping in axenic media Maximum encystment when MgCl$_2$ alone replaced mycological peptone	54 55, 58
Complex Media			
9.0% (v/v) Calf serum (Bvs) + 0.1% glc + 0.1% liver infusion + 0.016 mM MgSO$_4$ + 0.027 mM CaCl$_2$ + 1.0 mM Na$_2$HPO$_4$ + 1.0 mM KH$_2$PO$_4$ + 2.0 mM NaCl	*N. aerobia*[b] *N. fowleri*[b]	Maintenance 6 years without animal passage	22
5.0% (v/v) Serum + 0.5% casein + 0.5% YE + 5.0 mM KH$_2$PO$_4$, pH 7.0	*Entamoeba histolytica* *E. invadens*	Mass encystment within 30 hr	16
10% (v/v) Bvs + H-4 base containing 1% PP + 0.5% YE + 1% liver extract (LE) + 0.03 M glc + 10 mM PO$_4$ buffer, pH 6.5	*N. gruberi*	Generation time 8 hr; hemin supplement for serum as growth requirement	25

Table 1 (continued)
SUMMARY OF GROWTH OF SARCODINIDS ON VARIOUS MEDIA

Composition	Species	Assessment	Ref.
Complex Media (continued)			
Whole egg slant + dilute horse serum (Ehs) minus rice starch (RS)[f]	*E. histolytica* *E. moshkovskii* *E. terrapinae* *E. knowlesi* *E. ranarum* *E. invadens*	Survival 12 weeks Survival 36 weeks Survival 24 weeks Survival 24 weeks Survival 25 weeks Survival 7 weeks	27
Eagle's basal medium + 10% inactivated (Bvs) + 0.2% trypticase	*H. glebae,* *A. castellanii,* *H. rhysodes*	24–96-hr incubation period at 32°C required for growth[g]	35
Neff's medium and Neff + Serum + PP	*Acanthamoeba* sp., *Naegleria* sp.	8–9-hr generation time; population maximum 2.5–3.5 × 10^6 cells/ml	38
Part I: Buffered salt concentrate (40 g NaCl, 50 g K$_2$HPO$_4$, 6 g KH$_2$PO$_4$/l distilled H$_2$O) + 0.7 g YE/180 ml distilled H$_2$O[h] Part II: Ehs or Bvs + *Clostridium perfringens* + RS	*E. invadens*	Growth of trophozoites and cysts linear with time	36
Cerebrospinal fluid (CSF)	*Vahlkampfia jugasa,* *A. castellanii,* *H. vermiformis*	Growth after 24 hr[i]	40
Nutrient broth TTY and TTY-5-CEEM$_{2.5}$ diphasic media (Diamond)	*E. histolytica*[j]	Incubation period of 72 hr; axenic growth 2½–7 years	28
Filtered peptone + glc (PPGF) + DMG medium	*A. rhysodes,* *A. castellanii*	PPGF induction medium for amitotic division	29, 30
Balamuth's medium	*N. gruberi*	Growth 3 days	12, 21

f 4°C specified for all species.
g 32°C specified for all species.
h Final medium obtained by mixing Part I and Part II at 30°C.
i 30°C specified for all species.
j 35.5°C specified.

Table 1 (continued)
SUMMARY OF GROWTH OF SARCODINIDS ON VARIOUS MEDIA

Composition	Species	Assessment	Ref.
	Complex Media (continued)		
TYG and CAS media	N. gruberi	Serum + TYG and CAS required for growth	39
Whole egg slant + Ringer's solution + modified Lock's solution overlays	E. histolytica	Low-protein and -vitamin diets aggravate lesions	41
4.0 g NaCl + 2.0 g YE + 1.6 g KH_2PO_4 + 3.2 g K_2HPO_4 (encystation medium) + TPS-1 growth medium	E. invadens	Serum required for encystation	42
Diamond's TPS-1 medium minus vitamin mixture 107 and TP excystation medium, pH 7.0	E. histolytica[b]	Movement affected adversely in trophozoites exposed to excystation medium	43, 49
B-D medium with Lock solution overlay containing 500 units penicillin	E. polecki	Population maximum after 4 days	44
Adam's medium + DM-2 amino acid mix + vitamins + trace metals + balanced salt sol + 20 mg/ml glc; diluted 1:1 w gelatin, albumin, etc. pH 6.8	H. castellanii	High-molecular weight fractions inhibit cell adhesion, motility, and cytokinesis	45
Modified Chalkey's medium containing 80 mg NaCl + 4 mg NaHCO₃ + 4 mg KCl + 8 mg CaCl₂·2H₂O + 1.6 mg Ca(H₂PO₄)₂	Amoeba proteus, A. discoides	Mg, Zn, Fe levels varied with species	47
Shaffer-Frye medium + penicillin (CLG modification)	E. histolytica	Growth 0—24 hr; glc replaced by mal	53, 56

Table 2
MEDIA FOR EXTENDED CULTIVATION OF FLAGELLATES

Species	Components	Assessment	Ref.
	Marine		
Crypthecodinium cohnii	Basal medium: NaCl, MgSO$_4$, Ca succinate, L-glu, beatine, DL-ala, glycerol, Na$_2$ guanylate, suc, K$_3$ citrate, L-arg, L-asn, thiamine, folic acid, biotin + trace metal mix containing: Mn, Zn, Mo, Cu, V, Co, B, Ni, Cr	Heavy growth with salicyhydroxamic acid supplement	76
	Agar slant medium containing: K$_3$ citrate, KH$_2$PO$_4$, MgSO$_4$, KCl, NaCl, CaCO$_3$, NH$_4$Cl, L-glu, DL-ala, beatine HCl, suc, glc, nitrilotriacetic acid, biotin, thiamine HCl + trace elements	Biphasic cultures required for long-lasting inocula	82
	AXM medium + thiamine + biotin without acetate	Polyunsaturated fatty acids all *cis* 4,7,10,13, 16,19-docosahexaenoic acid biosynthesized	151
Chroomonds salina[a]	Sea water + tris buffer (8.3 mM), pH 7.9—8.1 + 0.25 M glycerol + 2.5 mM nitrate + 0.25 mM Pi + 0.5 mM m-silicate + vitamins (1.34 nM B$_{12}$, 1.48 μM thiamine, 4.1 nM biotin) + chelated trace metal ions (21.8 μM EDTA, 10 μM Fe^{3+}, 5 μM Mn^{2+}, 2 μM Zn^{2+}, 1 μM MoO$_4{}^{2-}$, 0.1 μM Cu^{2+}, 0.05 μM CO^{2+})	Maintenance 50 days	88
Gonyaulax catenella	Modified Provasoli's ASP 7 medium + PI metals and vitamin mix	Mitosis complete after 4–6 hr	101
G. tamerensis	750.0 ml filtered sea water + 2.29 ml NTA mix (Solution A) + 6.0 ml PII metal mix (Solution B) + 0.1 ml vitamin mix (Solution C) + 243.0 ml deionized distilled water	Biostimulation with NTA	119
	Freshwater Phytoflagellates		
Chlamydomonas acidophilia, *C. reinhardi*, *C. sphagonophila*	AM medium containing MgSO$_4$, KH$_2$PO$_4$, NH$_4$NO$_3$, NaCl, Ca, nitrilotriacetic acid, Fe, thiamine HCl, B$_{12}$, cyclo acid + trace metal mix 49A	Liver fraction "L" and Fe required for growth	70
Prymnesium parvum	SM salt medium, Sm minus NaNO$_3$ (SM-N) + 6 mM met or ethionine	Ethionine or met sole nitrogen source	175

[a] 20°C specified with continuous magnetic stirring.

Table 2 (continued)
MEDIA FOR EXTENDED CULTIVATION OF FLAGELLATES

Species	Components	Assessment	Ref.
	Freshwater Phytoflagellates (continued)		
Ochromonas sp.[b]	0.1% liver extract (LE) + 0.1% Bactotryptone + 0.1% suc	Increased acid phosphatase activity during logarithmic growth	111
O. malhamensis	Powdered whole liver + *Aerobacter aerogenes* + Brewer's yeast + Torula yeast	Prolonged cultivation at 5—8°C (dark)	77
	Base (g/100 ml): tripticase (BBL) or N-Z amine (Sheffield) 0.2 + yeast extract (YE) (Difco) or yeast autolysate (NBC) 0.2 + liver 1:20 (NBC) 0.01 + suc 1.0 + glycerol 0.5 + agar 4.0, pH 6.4	Survival 29—35 days with glycerol supplement	80
	0.1% LE + 0.1% Bactotryptone + 0.1% suc	Increased acid phosphatase activity during logarithmic growth	111
	Johnson's medium + B_{12} minus trp	Growth enhanced by trp	93
	Hutner's medium[c] + *A. aerogenes* + polystyrene latex particles	Particles required for digestive activity	98
O. danica	Powdered whole liver + *A. aerogenes* + Brewer's yeast + Torula yeast	Prolong cultivation at 5—8°C (dark)	77
	0.1% LE + 0.1% Bactotryptone + 0.1% suc	Increased acid phosphatase activity during logarithmic growth	111
	Aaronson and Scher defined medium (AG) minus arg, his, and nitrilotriacetic acid	Growth inhibition by ethionine reversed by met	165, 172
O. minute	0.1% glc + 0.1% tryptone + 0.1% YE + 1.5% LE	5—10 × 10⁶ organisms/ml in mid—late log-phase cultures 2—3 days after inoculation	100
	0.1% LE + 0.1% Bactotryptone + 0.1% suc	Increased acid phosphatase activity during logarithmic growth	111
Polytomella agilis	Modified Cramer-Meyers medium with OAc as carbon source + vitamins B_1 and B_{12}	Decrease in population with thiamine supplement[c]	87
Euglena gracilis[d] (green and colorless)	Modified Cramer-Meyer medium (M-CM) with 0.1% *n*-hexanoic acid as sole carbon source	Beta oxidation occurred in the dark with ethanol	89
	Cramer-Meyer medium (CM) + acetate; Lyman-Siegelman medium + CO_2; High salt medium + glu; Minimal culture medium + glu	Glu major free amino acid in cells grown in four different media	103
	Inorganic medium + OAc or glc	CO_2 required for catalase activity	120

b 18 to 20°C specified.
c 25°C specified.
d 28°C specified; maintained in the dark.

Table 2 (continued)
MEDIA FOR EXTENDED CULTIVATION OF FLAGELLATES

Freshwater Phytoflagellates (continued)

Species	Components	Assessment	Ref.
Euglena gracilis[d] (green and colorless) (continued)	CM medium, pH 3.0 and 8.1[e]	Variation in chloroplast structure with pH	135, 147
	Cambridge medium + 15 g glc + 3 g L-glu + 2 g DL-asp + 1 g DL-malate + 2.5 g gly + 0.47 g suc + distilled water to make 1.01	Early stationary phase after 5 days; population maximum 10^8 cells	136
	Edmund's modification of CM medium	Synchronization achieved by alternating periods of light (14 hr, 3500 lx) and darkness (10 hr)	145, 146
	CM medium + OAc, butyrate, glc, glu, pyruvate, succinate, butanol, ethanol, fumarate	Growth supported by butanol, ethanol, or fumarate	148
	Hunter's Medium A[c]	Chloramphenicol inhibits synthesis of chloroplastic 70s ribosomes	149
	M-CM medium with ferric citrate in place of ferrous sulfate	Highest hexokinase level in glc-grown cells	170
	CM medium + glc	Mutant produced with N-methyl-N-nitroso-*p*-toluenesulfonamide (MNTS)	171
	CM medium + 0.5% glc + 0.1% gly	Growth with glc as carbon source; gly decreased lag period	177
	5000 mg Na acetate + 200 mg $MgSO_4 \cdot 7H_2O$ + 645 mg Na_3 citrate$\cdot 2H_2O$ + 26.5 mg $CaCl_2 \cdot 2H_2O$ + 3.0 mg $Fe_2(SO_4)_3 \cdot x$ H_2O + 1.8 mg $MnCl_2 \cdot 4H_2O$ + 1.3 mg $CoCl_2 \cdot 6H_2O$ + 0.4 mg $ZnSO_4 \cdot 7H_2O$ + 0.2 mg $Na_2MoO_4 \cdot 2H_2O$ + 0.02 mg $CuSO_4 \cdot 5H_2O$ + 0.02 mg thiamine HCl + 0.01 mg B_{12}	Maximum cell density with 4—5 mg/ml inorganic phosphate	179
Volvox aureus	Darden and Sayers medium	Alternating light and dark cycle required for growth	117
Astrephomene gubernaculifera	MV medium, BT defined medium	Vitamin B_{12} required for growth	166

e $22°C$ specified; cultures bubbled with 5% CO_2, 95% air.

Table 2 (continued)
MEDIA FOR EXTENDED CULTIVATION OF FLAGELLATES

Species	Components	Assessment	Ref.
Freshwater Nonchlorophyll-bearing Species			
Monas	0.1–0.2% Cerophyl infusion + trace salts + *Escherichia coli*	Log-phase cultures of 1.5×10^7 organisms/ml after 2 or 3 days	100
M. sociabilis[b]	0.1% Bactotryptone + 0.1% suc + 0.1% LE	Acid phosphatase activity at pH 5.1	111
Parasitic			
Leptomonas *Crithidia fasciculata*	Stock culture medium containing (g/liter) liver infusion (oxoid), 5; proteose peptone, 7.5; brain heart infusion (BBL), 7.5; NaCl, 5; KCl, 2; ascorbic acid, 0.2; NaH_2PO_4 (anhydrous), 0.5; $MgSO_4 \cdot 7H_2O$, 0.5; suc, 2.5; morpholino-propane sulfonic acid, 2.5; and hemin (sigma "equine, type III"), 14 mg in 50% aqueous (v/v) $1,1',1'',1'''$ (ethylenedinitrilo)-tetra-2-propanol (J. T. Baker Chemical Co., Quadrol)	Leptomonas more susceptible than *C. fasciculata* to phenanthridines and diamidines; transfers at 3-week intervals	78
	Defined medium containing suc, hemin, Tween® 80, $MgSO_4$, NaCl, 16 amino acids, thiamine, uracil, orotic acid, adenine, guanine, beatine HCl, ethanolamine HCl, Na acetate, Na_2 fumarate, ferulic acid, biotin, thiamine nitrate, cystamine dichloride, choline hydrogen citrate, carnitine HCl, biopterin, folic acid, 5 vitamins, KCl, $Fe(NH_4)_2(SO_4)_2 \cdot 6H_2O$, $CaCO_3$, $MgCO_3$, KH_2PO_4, cyclo acid, nitrilo-triacetic acid + metal mix		
C. fasciculata	Defined medium containing suc, hemin, Tween® 80, $MgCO_3$, 14 amino acids, citric acid, malic acid, succinic acid, Ca succinate, KH_2PO_4, uracil, thymidine, 6 vitamins, biotin + metal mix	Growth response to biopterin concentration	79
	Low osmotic basal medium[f] containing suc, hemin, 11 amino acids, 6 vitamins, adenosine, Ca succinate, $MgCO_3$, KH_2PO_4, succinic acid,	Effective osmotic support with glycerophosphate, sorbitol, mannitol, gly, arg, KCl, NaCl	81

f   32°C specified.

Table 2 (continued)
MEDIA FOR EXTENDED CULTIVATION OF FLAGELLATES

Species	Components	Assessment	Ref.
Parasitic (continued)			
C. fasciculata (continued)	malic acid, citric acid, K₃ citrate, biotin, biopterin + trace metals, pH 6.3—6.5	Glycerol effective alone; growth without succinate equaled CHO	71
	Carbohydrate-free medium containing arg + pro + glu as substrates, pH 6.9—7.5	NaCl, suc, glc, or mannitol required to prevent change in shape	143
	Kidder and Dutta's medium	Cultivation for 3 years	105
	Brain-heart infusion medium (BHI)ᵍ + 10 mg/ml hemin		
	Kidder and Dutta's Medium I minus thr, met, and Tween® supplemented with Triton WR 1339 + L-cys	Cys and folate spare requirement for met and thr	164
C. deanei	Defined medium containing suc, glycerophosphate, KCl, K₃ citrate, citrate, malate, succinate, MgCO₃, CaCO₃, Fe, (NH₄)₂(SO₄)₂·6H₂O, L-met, L-tyr, L-phe, L-trp, nicotinamide, folic acid, thiamine HCl, biotin + trace elements	Met, tyr, folic acid, thiamine, biotin, and nicotinamide required for growth; no requirement for hemin or purine source	92
Blastocrithidia culicis	CAP medium	Growth of flagellate + symbiont in both media; chloramphenicol eliminated symbiotes	85
C. oncopelti	Grace's medium (GM) + 10% fetal bovine serum (FBS) or Trager's defined medium (TM)		
	Blood broth (BB) containing GN neopeptone broth (5.5%) + rabbit blood lysate (BL) 1:6 in distilled water (1:1); TM + 0.25% LE (TM + L); TM + L + BL; TM + L + 10% BL; TM minus hemin (TM-H); TM-H + L; TM-H + 0.25% LE	Exogenous hemin and LE required for growth	144
Trichomonas gallinae	Cysteine-peptone-liver (CPL) medium + mal; CPL-glc; CPL-gal; CPL-mal; CPL glc-mal + 5% (v/v) inactivated human serum	Utilization of sugars correlated with CHO type and exposure time	86
	Cysteine-tryptose-maltoseʰ medium (CTLM) + 5% (v/v) inactivated normal horse serum, pH 7.2	Log-phase cells at 24 hr	139

g 27°C specified; rotated bottles.
h 37°C specified.

Table 2 (continued)
MEDIA FOR EXTENDED CULTIVATION OF FLAGELLATES

Species	Components	Assessment	Ref.
Parasitic (continued)			
Trichomonas gallinae (continued)	CPL + 20 m*M* glc or 10 m*M* mal + 5% (v/v) human serum	Diauxic growth in STS medium + mal	122
	CPE glc medium minus agar, pH 7.0	Survival dependent on temperature and cell age	123
	Diamond's medium for growth of colonies on semi-solid medium		
T. vaginalis	Diamond's medium	Growth 3–6 weeks	109, 125
	Modified Bushby and Capp medium + activated horse serum (4:1) + nystatin 50 mg/ml	Cultivation 30 days	124
	CTLM + 5% (v/v)[h] inactivated horse serum, pH 6.0		139
	Eagle's MEM medium, Medium 199, Diamond's medium minus agar, Sprince and Kupferberg medium	Maintenance in axenic media	162
	6.5 g NaCl + 5 g glc + 25 g LE + 1 l distilled water + 80 ml inactivated horse serum + 500,000 units streptomycin + 1,000,000 units penicillin, pH 6.2	Bushby and FMM media supported growth	163
	Modified Bushby's medium + 20% calf serum, pH 6.0 Feinberg-Whittington medium (FMM)		
	30 g tryptone soy broth + 20 g glc + 18 g LE + calcium pantothenate (0.5% sol) + 1 l distilled water		
Tritrichomonas foetus	Modified Plastridge medium + 50% sterile in-activated ox serum	Growth 5–14 months	126
Trichomitus (Trichomonas) batrachorum	TYM medium + 0.05% (w/v) agar	Pseudocyst in TYM minus agar	137
Trypanosoma cruzi,[i] *T. lewisi*	Yaeger's LIT liquid medium	Maintenance for 1 year; antigenic differences demonstrated for various culture forms	83, 91, 128
T. cruzi	Eagle's Minimum Essential Medium (1×) + Earle's salts + L-gln + 2.2 g NaHCO$_3$ + Gibco 100× vitamin solution (10 ml) + nucleotide mixture + fetal calf serum (FCS)	Growth cycle culminated with 90% metacyclic forms in stationary phase	84

i   28° C specified.

Table 2 (continued)
MEDIA FOR EXTENDED CULTIVATION OF FLAGELLATES

Species	Components	Assessment	Ref.
T. cruzi (continued)	Diphasic medium containing (% w/v) nutrient agar, 1.3; agar, 1.5; NaCl, 0.8; liquid phase (% w/v) BHI (Difco), 2.8; glc, 1.0	5-day incubation for experimental cultures[j]	94
	Diphasic medium (solid phase): 1.0 g agar (Bacto) + 38 g BHI (Difco) + H_2O (1 l); overlay: 5 g YE + 5 g glc + 5 g neopeptone + 5 g NaCl + 80 ml inactivated calf serum + 1 l H_2O	O_2 uptake with pyruvate and OAc[i]	134
	Baker's diphasic medium + L4N liquid medium Transfer medium: Tissue Culture Medium 199 + 4 mM L-gln + 5 IU/ml, each penicillin and streptomycin (pH 7.2—7.4) + newborn calf serum Tissue Culture Medium same as transfer medium, pH unadjusted	Growth 3—4 weeks in liquid medium[d] 90% Trypomastigotes in preparations[l]	152 153
	T. (S.) cruzi Standard Liquid Medium (R. Wilson, unpublished): same as transfer medium except FCS replaced newborn calf serum Tissue Culture: African green monkey kidney cells 0.85% phosphate buffered saline (PBS) + Dulbecco's balanced salt solution (BSS) containing 0.25% trypsin, pH 7.8	Variable lag period prior to the onset of parasite reproduction	154
	RPMI 1640 medium containing 5% inactivated FBS + 100 units/ml penicillin and 100 mg/ml streptomycin, pH 7.0 or Eagle's complete medium containing 5% inactivated horse serum + 2 mmol/ml L-gln + antibiotics HeLa cell line (PPLO-free) maintained on RPMI 1640 + FBS + antibiotics		
	Primary bovine embryo skin and skeletal muscle (BESSM) cell culture + Eagle's Minimum Essential Medium containing 5% inactivated horse serum + 2 mmol/ml gln + 100 IU/ml penicillin + 100 μg/ml streptomycin Secondary BESM + HeLa cell lines + RPMI 1640 medium containing 5% inactivated FBS + 100 units/ml penicillin + 100 μg/ml streptomycin	Penetration rate in BESM cultures decreased exponentially with time	155

j 26° C specified.

Table 2 (continued)
MEDIA FOR EXTENDED CULTIVATION OF FLAGELLATES

Species	Components	Assessment	Ref.
Parasitic (continued)			
T. cruzi (continued)	BHI + 15% citrated human blood	Arg taken up by mediated transport system	168
	Basal medium containing Tissue Culture Medium 199 (10 ml), 0.8% glc (25 ml), 5.0% aqueous trypticase solution (10 ml), 2.8% sodium bicarbonate (1.5 ml), distilled water (53.5 ml) + hemin and FBS (0.5 mg/ml), (F-29 medium) 85:10:5, F-32 medium: F-29 + chicken plasma 85:5:5:5	Sevenfold increase of amastigotes in F-29 and F-32 after 4-day incubation at 35.5°C	169
	RPMI medium[k] containing 5% heat-inactivated FBS + 100 USP units/ml penicillin + 100 μg/ml streptomycin, pH 7.0	7—9 generations in BESM cell type culture	129
	TC199 calf serum medium + mouse peritoneal macrophages	Binary fission of amastigotes after 3 days	130
T. rotatorium	NN medium, BHI medium, Rugai Leishmanial medium, Diamond and Hermans SNB-9 medium + tryptone instead of neopeptone Parker's 199 medium	Growth in all media	131
T. ranarum	Diamond's SNB-9, Nicolle's modified NN	Maintenance 21 days	131
T. mega	Bone and Steinert's medium, Guttman's defined medium	Growth enhanced with human RBC Maintenance 1 year	131
	Diphasic blood agar medium	Nondividing stationary phase 8—9 days in supplemented medium	114
T. lewisi	Dusanic Medium (modified)	Nondividing stationary phase 15 days after inoculation	114
T. inopinatum	Ponselle's medium	Maintenance 1 year	131
T. maganorum	Blood culture medium-trypticase soy agar (TSA) VIM and NNN media containing 0.6% NaCl + defibrinated rabbit blood + 100 units penicillin + 0.1 mg streptomycin	NNN medium recommended as medium of choice	156
T. dionisii	Modified 4N medium containing 15.0 g proteose peptone + 2.5 g LE + 5.0 g YE + 5.0 g NaCl/1 + rabbit serum + 10% erythrocyte lysate	Strains maintained for 365 days[l]	116
T. vivax	Eagle's MEM-GBI-F15 dehydrated culture medium 0.98 g + 0.22 g NaHCO$_3$ + PIPES buffer, 0.30 g + BES buffer, 0.32 g/100 ml redistilled water, pH 6.8—7.0 + 20% sheep serum	Extensive multiplication of epimastigotes[l]	118

k 35°C specified; 95% air + 5% CO$_2$.

Table 2 (continued)
MEDIA FOR EXTENDED CULTIVATION OF FLAGELLATES

Species	Components	Assessment	Ref.
Parasitic (continued)			
T. vivax (continued)	Blood culture medium-TSA	Survival and multiplication in NNN medium[h]	156
	VIM and NNN media containing 0.6% NaCl + defibrinated rabbit blood + 100 units penicillin + 0.1 mg streptomycin		
T. gambiense	TCM-199 medium + 20% calf serum + 50 units/ml penicillin + 50 μg/ml streptomycin	Attachment and ingestion of parasite enhanced by homologous immune serum	127
T. congolense	Trager's medium minus sheep serum	Maintenance 30 days[i]	158
T. bufophlebotomi	NNN medium	Growth at 23°C, but not at 15 and 30°C	174
T. brucei	Bohringer's modified HX25 medium of Cross and Manning	Encapsulation in medium minus agar	110
	3.33 g NaCl + 5.40 g NaHPO$_4$ + 0.60 g Na$_2$ HPO$_4$ + 0.2 g KCl + 3.60 g nutrient broth + 5 ml glycerol + 900 ml blood lysate (9:1 v/v) with distilled water + 100 ml calf serum, pH 7.2–7.4	Maintenance 9 months[i]	115
	Tobie's medium + proline + inactivated defibrinated horse serum in solid phase; liquid phase contained (g/liter) KCl, 0.4; Na$_2$ HPO$_4$ · 12H$_2$O, 0.06; KH$_2$ PO$_4$, 0.06; CaCl$_2$, 0.14; MgSO$_4$·7H$_2$O, 0.1; MgCl$_2$ ·6H$_2$O, 0.1; NaCl, 8.0; L-pro, 1.0; pH 7	Respiration inhibited by *m*-chlorobenzhydroxamic acid (*m*-CLAM)	138, 140
	HXA medium minus crude serum	Defatted serum albumin essential for growth	157
	Tissue Culture Medium 199 + vitamins + amino acids + salts and other compounds	Casein requirement circumvented by vitamins and amino acids	157
	Trager's medium minus sheep serum	Maintenance 30 days[i]	158
T. cyclops[c]	Blood agar slants + Eagle's basal medium + Hanks salts + 5% (v/v) FCS	Pigmented bodies in cells grown in Hb-containing medium	99
T. evotomys[j]	Monolayer of hamster kidney cells (BHK) + RPMI 1640 HEPES + 10% FCS	8–15-day initial period as trypomastigote	113
Leptomonas	Composition (g/liter): N-Z amine AS, 3.0; yeast hydrolysate, 5.0; cane sugar, 5.0; MgSO$_4$ ·7H$_2$O, 0.1; hemin, 0.001; pH 7.4	Succinate selected as standard respiratory substrate	102
L. pessoai[d]	Tryptose, 0.5%; YE, 0.7%; glc, 1.5%; hemin, 2.0 mg%; folic acid, 0.2 mg%; pH 7.0 (*Leptomonas*)	Live cultures of *L. pessoai* protected mice against *T. cruzi* infection	97
	LIT medium, pH 7.2 (*T. cruzi*)		
Leptomonas	HY medium + 5% (w/v) glycerol	20–30 subcultures	108
L. pessoai, *Herpetomonas muscarum*	Yeager's "LIT" medium, Roitman et al. complex medium, Warren's medium, McConnell's medium	Man, rabbit, or cow serum required for growth in each medium	132
C. luciliae	NNN medium, pH range 5.6–8		

Table 2 (continued)
MEDIA FOR EXTENDED CULTIVATION OF FLAGELLATES

Species	*Blastocrithida* Components	Assessment	Ref.
H. megaseliae	Mansour's medium	Cultivation >3 days	133
Leishmania tarentolae[s]	BHI, Difco, pH 7.9	Late log phase 3—4 days	90
	BHI, modified defined medium C(CA) containing 20 μg/ml adenine in place of the original purine/pyrimidine mix	Prolonged cultivation in each medium	106, 178
	Low-phosphate neopeptone yeast extract medium (Neo YE), pH 7.9		
	Trager's Defined Medium C	Sodium sulfate as sole sulfur source; met sulfoxide used in place of L-met	167
	Sulfur-free Medium C	Asp aminotransferase inhibited by α-ketoglutarate concentration $>1.68 \times 10^{-2}\ M$	173
	Blood agar + Locke's solution overlay		
Leishmania sp.	NNN medium containing 1.4% (w/v) nutrient agar (Difco) + 0.6% (w/v) NaCl + 10% (v/v) defibrinated rabbit blood + 50 μ/ml each penicillin and streptomycin	Best growth on surface of agar base	161
	Locke's solution overlay containing 0.8% NaCl + 0.02% KCl + 0.02% CaCl$_2$ + 0.03% KH$_2$PO$_4$ + 0.25% glc, pH 7.2—7.4		
L. donovani[c]	Tanabe's medium + hemolyzed rabbit blood + 1.1% (w/v) NaCl + 1.33% sodium citrate	Lag phase 0—2 days; log growth through day 6	96
	NIH medium	O$_2$ uptake stimulated by L-pro, L-glc, and L-arg	141
	Ray's solid medium[l]	^{14}C succinate from 1-^{14}C glc much greater than from 6-^{14}C glc	159
L. enriettii	Blood agar medium + Hank's overlay	Log growth after 72 hr	121
	Modified blood agar base + Locke's solution overlay	Morphological abnormality increased with increased radiation	160
L. tropica	NIH blood agar overlayed with an equal amount of Locke's solution	Cultivation for 5 days	142
	NIH blood agar slants[m] overlayed with Locke's solution	O$_2$ uptake identical with glc and dGlc	95
	Blood agar medium + Hank's overlay	Log growth after 72 hr	121

l 22°C specified.
m 26°C specified, with shaking at 140 rpm.

Table 2 (continued)
MEDIA FOR EXTENDED CULTIVATION OF FLAGELLATES

Species	Components	Assessment	Ref.
	Blastocrithida (continued)		
L. tropica (continued)	Monophasic cell-free medium containing (per liter) bactotryptose, 15.0 g; YE, 10.0 g; glc, 2.0 g; NaCl, 4.0 g; KCl, 0.4 g; disodium hydrogen phosphate, 8.0 g; rabbit serum, 200 ml; Hb sol, 20 ml; + glass-distilled water to total 1 l	Growth inhibition with dGlc	107
Mycoplasma spp.	500 USP units penicillin + 0.025% thallium acetate + 15% equine serum	Maintenance >7 days in 5% CO_2 + nitrogen	112
Trichomonas fetus	Cysteine-peptone-liver medium (CPLM)		
Campylobacter fetus	Thioglycollate medium		
Monocercomonas sp.	Diamond's medium minus agar + 5% (v/v) heat inactivated horse serum	Mass cultures within 4 days	104
Histomonas meleagridis	Dwyer's modification of Lesser's medium, pH 7.8	5% chick embryo extract + 10% sheep or horse serum required for growth	150, 176

Table 3

GROWTH MEDIA AND IN VITRO UPTAKE OF SPECIFIC BIOCHEMICALS BY CILIATES

Species	Medium	Assessment	Ref.
	Free-living		
Tetrahymena pyriformis	Dewey's medium minus OAc, glc, and Tween® 85	OAc spared amino acids for glyconeogenesis	188
	20.0 g Oxoid Bactopeptone + 5.40 g KH_2PO_4 + 1.0 mg calcium pantothenate + 1.0 mg pyridoxine hydrochloride + 0.2 mg biotin + 0.2 mg thioctic acid + 10.0 mg $MgSO_4 \cdot 7H_2O$ + 10.0 mg $CaCl_2 \cdot 6H_2O$ + 10.0 mg $CuSO_4 \cdot 5H_2O$ + 5.0 mg $FeCl_3 \cdot 6H_2O$ + 5.0 mg $MnSO_4 \cdot 4H_2O$ and 2.0 mg each gua, adenine, urd, and cyd per liter deionized water, pH 6.0	Mean generation time 2.80 hr; 2×10^6 organisms/ml	190
	2% Proteose peptone broth (PP) + ado, cyd, guo, and glc each at 1.5 mM	Purines and pyrimidines required for growth	191
	1% PP + 0.5% liver extract (LE), pH 6.5	CoASAc used for lipogenesis	192
		Incorporation of 1-^{14}C OAc into glu induced by tolbutamide	193
		AMP reduced the rates of the TCA cycle, glyconeogenesis, and pyruvate utilization[a]	194
	2% PP + 0.5% (YE) + 0.87% glc	Complete cessation of DNA synthesis in medium minus amino acids	195
	Dewey's 2A medium + Tween® 80 + glc, guanylic acid, 1, 0.5, and 0.015%, respectively minus amino acids		
	2% PP + 1% LE, inorganic medium (IM) of Hamburger and Zeuthen	$7-8$ µg/ml cytochalasin B inhibited vacuole formation	196
	PP + 0.3 mM or 3.0 mM Ca + $CaCl_2$; PP + 0.3 mM or 3.0 mM Ca + $SrCl_2$	0.3 and 3 mM Ca depressed the mean Ca/P ratio and increased the Mg/P ratio of granules	197
	Casein-YE medium with $CaCl_2$ and $SrCl_2$ as per above defined medium		
	20 g PP + 1 g LE + 100 mg $MgSO_4 \cdot 7H_2O$ + 25 mg Fe$(NH_4)_2(SO_4)_2 \cdot 6H_2O$ + 0.5 mg $MnCl_2 \cdot 4H_2O$ + 0.05 mg $ZnCl_2$ + 50 mg $CaCl_2 \cdot 2H_2O$ + 5 mg $CuCl_2 \cdot 2H_2O$ + 1.25 mg $FeCl_3 \cdot 6H_2O$ per liter	Circadian period demonstrated at low temperature[b]	198

a   27°C specified; cultures aerated with sterile air containing 0.03% (v/v) CO_2.
b   10°C specified.

Table 3 (continued)
GROWTH MEDIA AND IN VITRO UPTAKE OF SPECIFIC BIOCHEMICALS BY CILIATES

Species	Medium — Free-living (continued)	Assessment	Ref.
Tetrahymena pyriformis (continued)	2% PP + 0.2% YE	Ethidium bromide inhibited mtDNA synthesis 95%	199
	2% PP + dephosphorylated YE + 7.5 g $MgSO_4 \cdot 7H_2O$ per 500 ml water, pH 8.0	RNA synthesis initiated 90 min after adding enriched nutrient medium	200
	2% PP + 0.1% liver fraction (LF) Inorganic medium containing 50 mM NaCl + 1 mM $MgCl_2$ + 5 mM KCl + 25 mM $NaHCO_3$ + 1 μM NaH_2PO_4, pH 7.4 PP + particles (ferric or aluminium hydroxides, heat-denatured egg albumin, polystyrene beads, or particles from heat sterilized PP)	40-hr generation time in particle-free medium; 6 hr in PP + particles	201, 217
	2% PP + cytochalasin, 5 mg/ml	Cytochalasin B reduced the rate of food vacuole formation	202
	2% PP + 0.1% LE + salts[c] IM medium of Hamburger and Zeuthen (starvation medium)	Dimethyl sulfoxide affected vacuole formation	204
	Levy-Hunt medium + 2 μg/ml $FeCl_3$	Specific activity of pyruvate kinase and lactate dehydrogenase increased in glc-supplemented medium	205
	2% PP + 0.1% YE + inorganic salts + penicillin + streptomycin (PPY + PS medium) PPY + PS + 10 μg/ml cycloheximide (CHX-10) PPY + PS + 30 μg/ml cycloheximide (CHX-30)	Evidence of macronuclear location of somatic mutations	206
	2% PP + 0.1% LE	Changes in nuclear acidic proteins correlated with DNA synthesis and cell division	207
	Inorganic buffer (starvation medium) 2% PP Dryl's Ca^{2+} free salt solution (starvation medium)	Trace amounts of Ca^{2+} required for conjugation	208
	Lipid-free synthetic medium[d] supplemented with 3 mg% (w/v) synthetic L-α-dipalmitoyl phosphatidyl choline + 3 mg% L-α-dipalmitoyl phosphatidyl ethanolamine	Lipids required for growth; structural changes induced by growth on saturated phospholipids at 40.1°C	209

c 28°C specified.
d 35°C specified.

Table 3 (continued)
GROWTH MEDIA AND IN VITRO UPTAKE OF SPECIFIC BIOCHEMICALS BY CILIATES

Species	Medium	Assessment	Ref.
	Free-living (continued)		
Tetrahymena pyriformis (continued)	PPY medium Rasmussen and Orias' medium Rasmussen and Orias' medium	10^6 cells/ml produced 10^7 cells/ml produced; higher yields in each by excess aeration	210
	Defined medium with uracil as pyrimidine source	Induction of TMP synthetase activity depended upon growth conditions[e]	211
	2.0% PP + 0.2% dex + 0.1 YE + 0.003% sequestrine	$5-7 \times 10^5$ cells/ml after 16–18 hr	213
	1% PP Dryl's salt solution (starvation medium)	Starvation produced pair formation and a series of nuclear events associated with conjugation	214
	10 g Peptone + 5 g glc + 2.5 g YE[c] + 4 drops aqueous silicone emulsion (BDH) + 50 mg $CaCl_2 \cdot 2H_2O$ + 5 mg $CuCl_2 \cdot 2H_2O$ + 100 mg $MgSO_4 \cdot 7H_2O$ + 25 mg $Fe(NH_4)_2 SO_4 \cdot 6H_2O$ + 0.5 mg $MnSO_4 \cdot 4H_2O$ + 0.05 mg $ZnSO_4$ per liter, pH 7.0	Behavior of contractile vacuole determined in axenic medium	226
	1% PP + 0.05% LE[f] + 0.02 M potassium phosphate buffer, pH 6.5	Growth inhibited by 4-pentenoic acid; OAc reversed growth inhibition	231
	Peptone-based medium[g]	Reduction in growth temperature produced alterations in fatty acid composition of the glycerophospholipids	232
	1% PP + 0.5% YE + 0.5% NaCl	Six major proteins demonstrated in isolated pellicles	233
	2% PP + 0.4% LF + salts in Medium A of Kidder and Dewey minus phosphates	Uridine transport inhibited by ribo- and deoxy-ribonucleosides	234
	1% PP + 0.05% LE in 0.02 M potassium phosphate buffer, pH 6.5	3 MPA inhibited appearance of label in glycogen from bicarbonate, OAc, pantanoate, octonate, and succinate, but not from glycerol or glc	235
		Transaldolase, transketolase, ribose 5-phosphate isomerase, ribulose-5-phosphate-3-epimerase, and ribokinase demonstrated via incorporation of label from $(1^{-14}C)$ ribose into CO_2 and glycogen	239
	1% Tryptone + 0.05% YE	Toxicity to 0.25 M suc and 1.4 M DMSO marked in late log- and stationary-phase cells	237
	Loefer's medium	Food vacuoles concentrate suspended particles	238

e   30°C specified.
f   26°C specified.
g   25°C specified.

Table 3 (continued)
GROWTH MEDIA AND IN VITRO UPTAKE OF SPECIFIC BIOCHEMICALS BY CILIATES

Species	Medium	Assessment	Ref.
Tetrahymena pyriformis (continued)		Free-living (continued)	
	1% (w/v) PP + 0.05%[h] LE in 0.02 M potassium phosphate, pH 6.5	Increase in acid phosphatase and decrease in α-glycosidase and ribonuclease with advancing culture age	240
	2% PP + 0.1% YE + 90 μM iron EDTA + 0.5% glc	Mg, cholesterol, and glc affected the ultimate distribution of RNA in cells suspended in non-nutrient buffer	241
	1% PP + 0.25% YE	Glc addition led to retention of a portion of purine, pyrimidine, and Pi by cells	242
		Fed cells showed a bimodal size distribution; starved cells were unimodal	243
	2% PP[c]	Doubling time reduced from 40 to 5 hr when PP was sterilized by autoclaving instead of filtration	250
	Basal medium containing (g/ml) Difco tryptone, 4.0; PP, 6.0; $K_2 HPO_4$, 0.1–0.5; glc, 0.0–2.5; and thiamine HCl, 0.001–0.005	Ca, Cu, Zn, and Mg reversed citrate inhibition; Ca, Al, Co, Cu, Fe, Mn, Ni, and Zn completely reversed EDTA growth inhibition	251
	Basal medium containing (g/l) PP, 15; tryptone, 5; YE, 2; glc, 4; $Na_3 PO_4$ (to pH 7.4), 3; $MgCl_2 \cdot 6H_2O$, 5; $FeSO_4 \cdot 7H_2O$, 30 mg; $Zn(NO_3)_2 \cdot 6H_2O$, 15 mg; $MnCl_2 \cdot 4H_2O$, 5 mg; $CuCl_2 \cdot 2H_2O$, 1 mg; $CoCl_2 \cdot 6H_2O$, 1 mg; and molybdic acid, 1 mg	Two periods of increased glyoxylate bypass in glc- or OAc-containing media	255
	2% PP + 0.1% YE Synthetic medium of Dewey et al.	Glc and OAc elevated free tyr pools; increased conversion of phe to tyr	260
	2% PP + Medium A of Kidder and Dewey, pH 7.2–7.4	24-hr incubation produced 5 × 10^5 cells/ml	261
	Maintenance medium containing (g/100 ml) dextrin (0.8) + Na acetate·$3H_2O$ (0.06) + yeast autolysate (0.5) + "Liver L" (0.06) + Hycase® SF (0.5) + agar (1.6), pH 6.8	Cultures remained vigorous for 1 month	262
	Substrate-free medium containing a 12-component metal mix + 12 amino acids + 6 vitamins + Na_2 guanylate·H_2O + uracil + folic acid + DL-thioctic acid + biotin, pH 6.4–6.8	L-ser essential; glycerol supplementation permitted growth equal to that with CHO supplement	
	Semidefined basal medium containing an 11-component trace metal mixture + uracil + adenosine + thymidine + 7 vitamins + folic acid + DL-lipoic (thioctic) acid + Casamino acids (Difco) + 4 amino acids + crude soy lecithin (Sigma), pH 5.4–5.6	Lecithin permitted prolonged survival	264

h   Shaking at 25°C specified.

Table 3 (continued)

GROWTH MEDIA AND IN VITRO UPTAKE OF SPECIFIC BIOCHEMICALS BY CILIATES

Species	Medium	Assessment	Ref.
Tetrahymena pyriformis (continued)	**Free-living (continued)**		
	Kidder and Dewey medium	Pro rapidly and completely oxidized to CO_2 + glutamate; Glc and OAc incorporated into cell components	267
	PP + YE + glc + Tween® 80 + PO_4 buffer, pH 6.8	Growth inhibition by 2-fluroadenine reversed by high levels of adenine	272
	Kidder and Dewey medium		
	1% PP + 0.05% LE + 0.2 M K phosphate, pH 6.5	5-Hydroxytryptophan increased cellular serotonin content	273
	2% PP + 0.1% YE + 90 μM iron EDTA, pH 7.0	Catabolic pathway for 5'AMP and 5'CMP suggested	274
	2% PP + 0.1% LE	Cultures synchronized by starving in a nonnutrient PO_4 buffer and refeeding with the growth medium	275
	2% PP, enriched PP medium	Extent to which endogenous lipids are degraded depends on culture age	278
	Tryptone-glc-vitamins-salts medium	Cells transferred from complete axenic medium to medium minus amino acids cease dividing and undergo oral replacement	280
	Medium minus amino acids, purines, and pyrimidines		
	0.5% Bactopeptone + 0.2% YE	Inhibition of polyssacharide utilization by ser	281
	0.5% Bactopeptone + 2% YE + 1% glc		
	0.2% YE + 0.9% dextrin		
	PP YE medium	DNase activity inhibited by acidic polymers and polyamines	283
	BM1 medium consisting of (g/100 ml) trypticase, 0.5; liver L, 0.04; yeast hydrolysate, 0.5; NaOAc, 0.01; dextrin, 0.05; pH 7.2 with Tris	Survival 16—22 weeks with Ionox® 330 or ascorbylpalmitate	285
	BM2 medium containing (g/100 ml) phytone, 0.05; yeast hydrolysate, 0.1, minus trypticase + other ingredients in BM1		
	2% PP + 0.1% YE + 0.2% NaOAc + mixture of divalent cations	Glc-grown cells with altered respiratory rate when exposed to culture medium containing this CHO	287
	2% PP + 0.1% LE + mineral salt mixture	Changes in lipid accumulation correlated with changes in cell size, shape, volume, and fragility	288
		Hydrolases demonstrated in food vacuoles of stationary phase cells	289

Table 3 (continued)

GROWTH MEDIA AND IN VITRO UPTAKE OF SPECIFIC BIOCHEMICALS BY CILIATES

Species	Medium (Free-living (continued))	Assessment	Ref.
Tetrahymena pyriformis (continued)	Defined medium containing 17 amino acids, 6 vitamins, guanosine-3'(2') phosphoric acid·2Na·H₂O, uracil, DL-thioctic acid, folic acid, 6 trace metals, citric acid·H₂O, K₂HPO₄·3H₂O, MgSO₄·7H₂O, CaCO₃, and glc	Supplementation with nucleic acid derivatives allowed full growth	290
	PP + LE medium	Total cellular lipid increased throughout log and early stationary phases	291
	2% PP	Activation of mitochondrial enzymes depends on the presence of ATP and Mg	294
	Medium W II containing (g/l) PP, 4; glc, 5; liver fraction, 1	Survival 128 days at −95°C	295
	Medium B containing (g/l) PP, 5; tryptone, 5; KH₂PO₄, 0.2		
	1% Bactotryptone + 1% YE	Appearance of acid phosphatase activity in food vacuoles dependent on vacuole formation and not on its contents	296
T. vorax	Loefer's medium	10,000 cells/ml produced	269
Paramecium aurelia	Synthetic medium containing 250 µg/ml cephalin + 0.5 µg/ml folic acid + 3.3 µg/ml dThd	High levels of folic acid required for growth	203
	Basal medium (mg/ml): PP (5) + trypticase (5) + yeast nucleic acid (1) + MgSO₄·7H₂O (500 µg/ml) + TEM-4T (100 µg/ml) + stigmasterol (5 µg/ml) + vitamins (µg/ml) + calcium pantothenate (5), nicotinamide (5), pyridoxamine HCl (2.5), riboflavin (5), folic acid (2.5), thiamine HCl (15), biotin (0.00125), DL-thioctic acid (0.05), pH 7.0	Requirements for fatty acids satisfied by oleic acid-containing lipids	249
	Lettuce infusion + PP + glc[i] + yeast autolysate (LPGY)	Change in nutrition affects ability to grow in axenic culture	254
	PP + trypticase + yeast nucleic acid + MgSO₄·7H₂O + stigmasterol + vitamin mixture + TEM-4T	Yeast fraction used previously replaced by trypticase + yeast nucleic acid + TEM-4T	256

i   Sterile cultures obtained by washing in 93–1300 µg/ml streptomycin + 787–1040 units/ml penicillin G.

Table 3 (continued)

GROWTH MEDIA AND IN VITRO UPTAKE OF SPECIFIC BIOCHEMICALS BY CILIATES

Species	Medium	Assessment	Ref.
	Free-living (continued)		
Paramecium aurelia (continued)	Synthetic medium consisting of amino acids, vitamins, purine, and pyrimidine derivatives, fatty acids, stigmasterol, sodium acetate, and salts	Arg, gly, his, ile, leu, lys, met, phe, pro, thr, trp, tyr, val, folic acid, nicotinamide, calcium pantothenate, pyridoxal, riboflavin, thiamine, DL-6-thioctic acid, guanosine, uridine (or cytidine), oleic acid, stigmasterol, Ca, and Mg essential nutrities	277
P. bursaria	Lettuce infusion + 0.5 mM tris-maleate buffer	Rate of directional change correlated w medium gradient	244
	Wheat straw-infusion medium[j]	Transfer of mal, glc, fru, and malate from endozoic sybiont to organism	245
	Glutaraldehyde-killed bacteria suspended in a 1:1 mixture of baked lettuce extract and Bristol's mineral medium	Growth improved with yeast nucleic acid, vitamins, and PP supplement	299
P. caudatum	Lettuce infusion[g] + 2.5–5 mM tris-maleate buffer, pH 7	Swimming velocity correlated with temperature and viscosity of medium	259, 268
P. tetraurelia	Phosphate-buffered cerophyl + 15–25 mM KCl + 1 mg% acriflavine	Conjugation induced with KCl + acriflavine and low Ca^{2+}	228
Glaucoma chattoni	2% PP + 1% Brewer's yeast powder + 0.5% anhydrous dextrose, pH 7.0	Growth improvement with increased concentrations of either amino or nucleic acids	276
	Modified defined medium of Lilly and Stillwell containing 17 amino acids, nucleic acid components, fatty acids, stigmasterol, sodium acetate, B vitamins, and inorganic salts		
Colpidium campylum, *Tetrahymena pyriformis,* *T. paravorax*	Chemically defined medium containing 16 amino acids + guanylic, adenylic, cytidylic, and uridylic acids + NaOAc + glc + stigmasterol + linoleic and oleic acids + 8 vitamins + 10 salts, pH 7.0	Lipids and proteins or peptides protect RNA from the action of chemical and physical agents	286
Stylonychia sp.	*Tetrahymena* as food source under bacteria-free conditions	High yield with protozoa as food source	216
S. mytilus	Pringsheim's medium + *Chlorogonium* sp.	Details of developing macronucleus determined	227
	Autoclaved pond water + rice grains	Encystment in response to increasing population density and food depletion	300

j   12/12 light-dark cycle specified.

Table 3 (continued)
GROWTH MEDIA AND IN VITRO UPTAKE OF SPECIFIC BIOCHEMICALS BY CILIATES

Species	Medium	Assessment	Ref.
	Free-living (continued)		
Opercularia coarctata	Defined medium containing 20 amino acids and 10 vitamins, an 8-component balanced salt solution, $Fe_2(SO_4)_3 \cdot (NH_4)_2 \cdot SO_4 \cdot 24H_2O$, Tween® 80, stigmasterol, a 7-component nucleic acid mixture, phenol red indicator, 2500 USP units/ml penicillin, thioctic acid, niacin, niacinamide, inositol, PABA, oleic acid, pH 7.0—7.2	Continuous cultivation 36—48 days	245
Telotrochidium henneguyi	20 mg% PP + 200 mg% cerophyl + 300 mg% wheat kernel broth (PPCW)k	Medium + *Bacillus cereus* required for growth	248
	Acid-hydrolyzed gelatin + aqueous LE + hydrolyzed yeast nucleic + glc + ser + EDTA + $CaCl_2$ + $FeCl_3$ + KCl + DL-β-hydroxybutyrate + $MgSO_4 \cdot 7H_2O$ + riboflavin + penicillin, pH 6.8	Maintenance 72 hr in axenic culture	
	Liver hydrolysate + yeast nucleic acid + EDTA + penicillin + 18 amino acids + 7 vitamins + 10 salts supplying trace metals + uridylic, cytidylic, guanylic, and adenylic acids + dTDP + NMN + choline, pH 6.8[l]	Continuous cultivation 36 days	
Vorticella convallaria	Egg yolk-lettuce extract + penicillin and streptomycin	Axenic growth 2 years	252
V. microstoma	Acid-hydrolyzed gelatin + aqueous LE + hydrolyzed yeast nucleic acid + glc + penicillin + PO_4 buffer, pH 6.4	Survival 96 hr	253
Tokophrya infusionum	YE medium	Reproduction stimulated by replacing old medium with fresh medium during stationary growth	189
Oxytricha bifaria	Lettuce medium	One generation per day	212
Euplotes crassus	Filtered sea water heated to 85°C twice + *Dunaliella* grown in Erd-Schreiber sea water medium as food source	8000 cells/ml produced after 48 hr	246
Pseudocohnilembus persalinus	Cerophyl + PP + trypticase + yeast nucleic acids + vitamins in 3% Osterout's saline	Growth stimulated by saturated and unsaturated fatty acids	229
Tokophrya infusionum	0.05% YE medium + sterile *Tetrahymena*	Weekly transfers required for axenic maintenance	258, 263

k   PPCW diluted 1:9 w distilled water.
l   23°C specified.

Table 3 (continued)
GROWTH MEDIA AND IN VITRO UPTAKE OF SPECIFIC BIOCHEMICALS BY CILIATES

Species	Medium	Assessment	Ref.
Free-living (continued)			
Uronema nigricans, Parauronema virginiatum, Miamiensis avidus, Miamiensis sp.	Cerophyl extract + PP + trypticase + yeast nucleic acid + biotin + Ca pantothenate + folic acid + nicotinamide + pyridoxal·HCl + riboflavin + thiamine·HCl + DL-thioctic acid in sea water, pH 7.2	Cerophyl replaced by a mixture of asolecithin + cephalin + Tween® 80	266
Blepharisma intermedium	Defined medium consisting of 18 amino acids, 5 purine derivatives, 8 vitamins, asolecithin, cephalin, and Tween® 80 in synthetic sea water, pH 7.2	$3{-}4 \times 10^6$ cells after 4—5 days	270
	Freshly killed or lyophilized and autoclaved bacteria + YE + lettuce infusion or stigmasterol + 6 B vitamins and PO_4 buffer	Division after 32 hr	279
	Unit standard balanced salt solution (USBSS)	Slow growth in media with high osmolal concentrations	284
	Hay infusion + Erd-Schreiber medium fortified with Horlick's malted milk[f]	Initial rise in temperature increases reproductive rate	271
Stentor coeruleus	Modified Peters' solution + axenic Tetrahymena pyriformis as food source	50% conjugation in axenic medium	282
Urostyla weissei	Chapman-Andersen's modification of Pringsheim's solution with axenic T. pyriformis as food source	Supplement of ground oven-dried yolk or hard boiled eggs required for growth	
Parasitic			
Epidinium ecaudatum caudatum, Polyplastron multivesiculatum	Wholemeal flour + dried grass + potassium phosphate-rich medium	Bacteria or other protozoa required for prolonged maintenance	215
Colpoda maupasi	7 Parts thioglycollate medium (29.8 g/l H_2O_2) + 2 parts beef heart infusion broth with 1% Pfanstiehl peptone + 3 parts distilled water	Weekly transfers to fresh medium required	257
Entodinium	Rumen fluid containing 0.2 mg/ml rice starch	Fluids with high concentrations of parasites are superior in supporting in vitro cultivation	293

Table 4
SUMMARY OF NUTRITIONAL REQUIREMENTS AND EFFECTS ON IN VITRO GROWTH AND MULTIPLICATION OF SPOROZOA

Species	Components Sporozoa	Assessment	Ref.
Plasmodium falciparum	Harvard and TC 199 media supplemented with human, fetal calf, or other sera	Mean multiplication rate × 3; IgG inhibited multiplication of East African parasites	305
	Modified Harvard medium + TES buffer	75% parasitemia in modified medium	307
	Modified Harvard medium consisting of inorganic salt solution + glc + 14 amino acids + vitamin mixture + ascorbic acid + purines and pyrimidines + TES supplement buffer	Red cell extract (RCE) beneficial for cultivation	312
	Modified Harvard medium with glycylglycine as supplement buffer instead of TES	BES, TES, and HEPES choice buffers; TES best growth	313
	Medium 199 + 200 units penicillin + 200 μg streptomycin supplemented with albumin and blood serum	Albumin stimulated motility and preserved viability of sporozoites in vitro[a]	340
	Trager's medium + host erythrocyte suspension	Antimalarial activity of pantothenic analogues inhibitory in vitro; compounds with R group (methoxyquinolyl) with activity approaching that of primaquine	353
P. coatneyi	Trager's duckling plasma + chick embryo extract medium for rocker dilution flask experiments (TRDF) + 10% heparinized rhesus monkey plasma	Both species completed 2 cycles of development without addition of fresh erythrocytes	354
P. falciparum	TRDF + 10% heparinized type B plasma		
P. knowlesi	Basal simplified medium (BSM) containing inorganic salt solution + glc + ile + met + PABA + biotin + pyrimidines + BES supplement buffer	Rhesus monkey RCE improved growth	312
	BSM + glycylglycin supplement buffer instead of BES	BES, TES, and HEPES choice buffers; BES best growth	313
P. berghei	Serum-free medium[b] consisting of 90% Eagle's minimum essential medium (MEM) + Eagle's salt solution without sodium bicarbonate or glc, pH 7.6	Ookinetes exhibited either gliding, snake-like, or spiral forward movements in vitro	308
	Medium 199 + 10% serum	Parasite number decreased at 31 and 37°C; multiplication greater than 1 at 15°C	309
	Trigg's medium	Parasites infective in vivo and metabolically active when leukocytes were removed from the blood	310

[a] 4°C specified.
[b] 18 to 21°C specified.

Table 4 (continued)
SUMMARY OF NUTRITIONAL REQUIREMENTS AND EFFECTS ON IN VITRO GROWTH AND MULTIPLICATION OF SPOROZOA

Species	Components	Assessment	Ref.
Sporozoa (continued)			
P. berghei (continued)	Medium 199 + radioactive glc	Glc converted to OAc and lactate; chloroquine-induced clumping of malaria pigment in vitro involved RNA and protein synthesis and required energy source	320
	TC medium 199 + 1000 units penicillin + 1 mg/ml streptomycin	Incorporation of adenosine and leu	333
P. relictum	Grace's insect culture medium + fetal bovine serum (FBS) in the presence of dividing cells of *Antheraea eucalypti*	Utilization of a mixture of arg, asn, glx, + glu + gly + his + lys + pro and ser	311
P. malariae	Modified Harvard medium with TES as supplement buffer	Parasites completed asexual cycle in medium but the number remained constant	314
P. vinckei chabaudi	Eagle's MEM medium + buffers + adenosine + vitamins	Lactate production and glc utilization correlated with development of parasite from ring to schizont stage	339
P. lophurae	Embryonic mouse liver monolayers	Mammalian cells supported growth comparable to that observed in avian cell cultures	315
	1% heparin + 0.85% NaCl + 0.2 ml organic mix (Anfinson) + erythrocyte suspension	Adenine-8-^{14}C and ^{14}C-orotic acid served as purine and pyrimidine nucleotide precursors, respectively	316
	Mature duck erythrocyte suspension	Glu, asp, cys, lys, and arg showed mediated entry into parasites	317
	Trager's buffer containing 1% bovine serum albumin fraction V (TBA) + ferritin	Low temperature inhibited ferritin uptake	336
	5 ml Medium 199 + 1 ml normal duck plasma + 1 ml duck blood infected with *P. lophurae* trophozoites + 1 ml saline solution of 800 µCi L-[^{3}H] his[c]	His, incorporated in vitro, functioned in penetration of merozoites	337
	7 ml of blood layered over 10 ml of 20% bovine albumen solution containing 0.6% NaCl	ATP with pyruvate or ADP with phosphoenolpyruvate favored in vitro cultivation	343
	Trager's duckling plasma + chick embryo extract medium (TRDF) + supplementary CoA, adenosine, triphosphate, Na pyruvate, and folinic acid	Incorporation of pro higher with ATP and pyruvate present	355

[c] 37°C in 5% CO_2, 95% air specified.

Table 4 (continued)
SUMMARY OF NUTRITIONAL REQUIREMENTS AND EFFECTS ON IN VITRO GROWTH AND MULTIPLICATION OF SPOROZOA

Species	Components	Assessment	Ref.
Sporozoa (continued)			
P. fallax	Embryonic mouse liver monolayers	Mammalian cells supported growth comparable to that observed in avian cell cultures	315
P. yoelii nigeriensis	Invertebrate tissue culture media	Exflagellation inhibited in vitro by temperature of 30°C and certain invertebrate tissue culture media	318
Babesia canis	Erythrocyte suspension + dextrose + citrate	Multiplication evident after 8 hr; marked after 24—33 hr	319
B. microti	TBA + ferritin	Low temperature inhibited ferritin uptake	337
Hepatozoon rarefaciens	*Culex pipiens* cells grown in equal parts of Hsu's, Schneider's, and Grace's insect tissue culture medium	Hsu's medium alone + FBS supported growth of parasite	321
Eimeria magna	Madin-Darby bovine kidney cells (MDBK) cultured in Eagle's MEM	Sporozoites readily entered ATP-depleted cells and cells treated with colchicine, colcemid, and vinblastine	306
	MDBK cultured in MEM with Eagle's BSS	No differences in structure between the microgamonts in cell cultures and those of the host	335
	MDBK cultured in MEM	Sporozoite caused invagination of host cell plasmalemma until parasite entered the cell	338
E. auburnensis	MDBK, EBTr, and EBS maintained in Eagle's MEM; human embryonic intestine (Int 407) in Eagle's Basal Medium (BME [E])	Mature first generation schizonts developed in MDBK, EBTr, and EBS cultures	348
E. meleagrimitis, *E. necatrix,* *E. acervulina,* *E. gallopavonis*	Monolayer cell cultures of bovine, ovine, porcine and human kidney maintained in BME(E) with Earle's BSS + 3% fcs, BME(E) + 5% fcs, BME(E) + 10% fcs, Medium 199 with Earle's BSS + 10% fcs, Earle's BSS + 2.5% lactalbumin hydrolysate (LAH) + 10% fcs, Earle's BSS + 20% LAH + 10% fcs, and Eagle's MEM with Hank's BSS + 5% fcs	Variations in the number of asexual generations, appearance, number, and size of mature schizonts with species	344
E. meleagrimitis	Monolayer cell cultures of primary embryonic turkey intestine (TEI) and EBK, maintained at 40.6 and 43°C in BME(E) + fcs, pH 7.0—7.4	Abundant merozoites liberated at 50, 52, 74, and 76 hr in bovine cultures	346
E. maxima	199 Medium containing 10% tryptose phosphate broth and 4% gelatin + sera 141 and 155	In vivo protection by serum 141 and decreased invasion of cultured cells by sporozoa	327
	Cultured kidney cells in phosphate buffer (PBS) containing 0.9% NaCl	Survival 27 hr; variation in survival time with other species	328

Table 4 (continued)
SUMMARY OF NUTRITIONAL REQUIREMENTS AND EFFECTS ON IN VITRO GROWTH AND MULTIPLICATION OF SPOROZOA

Species	Components	Assessment	Ref.
	Sporozoa (continued)		
E. ninakohlyakimova	Cell line cultures of OEK, embryonic ovine trachea (LeTr), embryonic ovine thymus (LETm), OET, and MDBK maintained in Eagle's MEM with Earle's BSS + 10% FBC	Sporozoites penetrated cells within 5 min after inoculation as well as 2 and 3 days after inoculation	350
E. adenoeides	EBK cell cultures maintained in BME(E) containing Hank's BSS + 10% fcs	Hyaluronidase did not increase the number of intracellular sporozoites	351
E. alabamensis	Monolayer primary cultures of BEInt, EBK, EBS, and bovine embryonic thyroid (BETy) and cell line cultures of EBTr, synovial cells (BESy) + established cell line cultures of MDBK, Int 407, and BHK suspended in Eagle's MEM containing 15% fcs + 50 units/ml penicillin G and 50 μg/ml dihydrostreptomycin	Sporozoites penetrated all cell types during the first 24 hr	352
E. callospermophili	EBTr, EBk, EBS, BESy, embryonic bovine thymus (BET), bovine intestinal (BE Int), Int 407, MDBK, BHK, and primary cells from whole embryos of ground squirrels (EGS-1) maintained in cell culture medium containing 3% FBS	Degenerative changes in host cells associated with intracellular parasitism seen throughout the period of infection	330
E bilamellata	EBK, spleen (EBS), EBTr, BESy, Int 407, embryonic lamb thyroid (LETh), Syrian hamster kidney (BHK), and MDBK maintained in cell culture medium containing 3% FBS		
E. bovis	Monolayer primary and secondary cultures of EBK, EBS, BE Int and testicle cells, and secondary cultures of BET maintained in LAH, Earle's BSS, and ovine serum	Sporozoites entered the cells of all cultures, but underwent development only in primary EBK and BE Int and secondary EBK, EBS, BET, BE Int, and testicle cells	345
	EBS, bovine embryonic ileum (BEI), EBK cell line cultures grown in LAH with Earle's BSS containing 10% lamb or calf serum + antibiotics; CEF, mouse macrophages (MM), bovine embryonic cecum (BEC), ovine embryonic kidney (OEK), ovine embryonic thyroid (OET), and EBTr cultured in Eagle's MEM containing 1% sodium pyruvate + 1% nonessential amino acids (aa) + 3% fetal calf serum (fcs) + 50 units penicillin + 50 μg/ml streptomycin	Number of merozoites produced increased daily reaching a peak 18—21 days after inoculation	347

Table 4 (continued)

SUMMARY OF NUTRITIONAL REQUIREMENTS AND EFFECTS ON IN VITRO GROWTH AND MULTIPLICATION OF SPOROZOA

Species	Components Sporozoa (continued)	Assessment	Ref.
E. brunetti	CK cells maintained in Hank's saline with 0.125% lactalbumin hydrolysate + 0.125% $NaHCO_3$ + 5% fcs[d]	First-generation schizonts found at 66 hr	331
	8% fresh bovine bile (FBB)[e]	Excystation levels above 90% in 50 and 10% dilutions in FBB	349
E. tenella	Chick embryo kidney cells + BUdR + Hank's BSS + thymidine (5 μg/ml)	Parasite unable to utilize exogenous thymidine	329
	Primary kidney cells maintained in 0.17% LAH + 5% fcs in modified Earle's BSS	In vitro development paralleled development of control strain	332
	Monolayer cultures of chick embryo kidney cells supplemented with tritiated nucleosides	Cytidine, thymidine and uridine incorporated into asexual stages	341
Toxoplasma gondii	HeLa cell cultures maintained in a medium containing lactalbumin hydrolysate + 10% calf serum + 100 units penicillin + 100 μg streptomycin/ml medium	Nicarbazin and robenziden depressed parasite in concentrations ranging from 0.01–1.0 mg/ml maintenance medium	322
	Detroit 532 cell line maintained in 5 ml MEM + nonessential amino acids + 2% FBS[c]	Viability of cultures, preserved by freezing, decreased rapidly with time from an initial 98 to 70% 12 days after infection	323
T. gondii	African green monkey kidney (Vero) cells suspended in Eagle's MEM containing Earle's BSS + 5% FBS	Penetration of 3 genera into cultured cells significantly inhibited by 0.1 mg quinine sulfate/ml in the inoculation medium	324
Besnoitia jellisoni	Embryonic bovine trachea (EBTr) cells + Eagle's basal medium containing Hank's BSS + 10% FBS		
Sarcocystis sp.	Embryonic bovine kidney (EBK) cells + Eagle's basal medium containing Hank's BSS + 10% FBS		
T. gondii	Rhesus monkey liver (RML), chick embryo fibroblast (CEF), rhesus monkey kidney (RMK), and vervet monkey kidney (VMK) cell lines maintained in MEMH growth medium containing Hank's BSS + 25,000 USP units penicillin and 25,000 μg/ml streptomycin, pH 7.5[f]	RH and Beverley strains produced plaques in all cell lines except WI-38; lincocin sulfisoxasole, cleocin, and aureomycin produced inactivation of parasite in high doses	325

[d] 41°C in an atmosphere of 5% CO_2, 95% air specified.

[e] Pretreatment with 50% CO_2/50% air in a medium containing 0.02 M cys·HCL for 12 hr at 39°C specified.

[f] 37°C in 4.5% CO_2 atmosphere specified.

Table 4 (continued)
SUMMARY OF NUTRITIONAL REQUIREMENTS AND EFFECTS ON IN VITRO GROWTH AND MULTIPLICATION OF SPOROZOA

Species	Components	Assessment	Ref.
	Sporozoa (continued)		
T. gondii (continued)	L929 cell cultures[g]	Pyrimidine precursors as orotic acid preferred over preformed pyrimidines as thymidine for DNA synthesis	356
Isospora felis	Int 407, esophageal epithelium (Minn EE), amnion (FL), diploid lung (WI-38), and HeLa cell lines maintained in Eagle's MEM with Hank's BSS + aa + fcs or human serum (hs) + LAH[c]	CHO utilized by sporozoites and replaced in the mature daughter organism	326
	Cnidosporida		
Myxosoma cerebralis	Eagle's MEM, Earle's BSS containing 10% FBS (MEM-10) + fish cartilage extract (FCE)	Incubation period 1–6 months	334

g Incubation for 20 hr at 37°C specified.

REFERENCES

1. **Kidder, G.,** Growth studies of ciliates. VII. Comparative growth characteristics of 4 species of sterile ciliates, *Biol. Bull.* (Woods Hole, Mass.), 80, 50–68, 1941.
2. **Kidder, G. W. and Dewey, V. C.,** The biochemistry of ciliates in pure culture, in *Biochemistry and Physiology of Protozoa,* Vol. 1, Lwoff, A., Ed., Academic Press, New York, 1951, 324–400.
3. **Lwoff, M.,** The nutrition of parasitic flagellates (Trypanosomidae, Trichomonadinae), in *Biochemistry and Physiology of Protozoa,* Vol. 1, Lwoff, A., Ed., Academic Press, New York, 1951, 129–176.
4. **Hutner, S. H.,** Plant animals as experimental tools for growth studies, *Bull. Torrey Bot. Club,* 88, 339–349, 1961.
5. **Hutner, S. H. and Provasoli, L.,** The Phytoflagellates, in *Biochemistry and Physiology of Protozoa,* Vol. 1, Lwoff, A., Ed., Academic Press, New York, 1951, 27–128.
6. **von Brand, T.,** Metabolism of Trypanosomidae and Bodonidae, in *Biochemistry and Physiology of Protozoa,* Vol. 1, Lwoff, A., Ed., Academic Press, New York, 1951, 177–234.
7. **Seaman, G. R.,** Metabolism of purines by extracts of *Tetrahymena, J. Protozool.,* 10, 87–91, 1963.
8. **Soldo, A. T. and van Wagtendonk, W. J.,** Nitrogen metabolism of *Paramecium aurelia, J. Protozool.,* 8, 41–54, 1961.
9. **Loefer, J. B. and Scherbaum, O. H.,** Amino acid composition of protozoa: comparative studies on *Tetrahymena, J. Protozool.,* 8, 184–191, 1961.
10. **Lwoff, M.,** Nutrition of parasitic amebae, in *Biochemistry and Physiology of Protozoa,* Vol. 1, Lwoff, A., Ed., Academic Press, New York, 1951, 235–250.
11. **Boss, H. J.,** Monoxenic and axenic cultivation of carrier and patient strains of *Entamoeba histolytica, Z. Parasitenkd.,* 47, 119–129, 1975.
12. **Visvesvara, G. S. and Balamuth, W.,** Comparative studies on related free living and pathogenic amebae with special reference to *Acanthamoeba, J. Protozool.,* 22, 245–256, 1975.
13. **Buro, N. C. and Weller, D. L.,** Purification and characterization of malic enzyme of *Entamoeba invadens:* evidence of isoenzymes, *J. Protozool.,* 21, 796–802, 1974.
14. **Diamond, L. S.,** Axenic cultivation of *Entamoeba histolytica, Science,* 134, 336–337, 1961.
15. **Diamond, L. S.,** Techniques of axenic cultivation of *Entamoeba histolytica* Schaudinn, 1903 and *E. histolytica*-like amoebae, *J. Parasitol.,* 54, 1047–56, 1968.
16. **Rengpien, S. and Bailey, G. B.,** Differentiation of *Entamoeba*: a new medium and optimal conditions for axenic encystation of *E. invadens, J. Parasitol.,* 61, 24–30, 1975.
17. **Dunnebacke, T. H. and Schuster, F. L.,** An infectious agent associated with amebas of the genus *Naegleria, J. Protozool.,* 21, 327–329, 1974.
18. **O'Dell, W. D. and Brent, M. M.,** Nutritional studies of three strains of *Naegleria gruberi, J. Protozool.,* 21, 129–133, 1974.
19. **Kadlec, V.,** The effect of some factors on the growth and morphology of *Naegleria sp.* and three strains of genus *Acanthamoeba, Folia Parasitol.* (Prague), 22, 317–321, 1975.
20. **Wang, L. T., Jen, G., and Cross, J. H.,** Axenic cultivation of four strains of *Entamoeba histolytica* from liver abscesses, *Southeast Asian J. Trop. Med. Public Health,* 5, 365–367, 1974.
21. **Balamuth, W.,** Nutritional studies on axenic cultures of *Naegleria gruberi, J. Protozool.,* 11(Suppl.), 19–20, 1964.
22. **Wong, M. M., Karr, S. L., Jr., and Balamuth, W. B.,** Experimental infections with pathogenic free-living amebae in primate hosts. A. A study on susceptibility to *Naegleria fowleri, J. Parasitol.,* 61, 199–208, 1975.
23. **Cerva, L., Serbus, C., and Skocil, V.,** Isolation of limax amoebae from the nasal mucosa of man, *Folia Parasitol.* (Prague), 20, 97–103, 1973.
24. **Chagla, A. H. and Griffiths, A. J.,** Growth and encystation of *Acanthamoeba castellanii, J. Gen. Microbiol.,* 85, 139–145, 1974.
25. **Brand, R. N. and Balamuth, W.,** Hemin replaces serum as a growth requirement for *Naegleria, Appl. Microbiol.,* 28, 64–65, 1974.
26. **Schuster, F. L.,** Ultrastructure of mitosis in the amoebaflagellate *Naegleria gruberi, Tissue Cell,* 7, 1–11, 1975.
27. **Neal, R. A.,** Survival of *Entamoeba* and related amoebae at low temperature. I. Viability of *Entamoeba* cysts at 4 degrees C, *Int. J. Parasitol.,* 4, 227–229, 1974.
28. **Weinbach, E. C. and Diamond, L. S.,** *Entamoeba histolytica.* I. Aerobic metabolism, *Exp. Parasitol.,* 35, 232–243, 1974.
29. **Band, R. N. and Mohrlok, S.,** Observations on induced amitosis in *Acanthamoeba, Exp. Cell Res.,* 79, 327–337, 1973.

30. **Band, R. N. and Mohrlok, S.,** The cell cycle and induced amitosis in *Acanthamoeba, J. Proto-zool.,* 20, 654–657, 1973.

31. **Rastogi, A. K., Sagar, P., and Agarwala, S. C.,** Role of riboflavin and certain amino acids in the excystment of *Schizopyrenus russelli, J. Protozool.,* 20, 453–455, 1973.

32. **Evans, D. A.,** Growth phase and the number of phosphorylation sites in the mitochondrial electron transport chain of *Acanthamoeba castellanii, J. Protozool.,* 20, 336–338, 1973.

33. **Raizada, M. K. and Krishna Murti, C. R.,** Synthesis of RNA, protein, cellulose, and mucopolysaccharide and changes in the chemical composition of *Hartmannella culbertsoni* during encystment uder axenic conditions, *J. Protozool.,* 19, 691–695, 1972.

34. **Sawyer, T. K.,** *Acanthamoeba griffini,* a new species of marine amoeba, *J. Protozool.,* 13, 650–654, 1971.

35. **McIntosh, A. H. and Shihman, C.,** A comparative study of 4 strains of Hartmannellid amoebae, *J. Protozool.,* 18, 632–636, 1971.

36. **Myer, D. F. and Morgan, R. S.,** Evidence for a link between division and differentiation in *Entamoeba invadens, J. Protozool.,* 18, 282–284, 1971.

37. **Childs, G. E.,** *Hartmannella culbertsoni:* enzymatic, ultrastructural, and cytochemical charac-teristics of peroxisomes in density gradient, *Exp. Parasitol.,* 34, 44–55, 1973.

38. **Stevens, A. R. and O'Dell, W. D.,** The influence of growth medium on axenic cultivation of virulent and avirulent *Acanthamoeba, Proc. Soc. Exp. Biol. Med.,* 143, 474–478, 1973.

39. **O'Dell, W. D. and Stevens, A. R.,** Quantitative growth of *Naegleria* in axenic culture, *Appl. Microbiol.,* 25, 621–627, 1973.

40. **Saygi, G., Warhurst, D. C., and Roome, A. P.,** A study of amoeba isolated from the Bristol cases of primary amoebic encephalitis, *Proc. R. Soc. Med.,* 66, 278–282, 1973.

41. **Rao, V. G. and Padma, M. C.,** Some observations on the pathogenicity of strains of *Entamoeba histolytica, Trans. R. Soc. Trop. Med. Hyg.,* 65, 606–616, 1971.

42. **Thepsuparungsikul, V., Seng, L., and Bailey, G. B.,** Differentiation of *Entamoeba:* encystation of *E. invadens* in monoxenic and axenic cultures, *J. Parasitol.,* 57, 1288–1292, 1971.

43. **Stringer, R. P.,** New bioassay system for evaluating percent survival of *Entamoeba histolytica* cysts, *J. Parasitol.,* 58, 306–310, 1972.

44. **McMillan, B. and Kelley, A.,** Attempts to cultivate *Entamoeba polecki* von Prowazek, 1912, *Trans. R. Soc. Trop. Med. Hyg.,* 66, 366–367, 1972.

45. **Pigon, A.,** Inhibition of movement, attachment, and cytokinesis by autogenous substances in the amoeba *Hartmannella, Exp. Cell Res.,* 73, 170–176, 1972.

46. **Bowers, B. and Olszewski, T. E.,** Pinocytosis in *Acanthamoeba castellanii:* kinetics and morphology, *J. Cell Biol.,* 53, 681–694, 1972.

47. **Friz, C. T.,** Ion levels of *Amoeba proteus, Amoeba discoides* and their heterospecific hybrids, *Exp. Cell Res.,* 71, 225–228, 1972.

48. **Jeon, K. W. and Danielli, J. F.,** Microsurgical studies with large free-living amebas, *Int. Rev. Cytol.,* 30, 49–89, 1971.

49. **Reeves, R. E., Lushbaugh, T. S., and Montalvo, F. E.,** Characterization of desoxyribonucleic acid of *Entamoeba histolytica* by casein chloride density centrifugation, *J. Parasitol.,* 57, 939–944, 1971.

50. **Singh, B. N. and Das, S. R.,** Studies on pathogenic and non-pathogenic small free-living amoebae and the bearing of nuclear division on the classification of the order amoebida, *Philos. Trans. R. Soc. Lond. Ser. B,* 259, 435–476, 1970.

51. **Wise, D. L.,** Effect of acetaldehyde on growth in succinate media and labeling RNA with ^{14}C succinate in *Polytomella caeca, J. Protozool.,* 17, 183, 1970.

52. **Raizada, M. K. and Krishna Murti, C. R.,** Changes in the activity of certain enzymes of *Hartmannella* (Culbertson strain A-1) during encystment, *J. Protozool.,* 18, 115–119, 1971.

53. **Sharma, N. H., Albach, R. A., and Shaffer, J. G.,** Autoradiographic studies of uridine-S-H^3 uptake in *Entamoeba histolytica* in the CLG medium, *J. Protozool.,* 16, 405–411, 1969.

54. **Band, N. and Mohrlok, S. H.,** An analysis of clumping in the soil amoeba *Mayorella palestinensis, J. Protozool.,* 16, 35–44, 1969.

55. **Griffiths, A. J. and Hughes, D. E.,** Starvation and encystment of soil amoeba *Hartmannella castellanii, J. Protozool.,* 15, 673–677, 1968.

56. **Charoenlarp, P., Reeves, R. E., and Warren, L. G.,** Carbohydrate utilization by *Entamoeba histolytica, Exp. Parasitol.,* 23, 205–211, 1968.

57. **Kjellstrand, P.,** The correspondence between growth, attachment and multinuclearity in *Acanthamoeba, Exp. Cell Res.,* 53, 37–43, 1968.

58. **Stratford, M. P. and Griffiths, A. J.,** Excystment of the amoeba *Hartmannella castellanii, J. Gen. Microbiol.,* 66, 247–249, 1971.

59. **Balamuth, W. and Thompson, P. E.,** Comparative studies on amebae and amebicides, in *Biochemistry and Physiology of Protozoa,* Vol. 2, Hutner, S. H. and Lwoff, A., Eds., Academic Press, New York, 1955, 277–345.

60. **Hall, R. P.,** *Protozoan Nutrition,* 1st ed., Blaisdell Publishing, New York, 1965, 32–69.

61. **Jamieson, A. and Anderson, K.,** A method for the isolation of *Naegleria* species from water samples, *Pathology,* 5, 55–58, 1973.

62. **Chapman, A. C.,** Biology of the large amoebae, *Annu. Rev. Microbiol.,* 25, 27–48, 1971.

63. **Svensson, R.,** *Entamoeba histolytica:* encystation and cultivation of several isolates, *Exp. Parasitol.,* 30, 270–283, 1971.

64. **Dunnebacke, T. H. and Schuster, F. L.,** Infectious agent from a free living soil amoeba, *Naegleri gruberi, Science,* 174, 516–518, 1971.

65. **Neal, R. A., Latter, V. S., and Richards, W. H. G.,** Survival of *Entamoeba* and related amoeba at low temperature. II. Viability of amoeba and cysts stored in liquid nitrogen, *Int. J. Parasitol.,* 4, 353–360, 1974.

66. **Hutner, S. H. and Provasoli, L.,** Comparative biochemistry of flagellates, in *Biochemistry and Physiology of Protozoa,* Vol. 2, Hutner, S. H. and Lwoff, A., Eds., Academic Press, New York, 1955, 18–41.

67. **Shorb, M. S.,** The physiology of Trichomonads, in *Biochemistry and Physiology of Protozoa,* Vol. 3, Hutner, S. H., Ed., Academic Press, New York, 1964, 384–459.

68. **Guttman, H. N. and Wallace, F. G.,** Nutrition and physiology of Trypanosomatidae, in *Biochemistry and Physiology of Protozoa,* Vol. 3, Hutner, S. H., Ed., Academic Press, New York, 1964, 460–494.

69. **Barker, S. A. and Bourne, E. J.,** Composition and synthesis of the starch of *Polytomella coeca,* in *Biochemistry and Physiology of Protozoa,* Vol. 2, Hutner, S. H. and Lwoff, A., Eds., Academic Press, New York, 1955, 45–56.

70. **Cassin, P. E.,** Isolation, growth, and physiology of acidophilic chlamydomonads, *J. Phycol.,* 10, 439–447, 1974.

71. **Tamburro, K. M. and Hutner, S. H.,** Carbohydrate free media for *Crithidia, J. Protozool.,* 18, 667–672, 1971.

72. **Kidder, G. W.,** Nitrogen: distribution, nutrition, and metabolism, in *Chemical Zoology,* Vol. 1, Florkin, M. and Scheer, B. T., Eds., Academic Press, New York, 1967, 93–159.

73. **Dewey, V. C.,** Lipid composition, nutrition, and metabolism, in *Chemical Zoology,* Vol. 1, Florkin, M. and Scheer, B. T., Eds., Academic Press, New York, 1967, 161–274.

74. **Lilly, D. M.,** Growth factors in protozoa, in *Chemical Zoology,* Vol. 1, Florkin, M. and Scheer, B. T., Eds., Academic Press, New York, 1967, 275–308.

75. **Hutner, S. H.,** Inorganic nutrition, *Annu. Rev. Microbiol.,* 26, 313–345, 1972.

76. **Levandowsky, M. and Hutner, S. H.,** Utilization of Fe^{3+} by the inshore colorless marine Dinoflagellate *Crypthecodinium cohnii, Ann. N.Y. Acad. Sci.,* 245, 16–25, 1975.

77. **Hutner, S. H.,** Maintaining protozoa and protozoan diversity in a culture collection, in *The Role of Culture Collections in the Era of Molecular Biology,* (ATCC 50th Anniversary Symposium), Colwell, R. R., Ed., American Society for Microbiology, Washington, D.C., 1975, 43–52.

78. **Bacchi, C. J., Lambros, C., Goldberg, B., Hutner, S. H., and de Carvalho, G. D. F.,** Susceptibility of an insect *Leptomonas* and *Crithidia* fasciculata to several established antitrypanosomatid agents, *Antimicrob. Agents Chemother.,* 6, 785–790, 1974.

79. **Baker, H., Frank, O., Bacchi, C. J., and Hutner, S. H.,** Biopterin content of human and rat fluids and tissue determined protozoologically, *Am. J. Clin. Nutr.,* 27, 1247–1253, 1974.

80. **Klein, S. R. S.,** Conservation of glycerolated cultures of *Ochromonas danica* and *O. malhamensis* at –10°C, *J. Protozool.,* 19, 140–143, 1972.

81. **Ellenbogen, B. B., Hutner, S. H., and Tamburro, K. M.,** Temperature-enhanced osmotic growth requirement of *Crithidia, J. Protozool.,* 19, 349–354, 1972.

82. **Keller, S. E., Hutner, S. H., and Keller, D. E.,** Rearing of the colorless marine Dinoflagellate *Cryptothecodinium cohnii* for use as a biochemical tool, *J. Protozool.,* 15, 792–795, 1968.

83. **Kloetzel, J., Camargo, M. E., and Giovannini, V. L.,** Antigenic differences among epimastigotes, amastigotes, and trypomastigotes of *Trypanosoma cruzi, J. Protozool.,* 22, 259–261, 1975.

84. **O'Daly, J. A.,** A new liquid medium for *Trypanosoma (Schizotrypanum) cruzi, J. Protozool.,* 22, 265–270, 1975.

85. **Chang, K. P.,** Reduced growth of *Blastocrithida culicis* and *Crithidia oncopelti* freed from intracellular symbiotes by chloramphenicol, *J. Protozool.,* 22, 271–276, 1975.

86. **Matthews, H. M. and Daly, J. J.,** Metabolic changes in *Trichomonas gallinae* resulting from growth in various carbohydrates, *J. Protozool.,* 22, 139–145, 1975.

87. **Cantor, M. H. and Burton, M. D.,** Effect of thiamine deprivation and replacement on the mitochondrion of *Polytomella agilis, J. Protozool.,* 22, 135–139, 1975.

88. **Cheng, J. Y., Don-Paul, M., and Antia, N. J.,** Isolation of an unusually stable *cis*-isomer of alloxanthin from a bleached autolyzed culture of *Chroomonas salina* grown photoheterotrophically on glycerol, observations on *cis-trans* isomerization of alloxanthin, *J. Protozool.,* 21, 761–768, 1974.

89.   Graves, L. B., Jr. and Becker, W. M., Beta-oxidation in glyoxysomes from *Euglena, J. Protozool.*, 21, 771–774, 1974.

90.   Braly, P., Simpson, L., and Kretzler, F., Isolation of kinetoplast-mitochondrial complexes from *Leishmania tarentolae, J. Protozool.*, 21, 782–790, 1974.

91.   Alves, M. J. M. and Colli, W., Agglutination of *Trypanosoma cruzi* by concanavalin A, *J. Protozool.*, 21, 575–578, 1974.

92.   Mundim, M. H., Roitman, I., Hermans, M. A., and Kitajima, E. W., Simple nutrition of *Crithidia deanei,* a Reduviid Trypanosomatid with an endosymbiont, *J. Protozool.*, 21, 518–521, 1974.

93.   Murdia, U. S. and Tamhane, D. V., Tryptophan metabolism in *Ochromonas malhamensis, J. Protozool.*, 21, 588–591, 1974.

94.   Segura, E. L., Cura, E. N., Paulone, I., Vasquez, C., and Cerisola, J. A., Antigenic makeup of subcellular fractions of *Trypanosoma cruzi, J. Protozool.*, 21, 571–574, 1974.

95.   Schaefer, F. W., Martin, E., and Mukkada, A. J., The glucose transport system in *Leishmania tropica* promastigotes, *J. Protozool.*, 21, 592–596, 1974.

96.   Giannini, M. S., Effect of promastigote growth phase, frequency of subculture, and host age on promastigote-initiated infections with *Leishmania donovani, J. Protozool.*, 21, 521–527, 1974.

97.   Souza, M. D. C., Reis, A. P., Dias Da Silva, W., and Brener, Z., Mechanism of acquired immunity induced by *Leptomonas pessoai* against *Trypanosoma cruzi* in mice, *J. Protozool.*, 21, 579–584, 1974.

98.   Dubowsky, N., Selectivity of ingestion and digestion in the Chrysomonad flagellate *Ochromonas malhamensis, J. Protozool.*, 21, 259–298, 1974.

99.   Heywood, P., Weinman, D., and Lipman, M., Fine structure of *Trypanosoma cyclops* in noncellular cultures, *J. Protozool.*, 21, 232–238, 1974.

100. Hill, F. G. and Outka, D. E., The structure and origin of mastigonemes in *Ochromonas minute* and *Monas* sp., *J. Protozool.*, 21, 299–312, 1974.

101. Tomas, R. N., Cell division in *Gonyaulax catenella* a marine catenate Dinoflagellate, *J. Protozool.*, 21, 316–321, 1974.

102. Goldberg, B., Lambros, C., Bacchi, C. J., and Hutner, S. H., Inhibition by several standard antiprotozoal drugs of growth and O_2 uptake of cells and particulate preparations of a *Leptomonas, J. Protozool.*, 21, 322–326, 1974.

103. Kempner, E. S. and Miller, J. H., The molecular biology of *Euglena gracilis.* IX. Amino acid pool composition, *J. Protozool.*, 21, 363–367, 1974.

104. Lindmark, D. G. and Muller, M., Biochemical cytology of Trichomonad flagellates. II. Subcellular distribution of oxidoreductases and hydrolases in *Monocercomonas* sp., *J. Protozool.*, 21, 374–378, 1974.

105. Simpson, A. M. and Simpson, L., Labeling of *Crithidia fasciculata* DNA with ^{3}H thymidine, *J. Protozool.*, 21, 379–382, 1974.

106. Simpson, L. and Berliner, J., Isolation of kinetoplast DNA of *Leishmania tarentolae, J. Protozool.*, 21, 382–393, 1974.

107. Mukkada, A. J., Schaefer, F. W., III, Simon, M. W., and Neu, C., Delayed in vitro utilization of glucose by *Leishmania tropica* promastigotes, *J. Protozool.*, 21, 393–397, 1974.

108. Bacchi, C. J., Lambros, C., Ellenbogen, B. B., Penkovsky, L. N., Sullivan, W., Eyinna, E. E., and Hutner, S. H., Drug-resistant *Leptomonas:* cross-resistance in trypanocide-resistant clones, *Antimicrob. Agents Chemother.*, 8, 688–692, 1975.

109. Nielsen, M. H., The ultrastructure of *Trichomonas vaginalis* Donn'e before and after transfer from vaginal secretion to Diamond's medium, *Acta Pathol. Microbiol. Scand.*, 83, 581–589, 1975.

110. Steiger, R. F., Ultracytochemistry of the surface coat pellicle complex in *Trypanosoma brucei, Acta Trop.*, 32, 152–158, 1975.

111. Lasman, M. and Kahan, D., Comparative study on hydrolases in five species of *Ochromonas* (Chrysomonadina), *Arch. Microbiol.*, 104, 185–188, 1975.

112. Stalheim, O. H. and Gallagher, J. E., Effect of *Mycoplasma* spp, *Trichomonas fetus,* and *Campylobacter fetus* on ciliary activity of bovine uterine tube organ cultures, *Am. J. Vet. Res.*, 36, 1077–80, 1975.

113. Hommel, M., Behavior of *Trypanosoma (Herpetosoma) evotomys* under experimental conditions, *Trans. R. Soc. Trop. Med. Hyg.*, 69, 3, 1975.

114. Lopez, T. and Melton, C. G., Jr., Cell volume and DNA content of *Trypanosoma lewisi* and *T. mega* in vitro, *J. Parasitol.*, 6, 209–212, 1975.

115. Hanas, J., Linden, G., and Stuart, K., Mitochondrial and cytoplasmic ribosomes and their activity in blood and culture from *Trypanosoma brucei, J. Cell Biol.*, 65, 103–111, 1975.

116. Baker, J. R., Green, S. M., Chaloner, L. A., and Gaborak, M., *Trypanosoma (Schizotrypanum) dionisii* of *Pipistrellus pipistrellus* (Chiroptera): intra- and extra-cellular development in vitro, *Parasitology,* 65, 251–263, 1972.

117. **Ely, T. H. and Darden, W. H.,** Concentration and purification of the male-inducing substance from *Volvox aureus* M-5, *Microbios,* 5, 51–58, 1972.

118. **Trager, W.,** On the cultivation of *Trypanosoma vivax:* a tale of two visits in Nigeria, *J. Parasitol.,* 61, 3–11, 1975.

119. **Yentsch, C. M., Yentsch, C. S., Owen, C., and Salvaggio, M.,** Stimulatory effect on growth and photosynthesis of the toxic red tide dinoflagellate, *Gonyaulax tamarensis,* with the addition of nitrilotriacetic acid (NTA)N (CH$_2$COOH), *Environ. Lett.,* 6, 231–238, 1974.

120. **Brody, M. and White, J. E.,** Environmental regulation of enzymes in the microbodies of mitochondria of dark-grown, greening and light grown *Euglena gracilis, Dev. Biol.,* 31, 348–361, 1973.

121. **Ardehali, S.,** The effect of homologous and heterologous rabbit antisera on the growth of *Leishmania enriettii,* and *L. tropica* major, and their differentiation by the quantitative Adler test, *Trans. R. Soc. Trop. Med. Hyg.,* 68, 266–267, 1974.

122. **Matthews, H. M. and Daly, J. J.,** Types of growth of *Trichomonas gallinae* in a maltose medium, *J. Parasitol.,* 60, 524–526, 1974.

123. **Matthews, H. M. and Daly, J. J.,** *Trichomonas gallinae:* use of solid medium to test survival under various environmental conditions, *Exp. Parasitol.,* 36, 288–298, 1974.

124. **Winston, R. M.,** The relationship between size and pathogenicity of *Trichomonas vaginalis, J. Obstet. Gynaecol. Br. Commonw.,* 81, 399–404, 1974.

125. **Forsgren, A. and Wallin, J.,** Trinidazole — a new preparation for *Trichomonas vaginalis* infection to laboratory evaluation, *Br. J. Vener. Dis.,* 50, 146–147, 1974.

126. **Clark, B. L., Parsonson, J. M., and Duffy, S. H.,** Experimental infection of bulls with *Tritrichomonas foetus, Aust. Vet. J.,* 50, 189–191, 1974.

127. **Takayanagi, T., Nakatake, Y., and Enriquez, G. L.,** Attachment and ingestion of *Trypanosoma gambiense* to the rat macrophage by special antiserum, *J. Parasitol.,* 60, 336–339, 1974.

128. **Dean, M. P. and Kloetzel, J.,** Lack of protection against *Trypanosoma cruzi* by multiple doses of *T. lewisi* culture forms: a discussion on some strains of "lewisi", *Exp. Parasitol.,* 35, 406–410, 1974.

129. **Luban, W. A. and Dvorak, J. A.,** *Trypanosoma cruzi:* interaction with vertebrate cells in vitro. III. Selection for biological characteristics following intracellular passage, *Exp. Parasitol.,* 36, 143–149, 1974.

130. **Behbehani, K.,** Developmental cycles of *Trypanosoma* (Schizotrypanum) *cruzi* (Chagas, 1909) in mouse peritoneal macrophages in vitro, *Parasitology,* 66, 343–53, 1973.

131. **Bardsley, J. E. and Harmsen, R.,** The trypanosomes of anura, *Adv. Parasitol.,* 11, 1–73, 1973.

132. **Carvalho, A. L. M. and Deane, M. P.,** Trypanosomatidae isolated from *Zelus leucogrammus* (Perty, 1834) (Hemiptera, Reduviidae), with a discussion on flagellates of insectivorous bugs, *J. Protozool.,* 21, 5–8, 1974.

133. **Janovy, J., Jr., Lee, K. W., and Brumbaugh, J. A.,** The differentiation of *Herpetomonas megaseliae:* ultrastructural observations, *J. Protozool.,* 21, 53–59, 1974.

134. **Deboiso, J. F. and Stoppani, A. O. M.,** The mechanism of acetate and pyruvate with *Trypanosoma cruzi, J. Protozool.,* 20, 673–678, 1973.

135. **Cook, J. R. and Li, T. C. C.,** Influence of culture pH on chloroplast structure in *Euglena gracilis, J. Protozool.,* 20, 652–653, 1973.

136. **Karn, R. C. and Hudock, G. A.,** A photorepressible isozyme of malic enzyme in *Euglena gracilis* strain Z, *J. Protozool.,* 20, 316–320, 1973.

137. **Mattern, C. F. T., Honigberg, B. M., and Daniel, W. A.,** Fine-structural changes associated with pseudocyst formation in *Trichomitus batrachorum, J. Protozool.,* 20, 222–229, 1973.

138. **Evans, D. A. and Brown, R. C.,** *m*-Chlorobenzhydroxamic acid — an inhibitor of cyanide-insensitive respiration in *Trypanosoma brucei, J. Protozool.,* 20, 157–160, 1973.

139. **Honigberg, B. M. and Mohn, F. A.,** An improved method for the isolation of highly polymerized native deoxyribonucleic acid from certain protozoa, *J. Protozool.,* 20, 146–150, 1973.

140. **Evans, D. A. and Brown, R. C.,** The utilization of glucose and proline by culture forms of *Trypanosoma brucei, J. Protozool.,* 19, 686–690, 1972.

141. **Krassner, S. M. and Flory, B.,** Proline metabolism in *Leishmania donovani, J. Protozool.,* 19, 682–685, 1972.

142. **Morales, N. M., Schaefer, F. W., Keller, S. J., and Myers, R. R.,** Effect of ethidium bromide and several acridine dyes on the kinetoplast DNA of *Leishmania tropica, J. Protozool.,* 19, 667–672, 1972.

143. **Maclean, F. I. and Amiro, E. R.,** The effect of solute concentration on the shape of the trypanosomatid flagellate *Crithidia fasciculata, Can. J. Microbiol.,* 19, 878–880, 1973.

144. **Chang, K. P. and Trager, W.,** Nutritional significance of symbiotic bacteria in two species of hemoflagellates, *Science,* 183, 531–532, 1974.

145. **Chotkowska, E. and Konopa, J.,** The influence of 1-nitro-9(3'-dimethyl-amino-propyl amino)-acridine on biosynthesis of DNA, RNA and protein in synchronized cultures of *Euglena gracilis, Arch. Immunol. Ther. Exp.,* 21, 767–774, 1973.

146. **Bertaux, O. and Valencia, R.,** Blocking of cell division and malformations induced by vitamin B₁₂ deficiency in synchronic cells of *Euglena gracilis* Z, *C. R. Acad. Sci. Ser. D,* 676, 753–756, 1973.

147. **Cook, J. R.,** Unbalanced growth and replication of chloroplast populations in *Euglena gracilis, J. Gen. Microbiol.,* 75, 51–60, 1973.

148. **Cook, J. R. and Kaiser, H., Jr.,** Factors affecting pH-dependent photoinhibition of division in *Euglena gracilis, J. Cell Physiol.,* 82, 489–495, 1973.

149. **Mo, Y., Harris, B. G., and Gracy, R. W.,** Triosephosphate isomerase and aldolases from light and dark grown *Euglena gracilis, Arch. Biochem. Biophys.,* 157, 580–587, 1973.

150. **McDougald, L. R. and Galloway, R. B.,** Blackhead disease: in vitro isolation of *Histomonas meleagridis* as a potentially useful diagnostic aid, *Avian Dis.,* 17, 847–850, 1973.

151. **Beach, D. H. and Holz, G. G., Jr.,** Environmental influences on the docosahexaenoate content of triacylglycerols and phosphatidyl-choline of a heterotrophic, marine dinoflagellate, *Crypthecodinium cohnii, Biochim. Biophys. Acta.,* 316, 56–65, 1973.

152. **Baker, J. R. and Price, J.,** Growth in vitro of *Trypanosoma cruzi* as amastigotes at temperatures below 37 degrees C, *Int. J. Parasitol.,* 3, 549–551, 1973.

153. **Stohlman, S. A., Kuwahara, S. S., and Kazan, B. H.,** Enzyme, protein, and nucleic acid content of two morphological forms of *Trypanosoma* (Schizotrypanum) *cruzi, Arch. Mikrobiol.,* 92, 301–311, 1973.

154. **Dvorak, J. A. and Hyde, T. P.,** *Trypanosoma cruzi:* interaction with vertebrate cells in vitro. I. Individual interactions at the cellular and subcellular levels, *Exp. Parasitol.,* 34, 268–283, 1973.

155. **Hyde, T. P. and Dvorak, J. A.,** *Trypanosoma cruzi:* interaction with vertebrate cells in vitro. II. Quantitative analysis of the penetration phase, *Exp. Parasitol.,* 34, 284–294, 1973.

156. **Kingston, N. and Morton, J.,** Trypanosomes from elk (*Cervus canadensis*) in Wyoming, *J. Parasitol.,* 56, 1132–1133, 1973.

157. **Cross, G. A. and Manning, J. C.,** Cultivation of *Trypanosoma brucei* spp. in semi-defined and defined media, *Parasitology,* 67, 315–331, 1973.

158. **Cunningham, I.,** Quantitative studies on trypanosomes in tsetse tissue culture, *Exp. Parasitol.,* 33, 34–45, 1973.

159. **Chatterjee, T. and Dutta, A. G.,** Anaerobic formation of succinate from glucose and bicarbonate in resting cells of *Leishmania donovani, Exp. Parasitol.,* 33, 138–146, 1973.

160. **Lemma, A. and Cole, L.,** *Leishmania enrietti:* radiation effects and evaluation of radio attenuated organisms for vaccination, *Exp. Parasitol.,* 35, 161–169, 1974.

161. **Stohlman, V., Mastright, G., and Kazan, B. H.,** Diffusion of nutrients in a biphasic medium for the cultivation of Trypanosomes, *Z. Parasitenkd.,* 41, 231–238, 1973.

162. **Cappuccinelli, P., Pizzoni, E., and Martinetto, P.,** Osservazioni sulla coltivazione di *T. vaginalis* in terreni privi di agar, *G. Batteriol. Virol. Immunol. Ann. Osp. Maria Vittoria Torino Parte I Sez. Microbiol.,* 65, 164–169, 1972.

163. **Cox, P. J. and Nicol, C. S.,** Growth studies of various strains of *T. vaginalis* and possible improvements in laboratory diagnosis of trichomoniasis, *Br. J. Vener. Dis.,* 49, 536–539, 1973.

164. **Kidder, G. W. and Dewey, V. C.,** Methionine or folate and phosphoenolpyruvate in the biosynthesis of threonine in *Crithidia fasciculata, J. Protozool.,* 19, 93–98, 1972.

165. **Hochberg, A., Pimstein, R., and Rahat, R.,** Properties of an ethionine-resistant mutant of *Ochromonas danica, J. Protozool.,* 19, 66–69, 1972.

166. **Brooks, A. E.,** The physiology of *Astrephoneme gubernaculifera, J. Protozool.,* 19, 195–199, 1972.

167. **DaCruz, F. S. and Krassner, S. M.,** Assimilatory sulfate reduction by the hemoflagellate *Leishmania tarentolae, J. Protozool.,* 18, 718–722, 1971.

168. **Hampton, J. R.,** Arginine transport in the culture form of *Trypanosoma cruzi, J. Protozool.,* 18, 701–703, 1971.

169. **Pan, C. T.,** Cultivation and morphogenesis of *Trypanosoma cruzi* in improved liquid media, *J. Protozool.,* 18, 556–560, 1971.

170. **Graves, L. B., Jr.,** Effect of different substrates on glucose uptake and hexokinase activity in *Euglena gracilis, J. Protozool.,* 18, 543–546, 1971.

171. **Mills, L. A. and McCalla, D. R.,** A *Euglena* mutant resistant to N-methyl-N-nitroso-p-toluenesulfonamide, *J. Protozool.,* 18, 538–543, 1971.

172. **Hochberg, A. and Rahat, M.,** Ethionine and methionine metabolism by the Chrysomonad flagellate *Ochromonas danica, J. Protozool.,* 18, 487–490, 1971.

173. **Fair, D. S. and Krassner, S. M.,** Alanine aminotransferase and aspartate aminotransferase in *Leishmania tarentolae, J. Protozool.,* 18, 441–444, 1971.

174. **Ayala, S. C.,** Trypanosomes in wild California sandflies, and extrinsic stages of *Trypanosoma bufophlebotomi, J. Protozool.,* 18, 433–436, 1971.

175. **Rahat, M. and Hochberg, A.,** Ethionine and methionine metabolism by the Chrysomonad flagellate *Prymnesium parvum, J. Protozool.,* 18, 378–382, 1971.

176. **Dwyer, D. M.,** Immunologic analysis by gel diffusion of effects of prolonged cultivation on *Histomonas meleagridis* (Smith), *J. Protozool.,* 18, 372–377, 1971.

177. **Hurlbert, R. E. and Bates, R. C.,** Glucose utilization by *Euglena gracilis* var. *bacillaris* at higher pH, *J. Protozool.,* 18, 298–306, 1971.

178. **Simpson, L. and Braly, P.,** Synchronization of *Leishmania tarentolae* by hydroxyurea, *J. Protozool.,* 17, 511–517, 1970.

179. **Buetow, D. E. and Schuit, K. E.,** Phosphorus and the growth of *Euglena gracilis, J. Protozool.,* 15, 770–773, 1968.

180. **Barna, I. and Weis, D. S.,** The utilization of bacteria as food for *Paramecium bursaria, Trans. Am. Microsc. Soc.,* 93, 434–440, 1973.

181. **van Wagtendonk, W. J.,** The nutrition of ciliates, in *Biochemistry and Physiology of Protozoa,* Vol. 2, Hutner, S. H. and Lwoff, A., Eds., Academic Press, New York, 1955, 57–84.

182. **Seaman, G. R.,** Metabolism of free living ciliates, in *Biochemistry and Physiology of Protozoa,* Vol. 2, Hutner, S. H. and Lwoff, A., Eds., Academic Press, New York, 1955, 91–158.

183. **Holz, G. G., Jr.,** Nutrition and metabolism of ciliates, in *Biochemistry and Physiology of Protozoa,* Vol. 3, Hutner, S. H., Ed., Academic Press, New York, 1964, 199–233.

184. **Holz, G. G., Jr., Wagner, B., Erwin, J., and Kessler, D.,** The nutrition of *Glaucoma chattoni* A., *J. Protozool.,* 8, 192–199, 1961.

185. **Tarantola, V. A. and van Wagtendonk, W. J.,** Further nutritional requirements of *Paramecium aurelia, J. Protozool.,* 6, 189–195, 1959.

186. **van Wagtendonk, W. J.,** *Paramecium – A Recent Survey,* Elsevier, Amsterdam, 1974.

187. **Hunter, N. W.,** Enzymes systems in *Stylonychia pustulata.* III. Hydrolysis of starches and glycogen, *Physiol. Zool.,* 33, 64–67, 1960.

188. **Mavrides, C.,** Regulation of glyconeogenesis from amino acids by acetate in *Tetrahymena pyriformis, Can. J. Biochem.,* 51, 323–331, 1973.

189. **Millecchia, L. L. and Rudzinska, M. A.,** Basal body replication and ciliogenesis in a suctorian, *Tokophrya infusionum, J. Cell Biol.,* 46, 553–563, 1970.

190. **Morrison, G. A. and Tomkins, A. L.,** Determination of mean cell size of *Tetrahymena* in growing cultures, *J. Gen. Microbiol.,* 77, 383–392, 1973.

191. **Rasmussen, L.,** On the role of food vacuole formation in the uptake of dissolved nutrients by *Tetrahymena, Exp. Cell Res.,* 82, 192–196, 1973.

192. **Raugi, G. J., Liang, T., and Blum, J. J.,** Structural organization of three pools of acetyl coenzyme A in *Tetrahymena, J. Biol. Chem.,* 248, 8064–8072, 1973.

193. **Liang, T., Raugi, G. J., and Blum, J. J.,** Effect of tolbutamide on the intracellular flow of acetyl coenzyme A in *Tetrahymena, J. Biol. Chem.,* 248, 8073–8078, 1973.

194. **Raugi, G. J., Liang, T., and Blum, J. J.,** Effect of adenosine monophosphate on intermediate metabolism and ribonucleic acid synthesis in *Tetrahymena, J. Biol. Chem.,* 248, 8079–8085, 1973.

195. **Watanabe, Y.,** Cessation of deoxyribonucleid acid polymerase activity in amino acid-starved *Tetrahymena pyriformis, Exp. Cell Res.,* 81, 8–14, 1973.

196. **Nilsson, J. R., Ricketts, T. R., and Zeuthen, E.,** Effect of cytochalasin B on cell division and vacuole formation in *Tetrahymena pyriformis* GL, *Exp. Cell Res.,* 79, 456–459, 1973.

197. **Coleman, J. R., Nilsson, J. R., Warner, R. R., and Batt, P.,** Effect of calcium and strontium on divalent ion content of refractive granules in *Tetrahymena pyriformis, Exp. Cell Res.,* 80, 1–9, 1973.

198. **Edmunds, L. N., Jr.,** Phasing effect of light on cell division in exponentially increasing grown at low temperatures, *Exp. Cell Res.,* 83, 367–379, 1974.

199. **Upholt, W. B. and Borst, P.,** Accumulation of replicative intermediates of mitochondrial DNA in *Tetrahymena pyriformis* grown in ethidium bromide, *J. Cell Biol.,* 61, 383–397, 1974.

200. **Stocco, D. M. and Zimmerman, A. M.,** Metabolism of acid-soluble nucleotides in starved-refed-synchronized *Tetrahymena pyriformis* GL, *Can. J. Biochem.,* 52, 310–318, 1974.

201. **Rasmussen, L.,** Food vacuole membrane in nutrient uptake by *Tetrahymena, Nature,* 250, 157–158, 1974.

202. **Hoffmann, E. K., Rasmussen, L., and Zeuthen, E.,** Cytochalasin B: aspects of phagocytosis in nutrient uptake in *Tetrahymena, J. Cell Sci.,* 15, 403–406, 1974.

203. **Soldo, A. T. and Godoy, G. A.,** A requirement for an unconjugated pteridine for the growth of *Paramecium aurelia, Biochim. Biophys. Acta,* 362, 521–526, 1974.

204. **Nilsson, J. R.,** Effect of DMSO on vacuole formation, contractile vacuole function, and nuclear division in *Tetrahymena pyriformis* GL, *J. Cell Sci.,* 16, 39–47, 1974.

205. **Levy, M. R.,** Synthesis of glycolytic and peroxisomal enzymes in *Tetrahymena* following a change in culture conditions, *J. Cell Physiol.,* 85, 41–45, 1975.

206. **Orias, E. and Newby, C. J.,** Macronuclear genetics of *Tetrahymena.* II. Macronuclear location of somatic mutations to cycloheximide resistance, *Genetics,* 80, 251–262, 1975.

207. **Jeter, J. R., Jr., Pavlat, W. A., and Cameron, J. L.,** Changes in the nuclear acidic proteins and chromatin structure in starved and refed *Tetrahymena, Exp. Cell Res.,* 93, 79–88, 1975.

208. **Allewell, N. M., Oles, J., and Wolfe, J.,** A physiochemical analysis of conjugation in *Tetrahymena pyriformis, Exp. Cell Res.,* 97, 394–405, 1976.

209. **Lo, H. K., Jasper, D., and Erwin, J. A.,** Ultrastructural alterations in *Tetrahymena pyriformis* induced by growth on saturated phospholipids at 40.1°C, *Tissue Cell,* 8, 19–32, 1976.

210. **Orias, E. and Bruns, P. J.,** Induction and isolation of mutants in *Tetrahymena, Methods Cell Biol.,* 13, 247–282, 1976.

211. **Dickens, M. S., Lucas-Leonard, J., and Roth, J. S.,** Induction of thymidylate synthetase activity in *Tetrahymena* by cyclic guanosine monophosphate, *Biochem. Biophys. Res. Commun.,* 67, 1319–1325, 1975.

212. **Ricci, N., Esposito, F., and Nobili, R.,** Conjugation in *Oxytricha bifaria:* cell interaction, *J. Exp. Zool.,* 192, 343–348, 1975.

213. **Gorovsky, M. A., Yao, M. C., Keevert, J. G., and Pleger, G. L.,** Isolation of micro and macronuclei of *Tetrahymena pyriformis, Methods Cell Biol.,* 9, 311–327, 1975.

214. **Bruns, P. J. and Brussard, T. B.,** Pair formation in *Tetrahymena pyriformis,* an inducible developmental system, *J. Exp. Zool.,* 188, 337–344, 1974.

215. **Coleman, G. S., Davies, J. I., and Cash, J. I.,** The cultivation of the rumen ciliates *Epidinium ecaudatum caudatum* and *Polyplastron multivesiculatum* in vitro, *J. Gen. Microbiol.,* 73, 509–521, 1972.

216. **Bostock, C. J. and Prescott, D. M.,** Evidence of gene diminution during the formation of the macronucleus in the protozoan, *Stylonychia, Proc. Natl. Acad. Sci. U.S.A.,* 69, 139–142, 1972.

217. **Rasmussen, L.,** Cell multiplication in *Tetrahymena* cultures after addition of particulate material, *J. Cell Sci.,* 12, 275–286, 1973.

218. **Hargitt, G. T. and Fray, W. W.,** *Paramecium* in pure cultures of bacteria, *J. Exp. Zool.,* 22, 421–454, 1917.

219. **Hall, R. P.,** *Protozoology,* 3rd ed., Prentice Hall, Englewood Cliffs, N.J., 1964, 428–505.

220. **Soldo, A.,** Cultivation of two strains of *Paramecium aurelia* in axenic medium, *Proc. Soc. Exp. Biol. Med.,* 105, 612, 1960.

221. **Johnson, W. H.,** Further studies on the sterile cultures of *Paramecium, Physiol. Zool.,* 25, 1–10, 1956.

222. **Johnson, W. H. and Miller, C. A.,** A further analysis of the nutrition of *Paramecium, J. Protozool.,* 3, 221–226, 1956.

223. **Seaman, G. R.,** Synthesis of B-vitamins by *Tetrahymena geleii, J. Physiol. Zool.,* 26, 22–28, 1953.

224. **Wu, C. and Hogg, J. F.,** The amino acid composition and nitrogen metabolism of *Tetrahymena geleii, J. Biol. Chem.,* 198, 753–764, 1952.

225. **Johnson, W. H.,** Nutrition of protozoa, *Annu. Rev. Microbiol.,* 10, 193–211, 1956.

226. **Patterson, D. and Sleigh, M. A.,** Behavior of contractile vacuole of *Tetrahymena pyriformis* W: a redescription with comments on terminology, *J. Protozool.,* 23, 410–417, 1976.

227. **Fedriani, C., Torres, A., and Perez-Silva, J.,** Application of coriphosphine staining to the study of macronuclear anlage of *Stylonychia mytilus, J. Protozool.,* 23, 417–420, 1976.

228. **Cronkite, D. L.,** A role of calcium ions in chemical induction of mating in *Paramecium tetraurelia, J. Protozool.,* 23, 431–433, 1976.

229. **Schwartz, J. R. L. and Hampton, J. R.,** Cultivation of *Pseudocohnilembus persalinus* in axenic medium, *J. Protozool.,* 23, 443–444, 1976.

230. **Hutner, S. H. and Corliss, J. O.,** Search for clues to the evolutionary meaning of ciliate phylogeny, *J. Protozool.,* 23, 48–56, 1976.

231. **Liang, T., Raugi, G. J., and Blum, J. J.,** Effect of 4-pentenoic acid on intermediate metabolism of *Tetrahymena, J. Protozool.,* 23, 186–193, 1976.

232. **Conner, R. L. and Stewart, B. Y.,** The effect of temperature on the fatty acid composition of *Tetrahymena pyriformis* WH-14, *J. Protozool.,* 23, 193–196, 1976.

233. **Vaudaux, P.,** Isolation and identification of specific cortical proteins in *Tetrahymena pyriformis, J. Protozool.,* 23, 458–464, 1976.

234. **Freeman, M. and Moner, J. G.,** The uptake of pyrimidine nucleosides in *Tetrahymena.* I. Uridine, *J. Protozool.,* 23, 465–472, 1976.

235. **Liang, T., Raugi, G. J., and Blum, J. J.,** Inhibition of P-enolpyruvate carboxykinase and glycogenesis in *Tetrahymena* by 3-mercaptopicolinic acid, *J. Protozool.,* 23, 473–477, 1976.

236. **Mishima, S.,** Effect of adenine on *Paramecium multimicronucleatum J. Protozool.,* 22, 443–447, 1975.

237. **Osborne, J. A. and Lee, D.,** Studies on the conditions required for optimum recovery of *Tetrahymena pyriformis* strain S (Phenoset A) after freezing to and thawing from −196°C, *J. Protozool.,* 22, 233–237, 1975.

238. **Rasmussen, L., Bushse, H. E., Jr., and Groh, K.,** Efficiency of filter feeding in two species of *Tetrahymena, J. Protozool.,* 22, 110–111, 1975.

239. **Eldan, M. and Blum, J. J.,** Presence of nonoxidative enzymes of the pentose phosphate shunt in *Tetrahymena, J. Protozool.,* 22, 145–149, 1975.

240. **Rothstein, T. L. and Blum, J. J.,** Lysosomal physiology in *Tetrahymena.* II. Effect of culture age and temperature on the extracellular release of three acid hydrolases, *J. Protozool.,* 21, 163–168, 1974.

241. **Koroly, M. J. and Conner, R. L.,** The fate of RNA degradation products in starved cultures of *Tetrahymena pyriformis, J. Protozool.,* 21, 169–177, 1974.

242. **Conner, R. L. and Koroly, M. J.,** Relationship of cellular energetics to RNA metabolism in *Tetrahymena pyriformis* W, *J. Protozool.,* 21, 177–182, 1974.

243. **Ricketts, T. R. and Rappitt, A. F.,** Determination of the volume and surface area of *Tetrahymena pyriformis,* and their relationship to endocytosis, *J. Protozool.,* 21, 549–551, 1974.

244. **Saji, M. and Osawa, F.,** Mechanism of photoaccumulation in *Paramecium bursaria, J. Protozool.,* 21, 556–561, 1974.

245. **McLaughlin, D., Johnson, G. R., and Bradley, C. J.,** Growth of the peritrich *Opercularia coarctata* in axenic culture, *J. Protozool.,* 21, 561–564, 1974.

246. **Miyake, A. and Noblish, R.,** Mating reaction and daily rhythm in *Euplotes crassus, J. Protozool.,* 21, 584–587, 1974.

247. **Brown, J. A. and Nielsen, P. J.,** Transfer of photosynthetically produced carbohydrate from endosymbiotic chlorellae to *Paramecium bursaria, J. Protozool.,* 21, 569–570, 1974.

248. **Finley, H. E. and McLaughlin, D.,** Cultivation of the peritrich *Telotrochidium henneguyi* in axenic and non-axenic media, *J. Protozool.,* 12, 41–47, 1965.

249. **Soldo, A. T. and van Wagtendonk, W. J.,** An analysis of the nutritional requirements for fatty acids of *Paramecium aurelia, J. Protozool.,* 14, 596–600, 1967.

250. **Rasmussen, L. and Kludt, T. A.,** Particulate material as a prerequisite for rapid cell multiplication in *Tetrahymena* cultures, *Exp. Cell Res.,* 59, 457–463, 1970.

251. **Hall, R. P.,** Effect of certain metal ions on growth of *Tetrahymena pyriformis, J. Protozool.,* 1, 74–79, 1954.

252. **Levine, L.,** Axenizing *Vorticella convallaria, J. Protozool.,* 6, 169–171, 1959.

253. **Finley, H. E., McLaughlin, D., and Harrison, D. H.,** Non-axenic and axenic growth of *Vorticella microstoma, J. Protozool.,* 6, 201–205, 1959.

254. **Burbanck, W. D. and Martin, V. L.,** Experimental microbial populations thirty-five years later: the influence of food on the symbiosis of *Paramecium aurelia* syngen 4, 51.7, *J. Protozool.,* 20, 135–138, 1973.

255. **Kemper, D. L., Thompson, S. H., and Parsons, J. A.,** Variation in glyoxylate bypass inducibility among strains of *Tetrahymena pyriformis, J. Protozool.,* 20, 467–470, 1973.

256. **Soldo, A. T., Godoy, G. A., and van Wagtendonk, W. J.,** Growth of particle-bearing and particle-free *Paramecium aurelia* in axenic culture, *J. Protozool.,* 13, 492–497, 1966.

257. **Rudzinska, M. A., Jackson, G. J., and Tuffrau, M.,** The fine structure of *Colpoda maupasi* with special emphasis on food vacuoles, *J. Protozool.,* 13, 440–459, 1966.

258. **Hascall, G. K.,** The stalk of the suctorian *Tokophrya infusionum:* histochemistry, biochemistry and physiology, *J. Protozool.,* 20, 701–704, 1973.

259. **Tawada, K. and Miyamoto, H.,** Sensitivity of *Paramecium* thermotaxis to temperature change, *J. Protozool.,* 20, 289–292, 1973.

260. **Mavrides, C., Whitlow, K. J., and D'Iorio, A.,** The conversion of phenylalanine to tyrosine in *Tetrahymena pyriformis, J. Protozool.,* 20, 342–344, 1973.

261. **Rifkin, J. L.,** The role of the contractile vacuole in the osmoregulation of *Tetrahymena pyriformis, J. Protozool.,* 20, 108–114, 1973.

262. **Cox, D., Frank, O., Hutner, S. H., and Baker, H.,** Growth of *Tetrahymena* in carbohydrate-free high-glutamate media, *J. Protozool.,* 15, 713–716, 1968.

263. **Millecchia, L. and Rudzinska, M. A.,** An abnormality in the life cycle of *Tokophrya infusionum, J. Protozool.,* 15, 665–673, 1968.

264. **Cox, D.,** Prolonged survival of *Tetrahymena* at 0–5°C in citrated, lecithinized defined media, *J. Protozool.,* 17, 150–152, 1970.

265. **Frank, O., Baker, H., and Hutner, S. H.,** Evaluation of protein quality with the phagotrophic protozoan *Tetrahymena,* in *Protein Nutritional Quality of Foods and Feeds,* Friedman, M., Ed., Marcel Dekker, New York, 1975, 203–209.

266. **Soldo, A. T. and Merlin, E. J.,** The cultivation of symbiote-free marine ciliates in axenic medium, *J. Protozool.,* 19, 519–524, 1972.

267. **Dewey, V. C. and Kidder, G. W.,** Proline metabolism in *Tetrahymena, J. Protozool.,* 19, 50–53, 1972.

268. **Tawada, K. and Osawa, F.,** Response of *Paramecium* to temperature change, *J. Protozool.,* 19, 53–57, 1972.

269. **Nicolette, J. A., Buhse, H. E., Jr., and Robin, M. S.,** The effect of 2-mercapto-1-(beta-4-pyridethyl) benzimidazole (MPB) on cell differentiation and RNA synthesis in the protozoon *Tetrahymena vorax, J. Protozool.,* 18, 87–90, 1971.

270. **Smith, S. G. and Giese, A. C.,** Axenic media for *Blepharisma intermedium, J. Protozool.,* 14, 649–654, 1967.

271. **Burchill, B. R.,** Conjugation in *Stentor coeruleus, J. Protozool.,* 14, 683–687, 1967.

272. **Hill, D. L., Straight, S., and Allan, P. W.,** Use of *Tetrahymena pyriformis* to evaluate the effects of purine and pyrimidine anologs, *J. Protozool.,* 17, 619–623, 1970.

273. **Brizzi, G. and Blum, J. J.,** Effect of growth conditions on serotonin content of *Tetrahymena pyriformis, J. Protozool.,* 17, 553–555, 1970.

274. **Conner, R. L. and Linden, C.,** Purine and pyrimidine 5′-nucleotide catabolism in *Tetrahymena pyriformis* W, *J. Protozool.,* 17, 659–662, 1970.

275. **Cameron, I. L. and Jeter, J. R., Jr.,** Synchronization of the cell cycle of *Tetrahymena* by starvation and refeeding, *J. Protozool.,* 17, 429–431, 1970.

276. **Meskill, V. P.,** Factors influencing the growth of *Glaucoma chatoni* in a chemically defined medium, *J. Protozool.,* 17, 104–107, 1970.

277. **Soldo, A. T. and van Wagtendonk, W. J.,** The nutrition of *Paramecium aurelia, J. Protozool.,* 16, 500–506, 1969.

278. **Thompson, G. A., Jr.,** The properties of an enzyme system degrading endogenous phospholipids of *Tetrahymena pyriformis, J. Protozool.,* 16, 397–400, 1969.

279. **Hilden, S. and Giese, A. C.,** Effect of salt concentration on regeneration rate in *Blepharisma* acclimated to high salt levels, *J. Protozool.,* 16, 419–422, 1969.

280. **Frankel, J.,** Participation of the undulating membrane in the formation of oral replacement primordia in *Tetrahymena pyriformis, J. Protozool.,* 16, 26–35, 1969.

281. **Reynolds, H.,** An apparent carbohydrate-amino acid interaction in *Tetrahymena pyriformis, J. Protozool.,* 16, 204–210, 1969.

282. **Jerka-Dziadosz, M. and Frankel, J.,** An analysis of the formation of ciliary primordia in the hypotrich ciliate *Urostyla weissei, J. Protozool.,* 16, 612–637, 1969.

283. **Holm, B. J.,** Inhibition of deoxyribonuclease activity in synchronized *Tetrahymena, J. Protozool.,* 16, 655–659, 1969.

284. **Kasturi, A. R. and Srihari, K.,** The effect of temperature on *Blepharisma intermedium, J. Protozool.,* 16, 738–743, 1969.

285. **Reid, R., Cox, D., Baker, H., and Frank, O.,** Phytosterols and other lipids as survival factors for *Tetrahymena* at 0–5°C, *J. Protozool.,* 16, 231–235, 1969.

286. **Stillwell, R. H.,** *Colpidium*-produced RNA as a growth stimulant for *Tetrahymena, J. Protozool.,* 14, 19–22, 1967.

287. **Conner, R. L. and Cline, S. G.,** Some factors governing respiration, glucose metabolism and iodoacetate sensitivity in *Tetrahymena pyriformis, J. Protozool.,* 14, 22–26, 1967.

288. **Allison, B. M. and Ronkin, R. R.,** Lipid cytochemistry and morphologic change in aging populations of *Tetrahymena, J. Protozool.,* 14, 313–319, 1967.

289. **Elliott, A. M. and Clemmons, G. L.,** An ultrastructural study of ingestion and digestion in *Tetrahymena pyriformis, J. Protozool.,* 13, 311–323, 1966.

290. **Rosenbaum, N., Erwin, J., Beach, D., and Holz, G. G., Jr.,** The induction of a phospholipid requirement and morphological abnormalities in *Tetrahymena pyriformis* by growth at supraoptimal temperatures, *J. Protozool.,* 13, 535–546, 1966.

291. **Everhart, L. P., Jr. and Ronkin, R. R.,** Changes in lipids of cells from aging population of *Tetrahymena pyriformis, J. Protozool.,* 13, 646–650, 1966.

292. **Holz, G. G., Jr.,** Is *Tetrahymena* a "plant?" *J. Protozool.,* 13, 2–4, 1966.

293. **Tompkin, R. B., Purser, D. B., and Weiser, H. H.,** Influence of rumen fluid source upon establishment and cultivation in vitro of the rumen protozoan *Entodinium, J. Protozool.,* 13, 55–58, 1966.

294. **Rosenbaum, J. L. and Holz, G. G., Jr.,** Amino acid activation in subcellular fractions of *Tetrahymena pyriformis, J. Protozool.,* 13, 115–123, 1966.

295. **Wang, G. T. and Marquardt, W. C.,** Survival of *Tetrahymena pyriformis* and *Paramecium aurelia* following freezing, *J. Protozool.,* 13, 123–128, 1966.

296. **Mueller, M., Rohlich, P., and Toro, I.,** Studies on feeding and digestion in protozoa. VII. Ingestion of polystyrene latex particles and its early effect of acid phosphatase in *Paramecium multimicronucleatum* and *Tetrahymena pyriformis, J. Protozool.,* 12, 27–34, 1965.

297. **Corliss, J. O.,** Guide to the literature on *Tetrahymena:* a companion piece to Elliott's "General Bibliography," *Trans. Am. Microsc. Soc.,* 92, 468–491, 1973.

298. **Corliss, J. O.,** *Tetrahymena* and some thoughts on the evolutionary origin of endoparasitism, *Trans. Am. Microsc. Soc.,* 91, 566–573, 1972.

299. **Weis, D. S.,** A medium for the axenic culture of chlorella bearing *Paramecium bursaria* in the light, *Trans. Am. Microsc. Soc.,* 94, 109–117, 1975.

300. **Walker, G. K., Maugel, T. K., and Goode, D.,** Some ultrastructural observations on encystment in *Stylonychia mytilus* (Ciliophora: Hypotrichida), *Trans. Am. Microsc. Soc.,* 94, 147–154, 1975.

301. **McKee, R. W.,** Biochemistry of *Plasmodium* and the influence of antimalarials, in *Biochemistry and Physiology of Protozoa,* Vol. 1, Lwoff, A., Ed., Academic Press, New York, 1951, 251–322.

302. **Goodwin, L. G. and Rollo, I. M.,** The chemotherapy of malaria, piroplasmosis, trypanosomiasis, and leishmaniasis, in *Biochemistry and Physiology of Protozoa,* Vol.2, Hutner, S. H. and Lwoff, A., Eds., Academic Press, New York, 1955, 225–276.

303. **Rollo, I. M.,** The chemotherapy of malaria, in *Biochemistry and Physiology of Protozoa,* Vol. 3, Hutner, S. H., Ed., Academic Press, New York, 1964, 525–561.

304. **Honigberg, B. M.,** Chemistry of parasitism among some protozoa, in *Chemical Zoology,* Vol. 1, Florkin, M. and Scheer, B. T., Eds., Academic Press, New York, 1967, 695–814.

305. **Mitchell, G. H., Butcher, G. A., Voller, A., and Cohen, S.,** The effect of human immune IgG on the in vitro development of *Plasmodium falciparum, Parasitology,* 72, 149–162, 1976.

306. **Jensen, J. B. and Edgar, S. A.,** Effects of antiphagocytic agents on penetration of *Eimeria magna* sporozoites in cultured cells, *J. Parasitol.,* 62, 203–206, 1976.

307. **Siddiqui, W. A., Schnell, J. V., and Richmond-Crum, S. M.,** In vitro cultivation of *Plasmodium falciparum* at high parasitemia, *Am. J. Trop. Med. Hyg.,* 23, 1015–1018, 1974.

308. **Rosales-Ronquillo, M. C. and Silverman, P. H.,** In vitro ookinete development of the rodent malaria parasite, *Plasmodium berghei, J. Parasitol.,* 60, 819–824, 1974.

309. **Smalley, M. E. and Butcher, G. A.,** The in vitro culture of blood stages of *Plasmodium berghei, Int. J. Parasitol.,* 5, 131–132, 1975.

310. **Williams, S. G. and Richards, W. H.,** Malaria studies in vitro. I. Techniques for the preparation and culture of leucocyte-free blood-dilution cultures of *Plasmodia, Ann. Trop. Med. Parasitol.,* 67, 169–178, 1973.

311. **Ball, G. H. and Chao, J.,** Use of amino acids by *Plasmodium relictum* oocysts in vitro *Exp. Parasitol.,* 39, 115–118, 1976.

312. **Siddiqui, W. A., Schnell, J. V., and Richmond-Crum, S. M.,** Effects of red cell extract on in vitro growth and multiplication of malaria parasites, *J. Parasitol.,* 61, 189–193, 1975.

313. **Siddiqui, W. A. and Schnell, J. V.,** Use of various buffers for in vitro cultivation of malaria parasites, *J. Parasitol.,* 61, 59, 516–519, 1973.

314. **Siddiqui, W. A., Schnell, J. V., and Geiman, Q. M.,** In vitro cultivation of *Plasmodium malariae, J. Parasitol.,* 58, 804, 1972.

315. **Beaudoin, R. L., Strome, C. P. A., and Clutter, W. G.,** Cultivation of avian malaria parasites in mammalian liver cells, *Exp. Parasitol.,* 36, 355–359, 1974.

316. **Booden, T. and Hull, R. W.,** Nucleic acid precursor synthesis by *Plasmodium lophurae* parasitizing chicken erythrocytes, *Exp. Parasitol.,* 34, 220–228, 1973.

317. **Sherman, I. W. and Tanigoshi, L.,** Incorporation of ^{14}C-amino acids by malarial parasite (*Plasmodium lophurae*). VI. Changes in the kinetic constants of amino acid transport during infection, *Exp. Parasitol.,* 35, 369–373, 1974.

318. **Sinden, R. E. and Croll, N. A.,** Cytology and kinetics of microgametogenesis and fertilization in *Plasmodium yoelii nigeriensis, Parasitology,* 70, 53–65, 1975.

319. **Chiodini, P. L.,** In vitro culture of *Babesia, Trans. R. Soc. Trop. Med. Hyg.,* 67, 27–28, 1973.

320. **Homewood, C. A., Atkinson, E. M., and Peters, W.,** I. Carbohydrate metabolism in *P. berghei:* preliminary observations. II. Chloroquine-induced pigment clumping in *P. berghei:* dependence on composition of the medium, *Trans. R. Soc. Trop. Med. Hyg.,* 67, 26–27, 1973.

321. **Ball, G. H. and Chao, J.,** The complete development of the sporogonous stages of *Hepatozoon rarefaciens* cultured in a *Culex pipiens* cell line, *J. Parasitol.,* 59, 513–515, 1973.

322. **Bedrnik, P.,** Antitoxoplasma activity of coccidiostatics, *Folia Parasitol.* (Prague), 19, 129–132, 1972.

323. **Bollinger, R. O., Musallam, N., and Stulberg, C. S.,** Freeze preservation of tissue culture propagated *Toxoplasma gondii, J. Parasitol.,* 60, 368–369, 1974.

324. **Fayer, R., Melton, M. L., and Sheffield, H. G.,** Quinine inhibition of host cell penetration by *Toxoplasma gondii, Besnoifia jellisoni,* and *Sarcocystis* sp. in vitro, *J. Parasitol.,* 58, 595–599, 1972.

325. **Roberts, C. O., Chaparas, S. D., and McLaughlin, D.,** The use of the plaque assay in chemotherapeutic and dermal hypersensitivity studies on *Toxoplasma gondii, Trans. Am. Microsc. Soc.,* 95, 470–482, 1976.

326. **Fayer, R. and Thompson, D. E.,** *Isospora felis:* development in cultured cells with some cytological observations, *J. Parasitol.,* 60, 160–168, 1974.

327. **Rose, M. E.,** Immunity to *Eimeria maxima:* reactions of antisera in vitro and protection in vivo, *J. Parasitol.,* 60, 528–530, 1974.

328. **Millard, B. J. and Long, P. L.,** The viability and survival of sporozoites of *Eimeria* in vitro, *Int. J. Parasitol.,* 4, 423–432, 1974.

329. **Quellette, C. A., Strout, R. G., and McDougald, L. R.,** Thymidylic acid synthesis in *Eimeria tenella* (Coccidia) cultured in vitro, *J. Protozool.,* 21, 398–400, 1974.

330. **Speer, C. A., Hammond, D. M., and Anderson, L. C.,** Development of *Eimeria callospermophili* and *E. bilamella* from the Uinta ground squirrel *Spermophilus armatus* in cultured cells, *J. Protozool.,* 17, 274–284, 1970.

331. **Ryley, J. F. and Wilson, R. G.,** The development of *Eimeria burnetti* in tissue culture, *J. Parasitol.,* 58, 660–663, 1972.

332. **McDougald, L. R. and Jeffers, T. K.,** Comparative in vitro development of precocious and normal strains of *Eimeria tenella* (Coccidia), *J. Protozool.,* 23, 530–534, 1976.

333. **Weiss, M. M. and Vanderberg, J. P.,** Studies of *Plasmodium* ookinetes. I. Isolation and concentration from mosquito midguts, *J. Protozool.,* 23, 547–551, 1976.

334. **Wolf, K. and Markiw, M.,** *Myxosoma cerebralis:* in vitro sporulation of the myxosporidan of salmonid whirling disease, *J. Protozool.,* 23, 425–427, 1976.

335. **Speer, C. A. and Danforth, H. D.,** Fine-structural aspects of microgametogenesis of *Eimeria magna* in rabbits and in kidney cell cultures, *J. Protozool.,* 23, 109–115, 1976.

336. **Langreth, S. G.,** Feeding mechanisms in extracellular *Babesia microti* and *Plasmodium lophurae, J. Protozool.,* 23, 215–223, 1976.

337. **Kilejian, A.,** Does a histidine-rich protein from *Plasmodium lophurae* have a function in merozoite penetration? *J. Protozool.,* 23, 272–277, 1976.

338. **Jensen, J. B. and Hammond, D. M.,** Ultrastructure of the invasion of *Eimeria magna* sporozoites into cultured cells, *J. Protozool.,* 22, 411–415, 1975.

339. **Coombs, G. H. and Gutteridge, W. E.,** Growth in vitro and metabolism of *Plasmodium vinckei chabaudi, J. Protozool.,* 22, 555–560, 1975.

340. **Vanderberg, J. P.,** Studies on the motility of *Plasmodium* sporozoites, *J. Protozool.,* 21, 527–537, 1974.

341. **Ouellette, C. A., Strout, R. G., and McDougald, L. R.,** Incorporation of radioactive pyrimidine nucleosides into DNA and RNA of *Eimeria tenella* (Coccidia) cultured in vitro, *J. Protozool.,* 20, 150–153, 1973.

342. **Manwell, R. D.,** The lesser haemosporidina, *J. Protozool.,* 12, 1–9, 1965.

343. **Trager, W.,** Adenosine triphosphate and the pyruvic and phosphoglyceric kinases of the malaria parasite *Plasmodium lophurae, J. Protozool.,* 14, 110–114, 1967.

344. **Doran, D. J. and Vetterling, J. M.,** Comparative cultivation of poultry coccidia in mammalian kidney cell cultures, *J. Protozool.,* 14, 657–662, 1967.

345. **Fayer, R. and Hammond, D. M.,** Development of first-generation Schizonts of *Eimeria bovis* in cultured bovine cells, *J. Protozool.,* 14, 764–772, 1967.

346. **Doran, D. J. and Vetterling, J. M.,** Survival and development of *Eimeria meleagrimitis* Tyzzer, 1929 in bovine kidney and turkey cell cultures, *J. Protozool.,* 15, 796–802, 1968.

347. **Hammond, D. M., Fayer, R., and Miner, M. L.,** Further studies on in vitro development of *Eimeria bovis* and attempts to obtain second-generation schizonts, *J. Protozool.,* 16, 298–302, 1969.

348. **Clark, W. and Hammond, D. M.,** Development of *Eimeria auburnensis* in cell cultures, *J. Protozool.,* 16, 646–654, 1969.

349. **Hibbert, L. E., Hammond, D. M., and Simmonds, J. R.,** The effect of pH, buffers, bile, and bile acids on excystation of sporozoites of various *Eimeria* species, *J. Protozool.,* 16, 441–444, 1969.

350. **Kelley, G. L. and Hammond, D. M.,** Development of *Eimeria ninakohlyakimova* from sheep in cell cultures, *J. Protozool.,* 17, 340–349, 1970.

351. **Fayer, R., Romanowski, R. D., and Vetterling, J. M.,** The influence of hyaluronidase and hyaluronidase substrates on penetration of cultured cells by eimerian sporozoites, *J. Protozool.,* 17, 432–436, 1970.

352. **Sampson, J. R., Hammond, D. M., and Ernst, J. V.,** Development of *Eimeria alabamensis* from cattle in mammalian cell cultures, *J. Protozool.,* 18, 120–128, 1971.

353. **Trager, W.,** Further studies on the effects of antipantothenates on malaria parasites, *J. Protozool.,* 18, 232–239, 1971.

354. **Trager, W.,** A new method for intraerythrocytic cultivation of malaria parasites, *J. Protozool.,* 18, 239–242, 1971.

355. **Trager, W.,** Malaria parasites (*Plasmodium lophurae*) developing extracellularly in vitro: incorporation of labeled precursors, *J. Protozool.,* 18, 392–399, 1971.

356. **Perrotto, J., Keister, D. B., and Gelderman, A. H.,** Incorporation of precursors into *Toxoplasma* DNA, *J. Protozool.,* 18, 470–473, 1971.

Plants

CULTURE MEDIA FOR BRYOPHYTES

D. V. Basile

Bryophytes, like flowering plants, are photoautotrophs. As such they theoretically need only be provided with CO_2, H_2O, and salt solutions containing the elements N, P, S, Ca, K, Mg, Fe, Mn, Mo, Cu, Co, Zn, B, and possibly Cr, Cl, V, Si, and Al, from which they can synthesize all else that is necessary to their nutrition. However, even though less than 0.5% of approximately 21,000 species of bryophytes have been cultured on chemically defined media, there is reason to believe that some bryophytes are not fully photoautotrophic, i.e., growth and development of some species are significantly slower on inorganic media than on mineral media supplemented with organics of various sorts; these species, then, can be considered auxotrophic. A few species investigated will grow but not develop a normal morphology on mineral media. Whether these latter would be considered auxotrophs or heterotrophs is a moot point.

Bryophytes have been successfully cultured for more than 70 years. In general, three methods of culture have been employed: (1) "clay pot cultures," (2) agnotobiotic cultures, and (3) axenic cultures. Clay pot cultures are cultures of plants on their natural substrata. This requires that the natural substratum be collected along with the plants to be cultivated or that their natural substrata be simulated as closely as possible.[1-3] Nutrients are presumably derived from the substratum, the atmosphere, and the water used to irrigate the cultures. Some inferences (based on the nature of the substrata) about bryophyte nutrition can be made from this method, but precise information is not possible. Agnotobiotic cultures are those in which a nonsterile "artificial" substratum such as sand, cinders, peat, filter paper, etc., is used to support the plants, and nutrients are provided as nonsterile solutions of known composition.[4,5] Useful information regarding bryophyte nutrition has been obtained using this procedure, but the influence of other organisms which invariably contaminate the system limits the use of this culture method for rigorous studies on bryophyte nutrition. Axenic cultures are those in which the culture vessels, supporting substratum, nutrient medium, and plants to be cultured are freed of any contaminants. This method has the greatest potential for obtaining precise information about the nutritional requirements of individual species of bryophytes.

A relatively large number of mineral media formulations have been successfully used to culture bryophytes over the past 72 years. These media, both macro- and micronutrient, have been adapted for bryophyte culture with little or no change from media originally developed for algae or flowering plants. It is not possible to state with confidence which is the best formulation for any particular bryophyte. Extensive comparative studies have not been reported. What follows are formulations of different components (i.e., macronutrients, micronutrients, and organic supplements) of media that have been used to culture a variety of bryophytes. Most of the mineral media, adjusted to a suitable pH, would probably support growth of fully autotrophic bryophytes in either agnotobiotic or axenic culture. Which would best support a sufficiently rapid, characteristic mode of development can only be determined empirically.

MACRONUTRIENT MEDIA

The salt compositions of 18 macronutrient media are given in Table 1. When iron salts are not indicated it is because they were supplied as "trace" elements or not listed at all. Iron will be considered again in the section on micronutrients.

Table 1
SALT COMPOSITION OF SOME MACRONUTRIENT MEDIA USED TO CULTURE BRYOPHYTES

Salt	Medium																	
	A[6]	B[7]	C[8]	D[9]	E[10]	F[11]	G[12]	H[13]	I[14]	J[15]	K[16]	L[17]	M[18]	N[19]	O[20]	P[21]	Q[22]	R[23]
NH_4NO_3	0.2	0.5	1.0	1.0	0.5	—	—	—	—	—	—	—	—	—	—	—	—	—
$(NH_4)_2SO_4$	—	—	—	—	—	0.5	—	—	—	—	—	—	—	—	—	—	—	—
$(NH_4)_2HPO_4$	—	—	—	—	—	—	0.02	—	—	—	—	—	—	—	—	—	—	—
$(NH_4)_3PO_4$	—	—	0.5	0.5	—	—	—	—	—	—	—	—	—	—	—	—	—	—
$Mg(NO_3)_2 \cdot 6H_2O$	—	—	—	—	—	—	—	0.15	—	—	—	—	—	—	—	—	—	—
$Ca(NO_3)_2 \cdot 4H_2O$	—	—	—	—	0.25	1.0	—	0.165	0.5	1.0	0.12	0.26	0.03	1.0	0.5	—	—	—
KNO_3	—	—	—	—	—	—	0.1	0.08	0.125	0.25	1.2	0.08	—	—	—	1.2	—	—
$NaNO_3$	—	—	—	—	—	—	—	—	—	—	—	—	—	—	—	—	0.375	0.25
$CaCl_2$	0.1	0.1	—	—	—	—	0.05	—	—	—	—	—	—	—	—	—	0.125	0.025
$CaSO_4$	—	—	0.5	0.5	—	—	—	—	—	—	—	—	—	—	—	—	—	—
$MgSO_4 \cdot 7H_2O$	0.1	0.2	0.5	0.5	0.25	0.25	0.02	0.196	0.125	0.2	0.24	0.735	0.075	0.25	0.125	0.3	0.125	0.075
$MgCl_2 \cdot 6H_2O$	—	—	—	—	—	—	—	—	—	—	—	—	0.075	—	0.125	0.3	—	—
K_2HPO_4	0.1	0.2	0.5	—	—	—	—	—	—	—	—	—	—	—	—	0.8	—	0.075
KH_2PO_4	—	—	—	0.20	—	0.25	—	0.054	0.125	0.25	0.12	—	—	0.25	0.125	—	0.125	0.175
KCl	—	—	—	—	—	—	—	—	—	—	0.12	0.06	0.0036	0.12	—	—	0.06	—
K_2SO_4	—	0.5	0.5	0.5	0.25	—	—	—	—	—	—	—	—	—	—	—	—	—
$NaCl$	—	—	—	—	—	—	—	—	—	—	—	—	—	—	—	—	—	0.025
$Na_2HPO_4 \cdot 2H_2O$	—	—	—	0.05	—	—	—	—	—	—	—	0.165	—	—	—	—	—	—
NaH_2PO_4	—	—	—	—	—	—	—	—	—	—	—	0.20	—	—	—	—	—	—
Na_2SO_4	—	—	—	—	—	—	—	—	—	0.2	—	—	—	—	—	—	—	—
$Fe_3(PO_4)_2$	0.004	—	—	—	—	—	—	—	—	—	—	—	—	—	—	—	—	—
$FeCl_3 \cdot 6H_2O$	—	—	0.01	—	0.001	—	—	—	0.01	—	—	—	—	—	—	—	—	—
$Fe_2(SO_4)_3 \cdot 7H_2O$	—	—	—	0.01	—	—	—	—	—	—	—	—	—	—	—	—	—	—
Total salt concentration (g/l)	0.448	0.873	2.639	2.638	1.286	1.567	0.170	0.431	0.658	1.663	1.519	1.044	0.137	1.187	0.591	2.386	0.640	0.587

From Basile, D. V., *Bryologist*, 78, 401—413, 1975. With permission.

The grouping of media formulations was determined primarily by the manner in which nitrogen is supplied. Media A through G supply nitrogen as both ammonium and nitrate while media H through R supply it only as nitrate. Within these two categories the media are ordered according to whether nitrogen is supplied by a single salt (A to C and M to R), by two salts (D to G and I to L), or three salts (H). Finally, within each subcategory, the media are grouped according to their greatest similarity in overall composition.

Ionic Composition of Media

The ionic compositions of the 23 salts listed in Table 1 are given in Table 2. Concentrations of each ion for each medium is expressed in three ways: parts per million, milliequivalents, and millimoles. For consideration of ionic balances and interactions, milliequivalents is the most useful expression of concentration. Careful study of the milliequivalent ratios of ions known to have physiological relationship (e.g., NH_4^+ and NO_3^-, Ca^{++} and Mg^{++}, Ca^{++} and K^+, K^+ and Na^+) reveal that they can be very different among the 18 media. It cannot be predicted, however, what influence the different ionic ratios would have on the growth and development of a given bryophyte.

Hydrogen Ion Concentration of Media

The unadjusted pH values of 17 of the macronutrient media are given in Table 3. The pH of the media can be seen to range from about 4.4 to 7.9. The pH range of natural substrata for bryophytes is between 3.0 and 8.0.[24,25] Although many bryophytes can tolerate a relatively wide range of pH, others seem to have relatively narrow tolerances. Therefore, the pH of each medium would have to be adjusted to correspond to the tolerances of the particular bryophyte to be cultured. Normal solutions of NaOH or HCl are commonly used to adjust the pH of culture media. Marked differences in pH change can occur when a fixed amount of N NaOH or N HCl are added to each of the media as indicated in Table 3. The difference in the magnitude of response to added base or acid may be indicative of different internal buffering capacities of the media. This can be an important consideration for many experiments, especially long-term ones. Buffers such as 2-(N-morpholino) ethansulfonic acid (MES) have been used in bryophyte culture media.[11] There are some disadvantages to this practice, however. Relatively large amounts of NaOH may be needed to adjust media upward from an initially low pH, and media containing MES must be cold-sterilized through membrane filters having pore diameters of 0.45 μm or less.

Some Suggested Macronutrient Media

While all the media listed in Table 1 have been found to be suitable sources of macronutrients for some bryophytes, probably none would be ideal for all bryophytes. It would be impractical, however, to test the suitability of each of 18 macronutrient solutions for culturing a particular bryophyte. It is suggested that at least three distinctly different macronutrient formulations be tried when attempting to choose one that will best support rapid growth and a characteristic pattern of development for a specific bryophyte: one medium with ammonium as a sole nitrogen source, one with nitrate as a sole source, and one with both amonium andnitrate. At this point, however, there are insufficient data for specifying which of the 18. Table 4 provides formulations of seven media from which a choice can be made. The first medium (F′) was modified from medium F[26] to provide ammonium as the sole source of nitrogen. Any of the media F, G′, and A will supply both ammonium and nitrate ions. A medium with nitrate as a sole nitrogen source can be selected from media I, L, and R. Table 5 gives the ionic composition of these seven selected macronutrient media.

Table 2
IONIC COMPOSITION OF SOME MACRONUTRIENT MEDIA USED FOR BRYOPHYTE CULTURE

Ion		A	B	C	D	E	F	G	H	I	J	K	L	M	N	O	P	Q	R
											Medium[a]								
NH_4^+	ppm[b]	45.0	112.5	225.0	406.0	112.5	136.4	5.45	—	—	—	—	—	—	—	—	—	—	—
	meq[c]	2.5	6.25	12.5	22.5	6.25	7.57	0.30	—	—	—	—	—	—	—	—	—	—	—
	mmol[d]	2.5	6.25	12.5	22.5	6.25	7.57	0.30	—	—	—	—	—	—	—	—	—	—	—
NO_3^-	ppm	155.0	387.5	775.0	775.0	518.8	525.2	61.39	208.5	339.4	909.6	799.7	185.7	15.76	525.4	262.7	736.6	273.5	182.4
	meq	2.5	6.25	12.5	12.5	8.37	8.47	0.99	3.36	5.47	14.67	12.89	2.99	0.25	8.47	4.24	11.88	4.41	2.94
	mmol	2.5	6.25	12.5	12.5	8.37	8.47	0.99	3.36	5.47	14.67	12.89	2.99	0.25	8.47	4.24	11.88	4.41	2.94
Ca^{++}	ppm	36.36	27.39	116.5	116.5	42.45	172.9	11.19	28.52	84.92	243.9	20.34	44.16	5.08	169.8	84.92	69.91	7.56	9.09
	meq	1.82	1.36	5.82	5.82	2.12	8.64	0.56	1.42	4.24	12.18	1.02	2.21	0.26	8.50	4.24	3.50	0.38	0.46
	mmol	0.91	0.68	2.91	2.91	1.06	4.32	0.28	0.71	2.12	6.09	0.51	1.10	0.13	4.25	2.12	1.75	0.19	0.23
Mg^{++}	ppm	9.88	19.76	49.40	49.40	24.31	24.71	1.95	33.59	12.36	4.00	23.41	72.66	7.41	24.62	27.31	29.66	12.19	7.32
	meq	0.82	1.62	4.06	4.06	2.0	2.04	0.16	2.76	1.02	3.30	1.92	5.98	0.60	2.02	2.24	2.44	1.0	0.60
	mmol	0.41	0.81	2.03	2.03	1.0	1.02	0.08	1.38	0.51	1.65	0.96	2.99	0.30	1.01	1.12	1.22	0.5	0.30
$H_2PO_4^-$	ppm	69.85	139.7	—	—	142.6	174.6	—	37.72	87.31	178.3	83.82	130.6	—	174.6	87.32	—	87.31	124.8
	meq	0.74	1.47	—	—	1.47	1.84	—	0.39	0.92	1.84	0.88	1.38	—	1.84	0.92	—	0.92	1.29
	mmol	0.74	1.47	—	—	1.47	1.84	—	0.39	0.92	1.84	0.88	1.38	—	1.84	0.92	—	0.92	1.29
HPO_4^{--}	ppm	—	—	275.9	—	26.8	—	13.97	—	—	—	—	—	41.37	—	—	441.4	—	41.4
	meq	—	—	5.74	—	0.56	—	0.30	—	—	—	—	—	0.86	—	—	9.18	—	0.86
	mmol	—	—	2.87	—	0.28	—	0.15	—	—	—	—	—	0.43	—	—	4.59	—	0.43
SO_4^{--}	ppm	39.02	78.04	757.2	757.2	235.5	461.2	35.71	76.49	48.78	65.0	93.66	422.0	29.27	97.56	48.78	284.5	48.78	29.26
	meq	0.82	1.62	15.78	15.78	4.90	9.60	0.74	1.58	1.02	1.36	1.96	8.79	0.60	2.04	1.02	5.92	1.02	0.60
	mmol	0.41	0.81	7.89	7.89	2.45	4.80	0.37	0.79	0.51	0.68	0.98	4.39	0.30	1.02	0.51	2.96	0.51	0.30
K^+	ppm	28.67	57.5	449.4	224.7	168.1	71.87	38.71	46.38	84.33	168.2	497.7	62.67	35.61	135.3	35.94	824.1	67.47	83.80
	meq	0.74	1.47	11.49	5.76	4.31	1.84	0.99	1.19	2.16	4.31	12.76	1.61	0.91	3.47	0.92	21.13	1.73	2.15
	mmol	0.74	1.47	11.49	5.76	4.31	1.84	0.99	1.19	2.16	4.31	12.76	1.61	0.91	3.47	0.92	21.13	1.73	2.15
Na^+	ppm	—	—	—	—	12.85	—	—	—	—	—	—	96.37	—	—	—	—	101.4	77.56
	meq	—	—	—	—	0.56	—	—	—	—	—	—	4.07	—	—	—	—	4.41	3.37
	mmol	—	—	—	—	0.56	—	—	—	—	—	—	4.07	—	—	—	—	4.41	3.37

Table 2 (continued)
IONIC COMPOSITION OF SOME MACRONUTRIENT MEDIA USED FOR BRYOPHYTE CULTURE

Ion		Medium[a]																		
		A	B	C	D	E	F	G	H	I	J	K	L	M	N	O	P	Q	R
Cl^-	ppm	63.63	47.95	—	—	0.40	—	—	—	3.90	—	—	28.38	1.70	56.76	43.32	—	41.64	30.99
	meq	1.82	1.37	—	—	0.01	—	—	—	0.11	—	—	0.81	0.05	1.62	1.24	—	1.19	0.89
	mmol	1.82	1.37	—	—	0.01	—	—	—	0.11	—	—	0.81	0.05	1.62	1.24	—	1.19	0.89
Fe^{++} or	ppm	0.80	—	2.79	2.79	0.20	—	—	—	2.08	93.8	—	—	—	—	—	—	—	—
Fe^{+++}	meq	0.03	—	0.15	0.15	0.012	—	—	—	0.12	3.36	—	—	—	—	—	—	—	—
	mmol	0.01	—	0.05	0.05	0.004	—	—	—	0.04	1.68	—	—	—	—	—	—	—	—

[a] See Table 1 for references to media.
[b] Parts per million.
[c] Milliequivalents.
[d] Millimoles.

From Basile, D. V., *Bryologist*, 78, 401—413, 1975. With permission.

Table 3
UNADJUSTED pH VALUES OF SOME MACRONUTRIENT MEDIA USED TO CULTURE BRYOPHYTES[a]

Medium[b]

pH	A	B	C	D	E	F	G	H	I	J	K	L	M	N	O	P	Q	R
$\bar{m}$[c]	5.03	4.95	7.32	—	6.13	4.80	7.42	5.25	4.93	4.42	4.97	4.36	7.88	4.83	4.90	7.50	4.99	6.34
s[d]	0.09	0.02	0.11	—	0.12	0.02	0.03	0.21	0.11	0.05	0.07	0.05	0.13	0.03	0.06	0.08	0.07	0.04
Δ	2.05	1.61	0.28		0.56	1.40	2.38	4.06	1.94	1.33	1.90	0.53	2.42	1.37	1.88	0.24	1.85	0.54

[a] The change in pH (ΔpH) in response to 0.5 ml N NaOH is given as an *indication* of the buffering capacity of each medium.

[b] See Table 1 for references to media.

[c] Mean of three measurements.

[d] Standard deviation.

From Basile, D. V., *Bryologist*, 78, 401–413, 1975. With permission.

Table 4

SALT COMPOSITION OF SEVEN SELECTED MACRONUTRIENT MEDIA FOR BRYOPHYTE CULTURE

Salt	F'^b	F	G'^b	A	I	L	R
			Medium[a]				
NH_4NO_3	—	—	—	0.2	—	—	—
$(NH_4)_2SO_4$	0.5	0.5	—	—	—	—	—
$(NH_4)_2HPO_4$	—	—	0.2	—	—	—	—
$Ca(NO_3)_2 \cdot 4H_2O$	—	1.0	—	—	0.5	0.26	—
KNO_3	—	—	0.1	—	0.125	0.08	—
$NaNO_3$	—	—	—	—	—	—	0.25
$CaCl_2 \cdot 2H_2O$	0.5	—	0.05	0.1	—	—	0.025
$MgSO_4 \cdot 7H_2O$	0.5	0.25	0.05	0.1	0.125	0.735	0.075
K_2HPO_4	—	—	—	—	—	—	0.075
KH_2PO_4	0.25	0.25	—	0.1	0.125	—	0.175
KCl	—	—	—	—	—	0.06	—
NaCl	—	—	—	—	—	—	0.025
NaH_2PO_4	—	—	—	—	—	0.165	—
Na_2SO_4	—	—	—	—	—	0.02	—
Total salt concentration (g/l)	1.362	1.565	0.360	0.448	0.658	1.044	0.587

[a] See Table 1 for references to media.

[b] F' and G' are modifications of F and G, respectively.

From Basile, D. V., *Bryologist,* 78, 401–413, 1975. With permission.

MICRONUTRIENTS OR "TRACE ELEMENTS"

The micronutrient requirements of bryophytes have not been well studied;[27] indeed, many, if not most, earlier workers did not consciously supply micronutrients other than iron (and sometimes not even iron). The practice of not supplying micronutrients was then possible, and to some extent still is, because they are often present in sufficient amounts as contaminants of the macronutrient salts, glassware, water, and/or agar used. With the use of highly purified reagents, scrupulously cleaned culture vessels, washed agar, and redistilled, deionized water, the addition of micronutrients becomes necessary.

Table 6 provides formulations of eight trace element supplements which may be used for culturing bryophytes. None of these were specifically designed for bryophytes, and probably any balanced mixture of micronutrients found suitable for culturing plants or plant parts would suffice. In the absence of detailed comparative studies on which to base a selection, the choice of micronutrient supplements remains somewhat arbitrary. Nevertheless, the most sophisticated supplement given in Table 6 is Medium II. (Medium II is a refinement of the trace element supplement, Medium I, successfully used in my laboratory for almost 15 years; both formulations are the products of extensive research by Seymore Hutner and his collaborators.) The amounts given are for preparing liter quantities of stock solutions at 1000X concentration (i.e., 1 ml stock/liter culture medium). It may prove desirable to use somewhat more or less than a milliliter per liter for some solutions and some bryophytes.[17,27]

Table 7 lists the concentrations of trace elements that are supplied by the salts listed in Table 6. It is obvious from this table that none of the formulations supply all the trace elements known or suspected to be essential to growth and development of bryophytes. Medium VIII supplies the fewest and Medium II supplies the most since in Medium III the essentiality of Be and Ti is doubtful. Equally apparent is the considerable difference in

Table 5
IONIC COMPOSITION OF SEVEN SELECTED MACRONUTRIENT MEDIA FOR BRYOPHYTE CULTURE

Ion		Medium[a]						
		F'[b]	F	G'[b]	A	I	L	R
NH_4^+	ppm[c]	136.4	136.4	54.55	45.0	—	—	—
	meq[d]	7.57	7.57	3.03	2.5	—	—	—
	mmol[e]	7.57	7.57	3.03	2.5	—	—	—
NO_3^-	ppm	—	525.2	61.39	155.0	339.4	185.7	182.4
	meq	—	8.47	0.99	2.5	5.47	2.99	2.94
	mmol	—	8.47	0.99	2.5	5.47	2.99	2.94
Ca^{++}	ppm	136.9	172.9	13.69	36.36	84.92	44.16	9.09
	meq	6.84	8.64	0.68	1.82	4.24	2.21	0.46
	mmol	3.42	4.32	0.34	0.91	2.12	1.10	0.23
Mg^{++}	ppm	47.46	24.71	4.75	9.88	12.36	72.66	7.32
	meq	3.91	2.04	0.39	0.82	1.02	5.98	0.60
	mmol	1.95	1.02	0.19	0.41	0.51	2.99	0.30
$H_2PO_4^-$	ppm	178.6	178.6	—	69.85	87.31	130.6	124.8
	meq	1.84	1.84	—	0.74	0.92	1.38	1.29
	mmol	1.84	1.84	—	0.74	0.92	1.38	1.29
HPO_4^{--}	ppm	—	—	145.5	—	—	—	41.4
	meq	—	—	3.04	—	—	—	0.86
	mmol	—	—	1.52	—	—	—	0.43
SO_4^{--}	ppm	551.1	461.2	18.75	39.02	48.78	422.0	29.26
	meq	11.48	9.60	0.39	0.82	1.02	8.79	0.60
	mmol	5.74	4.80	0.19	0.41	0.51	4.39	0.30
K^+	ppm	71.87	71.87	38.71	28.67	84.33	62.67	83.80
	meq	1.84	1.84	0.99	0.74	2.16	1.61	2.15
	mmol	1.84	1.84	0.99	0.74	2.16	1.61	2.15
Na^+	ppm	—	—	—	—	—	96.37	77.56
	meq	—	—	—	—	—	4.07	3.37
	mmol	—	—	—	—	—	4.07	3.37
Cl^-	ppm	239.7	—	23.97	63.63	—	28.38	30.99
	meq	6.85	—	0.68	1.82	—	0.81	0.89
	mmol	6.85	—	0.68	1.82	—	0.81	0.89

[a] See Table 1 for references to media.
[b] F' and G' are modifications of F and G, respectively.
[c] Parts per million.
[d] Milliequivalents.
[e] Millimoles.

From Basile, D. V., *Bryologist*, 78, 401—413, 1975. With permission.

Table 6
SALT COMPOSITION (GRAMS) PER LITER OF MICRONUTRIENT SUPPLEMENTS FOR BRYOPHYTE CULTURE MEDIA[a]

Salt	Gravimetric number[c]	I[d]	II[d]	III	IV	V	VI	VII	VIII
$Fe(NH_4)_2SO_4 \cdot 6H_2O$	7.02	7.0	4.2	—	—	—	—	—	—
$FeCl_3 \cdot 6H_2O$	4.84	—	—	—	—	1.0	—	—	—
$FeC_6H_5O_7 \cdot 7H_2O$	6.64	—	—	—	2.5[f]	—	—	—	—
$Fe_2(SO_4)_3$	7.16	—	—	5.0	—	—	—	—	—
$FeSO_4 \cdot 7H_2O$	4.98	—	—	—	—	—	—	0.15[g]	0.02
$Fe_2(C_4H_4O_6)_3$	4.97	—	—	—	—	—	1.825	—	—
H_2SO_4	—	—	—	1 ml	0.5 ml	—	—	—	—
$MnCl_2 \cdot 4H_2O$[e]	3.60	—	—	—	—	—	—	0.905	—
$MnSO_4 \cdot H_2O$[e]	3.08	3.04	1.55	2.24	3.0	0.076	0.5	—	0.20
$CrK(SO_4)_2 \cdot 12H_2O$	9.6	—	0.096	—	—	—	—	—	—
$ZnCl_2$	2.08	—	—	—	—	—	0.625	—	0.20
$ZnSO_4 \cdot 7H_2O$	4.4	2.2	2.2	0.10	0.5	1.0	—	0.11	—
$BeSO_4$	11.66	—	—	0.10	—	—	—	—	—
$(NH_4)_6(Mo_7)_{24} \cdot 4H_2O$	12.88	0.0736	0.36	—	—	—	—	—	—
$H_2MoO_4 \cdot H_2O$	1.86	—	—	—	—	—	—	0.045	—
$NaMoO_4 \cdot 2H_2O$	2.52	—	—	—	0.025	—	0.252	—	—
$TiSO_4$	2.2	—	—	0.20	—	—	—	—	—
$CuCl_3 \cdot 2H_2O$	2.68	—	—	—	—	—	0.268	—	—
$CuSO_4$[e]	2.51	0.394	0.10	0.05	0.016	0.019	—	0.026	—
$CoCl_2 \cdot 6H_2O$[e]	4.04	—	—	0.092	0.046	—	—	—	—
$CoSO_4 \cdot 7H_2O$	4.77	0.238	0.048	—	—	—	—	—	—
H_3BO_3	5.72	0.0572	0.057	0.05	0.5	1.0	0.57	1.43	—
$Na_2B_4O_7$	18.6	—	—	—	—	—	—	—	0.20
$AlCl_3$	4.94	—	—	—	—	0.03	—	—	—
$NiCl_2 \cdot 6H_2O$	4.05	—	—	0.092	—	0.03	—	—	—
$NiSO_4 \cdot 6H_2O$	4.48	—	0.045	—	—	—	—	—	—
KI	1.31	—	—	0.5	—	0.01	—	—	—
NH_4VO_3	2.3	—	0.046	—	—	—	—	—	—
$Na_3VO_4 \cdot 16H_2O$	9.27	0.0462	—	—	—	—	—	—	—

[a] Normally, 1 ml of micronutrient solution is added to one liter of macronutrient medium.

[b] References for the numbered media: I,[11] II,[28] III,[29] IV,[17] V,[10] VI,[7] VII,[30] VIII.[13,31]

[c] Formula weight salt/formula weight trace element.

[d] To prepare, make up 1% solution (w/v) of total salts in 0.05% aqueous 5-sulfosalicylic acid and heat to dissolve. Bring up to liter volume with distilled/deionized water.

[e] Adjusted to compensate for different waters of hydration of salts used by different investigators.

[f] Prepared separately by adding 2.5 g to one liter water and heating to dissolve (see Footnote g, below).

[g] Prepared separately as a 0.5% solution in 0.4% aqueous tartaric acid. A suggested alternative to this solution and of above is 2.78 g $FeSO_4 \cdot 7H_2O$ and 3.73 g Na_2 EDTA dissolved with heating in a liter of distilled/deionized water; 1 ml/l supplies Fe at about 0.5 ppm.

Table 7
TRACE ELEMENT CONCENTRATION, IN PARTS PER MILLION, OF MICRONUTRIENT SUPPLEMENTS FOR BRYOPHYTE CULTURE MEDIA

	Medium[a]							
	I	II	III	IV	V	VI	VII	VIII
Fe	1	0.6	0.7	0.4	0.2	0.4	0.03	0.004
Mn	1	0.5	0.7	1	0.025	0.2	0.25	0.07
Zn	0.5	0.5	0.02	0.1	0.2	0.3	0.025	0.1
Cu	1.6	0.04	0.02	0.006	0.008	0.1	0.01	—
B	0.01	0.01	0.009	0.009	0.2	0.1	0.25	0.01
Co	0.05	0.01	0.02	0.01	—	—	—	—
Mo	0.006	0.03	—	0.01	—	0.1	0.02	—
Ni	—	0.01	0.02	—	0.007	—	—	—
V	0.005	0.02	—	—	—	—	—	—
I	—	—	0.4	—	0.008	—	—	—
Cr	—	0.01	—	—	—	—	—	—
Al	—	—	—	—	0.006	—	—	—
Be	—	—	0.009	—	—	—	—	—
Ti	—	—	0.09	—	—	—	—	—

[a] See Table 6 for references to media.

concentration of a particular trace element supplied by the different supplements. For some purposes it may be more desirable to adjust the concentrations of specific elements rather than the concentration of the entire mixture. The adjusted amount of salt needed to make up the stock solution can easily be determined as follows: salt concentration (g/l) = trace element concentration (ppm) × gravimetric number × 10^3. The gravimetric number (formula weight salt/formula weight element) for each salt is given in Table 6.

ORGANIC NUTRITION

Over 100 years ago, the ineffable Kerner Von Marilaun[32] observed that many mosses and liverworts were partial saprophytes, taking nutriment in the form of organic compounds from products of decay in their substrata. The exceedingly limited data that can be gleaned from a scattered literature tend to support Kerner's contention. No systematic investigation as to which of the organic compounds present in bryophyte substrata are utilized have been performed, although Killian[9] began such an inquiry.

Table 8 provides some information on organic additives to bryophyte culture media which have reportedly stimulated growth and/or development of one or more species of bryophyte. These data do not represent an exhaustive search of the literature, but are only meant to be indicative and to provide a basis for designing better media. The reports of stimulation were not always based on carefully controlled, quantitative experiments, so generalizations can only be made with reservations.

Sugars seem to be the most widely used type of organic additive to bryophyte culture media. Probably most bryophytes would be stimulated by glucose or sucrose, and at least one *Sphagnum* species will not grow in culture without an exogenous carbon source such as sucrose or glucose.[36] Hatcher[17] cultured 58 species in 36 genera of hepatics on a medium containing 0.5% sucrose. Basile (unpublished) cultured 42 species in 32 genera on several media containing 1% glucose. Concentrations of these sugars between 0.5 and 1% are generally suitable, but some bryophytes are optimally stimulated at higher

Table 8
ORGANIC ADDITIVES TO NUTRIENT MEDIA
REPORTED STIMULATORY TO GROWTH AND/OR DEVELOPMENT
OF ONE OR MORE BRYOPHYTES

Compounds	Concentration range	Ref.
Sugars		
Sucrose	0.5–1%(–2%)	17, 28, 31, 33–37
Glucose	0.5–1%(–2%)	9, 28, 33, 36–42
Fructose	0.5–1%(–2%)	28, 33, 36, 37
Amino Acids, Singly or Defined Groups		
L-Alanine	$10^{-6}M{-}10^{-3}M$	16, 43, 44
L-Arginine	$10^{-6}M{-}10^{-3}M$	7, 42, 45
L-Asparagine	$10^{-6}M{-}10^{-3}M$	9, 16, 21, 43
L-Aspartic acid	$10^{-6}M{-}10^{-3}M$	44
L-Cysteine	$10^{-6}M{-}10^{-3}M$	45
L-Glutamic acid	$10^{-6}M{-}10^{-3}M$	42, 44
L-Glycine	$10^{-6}M{-}10^{-3}M$	7, 16, 43, 44
L-Histidine	$10^{-6}M{-}10^{-3}M$	42
Hydroxy-L-proline	$10^{-6}M{-}10^{-3}M$	44
L-Leucine	$10^{-6}M{-}10^{-3}M$	16
L-Proline	$10^{-6}M{-}10^{-3}M$	7, 44–47
L-Tyrosine	$10^{-6}M{-}10^{-3}M$	7
Amino Acid Complexes		
Casein hydrolysate, vitamin free	0.01–0.05%	44
Peptone	0.01–0.05%	9, 16, 38, 39
Tryptone	0.01–0.05%	31, 41, 44
Yeast extract	0.01–0.05%	16, 31, 41, 48

concentrations. On the other hand, exogenous sugar has been found to be incompatible with normal growth and development of some bryophytes.

Amino acids have been relatively infrequently used but in a few cases have profoundly stimulated growth and/or development. Because they can interfere with each others' uptake and/or metabolism, they should be tested singly at first and over a range of concentrations — between 10^{-3} M and 10^{-6} M in my judgment.

Complex mixtures of amino acids and other nutrients from protein hydrolysates such as peptone, tryptone, casein hydrolysate, or from yeast extract have proven stimulatory to some bryophytes. These complexes, like the individual amino acids, should be tried over a range of concentrations. Concentrations between 0.2 and 0.04% w/v are indicated in the literature, but these levels may be too high. Probably levels between 0.05 and 0.01% w/v would be more reasonable.

It is likely that other organic compounds from products of decay are utilized by some bryophytes and conceivably play an important role in their nutrition. The subject of bryophyte nutrition warrants much more study, not only for its relevance to plant nutrition *per se* but also for its relationship to plant development and plant ecology.

REFERENCES

1. **Richards, P. W.,** *Trans. Br. Bryol. Soc.,* 1, 1–3, 1947.
2. **Berrie, G. K.,** *Trans. Br. Bryol. Soc.,* 1, 485, 1951.
3. **Matzke, E. B.,** *Bryologist,* 67, 136–141, 1964.
4. **Schelpe, E. A.,** *Trans. Br. Bryol. Soc.,* 2, 216–219, 1952.
5. **Schneider, M. J., Voth, P. D., and Troxler, R. F.,** *Bot. Gaz.,* 128, 169–174, 1967.
6. **Beneke, W.,** *Bot. Zeitung.,* 61, 19–46, 1903.
7. **Burkholder, P. R.,** *Bryologist,* 62, 6–15, 1959.
8. **Marchal, E. and Marchal, E.,** *Bull. Soc. R. Bot. Belg.,* 43, 115–214, 1906.
9. **Killian, C.,** *C. R. Soc. Biol. Strasbourg (Paris),* 88, 746–748, 1923.
10. **Kofler, L.,** *Rev. Bryol. Lichenol.,* 28, 1–202, 1959.
11. **Basile, D. V.,** *Bull. Torrey Bot. Club,* 100, 350–352, 1973.
12. **Waris, H.,** *Physiol. Plant.,* 6, 538–543, 1953.
13. **Voth, P. D.,** *Bot. Gaz.,* 104, 591–601, 1943.
14. **Hurey-Py, G.,** *C. R. Acad. Sci.,* 227, 1256–1258, 1948.
15. **Meyer, S. L.,** *Bryologist,* 51, 213–217, 1948.
16. **Pringsheim, E. G. and Pringsheim, O.,** *Jahrb. Wiss. Bot.,* 82, 310–332, 1935.
17. **Hatcher, R. E.,** *Bryologist,* 68, 227–231, 1965.
18. **Dachnowski, A.,** *Jahrb. Wiss. Bot.,* 44, 254–286, 1907.
19. **Bopp, M.,** *Z. Bot.,* 40, 119–152, 1952.
20. **Detmer, W.,** *Das Pflanzenphysiologie Praktikum,* G. Fisher Verlag, Jena, E. Germany, 1912.
21. **Buch, H.,** *Overs. Fin. Vetensk. Soc. Forh.,* 62, 1–46, 1920.
22. **Nehira, K.,** *Hikobia J. Hiroshima Bot. Club,* 2, 185–189, 1961.
23. **Bold, H.,** *Bull. Torrey Bot. Club,* 76, 101–108, 1949.
24. **Apinis, A. and Dioguės, A. M.,** *Acta Hortic. Bot. Univ. Latviensis,* 8, 1–19, 1933.
25. **Apinis, A. and Lacis, L.,** *Acta Hortic. Bot. Univ. Latviensis,* 9–10, 1–100, 1934–1935.
26. **Basile, D. V.,** *Bryologist,* 78, 401–413, 1975.
27. **Hoffmann, G. R.,** *Bryologist,* 69, 182–192, 1966.
28. **Hutner, S. H.,** *Ann. Rev. Microbiol.,* 26, 313–345, 1972.
29. **Allsopp, A. and Ilahi, I.,** *Phytomorphology,* 19, 242–253, 1969.
30. **Yang, B.,** *Taiwania,* 15, 301–318, 1970.
31. **Iverson, G. B.,** *Bryologist,* 60, 348–358, 1957.
32. **Kerner von Marilaun, A. J.,** *The Natural History of Plants,* (Oliver, F. W. et al., transl.), Blackie & Son, London, 1895.
33. **Robbins, W. J.,** *Bot. Gaz.,* 65, 542-551, 1918.
34. **Machlis, L.,** *Plant Physiol.,* 15, 354–362, 1962.
35. **Belkengren, R. O.,** *Am. J. Bot.,* 49, 567–571, 1962.
36. **Simola, L. K.,** *Physiol. Plant.,* 22, 1079–1084, 1969.
37. **Chopra, R. N. and Sood, S.,** *Phytomorphology,* 23, 230–243, 1973.
38. **Servettaz, C.,** *Ann. Sci. Nat. Bot. IXe,* 17, 111–224, 1913.
39. **Von Ubisch, G.,** *Ber. Dtsch. Bot. Ges.,* 31, 543–552, 1913.
40. **Fries, N.,** *Bot. Not.,* 1945, 417–424, 1945.
41. **Diller, V. M., Fulford, M., and Kersten, H. J.,** *Bryologist,* 58, 173–192, 1955.
42. **Basile, D. V.,** *Am. J. Bot.,* 52, 443–454, 1965.
43. **Keilova-Kleckova, V.,** *Preslia,* 31, 166–178, 1959.
44. **Sharma, K. K., Diller, V. M., and Fulford, M.,** *Bryologist,* 63, 203–212, 1960.
45. **Lockwood, L.,** *Am. J. Bot.,* 62, 893–900, 1975.
46. **Basile, D. V.,** *Bull. Torrey Bot. Club,* 95, 127–134, 1968.
47. **Yang, B.,** *Taiwania,* 15, 199–209, 1970b.
48. **Sironal, C.,** *Bull. Soc. R. Bot. Belg.,* 79, 48–78, 1947.

CULTURE MEDIA (NATURAL AND SYNTHETIC): PTERIDOPHYTES

J. D. Caponetti

INTRODUCTION

The pteridophytes are a group of vascular plants which do not bear seeds. They are commonly referred to as the lower vascular plants, and include the so-called "fern allies" and the true ferns. The fern allies are the homosporous whisk ferns (Psilotaceae); club mosses, ground pines, or trailing evergreens (Lycopodiaceae); horsetails, scouring rushes, or pipes (Equisetaceae); the small heterosporous spike mosses (Selaginellaceae); and quillworts (Isoetaceae). The true ferns include the homosporous adder's tongue ferns and grape ferns (Ophioglossaceae); the marattiaceous ferns (Marattiaceae); the polypodiaceous ferns (Osmundaceae, Gleicheniaceae, Schizaeaceae, Polypodiaceae, etc.); the filmy ferns (Hymenophyllaceae); the tree ferns (Cyatheaceae); and the heterosporous water ferns (Marsileaceae, Salviniaceae).

Since pteridophytes do not bear seeds, their dispersal unit is usually the spore. Therefore, spores are the most practical starting point for culture. In many cases, the gametophytes that develop from germinated spores can be propagated by fragmentation (pieces of the gametophyte). The sporophytes that develop on gametophytes after fertilization or by apogamy can also be propagated by fragmentation (stem cuttings or bulbils, for example).

The exogenous nutritional requirements of pteridophytes are rather simple and uncomplicated. Like seeds, spores generally contain an adequate amount of stored nutrients to permit germination when exposed to sufficient water. Those resulting gametophytes that contain chlorophyll can supply their own organic nutrition by the process of photosynthesis; they need only be supplied with the inorganic mineral elements that are common to plants. The same is true for sporophytes. Gametophytes that contain no chlorophyll must have an exogenous source of organic nutrition available besides the inorganic mineral elements and water. These non-green gametophytes are usually subterranean and require organic substrates. Certain species of the Psilotaceae, Lycopodiaceae, Selaginellaceae, Isoetaceae, Equisetaceae, and the Ophioglossaceae have non-green gametophytes. In general, most species of the Marattiaceae and other ferns have chlorophyll-containing gametophytes. The sporophytes of all pteridophytes contain chlorophyll.

NUTRITIONAL MEDIA FOR PTERIDOPHYTES

Natural

1. Moist sandy loam with or without peat moss, leaf compost, or bark mulch for spores, gametophytes, and sporophytes.

2. Pieces of cement block, brick, limestone, or clay pot partially submerged in water in suitable, covered, clear glass or plastic containers for spores and subsequent gametophytes.

Synthetic (Chemically Defined)

1. Sterilized or nonsterilized liquid media containing mineral elements in special culture dishes (e.g., petri dishes) or covered, clear glass or plastic containers for floating spores and subsequent gametophytes.

2. Sterilized agar-solidified media containing mineral elements in special culture dishes (e.g., petri dishes) or culture tubes for spores, subsequent gametophytes, and small sporophytes.

The members of all pteridophyte families can be propagated on natural nutritional media in suitable containers under sterile or nonsterile conditions. However, for experimental purposes under controlled conditions, media of known composition are preferred. The essential mineral elements for successful pteridophyte propagation, along with their occurrence in the three most commonly employed synthetic nutritional media, are shown in Table 1.

Table 1
MINERAL ELEMENTS FOR CULTURING PTERIDOPHYTES

Element desired	Compound name	Compound formula	Occurence in media[a] Knop's	Knudson's	Moore's
A. Calcium and nitrogen	1. Calcium nitrate	$Ca(NO_3)_2 \cdot 4H_2O$	+	+	−
	2. Calcium chloride	$CaCl_2 \cdot 2H_2O$	−	−	+
	3. Potassium nitrate	KNO_3	+	−	−
	4. Ammonium nitrate	NH_4NO_3	−	−	+
	5. Ammonium sulfate	$(NH_4)_2SO_4$	−	+	−
B. Magnesium and sulfur	1. Magnesium sulfate	$MgSO_4 \cdot 7H_2O$	+	+	+
C. Potassium and phosphorus	1. Monobasic potassium phosphate	KH_2PO_4	+	−	−
	2. Dibasic potassium phosphate	K_2HPO_4	−	+	+
D. Iron	1. Ferric citrate	$FeC_6H_5O_7 \cdot 5H_2O$	+	+	+
E. Manganese	1. Manganous chloride	$MnCl_2 \cdot 4H_2O$	+	+	−
	2. Manganous sulfate	$MnSO_4 \cdot H_2O$	−	−	+
F. Boron	1. Boric acid	H_3BO_3	+	+	+
G. Zinc	1. Zinc sulfate	$ZnSO_4 \cdot 7H_2O$	+	+	+
H. Cobalt	1. Cobalt chloride	$CoCl_2 \cdot 6H_2O$	+	+	+
I. Copper	1. Cupric chloride	$CuCl_2 \cdot 2H_2O$	+	+	−
	2. Cupric sulfate	$CuSO_4 \cdot 5H_2O$	−	−	+
J. Molybdenum	1. Sodium molybdate	$Na_2MoO_4 \cdot 2H_2O$	+	+	+

[a] +: present; −: absent.

The precise composition of the three media is outlined below in such a way as to show appropriate stock solutions and appropriate mixing sequence. In actual practice, it is more convenient to work with prepared stock solutions of certain combinations of minerals in order to avoid repetitive weighings. Moreover, each ingredient and subsequently each stock solution should be dissolved or mixed in about seven eighths of the volume of water one at a time in the order given so as to prevent the formation of insoluble complexes. Also, the previous ingredient should be completely dissolved or mixed before adding the next ingredient. The final mineral solution should be brilliantly clear. Stock solutions and final composition of the three media in grams or milliliters per liter are as follows:

I. Knop's Medium
A. Macroelements stock solution — double strength

Ingredient	Amount
(1) $Ca(NO_3)_2 \cdot 4H_2O$	1.000
(2) KNO_3	0.250
(3) $MgSO_4 \cdot 7H_2O$	0.250
(4) KH_2PO_4	0.250
(5) Water[a] to make	1000.0

[a] Should be glass distilled.

B. Microelements stock solution

Ingredient	Amount
(1) H_2SO_4, concentrated[a]	0.500
(2) $MnCl_2 \cdot 4H_2O$	2.500
(3) H_3BO_3	2.000
(4) $ZnSO_4 \cdot 7H_2O$	0.050
(5) $CoCl_2 \cdot 6H_2O$	0.030
(6) $CuCl_2 \cdot 2H_2O$	0.015
(7) $Na_2MoO_4 \cdot 2H_2O$	0.025
(8) Water[b] to make	1000.0

[a] Included to acidify the solution in order to maintain clarity.
[b] Should be glass distilled.

C. Ferric citrate stock solution

Ingredient	Amount
(1) $FeC_6H_5O_7 \cdot 5H_2O$	2.500
(2) Water to make	100.0

D. Final medium

Ingredient	Amount
(1) Macroelements solution X2	500.0
(2) Water[a]	400.0
(3) Microelements solution	0.5
(4) Ferric citrate solution	0.4
(5) Water[a] to make	1000.0

[a] The final medium should be brought to accurate volume. Compensate for the addition of organic nutrients. Adjust pH to 5.5. Media can be solidified with purified agar in a concentration range of 0.5 to 1% with 0.8% being the usual concentration.

II. Knudson's Medium

A. Macroelements stock solution — quadruple strength

Ingredient	Amount
(1) $Ca(NO_3)_2 \cdot 4H_2O$	2.000
(2) $(NH_4)_2SO_4$	1.000
(3) $MgSO_4 \cdot 7H_2O$	0.500
(4) Water to make	1000.0

B. Phosphate stock solution

Ingredient	Amount
(1) K_2HPO_4	25.000
(2) Water to make	100.0

C. Microelements stock solution: Same as for Knop's Medium
D. Ferric citrate stock solution: Same as for Knop's Medium
E. Final medium

Ingredient	Amount
(1) Macroelements solution X4	250.0
(2) Water	700.0
(3) Phosphate solution	0.5
(4) Microelements solution	0.5
(5) Ferric citrate solution	0.4
(6) Water to make	1000.0

III. Moore's Medium

A. Macroelements stock solution — quadruple strength

Ingredient	Amount
(1) $CaCl_2 \cdot 2H_2O$	0.300
(2) NH_4NO_3	2.000
(3) $MgSO_4 \cdot 7H_2O$	0.800
(4) K_2HPO_4	0.800
(5) Water to make	1000.0

B. Microelements stock solution

Ingredient	Amount
(1) H_2SO_4, concentrated	0.500
(2) $MnSO_4 \cdot H_2O$	3.000
(3) H_3BO_3	0.500
(4) $ZnSO_4 \cdot 7H_2O$	0.500
(5) $CoCl_2 \cdot 6H_2O$	0.025
(6) $CuSO_4 \cdot 5H_2O$	0.025
(7) $Na_2MoO_4 \cdot 2H_2O$	0.025
(8) Water to make	1000.0

C. Ferric citrate stock solution: Same as for Knop's Medium

D. Final medium

Ingredient	Amount
(1) Macroelements solution X4	250.0
(2) Water	700.0
(3) Microelements solution	1.0
(4) Ferric citrate solution	0.4
(5) Water to make	1000.0

Any one of the three media could be used to culture pteridophyte units. However, investigators have personal preferences based on successful results of culture during experimental investigations. Other media are available, but these three have been the most popular. Organic nutrients such as sucrose, dextrose, amino acids, vitamins, hormones, etc. are added only under special experimental circumstances such as the culture of subterranean spores and gametophytes, e.g., those of the Psilotaceae. In such cases, the medium should include at least dextrose or sucrose in a concentration range of 0.25 to 1.0%.

A few examples will serve to illustrate the use of the three media in pteridophyte nutrition. Knop's Medium has been successfully employed in the culture of spores and gametophytes of polypodiaceous ferns;[1,2] gametophytes and callus-like tissues of gametophytes of polypodiaceous ferns,[3] including the use of Knudson's Medium;[4] sporophytes of Lycopodiaceae, Selaginellaceae, Equisetaceae, and polypodiaceous ferns;[5] sporophytes of water ferns;[6,7] isolated sporophyte leaves of polypodiaceous ferns,[8-10] including leaf tips[11] and the use of Knudson's Medium.[12,13]

Knudson's Medium has been used in the culture of spores and gametophytes of Psilotaceae;[14] Ophioglossaceae;[15] spores and gametophytes of polypodiaceous ferns[16,17] including sporophytes;[18,19] plus embryos,[20] isolated leaves,[21] and leaf fragments.[22]

Moore's Medium has been used to culture spores and gametophytes of Lycopodiaceae,[23] including callus-like tissues,[24] and spores and gametophytes of polypodiaceous ferns.[25]

An example of natural medium for the culture of spores and gametophytes of Psilotaceae is soil.[26]

ENVIRONMENTAL FACTORS RELATED TO NUTRITION

To effect the developmental and growth potential of the several chlorophyll-containing pteridophyte units, it is necessary to expose cultures to the proper physical environmental conditions. In general, successful cultures can be attained at a light intensity range of 2500 to 5000 lx at the level of the cultures, a temperature range of 23 to 27°C, and a relative humidity of 50 to 75%. If daylight is used, exposure of the cultures to the diffuse light of a north window is preferred. Exposure to a south window may be detrimental to pteridophyte cultures because the intensity could be too high and the temperature might excede 27°C. If cultures are to be exposed to artificial lighting in the interior of a room or in special environmental chambers, the light color spectrum supplied should be broad enough to include red and blue. This can be accomplished by using a combination of white incandescent bulbs and white fluorescent tubes, or by using special broad-spectrum fluorescent tubes. Unless otherwise specified, the light duration time is usually set at 12/12 (12 hr of light and 12 hr of darkness).

REFERENCES

1. Miller, J. H. and Miller, P. M., *Am. J. Bot.,* 48, 154–159, 1961.
2. Raghavan, V., *Am. J. Bot.,* 52, 900–910, 1965.
3. Sussex, I. M. and Steeves, T. A., *Ann. Bot. London,* 17, 395–401, 1953.
4. Steeves, T. A., Sussex, I. M., and Partanen, C. R., *Am. J. Bot.,* 42, 232–245, 1955.
5. Wetmore, R. H., Abnormal and pathological plant growth, *Brookhaven Symp. Biol.,* No. 6, pp. 22–40, 1954.
6. Allsopp, A., *Ann. Bot. London,* 16, 165–183, 1952.
7. White, R. A., *Am. J. Bot.,* 53, 158–165, 1966.
8. Steeves, T. A. and Sussex, I. M., *Am. J. Bot.,* 44, 665–673, 1957.
9. Sussex, I. M., *Phytomorphology,* 8, 96–107, 1958.
10. Caponetti, J. D. and Steeves, T. A., *Can. J. Bot.,* 41, 545–556, 1963.
11. Caponetti, J. D., *Bot. Gaz. Chicago,* 133, 331–335, 1972.
12. Sussex, I. M. and Steeves, T. A., *Bot. Gaz. Chicago,* 119, 203–208, 1958.
13. Harvey, W. H. and Caponetti, J. D., *Can. J. Bot.,* 50, 2673–2682, 1972.
14. Whittier, D. P., *Can. J. Bot.,* 51, 2000–2001, 1973.
15. Whittier, D. P., *Bot. Gaz. Chicago,* 133, 336–339, 1972.
16. Whittier, D. P. and Steeves, T. A., *Can. J. Bot.,* 40, 1525–1531, 1962.
17. Sobota, A. E. and Partanen, C. R., *Can. J. Bot.,* 44, 497–506, 1966.
18. Munroe, M. H. and Sussex, I. M., *Can. J. Bot.,* 47, 617–621, 1969.
19. DeMaggio, A. E., *Am. J. Bot.,* 55, 915–922, 1968.
20. DeMaggio, A. E. and Wetmore, R. H., *Am. J. Bot.,* 48, 551–565, 1961.
21. Kuehnert, C. C., *Can. J. Bot.,* 45, 2109–2113, 1967.
22. Kuehnert, C. C. and Steeves, T. A., *Nature,* 196, 187–189, 1962.
23. Freeberg, J. A. and Wetmore, R. H., *Phytomorphology,* 7, 204–217, 1957.
24. DeMaggio, A. E., *Proc. Natl. Acad. Sci. U.S.A.,* 52, 854–859, 1964.
25. Voeller, B. R., *Régulateurs Naturels de la Croissance Végétale,* Editions du Centre National de la Recherche Scientifique, Paris, 1964, 665–684.
26. Bierhorst, D. W., *Am. J. Bot.,* 40, 649–658, 1953.

NATURAL AND SYNTHETIC CULTURE MEDIA
FOR SPERMATOPHYTES

C. J. Asher

INTRODUCTION

A satisfactory culture medium* for spermatophytes must provide adequate amounts of water, essential mineral nutrient elements, and, except in certain cases, oxygen. These special cases involve species possessing well-developed structures for the supply of atmospheric oxygen to the root system via shoots or pneumatophores. For terrestrial and semiaquatic spermatophytes, the medium must also provide a secure anchorage for the plant unless some alternative means of mechanical support is to be provided.

The essential mineral elements to be supplied by the medium include (in approximately descending order of quantitative requirement) nitrogen, potassium, calcium, magnesium, phosphorus, sulfur, chlorine, iron, boron, manganese, zinc, copper, and molybdenum. Species which form symbiotic associations with nitrogen-fixing microorganisms can be grown in media lacking nitrogen, provided cobalt is present.[1-3] Additional mineral elements may be essential for the growth of particular plant species, e.g., sodium for *Atriplex vesicaria*[4] and *Halogeton glomeratus*[5] and silicon for *Equisetum arvense*.[6] The presence of sodium has been reported to have beneficial effects on the growth of a number of species including *Beta vulgaris* and *Apium graveolens*,[7] although essentiality for these species has not been established. Similarly beneficial effects of silicon have been reported on the growth of *Saccharum* spp.[8] and *Oryza sativa*,[9] but, again, evidence of essentiality is lacking. In *Canavalia ensiformis*, nickel has been shown to be a constituent of the enzyme urease (EC.3.5.1.5),[10] which is involved in the nitrogen metabolism of young legume seedlings,[11] but direct evidence of a nickel requirement for the growth of legumes has not been presented to date. However, it seems likely that with improvements in techniques for purifying synthetic culture media, further additions will have to be made to the current list of essential elements.

LIMITING ION CONCENTRATIONS FOR PLANT GROWTH

There is a scarcity of reliable data on concentrations of essential mineral elements and other elements present in the root environment that are adequate or limiting for plant growth. Much of the published information in this area is suspect in that it comes from experimental systems in which substantial changes in the composition of the culture medium could have occurred during the course of the experiment. The extent of such changes in composition can be reduced greatly by the use of continuously flowing liquid media[12] or large volumes of a well-stirred medium.[13] However, it is important that the flow rate of the medium past the roots is high enough to minimize problems due to unstirred layers in contact with root surfaces.[14] The limiting values given in Table 1 are primarily from flowing culture experiments in which the concentrations of ions in the root environment were controlled closely or other experiments in which the absence of serious concentration changes in the medium was established by actual measurement. It should be stressed that each set of values relates to a specific set of experimental conditions and the limiting concentrations for specific ions may vary with changes in pH or the concentration of other ions. For example, with *Medicago tornata* at pH 7.0 a manganese concentration of 0.3 μM was found to be toxic if the calcium concentration

---

* The term "culture medium" is taken to exclude the shoot environment.

Table 1
APPROXIMATE LIMITING ION CONCENTRATIONS FOR THE GROWTH OF VARIOUS SPERMATOPHYTES UNDER CONDITIONS OF CONSTANT OR NEAR-CONSTANT NUTRIENT SUPPLY

Concentration (μM)	Deficient	Concentration (μM)	Adequate	Concentration (μM)	Toxic	Ref.
			Nitrogen (As Nitrate)			
≤10	Zingiber officinale	>20	Zingiber officinale	>5000	Zingiber officinale[a]	17
	Zea mays		Zea mays			18
		>50	Gossypium hirsutum cvs Rode, DP16, RGSS; Gossypium barbadense cv Pima S.1			19
≤35	Cucurbita pepo[b]	>70	Citrus sinensis			20, 23
≤50	Triticum aestivum	>100	Triticum aestivum			21, 23
		>140	Pisum sativum[b]			22, 23
		>200	Triticum aestivum			23
		>360	Pisum sativum,[c] Cucurbita pepo[b]	>1430	Cucurbita pepo[b]	18
≤50	Manihot esculenta cvs Ceiba, Seda, CUQ4; Sorghum bicolor; Helianthus annuus	>500	Manihot esculenta cvs Ceiba, Seda, CUQ4; Sorghum bicolor; Helianthus annuus			19
	Gossypium hirsutum cvs Rex Bar 7/8, Deltaphine; Gossypium barbadense cv Sea Island; Gossypium arboreum		Gossypium hirsutum cvs Rex, Bar 7/8, Deltapine; Gossypium barbadense cv Sea Island; Gossypium arboreum			
≤70	Zea mays[c]	>710	Zea mays[c]			23
		>1790	Zea mays[b]			23
≤140	Cucurbita pepo[c]	>3500	Cucurbita pepo[c]			23
≤500	Gossypium barbadense cv Pima 32, Gossypium herbaceum, Gossypium thurberi, Gossypium sturtii	>5000	Gossypium barbadense cv Pima 32, Gossypium herbaceum, Gossypium thurberi, Gossypium sturtii			19
	Gossypium hirsutum, Manihot esculenta[d]		Gossypium hirsutum, Manihot esculenta[d]			18
≤710	Zea mays[b]			>14,300	Pisum sativum[c]	23
			Nitrogen (As Ammonium)			
≤0.5	Zingiber officinale	>2	Zingiber officinale	>30	Zingiber officinale	17
≤2	Zea mays, Manihot esculenta cv Mameya	>7	Zea mays; Manihot esculenta cv Mameya	>30	Manihot esculenta cv CUQ4	18
		>14	Pisum sativum[c]			23
≤7	Gossypium hirsutum; Helianthus annuus; Manihot esculenta cvs Ceiba, Nina, CUQ4, CUQ6	>30	Gossypium hirsutum; Helianthus annuus; Manihot esculenta cvs Ceiba, Nina, CUQ4, CUQ6	>120	Helianthus annuus	18

a   Above-optimum nitrate concentrations caused scorching of leaves, but yield reductions were small and not statistically significant.
b   At pH 7 to 8.
c   At pH 4 to 5.

Table 1 (continued)
APPROXIMATE LIMITING ION CONCENTRATIONS FOR THE GROWTH OF VARIOUS SPERMATOPHYTES UNDER CONDITIONS OF CONSTANT OR NEAR-CONSTANT NUTRIENT SUPPLY

Concentration (μM)	Deficient	Concentration (μM)	Adequate	Concentration (μM)	Toxic	Ref.
			Nitrogen (As Ammonium) (continued)			
≤10	Triticum aestivum	≥30	Triticum aestivum	≥70	Triticum aestivum	21
		≥36	Pisum sativum,b Cucurbita pepo c	≥140	Pisum sativum c	23
≤10	Gossypium hirsutum cvs Rex, Rode, DP16, RGSS; Gossypium barbadense cvs Sea Island, Pima S.1; Gossypium herbaceum; Gossypium thurberi; Gossypium sturtii	≥50	Gossypium hirsutum cvs Rex, Rode, DP16, RGSS: Gossypium barbadense cvs Sea Island, Pima S.1; Gossypium herbaceum; Gossypium thurberi; Gossypium sturtii	≥500	Gossypium hirsutum cvs Rex, DP16, RGSS; Gossypium sturtii	19
≤14	Cucurbita pepo b	≥70	Zea mays c	≥710	Zea mays c	23
≤30	Manihot esculenta cvs CUQ3, CUQ5	≥120	Manihot esculenta cvs CUQ3, CUQ5			18
≤50	Gossypium hirsutum cvs Bar 7/8, Deltapine; Gossypium barbadense cv Pima 32; Gossypium arboreum	≥360	Cucurbita pepo b	≥1430	Cucurbita pepo b,c	23
		≥500	Gossypium hirsutum cvs Bar 7/8, Deltapine; Gossypium barbadense cv Pima 32; Gossypium arboreum			19
≤120	Sorghum bicolor; Manihot esculenta cvs Seda, Amarillo, CUQ1, CUQ2	≥500	Sorghum bicolor; Manihot esculenta cvs Seda, Amarillo, CUQ1, CUQ2			18
≤360	Zea mays b	≥1790	Zea mays b	≥7140	Zea mays b	23
			Potassium (In the Presence of Ammonium) e			
≤10	Vulpia (Festuca) myuros, Lolium rigidum, Bromus rigidus, Ornithopus sativus, Avena sativa, Pisum arvense; Vicia sativa, Medicago truncatula	≥25	Vulpia (Festuca) myuros, Lolium rigidum, Bromus rigidus, Ornithopus sativus, Avena sativa, Pisum arvense; Vicia sativa, Medicago truncatula			25
≤25	Trifolium hirtum, Trifolium subterraneum, Erodium botrys, Ehrharta longifolia, Arctotheca calendula	≥100	Trifolium hirtum, Trifolium subterraneum, Erodium botrys, Ehrharta longifolia, Arctotheca calendula			26
≤100	Hordeum vulgare	≥1000	Hordeum vulgare			26
≤150	Arachis hypogaea	≥250	Arachis hypogaea			26

d Eight cultivars: Nina, Mameya, Amarillo, CUQ1, CUQ2, CUQ3, CUQ5, CUQ6.

e Ammonium has been shown to act as a powerful inhibitor of potassium uptake.[24]

Table 1 (continued)
APPROXIMATE LIMITING ION CONCENTRATIONS FOR THE GROWTH OF VARIOUS SPERMATOPHYTES UNDER CONDITIONS OF CONSTANT OR NEAR-CONSTANT NUTRIENT SUPPLY

Concentration (μM)	Deficient	Concentration (μM)	Adequate	Concentration (μM)	Toxic	Ref.
		Potassium (In the Absence of Ammonium)				
<1	Dactylis glomerata, Anthoxanthum odoratum, Trifolium pratense	>3	Dactylis glomerata, Anthoxanthum odoratum, Trifolium pratense			27
<3	Zea mays, Manihot esculenta[f]	>10	Zea mays, Manihot esculenta[f]			28
	Medicago lupulina		Medicago lupulina			27
<10	Hordeum vulgare	>25	Hordeum vulgare			13
	Manihot esculenta,[g] Helianthus annuus		Manihot esculenta,[g] Helianthus annuus			28
<100	Beta vulgaris, Brassica oleracea capitata	>125	Beta vulgaris, Brassica oleracea capitata			29
	Manihot esculenta[g]		Manihot esculenta[g]			28
		Calcium[h]				
<0.5	Zea mays	>3	Zea mays			30
<1.0	Lupinus albus, Ornithopus sativus, Lolium perenne, Zea mays		Lupinus albus, Ornithopus sativus, Lolium perenne, Zea mays			31
<3	Lupinus angustifolius, Trifolium subterraneum cv Mt. Barker, Secale cereale, Triticum aestivum cv Gabo	>10	Lupinus angustifolius, Trifolium subterraneum cv Mt. Barker, Secale cereale, Triticum aestivum cv Gabo			31
	Oryza sativa, Lupinus angustifolius, Triticum aestivum cv Gatcher		Oryza sativa, Lupinus angustifolius, Triticum aestivum cv Gatcher			30
<25		>25	Triticum aestivum			22
<10	Lycopersicon esculentum; Lupinus luteus; Medicago sativa cv Du Puits; Medicago scutellata; Ornithopus compressus; Trifolium subterraneum cvs Clare, Bacchus Marsh, Yarloop; Lolium rigidum; Avena sativa, Triticum aestivum cv Wongoondy	>100	Lycopersicon esculentium; Lupinus luteus; Medicago sativa cv Du Puits; Medicago scutellata; Ornithopus compressus; Trifolium subterraneum cvs Clare, Bacchus Marsh, Yarloop; Lolium rigidum; Avena sativa, Triticum aestivum cv Wongoondy			31
	Phaseolus vulgaris; Zingiber officinale; Sorghum bicolor; Manihot esculenta cv CUQ5; Helianthus annuus		Phaseolus vulgaris; Zingiber officinale; Sorghum bicolor; Manihot esculenta cv CUQ5; Helianthus annuus			30

f   Of the 10 cultivars, 9 showed yield depression due to potassium-induced magnesium deficiency above 125 μM K.

g   One cultivar; it developed potassium-induced magnesium deficiency above 125 μM K.

h   No evidence of direct toxic effects of high calcium concentrations found, but in studies in which CaCl$_2$ has been used as the calcium source, numerous species show yield depressions at 1000 μM or 3000 μM Ca attributable to chloride injury.

Table 1 (continued)
APPROXIMATE LIMITING ION CONCENTRATIONS FOR THE GROWTH OF VARIOUS SPERMATOPHYTES UNDER CONDITIONS OF CONSTANT OR NEAR-CONSTANT NUTRIENT SUPPLY

Concentration (μM)	Deficient	Concentration (μM)	Adequate	Concentration (μM)	Toxic	Ref.
	Calcium[h]		Calcium[h] (continued)			
≤100	Gossypium hirsutum, Carthamus tinctoris, Glycine max, Manihot esculenta cv Nina, Medicago truncatula	≥1000	Gossypium hirsutum, Carthamus tinctoris, Glycine max, Manihot esculenta cv Nina, Medicago truncatula			30
	Lupinus cosentini, Medicago falcata, Medicago sativa cv Hunter River, Medicago littoralis, Medicago truncatula, Bromus rigidus, Hordeum vulgare		Lupinus cosentini, Medicago falcata, Medicago sativa cv Hunter River, Medicago littoralis, Medicago truncatula, Bromus rigidus, Hordeum vulgare			29
≤600	Beta vulgaris[i]		Beta vulgaris[i]			32
	Magnesium					
		≥10	Triticum aestivum			22
		≥40	Arachis hypogaea			33
	Phosphorus					
≤0.1	Stylosanthes guyanensis	≥1.0	Stylosanthes guyanensis	≥1.0	Stylosanthes guyanensis	34
	Glycine max, Zea mays	≥3	Glycine max, Zea mays	≥3	Glycine max	35
≤0.2	Vulpia myuros	≥5	Vulpia myuros	≥5	Vulpia myuros	36
	Stylosanthes humilis, Glycine wightii	≥8	Stylosanthes humilis, Glycine wightii, Macroptilium atropurpureum, Phaseolus lathyroides, Centrosema pubescens	≥8	Glycine wightii	34
≤1.0	Macroptilium atropurpureum, Phaseolus lathyroides, Centrosema pubescens, Desmodium intortum	>25	Triticum aestivum	>25	Stylosanthes humilis	34
≤5	Trifolium subterraneum, Erodium botrys, Lupinus cosentinii, Bromus rigidus, Arctotheca calendula	>5	Trifolium subterraneum, Erodium botrys, Lupinus cosentinii, Bromus rigidus, Arctotheca calendula		Trifolium subterraneum	22, 36
≤5	Medicago truncatula, Hypochoeris radicata	>25	Medicago truncatula, Hypochoeris radicata			36
≤12	Manihot esculenta[j]	>50	Manihot esculenta[j]	>130	Zea mays[k]	35

i Experiment conducted at relatively high Mg concentration (2000 μM).

j Twelve cultivars.

k Attributed to phosphorus-induced iron deficiency.

Table 1 (continued)
APPROXIMATE LIMITING ION CONCENTRATIONS FOR THE GROWTH OF VARIOUS SPERMATOPHYTES UNDER CONDITIONS OF CONSTANT OR NEAR-CONSTANT NUTRIENT SUPPLY

Concentration (μM)	Deficient	Concentration (μM)	Adequate	Concentration (μM)	Toxic	Ref.
			Sulfur			
		≥4	Triticum aestivum			22
			Manganese			
				≥0.3	Medicago truncatula,[l] Medicago tornata[l]	15
≤0.01	Medicago truncatula; Trifolium subterraneum,[m] Avena sativa cvs Avon, Mulga	≥0.01 ≥0.05	Medicago truncatula Trifolium subterraneum,[m] Avena sativa cvs Avon, Mulga	≥6	Medicago truncatula; Medicago sativa Trifolium subterraneum;[m] Triticum aestivum, Avena sativa cvs Avon, Mulga	37
≤0.05	Medicago sativa, Triticum aestivum, Avena sativa cv Algerian	≥0.25	Medicago sativa, Triticum aestivum, Avena sativa cv Algerian			37
≤0.3	Sorghum sudanense[n]	≥4.5	Sorghum sudanense[n]	≥18	Sorghum sudanense[n]	38
			Zinc			
≤0.01	Trifolium subterraneum,[m] Medicago sativa, Triticum aestivum, Avena sativa[o]	≥0.05	Trifolium subterraneum,[m] Medicago sativa, Triticum aestivum, Avena sativa[o]	≥3	Medicago sativa, Medicago truncatula, Trifolium subterraneum cv Bacchus Marsh	39
≤0.05	Medicago truncatula	≥0.25	Medicago truncatula	≥6	Trifolium subterraneum cv Clare	39
			Copper			
≤0.01	Chrysanthemum morifolium[p]	≥0.02	Chrysanthemum morifolium[p]			40
			Aluminium			
				≥20	Manihot esculenta cv CUQ5, Zea mays	41
				≥40	Ipomoea batatus cvs SUQ1, SUQ2; Glycine max	41
				≥80	Ipomoea batatus cv SUQ3, Colocasia esculenta, Zingiber officinale, Manihot esculenta cv Nina	41

l Toxic at 250 μM Ca and pH 7 but not at 2500 μM or pH 5.4.

m Two cultivars: Bacchus Marsh, Clare.

n Critical concentrations may have been overestimated, as solutions were analyzed and brought up to strength only every 3 days.

o Three cultivars: Avon, Mulga, and Algerian.

p Critical concentrations may have been overestimated as solutions were replaced only every 7 days.

was 250 μM, but not if the calcium concentration was 2500 μM.[15] At 250 μM calcium, the toxicity could also be overcome by lowering the pH to 5.4. Again, recent research on *Beta vulgaris* suggests that limiting concentrations of calcium may be influenced to a substantial degree by the concentration of magnesium in the medium.[16]

EFFECTS OF pH ON PLANT GROWTH

The pH of culture media can have large effects on plant growth. These effects are often complex, so that it is difficult to partition the total observed effect among the various factors likely to be involved in the case of a particular plant species and a particular medium. In liquid culture media the following effects may be recognized:

1. At very low pH (below pH 4) development of both main root axes and laterals may be severely depressed (Figure 1).[30]
2. At low pH the ability of roots to take up ions may be impaired (especially at low calcium concentrations),[42] and H^+ ion injury may cause soluble constituents (including nutrient ions) to be lost from the cells.[43,44]
3. Within the pH range of approximately 4 to 8, increasing pH increases the ability of the roots to absorb zinc[45] and manganese[15] ions, but decreases the ability to absorb phosphorus[46-48] and molybdenum.[49]
4. For some elements, the predominant ionic species present changes markedly with pH. Effects of pH on the relative proportions of orthophosphate and aluminum ions (the latter not usually added to liquid culture media, but present in soil solutions) are shown in Figure 2. Other biologically important elements in which appreciable changes in ionic species occur with changes in pH include molybdenum, boron, selenium, and arsenic. In the case of phosphorus, effects of pH on plant uptake have been found to closely parallel changes in the concentration of the element present as the $H_2PO_4^-$ ion.[46,48]
5. In aerated liquid culture media it is often difficult to maintain adequate concentrations of Fe^{2+} ions in solution, due to oxidation to Fe^{3+} and precipitation as $Fe(OH)_3$. The problem becomes more severe as the pH increases (see section on composition of nutrient solution).
6. Important effects of pH on root nodule formation have been reported in legumes such as *Trifolium subterraneum*,[51] *Medicago truncatula*,[52] and *M. sativa*,[53] and nonlegumes such as *Myrica gale*[54] and *Casuarina cunninghamiana*.[55] In the legume species, the effect of pH on nodulation is markedly affected by the level of calcium supply.[51,52,56] Effects of pH on nodulation in *T. subterraneum* and *C. cunninghamiana* are illustrated in Figure 3.

It is to be expected that individual plant species will differ in the extent to which their growth is affected by the various factors listed above and, hence, in their pH optima. However, results of experiments in which close attention has been paid to the control of pH and to the supply of phosphate, iron, and other nutrient elements[57,58] suggest, for many species, an optimum pH in the region of pH 5 to 6 (Figure 4).

However, pH optima for plants in liquid culture media may not coincide with those in solid media because of substantial effects of pH on the solubility of many biologically important elements associated with solid phase components. Such elements include aluminum, manganese, iron, copper, zinc, and molybdenum.[59-61]

Another problem when considering pH optima for culture media is that differential uptake of anions and cations by the plants may cause substantial shifts in pH, especially in weakly buffered solution culture and sand culture systems. In general, an excess of cation over anion uptake leads to a decrease in pH, whereas an excess of anion over cation

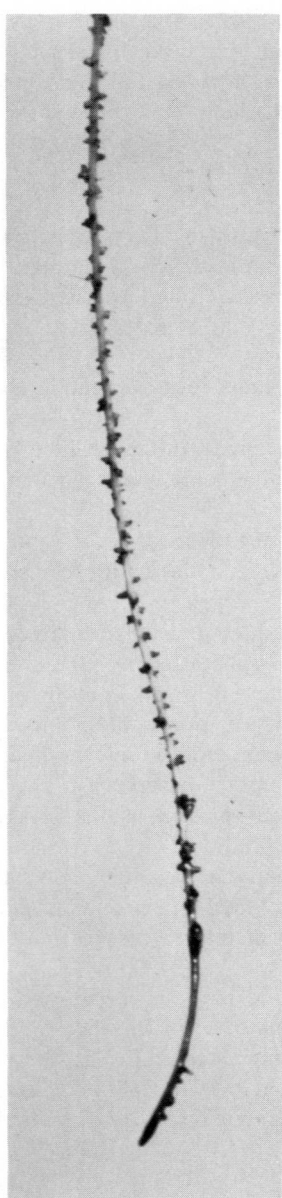

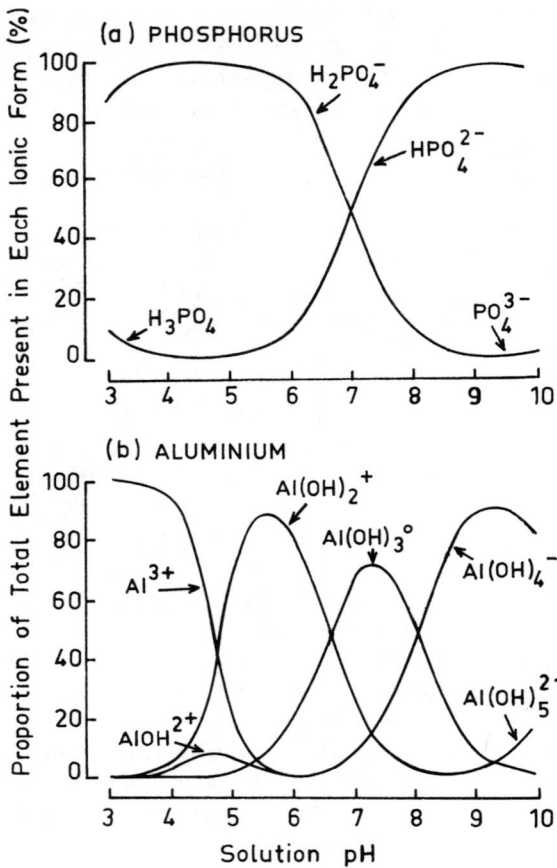

FIGURE 2.   Effects of solution pH on the proportions of (a) phosphorus (assuming an ionic strength of 0.01 *M*) and (b) aluminum (assuming an ionic strength of 0.1 *M*) present in various ionic forms. ([b] From Marion, G. M., Hendrix, D. M., Dutt, G. R., and Fuller, W. H., *Soil Sci.*, 121, 76–85, 1976. With permission.)

FIGURE 1.   Effect of a constant low pH (3.3) on the growth of a root of *Zea mays* in a complete nutrient solution containing an adequate concentration of calcium for plant growth (250 μM). Note necrosis of the tip of main axis and of the greatly fore-shortened laterals.

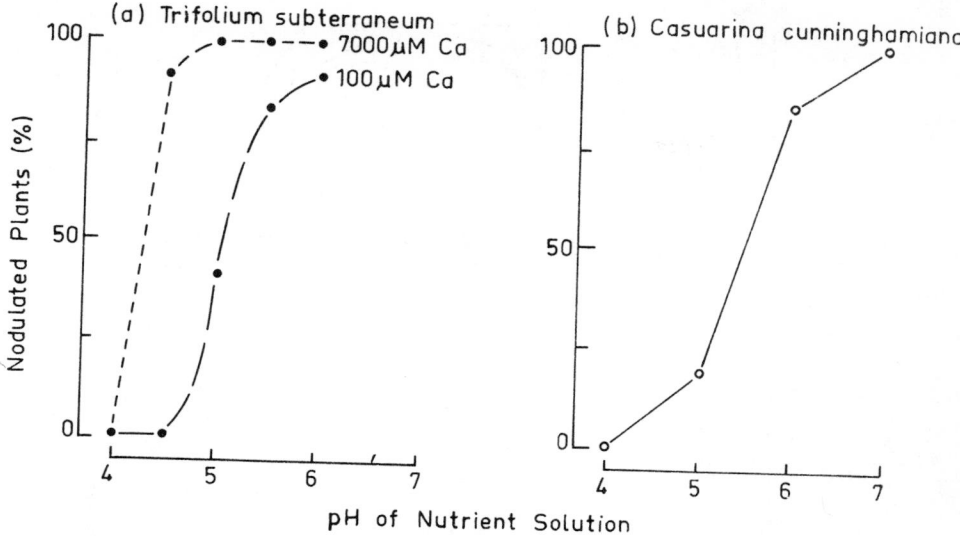

FIGURE 3.   Effects of pH on the nodulation of (a) the legume *Trifolium subterraneum* and (b) the nonlegume *Casuarina cunninghamiana*. ([a] From Loneragen, J. F. and Dowling, E. J., *Aust. J. Agric. Res.*, 9, 464—472, 1958. With permission. [b] From Bond, G., *Ann. Bot.* (London), 21, 373—380,. 1957. With permission.)

uptake leads to an increase in pH. As nitrogen (an element required in large quantities for healthy plant growth) may be supplied either as a cation (NH_4^+) or an anion (NO_3^-), the ratio of these two forms of nitrogen in the culture medium can have large effects on both the rate and direction of pH changes with time. Work of Trelease and Trelease[62] with *Triticum aestivum* suggests that for pH stability in liquid culture media, approximately 80 to 90% of the nitrogen should be supplied in the nitrate form (Figure 5).

For a discussion of effects of pH in soils see reviews by Pearson.[63,64]

TYPES OF CULTURE MEDIA

A distinction has already been made between liquid and solid media. Some properties of various types of liquid and solid media are discussed further in the following sections.

Solid Media
Natural
Soil

Soils are complex, heterogeneous materials differing widely in chemical and physical composition depending on age, parent material, climatic factors, and biological factors including previous use by man. In soils, the liquid phase, or soil solution, provides the major immediate source of water and mineral nutrient elements to nonepiphytic terrestrial spermatophytes. Usually only a very small portion of the total mineral nutrient content of the soil is present in the soil solution at any point in time. Replenishment of the soil solution comes from dynamic equilibria between the liquid and solid phases of the soil and from decomposition of organic matter. Data on soil solution concentrations of several biologically important elements are presented in Figure 6. A comparison of the data in Figure 6 with those in Table 1 indicates that equilibrium soil solution concentrations are frequently somewhat higher than the minimum concentrations reported to permit unrestricted plant growth. However, with actively growing plants, concentrations of many elements will often be lower at the soil/root interface than in the bulk of the soil solution.

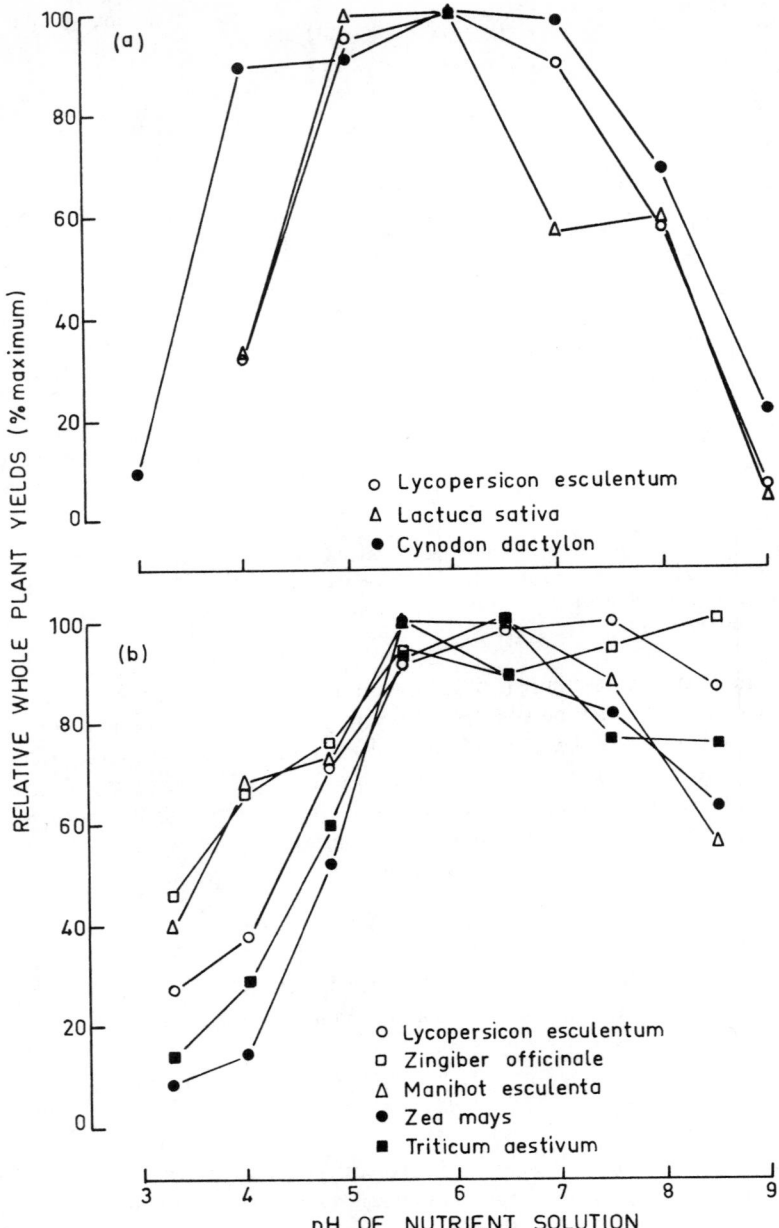

FIGURE 4.   Response of seven plant species to the pH of the nutrient solution.
([a] From Arnon, D. I. and Johnson, C. M., *Plant Physiol.,* 17, 525–539, 1942.
With permission. [b] From data of Islam et al.[58]

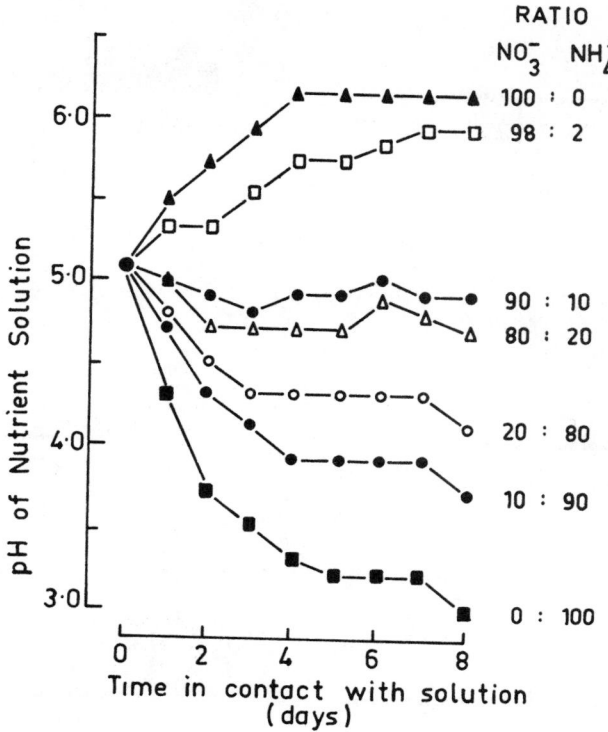

FIGURE 5.   Effect of the ratio of nitrate to ammonium nitrogen on the rate and direction of pH changes in nutrient solutions in contact with the roots of *Triticum aestivum* plants. (Modified from Trelease, S. F. and Trelease, H. M., *Am. J. Bot.,* 22, 520–542, 1935. With permission.)

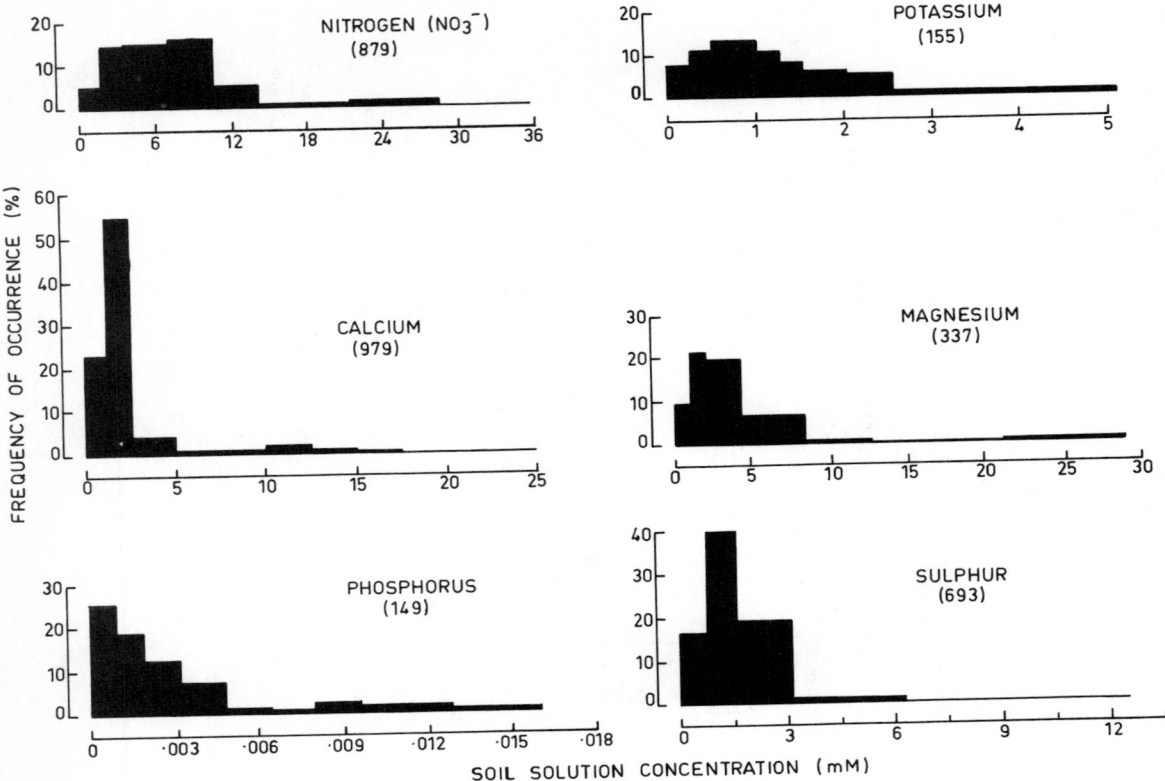

FIGURE 6.   Frequency distribution diagrams for the concentrations of six essential elements in soil solutions. (Based on data compiled by Altman;[65] numbers of soils indicated in parentheses.)

Physical attributes of the soil can have substantial effects on germination and subsequent growth of plants. Important factors affecting physical conditions are particle size distribution of the mineral fraction, type and cation status of predominant clay minerals, and organic matter content. For container-grown plants, surface-tension effects at air/water interfaces in the soil at the base of the container serve to impede drainage. Consequently, the best results tend to be obtained with free-draining soils such as sandy loams, loams, and silt loams rather than finer textured soils such as clay loams or clays. Typical relationships between the principal constituents of a silt loam soil in good condition for plant growth are given in Figure 7. Properties of five silt loam soils are given in Table 2.

Synthetic Mixtures
Standard Potting Mixtures
Mixtures of a relatively small number of ingredients can be combined to produce materials with physical and chemical characteristics similar to a fertile natural soil, but with a more uniform batch composition. The mixtures listed below have been used widely for the growth of nonepiphytic terrestrial spermatophytes.

John Innes Seed Compost* — This is used for propagation from seed or cuttings. Each cubic meter contains

* Metric conversions of quantities given in Imperial units by Darlington.[68]

Sterilized composted meadow turf	0.5 m³
Peat (particles <9.5 mm, 3 mm to 4 mm predominating)	0.25 m³
Coarse sand (60 to 70% of particles >1.6 mm < 3.2 mm)	0.25 m³
Ground limestone or chalk	593 g
Single superphosphate	1187 g

John Innes Potting Compost No.1* — This is used for established seedlings or rooted cuttings. Each cubic meter contains

Sterilized composted meadow turf	0.58 m³
Peat (particles <9.5 mm, 3 mm to 4 mm predominating)	0.25 m³
Coarse sand (60 to 70% of particles >1.6 mm < 3.2 mm)	0.17 m³
Ground limestone or chalk	593 g
Single superphosphate[a]	1187 g
Hoof and horn (13% N, 3.2 mm grist)[a]	1187 g
Potassium sulfate[a]	593 g

[a] For composts Nos. 2 and 3 these amounts are respectively doubled or tripled.

University of California Potting Mixes — These consist of mixtures of fine sand (particles > 0.05 < 0.5 mm) and sphagnum or hypnum peat moss to which one of six fertilizer mixtures may be added, giving a total of 30 combinations. These mixes differ importantly from the John Innes composts in that their preparation does not involve the procurement and composting of meadow turf, a material likely to vary substantially from location to location. The composition of the 30 combinations is given in Table 3. Full details concerning the preparation and use of these mixtures are given by Baker.[69]

Sand Culture

Plants are grown in coarse sand or fine gravel which is moistened at suitable time

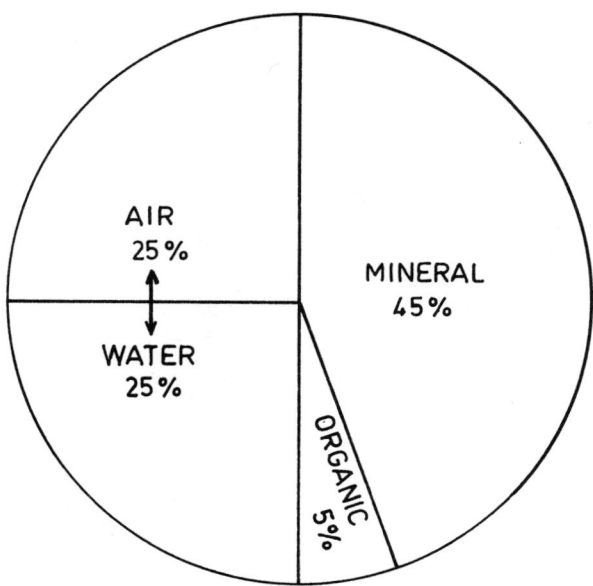

FIGURE 7. Volume composition of a silt loam in good condition for plant growth. (From Buckman, H. O. and Brady, N. C., *The Nature and Property of Soils*, 6th ed., Macmillan, New York, 1965, 9. With permission.)

* Metric conversions of quantities given in Imperial units by Darlington.[68]

Table 2
PHYSICAL PROPERTIES REPORTED FOR VIRGIN
AND CULTIVATED LOAMS AND SILT LOAMS

Soil	Condition	Bulk density (g/cm^3)	Aeration porosity[a] (%)	Available water[b] (% by volume)
Webster loam	Virgin	0.91	19.3	12.5
	Cultivated	1.14	7.9	15.6
Grundy silt loam	Virgin	1.02	7.2	15.7
	Cultivated	1.20	5.4	16.9
Edina silt loam	Virgin	1.04	9.0	15.5
	Cultivated	1.36	7.1	16.9
Marshall silt loam	Virgin	1.08	7.1	15.0
	Cultivated	1.08	9.1	14.4
Carrington silt loam	Virgin	1.13	9.5	15.8
	Cultivated	1.13	10.3	16.3

[a] Pores drained at a tension of 40 cm water.
[b] Difference between moisture equivalent and wilting point.

Modified from Anderson, M. A. and Browning, G. M., *Soil Sci. Am. Proc.,* 14, 370—374, 1949.
By permission of the Soil Science Society of America.

Table 3

COMPOSITION OF UNIVERSITY OF CALIFORNIA MIXES A TO E AND RECOMMENDED FERTILIZER MIXTURES (I TO VI) FOR USE WITH EACH MIX (FERTILIZER UNITS ARE g/m³, OBTAINED BY METRIC CONVERSION OF IMPERIAL UNITS GIVEN BY BAKER[69]

U.C. Mix A:[a] 100% Fine Sand + One of the Following Fertilizer Mixtures; Bulk Density 1.43 g/cm³, Maximum Water Retention 43% by Volume, pH After Fertilizer Addition 6.8

	Fertilizer mixture[b]					
Ingredient	I	II	III	IV	V	VI
Hoof and horn[c]	—	868	1735	—	868	1735
Potassium nitrate	173	173	173	—	—	—
Potassium sulfate	87	87	87	260	260	260
Single superphosphate	868	868	868	868	868	868
Dolomite	521	521	521	521	521	521
Gypsum	868	868	868	868	868	868

U.C. Mix B:[d] 75% Fine Sand + 25% Peat Moss + One of the Following Fertilizer Mixtures; Bulk Density 1.22 g/cm³, Maximum Water Retention 46% by Volume, pH After Fertilizer Addition 6.8

	Fertilizer mixture[b]					
Ingredient	I	II	III	IV	V	VI
Hoof and horn[c]	—	868	1735	—	868	1735
Potassium nitrate	130	130	130	—	—	—
Potassium sulfate	87	87	87	217	217	217
Single superphosphate	868	868	868	868	868	868
Dolomite	1562	1562	1562	1562	1562	1562
Gypsum	434	434	434	434	434	434
Calcium carbonate	434	434	434	434	434	434

U.C. Mix C:[e] 50% Fine Sand + 50% Peat Moss + One of the Following Fertilizer Mixtures; Bulk Density 1.01 g/cm³, Maximum Water Retention 48% by Volume, pH After Fertilizer Addition 6.5

	Fertilizer mixture[b]					
Ingredient	I	II	III	IV	V	VI
Hoof and horn[c]	—	868	1735	—	868	1735
Potassium nitrate	87	87	87	—	—	—
Potassium sulfate	87	87	87	173	173	173
Single superphosphate	868	868	868	868	868	868
Dolomite	2603	2603	2603	2603	2603	2603
Calcium carbonate	868	868	868	868	868	868

U.C. Mix D:[f] 25% Fine Sand + 75% Peat Moss + One of the Following
Fertilizer Mixtures; Bulk Density 0.55 g/cm^3, Maximum Water
Retention 51% by Volume, pH After Fertilizer Addition 6.0

	Fertilizer mixture[b]					
Ingredient	I	II	III	IV	V	VI
Hoof and horn[c]	–	868	1735	–	868	1735
Potassium nitrate	87	87	87	–	–	–
Potassium sulfate	87	87	87	173	173	173
Single superphosphate	694	694	694	694	694	694
Dolomite	1735	1735	1735	1735	1735	1735
Calcium carbonate	1388	1388	1388	1388	1388	1388

U.C. Mix E:[g] 100% Peat Moss + One of the Following
Fertilizer Mixtures; Bulk Density 0.11 g/c, Maximum Water
Retention 59% by Volume pH After Fertilizer Addition 5.7

	Fertilizer mixture[b]					
Ingredient	I	II	III	IV	V	VI
Hoof and horn[c]	–	868	1735	–	868	1735
Potassium nitrate	130	130	130	–	–	–
Potassium sulfate	–	–	–	130	130	130
Single superphosphate	347	347	347	347	340	340
Dolomite	868	868	868	868	868	868
Calcium carbonate	1735	1735	1735	1735	1735	1735

[a] Seldom used; has highest bulk density of poorest retention of nutrients of the five mixes.

[b] Mixes containing fertilizer mixtures I and IV may be stored indefinitely, all others should be planted within one week of preparation. Fertilizer mixtures I, II, and III contain moderate amounts of readily available nitrogen as nitrate. I contains no reserve of organic nitrogen, II contains a moderate reserve, and III contains a large reserve to support later plant growth. Fertilizer mixtures IV, V, and VI contain no added nitrate and IV contains no organic nitrogen, either. Mixtures V and VI contain moderate and large reserves of organic nitrogen, respectively.

[c] Blood meal may be substituted.

[d] Commonly used; has good physical properties.

[e] Commonly used; has excellent physical properties.

[f] A lightweight mix with good aeration.

[g] A very light weight, slightly acid mix used in commercial horticulture for *Azalea, Gardenia,* and *Camellia* sp.

intervals with a nutrient solution. The method has many variations (see References 70 and 71) and permits plants to be grown in a more chemically defined medium than is the case with natural soils or synthetic potting mixtures containing organic materials.

Properties of Sands

Sand is usually regarded as a chemically inert supporting medium which does not contribute to the supply of mineral nutrients. However, the micronutrient content of sands may be appreciable, necessitating acid extraction before use in studies concerned with effects of micronutrient supply on plant growth. Apparatus for extraction of silica sand with steam heated 15 to 18% hydrochloric acid plus 1% oxalic acid is described in detail by Hewitt.[71] Data on contamination levels are given in Tables 4 and 5.

Particle size distribution of sands is a major determinant of water retention, aeration, and ease of penetration by the roots. In general, fine sands have high water retention, but inferior aeration and root penetration characteristics. Coarse sands and gravels are easily penetrated by roots and have good aeration, but may require frequent applications of nutrient solution to ensure an adequate supply of water. Particle size distributions for sands used by various workers are given in Table 6.

Composition of Nutrient Solutions

The nutrient solutions commonly used for sand and nonrenewed or intermittently renewed solution culture systems have been developed by empirical means and for the most part without the benefit of precise information regarding limiting solution concentrations for the growth of particular species (see Table 1). As there are no obvious differences in composition between the nutrient solutions used for sand culture and solution culture and some (such as Hoagland's solution[113]) have been used for both purposes, the following discussion applies to solutions used for either purpose.

Sand culture and solution culture systems differ importantly from natural soils and most synthetic potting mixtures in that the solution in contact with the roots is not continuously replenished in essential elements by the decomposition of organic matter, dissolution of relatively insoluble minerals, and adsorption/desorption reactions with clay minerals and other solid-phase components. Consequently, the concentrations of essential elements in the nutrient solutions tend to fall continuously as a result of uptake by the plant roots. Where well-grown plants are involved and the volume of solution available per plant is relatively small, rates of depletion may be very rapid (Figure 8).

One potentially valuable method of stabilizing the composition of nutrient solutions with respect to time would be to introduce suitable natural or synthetic ion exchange materials into the culture medium. Although some degree of success has been achieved in supplying essential elements through the addition of ion exchange materials to culture media,[115-117,120,121] few workers have studied the maintenance of appropriate concentrations of one or more ions in the solution phase of the system.[118-120] Further research is necessary before the full potential of this method can be realized.

Because of the magnitude of the nutrient depletion problems encountered in conventional sand culture and solution culture methods, nutrient solutions are usually employed which provide initial nutrient concentrations substantially higher than those commonly encountered in soil solutions (Figure 6) or needed at root surfaces for healthy plant growth (Table 1). In some cases the stated initial concentrations are so high that the solubility product of some ion pairs is exceeded, resulting in precipitation or incomplete dissolution of the component salts.[121-124] Users of such "solutions" may either filter off the supernatant[121] or suspend the insoluble material before use.[122-124]

In the case of some essential elements, initial concentrations may fall within the toxic range for some species. Although such effects are occasionally discussed in the literature,[125-127] the toxicity may be transitory and pass unnoticed if uptake by the

Table 4
MINERAL IMPURITIES IN SILICEOUS SANDS USED IN SAND CULTURE EXPERIMENTS BY VARIOUS WORKERS

Sand	Fraction and extraction method	SiO_2 (%)	N (ppm)	P_2O_5 (ppm)	CaO (ppm)	MgO (ppm)	K_2O (ppm)	Fe (ppm)	Al (ppm)	Mn (ppm)	Cu (ppm)	Zn (ppm)	B (ppm)	Mo (ppm)	V (ppm)	Ref.
Allahabad, Uttar Pradesh, India	HCl							White, 75; tinted, 1,440								72
Hohenbockaer glass	Water soluble										0.0042					73
	Total										0.09					73
Silica	Water soluble (pH 7.4)		0.5	0.15	61	7	9.5									74
	Exchangeable (0.01 M NaCl, pH 7.1)		5	0.15	96	7	9.5									74
	Acid soluble (0.01 M HCl, pH 2.3)		5	7.5	130	13	13									74
Pure Ottawa	Acid soluble (0.01 M HCl, pH 2.3)				20	4										74
Pale beach				40	600	2,800	Trace	4,000	2,500							75
Pale brown river, Kuala Lumpur	Acid soluble in 0.5—1.0 mm fraction			38	3.5	16	50	953		28	2.3	4.6				76
	Biological assay: sunflower												0.02—0.05			77
Silica		98.84		500	1,000	400	700	4,500 (as Al_2O_3 + Fe_2O_3)								78
		94.0					20,000	15,000 (as Al_2O_3)	800							79
Ottawa silica	HCl				ND[a]	Trace		180								80
Coarse river														0.0046	10	81
High-grade silica	Cold H_2SO_4									20			ND[a]			82
Hohenbockaer glass	HCl		70	20	190	200	150									83
Benin	Water washed				20	6	6									84
Blackhawk Sand Blast	Total	99.8			60	31		130	250							85
Leached river		99.55	43	70—90				470	1,050							87
Fine white silica		99.71						346								88
					120	60										89

[a] Not detected.

Adapted from Hewitt, E. J., *Commonw. Bur. Hortic. Plant Crops G. B. Tech. Commun.*, No. 22 (revised 2nd ed.), 1966. With permission.

Table 5

MICRONUTRIENT IMPURITIES IN SILICA SANDS EXAMINED AT UNIVERSITY OF BRISTOL AGRICULTURAL AND HORTICULTURAL RESEARCH STATION, LONG ASHTON, U.K.

Sand	Fraction	Fe[a] (ppm)	Mn[a] (ppm)	Cu[a] (ppm)	Zn[a] (ppm)	B[a] (ppm)	Mo[a] (ppm)	Mo[b] (ppm)
Leighton Buzzard, Double Arches Pits[c]								
1943	Unsieved	360	0.28	—	—	3.5	—	—
1944	Sieved through 24-mesh copper gauze	240	0.25	—	—	0.6	—	—
1945	Sieved through 24-mesh copper gauze	170	0.15	—	—	0.7	—	—
1946	Sieved through 24-mesh copper gauze	150	0.29	0.2	—	0.4	0.010	—
1948	Retained 24-mesh	212	0.8	0.7	2.7	2.25	—	—
	Passed 24-mesh	142	0.15	0.7	1.6	1.55	—	—
1949	Retained 24-mesh	463	0.6	2.3	—	1.3	0.012	0.00135
	Passed 24-mesh	196	0.3	1.0	2.1	4.0	0.032	0.00285
Fragments of iron sandstone and calcareous concretions on 24-mesh sieve		72,000	295	30	—	33	0.25	—
Heavily iron-stained sample		3,802	16	1.9	13.5	3.7	0.015	0.0028
Pink-stained profile		488	0.76	0.39	0.82	4.0	0.017	0.0027
Exface 1949 (Front Pit)		136	0.1	0.27	1.75	1.0	ND[d]	0.00065
Exface 1949 (Big Pit)		257	0.4	0.25	0.55	4.7	0.006	0.0009
Average values (Big Pit) assessed 1943–50 (as used)		216	0.26	0.55	1.4	2.2	0.018	—
Aylesbury Pit "60"		296	1.3	0.16	2.25	3.1	0.047	0.00215
Aylesbury Pit "60"	Carbonate boiled	320	10.7	0.07	0.7	1.0	0.051	0.0036
Loch Aline (N.W. Scotland), 98% SiO_2	White sand retained by 1 mm (24-mesh)	13,410	317	7.8	—	12.5	0.32	—
Loch Aline (N.W. Scotland), 99.9% SiO_2	Passed by 1 mm (24-mesh)	125	1.6	0.9	3.7	0.60	0.015	0.0035
Leziate Sands, 99.8% SiO_2								
Washed No. 1		130	0.3	0.4	1.0	1.4	0.043	0.0012
Fine		222	0.6	0.6	ND[d]	2.9	0.019	0.0011
Coarse		212	0.3	0.35	2.5	3.1	0.044	0.0011
Norwegian crushed quartz (Rakestadt)	24–50 mesh	—	—	—	—	—	—	0.0026–0.005

[a] Extracted with boiling 6N HCl.

[b] Aspergillus niger bioassay.

[c] Samples contained 99.8% SiO_2 and approximately 40 ppm Ca.

[d] Not detected.

Adapted from Hewitt, E. J., Commonw. Bur. Hortic. Plant. Crops G. B. Tech. Commun., No. 22 (revised 2nd ed.), 1966. With permission.

Table 6
PARTICLE SIZE DISTRIBUTION OF SANDS USED AS THE SUPPORTING MEDIUM WHEN GROWING VARIOUS SPECIES IN SAND CULTURE

Percentage of particles falling in each size class[a]

>2 mm	2–1 mm	1–0.5 mm	0.5–0.4 mm	0.4–0.3 mm	0.3–0.25 mm	0.25–0.2 mm	0.2–0.15 mm	0.15–0.125 mm	0.125–0.1 mm	<0.1 mm	Species grown	Remarks	Ref.
				100							Several species including Hordeum vulgare, Avena sativa, Raphanus sativa, Brassica oleraceus, Oryza sativa, Beta vulgaris	25-cm clay pots and 51 urns	71
		0.8	99	0.3							Rosa spp.	Cultures 15 and 30 cm deep, intermittent surface application of nutrients	90
							100				Several species, details not given	Beds 28 cm deep, surface irrigated	91
		2			25	54	13	6			Several species, including Zea mays	Beds 33 cm deep	92
				85							Fruit trees (unspecified)	Beds 1.9 m deep	92
					50						Not specified	Surface-irrigated pot cultures	93
											Frageria sp., Trifolium repens, Medicago sativa, Phaseolus vulgaris, Linum usitatissimum, Parthenium argentatum	Surface-irrigated pot cultures	94
27	68										Triticum aestivum, Oryza sativa, Brassica alba, Gossypium hirsutum, Helianthus annuus	Plants prone to wilting	95
	16	64		17								Satisfactory	
							84			13		Too compact	
0.15	1.0	86.6				12.3					Not specified	Pot cultures 20 cm deep	96
	5.0	90.5				4.5					Not specified		97, 98
0.2	12.4	83.7	0.63		2.4	0.4	0.2				Lycopersicon esculentum, Beta vulgaris, Avena sativa, Solanum tuberosum		99
0.1	0.2	4.4	2.0		85.3	5.0	2.8						
	0.3	32.3 (<20)		41.1	21.1		4.9				Lycopersicon esculentum, Brassica oleracea botrytis, Cucurbita pepo var. medullosa	5 or 101 buckets, irrigated 3 times daily	100
25		30			40		5				Dianthus caryophyllus, Lycopersicon esculentum, Chrysanthemum sp.		101
	0.14					60	26.7	9.7		3.5	Hordeum vulgare	Undrained vessels, excess liquid removed by siphon	102
		48.6		26.4			22.9			1.5	Triticum aestivum	Undrained vessels, 1.5-kg capacity	103
	35.0	63.7	1.3								Citrus auranteum		104
2.3	41.7	53.3	2.7								Lycopersicon esculentum	Drip cultures 20—22 cm deep	105
		0.3			7.85	54.1	21.6		16.1				

a   — x — indicates x% between the limits shown.

Table 6 (continued)
PARTICLE SIZE DISTRIBUTION OF SANDS USED AS THE SUPPORTING MEDIUM WHEN GROWING VARIOUS SPECIES IN SAND CULTURE

Percentage of particles falling in each size class[a]

>2 mm	2—1 mm	1—0.5 mm	0.5—0.4 mm	0.4—0.3 mm	0.3—0.25 mm	0.25—0.2 mm	0.2—0.15 mm	0.15—0.125 mm	0.125—0.1 mm	<0.1 mm	Species grown	Remarks	Ref.
5.1	29.6	51.1	7.7	7.7				6.5			*Lycopersicon esculentum, Avena sativa*	Drip cultures 20—22 cm deep	106
		43	31	16			9				*Aleurites moluccana*	141 pots, intermittent application	107
0.6	5.6	30.8	53.9	53.9	7.36			1.7			*Fagopyrum esculentum*		108
		0.6—0.5 100									*Medicago sativa, Hordeum vulgare, Triticum aestivum*	Undrained pots maintained at 15% moisture	109
7.8	7.8			90.1	90.1	2.1		2.1			Various fruit crops	Cultures 75 cm deep, in air conditioned green houses	110
					2.1	22.5	44.1	24.7	2.9	2.7	*Lactuca sativa*		111, 112

Adapted from Hewitt, E. J., *Commonw. Bur. Hortic. Plant. Crops G. B. Tech. Commun.*, No. 22 (revised 2nd ed.), 1966. With permission.

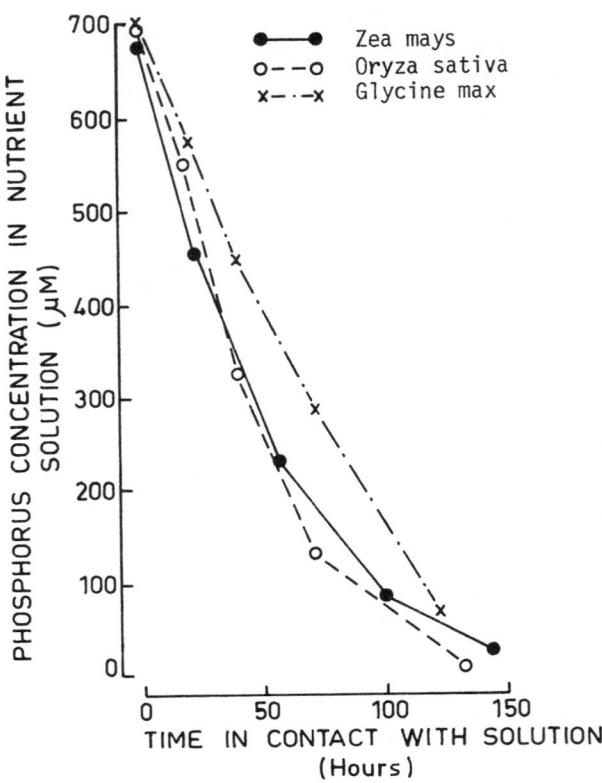

FIGURE 8.   Removal of phosphorus from a nutrient solution by 4- to 6-week-old plants of three species. Volume of solution per plant was approximately 250 ml. (From Franco, C. M. and Loomis, W. E., *Plant Physiol.*, 22, 627–634, 1947. With permission.)

plants quickly reduces the solution concentration to nontoxic levels. Again, growth reductions due to potentially toxic concentrations of an element may fail to eventuate if uptake of the element in question is inhibited by the presence of a second element also at a very high concentration.

Many nutrient solutions have been devised over the years for the growth of spermatophyte species (see Reference 71). In the majority of cases the approach has been wholly empirical, and the initial concentrations have been selected without consideration of limiting concentrations for deficiency or toxicity in the species to be grown or the changes likely in the composition of the nutrient solution as growth proceeds. Inasmuch as deficiency and toxicity thresholds differ widely among species (Table 1) and changes in solution composition with time depend on several factors (including the volume of solution provided per plant, the rate of nutrient uptake [which is largely a function of absolute growth rate] and the frequency of solution renewal), it follows that there can be no single nutrient solution which can be expected to give satisfactory results under all circumstances. Nevertheless, some solutions (such as the Long Ashton solution[71] and the Hoagland Arnon solution[113]) are widely used and have been found to work well in experiments with many species. Details of the composition of these and several other nutrient solutions are given in Tables 7 and 8.

It is clear from Table 7 that a wide range of salts can be used in the preparation of nutrient solutions, and that large differences in both the total concentration of ions and the concentrations of individual ions exist among solutions reportedly giving satisfactory results (Table 8).

Table 7

COMPOSITION OF SELECTED NUTRIENT SOLUTIONS USED FOR SAND CULTURE OR SOLUTION CULTURE EXPERIMENTS WITH VARIOUS SPERMATOPHYTE SPECIES

Milligrams Salt per Liter of Solution

Salts used	Nitrogen-free solutions		Solutions with N as NO_3^-		Solutions with N as NH_4^+		Solutions with N as both NO_3^- and NH_4^+		
	Bond[54]	Norris and Date[128]	Hoagland's No. 1[113]	Long Ashton[71]	Sideris et al.[128]	Addom's "B'''"[128]	Hoagland's No. 2[113]	Mulder[130]	Trelease and Trelease[131]
$(NH_4)_2SO_4$	750				132	734		150	78
$NH_4H_2PO_4$							115		
KNO_3			505	404			607	350	677
$Ca(NO_3)_2 \cdot 4H_2O$			1181				945		
$Ca(NO_3)_2$				656					
KCl		74.5							
KH_2PO_4			136		68.1	870		200	347
K_2HPO_4		174						125	13.7
K_2SO_4					87.2				
$NaH_2PO_4 \cdot 2H_2O$				208					
$CaSO_4 \cdot 2H_2O$	500	344							
$CaCl_2$	250				111	324			437
$Ca_3(PO_4)_2$									
$MgSO_4 \cdot 7H_2O$	500	247	493	368			493		
$MgSO_4$					120	167		250	317
$Na_2CO_3 \cdot 10H_2O$								1.25	
$NaCl$	250			5.9					
$Fe_3(PO_4)_2 \cdot 8H_2O$					27.8[a]				
$FeCl_3 \cdot 6H_2O$								25	
$FeCl_3$						2			
$FeSO_4 \cdot 7H_2O$[a]									13.9[a,b]
Fe tartrate			5[c]				5[c]		
Fe citrate $5H_2O$	1.8	0.7		33.5					
H_3BO_3	2.86		2.86	3.1	d	1	2.86		d
$MnCl_2 \cdot 4H_2O$	1.81		1.81		d		1.81	0.25	d

[a] Both authors state the molar concentration and give the formula of the salt as "$FeSO_4$." However, it seems more likely that one of the hydrated salts was used. Weights given are on the assumption that the salt employed was the heptahydrate.

[b] An equimolar concentration of potassium citrate was also added.

[c] Renewed twice weekly.

[d] Salts used to supply micronutrients other than iron not specified.

[e] Included only in N fixation studies.

Table 7 (continued)
COMPOSITION OF SELECTED NUTRIENT SOLUTIONS USED FOR SAND CULTURE OR SOLUTION CULTURE EXPERIMENTS WITH VARIOUS SPERMATOPHYTE SPECIES

Milligrams Salt per Liter of Solution

Salts used	Nitrogen-free solutions		Solutions with N as NO_3^-		Solutions with N as NH_4^+		Solutions with N as both NO_3^- and NH_4^+		
	Bond[54]	Norris and Date[128]	Hoagland's No. 1[113]	Long Ashton[71]	Sideris et al.[128]	Addom's "B"[128]	Hoagland's No. 2[113]	Mulder[130]	Trelease and Trelease[131]
$MnSO_4 \cdot 4H_2O$	0.22	1			d				d
$MnSO_4$				2.23	d	1		1	d
$ZnSO_4 \cdot 7H_2O$	0.08	0.11	0.22	0.29	d		0.22	0.25	d
$CaSO_4 \cdot 5H_2O$		0.04	0.08	0.25	d		0.08	0.25	d
$(NH_4)_6Mo_7O_{24} \cdot 4H_2O$		0.005							
$Na_2MoO_4 \cdot 2H_2O$	0.02			0.12	d			0.1	d
$H_2MoO_4 \cdot H_2O$					d				d
$CoSO_4 \cdot 6H_2O$				0.05e	d				d

d Salts used to supply micronutrients other than iron not specified.

e Included only in N fixation studies.

Table 8

APPROXIMATE INITIAL CONCENTRATIONS[a] OF INDIVIDUAL ELEMENTS IN NUTRIENT SOLUTIONS USED FOR SAND CULTURE AND SOLUTION CULTURE EXPERIMENTS

Concentrations in Micromoles per Liter

Nutrient element	Nitrogen-free solutions		Solutions with N as NO_3^-		Solutions with N as NH_4^+		Solutions with N as both NO_3^- and NH_4^+		
	Bond[54]	Norris and Date[123][b]	Hoagland's No. 1[113]	Long Ashton[71]	Sideris et al.[128][c]	Addom's "B"[129]	Hoagland's No. 2[113]	Mulder[130]	Trelease and Trelease[131]
Nitrogen									
NO_3^-	10,060		15,000	12,000			14,000	3,460	6,700
NH_4^+					2,000	11,110	1,000	2,270	1,180
Potassium	5,320	3,000	6,000	4,000	1,500	6,390	6,000	6,370	9,410
Calcium	2,030	2,000	5,000	4,000	1,000	2,920	4,000	1,453	3,940
Magnesium	2,610	1,000	2,000	1,500	1,000	1,385	2,000	1,014	2,630
Phosphorus	4,930	1,000	1,000	1,330	500	6,390	1,000	2,190	2,630
Sulfur	10,075	3,005	2,000	1,510	1,500	6,950	2,000	3,610	3,270
Chlorine	1,495	1,000	18	100	2,000		18	279	7,880
Iron	46	5.4	25[d]	100	100	12	25[d]	93	50
Boron	9	12	46	50	46	16	46	4	4.6
Manganese	0.8	4.6	9	10	9	7	9	4.5	2
Zinc	0.3	0.4	0.8	1	1.5		0.8	0.9	1.1
Copper	0.16	0.16	0.3	1	1.6		0.3	1	0.03
Molybdenum	0.1	0.03	0.1	0.5			0.1	0.4	
Other elements				Na, 1,430; Co, 0.2	I, 0.8			Na, 9	Na, 5.2; As, 0.1; Si, 3.6; Ni, 0.2; Co, 0.2; I, 0.08; Li, 0.7; Al, 3.3
Total nutrients	36,576	11,027	31,099	26,033	9,659	35,180	30,099	20,759	37,711

[a] Concentrations calculated on the assumption that all nutrient salts dissolve completely and that no precipitation occurs. These assumptions would not fully be met in the case of Bond's solution.[54]

[b] A relatively dilute nutrient solution is now commonly used in place of the more concentrated Norris' solution,[122] which has been found to inhibit the growth and nodulation of some legume species.[123]

[c] A relatively dilute nutrient solution used for experiments with Ananas comosus.

[d] Renewed twice weekly.

Liquid Media

Natural

Lake and river waters provide a natural liquid culture medium for a variety of free-floating spermatophytes such as *Lemna minor, Wolffia australiana, Spirodela oligorrhiza, Eichhornia crassipes, Aldrovanda vesiculosa,* and *Utricularia australis.* These waters also provide the nutrient medium for some water plants anchored to rocks by means of a holdfast, e.g., *Torrenticola queenslandica* and *Tristicha trifaria.*

Relatively little is known about the mineral nutrient requirements of such species or of their tolerance to high concentrations of specific ions. However, it is evident that, depending on factors such as climate, geology, and distance from source, continental natural waters may vary greatly in mineral composition. Some mean values for major constituents in river waters are given in Table 9. For a more comprehensive coverage of this aspect see Reference 65.

Synthetic

Conventional Solution Culture*

Plants are commonly grown with their roots immersed in relatively small volumes (a few hundred milliliters to a few liters per plant) of relatively concentrated nutrient solution (see Table 8 and section entitled "Composition of Nutrient Solutions"). The principal features of a typical solution culture system are shown in Figure 9. Although the nutrient solutions used are essentially the same as in sand culture systems, this method differs in several other respects. Firstly, in the absence of regular wetting and drying cycles which serve to draw atmospheric oxygen into solid rooting media, artificial aeration of liquid media is required to ensure an adequate supply of oxygen to the roots. Aeration is probably also valuable as a means of stirring the solution, and the bursting of bubbles at the solution surface helps to keep roots moist where they cross an air gap between the solution surface and the lid of the culture vessel.

Another difference concerns the need for mechanical support of plants grown in solution culture systems. This may be achieved for some species by wrapping a wad of dacron wool or soft polyurethane foam round the base of the stem and forcing this into

Table 9
APPROXIMATE MEAN CONCENTRATIONS OF MAJOR CONSTITUENTS OF RIVER WATERS FROM VARIOUS REGIONS OF THE WORLD

Mean concentrations reported for various regions (μM)

Element or ion	Europe	North America	South America	Asia	Africa	Australia and New Zealand
Nitrogen (NO_3^-)	59.7	16.1	11.3	11.3	12.9	0.8
Potassium	44	36	51	238	—	36
Calcium	776	524	180	459	312	97
Magnesium	230	206	62	230	156	111
Sulphur	250	208	50	4	141	27
Chlorine	195	226	138	245	341	28
Iron	14	29	25	2	23	5
Sodium	235	392	174	405	479	126
Silicon	99	118	156	154	305	51
Bicarbonate	1557	1115	508	1295	705	518

Adapted from Altman, P. L., Ed., *Environmental Biology,* Federation of the American Societies for Experimental Biology, Bethesda, Md., 1966, 507–508. With permission.

* Small-volume nonrenewed or intermittently renewed liquid culture systems.

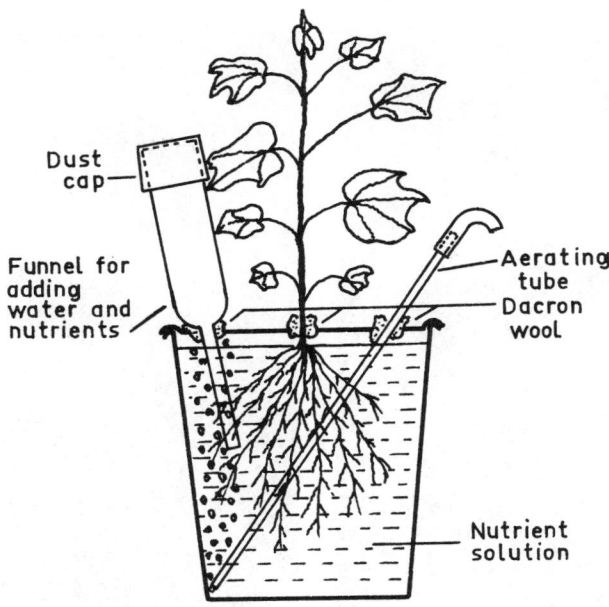

FIGURE 9. Diagram of a conventional small-volume inter-mittently renewed solution culture system. (From Epstein, E., *Mineral Nutrition of Plants: Principles and Perspectives,* John Wiley & Sons, New York, 1972, 37. With permission.)

a hole in the lid of the culture vessel, as illustrated in Figure 9. Alternatively, small plants may be supported by passing them through small slits in a piece of soft polyethylene film secured to the lid of the container.[12] However, this method does not work well with species which form tillers at the base of the main shoot axis.

A highly versatile plant-supporting system which has been used extensively in solution culture studies at the University of Queensland involves the use of plant support baskets with rigid polyethylene sides and a base made of nylon, polyethylene, or nontoxic polyvinyl chloride mesh. Use of mesh in which the fibers are not bonded at points of intersection allows the roots to push the fibers apart, providing a firm anchorage without any restriction in radial growth of the roots. The baskets are usually filled with a lightweight, chemically inert medium such as the polyethylene beads used in injection molding processes. Use of black beads helps to exclude light from the culture solution below and, thus, to suppress algal growth. For short-term experiments with small plants, baskets as small as 2 to 3 cm may be used (Figure 10). For experiments in which plants are to be grown to an advanced stage of development, larger baskets will be needed to provide a stable anchorage and adequate room for tiller development. Some baskets used in solution culture studies at the University of Queensland are illustrated in Figure 11.

Maintenance of adequate concentrations of inorganic iron in solution in aerated solution culture systems is difficult because of oxidation of Fe^{2+} to Fe^{3+} and precipitation of $Fe(OH)_3$, which is highly insoluble, especially within the pH range of 6.5 to 8.0.[61] The problem may be overcome by periodically adding iron salts to the nutrient solution,[113] spraying dilute solutions of iron salts on the leaves, or using natural iron chelates (e.g., citrate or tartrate) or synthetic iron chelates (e.g., EDTA, EDDHA, HEEDTA, DTPA). Much information on the use of synthetic iron chelates for this purpose has been assembled by Wallace.[133]

Continuously Flowing Solution Cultures

Nutrient depletion effects in sand culture and conventional solution culture systems have already been discussed. In continuously flowing solution culture systems, an attempt

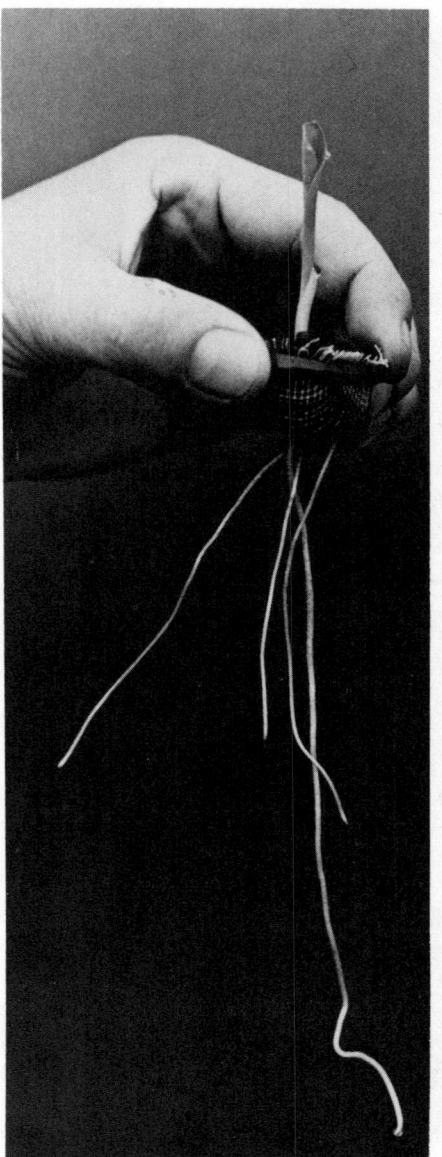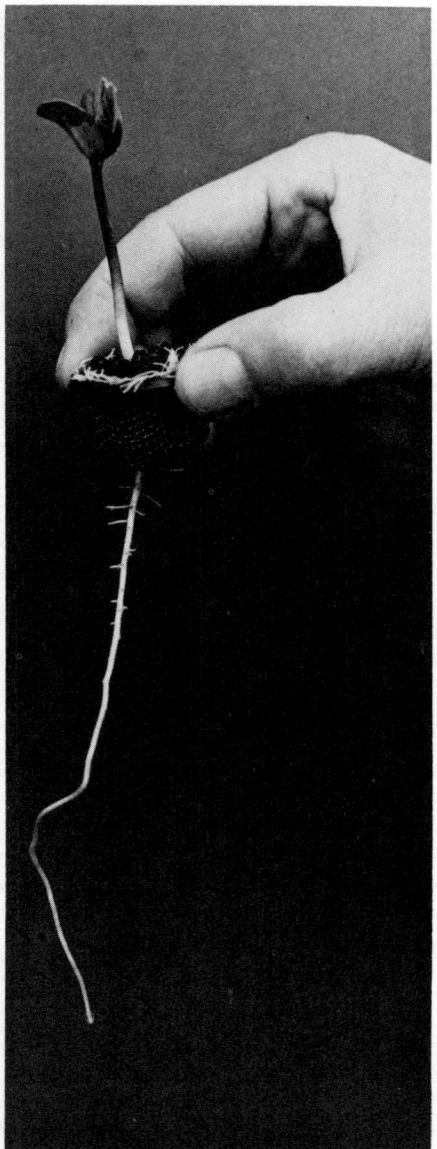

FIGURE 10.   Small polyethylene plant support baskets with seedlings of *Zea mays* (left) and *Glycine max* (right) used in short-term plant nutrition experiments at the University of Queensland.

is made to minimize the depletion problem by growing plants with their roots in a continuously flowing stream of nutrient solution.

The earliest such system which appears to have been successful in achieving good control of nutrient ion concentrations was that of van den Honert,[46] a thorough description of which was given by Becking.[134] In this system up to four nutrient solutions are fed in to each pot and the rates of flow are adjusted so that the rate of addition of the element under test exactly equals the rate of plant uptake, thereby maintaining a constant concentration somewhat lower than the concentration of the incoming solution. However, because the size and uptake rates of young plants tend to

FIGURE 11.   Polyethylene plant support baskets varying from 2.5 to 25 cm in diameter. In the largest basket note the presence of a perforated polyethylene sheet under the mesh to give the basket the necessary strength to support the weight of a large plant.

increase exponentially with time, frequent analyses of solution samples from each pot and corresponding adjustments to solution flow rates are necessary to maintain accurate control of concentration. Solutions are discarded after passage through the pots, but consumption of nutrient solution tends to be moderate because of the relatively low flow rates that are used (of the order of hundreds of milliters per pot per hour). Because of the laborious nature of the technique, its use has been restricted to experiments with small numbers of plants.

An alternative method which has gained some popularity in recent years is rapid circulation of a single dilute nutrient solution through a group of pots arranged in parallel with respect to solution flow, with provision made for frequent monitoring and adjustment of the composition of this common solution.[12,135] In this way, much larger numbers of plants can be studied in a given experiment, and comparisons between species and varieties become practical (the flowing culture facility at the University of Queensland involves nine 60-pot units, allowing control of solution composition to be maintained simultaneously in a total of 540 pots). Solution temperature and pH are usually continuously monitored and controlled.[12,135] In such systems, high flow rates are usually needed to prevent appreciable depletion of the solution in contact with the roots (of the order of liters per pot per minute), especially when dealing with large plants and low nutrient ion concentrations.[14]

The composition of some modern flowing culture solutions is given in Table 10. These nutrient solutions are usually much more dilute than those used in conventional culture systems (see Tables 8 and 10), and it is now clear that many species will grow vigorously in solutions in which the total nutrient concentration is less than 2000 μM.[17,21,27,36,39]

Table 10
APPROXIMATE COMPOSITION OF SOME MODERN FLOWING CULTURE SOLUTIONS USED IN STUDIES OF THE RESPONSE OF VARIOUS SPERMATOPHYTE SPECIES TO VARIATION IN THE CONCENTRATION OF NITROGEN, POTASSIUM, CALCIUM, PHOSPHORUS, AND ZINC

Concentrations in Micromoles per Liter

Element under study

Nutrient element	Nitrogen (NO_3^-) Lee[17]	Nitrogen (NH_4^+) Lee[17]	Nitrogen (NH_4^+) Cox and Reisenauer[21]	Nitrogen (NO_3^- and NH_4^+) Malik[19]	Potassium Asher and Ozanne[25]	Potassium Fageria[26]	Calcium Wilde et al.[27]	Calcium Loneragan et al.[31]	Calcium Islam[30]	Phosphorus Asher and Loneragan[36]	Phosphorus Jintakanon[35]	Zinc Carroll and Loneragan[39,a]
Nitrogen NO_3^-	0.5–5,000	—		50–5,000	2,400–3,025	250	700	350	250	750	1,000	750
NH_4^+	—	0.5–500	0–100	10–500	300	454	—	100	—	100	—	100
Potassium	250–1,566	250	110	250–1,566	1–1,000	51–512	1–33	250	250	250	250–2,060	250
Calcium	250–1,566	250	100	500–1,566	1,000	125	420–470	0.3–1,000	0.3–3,000	250	1,000	250
Magnesium	100–626	100	25	200–625	400	41	100	100	10	100	200	100
Phosphorus	15	15	5	15	200–325	32	50	10	15	0.04–25	0.05–810	5
Sulphur	218	218–418	170	825–950	200–325	268	100	100	10	100	825	100
Chlorine	500	506	15	15	100–225	21–482	125	0.6–2,000	0.6–6,000	100	15	100
Iron	20[b]	20[b]	2[c]	20[b]	9[c]	9.5[c]	5[d]	3[e]	2[e]	2[c]	20[e]	10[c]
Boron	3	3	1	3	13	9.7	2.5	3	3	3	3	5
Manganese	0.25	0.25	0.1	0.25	5	2	0.5	1	0.25	1	0.25	0.25
Zinc	0.5	0.5	0.01	0.5	2	0.2	0.05	0.5	0.5	0.5	0.5	0.01–3.2
Copper	0.1	0.1	0.01	0.1	0.5	0.2	0.02	0.1	0.1	0.1	0.1	0.1
Molybdenum	0.02	0.02	0.005	0.02	0.1	NS[g]	0.005	0.02	0.02	0.02	0.02	0.02
Other elements/ions	Na, 100[h]; Si, 10; Co, 0.04	Na, 100[h]; Si, 10; Co, 0.04	Na, 5; HCO_3^-, 0–100[f]	Na, 115[h]; Si, 10; Co, 0.04	Co, 0.2	Na, 9.5	—	Na, 10; Co, 0.04	Na, 25; Si, 5; Co, 0.04	Co, 0.04	Na, 35; Co, 0.04	Co, 0.04
Total nutrients	1,467–9,625	1,473–2,119	1,423–1,523	2,014–10,386	4,330–6,330	1,273–2,195	1,454–1,536	937–3,927	572–9,571	1,658–1,682	3,349–4,969	1,670–1,674

[a] Solution used in third experiment.

[b] As ferric citrate.

[c] As Fe EDTA.

[d] Iron source not stated.

[e] As Sequestrene® 138. Batch used contained 6.7% Fe; molar ratio of Fe: Na was 1:4.

[f] HCO_3^- not included in nutrient total.

[g] Not stated.

[h] 80 μM from Sequestrene® 138; remainder from other sources.

REFERENCES

1. **Ahmed, S. and Evans, H. J.,** Cobalt: a micronutrient element for the growth of soybean plants under symbiotic conditions, *Soil Sci.,* 90, 205–210, 1960.
2. **Hallsworth, G. E., Wilson, S. B., and Greenwood, E. A. N.,** Copper and cobalt in nitrogen fixation, *Nature,* 187, 79–80, 1960.
3. **Reisenauer, H. M.,** Cobalt in nitrogen fixation by a legume, *Nature,* 186, 375–376, 1960.
4. **Brownell, P. F. and Wood, J. G.,** Sodium as an essential micronutrient element for *Atriplex vesicaria,* Heward, *Nature,* 179, 635–636, 1957.
5. **Williams, M. C.,** Effect of sodium and potassium salts on growth and oxalate content of halogeton, *Plant Physiol.,* 35, 500–505, 1960.
6. **Chen, C. H. and Lewin, J. C.,** Silicon as a nutrient element for *Equisetum arvense, Can. J. Bot.,* 47, 125–131, 1969.
7. **Lehr, J. J.,** *J. Sci. Food Agric.,* 4, 460–471, 1953.
8. **Fox, R. L., Silva, J. A., Young, O. R., Plucknett, D. L., and Sherman, G. D.,** Soil and plant silicon and silicate response by sugar cane, *Soil Sci. Soc. Am. Proc.,* 31, 755–779, 1967.
9. **Mitsui, S. and Takatoh, H.,** I. Nutritional study of silicon in graminaceous crops. II. Study of silicon, *Soil Sci. Plant Nutr. (Tokyo),* 9, 49–58, 1963.
10. **Dixon, N. E., Gazzola, C., Blakeley, R. L., and Zerner, B.,** Jack bean urease (E.C. 3.5.1.5). A metalloenzyme. A simple biological role for nickel? *J. Am. Chem. Soc.,* 97, 4131–4133, 1975.
11. **Dixon, N. E., Gazzola, C., Blakeley, R. L., and Zerner, B.,** Metal ions in enzymes using ammonia or amides, *Science,* 191, 1144–1150, 1976.
12. **Asher, C. J., Ozanne, P. G., and Loneragan, J. F.,** A method for controlling the ionic environment of plant roots, *Soil Sci.,* 100, 149–156, 1965.
13. **Williams, D. E.,** The absorption of potassium as influenced by its concentration in the nutrient medium, *Plant Soil,* 15, 387–399, 1961.
14. **Edwards, D. G. and Asher, C. J.,** The significance of solution flow rate in flowing culture experiments, *Plant Soil,* 41, 161–175, 1974.
15. **Robson, A. D. and Loneragan, J. F.,** Sensitivity of annual *Medicago* species to manganese toxicity as affected by calcium and pH, *Aust. J. Agric. Res.,* 21, 223–232, 1970.
16. **Mostafa, M. A. E. and Ulrich, A.,** Interaction of calcium and magnesium in nutrition of intact sugarbeets, *Soil Sci.,* 121, 16–20, 1976.
17. **Lee, M. T.,** Ph.D. thesis, University of Queensland, 1974.
18. **Forno, D. A.,** Ph.D. thesis, University of Queensland, 1977.
19. **Malik, M. N. A. A.,** Ph.D. thesis, University of Queensland, 1976.
20. **Chapman, N. D. and Liebig, G. F.,** Nitrate concentration and ion balance in relation to citrus nutrition, *Hilgardia,* 13, 14–173, 1940.
21. **Cox, W. J. and Reisenauer, H. M.,** Growth and ion uptake by wheat supplied nitrogen as nitrate, or ammonium, or both, *Plant Soil,* 38, 363–380, 1973.
22. **Reisenauer, H. M.,** 60th Annu. Meet. American Society of Agronomy, New Orleans, November 1968, *Agron. Abstr.,* p. 108, 1968.
23. **Pirschle, K.,** Nitrate-used Ammonsalze als Stickstoffquellen für höhere Pflanzen, bei konstanter Waesser's offionenkronzentra Vion, *Planta,* 14, 583–676, 1931.
24. **Tromp, J.,** Interactions in the absorption of ammonium, potassium, and sodium ions by wheat roots, *Acta Bot. Neerl.,* 11, 147–192, 1962.
25. **Asher, C. J. and Ozanne, P. G.,** Growth and potassium content of plants in solution cultures maintained at constant potassium concentrations, *Soil Sci.,* 103, 155–161, 1967.
26. **Fageria, N. K. M.,** Uptake of potassium and its influence on growth and magnesium uptake by ground nut (*Arachis hypogaea* L.) plants, *Biol. Plant.,* 16, 210–214, 1974.
27. **Wild, A., Skarlou, V., Clement, C. R., and Snaydon, R. W.,** Comparison of potassium uptake by four plant species grown in sand and flowing solution culture, *J. Appl. Ecol.,* 11, 801–812, 1974.
28. **Spear, S., Asher, C. J., and Edwards, D. G.,** manuscript in preparation.
29. **Freeman, G. G.,** *J. Sci. Food Agric.,* 18, 121–126, 1970.
30. **Islam, A. K. M. S.,** unpublished.
31. **Loneragan, J. F., Snowball, K., and Simmons, W. J.,** Response of plants to calcium concentration in solution culture, *Aust. J. Agric. Res.,* 19, 845–857, 1968.
32. **Berry, W. L. and Ulrich, A.,** Cation absorption from culture solution by sugar beets, *Soil Sci.,* 106, 303–308, 1968.
33. **Fageria, N. K.,** Absorption of magnesium and its influence on the uptake of phosphorus, potassium and calcium by intact groundnut plants, *Plant Soil,* 40, 313–320, 1974.
34. **Chantkam, S.,** unpublished.
35. **Jintakanon, S.,** unpublished.

36. **Asher, C. J. and Loneragan, J. F.,** Response of plants to phosphate concentration in solution culture. I. Growth and phosphorus content, *Soil Sci.,* 103, 225–233, 1967.

37. **Carroll, M. D.,** unpublished.

38. **Brown, J. E.,** Manganese-silicon interaction and its effect on growth of Sudan grass, *Plant Soil,* 37, 577–588, 1972.

39. **Carroll, M. D. and Loneragan, J. F.,** Response of plant species to concentrations of zinc in solution. I. Growth and zinc content of plants, *Aust. J. Agric. Res.,* 19, 859–868, 1968.

40. **Graves, C. J. and Sutcliffe, J. F.,** An effect of copper deficiency on the initiation and development of flower buds of chrysanthemum morifolium grown in solution culture, *Ann. Bot.* (London), 38, 729–738, 1974.

41. **Gunatilaka, A.,** M. Agric. Sci. thesis, University of Queensland, 1977.

42. **Jacobson, L., Moore, D. P., and Hannapel, R. J.,** Role of calcium in absorption of monovalent cations, *Plant Physiol.,* 35, 352–358, 1960.

43. **Jacobson, L., Overstreet, R., King, H. M., and Handley, R.,** A study of potassium absorption by barley roots, *Plant Physiol.,* 25, 639–647, 1960.

44. **Marschner, H., Handley, R., and Overstreet, R.,** Potassium loss and changes in the fine structure of corn root tips induced by H- ion, *Plant Physiol.,* 41, 1725–1735, 1966.

45. **Chaudhry, F. M. and Loneragan, J. F.,** Zinc absorption by wheat seedlings. II. Inhibition by hydrogen ions and by micronutrient cations, *Soil Sci. Soc. Am. Proc.,* 36, 327–331, 1972.

46. **van den Honert, T. H.,** *Meded. Proefstn. Groenteteelt Vollegrond Ned.,* 23, 1120–1156, 1933.

47. **Hagen, C. E. and Hopkins, H. T.,** Ionic species in orthophosphate absorption by barley roots, *Plant Physiol.,* 30, 193–199, 1955.

48. **Hendrix, J. E.,** The effect of pH on the uptake and accumulation of phosphate and sulphate ions by bean plaslets, *Am. J. Bot.,* 54, 560–564, 1967.

49. **Stout, P. R., Meagher, W. R., Pearson, G. A., and Johnson, C. M.,** Molybdenum nutrition of crop plants. I. The influence of phosphate and sulphate on the absorption of molybdenum from soils and solution cultures, *Plant Soil,* 3, 51–87, 1951.

50. **Marion, G. M., Hendrix, D. M., Dutt, G. R., and Fuller, W. H.,** *Soil Sci.,* 121, 76–85, 1976.

51. **Loneragan, J. F. and Dowling, E. J.,** The interaction of calcium and hydrogen ions in the nodulation of subterranean clover, *Aust. J. Agric. Res.,* 9, 464-472, 1958.

52. **Robson, A. D. and Loneragan, J. F.,** Nodulation and growth of *Medicago truncatula* on acid soils, *Aust. J. Agric. Res.,* 21, 427–434, 1970.

53. **Munns, D. N.,** Nodulation of *Medicago sativa* in solution culture. I. Acid-sensitive steps, *Plant Soil,* 28, 129–146, 1968.

54. **Bond, G.,** The fixation of nitrogen associated with the root nodules of *Myrica gale* L., with special reference to its pH relation and ecological significance, *Ann. Bot. New Ser.* (London), 15, 447–459, 1951.

55. **Bond, G.,** The development and significance of the root nodules of *Casuarina, Ann. Bot. New Ser.* (London), 21, 373–380, 1957.

56. **Munns, D. M.,** Nodulation of *Medicago sativa* in solution culture. V. Calcium and pH requirements during infection, *Plant Soil,* 32, 90–102, 1970.

57. **Arnon, D. I. and Johnson, C. M.,** Influence of hydrogen ion concentration on the growth of higher plants under controlled conditions, *Plant Physiol.,* 17, 525–539, 1942.

58. **Islam, A. K. M. S., Forno, D. A., Edwards, D. G., and Asher, C. J.,** manuscript in preparation.

59. **Black, C. A.,** *Soil-Plant Relationships,* 2nd ed., John Wiley & Sons, New York, 1968.

60. **Magistad, O. C.,** *Soil Sci.,* 20, 181–225, 1925.

61. **Lindsay, W. L.,** in *Micronutrients in Agriculture,* Mortvedt, J. J., Giordano, P. M., and Lindsay, W. L., Eds., Soil Science Society of America, Madison, Wis., 1972, 41–57.

62. **Trelease, S. F. and Trelease, H. M.,** Changes in hydrogen-ion concentration of culture solutions containing nitrate and ammonium nitrogen, *Am. J. Bot.,* 22, 520–542, 1935.

63. **Pearson, R. W. and Adams, F., Eds.,** *Soil Acidity and Liming,* Agron. Monogr. 12, American Society of Agronomy, Madison, Wis., 1967.

64. **Pearson, R. W.,** *Soil Acidity and Liming in the Humid Tropics,* Cornell Univ. Int. Agric. Devel. Bull. No. 30, Cornell University, Ithaca, N.Y., 1975.

65. **Reisenauer, H. M., Goldman, C. R., and Wetzel, R. C.,** in *Environmental Biology,* Altman, P. L., Ed., Federation of the American Society for Experimental Biology, Bethesda, Md., 1966, 508–510.

66. **Buckman, H. O. and Brady, N. C.,** *The Nature and Properties of Soils,* 6th ed., Macmillan, New York, 1965, 9.

67. **Anderson, M. A. and Browning, G. M.,** Some physical and chemical properties of six virgin and six cultivated Iowa soils, *Soil Sci. Soc. Am. Proc.,* 14, 370–374, 1949.

68. **Darlington, C. D.,** *The Fruit, the Seed and the Soil,* Oliver and Boyd, London, 1949.

69. **Baker, K. F., Ed.,** The U.C. System for Producing Healthy Container-Grown Plants, Calif. Agric. Exp. Stn. Ext. Serv. Man. No. 23, University of California Division of Agricultural Science, Davis, 1957, 1–332.

70. **Miles, R. O.,** *The Culture of Plants in Sand and in Aggregate,* Jealott's Hill Res. Stn. Bull. No. 2 (revised), Imperial Chemical Industries, London, 1928.

71. **Hewitt, E. J.,** Sand and water culture methods used in the study of plant nutrition, *Commonw. Bur. Hortic. Plant Crops G.B. Tech. Commun.* No. 22, (revised 2nd ed.), 1966.

72. **Agarwala, S. C. and Sharma, C. P.,** *Curr. Sci.,* 30, 427–428, 1961.

73. **Arnd, T. H. and Hoffman, W.,** Spurenelemente und ihre Wirkung auf das Pflanzenwachstum unter besonderer Berucksichtigung von Versuchsergebnissen mit Kupfer, *Landwirtsch. Vers. Stn.,* 129, 71–79, 1937.

74. **Arnon, D. I. and Meagher, W. R.,** Factors influencing availability of plant nutrients from synthetic ion-exchange materials, *Soil Sci.,* 64, 213–221, 1947.

75. **Ayres, A. H.,** Influence of the composition and concentration of the nutrient solution on plants grown in sand cultures, *Univ. Calif. Berkeley Publ. Agric. Sci.,* 1, 341–394, 1917.

76. **Bolle-Jones, E. W.,** Nutrition of *Hevea brasiliensis.* I. Experimental methods, *J. Rubber Res. Inst. Malays.,* 14, 183–207, 1954.

77. **Colwell, W. E.,** Effect of chloride and sulphate salts on the growth and development of the Elberta peach on shalil and lovell rootstocks, *Soil Sci.,* 56, 71–94, 1943.

78. **Davis, M. B.,** *Sci. Agric.,* 8, 41–55, 1927.

79. **Hayward, H. E., Long, E. M., and Uhvits, R.,** *U.S. Dep. Agric., Tech. Bull.* No. 922, 1946.

80. **Hobbs, C. H.,** Studies on mineral deficiency in pine, *Plant Physiol.,* 19, 590-602, 1944.

81. **Jensen, H. L. and Betty, R. L.,** Nitrogen fixation in leguminous plants. III. The importance of molybdenum in symbiotic nitrogen fixation, *Proc. Linn. Soc. N.S.W.,* 68, 1–8, 1943.

82. **Muckenhirn, R. J.,** Response of plants to boron, copper, and manganese, *J. Am. Soc. Agron.,* 28, 824–842, 1936.

83. **Scharrer, K. and Schropp, W.,** Sand und Wasserkulturoersuch über die Wirkung des Kupferions, *Z. Pflanzenernaehr. Dueng.,* 32A, 184–200, 1933.

84. **Smilde, K. W.,** *10th Annu. Rep. West African Institute of Oil Palm Research,* 1961–62, W.A.I.F.O.R., Benin, Nigeria, 1962, 71.

85. **Swan, H. S. D.,** *Pulp Pap. Res. Inst. Can. Tech. Rep.,* No. 168, Pulp and Paper Research Institute of Canada, Pointe Claire, Quebec, 1960.

86. **Trumble, H. C. and Strong, T. H.,** Investigations on the associated growth of herbage plants. I. On the nitrogen accretion of pasture grasses when grown in association with legumes, *Bull. Counc. Sci. Ind. Res. Aust.,* 105, 11, 1937.

87. **Trumble, H. C. and Shapter, R. E.,** Investigation on the associated growth of herbage plants. II. The influence of nitrogen and phosphorus treatments on the yield and chemical composition of Wimmera rye grass and subterranean clover grown separately and in association, *Bull. Counc. Sci. Ind. Res. Aust.,* 105, 25, 1937.

88. **Waltman, C. S.,** Effect of hydrogen ion concentration on the growth of strawberries in sand and soil, *K. Agric. Exp. Stn. Bull.* No. 321, 1931.

89. **Weinberger, J. H. and Cullinan, F. P.,** Symptoms of some mineral deficiencies in one-year Elberta peach trees, *Proc. Am. Soc. Hortic. Sci.,* 34, 249–254, 1936.

90. **Davidson, O. W.,** Large-scale soilless culture for plant research, *Soil Sci.,* 62, 71–86, 1946.

91. **Eaton, F. M.,** Large scale land culture apparatus, *Soil Sci.,* 31, 235–241, 1931.

92. **Eaton, F. M.,** Automatically operated sand-culture equipment, *J. Agric. Res.,* 53, 433–444, 1936.

93. **Eaton, F. M.,** Plant culture equipment, *Plant Physiol.,* 16, 385–392, 1941.

94. **Gauch, H. G. and Wadleigh, C. H.,** A new type of intermittently-irrigated land culture equipment, *Plant Physiol.,* 18, 543–547, 1943.

95. **Gile, P. L. and Feustel, I. C.,** The effect of soil and peat admixtures on the growth of plants in quartz sand, *J. Agric. Res.,* 66, 49–65, 1943.

96. **Hewitt, E. J.,** Experiments in mineral nutrition. III. The visual syptoms of mineral deficiencies in crop plants grown in sand cultures, *Annu. Rep. Long Ashton Agric. Hortic. Res. Stn. 1944* pp. 50–60, 1945.

97. **Hewitt, E. J.,** The resolution of the factors in soil acidity, *Annu. Rep. Long Ashton Agric. Hortic. Res. Stn. 1945,* pp. 44–51, 1946.

98. **Hewitt, E. J.,** Experiments in mineral nutrition. II. The visual symptoms of mineral deficiencies in crop plants grown in sand cultures, *Annu. Rep. Long Ashton Agric. Hortic. Res. Stn. 1945,* pp. 51–60, 1946.

99. **Hewitt, E. J.,** unpublished.

100. **Hewitt, E. J.,** unpublished.

101. **Hicks, F. and Tincker, M. A. H.,** A simple method of growing carnations and other plants in sand, *J. R. Hortic. Soc.,* 69, 112–116, 1944.

102. **Hoagland, D. R.,** Relation of concentration and reaction of the nutrient Medium to the growth and absorption of the plant, *J. Agric. Res.,* 18, 73–117, 1919.

103. **McCall, A. G.,** Physiological balance of nutrient solutions for plants in sand cultures, *Soil Sci.,* 2, 207–253, 1916.

104. **Reed, H. S. and Haas, A. R. C.,** Growth and composition of orange trees in sand and soil cultures, *J. Agric. Res.,* 24, 801–814, 1923.

105. **Robbins, W. R.,** Relation of nutrient salt concentration to growth of the tomato and to the incidence of blossom-end rot of the fruit, *Plant Physiol.,* 12, 21–50, 1937.

106. **Robbins, W. R.,** Growing plants in sand cultures for experimental work, *Soil Sci.,* 62, 3–22, 1946.

107. **Shear, C. B., Crane, H. L., and Myers, A. T.,** Nutrient element balance response of tung trees grown in sand culture to potassium, magnesium, calcium, and their interactions, *U.S. Dep. Agric. Tech. Bull.* No. 1085, 1953.

108. **Shive, J. W.,** Some availability studies with ammonium phosphate and its chemical and biological effects upon the soil, *Soil Sci.,* 6, 1–32, 1918.

109. **Thomas, M. D., Hendricks, R. H., Ivie, J. D., and Hill, G. R.,** An installation of large sand-culture beds surmounted by individual air-conditioned greenhouses, *Plant Physiol.,* 18, 334–344, 1943.

110. **Wallace, T.,** Experiments on the manuring of fruit trees, *J. Pomol.,* 4, 117–140, 1924.

111. **Woodman, R. M.,** Pure silicon sand as a basis for phosphate deficiency tests in lettuce, *Sands Clays Miner.,* 3, 22, 1936.

112. **Woodman, R. M.,** *J. Agric. Sci.,* 29, 229–248, 1939.

113. **Hoagland, D. R. and Arnon, D. I.,** *The Water Culture Method for Growing Plants Without Soil,* Calif. Agric. Exp. Stn. Circ. No. 347, College of Agriculture, University of California, Berkeley, 1950, 1–32.

114. **Franco, C. M. and Loomis, W. E.,** The absorption of phosphorus and iron from nutrient solutions, *Plant Physiol.,* 22, 627–634, 1947.

115. **Arnon, D. I. and Grossenbacher, K. A.,** Nutrient culture of crops with the use of synthetic ion-exchange materials, *Soil Sci.,* 63, 159–180, 1947.

116. **Arnon, D. I. and Meagher, W. R.,** Factors influencing availability of plant nutrients from synthetic ion-exchange materials, *Soil Sci.,* 64, 213–221, 1947.

117. **Handley, M. F., Cozart, E. R., Baggett, P., and Seymour, K, G.,** *Down Earth,* pp. 1–7, Winter 1961.

118. **Parr, J. F. and Norman, A. G.,** pH control in nitrate uptake studies with excised roots, *Plant Soil,* 21, 185–190, 1964.

119. **Skogley, E. O. and Dawson, J. E.,** Synthetic ion-exchange resins as a medium for plant growth, *Nature,* 198, 1328–1329, 1963.

120. **Skogley, E. O.,** The "Donnan theory" in development of plant growth media from ion-exchange resins, *Agron. J.,* 61, 317–322, 1969.

121. **Bond, G. and MacKintosh, A. H.,** Effect of nitrate-nitrogen on the nodule symbioses of *Coriaria* and *Hippophae, Proc. R. Soc. London Ser. B,* 190, 199–209, 1975.

122. **Norris, D.,** Some concepts and methods in sub-tropical pasture research, *Commonw. Bur. Pastures Field Crops Hurley Berkshire Bull.,* 47, 186–198, 1964.

123. **Norris, D. O. and Date, R. A.,** in *Tropical Pasture Research: Principles and Methods,* Shaw, N. H. and Bryan, W. W., Eds., Bull. 51, Commonwealth Bureau of Agriculture, 1976, 134–174.

124. **Zinzadze, C.,** Nutrition artificielle des plantes cultivées. I. Melanges nutritifs á pH stabile, *Ann. Agron.,* 2, 809–853, 1932; as cited by Franco, C. M. and Loomis, W. E., *Plant Physiol.,* 22, 627–634, 1947.

125. **Shive, J. W.,** Toxicity of monobasic phosphates towards soybeans growers in soil- and solution-cultures, *Soil Sci.,* 5, 87–122, 1918.

126. **Williams, D. E. and Vlamis, J.,** Manganese toxicity in standard culture solutions, *Plant Soil,* 8, 183–193, 1957.

127. **Williams, D. E. and Vlamis, J.,** Manganese and boron toxicities in standard culture solutions, *Soil Sci. Soc. Am. Proc.,* 21, 205–209, 1957.

128. **Sideris, C. P., Young, H. Y., and Krauss, B. H.,** Effects of iron on the growth and ash constituents of *Ananas comosus* (L.) merr, *Plant Physiol.,* 18, 608–632, 1943.

129. **Addoms, R. M.,** Nutritional studies on loblolly pine, *Plant Physiol.,* 12, 199–205, 1937.

130. **Mulder, E. G.,** Importance of molybdenum in the nitrogen metabolism of microorganisms and higher plants, *Plant Soil,* 1, 94–119, 1948.

131. **Trelease, S. F. and Trelease, H. M.,** Physiologically balanced culture solutions with stable hydrogen-ion concentration, *Science,* 78, 438–439, 1933.

132. **Epstein, E.,** *Mineral Nutrition of Plants: Principles and Perspectives,* John Wiley & Sons, New York, 1972, 37.
133. **Wallace, A., Ed.,** *A Decade of Synthetic Chelating Agents in Inorganic Plant Nutrition,* A. Wallace, Los Angeles, 1962.
134. **Becking, J. H.,** On the mechanism of ammonium ion uptake by maize roots, *Acta Bot. Neerl.,* 5, 1–79, 1956.
135. **Clement, C. R., Hopper, M. J., Canaway, R. J., and Jones, L. H. P.,** A system for measuring the uptake of ions by plants from flowing solutions of controlled composition, *J. Exp. Bot.,* 25, 81–99, 1974.

Index

INDEX

A

L

N

P

X

Y

Z

CRC PUBLICATIONS OF RELATED INTEREST

CRC HANDBOOKS:

CRC FENAROLI'S HANDBOOK OF FLAVOR INGREDIENTS, 2nd Edition
Edited, translated, and revised by: **Thomas E. Furia** and **Nicolo Bellanca**, Dynapol, Palo Alto, California.
This two-volume Handbook is an update of the 1st Edition and includes comprehensive review chapters by recognized experts in the field. It is extensively indexed for easy use.

CRC HANDBOOK OF FOOD ADDITIVES, 2nd Edition
Edited by **Thomas E. Furia**, Dynapol, Palo Alto, California.
Nearly 1000 pages of pertinent food additive information is offered, reflecting the important changes that have occurred in recent years in the area of food additives.

CRC UNISCIENCE PUBLICATIONS:

FOOD ANALYSIS: Analytical Quality Control Methods for the Manufacturer and Buyer, 3rd Edition
This book brings together methods of analysis which are of most value to the factor control chemist.

LOW CALORIE AND DIETETIC FOODS
Edited by **Basant K. Dwivedi, Ph.D.**, Estee Candy Company, Inc.
This book includes all aspects of low calorie and dietetic foods including discussions on fructose, applications and commercial potential of sweetening agents, and the present status of food products for people with special dietary requirements.

MAN, FOOD AND NUTRITION
Edited by **Miloslav Rechcigl, Jr., M.N.S., Ph.D., F.A.A.A.S., F.A.I.C., F.W.A.S.**, Agency for International Development, U.S. Department of State.
This interdisciplinary treatise offers a comprehensive and integrated critical review of the nature and the scope of the world food problem. It presents strategies and discusses various technological approaches to overcoming world hunger and malnutrition.

TOXICITY OF PURE FOODS
By **Eldon M. Boyd, Ph.D.** (deceased), Queen's University, Kingston, Ontario. Edited by **Carl E. Boyd, M.D.**, Health and Welfare, Canada.
A systematic study of the toxicity of pure foods is described in this volume.

WORLD FOOD PROBLEM
Edited by **Miloslav Rechcigl, Jr., M.N.S., Ph.D., F.A.A.A.S., F.A.I.C., F.W.A.S.**, Agency for International Development, U.S. Department of State.
This is a comprehensive and up-to-date bibliography on all important facets of the world food problem, encompassing such areas as the availability of natural resources and the present future sources of energy.

CRC MONOTOPIC REPRINTS:

FLEXIBLE PACKAGING OF FOODS
By **Aaron L. Brody, S. B., M.B.A., Ph.D.**, Arthur D. Little, Inc.
The aim of this book is to describe food products employing flexible packaging, the requirements dictating the flexible packaging being used and the current state of flexible packaging in the food industry.

FREEZE-DRYING FOODS
By **C. Judson King, B.E., E.M., Sc.D.**, University of California.
This review concentrates on several papers published in recent years, relates them to the rest of the field, gives an evaluation of their findings, and the conclusions they draw.